THE ASTROPHYSICS OF PLANETARY SYSTEMS:
FORMATION, STRUCTURE, AND DYNAMICAL EVOLUTION

IAU SYMPOSIUM No. 276

COVER ILLUSTRATION:

The cover illustration is an image realized by *Effetti*. The upper part of the image is an artist's view of an extrasolar planetary system containing a potentially habitable Super Earth (credits: David A. Hardy's AstroArt). The lower part of the image is a view of 2006 Winter Olympics host city Torino, with the snow-capped Alps in the background. The tall building in the foreground is the Mole Antonelliana, the major landmark of Torino. It is named for the architect who built it, Alessandro Antonelli. Construction began in 1863 and was completed 26 years later, after the architect's death. Initially conceived to be a synagogue, today it houses the National Museum of Cinema. With its 167 m vertical size, it is 0.49 m higher than the Philadelphia City Hall in Pennsylvania (USA), thus holding the record for the tallest masonry building in the world.

The image, ideally representing both the location and the scientific topics addressed at the meeting, was chosen to be the official poster of IAU Symposium 276.

INTERNATIONAL ASTRONOMICAL UNION
UNION ASTRONOMIQUE INTERNATIONALE

THE ASTROPHYSICS OF PLANETARY SYSTEMS: FORMATION, STRUCTURE, AND DYNAMICAL EVOLUTION

PROCEEDINGS OF THE 276th SYMPOSIUM OF THE INTERNATIONAL ASTRONOMICAL UNION
HELD IN TORINO, ITALY
OCTOBER 10–15, 2010

Edited by

ALESSANDRO SOZZETTI
INAF - Osservatorio Astronomico di Torino, Italy

MARIO G. LATTANZI
INAF - Osservatorio Astronomico di Torino, Italy

and

ALAN P. BOSS
Carnegie Institution of Washington, DC, USA

CAMBRIDGE
UNIVERSITY PRESS

University Printing House, Cambridge CB2 8BS, United Kingdom

One Liberty Plaza, 20th Floor, New York, NY 10006, USA

477 Williamstown Road, Port Melbourne, VIC 3207, Australia

314-321, 3rd Floor, Plot 3, Splendor Forum, Jasola District Centre, New Delhi - 110025, India

103 Penang Road, #05-06/07, Visioncrest Commercial, Singapore 238467

Cambridge University Press is part of the University of Cambridge.

It furthers the University's mission by disseminating knowledge in the pursuit of education, learning and research at the highest international levels of excellence.

www.cambridge.org
Information on this title: www.cambridge.org/9780521196529

First published 2011

A catalogue record for this publication is available from the British Library

ISBN 978-0-521-19652-9 Hardback
ISSN 1743-9213

Table of Contents

Part 1. PLANET FORMATION

Chairs: A. Sozzetti, G. Marcy, Y. Alibert, A. Boss, W. Kley & M.G. Lattanzi

Part 2. STRUCTURE AND ATMOSPHERES

Chairs: D. Latham, T. Mazeh, W. Borucki & D. Queloz

Part 3. INTERACTIONS

Chairs: D. Charbonneau, S. Udry, S. Seager & T. Guillot

Part 4. THE NEXT DECADE

Chairs: S. Raymond, R. Gratton, W. Traub & A. Sozzetti

Part 5. POSTER PAPERS

Section A: Planet Formation

Section B: Structure & Atmospheres

Section C: Interactions

Section D: The Next Decade

Preface

More than 500 planets are now known to orbit main-sequence stars in the neighborhood of our Sun, discovered and characterized using a variety of techniques, both from the ground and in space. On the one hand, the observational data on extrasolar planets show striking properties indeed, likely evidence of the complexity of the process of planet formation and evolution. On the other hand, the large flow of empirical information gathered on extrasolar planets in the Solar neighbourhood is such that in-depth studies are now possible, which allow us to reach a deeper understanding of the mechanisms regulating their formation processes, their internal structure and atmospheres, and their long-term dynamical evolution. Next-generation observatories (both from the ground and in space) and new methods of data analysis have reached a degree of ripeness that the discovery of planets similar to our Earth, for which it might be possible to establish the degree of habitability, appears to be behind the corner. Fifteen years after the first announcement of a Jupiter-mass companion orbiting a normal star other than the Sun, the formation and evolution of planetary systems is now emerging as a new, quickly expanding interdisciplinary research field.

When the vast breadth of exoplanets research is taken as a whole, one then realizes how we're now witnessing the beginning of a new era of comparative planetology, in which our Solar System can finally be put in the broader context of the astrophysics of planetary systems. To this end, help from future data obtained with a variety of techniques will prove invaluable. Planet search surveys, initially focused solely on planet discovery, are now being designed to put the emerging properties of planetary systems on firm statistical grounds and thus thoroughly test the theoretical models put forth to explain their existence. Furthermore, both NASA and ESA are now formulating strategies to establish a logical sequence of missions and telescope construction to optimize the pace and quality of exoplanet discoveries (with both direct and indirect techniques) and address key questions on the physical characterization and architecture of planetary systems.

With the approaching 15th anniversary of the 51 Pegb discovery announcement, and considering the quickening pace of development of the exoplanet field, a preliminary program was drafted in 2009 by members of the Scientific Organizing Committee (SOC) for an IAU Symposium focused on addressing two main questions: Where do we stand? What's next? The 276th IAU Symposium (IAUS 276) was held in Torino during the week of October 10-15, 2010. At the time of definition of the final scientific program, the broad range of issues in the astrophysics of planetary systems selected to provide answers to these questions was divided in to four main topical sessions: *Planet Formation*, *Internal Structure and Atmospheres*, *Interactions*, and *The Next Decade*. The first three sessions allowed for vibrant confrontations between theory and observations. Datasets of the highest quality, state-of-the-art numerical tools, and increasingly sophisticated theoretical models showed the impressive progress being made in our understanding of planet formation and evolution. The last session provided a forward look into strategic planning exercises of both community and agencies and into ongoing preparations and developments of future ground-based and space-borne observatories devoted to exoplanetary sciences. We warmly acknowledge all the SOC members who actively contributed with their suggestions to shape a strong and attractive scientific program (while making sometimes difficult choices given the large number of talk requests). One major objective achieved during the Symposium was indeed that of connecting scientific results obtained by ground-based and space-borne research programs for the detection and characterization of extrasolar planets with the grand projects that will contribute to move forward the frontier of research in the field during the next decade. The most recent, exciting discoveries of transiting rocky planets ("Super Earths") by the Kepler and CoRoT space telescopes were discussed in parallel to unprecedented results obtained with large ground-based facilities, such as the VLT and the Keck Observatories, regarding the characterization of the chemical composition of the atmospheres of nearby exoplanets. From the ground, ambitious project to search for Earth analogs around the nearest stars with the HARPS spectrograph were discussed in the context of the science potential of next generation instruments that will come online during the next decade, such as ESPRESSO on the VLT, or CODEX on the 42-m E-ELT. From space, the heritage of the great results obtained by the Hubble and Spitzer space telescopes (at visible and infrared wavelengths) on the characterization of the structural and atmospheric properties of extrasolar gas giants was shown to form the basis for the design of new challenging exoplanet characterization programs with the next generation of space observatories, such NASA's JWST and ESA's Gaia.

The community answered even more enthusiastically than we could hope for. The great interest in the Symposium can be easily quantified in terms of its sheer numbers: 12 invited review talks ("(IR)" in the table of contents), 27 invited talks ("(I)" in the table of contents), 39 oral contributions ("(C)" in the table of contents), and some 120 posters, whose authors had the opportunity to illustrate with 2-minute presentations within five dedicated daily poster popups sessions (a significant fraction of this volume is devoted to the poster papers, arranged according to their topic in a sequence echoing that of the oral sessions). Overall, the Symposium entertained 218 astronomers (of which 27% were female) from 27 countries. The enthusiasm and professionalism of the participants crucially helped in making IAUS 276 an overwhelming success.

The choice of Torino as the Symposium venue was deemed timely as the conference would fall during a particular period of large-scale celebrations: the Torino Astronomical Observatory concludeed in 2010 the celebrations of the 250th anniversary from its foundation, Torino was the 2010 European Capital of Science, hosting the Euroscience Open Forum, and significant preparations started in 2010 for the great celebrations of the 150th anniversary of the Unification of Italy in 2011 (Torino having been the first capital of unified Italy). While not being a famous beach or ski resort, Torino has been rediscovered in recent years as an important tourist destination. After hosting the 2006 Winter Olympics, and at the end of a 20-year long redevelopment plan which is unparalleled in Europe since the one carried out by the city Barcelona in the 1980's, Torino is today in the top ten of the most visited cities in Italy. It was a cause of major satisfaction to hear the impressions of many of the participants (and their accompanying guests), who confessed to having thoroughly enjoyed the unexpected beauties of Torino.

The success of IAUS 276 was not only *scientific*, but also *logistic*. The smoothness of all activities related to the Symposium, and the virtually non-existent organizational "glitches" is the result of the extraordinary joint efforts of a large number of people. The Torino Astronomical Observatory members of the Local Organizing Committee (LOC) worked very hard to make this conference both enjoyable and highgly memorable. Particular thanks to Maria Sarasso, Ummi Abbas, Tullia Carriero, Richard Smart, and Roberto Silvotti for their dedication in taking care of all organizational aspects with lucidity, calmness, and professionalism. An excellent team of people helped the LOC in coordinating the daily activities at the Torino Incontra Conference Center (Roberta Ghiringhelli,

Alessandra Quaranta, Maurizio Pesce, Deborah Busonero, Sebastiano Ligori, Alberto Riva). The schedule of the meeting was quite compressed, and the fact that we could always end the sessions in perfect time is particularly due to the professionalism and efficiency of Roberto Morbidelli (Torino Astronomical Observatory), Massimo d'Ambrosio and Marco Gonzatto (Torino Incontra), who chased speakers and poster presenters across the whole of the conference venue, made sure all presentations worked correctly, and ran smoothly all display operations from inside the slide room. Alessandro Spagna (Torino Astronomical Observatory) is to be warmly thanked for providing all the exquisite Symposium pictures (available at `http://iaus276.oato.inaf.it/IAUS_276/index.htm`), some of which are included in this volume. We gratefully acknowledge David Charbonneau, who failed to assign the prize (a large bottle of Canadian maple syrup of the highest quality) for the 100th transiting planet discovered during the Symposium, but generously left it behind for us to enjoy it thoroughly.

Finally, a special thought goes to Ummi, for always being there and providing crucial support at any time.

Extrasolar planets and the search for life in the Universe are topics of particular appeal for the general public. It was thus natural to offer, in parallel to the Symposium science activities, a strong public outreach program to the wider community. Two scientific lectures open to the public were scheduled at the Planetarium of Torino during the time of IAUS 276, delivered by prominent actors in the exoplanet arena (Dr. David W. Latham and Prof. Sara Seager). The public lecture given by Prof. David Charbonneau at the historic Gobetti Theater in downtown Torino proved a very successful means of dissemination of the latest hot results in the field to the greater public. In addition, a long reportage on the Symposium with interviews to David Charbonneau, Sara Seager, Bill Borucki, and the SOC Chair by Dr. Silvia Rosa Brusin was broadcasted on the public (RAI) national TV channels in October 2010, and more than 25 articles covering IAUS 276 appeared on local, national, and international news media during the same timeframe.

It is a great pleasure to acknowledge the patronage and generous financial contributions of the public and private sponsors listed on page *xvii* of these Proceedings. Their support made the idea of IAUS 276 come true. Essential travel sponsorship for young graduate students and early-stage researchers was generously provided by IAU.

Very special thanks go to the Regione Piemonte and Thales Alenia Space S.p.A. for providing the funds that made the realization of this volume possible.

The field of exoplanet science is now moving so fast that, just a few months after the end of IAUS 276, spectacular new discoveries are already looming on the horizon. This Proceedings volume serves two important purposes: It provides a detailed still picture of the state-of-the-art of the field fifteen years after the first discovery announcement, and with the breadth of its scope, it constitutes a tribute to the extraordinary diversity and dynamism of research in planetary systems astrophysics. As much as we enjoyed assembling it, we trust the readers will enjoy perusing this volume, and find the motivation and inspiration for the next Symposium on exoplanets astrophysics.

Alessandro Sozzetti (SOC Chair, Lead Editor), Mario G. Lattanzi (LOC Chair), and Alan P. Boss
Torino, Italy, and Washington D.C., USA, January 2011

THE ORGANIZING COMMITTEE

Scientific

Y. Alibert (Switzerland)
A. Boss (USA)
D. Charbonneau (USA)
D. Fischer (USA)
K. Gozdziewski (Poland)
R. Gratton (Italy)
T. Guillot (France)
J. Laskar (France)
D. Latham (USA)
T. Mazeh (Israel)
T. Michtchenko (Brazil)
R. Nelson (UK)
N. Santos (Portugal)
D. Queloz (Switzerland)
H. Rauer (Germany)
I. Ribas (Spain)
A. Sozzetti (Chair, Italy)
G. Tinetti (UK)
W. Traub (USA)
J.-L. Zhou (China)

Local

U. Abbas
T. Carriero
M.G. Lattanzi (Chair)
R. Morbidelli
M. Sarasso
R. Silvotti
R.L. Smart
A. Spagna

Acknowledgements

The Symposium was sponsored and supported by the International Astronomical Union (IAU), Division III (Planetary Systems Sciences), and by the IAU Commissions No. 51 (Bio-Astronomy), and No. 53 (Extrasolar Planets).

The Local Organizing Committee operated under the auspices of the Istituto Nazionale di Astrofisica, Osservatorio Astronomico di Torino.

Funding and patronage by the
International Astronomical Union,
Istituto Nazionale di Astrofisica,
Osservatorio Astronomico di Torino,
Università degli Studi di Torino,
Agenzia Spaziale Italiana (ASI),
Advanced Logistics Technology Engineering Center S.p.A. (ALTEC),
Thales Alenia Space S.p.A.,
Planetario di Torino (INFINI-TO),
Camera di Commercio di Torino,
UNESCO,
Città di Torino,
Provincia di Torino,
and
Regione Piemonte,
are gratefully acknowledged.

CONFERENCE PHOTOGRAPHS

Figure 1. IAUS 276 Group Picture in Piazzale Valdo Fusi, in front of the Symposium venue.

Figure 2. Top: the Scientific Organizing Committee. Bottom: the Local Organizing Committee and the SOC Chair.

Figure 3. Top: the participants attending the opening session of IAUS 276 in the Cavour Hall of the Torino Incontra Conference Center. Bottom: Professor Geoff Marcy addressing the audience during the opening review talk of the Symposium.

Figure 4. Top: Dr. David Latham coordinating the discussion time as Chairman of one of the sub-Sessions on transiting planets. Bottom: a bird's eye view of one of the poster popup sessions.

Figure 5. Top: interaction between participants was frequent at the posters viewing area, in which they were exposed for the duration of IAUS 276 in the Torino Hall. Bottom: participants had no trouble standing in line during coffee breaks in order to savor the highly praised *espresso*.

Figure 6. Top: Prof. David Charbonneau during his Public Lecture at the Gobetti Theater in downtown Torino. Bottom: the inside of the historic theater, once the Royal Savoy family's private entertainment place.

Figure 7. Top: a view of the conference dinner in the Senate Hall of Palazzo Madama. Bottom: SOC and LOC Chairs

Figure 8. Women in Astronomy: a significant fraction (27%) of the participants to IAUS 276 were female researchers at either an early or advanced stage of their careers. Women represented $\sim 40\%$ of the LOC, 20% of the SOC, $\sim 20\%$ of the (invited or contributing) speakers, $\sim 40\%$ of the poster presenters, and $\sim 40\%$ of the grant recipients.

Figure 9. Top: one of the groups of participants gathering at the entrance of the Venaria Reale royal residence, a World Heritage site just outside Torino. Bottom: a group of participants enjoying the Great Gallery, or 'Galleria di Diana'

Participants

Ummi **Abbas**, INAF - Astronomical Observatory of Torino, Italy abbas@oato.inaf.it
Yann **Alibert**, University of Bern, Switzerland alibert@space.unibe.ch
Roi **Alonso**, Geneva Observatory, Switzerland roi.alonso@unige.ch
Leonardo **Andrade de Almeida**, National Institute for Space Research, Brazil leonardo@das.inpe.br
Daniel **Angerhausen**, German SOFIA Institute, Germany daniel.angerhausen@gmail.com
Alberto **Anselmi**, Thales Alenia Space SpA, Torino, Italy Alberto.Anselmi@thalesaleniaspace.com
Serena **Arena**, CRAL/ENS, Lyon, France serena.arena@ens-lyon.fr
Paola **Ballerini**, INAF - Astronomical Observatory of Catania, Italy pballerini@astropa.unipa.it
Mauro **Barbieri**, INAF - Astronomical Observatory of Padova, Italy mauro.barbieri@oapd.inaf.it
Richard K. **Barry**, NASA-GSFC, Greenbelt MD, USA richard.k.barry@nasa.gov
Jean-Philippe **Beaulieu**, Institut d'Astrophysique de Paris, France beaulieu@iap.fr
Andreas **Becker**, University of Rostock, Germany andreas.becker@uni-rostock.de
Bjoern **Benneke**, MIT MA, USA bbenneke@mit.edu
David **Bennett**, University of Notre Dame IN, USA bennett@nd.edu
Willy **Benz**, University of Bern, Switzerland wbenz@space.unibe.ch
Carolina **Bergfors**, Max Planck Institute for Astronomy, Germany bergfors@mpia.de
Andrea Ettore **Bernagozzi**, Astron. Obs. of the Aosta Valley, Italy andrea.bernagozzi@gmail.com
Enzo **Bertolini**, Astronomical Observatory of the Aosta Valley, Italy direttore@oavda.it
Bertram **Bitsch**, Universität Tübingen, Germany bertram.bitsch@uni-tuebingen.de
Isabelle **Boisse**, Institut d'Astrophysique de Paris, France iboisse@iap.fr
Aaron **Boley**, University of Florida, Gainesville FL, USA aaron.boley@gmail.com
Emeline **Bolmont**, Laboratoire d'Astrophysique de Bordeaux, France bolmont@obs.u-bordeaux1.fr
Mariangela **Bonavita**, University of Toronto, Canada bonavita@utoronto.ca
Aldo Stefano **Bonomo**, Laboratoire d'Astrophysique de Marseille, France aldo.bonomo@oamp.fr
William **Borucki**, NASA Ames Research Center CA, USA William.J.Borucki@nasa.gov
Alan **Boss**, Carnegie Institution of Washington DC, USA boss@dtm.ciw.edu
David **Brown**, University of St Andrews, UK djab@st-andrews.ac.uk
Joanna **Bulger**, University of Exeter, UK joanna@astro.ex.ac.uk
Adam **Burgasser**, UCSD, La Jolla CA, USA aburgasser@ucsd.edu
Deborah **Busonero**, INAF - Astronomical Observatory of Torino, Italy busonero@oato.inaf.it
Susana Cristina **Cabral de Barros**, Queen's University Belfast, UK s.barros@qub.ac.uk
Juan **Cabrera**, DLR German Aerospace Center, Germany juan.cabrera@dlr.de
Elena **Carolo**, INAF - Astronomical Observatory of Padova, Italy elena.carolo@oapd.inaf.it
Ludmila **Carone**, University of Köln, Germany ludmila.carone@uni-koeln.de
Tullia **Carriero**, INAF - Astronomical Observatory of Torino, Italy carriero@oato.inaf.it
Marco **Castronuovo**, ASI Headquarters, Roma, Italy marco.castronuovo@asi.it
Alberto **Cellino**, INAF - Astronomical Observatory of Torino, Italy cellino@oato.inaf.it
Stefano **Cesare**, Thales Alenia Space SpA, Torino, Italy Stefano.Cesare@thalesaleniaspace.com
Gilles **Chabrier**, CRAL/ENS, Lyon, France chabrier@ens-lyon.fr
David **Charbonneau**, Harvard-Smithsonian CfA MA, USA dcharbon@cfa.harvard.edu
Sourav **Chatterjee**, Northwestern University, IL, USA souravchatterjee2010@u.northwestern.edu
Armando **Ciampolini**, ALTEC SpA, Torino, Italy armando.ciampolini@altecspace.it
Andrew **Collier Cameron**, University of St Andrews, UK acc4@st-and.ac.uk
Alexandre **Correia**, Universidade de Aveiro, Portugal correia@ua.pt
Vincent **Coudé du Foresto**, LESIA, Paris Observatory, France vincent.foresto@obspm.fr
Elisabeth **Crespe**, CRAL/ENS, Lyon, France elisabeth.crespe@ens-lyon.fr
Bryce **Croll**, University of Toronto, Canada croll@astro.utoronto.ca
Szilard **Csizmadia**, DLR German Aerospace Center, Germany szilard.csizmadia@dlr.de
Mario **Damasso**, Astronomical Observatory of the Aosta Valley, Italy mario.damasso@studenti.unipd.it
Cilia **Damiani**, INAF - Astronomical Observatory of Catania, Italy damiani@oact.inaf.it
Melvyn **Davies**, Lund Observatory, Sweden mbd@astro.lu.se
Ernst **de Mooij**, Leiden Observatory, Netherlands demooij@strw.leidenuniv.nl
Magali **Deleuil**, Laboratoire d'Astrophysique de Marseille, France magali.deleuil@oamp.fr
Elisa **Delgado Mena**, Instituto de Astrofisica de Canarias, Spain edm@iac.es
Annalisa **Deliperi**, INAF - Astronomical Observatory of Torino, Italy deliperi@oato.inaf.it
Brice-Olivier **Demory**, MIT, MD, USA demory@mit.edu
Xavier **Dumusque**, Geneva Observatory, Switzerland xavier.dumusque@unige.ch
Natalia **Dzyurkevich**, Max-Planck Institute for Astronomy, Germany natalia@mpia.de
Jason **Eastman**, The Ohio State University, OH, USA jdeast@astronomy.ohio-state.edu
Anne **Eggenberger**, Grenoble Observatory, France anne.eggenberger@obs.ujf-grenoble.fr
Carlos **Eiroa**, Universidad Autónoma de Madrid, Spain carlos.eiroa@uam.es
Sebastian **Elser**, University of Zurich, Switzerland sebastian.elser@uzh.ch
Becky **Enoch**, University of St Andrews, UK bel2@st-andrews.ac.uk
Anders **Erikson**, DLR Institute of Planetary Research, Germany anders.erikson@dlr.de
Daniel **Fabrycky**, University of California, Santa Cruz, CA, USA daniel.fabrycky@gmail.com
Francesca **Faedi**, Queen's University Belfast, UK f.faedi@qub.ac.uk
Fabio **Favata**, ESA, Netherlands Fabio.Favata@rssd.esa.int
Mario **Flock**, Max-Planck Institute for Astronomy, Germany flock@mpia.de
Eric **Ford**, University of Florida, Gainesville, FL, USA eford@astro.ufl.edu
Andrea **Fortier**, University of Bern, Switzerland andrea.fortier@space.unibe.ch
Luca **Fossati**, The Open University, UK l.fossati@open.ac.uk
Richard **Freedman**, SETI Institute, NASA Ames Research Center, CA, USA freedman@darkstar.arc.nasa.gov
Misato **Fukagawa**, Osaka University, Japan misato@iral.ess.sci.osaka-u.ac.jp
Pavel **Gabor**, Vatican Observatory, Vatican City p.gabor@jesuit.cz
Mario **Gai**, INAF - Astronomical Observatory of Torino, Italy gai@oato.inaf.it
Marina **Galvagni**, University of Zurich, Switzerland galva@physik.uzh.ch
Daniele **Gardiol**, INAF - Astronomical Observatory of Torino, Italy gardiol@oato.inaf.it
Nikolaos **Georgakarakos**, ATEI of Western Macedonia, Greece georgakarakos@hotmail.com
Paolo **Giacobbe**, University of Trieste, Italy paologiacobbe85@gmail.com
Neale **Gibson**, University of Oxford, UK Neale.Gibson@astro.ox.ac.uk

Michaël **Gillon**, University of Liège, Belgium michael.gillon@ulg.ac.be
Roberto **Gilmozzi**, ESO, Germany Roberto.Gilmozzi@eso.org
Vincenzo **Giorgio**, Thales Alenia Space SpA, Torino, Italy Vincenzo.Giorgio@thalesaleniaspace.com
Jonay I. **Gonzalez Hernandez**, Instituto de Astrofisica de Canarias, Spain jonay@iac.es
Raffaele **Gratton**, INAF - Astronomical Observatory of Padova, Italy raffaele.gratton@oapd.inaf.it
John Lee **Grenfell**, Technische Universität, Berlin, Germany lee.grenfell@dlr.de
Olivier **Gressel**, Queen Mary University, London, UK o.gressel@qmul.ac.uk
Sheng-hong **Gu**, Yunnan Astronomical Observatory, China shenghonggu@ynao.ac.cn
Octavio Miguel **Guilera**, Universidad Nacional de La Plata, Argentina oguilera@fcaglp.unlp.edu.ar
Tristan **Guillot**, Observatoire de la Cote d'Azur, France tristan.guillot@oca.eu
Nader **Haghighipour**, University of Hawaii, HI, USA nader@ifa.hawaii.edu
Yasuhiro **Hasegawa**, McMaster University, Canada hasegay@physics.mcmaster.ca
Artie **Hatzes**, Thueringerr Landessternwarte Tautenburg, Germany artie@tls-tautenburg.de
Mathieu **Havel**, Observatoire de la Cote d'Azur, France mathieu.havel@oca.eu
Ravit **Helled**, UCLA, CA, USA rhelled@ucla.edu
Coel **Hellier**, Keele University, UK ch@astro.keele.ac.uk
Teruyuki **Hirano**, University of Tokyo, Japan hirano@utap.phys.s.u-tokyo.ac.jp
Matthew **Holman**, Harvard-Smithsonian CfA, MA, USA mholman@cfa.harvard.edu
Douglas **Hudgins**, NASA Headquarters, Washington DC, USA Douglas.M.Hudgins@nasa.gov
Nawal **Husnoo**, University of Exeter, UK nawal@astro.ex.ac.uk
Shigeru **Ida**, Tokyo Institute of Technology, Japan ida@geo.titech.ac.jp
Ray **Jayawardhana**, University of Toronto, Canada rayjay@astro.utoronto.ca
Sheng **Jin**, Purple Mountain Observatory, China qingxiaojin@gmail.com
Anders **Johansen**, Lund Observatory, Sweden anders@astro.lu.se
Paul **Kalas**, University of California, Berkeley CA, USA kalas@berkeley.edu
Lisa **Kaltenegger**, Max-Planck Institute for Astronomy, Germany lkaltene@cfa.harvard.edu
David **Kirsh**, McMaster University, Canada kirshdr@mcmaster.ca
Laszlo **Kiss**, Konkoly Observatory, Hungary kiss@konkoly.hu
Rainer **Klement**, Max-Planck Institute for Astronomy, Germany klement@mpia.de
Wilhelm **Kley**, Universität Tübingen, Germany wilhelm.kley@uni-tuebingen.de
Ludwik **Kostro**, University of Gdansk, Poland fizlk@univ.gda.pl
Ulrike **Kramm**, University of Rostock, Germany ulrike.kramm2@uni-rostock.de
Nobuhiko **Kusakabe**, National Astronomical Observatory of Japan, Japan nb.kusakabe@nao.ac.jp
Pierre-Olivier **Lagage**, CEA-IRFU, France pierre-olivier.lagage@cea.fr
Anne-Marie **Lagrange**, Laboratoire d'Astroph. de Grenoble, France lagrange@obs.ujf-grenoble.fr
Dong **Lai**, Cornell University, NY, USA dong@astro.cornell.edu
David **Latham**, Harvard-Smithsonian CfA, MA, USA dlatham@cfa.harvard.edu
Mario G. **Lattanzi**, INAF - Astronomical Observatory of Torino, Italy lattanzi@oato.inaf.it
Jeremy **Leconte**, CRAL/ENS, Lyon, France jeremy.leconte@ens-lyon.fr
Alain **Léger**, IAS, Paris, France alain.leger@ias.fr
Monika **Lendl**, Geneva Observatory, Switzerland monika.lendl@unige.ch
Sebastiano **Ligori**, INAF - Astronomical Observatory of Torino, Italy ligori@oato.inaf.it
Giuseppe **Lodato**, University of Milano, Italy giuseppe.lodato@unimi.it
Christophe **Lovis**, Geneva Observatory, Switzerland christophe.lovis@unige.ch
Wladimir **Lyra**, American Museum of Natural History, NY, USA wlyra@amnh.org
Avi M. **Mandell**, NASA-GSFC, Greenbelt, MD, USA Avi.Mandell@nasa.gov
Geoffrey W. **Marcy**, University of California, Berkeley CA, USA bhovers@astro.berkeley.edu
Rosemary **Mardling**, Monash University, Australia mardling@sci.monash.edu.au
Michele **Martino**, ALTEC SpA, Torino, Italy michele.martino@altecspace.it
Soko **Matsumura**, University of Maryland, MD, USA soko@astro.umd.edu
Anne-Sophie **Maurin**, Laboratoire d'Astrophysique de Bordeaux, France maurin@obs.u-bordeaux1.fr
Satoshi **Mayama**, Graduate Univ. for Advanced Studies, Japan mayama_satoshi@soken.ac.jp
Tsevi **Mazeh**, Wise Observatory, Tel Aviv University, Israel mazeh@post.tau.ac.il
Farzana **Meru**, University of Exeter, UK farzana@astro.ex.ac.uk
Yamila **Miguel**, Universidad Nacional de La Plata, Argentina ymiguel@fcaglp.unlp.edu.ar
Eliza **Miller-Ricci Kempton**, University of California, Santa Cruz, CA, USA elizamr@ucolick.org
Roberto **Morbidelli**, INAF - Astronomical Observatory of Torino, Italy morbidelli@oato.inaf.it
Christoph **Mordasini**, Max Planck Institute for Astronomy, Germany mordasini@mpia.de
Amaya **Moro-Martin**, Centro de Astrobiologia (CSIC-INTA), Spain amaya@cab.inta-csic.es
Andres **Moya**, Centro de Astrobiologia (CSIC-INTA), Spain amoya@cab.inta-csic.es
Alexander **Mustill**, University of Cambridge, UK ajm233@ast.cam.ac.uk
Smadar **Naoz**, Northwestern University, IL, USA snaoz@northwestern.edu
Norio **Narita**, National Astronomical Observatory of Japan, Japan norio.narita@nao.ac.jp
Marie-Eve **Naud**, University of Montreal, Canada naud@astro.umontreal.ca
Sergei **Nayakshin**, University of Leicester, UK sn85@astro.le.ac.uk
Richard **Nelson**, Queen Mary University, London, UK R.P.Nelson@qmul.ac.uk
Vasco **Neves**, Universidade do Porto, Portugal vasco.neves@astro.up.pt
Andrzej **Niedzielski**, N. Copernicus University, Torun, Poland aniedzi@astri.uni.torun.pl
Anna **Nobili**, University of Pisa, Italy nobili@dm.unipi.it
Aake **Nordlund**, Niels Bohr Institute, Copenhagen, Denmark aake@nbi.dk
Claudia **Orlando**, Liceo Scientifico "Leonardo da Vinci", Pescara, Italy orlando.jetlag@gmail.com
Mahmoudreza **Oshagh**, Universidade do Porto, Portugal moshagh@astro.up.pt
Fabio **Pagan**, SISSA, Trieste, Italy pagan@sissa.it
Enric **Palle**, Instituto de Astrofisica de Canarias, Spain epalle@iac.es
Olja **Panic'**, ESO, Germany opanic@eso.org
Neil **Parley**, University of St Andrews, UK neil.parley@st-andrews.ac.uk
Karla **Peña Ramírez**, Instituto de Astrofsica de Canarias, Spain karla@iac.es
Francesco **Pepe**, Geneva Observatory, Switzerland Francesco.Pepe@unige.ch
Hagai **Perets**, Harvard-Smithsonian CfA, MA, USA hperets@cfa.harvard.edu
Giovanni **Picogna**, University of Padova, Italy giovanni.picogna@studenti.unipd.it
Elke **Pilat-Lohinger**, University of Wien, Austria elke.pilat-lohinger@univie.ac.at
Ennio **Poretti**, INAF - Astronomical Observatory of Brera, Italy ennio.poretti@brera.inaf.it

Loredana **Prisinzano**, INAF - Astron. Obs. of Palermo, Italy loredana@astropa.inaf.it
Didier **Queloz**, Geneva Observatory, Switzerland didier.queloz@unige.ch
Andreas **Quirrenbach**, Heidelber University, Germany A.Quirrenbach@lsw.uni-heidelberg.de
Natalie **Raettig**, Max-Planck Institute for Astronomy, Germany raettig@mpia.de
Heike **Rauer**, DLR Institute of Planetary Research, Germany heike.rauer@dlr.de
Sean **Raymond**, Laboratoire d'Astrophysique de Bordeaux, France rayray.sean@gmail.com
Zsolt **Regaly**, Konkoly Observatory, Hungary regaly@konkoly.hu
Martin **Reidemeister**, AIU Jena, Germany martin.reidemeister@astro.uni-jena.de
Alberto **Riva**, INAF - Astronomical Observatory of Torino, Italy riva@oato.inaf.it
Adrian **Rodríguez Colucci**, Universidade de São Paulo, Brazil adrian@astro.iag.usp.br
Leslie **Rogers**, MIT, MA, USA larogers@mit.edu
Cristoforo **Romanelli**, ALTEC SpA, Torino, Italy cristoforo.romanelli@altecspace.it
Johannes **Sahlmann**, Geneva Observatory, Switzerland Johannes.Sahlmann@unige.ch
Roberto **Sanchis-Ojeda**, MIT, MA, USA rsanchis@mit.edu
Esther **Sanroma Ramos**, Instituto de Astrofisica de Canarias, Spain mesr@iac.es
Alexandre **Santerne**, LAM, France alexandre.santerne@oamp.fr
Maria **Sarasso**, INAF - Astronomical Observatory of Torino, Italy sarasso@oato.inaf.it
Sara **Seager**, MIT, MA, USA seager@MIT.EDU
Damien **Ségransan**, Geneva Observatory, Switzerland Damien.Segransan@unige.ch
Franck **Selsis**, Laboratoire d'Astroph. de Bordeaux, France franck.selsis@obs.u-bordeaux1.fr
Eugene **Serabyn**, JPL, CA, USA gene.serabyn@jpl.nasa.gov
Johny **Setiawan**, Max-Planck Institute for Astronomy, Germany setiawan@mpia.de
Michael **Shao**, JPL, CA, USA michael.shao@jpl.nasa.gov
Avi **Shporer**, LCOGT, University of California, Santa Barbara, CA, USA ashporer@lcogt.net
Roberto **Silvotti**, INAF - Astronomical Observatory of Torino, Italy silvotti@oato.inaf.it
Richard **Smart**, INAF - Astronomical Observatory of Torino, Italy smart@oato.inaf.it
Ignas **Snellen**, Leiden Observatory, Netherlands snellen@strw.leidenuniv.nl
Frank **Sohl**, DLR Institute of Planetary Research, Germany frank.sohl@dlr.de
Filomena **Solitro**, ALTEC SpA, Torino, Italy filomena.solitro@altecspace.it
Alessandro **Sozzetti**, INAF - Astronomical Observatory of Torino, Italy sozzetti@oato.inaf.it
Alessandro **Spagna**, INAF - Astronomical Observatory of Torino, Italy spagna@oato.inaf.it
Vlada **Stamenkovic**, DLR, Germany Vlada.Stamenkovic@dlr.de
Rachel **Street**, Las Cumbres Observatory, CA, USA rstreet@lcogt.net
Mark **Swain**, JPL, CA, USA swain@s383.jpl.nasa.gov
Yuhei **Takagi**, Kobe University, Japan takagi@stu.kobe-u.ac.jp
Yasuhiro **Takahashi**, Grad. Univ. for Advanced Studies, Japan yasuhiro.takahashi@nao.ac.jp
Stuart F. **Taylor**, National Tsing Hua University, Taiwan astrostuart@gmail.com
Giovanna **Tinetti**, University College, London, UK g.tinetti@ucl.ac.uk
Wesley **Traub**, JPL, CA, USA wtraub@jpl.nasa.gov
Amaury **Triaud**, Geneva Observatory, Switzerland Amaury.Triaud@unige.ch
Stephane **Udry**, Geneva Observatory, Switzerland stephane.udry@unige.ch
Ana L. **Uribe**, Max Planck Institute for Astronomy, Germany uribe@mpia.de
Diana **Valencia**, Observatoire de la Cote d'Azur, France valencia@oca.eu
Sylvie **Vauclair**, LATT/OMP, France sylvie.vauclair@ast.obs-mip.fr
Allona **Vazan Shukrun**, Tel Aviv University, Israel allonava@post.tau.ac.il
Ernesto **Vittone**, ALTEC SpA, Torino, Italy ernesto.vittone@altecspace.it
Eduard **Vorobyov**, Southern Federal University, Russian Federation vorobyov@astro.uwo.ca
Frank W. **Wagner**, DLR Institute of Planetary Research, Germany frank.wagner@dlr.de
Xiao-bin **Wang**, Yunnan Astronomical Observatory, China wangxb@ynao.ac.cn
Joshua **Winn**, MIT, MA, USA jwinn@mit.edu
Paul **Withers**, Boston University, MA, USA withers@bu.edu
Patricia **Wood**, Keele University, UK p.wood@epsam.keele.ac.uk
Günther **Wuchterl**, Thüringer Landessternwarte, Germany gwuchterl@TLS-Tautenburg.de
Chao-Chin **Yang**, American Museum of Natural History, NY, USA cyang@amnh.org
Olga **Zakhozhay**, Main Astronomical Observatory NAS, Kyiv, Ukraine zkholga@mail.ru
Maria Rosa **Zapatero Osorio**, CSIC-INTA, Spain mosorio@cab.inta-csic.es

PART 1:
PLANET FORMATION

The Astrophysics of Planetary Systems: Formation, Structure, and Dynamical Evolution
Proceedings IAU Symposium No. 276, 2010
A. Sozzetti, M. G. Lattanzi, & A. P. Boss, eds.

doi:10.1017/S1743921311019867

The occurrence and the distribution of masses and radii of exoplanets

Geoffrey W. Marcy[1], Andrew W. Howard[1] and the *Kepler* Team

[1]Astronomy Dept., University of California, Berkeley
Berkeley, CA, USA, 94720 email: gmarcy@berkeley.edu and howard@astro.berkeley.edu

Abstract. We analyze the statistics of Doppler-detected planets and *Kepler*-detected planet candidates of high integrity. We determine the number of planets per star as a function of planet mass, radius, and orbital period, and the occurrence of planets as a function of stellar mass. We consider only orbital periods less than 50 days around Solar-type (GK) stars, for which both Doppler and *Kepler* offer good completeness. We account for observational detection effects to determine the actual number of planets per star. From Doppler-detected planets discovered in a survey of 166 nearby G and K main sequence stars we find a planet occurrence of 15^{+5}_{-4}% for planets with $M \sin i$ = 3–30 M_E and $P < 50$ d, as described in Howard *et al.* (2010). From *Kepler*, the planet occurrence is 0.130 ± 0.008, 0.023 ± 0.003, and 0.013 ± 0.002 planets per star for planets with radii 2–4, 4–8, and 8–32 R_E, consistent with Doppler-detected planets. From *Kepler*, the number of planets per star as a function of planet radius is given by a power law, $df/d\log R = k_R R^{\alpha}$ with $k_R = 2.9^{+0.5}_{-0.4}$, $\alpha = -1.92 \pm 0.11$, and $R = \mathrm{R}_P/R_E$. Neither the Doppler-detected planets nor the *Kepler*-detected planets exhibit a "desert" at super-Earth and Neptune sizes for close-in orbits, as suggested by some planet population synthesis models. The distribution of planets with orbital period, P, shows a gentle increase in occurrence with orbital period in the range 2–50 d. The occurrence of small, 2–4 R_E planets increases with decreasing stellar mass, with seven times more planets around low mass dwarfs (3600–4100 K) than around massive stars (6600–7100 K).

Keywords. planetary systems: formation, stars: statistics

---

1. Introduction

The occurrence of gas-giant exoplanets has been quantitatively studied from careful counting of planets detected by the Doppler technique within well-defined samples of stars. Cumming *et al.* (2008) found that 10.5% of Solar-type stars host a gas-giant planet in the mass range, $M \sin i$ = 100–3000 M_E, and orbital period range, P = 2 d – 5.5 yr. These gas-giants occur around solar-type stars with a frequency that depends on planet mass and orbital period as, $df \propto M^{-0.31\pm0.2} P^{0.26\pm0.1}\, d\log M\, d\log P$. The number of planets rises with smaller masses and larger orbital distances (in logarithmic intervals). This distribution as a function of planet mass and orbital-period reveals important information about planet formation and migration as shown by Ida & Lin (2010), Mordasini *et al.* (2011), Raymond *et al.* (2011), Bromley & Kenyon (2011) and Wittenmyer *et al.* (2011). Similarly, the clear dependence of planet occurrence on stellar mass and metallicity, shown for example by Johnson *et al.* (2010), is consistent with formation by rocky-core-nucleated accretion of H and He gas in a protoplanetary disk as shown by Ida & Lin (2008b), Mordasini *et al.* (2011) and Alibert *et al.* (2011).

Here we present two recent studies of the planet occurrence for smaller planets that have masses and radii less than those of Saturn. We first consider the small planets discovered by the Doppler technique, which is sensitive to planets having masses as low

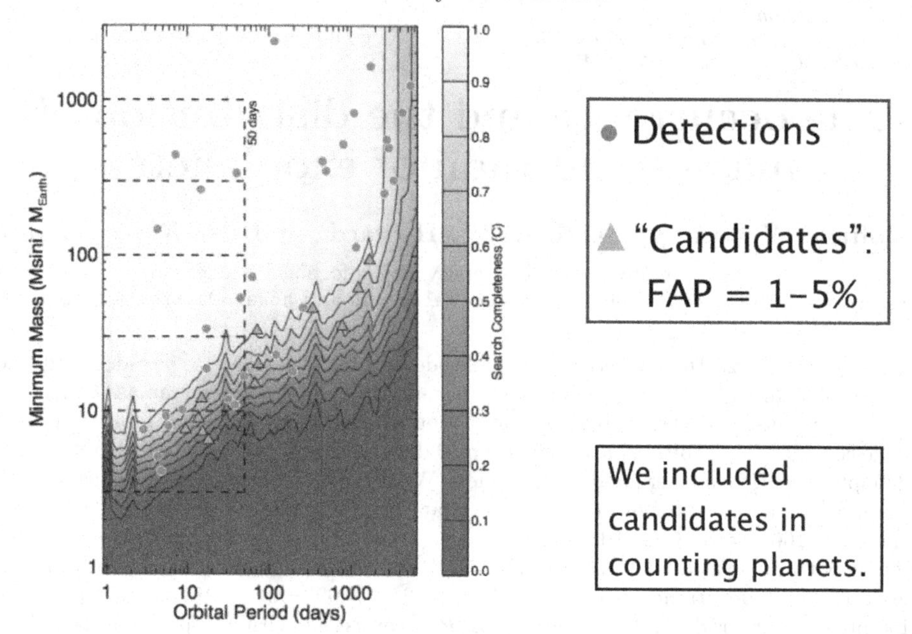

Figure 1. Doppler-detected planets (green circles) and candidate planets (orange triangles) from the survey by Howard *et al.* (2010), in a two-parameter space of orbital period and minimum mass. We consider five mass domains of 3–10, 10–30, 30–100, 100–300, 300-1000 Earth-masses, and orbital periods less than 50 days, as marked by dashed lines. The fraction of stars with sufficient measurements to rule out planets in circular orbits of a given minimum mass and orbital period is shown as blue contours from 0.0 to 1.0 in steps of 0.1. The white regions have 100% detectability and the blue regions have low detectability. Our occurrence statistic corrects for this detectability in each domain, described in Howard *et al.* (2010).

as a few Earth masses for close-in orbits with periods under 50 days, as described by Howard *et al.* (2010). We also carry out an analysis of the epochal *Kepler* results for transiting planet candidates from Borucki *et al.* (2011) with a careful treatment of the completeness. We focus attention on the planets with orbital periods less than 50 days to match the period range for which the Doppler surveys are robust, as described by Howard *et al.* (2011b). Here, we review planet occurrence as a function of planet masses and radii, restricting our attention to planets within 0.25 AU of G and K-type main sequence stars, i.e. solar-type stars. We also consider planet occurrence as a function of stellar mass.

2. The Occurrence of Small Planets from Doppler Surveys

We have conducted a sensitive Doppler survey for planets during the past five years, beginning with a blind sample of 166 G & K-type main sequence stars that are chromospherically quiet. From this sample we detected many planets from our Doppler measurements, and some planets were found by others. We have included all planets here, to be complete. In Fig. 1 we show all of the detected planets in this survey by Howard *et al.* (2010). The Doppler-detected planets are shown as green circles, and those with a false-alarm probability 1-5% are deemed "candidate planets" and shown as orange triangles. We exhibit all planet detections in a two-parameter space of orbital period and minimum mass. We divide this space into five domains of minimum planet mass ($M \sin i$) of 3–10, 10–30, 30–100, 100–300, 300-1000 Earth-masses. We only consider orbital periods less

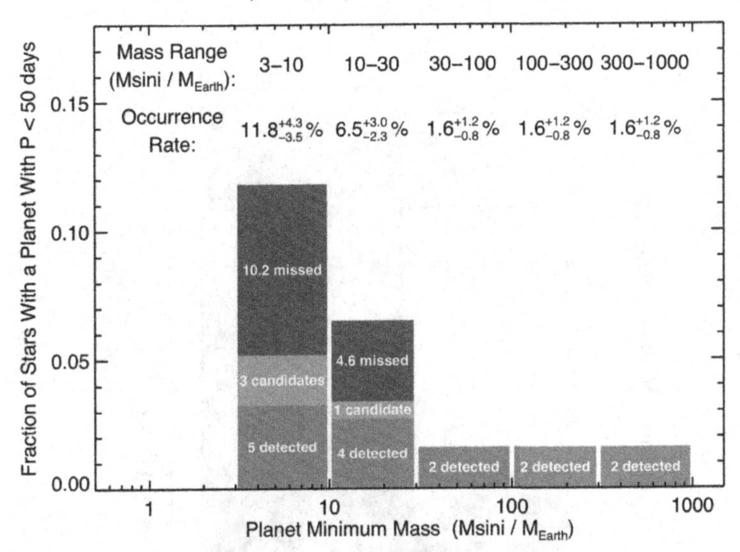

Figure 2. The fraction of solar-type stars with a planet as a function of planet (minimum) mass, Msin*i*. Only periods less than 50 d are included. The detected (green), candidate (orange), and missed (blue) planet are shown separately. Missed planets represents the correction for detectability due to inadequate sensitivity, based on those stars that did have adequate sensitivity. (From Howard *et al.* (2010).)

than 50 days for which Doppler detection remains strong down to masses of 3 M_E, as marked by dashed lines. The fraction of stars with sufficient measurements to rule out planets in circular orbits of a given minimum mass and orbital period is shown as blue contours from 0.0 to 1.0 in steps of 0.1. The white regions have 100% detectability of planets and the blue regions have low detectability. Our occurrence statistic corrects for this detectability in each domain.

From the knowledge of the 166 GK stars in the original Doppler survey, one may compute the planet occurrence. We exhibit the planet occurrence as a function of minimum-planet mass ($M \sin i$) in Fig. 2. The occurrence of planets rises rapidly toward smaller masses. For planets having minimum masses 3–10 M_E, the occurrence is 11.8%, and for planets of minimum masses 10–30 M_E, the occurrence is 6.5%. These two mass bins give the occurrence of "super-Earths" and "exo-Neptunes" for periods less than 50 days around solar type stars, a total of 18.3%.

In Fig. 3 we exhibit the Doppler-detected planets again, but this time overplotted with the planets predicted from population syntheses. The small black dots represent those predicted by the theory of Mordasini *et al.* (2009) for which the results are similar to those from Ida & Lin (2010). We note that the observed planet distribution differs from those of theory in two ways. The predicted planet desert from 3 to 10 M_E is in fact well populated with planets. The desert doesn't actually exist. And the predicted uniform distribution of planets with mass from 100 down to 10 M_E actually is populated by an increasing number of planets with decreasing mass. The observed increase in planet number with decreasing mass was not predicted by the simulations. Thus, there seems to be a missing physics in either the planet formation or migration included in the simulations.

3. The Occurrence of Small Planets from *Kepler*

From all 1235 planet candidates announced in Borucki *et al.* (2011), we consider only a subset of target stars and the associated planets. We consider only stars having surface

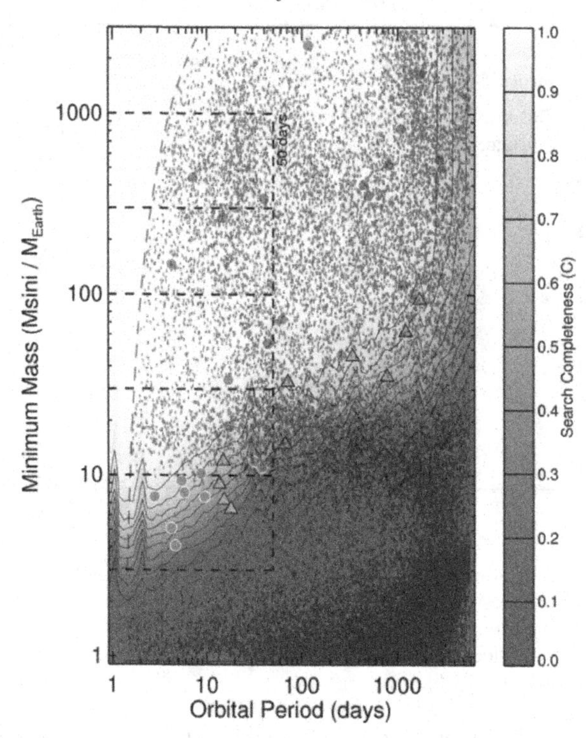

Figure 3. Same as Figure 1, but overplotted is the synthetic population of planets from Mordasini *et al.* (2009) and Ida & Lin (2010). The predicted "desert" of planets from 3-10 Earth–masses and P <50 d from theory is quite well populated with actual planets. Thus the theory of planet formation and migration is missing a key ingredient that actually populates the desert as noted by Howard *et al.* (2010).

temparature $T_{\rm eff}$= 4100–6100 K, surface gravity $\log g$= 4.0–4.9, and *Kepler* magnitude <15 mag. This restriction reduces the number of *Kepler* target stars under consideration to ~58,000. We construct a two-dimensional space of orbital period and planet radius, as shown in Fig. 4. We divide this space into small cells of specified increments in orbital period and planet radius, and we carefully determine the subset of target stars for which the transit depths of planets in that domain would have a signal-to-noise ratio, SNR >10. In that way, each domain of orbital period and planet size (or mass) has its own subsample of target stars (typically 58,000 or less) that are selected *a priori*, within which the detected planets can be counted and compared to that number of stars. The occurrence of planets within each cell is a simple ratio. In the numerator is the number of planets detected in that cell, with each planet multiplied by the correction for orbital inclination, $a/R_{\rm Star}$, to account for inclined, non-transiting orbits. In the denominator is the total number of stars for which such a detection would have been possible with SNR >10.

Thus, planet occurrence is simply the number of detected planets having some set of properties (radius and orbital periods) compared to the set of stars from which planets with those properties could have been reliably detected. We include only planet candidates found in three *Kepler* data segments ("Quarters") labeled Q0, Q1, and Q2, for which all photometry is published in Borucki *et al.* (2011). Planet radii stem from stellar radii which are estimated from $T_{\rm eff}$ and $\log g$ and carry an uncertainty of 35% rms as measured by Brown *et al.* (2011).

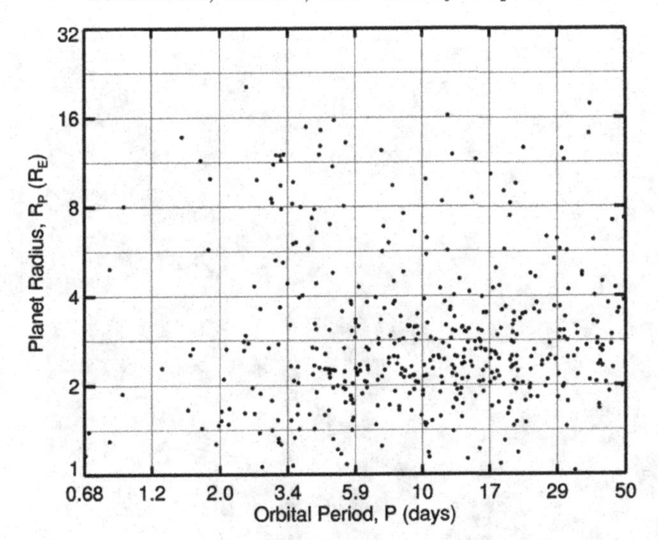

Figure 4. *Kepler* planet candidates plotted in a two-parameter space of orbital period and planet radius from Borucki *et al.* (2011). The space is divided into a grid of equal logarithmically spaced cells within which planet occurrence is computed. The planet candidates detected by *Kepler* having SNR > 10 are shown as black dots. We compute planet occurrence within each cell individually. In each cell there are a certain number of *Kepler* target stars with low enough noise to permit detection of that cell's planets with SNR > 10. The planet occurrence is the number of detected planets, multiplied by factor that accounts for the missed planets due to inclination, namely $a/R_{\rm Star}$, divided by the number of stars amenable to such detection. The method is described in detail in Howard *et al.* (2011b).

The determination of planet occurrence is carried out *only* among those stars having photometric quality so high that the transit signals stand out easily. We adopt the metric of the signal-to-noise ratio (SNR) of the transit signal integrated over a 90 day photometric time series, setting a threshold, SNR > 10. Fig. 5 shows the parameter space again, but with the number of surviving target stars written in white within each cell. We restrict our study to orbital periods under 50 days.

We adopt the *Kepler* planet candidates and their orbital periods and planet radii from Table 2 of Borucki *et al.* (2011). Morton & Johnson (2011) note that the false positive probability depends on transit depth, galactic latitude, and *Kepler* magnitude. Using their model we estimate that 22 planet candidates are actually false positives. The resulting false positive rate is 5–10% for planet radii, $R_P > 2\ R_E$.

The above description defines planet occurrence, f, as the number of planets detected within a cell in Fig. 5 augmented by the factor that accounts for all orbital inclinations, $a/R_{\rm Star}$, divided by the number of a target stars within that same cell in Fig. 5. The planet occurrence within a cell is given by

$$f_{\rm cell} = \sum_{j=1}^{n_{\rm pl,cell}} \frac{1/p_j}{n_{\star,j}}, \tag{3.1}$$

where the sum is over all detected planets within the cell that have SNR > 10. In the numerator, $p_j = (R_\star/a)_j$ is the probability of a transiting orientation of the orbital plane for planet j. Thus each detected planet is augmented in its contribution to the planet count by a factor of $a/R_\star$ to account for the number of planets with similar radii and periods that are not detected because of non-transiting geometries. For each planet, its specific value of $(a/R_\star)_j$ is used, not the average $a/R_\star$ of the cell in which it resides.

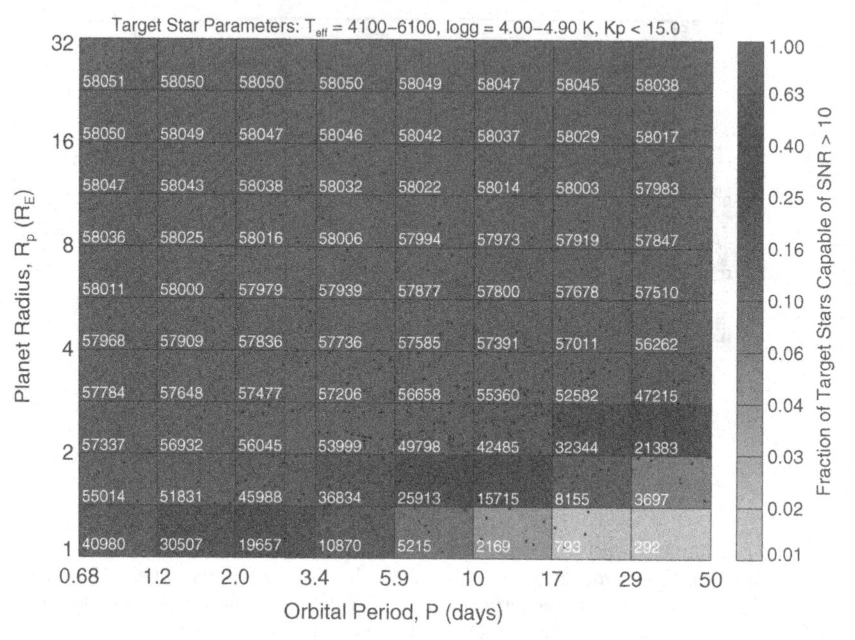

Figure 5. The two-parameter space of orbital period and planet radius. The white number in each cell is the number of *Kepler* target stars under consideration within the cell. The *Kepler* target stars must satisfy these criteria: $T_{\rm eff} = 4100$–6100 K, $\log g = 4.0$–4.9, *Kepler* magnitude <15. Also we used the observed photometric noise to include only those stars quiet enough so that planets in that cell would be detected with SNR > 10. The number of stars shown in each cell (typically $\sim$58,000 for the "red" cells) is the denominator of the planet occurrence calculation. The color code indicates the fraction of target stars capable of achieving SNR >10, indicating low detectability for small radii and long periods (fewer transits). (See Howard *et al.* (2011b).)

Note that each scaled semi-major axis $(a/R_\star)_j$ is measured directly from *Kepler* photometry and is not the ratio of two quantities, a_j and $R_{\star,j}$, separately measured with lower precision. In the denominator, $n_{\star,j}$ is the number of stars for which a planet of radius $R_{\mathrm{p},j}$ and period P_j would have been detected with SNR > 10.

We computed planet occurrence as a function of planet radius by integrating the planet occurrence over all orbital periods with $P < 50$ days. The resulting distribution of occurrence with planet radius is shown in Fig. 6, which shows a clear increase with decreasing planet radius. Smaller planets are far more numerous than large planets, for solar-type stars and periods less than 50 d.

We fit the data with a power law, finding:

$$\frac{\mathrm{d}f(R)}{\mathrm{d}\log R} = k_R R^\alpha . \tag{3.2}$$

Here $\mathrm{d}f(R)/\mathrm{d}\log R$ is the mean number of planets per star having $P < 50$ days in a $\log_{10}$ radius interval centered on R (in R_E), with $k_R = 2.9^{+0.5}_{-0.4}$, $\alpha = -1.92 \pm 0.11$, and $R = \mathrm{R}_P/R_E$. For comparison, Howard *et al.* (2010) found a power law planet *mass* function, $\mathrm{d}f/\mathrm{dlog}M = k' M^{\alpha'}$, with $k' = 0.39^{+0.27}_{-0.16}$ and $\alpha' = -0.48^{+0.12}_{-0.14}$ for periods $P < 50$ days and masses $M \sin i = 3$–$1000\ M_E$.

We also computed planet occurrence as a function of orbital period. Fig. 7 shows that planet occurrence increases slowly with increasing orbital period for the logarithmic binning used here, and the increase is fastest for the smallest planets with $\mathrm{R}_P = 2$–$4\ R_E$ (shown in yellow). Finally, we computed the planet occurrence as a function of

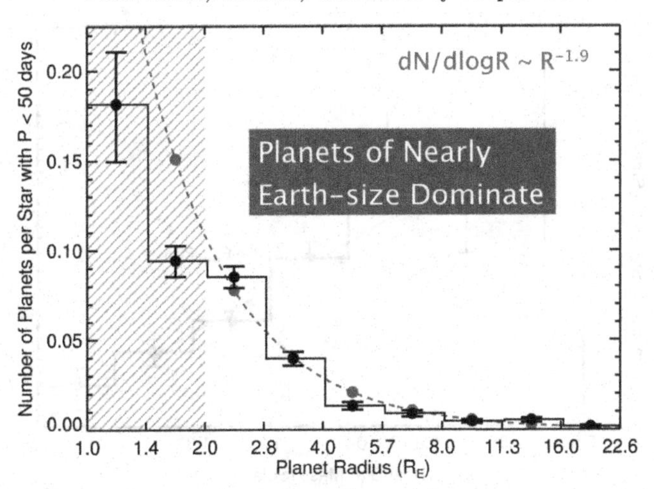

Figure 6. The number of planets within a bin of radius divided by the number of target stars that could have detected such planets by *Kepler*, as a function of planet radius. We include only orbital periods of $P < 50$ days (black filled circles and histogram). We also only include GK main sequence stars consistent with the selection criteria in Figure 5 to compute planet occurrence. A power law fit is shown in red. The estimates of planet occurrence are incomplete for radii below 2 R_E, shown hatched. Error bars indicate statistical uncertainties and do not include effects of the poorly known stellar radii (35% uncertainty). (From Howard *et al.* (2011b).)

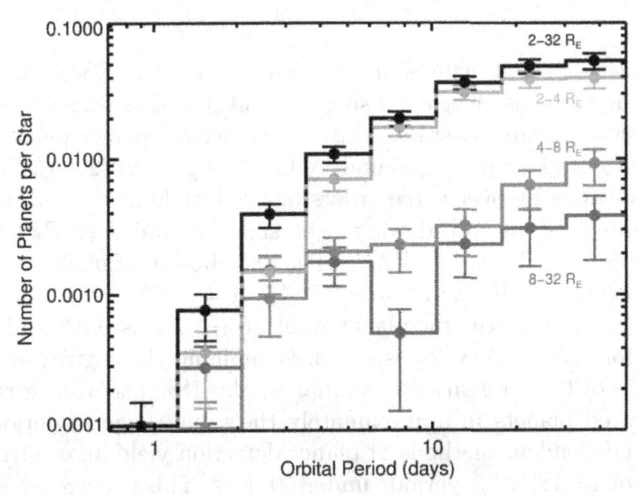

Figure 7. Planet occurrence as a function of orbital period from *Kepler* , for different planet sizes designated by color. Only stars consistent with the target selection in Figure 5 (and their associated planets) were included, namely solar type stars brighter than 15th magnitude. (See Howard *et al.* (2011b).)

stellar effective temperature and converted them to stellar masses (approximately). As the targets were restricted to main sequence stars this conversion is straightforward. Fig. 8 shows that planet occurrence of the smallest planets, 2-4 R_E, increases with decreasing stellar mass. Apparently the K and M dwarfs have small planets more commonly than the G and F main sequence stars for orbital periods under 50 days considered here.

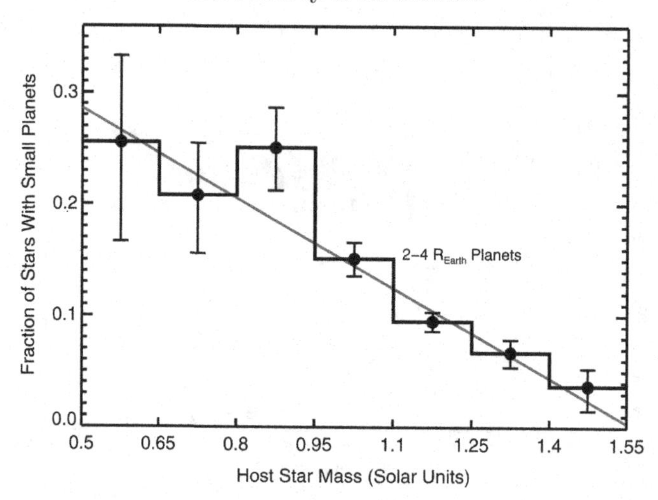

Figure 8. Occurrence of small planets, 2-4 R_E, as a function of stellar mass on the main sequence. Stellar masses were assigned based in the effective temperatures, $T_{\rm eff}$, in the *Kepler* Input Catalog (uncertain by ~135K), yielding stellar masses uncertain by ~20%. We consider only planets with $P < 50$ days and host main sequence stars having $T_{\rm eff}$ = 3600–7100 K. The occurrence of small planets of 2–4 R_E rises substantially toward lower mass stars. A similar result is shown in Figure 13 in Borucki *et al.* (2011) for the Neptune-size planets (2-6 R_E). The best-fit linear occurrence model for these small planets is shown as a red line.

4. Summary

From the Doppler-detected planets in the Solar neighborhood, we find a planet occurrence of 15^{+5}_{-4}% for the mass range, $M \sin i$ = 3–30 M_E and period range, $P < 50$ d, around main sequence G and K stars. This occurrence of smaller planets is continuous with the occurrence of giant planets found by Cumming *et al.* (2008). From *Kepler* the planet occurrence varies by over three orders of magnitude in the radius-orbital period plane and increases substantially down to the smallest radius (2 R_E) and out to the longest orbital period (50 days, ~0.25 AU). The distribution of planet radii is given by a power law, $df/d\log R = k_R R^{\alpha}$ with $k_R = 2.9^{+0.5}_{-0.4}$, $\alpha = -1.92 \pm 0.11$, and $R = R_P/R_E$. The number of planets per star for planet radii of 2–4 R_E is 0.13, and the number of planets per star for radii of 4–8 R_E is 0.023. Combining these gives an occurrence for planets of 2–8 R_E of 0.15, remarkably similar to the 18% occurrence rate found from the Doppler-detected planets in approximately the same range of periods and for GK stars. Thus two independent methods of planet detection yield an occurrence of planets between 2–8 R_E of 15–18%, for periods under 50 days. This occurrence stands as a test benchmark for the theory of planet formation, migration, and planet-planet interactions, such as those of Ida & Lin (2010), Mordasini *et al.* (2011), Schlaufman *et al.* (2010) and Wu & Lithwick (2011).

Acknowledgements

We thank the entire *Kepler* team and Ball Aerospace Corp. for their tireless work making the *Kepler* mission and data possible. We thank Howard Isaacson, John Johnson, Debra Fischer and the Keck Observatory staff for excellent help with the Keck Observatory data and analysis. We thank E. Chiang and H. Knuston for helpful conversations. We thank the W. M. Keck Observatory, and both NASA and the University of California for use of the Keck telescope. We extend special thanks to those of Hawai'ian ancestry on whose sacred mountain of Mauna Kea we are privileged to be guests. G. M. acknowledges NASA grant NNX06AH52G. Funding for the *Kepler* Discovery mission is provided by NASA's Science Mission Directorate.

References

Adams, E. R., Seager, S., & Elkins-Tanton, L. 2008, *ApJ*, 673, 1160

Alibert, Y., Mordasini, C., & Benz, W. 2011, *A&A*, 526, A63

Borucki, Koch, Basri, Batalha, Brown, Caldwell, Caldwell, Christensen-Dalsgaard, Cochran, DeVore, Dunham, Dupree, Gautier, Geary, Gilliland, Gould, Howell, Jenkins, Kondo, Latham, Marcy, Meibom, Kjeldsen, Lissauer, Monet, Morrison, Sasselov, Tarter, Boss, Brownlee, Owen, Buzasi, Charbonneau, Doyle, Fortney, Ford, Holman, Seager, Steffen, Welsh, Rowe, Anderson, Buchhave, Ciardi, Walkowicz, Sherry, Horch, Isaacson, Everett, Fischer, Torres, Johnson, Endl, MacQueen, Bryson, Dotson, Haas, Kolodziejczak, Van Cleve, Chandrasekaran, Twicken, Quintana, Clarke, Allen, Li, Wu, Tenenbaum, Verner, Bruhweiler, Barnes, Prsa. 2011, *ApJ*, 736, id.19

Brown, T. M., Latham, D. W., Everett, M. E., & Esquerdo, G. A. 2011, *AJ* submitted, arXiv:1102.0342

Bromley, B. C. & Kenyon, S. J. 2011, *ApJ*, 731, 101

Chatterjee, S., Ford, E. B., Matsumura, S., & Rasio, F. A. 2008, *ApJ*, 686, 580

Chatterjee, S., Ford, E. B., & Rasio, F. A. 2011, this volume

Cox, A. N., ed. 2000, *Allen's Astrophysical Quantities* (Springer)

Cumming, A., Butler, R. P., Marcy, G. W., Vogt, S. S., Wright, J. T., & Fischer, D. A. 2008, *PASP*, 120, 531

Fischer, D. A. & Valenti, J. 2005, *ApJ*, 622, 1102

Ford, E. B., Lystad, V., & Rasio, F. A. 2005, *Nature*, 434, 873

Ford, E. B. & Rasio, F. A. 2008, *ApJ*, 686, 621

Fortney, J. J., Marley, M. S., & Barnes, J. W. 2007a, *ApJ*, 668, 1267

—. 2007b, *ApJ*, 659, 1661

Gilliland, R. L., *et al.* 2000, *ApJL*, 545, L47

Howard, A. W., *et al.* 2010, *Science*, 330, 653

—. 2011a, *ApJ*, 726, 73

Howard, A. W & Marcy, G., The *Kepler* Team 2011b, *ApJ* submitted (arXiv:1103.2541)

Ida, S., & Lin. 2008a, *ApJ*, 673, 487

—. 2008b, *ApJ*, 685, 584

—. 2010, *ApJ*, 719, 810

Jenkins, J. M., *et al.* 2010a, *ApJ*, 724, 1108

—. 2010b, *ApJL*, 713, L120

—. 2010c, *ApJL*, 713, L87

Johnson, J. A., Aller, K. M., Howard, A. W., & Crepp, J. R. 2010, *PASP*, 122, 905

Johnson, J. A., Winn, J. N., Albrecht, S., Howard, A. W., Marcy, G. W., & Gazak, J. Z. 2009, *PASP*, 121, 1104

Kepler Mission Team. 2009, VizieR Online Data Catalog, 5133

Koch, D. G., *et al.* 2010a, *ApJL*, 713, L131

—. 2010b, *ApJL*, 713, L79

Lissauer, J. J., Hubickyj, O., D'Angelo, G., & Bodenheimer, P. 2009, *Icarus*, 199, 338

Lissauer, J. J., *et al.* 2011a, *Nature*, 470, 53

—. 2011b, *ApJ* submitted (arXiv:1102.0543)

Lithwick, Y. & Wu, Y. 2010, *ApJ* submitted (arXiv:1012.3706)

Marcy, G., Butler, R. P., Fischer, D., Vogt, S., Wright, J. T., Tinney, C. G., & Jones, H. R. A. 2005a, *Progress of Theoretical Physics Supplement*, 158, 24

Marcy, G. W., Butler, R. P., Vogt, S. S., Fischer, D. A., Henry, G. W., Laughlin, G., Wright, J. T., & Johnson, J. A. 2005b, *ApJ*, 619, 570

Marcy, G. W., *et al.* 2008, *Physica Scripta*, 130, 014001

Mayor, M., *et al.* 2009, *A&A*, 493, 639

Moorhead, A. V., *et al.* 2011, *ApJ* submitted (arXiv:1102.0547)
Mordasini, C., Alibert, Y., Benz, W., & Naef, D. 2009, *A&A*, 501, 1161
Mordasini, C., Alibert, Y., Benz, W., & Klahr, H. 2011, *EPJ Web of Conferences*, 11, 04001
Morton, T. D. & Johnson, J. A. 2011, *ApJ* in press (arXiv:1101.5630)
Raymond, S. N., *et al.* 2011, *A&A*, 530, A62
Robin, A. C., Reylé, C., Derrière, S., & Picaud, S. 2003, *A&A*, 409, 523
Schlaufman, K. C., Lin, D. N. C., & Ida, S. 2010, *ApJL*, 724, L53
Udry, S. & Santos, N. C. 2007, *ARA&A*, 45, 397
Wittenmyer, R. A., Tinney, C. G., Butler, R. P., O'Toole, S. J., Jones, H. R. A., Carter, B. D., Bailey, J., & Horner, J. 2011, *ApJ* submitted (arXiv:1103.4186)
Wright, J. T., Upadhyay, S., Marcy, G. W., Fischer, D. A., Ford, E. B., & Johnson, J. A. 2009, *ApJ*, 693, 1084
Wu, Y. & Lithwick, Y. 2011, *ApJ*, 735, id.109

The Astrophysics of Planetary Systems: Formation, Structure, and Dynamical Evolution
Proceedings IAU Symposium No. 276, 2010
A. Sozzetti, M. G. Lattanzi & A. P. Boss, eds.

doi:10.1017/S1743921311019879

Hunting for the lowest-mass exoplanets

Francesco Pepe[1], Michel Mayor[1], Christophe Lovis[1], Willy Benz[2], François Bouchy[3], Xavier Dumusque[1], Didier Queloz[1], Nuno C. Santos[4], Damien Ségransan[1] and Stéphane Udry[1]

[1]Observatoire Astronomique de l'Université de Genève, 51, ch. des Maillettes, CH-1290 Versoix, Switzerland, email: Francesco.Pepe@unige.ch
[2]Physikalisches Institut der Universität Bern, Sidlerstrasse 5, CH-3012 Bern, Switzerland
[3]Institut d'Astrophysique de Paris, 98bis, bd Arago, F-75014 Paris, France
[4]Centro de Astrofísica da Universidade do Porto (CAUP), Rua das Estrelas, P-4150-762 Porto, Portugal

Abstract. In order to understand general planet characteristics and constrain formation models it is necessary to scan over the widest possible parameter range of discovered systems. Due to detection biases, the domain of very-low mass planets had remained poorly explored. Only with improving measurement precision it has been possible to enter in the sub-Neptune mass range. The HARPS planet search program has been particularly efficient in detecting such ice giants and super earths. The present talk will summarize the obtained results and the characteristics of the low-mass population of exoplanets.

Keywords. techniques: radial velocities, planetary systems

---

1. The discovery of a rich population of low-mass planets on tight orbits

Today, more than 500 extrasolar planets have been discovered. Most of the detected exoplanets have been found by using precise measurements of stellar radial velocities. The planetary mass estimate from Doppler measurements is directly proportional to the amplitude of the stellar reflex motion. Our progress to detect very-low-mass planets are directly related to the progress done to improve the sensitivity and stability of spectrographs. In 1989, the detection of HD 114762 b, a companion of 11 Jupiter masses to a metal deficient F- star was obtained with spectrographs allowing Doppler measurements with a precision of some 300 m/s (Latham *et al.* 1989). Fifteen years ago, the precision achieved by any team searching for exoplanets was of the order of 15 m/s. Today, the instrumental precision achieved with the HARPS spectrograph at La Silla Observatory is better than 0.5 m/s (Mayor *et al.* 2003). At this level of precision we are mostly limited by the intrinsic variability of stellar velocities induced by diverse phenomena (acoustic modes, granulation, magnetic activity). However, by adopting an improved observing strategy, we have already some indications that planetary signals as small as a tiny fraction of a meter per second are detectable.

This progress in instrumentation and observing strategy have made possible the discovery of a rich population of super-Earths and Neptune-mass planets in tight orbits around solar-type stars (Mayor & Udry 2008).

The name "super-Earth" is used to qualify planets more massive than the Earth but with masses smaller than 10 Earth masses, a category of planets absent in the solar system. We mention here a few landmark discoveries of these low-mass planets orbiting solar-type stars. Limiting ourself to planets in the super-Earth range we can mention: μ Ara c with a mass of 10.5 $M_{\oplus}$ and a period of 9.7 days (Santos *et al.* 2004, revised in

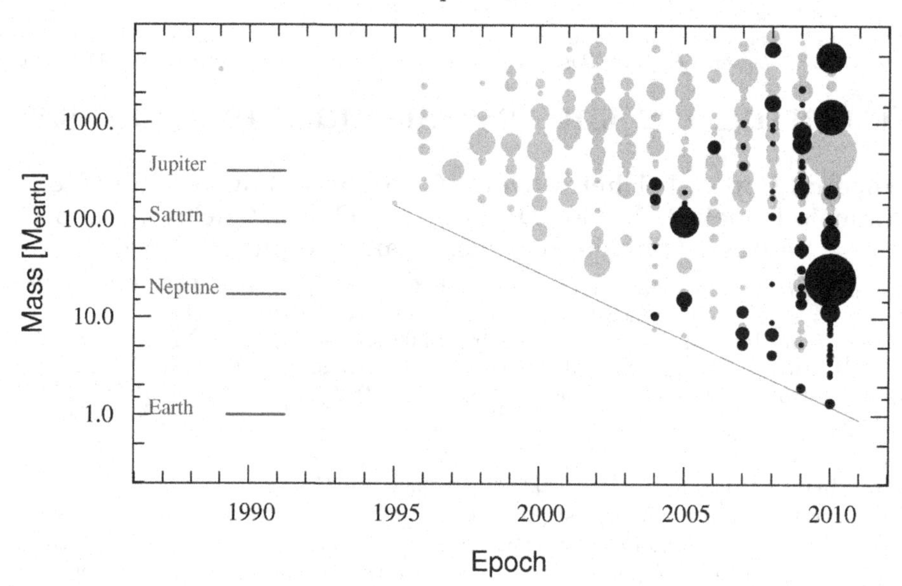

Figure 1. Minimum mass of planets detected by Doppler spectroscopy as a function of epoch of discovery. This figure illustrates the impressive progress made in detection sensitivity over the past 15 years. Black symbols indicate HARPS discoveries. Dot size is related to orbital semi-major axis.

Pepe *et al.* 2007), HD 69830 b with a mass of 10.2 $M_\oplus$ and a period of 8.7 days (Lovis *et al.* 2006), HD 40307 b, c, d, a system with three super-Earths with masses comprised between 4 and 9 $M_\oplus$ and periods from 4 to 20 days (Mayor *et al.* 2009b). We also have to mention the exceptional system around HD 10180, with 7 planets of which one with a mass as small as 1.4 $M_\oplus$ on a tight orbit with a period of 1.17 day (Lovis *et al.* 2011). In addition to these early detections of super-Earths orbiting solar-type stars, we also have to mention the discoveries of super-Earths hosted by M-dwarfs: GJ 876 d, a planet with a mass of 5.9 $M_\oplus$ and a period of 1.94 day (Rivera *et al.* 2005; Correia *et al.* 2010), GJ 581 c, d, e with masses of 5, 7 and 1.9 $M_\oplus$ (Udry *et al.* 2007; Mayor *et al.* 2009a). It is impressive to see that all these super-Earths are part of rich multi-planetary systems with 3 to 7 planets per system. The remarkable progress of instrumentation in the last 15 years is obvious in Fig. 1. The masses of planetary companions are plotted as a function of the epoch of their discovery. The mass of HD 10180 b (Lovis *et al.* 2011) is a factor 100 smaller than the mass of 51 Peg b (Mayor & Queloz 1995).

2. Reaching sub-meter-per-second precision

There is no doubt that successful detections depend on the observational precision, in particular when searching for signals with semi-amplitudes below 1-2 $\mathrm{m\,s^{-1}}$, as it is the case for the presented program. The instrument used for our program, ESO's HARPS, is recognized as the most precise RV instrument worldwide. Five years of observations have proved that, on quiet and bright dwarf stars, a radial-velocity precision well below 1 $\mathrm{m\,s^{-1}}$ can be achieved on both short-term (night) *and* long-term (years) timescales. A direct confirmation is offered by the raw RV dispersions measured for the candidates discovered with HARPS. Furthermore, when binning the data-points over several days to remove meso-granulation effects of the stellar surface, the radial-velocity dispersion can be reduced below 30 cms over over observational period of several years (see e.g. Lovis *et al.* (2006) describing the detection of three Neptunes orbiting HD 69830).

The recent impressive HARPS discoveries made us aware of the fact that discovering Earth analogs is already within the reach of HARPS, although important questions still remain open: How frequent are low-mass planets? At what level is the detection bias? Where is the precision limit set by stellar noise? Can we detect low-mass planets in the habitable zone of their parent star if sufficient observations on a high-precision instrument are invested? Driven by the very encouraging results of the HARPS High-Precision Program (Udry & Mayor 2008) and the hints for high planet occurrence announced by Lovis *et al.* (2009), we have decided to investigate these questions in further detail. For this purpose, we have defined a specific program on HARPS for the search of rocky planets in the habitable zone HZ based on Guarantee Time Observations (GTO) awarded to us by ESO for the upgrade and maintenance of the HARPS spectrograph. In this program we have focused on 10 quiet, non-rotating, non-active stars of the original HARPS high-precision program and the observational strategy was optimized. The program is being carried out presently and the first results are being obtained.

In order to illustrate the kind of objects we are dealing with and the level of precision which can be attained with HARPS, we shall focus a moment on HD 10700, also known as Tau Ceti. This is the most quiet star in our sample and shows a dispersion of only $0.92\,\mathrm{m\,s^{-1}}$, despite the large number of data points spanning 6 years of observations. What is even more remarkable it that none of the parameters shown in Fig. 2, i.e. the radial velocity, the chromospheric activity indicator $\log(R'_{\mathrm{HK}})$, and the average line bisector of the cross-correlation function BIS, show any trend over the 6 years of observations. Most amazing is the line bisector, which has a dispersion of about $0.5\,\mathrm{m\,s^{-1}}$, confirming the extreme stability of HARPS' instrumental profile (IP).

One may also wonder how strongly stellar activity may influence the radial-velocities. We refer to Dumusque *et al.* (2011a) and references therein for a detailed discussion. In the specific case of Tau Ceti we observe only little RV jitter despite the fact that the $\log(R'_{\mathrm{HK}})$seems to vary significantly at some periods, while it remains stable over other periods. As an example, Fig. 3 shows the radial velocity and the $\log(R'_{\mathrm{HK}})$over a single seasons. Significant variations (in terms of photon noise) of the $\log(R'_{\mathrm{HK}})$can bee seen, while no correlated variation is detected in the radial velocity. This is better illustrated in Fig. 4 where the correlation between the two quantities is plotted. We deduce that in the specific case of Tau Ceti, the radial velocity is marginally affected by stellar activity. This behavior is expected for late G and K dwarfs (Dumusque *et al.* 2011a).

3. The case of Gl 581

Recently, Vogt *et al.* (2010) have announced the discovery of a 5th and 6th planet around the star Gl 581. This detection resulted from the combination of HARPS and HIRES data. When fitting a given model to the HIRES and HARPS data equally weighted, the residuals of the HARPS data are of the order of $1\,\mathrm{m\,s^{-1}}$, while the for the HIRES data they are significantly higher than $2\,\mathrm{m\,s^{-1}}$. From this fact we concluded that the HIRES data where not sufficiently precise to allow the detection of a planet with $\mathrm{m\,s^{-1}}$amplitude. On the other hand, the high number of HARPS data and the good time sampling, would allow to see a potential 5th and 6th signal with HARPS data only. Therefore, we have investigated the presence of possible related signals in the HARPS data only. Fig. 5 shows the generalized Lomb-Scargle periodogram (GLS) of the residuals after fitting the four already know planets. It appear evident, that none of the residual peaks is significant, being all of them well below the 10% false-alarm probability (FAP)

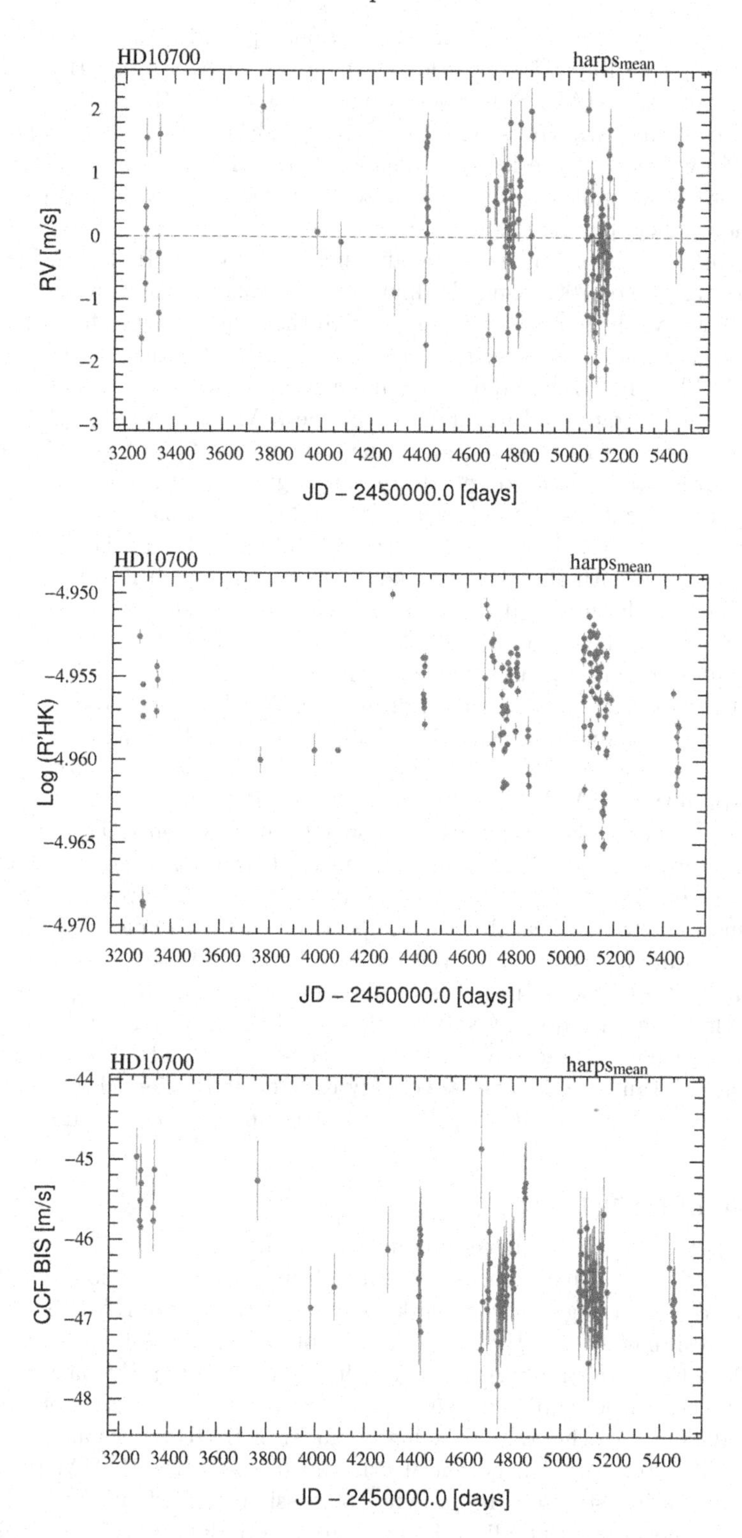

Figure 2. Radial velocities, chromospheric activity indicator $\log(R'_{\mathrm{HK}})$, and line bisector $CCFBIS$ of HD 10700 versus time.

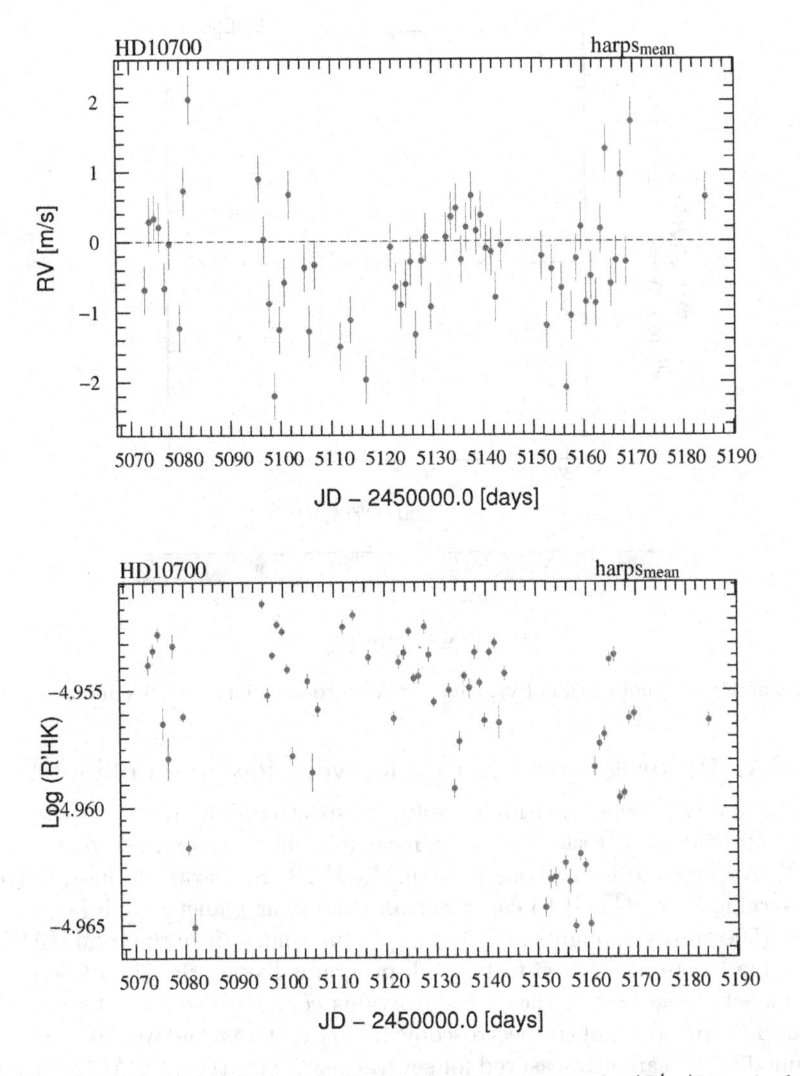

Figure 3. Radial velocitiy and chromospheric activity indicator $\log(R'_{\rm HK})$in a single season.

threshold. It is however true that, if the eccentricity of the planet at $P = 66$ days is fixed at 0, then a peak at $P = 33$ days appears. However, the resulting 'signal' has a FAP of several percent, and in particular, its period is significantly different from the period of 37 days of the f-planet announced by Vogt *et al.* The same discussion applies for the g-planet. We conclude from this that the more precise HARPS data show no evidence for the existence of potential 5th and 6th planets around Gl 581 with orbital parameters as announced by Vogt *et al.* We would like to refer at this point to two independent studies carried out by Gregory (2011) and Andrae *et al.* (2010) and published shortly after the IAUS 276 Symposium. They proof that the signals detected by Vogt *et al.* are statistically not relevant. Both studies indicate that, using the same data sample on which Vogt *et al.* based their analysis, the most likely model is the one with 4 planets. Furthermore they show that HIRES data only would degrade the quality of the found solutions.

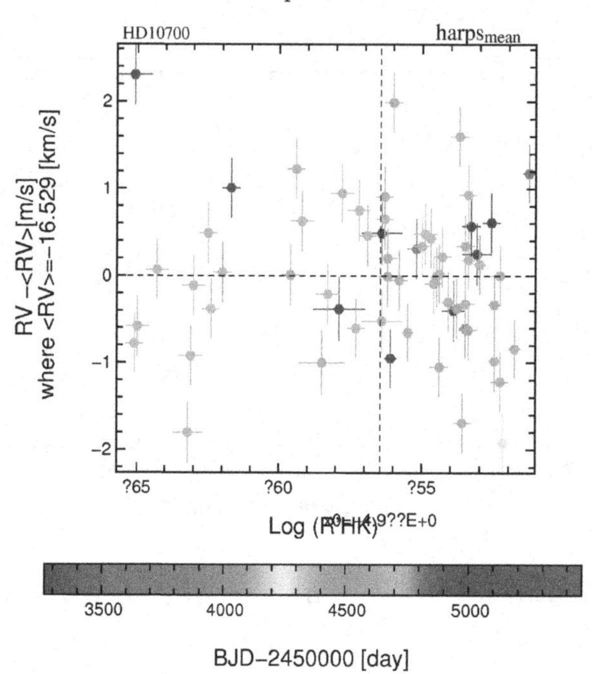

Figure 4. Correlation plot of radial velocity versus chromospheric activity indicator $\log(R'_{\rm HK})$.

4. The HARPS program to search for very low mass planets

HARPS is a vacuum-operated high-resolution spectrograph (R = 115,000), fiber-fed, optimized to provide stellar radial-velocity measurements with extreme precision (Mayor *et al.* 2003). As a reward for its construction, the HARPS consortium has received guaranteed observing time (GTO) to carry out an extrasolar planet search in the southern hemisphere (500 observing nights over 5 years). More than 60% of the total HARPS GTO observing time has been devoted to two sub-programs having the aim of detecting very low-mass planets. The first of these sub-programs comprises some 400 stars which are non-active, slow rotators, not in spectroscopic binary systems, and were selected from the large volume-limited sample measured for several years with the CORALIE spectrograph on the 1.2 m-Euler telescope at la Silla Observatory. The second sub-program consists of a volume-limited sample of about 120 M dwarfs at the bottom of the main sequence, also selected to be slow rotators and not members of spectroscopic binary systems.

What are the limits presently achieved in terms of radial velocity precision? Several sources of noise can be identified:

- As a result of the efficiency of the cross-correlation technique, a photon noise level of only a fraction of a meter per second is achieved in a few minutes for most of our targets. Sometimes the exposure time is shorter than the typical periods of stellar acoustic modes. In a few minutes, the full amplitude of the stellar velocity variations resulting from acoustic modes could be as large as several meters per second. Long integrations compared to acoustic mode periods are sufficient to have acoustic noise residuals smaller than 0.2 m/s (rms). For most stars, integrations of 15 minutes are sufficient.
- Dumusque *et al.* (2011a) have shown that stellar granulation in solar-type stars can induce radial velocity variability comparable to or larger than 1 m/s on longer timescales compared to acoustic modes. Several measurements spanning several hours are requested to damp the granulation noise.

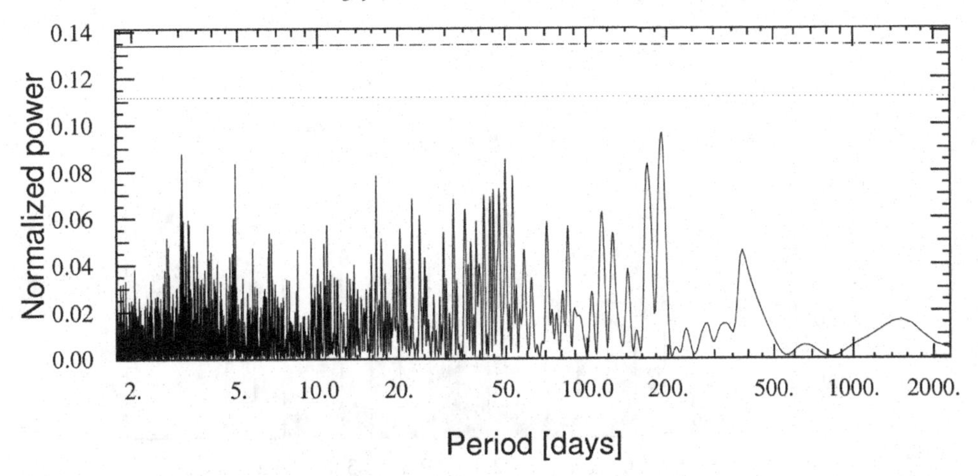

Figure 5. Generalized Lomb-Scargle periodogram of the HARPS radial-velocities of Gl 581. The horizontal line mark the 10%, respectively 1% FAP limits.

• Any anisotropies in stellar atmospheres related to magnetic activity will induce radial velocity variations at the stellar rotation period. The amplitude of the radial-velocity jitter is related with stellar chromospheric activity. If we want to search for very low mass planets we need to carefully select "non-active" stars . The reemission in the core of the calcium H and K lines is a good indicator of the chromospheric activity and has been used for the selection of the stellar sample.

• The analysis of the radial velocity variations of several solar-type stars has recently revealed well-defined variations of several m/s on rather long periods (more than five years). These velocity variations are strongly correlated with the mean shape of absorption lines and chromosperic indicators like Ca II H & K core emission. These variations are related to the stellar analogs of the solar magnetic cycle. This effect has been observed in stars with rather modest chromospheric activity levels (e.g. $\log R'_{\rm HK}$ around -4.90, see Lovis *et al.* 2011, in prep.). Any long-term drift in stellar radial velocities cannot be a priori attributed to long period planets if a careful check of the long-term behavior of the line bisector and other activity indicators has not been performed.

• Finally, we still have instrumental noise. Lovis & Pepe (2007) have considerably improved the precision of the wavelength of thorium lines as well as the number of lines to be used for the calibration of the spectrograph. Pressure changes in the plasma with the aging of the ThAr calibration lamp induce a very small shift in the wavelengths. As this effect is smaller for thorium lines than argon, we can use this differential effect to correct the aging effect. Long term drifts have thus been reduced below 0.3 m/s over timescales of several years. The scrambling effect in optical fibers is excellent... but not perfect and some sub-meter per second error could result from imperfect guiding.

The global budget of all these errors is difficult to determine. The best estimation of the lower limit of the quadratic sum of the different components of the noise is provided by the residuals observed around fitted radial velocity curves. Several stars with a very large number of velocity measurements spanning several years have residuals with a dispersion as low as 0.6 m/s (when binning the data over a few days). For stars with larger chromospheric activity, we can obviously have larger residuals.

This is the precision presently achieved for the HARPS program, for which we have derived preliminary results for the population of low mass planets, as discussed in the next section. If we are searching for low-mass planets on rather long periods, it could

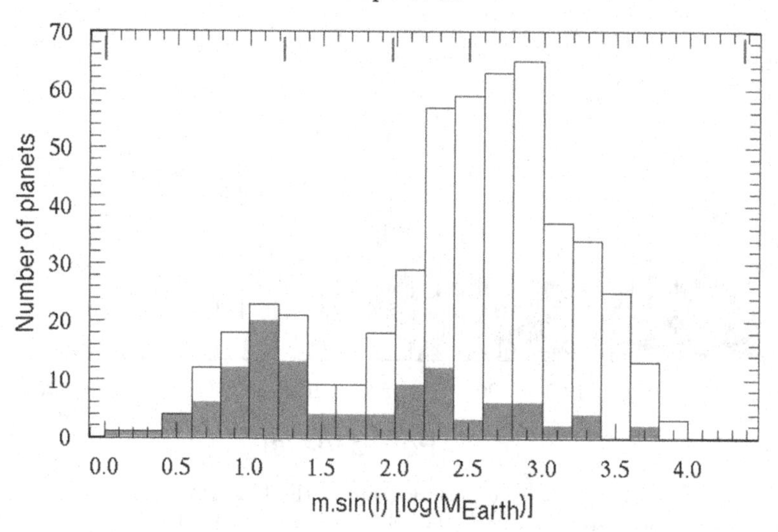

Figure 6. Mass distribution of all detected planets. The contribution of the HARPS program (solid histogram) for the detection of very low mass planets is evident.

be useful to bin the measurements done on N consecutive nights. This procedure could help to damp the noise induced by chromospheric activity, with a time scale comparable to the stellar rotation period. First experiments done on stars with a large number of measurements have shown that the residuals decrease to 0.3–0.5 m/s after binning over ten consecutive nights.

5. Emerging characteristics of low-mass planets and their host star

We are still far from having a detailed and unbiased view of the population of planets with masses in the range of super-Earths and Neptunes. Nevertheless, we can already notice a few emerging properties. The study of planet hosts themselves also provide additional information to constrain planet formation. In particular the metallicity of the parent stars seems to be of prime importance for models of planetary formation.

5.1. *The mass distribution*

The mass distribution of all detected planets is illustrated in Fig. 6. In this plot the contribution of the HARPS program for the detection of very low mass planets is evident. Due to the better detection sensitivity of Doppler spectroscopy for massive and/or short period planets, we still have a strong bias against the detection of low-mass planets, especially if they are on long-period orbits.

The bimodal aspect of the mass distribution is a clear indication that the decrease of the distribution for masses less than about one mass of Jupiter is not the result of a detection bias, but is real. The extrapolation by a power-law distribution, as for example $f(m) \sim m^{-1}$, to estimate the number of planets with a mass smaller than the mass of Jupiter is certainly not justified. The observed bimodal shape of the mass distribution from gaseous giant planets to the super-Earth regime provides an interesting constraint for planetary formation scenarios. The planetary formation simulations carried out by Mordasini *et al.* (2009a,b) also predict a bimodal distribution for that range of planetary mass. In addition these simulations also predict a sharp rise in the mass distribution at a few Earth masses and below. This domain of mass is still at the limit of present instrumental sensitivity. Nevertheless, once again the expected shape of this theoretical

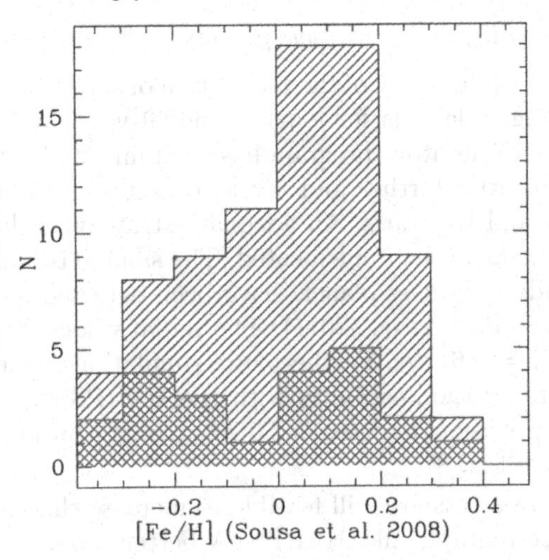

Figure 7. Correlation of the frequency of planets as a function of host star metallicity. Planetary systems with masses smaller than 20 $M_\oplus$ are frequently hosted by metal-deficient stars (cross-hatched histogram), at the opposite of the correlation observed for gaseous giant planets (hatched histogram).

mass distribution from 10 down to 1 $M_\oplus$ is clearly not an exponential and any estimate of the frequency of Earth-twins based on an exponential extrapolation is completely unjustified.

5.2. *The frequency of low-mass multiplanet systems*

With the HARPS data presently available from the high-precision sample, we have 48 stars with well-characterized planetary systems. More than 50% of these systems are multiplanetary. Four of them have 4 planets and the amazing system HD 10180 is the host of 7 planets (Lovis *et al.* 2011), one of them having a mass as small as 1.5 $M_\oplus$.

5.3. *The correlation with the metallicity of host stars*

The correlation between the occurrence of gaseous giant planets and the metallicity of host stars is striking. Based on large planetary surveys this correlation is well established by independent teams (Santos, Israelian & Mayor 2001, 2004; Fischer & Valenti 2005). We have a completely different result if we examine the metallicity of host stars for systems having all planets less massive than 40 $M_\oplus$. We do not have any correlation between the presence of these low-mass planets and the host star metallicity (see Fig. 7), a result already mentioned by Udry *et al.* (2006) and Sousa *et al.* (2008), based at that time on a very limited number of stars. With the present study, this lack of correlation with the host star metallicity is robust. The mean metallicity of the 28 planetary systems with planets less massive than 40 $M_\oplus$ is [Fe/H] = -0.12, a metallicity not so different from the mean metallicity of stars in the solar neighborhood.

5.4. *The occurrence of low-mass planets orbiting solar-type stars*

The occurence of low-mass planets on tight orbits has been estimated by Lovis *et al.* (2009). For planets with masses between ~5 and 50 $M_\oplus$ and periods shorter than 100 days, we have detected low-mass planets orbiting about 30% of the stars in the HARPS sample. A more complete estimate is currently in progress, based on the present, more complete survey.

5.5. *Searching for Earth-type planets in the habitable zone*

The programme devoted to the study of the population of super-Earths and Neptune-type planets is still continuing at la Silla for four additional years after the end of the GTO time. In addition, a new exploratory program has been initiated with the goal of pushing the HARPS precision a little further and try to detect super-Earths in the habitable zone of very nearby G and K dwarfs. An adequate strategy to damp the acoustic and granulation noise sources has been implemented. The sample is limited to only 10 bright non-active stars. Already, low-mass planets have been detected on three stars members of that small sample, see Pepe *et al.* (2011). The radial velocity signal for one of these planets is as small as K = 0.6 m/s. Furthermore, simulations done by Dumusque *et al.* (2011b) have demonstrated the possibility with the HARPS spectrograph, the present observing strategy and precision, to detect a 2.5-$M_{\oplus}$ planet orbiting a non-active K-dwarf in its habitable zone (see Fig. 8).

Some technical improvements are still feasible to increase the sensitivity and stability of cross-correlation spectrographs like HARPS. A better scrambling of the input beam could be achieved by new optical fibers with octogonal cross sections. These new fibers will strongly diminish the already very small effect of input conditions (guiding errors, variable seeing and focus) on the spectrograph illumination, a mandatory condition to achieve 0.1 m/s precision. To secure the stability of radial velocity measurements over a span of several years at the level of 0.1 m/s or better, we must have a calibration device better than the existing ThAr lamps. Developments of laser frequency combs adapted to the resolution and wavelength coverage of HARPS will provide the requested stability (Wilken *et al.* 2010).

A photon noise on the Doppler signal at the level of 0.1 m/s requires a rather large telescope size to achieve the needed signal-to-noise ratio in a reasonable exposure time. The ESPRESSO project, presently in development, to be implemented on the 8.2-m VLT telescope at Paranal is designed to achieve the 0.1 m/s Doppler precision and stability on the long term (Pepe *et al.* 2010). The ESPRESSO project can also be seen as a precursor for an even more ambitious stable spectrograph, the CODEX project presently at the study phase level for the 42-m E-ELT telescope, to be implemented by ESO at Cerro Armazones (Chile) in the next decade (Pasquini *et al.* 2010).

We have to keep in mind that for stars with the lowest chromospheric activity, we still do not know the true level of radial velocity jitter. Analysis of the radial velocity scatter of HARPS measurements for non-active stars suggest a minimum jitter of 0.5 m/s or less. This stellar variability, depending on the changing number and phase of magnetic spots (or other features) will be difficult to model. Preliminary studies show that non-active K dwarfs will be the most suitable targets to search for Earth twins. A large number of Doppler measurements has the potential to overcome the effects of the stellar intrinsic variability and permit detections of planetary signals of 0.2 m/s or less.

The discovery of radial velocity variations associated with solar cycle analogues with full amplitude as large as 10 m/s seems a priori to be casting doubts on our ability to detect Earth analogues in the habitable zone. However, using parameters of the cross-correlation function it has been possible to correct the magnetic cycle effects to less than 1 m/s. In addition, for some domain of stellar masses (K-dwarfs), we observe that the amplitude of the radial velocity effect is vanishing despite quite noticeable magnetic cycles. Finally, we notice that the periods of magnetic cycles are much longer (about a factor 10) than the expected periods of habitable planets orbiting K dwarfs. We are thus still convinced that Doppler spectroscopy has the potential to detect rocky planets in the habitable zone of K dwarfs.

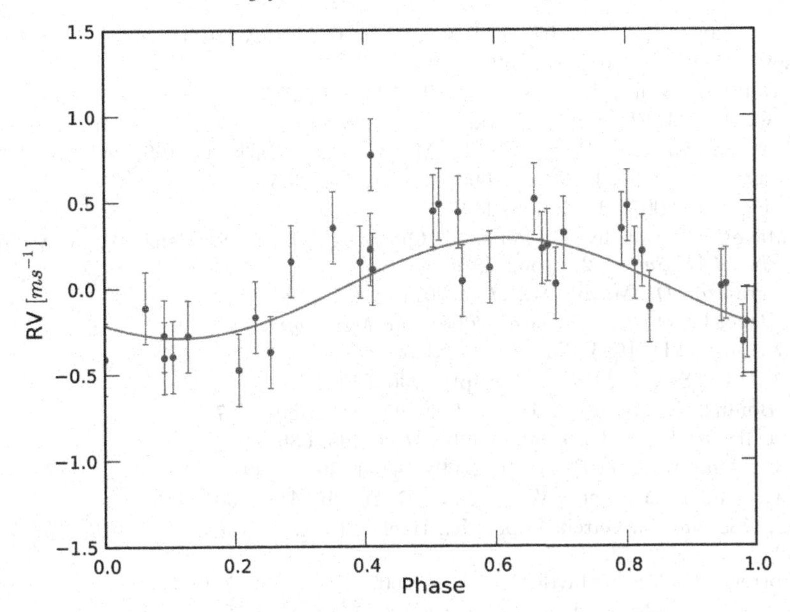

Figure 8. Simulation for the detection of a super-Earth of 2.5 $M_{\oplus}$ in the habitable zone of an inactive K-dwarf (from Dumusque *et al.* 2011b).

The medium- or long-term scientific goal to search for chemical signatures of life in the atmospheric spectra of Earth twins via space experiments as the ESA-DARWIN concept will first require identification of targets. It seems that at the moment Doppler spectroscopy is the only method with the potential to detect Earth-type planets in the habitable zone of stars as close as possible to the Sun. The last condition is mandatory, if we want to have a star-planet angular separation large enough for the need of planetary atmosphere spectroscopy, as well as bright enough targets to maximize the signal-to-noise ratio.

From Doppler surveys we know that super-Earths on tight orbits are frequent. We have first hints from microlensing searches that super-Earths could also be frequent at large semi-major axis (Gould *et al.* 2010). But we do not have any estimate of the frequency of Earth-twins in the habitable zone of solar-type stars and no ideas on their orbital eccentricity distribution. The orbital eccentricity of Earth-twins is also relevant in the frame of life-search experiments. The ESA-PLATO space project is, in that context, the most interesting experiment, complementary to Doppler surveys to explore the domain of Earth-type planets orbiting relatively close stars.

Acknowledgements

We would like to thank the Swiss National Science Foundation for its continuous support.

References

Andrae, R., Schulze-Hartung, T., & Melchior, P. 2010, arXiv:1012.3754v1

Correia, A., Udry, S., Mayor, M., *et al.* 2009, *A&A*, 496, 521

Correia, A. C. M.., Couetdic, J., Laskar, J., *et al.* 2010, *A&A*, 511, A21

Dumusque, X., Udry, S., Lovis, C., Santos, N. C., & Monteiro, M. J. P.. F. G.. 2011a, *A&A*, 525, A140

Dumusque, X., Santos, N. C., Udry, S., Lovis, C., & Bonfils, X. 2011b, *A&A*, 527, A82
Fischer, D. & Valenti, J. 2005, *ApJ*, 622, 1102
Gould, A., Dong, S., Gaudi, B. S., *et al.* 2010, *ApJ*, 720, 1073
Gregory, P. C. 2011, *MNRAS* in press (arXiv:1101.0800)
Latham, D. W., Stefanik, R. P., Mazeh, T., Mayor, M., & Burki, G. 1989, *Nature*, 339, 38
Lovis, C., Mayor, M., Pepe, F., *et al.* 2006, *Nature*, 441, 305
Lovis, C. & Pepe, F. 2007, *A&A*, 468, 1115
Lovis, C., Mayor, M., Bouchy, F., Pepe, F., Queloz, D., Udry, S., Benz, W., & Mordasini, C. 2009, *Proc. IAU Symp.*, 253, 502
Lovis, C., Ségransan, D., Mayor, M., *et al.* 2011, *A&A*, 528, A112
Mayor, M., Pepe, F., Queloz, D., *et al.* 2003, *The Messenger*, 114, 20
Mayor, M. & Queloz, D. 1995, *Nature*, 378, 355
Mayor, M. & Udry, S. 2008, Physica Scripta, 130, 014010
Mayor, M., Bonfils, X., Forveille., T., *et al.* 2009a, *A&A*, 507, 487
Mayor, M., Udry, S., Lovis, C., *et al.* 2009b, *A&A*, 493, 639
Mordasini, C., Alibert, Y., & Benz, W. 2009a, *A&A*, 501, 1139
Mordasini, C., Alibert, Y., Benz, W., & Naef, D. 2009b, *A&A*, 501, 1161
Pasquini, L., Cristiani, S., Garcia-Lopez, R., Haehnelt, M., & Mayor, M. 2010, *The Messenger*, 140, 20
Pepe, F., Correia, A. C. M., Mayor, M., *et al.* 2007, *A&A*, 462, 769
Pepe, F., Cristiani, S., Rebolo Lopez, R., *et al.* 2010, *Proc. SPIE*, 7735, 14
Pepe, F., *et al.* 2011, *A&A*, submitted
Rivera, E. J., Lissauer, J. J., Butler, R. P., *et al.* 2005, *ApJ*, 634, 625
Santos, N. C., Israelian, G., & Mayor, M. 2001, *A&A*, 373, 1019
Santos, N. C., Israelian, G., & Mayor, M. 2004, *A&A*, 415, 1153
Santos, N. C., Bouchy, F., Mayor, M., *et al.* 2004, *A&A*, 426, L19
Sousa, S. G., Santos, N. C., Mayor, M., *et al.* 2008, *A&A*, 487, 373
Udry, S., Mayor, M., Benz, W., *et al.* 2006, *A&A*, 447, 361
Udry, S., Bonfils, X., Delfosse, X., *et al.* 2007, *A&A*, 469, L43
Udry, S. & Mayor, M. 2008, *ASP Conf. Ser.*, 398, 13
Vogt, S. S., Butler, R. P., Rivera, E. J., Haghighipour, N. H., Gregory, W. W., & Michael, H. 2010, *ApJ*, 723, 954
Wilken, T., Lovis, C., Manescau, A., *et al.* 2010, *MNRAS*, 405, L16

The Astrophysics of Planetary Systems: Formation, Structure, and Dynamical Evolution
Proceedings IAU Symposium No. 276, 2010
A. Sozzetti, M. G. Lattanzi & A. P. Boss, eds.

doi:10.1017/S1743921311019880

Chemical clues on the formation of planetary systems

Elisa Delgado Mena[1,2], Garik Israelian[1,2], Jonay I. González Hernández[1,2], Jade C. Bond[3], Nuno C. Santos[4,5], Stéphane Udry[6] and Michel Mayor[6]

[1]Instituto de Astrofísica de Canarias, E-38200 La Laguna, Tenerife, Spain. email: edm@iac.es

[2]Departamento de Astrofísica, Universidad de La Laguna, 38205 La Laguna, Tenerife, Spain.

[3]Planetary Science Institute, 1700 E. Fort Lowell, Tucson, AZ 85719, USA.

[4]Centro de Astrofísica, Universidade do Porto, Rua das Estrelas, 4150-762 Porto, Portugal.

[5]Departamento de Física e Astronomia, Faculdade de Ciências, Universidade do Porto, Portugal.

[6]Observatoire de Genève, 51 ch. des Maillettes, CH-1290 Sauverny, Switzerland.

Abstract. Theoretical studies suggest that C/O and Mg/Si are the most important elemental ratios in determining the mineralogy of terrestrial planets. The C/O ratio controls the distribution of Si among carbide and oxide species, while Mg/Si gives information about the silicate mineralogy. We find mineralogical ratios quite different from those of the Sun, showing that there is a wide variety of planetary systems which are not similar to Solar System. Many of planetary host stars present a Mg/Si value lower than 1, so their planets will have a high Si content to form species such as $MgSiO_3$. This type of composition can have important implications for planetary processes like plate tectonics, atmospheric composition or volcanism. Moreover, the information given by these ratios can guide us in the search of stars more probable to form terrestrial planets.

Keywords. stars: abundances, stars: atmospheres, stars: fundamental parameters, planetary systems, planetary systems: formation

---

1. Introduction

The study of the photospheric stellar abundances of planet host stars is the key to understand how and which of the protoplanetary clouds form planets and which do not. These studies also help us to investigate the internal and atmospheric structure and composition of extrasolar planets.

One remarkable characteristic of planet host stars is that they are considerably metal rich when compared with single field dwarfs (Gonzalez 1998, Gonzalez *et al.* 2001, Santos *et al.* 2000, 2001, 2004, Fischer & Valenti 2005). Two main explanations have been suggested to clarify this difference. The first of these is that the origin of this metallicity excess is primordial, so the more metals you have in the proto-planetary disk, the higher should be the probability of forming a planet. On the other hand, this excess might be produced by accretion of rocky material by the star some time after it reached the main-sequence. Recent studies on chemical abundances in stars with and without planets showed no important differences in [X/Fe] vs. [Fe/H] trends between both groups of stars (Takeda 2007, Bond *et al.* 2008, Neves *et al.* 2009, González Hernández *et al.* 2010). However, other works have reported less statistically significant enrichments in other species such as C, Na, Si, Ni, Ti, V, Co, Mg and Al (Santos *et al.* 2000, Gonzalez *et al.* 2001, Sadakane *et al.* 2002, Bodaghee *et al.* 2003, Fischer & Valenti 2005, Beirão *et al.* 2005, Gilli *et al.* 2006, Bond *et al.* 2006, Gonzalez & Laws 2007).

These results have important implications for models of giant planet formation and evolution. There are two major planet formation models: the core accretion model (Pollack *et al.* 1996), more likely to form planets in the inner disk, and the disk instability model (Boss 1997), which is in better agreement with the conditions in the extended disk. In the core accretion model, planet formation is dependent on the dust content of the disk (Pollack *et al.* 1996) while in the disk instability model it is not so clear (Boss 2002, Cai *et al.* 2006). Present observations are thus more compatible with core accretion model although they do not exclude disk instability.

Theoretical studies suggest that C/O and Mg/Si are the most important elemental ratios in determining the mineralogy of terrestrial planets and they can give us information about the composition of these planets. The C/O ratio controls the distribution of Si among carbide and oxide species, while Mg/Si gives information about the silicate mineralogy (Bond *et al.* 2010a, Bond *et al.* 2010b). Bond *et al.* (2010b) carried out simulations of planet formation where the chemical composition of the protoplanetary cloud was taken as an input parameter. Terrestrial planets were found to form in all the simulations with a wide variety of chemical compositions so these planets might be very different from the Earth. In order to investigate the mineralogical characteristics of those systems we will present C/O and Mg/Si ratios in a sample of 71 and 380 stars with and without detected planets, respectively, using new high quality spectra from the HARPS GTO sample (Mayor *et al.* 2003) and very precise stellar parameters (Sousa *et al.* 2008).In addition we use high quality spectroscopic observations for 42 stars hosting planets from the CORALIE survey, using the same spectral tools to determine their stellar parameters (Santos *et al.* 2004, 2005), and thus ensuring that the final sample is homogeneous.

2. Abundances

For all the elements we performed a standard LTE analysis with the 2002 revised version of the spectral synthesis code MOOG (Sneden 1973) and a grid of Kurucz ATLAS9 atmospheres with overshooting (Kurucz 1993), by measuring the equivalent width (EW) of the different lines with the ARES program† (Sousa *et al.* 2007).

Mg and Si abundances were calculated using the line list of Neves *et al.* (2009), adding a Mg line at λ 6318.72 Å. On the other hand, C and O abundances were determined by measuring the equivalent widths of *CI* lines at λ 5380.3 Å and λ 5052.2 Å and *OI* forbidden line at λ 6300 Å. We removed from the sample stars with $T_{\rm eff} < 5100$ K since C abundance is not reliable for those stars. In addition, the spectral region around the forbidden line has telluric lines which can be blended with the *OI* line in some stars. So we made a detailed observation of the spectra to remove these objets from the sample in order to avoid wrong values of the O abundance. This, together with the limitation on $T_{\rm eff}$, makes a final sample of 69 and 270 stars with and without detected planets from HARPS, and 31 stars with planets from other surveys.

In Fig. 1 [X/Fe] ratios as a function of metallicity are plotted. There is an average overabundance in the total planet-host stars with respect to the comparison sample stars for all the elements. Since targets with planets are on average more metal-rich than the stars of comparison sample, their abundance distributions correspond to the extensions of the comparison sample trends at high metallicity. Such a trend supports the primordial

† The ARES code can be downloaded at http://www.astro.up.pt/ sousasag/ares/

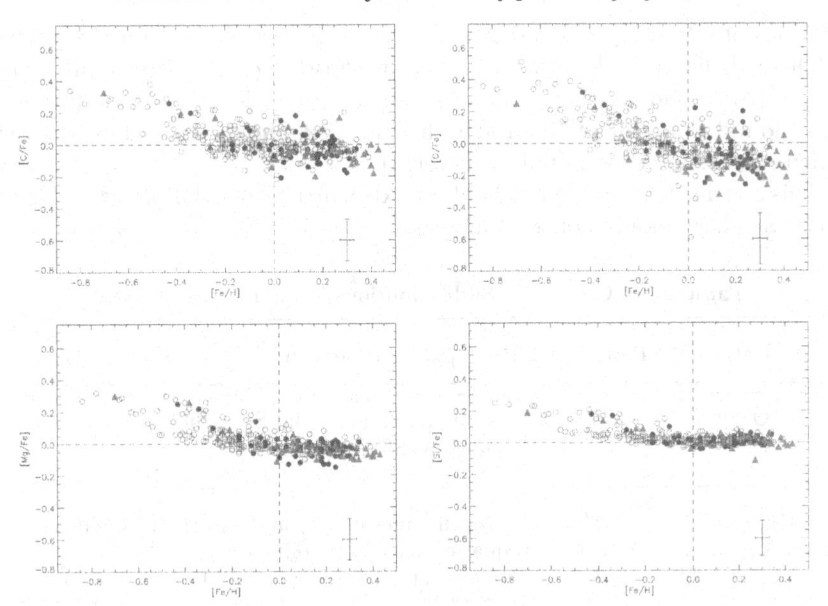

Figure 1. [X/Fe] ratios as a function of metallicity for C, O, Mg and Si.

scenario as an explanation of the overmetallicity of planet-host stars. The samples of stars both with and without detected planets behave quite similarly except for Mg. At subsolar metallicities all stars present high Mg abundances irrespective of $T_{\rm eff}$. However, this is not the case for [Fe/H] $\geqslant 0$, where stars without detected planets have higher Mg abundances. Nevertheless, this effect dissapears when we take into account only solar analogs, with $5600 < T_{\rm eff} < 5950$ K, perhaps due to the low number of stars with planets in this group. Therefore, it might be an effect in Mg abundances due to the presence of planetary companions.

3. C/O vs Mg/Si

In Fig. 2, C/O ratios as a function of Mg/Si are presented for different temperature ranges. These ratios are calculated as:

$$\mathrm{A/B} = N_{\mathrm{A}}/N_{\mathrm{B}} = 10^{\log\epsilon(\mathrm{A})}/10^{\log\epsilon(\mathrm{B})} \qquad (3.1)$$

where $\log\epsilon(\mathrm{A})$ and $\log\epsilon(\mathrm{B})$ are the absolute abundances, so they are not dependent on solar reference abundances.

In our sample, 34% of stars with known planets have C/O values greater than 0.8 (see Table 1), which means that under the assumption of equilibrium those systems will contain carbide-rich phases (such as graphite, SiC and TiC) in the innermost regions of the disk. Metallic Fe and Mg-silicates such as olivine (Mg_2SiO_4) and pyroxene ($MgSiO_3$) are also present and are located further from the host star. Terrestrial planets forming in these planetary systems are expected to be C-rich, containing significant amounts of C in addition to Si, Fe, Mg and O. Those systems may possess an alternative mass distribution profile for solid material, potentially making it easier either for giant planets to form closer to the host star than previously expected or for terrestrial planets to form in the inner regions of the disk (Bond *et al.* 2010b). However, we find no evidence of any trends with C/O values for either planetary period, semi-major axis or mass. As such, it appears that any effects of an alternative solid mass distribution due to high concentrations of refractory C-rich material are not preserved in the architecture of the system. This is believed to be due to the fact that Bond *et al.* (2010b) only considered equilibrium-driven condensation and did not include the effects of disequilibrium or the migration and radial mixing of material within the disk. Simulations addressing this issue are in progress. It should be noted, however, that we are still only able to detect giant planets. This conclusion may be not hold for extrasolar terrestrial planets which require significantly smaller amounts of solid material.

Table 1. C/O and Mg/Si distributions for stars with planets.

Ratio	Percentage	Principal Composition
C/O > 0.8	34%	graphite, TiC and solid Si as SiC
C/O < 0.8	66%	solid Si as SiO_4^{4-} or SiO_2
Mg/Si < 1	56%	pyroxene, metallic Fe and excess Si as feldspars
1 < Mg/Si < 2	44%	equal pyroxene and olivine
Mg/Si >2	0%	olivine and excess Mg as MgO

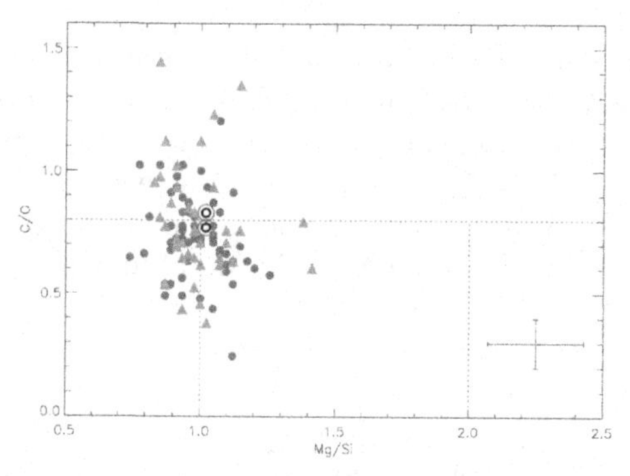

Figure 2. C/O vs Mg/Si for stars with planets from the HARPS GTO sample (red filled circles). Green triangles are stars with planets from other surveys. Solar abundances are calculated with the Kurucz Solar Atlas and the Harps daytime spectrum.

On the other hand, 66% of stars with known planets have C/O values lower than 0.8 and Si will be present in the solid form as SiO_4^{4-} or SiO_2, predominantly forming Mg-silicates. Silicate mineralogy will be controlled by Mg/Si ratio. For systems with a Mg/Si value between 1 and 2, the silicates present are predominately olivine and pyroxene in a condensation sequence closely resembling Solar. This is expected to result in the production of terrestrial planets similar in composition to that of Earth (in that their composition will be dominated by O, Fe, Mg and Si, with small amounts of Ca and Al also present). However, 56% of all planetary host stars in this study have a Mg/Si value less than 1. For such a composition, the solid component of the disk is dominated by approximately equal amounts of pyroxene and metallic Fe with minimal amounts of olivine present. Feldspars are also likely to be present as all available Mg is partioned into pyroxene, leaving excess Si available to form other silicate species.

Such an excess of Si content is predicted to produce a quartz-feldspar rich terrestrial planet with a composition more like that of Earth's continental crust material than that of Earth's olivine-dominated mantle. A composition such as this can have drastic implications for planetary processes such as plate tectonics and atmospheric composition. For example, volcanism on a Si-rich planet is expected to be intermediate to felsic in composition (i.e. >52% silica by weight) due to the potentially high SiO_2 content of the planet itself, producing igneous species such as andesite, rhyolite and granite. Eruptions may also be more explosive in nature due to the high viscosity of SiO_2-rich magma trapping volatiles within the magma.

Although we also found stars very similar to our Sun, it is clear that a wide variety of planets will probably exist within extrasolar planetary systems. These results can give us hints of what type of terrestrial planets we could find in different stars and help to guide the future surveys of low-mass planets.

References

Beirão, P., Santos, N. C., Israelian, G., & Mayor, M. 2005, *A&A*, 438, 251
Bodaghee, A., Santos, N. C., Israelian, G., & Mayor, M. 2003, *A&A*, 404, 715
Bond, J. C., Tinney, C. G., Butler, R. P., Jones, H. R. A., Marcy, G. W., Penny, A. J., & Carter, B. D. 2006, *MNRAS*, 370, 163
Bond, J. C., *et al.* 2008, *ApJ*, 682, 1234
Bond, J. C., Lauretta, D. S., & O'Brien, D. P. 2010a, *Icarus*, 205, 321
Bond, J. C., O'Brien, D. P., & Lauretta, D. S. 2010b, *ApJ*, 715, 1050
Boss, A. P. 1997, *Science*, 276, 1836
Boss, A. P. 2002, *ApJ*, 567, 149
Cai, K., Durisen, R. H., Michael, S., Boley, A. C., Mejía, A. C., Pickett, M. K., & D'Alessio, P. 2006, *ApJL*, 636, L149
Gilli, G., Israelian, G., Ecuvillon, A., Santos, N. C., & Mayor, M. 2006, *A&A*, 449, 723
Gonzalez, G. 1998, *A&A*, 334, 221
Gonzalez, G., Laws, C., Tyagi, S., & Reddy, B. E. 2001, *AJ*, 121, 432
Gonzalez, G. & Laws, C. 2007, *MNRAS*, 378, 1141
González Hernández, J. I., Israelian, G., Santos, N. C., Sousa, S. G., Delgado Mena, E., Neves, V., & Udry, S. 2010, *ApJ*, 720, 1592
Fischer, D. A. & Valenti, J. 2005, *AJ*, 622, 1102
Kurucz, R. L. 1993, *CD-ROMs, ATLAS9 Stellar Atmospheres Programs (Cambridge: Smithsonian Astrophys. Obs.)*
Mayor, M. & Queloz, D., *et al.*2003, *The Messenger*, 114,20
Neves, V., Santos, N. C., Sousa, S. G., Correia, A. C. M., & Israelian, G. 2009, *A&A*, 497, 563
Pollack, J. B., Hubickyj, O., Bodenheimer, P., Lissauer, J. J., Podolak, M., & Greenzweig, Y. 1996, *Icarus*, 124, 62
Sadakane, K., Ohkubo, M., Takeda, Y., Sato, B. Kambe, E., & Aoki, W. 2002, *PASJ*, 54, 911
Santos, N. C., Israelian, G., & Mayor, M. 2000, *A&A*, 363, 228
Santos, N. C., Israelian, G., & Mayor, M. 2001, *A&A*, 373, 1019
Santos, N. C., Israelian, G., & Mayor, M. 2004, *A&A*, 415, 1153
Santos, N. C., Israelian, G., Mayor, M., Bento, J. P., Almeida, P. C., Sousa, S. G., & Ecuvillon, A., 2005, *A&A*, 437, 1127
Sousa, S. G., Santos, N. C., Israelian, G., Mayor, M., & Monteiro, M. J. P. F. G. 2007, *A&A*, 469, 783
Sousa, S. G., Santos, N. C., Mayor, M., Udry, S., Casagrande, L., Israelian, G., Pepe, F., Queloz, D., & Monteiro, M. J. P. F. G. 2008, *A&A*, 487, 373
Sneden, C. 1973 Ph.D Thesis, University of Texas.
Takeda, Y. 2007, *PASJ*, 59, 335

The Astrophysics of Planetary Systems: Formation, Structure, and Dynamical Evolution
Proceedings IAU Symposium No. 276, 2010
A. Sozzetti, M. G. Lattanzi & A. P. Boss, eds.

doi:10.1017/S1743921311019892

Precise characterisation of exoplanet-host stars parameters

Sylvie Vauclair[1]

[1]Institut de Recherches en Astrophysique et Planetologie,
14 avenue Edouard Belin, 31400, Toulouse, France
email: sylvie.vauclair@ast.obs-mip.fr

Abstract. Studying the internal structure of exoplanet-host stars compared to that of similar stars without detected planets is particularly important for the understanding of planetary formation. In this framework, asteroseismic studies represent an excellent tool for a better characterization of stars and for a precise determination of the stellar parameters like mass, radius, gravity, effective temperature. The detection of stellar oscillations is obtained with the same instruments as used for the discovery of exoplanets, both from the ground and from space, although the time scales are very different. Here I discuss some details about the characterization of exoplanethost stars from seismology and the importance of the helium and heavy element abundances in this respect.

Keywords. stars: abundances, stars: oscillations, stars: interiors

1. Introduction

Detailed studies of exoplanets need precise knowledge of the central stars of planetary systems. In this framework, the best way to obtain precise values of the stellar parameters is asteroseismology combined with spectroscopic observations.

The observations for stellar oscillations and exoplanet searches are done with the same instruments. In some cases, the same observations, analysed on different time scales, can lead to both planet detection and seismic studies. This was the case for the star μ Arae, observed with HARPS during eight nights in June 2004: these observations, aimed for asteroseismology, lead to the discovery of the exoplanet μ Arae d (Santos *et al.* 2004). Another case is that of the planet discovered around the extreme horizontal branch star V391 Peg (Silvotti *et al.* 2007), which was detected owing to the seismic period variations (the so-called "time delay method").

Generally speaking, the goals of the asteroseismology of exoplanets-host stars are to derive precisely their masses, radii, ages, evolutionary stages, and internal structure. At the present time, with a detailed seismic analysis, it is possible to obtain the masses and radii of solar-type stars with a precision of a few 1 percent, from both ground and space observations (C.f. Vauclair *et al.* 2008; Metcalfe *et al.* 2010).These precise values are important for the determination of the parameters of the planets. They can also be compared to those of the Sun and stars without detected planets, to obtain hints about the theories of planetary formation and migration.

2. Basics of asteroseismology for slowly rotating solar-type stars

In solar-type stars, acoustic waves are permanently created by the motions which occur in their outer layers, induced by convection and related processes. The waves are damped, but as other waves always appear, the stellar sphere globally behaves as a resonant cavity, and the oscillations can be treated, with a very good approximation, as standing waves.

Each mode can be characterized with three numbers: the number of nodes in the radial direction, n, and the two tangential numbers ℓ and m, which appear in the development of the waves on the spherical harmonics:

$$Y_l^m(\theta,\phi) = (-1)^m C_{l,m} P_l^m(cos\theta)exp(im\theta) \tag{2.1}$$

Several combinations of the oscillation frequencies are used to obtain more precise constraints on the stellar internal structure, like the "large separations", which are defined as the differences between two consecutive modes of the same ℓ number, and the "small separations", defined as:

$$\delta\nu = \delta\nu_{n,l} - \nu_{n-1,l+2} \tag{2.2}$$

For acoustic modes in a given star, the large separations are nearly constant, and their average value is approximately half the inverse of the stellar acoustic time, i.e. the time needed for the $l = 0$ waves to travel along the whole radius. However detailed studies of these frequency separations as a function of the frequency itself, give evidence of characteristic modulations. They are related to the regions of partial reflection of the waves inside the stars, which occur for rapid variations of the sound velocity. The modulations of the small separations are directly related to the structure of the stellar cores (Roxburgh & Vorontsov 1994; Roxburgh 2007; Soriano *et al.* 2007; Soriano & Vauclair 2008), whereas those of the large separation give information on the stellar upper layers, like the regions of helium ionization, the bottom of the convective zones or the diffusion-induced helium gradient (Vauclair & Théado 2004; Castro & Vauclair 2006).

3. Asteroseismology of exoplanet-host stars

Aseroseismology of exoplanet-host stars is a useful tool to determine their internal structure and their behaviour with respect to stars without detected planets. Let us recall that, due to the present observation bias, many stars may very well host undetected planets. In this framework, the special characteristics of exoplanet-host stars compared to the other ones are more related to the migration process of the planets than to their mere existence.

The most important goals at the present time are the determination of masses, radii and ages of exoplanet-host stars with the best possible precision. When comparing the observed frequencies and frequency differences with the computed ones, one must take into account the details of the chemical composition.

Most exoplanet-host stars are overmetallic, compared to the other stars and to the Sun (e.g. Santos *et al.* 2005). This overmetallicity is not due to accretion of planetary material, as the accreted heavy matter falls down due to thermohaline convection induced by the unstable μ-gradient (Vauclair 2004). It was present in the instersellar cloud out of which the stellar system formed, and this must be taken into account in the models.

A crucial unknown parameter, very important for the determination of the stellar characteristics, is the helium abundance. Unfortunately, helium is not directly observable in the spectra of solar-type stars. Contrary to what is often admitted, it cannot be safely related to the overmetallicity, as the helium versus metal enrichment may strongly vary inside a galaxy. For the Hyades, for example, the helium content is much smaller than could be expected from the metal enrichment (Lebreton *et al.* 2001).

When comparing the observed frequencies and frequency differences for a given star with those of different models computed with different helium contents, one find the same gravities (log g) with a precision better than 1 percent, very close radii and masses. On

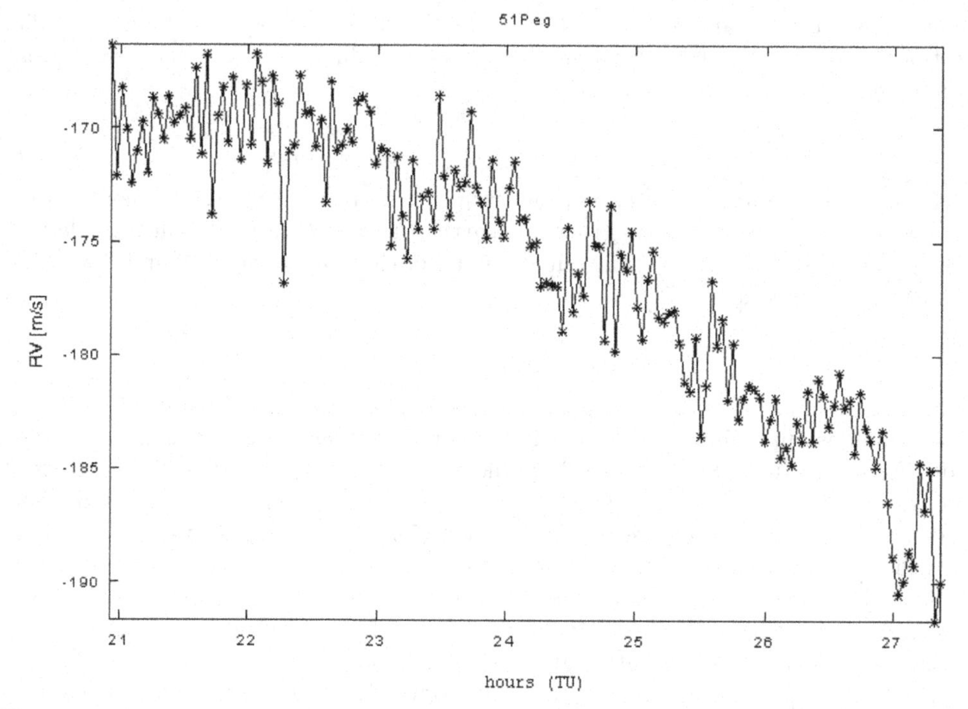

Figure 1. The star 51 Peg has been observed for asteroseismology with the SOPHIE spectrometer, at Haute Provence Observatory, in August 2007 and 2010. The graph presents part of the radial velocity curve, obtained during one night. The solar-like oscillations are clearly visible. The overall decrease of the radial velocity during the night is the signature of the exoplanet discovered by Mayor and Queloz 1995

the other hand, ages may differ by large values. Detailed seismic analysis can help solving this question, with the help of spectroscopy, as has been proved for ι Hor (Vauclair *et al.* 2008) and μ Arae (Soriano & Vauclair 2010).

Up to now four exoplanet-host stars have been observed and deeply analyzed from the ground on relatively long periods (8 or 9 consecutive nights) with the HARPS and SOPHIE spectrometers. They are: μ Arae (HARPS, 2004) (Bazot *et al.* 2005; Soriano & Vauclair 2010); ι Hor (HARPS, 2006) (Vauclair *et al.* 2008), 94 Cet (HARPS, 2007) and 51 Peg (SOPHIE, 2007 and 2010) (Bazot *et al.* 2011 in preparation, Figure 1). Several of them have been observed from space, but the most detailed analysis is that of the main COROT target HD52265 (Ballot *et al.* 2011).

The asteroseismology of exoplanets-host stars is a new field which will become more and more important in the near future.

4. Conclusion

From the few examples already available, asteroseismology has proved to be a powerful tool for determining stellar parameters, in particular for exoplanet-host stars. The scientific community is well prepared for future observations with on-going or planned projects. Space missions like KEPLER, and later on PLATO, are expected to give a large amount of new data for seismology, besides planet searches. Recent studies of solar-type stars with Kepler lead to radii determinations with a precision of 1 percent (Metcalfe *et al.* 2010). Meanwhile ground based instruments devoted to exoplanets like HARPS or SOPHIE can be used for seismology. All these observations will give the possibility of using the seismic tests with a precision never obtained before. They will lead to better parameters for exoplanet-host stars and help for a better understanding of planetary formation and evolution.

References

Bazot, M. & Vauclair, S. 2004, *A&A*, 427, 965
Bazot, M., Vauclair, S., Bouchy, F., & Santos, N. 2005, *A&A*, 440, 615
Ballot, J., Gizon, L., *et al.* 2011 in preparation
Bouchy, F., Bazot, M., Santos, N., Vauclair, S., & Sosnowska, D. 2005, *A&A*, 440, 609
Castro, M. & Vauclair, S. 2006, *A&A*, 456, 611
Lebreton, Y., Fernandes, J., & Lejeune, T. 2001, *A&A*, 374, 540
Mayor, M. & Queloz, D. 1995, *Nature*, 78, 355
Metcalfe, T. S., Monteiro, M. J. P.. F. G.., Thompson, M. J. *et al.* 2010, *ApJ*, 723, 1583
Roxburgh, I. W., & Vorontsov, S. V. 1994, *MNRAS*, 267, 297
Roxburgh, I. W. 2007, *MNRAS*, 379, 801
Santos, N. C., Israelian, G., & Mayor, M. 2004, *A&A*, 415, 1153
Santos, N. C., Israelian, G., Mayor, M., Bento, J. P., Almeida, P. C., Sousa, S. G., & Ecuvillon, A. 2005, *A&A*, 437, 1127
Silvotti, R., Schuh, S., Janulis, R., *et al.* 2007, *Nature*, 449, 189
Soriano, M., Vauclair, S., Vauclair, G., & Laymand, M. 2007, *A&A*, 471, 885
Soriano, M. & Vauclair, S. 2008, *A&A*, 488, 975
Soriano, M. & Vauclair, S. 2010, *A&A*, 513, 49
Vauclair, S. 2004, *ApJ*, 605, 874
Vauclair, S. & Théado, S. 2004, *A&A*, 425, 179
Vauclair, S., Laymand, M., Bouchy, F., Vauclair, G., Hui Bon Hoa, A., Charpinet, S., & Bazot, M. 2008, *A&A*, 482, L5

The Astrophysics of Planetary Systems: Formation, Structure, and Dynamical Evolution
Proceedings IAU Symposium No. 276, 2010
A. Sozzetti, M. G. Lattanzi & A. P. Boss, eds.

doi:10.1017/S1743921311019909

Kepler mission highlights

William J. Borucki[1], David G. Koch[1] and the *Kepler* Team

[1]Space Science Directorate, Mail Stop 244-30, NASA Ames Research Center Moffett Field, CA 94035
email: William.J.Borucki@NASA.gov

Abstract. During the first 33.5 days of science-mode operation of the Kepler Mission, the stellar flux of 156,000 stars were observed continuously. The data show the presence of more than 1800 eclipsing binary stars, over 700 stars with planetary candidates, and variable stars of amazing variety. Analyses of the commissioning data also show transits, occultations and light emitted from the known exoplanet HAT-P7b. The depth of the occultation is similar in amplitude to that expected from a transiting Earth-size planet and demonstrates that the Mission has the precision necessary to detect such planets. On 15 June 2010, the Kepler Mission released most of the data from the first quarter of observations. At the time of this data release, 706 stars from this first data set have exoplanet candidates with sizes from as small as that of the Earth to larger than that of Jupiter. More than half the candidates on the released list have radii less than half that of Jupiter. Five candidates are present in and near the habitable zone; two near super-Earth size, one similar in size to Neptune, and two bracketing the size of Jupiter. The released data also include five possible multi-planet systems. One of these has two Neptune-size (2.3 and 2.5 Earth-radius) candidates with near-resonant periods as well as a super-Earth-size planet in a very short period orbit.

Keywords. space vehicles, telescopes, planetary systems, techniques: photometric

1. Introduction

Kepler is a Discovery-class mission designed to determine the frequency of Earth-size planets in and near the habitable zone (HZ) of solar-type stars. In Sections 2 and 3, the Mission is summarized and the data described. Discoveries are described in Sections 4 through 9. Sections 10 and 11 discuss the availability of the data and provide a summary.

2. Kepler Mission description

The instrument is a wide field-of-view photometer designed to monitor over 150,000 stars for at least 3.5 years with enough precision to find Earth-size planets in the habitable zone. The instrument design is shown in Figure 1. The instrument has a 0.95 meter aperture and a spectral bandpass with a throughput greater than 40% from 430 to 820 nm and a 5% response cutoff at 423 and 897 nm. The shortwave cutoff reduces stellar variability due to Calcium H and K lines. The instrument monitors a single large area of the sky centered on R.A. = $19^h 22^m 40^s$ and decl. = $44°30^{min}00^{sec}$ and is in a 53-week long heliocentric orbit to allow continuous observations of the targets to avoid missing transits

The focal plane of the Schmidt-type telescope contains 42 CCDs with a total of 95 megapixels that cover 115 square degrees of sky. Kepler was launched into an Earth-trailing heliocentric orbit on 6 March 2009, finished its commissioning on 12 May 2009, and is now in science operations mode. Further details of the Kepler Mission and instrument can be found in Koch *et al.* (2010a), Jenkins *et al.* (2010a), and Caldwell *et al.* (2010).

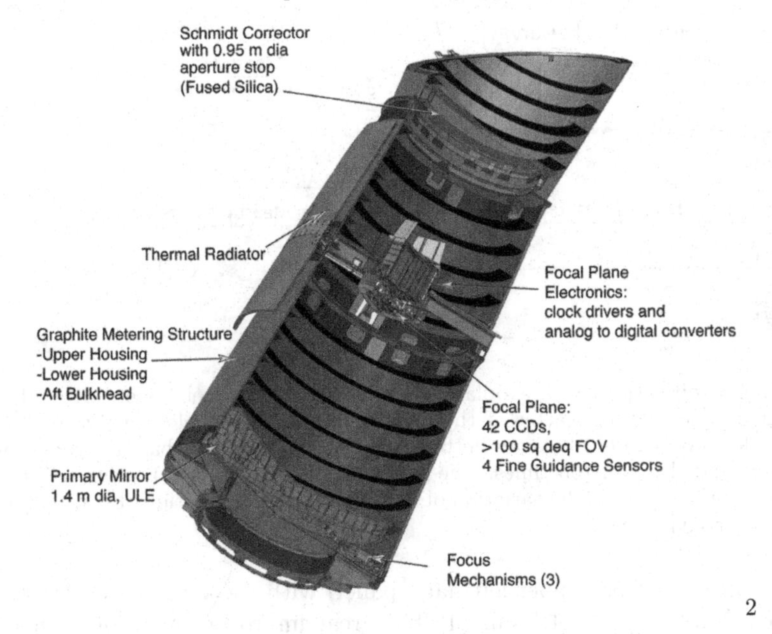

2

Figure 1. Diagram of the *Kepler* Instrument

3. Data description

During the first 33.5 days of science-mode operation, 156, 097 stars were observed. Five new exoplanets with sizes between 0.3 and 1.6 Jupiter radii and orbital periods from 3.2 to 4.9 days were confirmed by radial velocity observations (Borucki *et al.* 2010; Koch *et al.* 2010a; Dunham *et al.* 2010; Jenkins *et al.* 2010b; Latham *et al.* 2010). Later, three additional planets were confirmed using a combination of RV observations and variations in the times of transits (Holman *et al.* 2010; Torres *et al.* 2011). Most of the results shown in this paper are based on the first data segment taken at the beginning of science operations on 13 May 2009 UT and finished on 15 June 2009 UT; a 33.5-day segment (labeled Q1). Additional data were used to confirm the discovery of three planets detected by transit timing variations and 9.7 days of data (labeled *Q*0) taken during commissioning of spacecraft and instrument were used to analyze the exoplanets known to be in the *Kepler* field-of-view (FOV).

The target stars are primarily main sequence dwarfs chosen from the Kepler Input Catalog (KIC). These were chosen to maximize the number of stars that were both bright and small enough to show detectable transit signals for small planets in and near the HZ (Batalha *et al.* 2010). Most stars were in the magnitude range $9 < Kp < 16$. Data for all stars are recorded at a cadence of one per 29.4 minutes (hereafter, long cadence, or LC). Data for a subset of 512 stars are also recorded at a cadence of one per 58.5 seconds (hereafter, short cadence or SC), sufficient to conduct asteroseismic observations needed for measurements of the stars' size, mass, and age. The results presented here are based only on LC data. For a full discussion of the LC data and their reduction, see Jenkins *et al.* (2010b,c). See Gilliland *et al.* (2010) for a discussion of the SC data.

4. Detection of Radiation from HAT-P-7b

The Q0 data were analyzed to investigate the transit characteristics of the previously known giant transiting exoplanet HAT-P-7b (Pal *et al.* 2008). A comparison of the

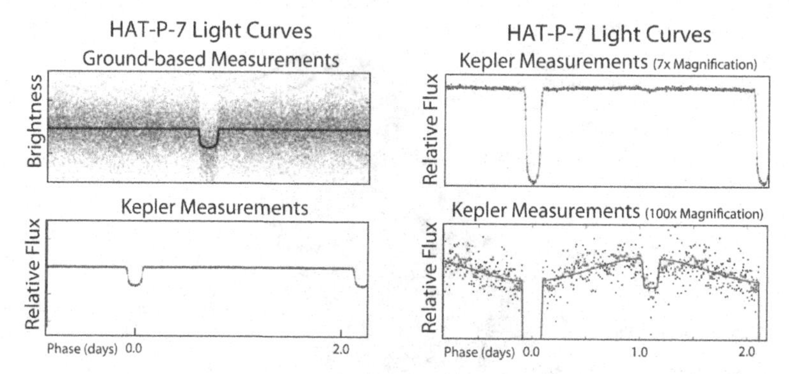

Figure 2. Kepler observations of exoplanet HAT-P-7b. Left panel upper: Ground based observations. Left panel lower: Scatter of the data points in the Kepler data is within the line thickness. Kepler precision is 100 times better than that from ground-based observations. Right panel; Scale expanded 7 and 100 times. The transit is off scale on the bottom half of the right panel, but the occultation and the variation of light from the combination of starlight and planet emission are clearly visible.

ground based observations (upper left hand panel) with those from *Kepler* is presented in the lower left hand panel of Figure 1. The great improvement in photometric precision obtained for space-based operation compared to the ground-based measurement is evident in the left hand panel. In the right hand panel, the *Kepler* data for HAT-P-7b show a smooth rise and fall of light from the planet as it orbits its star, punctuated by a drop of 130 ± 11 ppm in flux when the planet passes behind its star. We interpret this as the phase variation of the dayside thermal emission plus reflected light from the planet as it orbits its star and is occulted. The depth of the occultation is similar in significance to the detection of a transiting Earth-size planet for which the mission was designed.

Kepler's photometric detection of the optical phase curve and occultation of HAT-P-7b confirms the prediction of Raymond *et al.* (2006). The depth of the occultation and the shape and amplitude of the phase curve indicate that HAT-P-7b could have a strongly absorbing atmosphere and inhibited advection to the night side. If the planet has a completely absorbing atmosphere, its dayside temperature is estimated to be $2650{\pm}100K$. The position in phase of the occultation is consistent with zero orbital eccentricity, as expected from the radial velocity variation.

5. Validation of Discoveries

After the beginning of science operations, several thousand candidates were recognized, but most of these were quickly shown to be false positive events, usually energetic-particle events or data breaks. A comprehensive program was initiated to confirm the candidates as planets before announcing them to the public.

The search for planets starts with a search of the time series of each star for a pattern that exceeds a detection threshold commensurate with a non-random event. Observed patterns of transits consistent with those from a planet transiting its host star are labeled "planetary candidates." Those that were at one time considered to be planetary candidates but subsequently failed some consistency test are labeled "false positives". After passing the consistency tests listed below, and only after a review of all the evidence by the entire *Kepler* Science Team, does the candidate become a *confirmed* or *validated* exoplanet.

List of Tests Conducted Prior to Radial Velocity Observations;

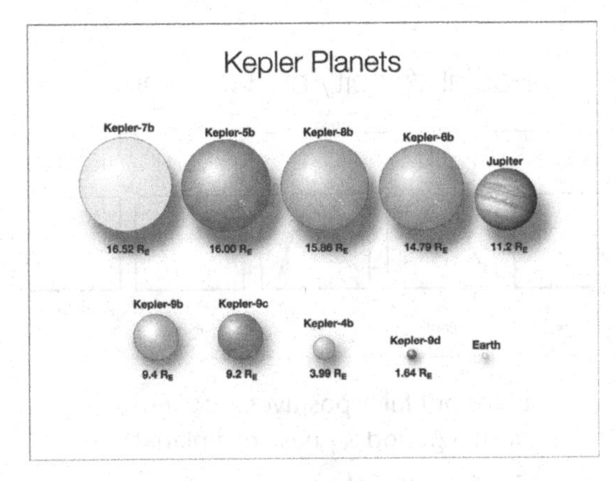

Figure 3. Planets discovered and confirmed by the Kepler Mission in 2010.

- SNR $> 7.1\sigma$ to rule out statistical fluctuations
- At least three transits to confirm orbital periodicity
- Light curve depth, shape, and duration must be consistent with a Keplerian orbit
- Search for secondary eclipses
- Centroid analysis to identify signals from background stars

6. Confirmation of Early Discoveries

After the first month of data were transmitted to ground and analyzed, only a few weeks remained to make radial velocity (RV) measurements before the star field set for ground-based observations. There was sufficient time during the Fall of 2009 to make the necessary RV observations to confirm five short-period planets; *Kepler* 4 through 8. See top row of Figure 3.

Kepler$-5b$, $-6b$, $-7b$, and $-8b$ are larger and less dense than Jupiter. Kepler$-8b$ with a density of 0.17 gr/cc is one of the lowest density planets ever discovered. Kepler$-4b$ is similar in mass and density (1.91 gr/cc) to Neptune (1.67 gr/cc), but with a stellar flux nearly a million times larger because it is only 0.046 AU from it star. Its radiative equilibrium temperature is 1650 K. Because of such high irradiation levels, such planets were expected to have a much lower density (Fortney 2008; Chabrier *et al.* 2009).

Although RV observations have been critical to the confirmation of most of these planets and to the determination of their masses and densities, the present-day RV precision of 1m/s for bright stars with sharp spectral lines is insufficient for low-mass planets in long period orbits around dim stars; i.e., for most of the planetary candidates found by the Kepler Mission.

The relative sizes of the measurement error bars in Figure 4 shows the increasing difficulty of confirming photometric detections as the planet mass decreases. For longer period orbits, less massive planets, and dimmer stars, the precision of the RV observations will usually be too low to confirm the photometrically discovered planets and therefore other methods must be employed for confirmation and validation.

For the case of a low-mass planet, the reflex orbital motion of the star will often be below the limit of detectability. However, under favorable circumstances, the distortion of stellar spectral lines caused by a transiting planets successively blocking Doppler shifted portions of the stellar surface during a transit (the Rossiter-McLaughlin effect) may produce an anomalous Doppler signal that is detectable even for the transit of an Earth-size planet (Gaudi *et al.* 2007; Welsh & Orosz 2007). See Figure 5. Although the signal

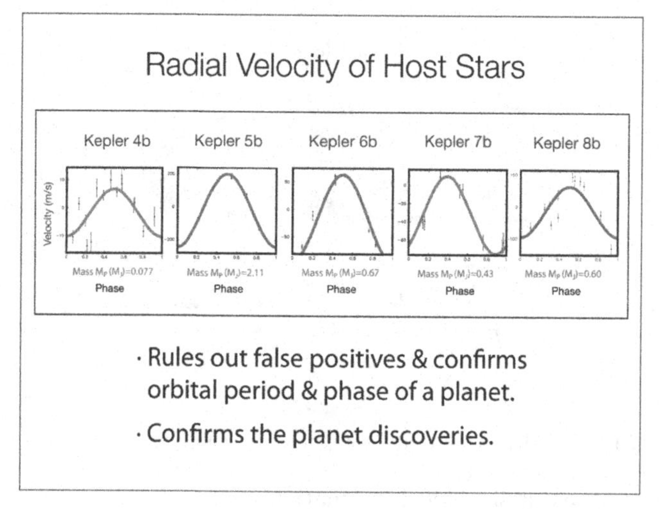

Figure 4. RV confirmation of first five Kepler planets.

cannot be used to derive the mass of the planet, it does rule out the possibility that a background binary is the cause of the signal. The Rossiter-McLaughlin effect (Winn *et al.* 2005) helped to confirm Kepler-8b (Jenkins *et al.* 2010b).

Another method used to confirm planets is the mutual gravitational interaction that causes variations in the transit times. The interactions are proportional to the planet masses and are strongest for resonant or near-resonant orbits. This situation allowed a three-planet system to be announced; Kepler$-9b$,$-c$, and $-d$ (Holman *et al.* 2010; Torres *et al.* 2011). The size distribution of the three planets is shown in the bottom panel of Figure 3. Only the two Saturn size planets are in near-resonant orbits and thus show strong transit timing variations. See Figure 6. The third planet Kepler$-9d$ is in a 1.59 day period and not significantly perturbed by the two outer planets (Torres *et al.* 2011).

A comparison of the distributions shown in Figure 7 indicates that the majority of the candidates discovered by *Kepler* are Neptune-size (i.e., $3.8R_{\oplus}$) and smaller; in contrast, the planets discovered by the transit method and listed in the Extrasolar Planets Encyclopedia (EPE) are typically Jupiter-size (i.e., $11.2R_{\oplus}$) and larger. This difference is understandable because of the difficulty in detecting small planets when observing through the Earth's atmosphere and because of the day/night cycle which causes most transits to be missed.

The *Kepler* results shown in Figure 7 imply that small candidate planets with periods less than 33.5 days are much more common than large candidate planets with periods less than 33.5 days and that the ground-based discoveries are potentially sampling only the upper tail of the size distribution (Gaudi *et al.* 2005; Gould *et al.* 2003). Note that for a substantial range of planet sizes, an R^{-2} curve fits the *Kepler* data well. Because it is much easier to detect larger candidates than smaller ones, the result is robust; it implies that the frequency of planets decreases with the area of the planet, assuming that the false positive rate and other biases are independent of planet size for planets larger than 2 Earth radii.

7. Characteristics of the Detected planets

Most of the planets shown in Figure 8 have short orbital periods and are therefore quite hot; hotter than molten lava and self-luminous. As the mission duration increases,

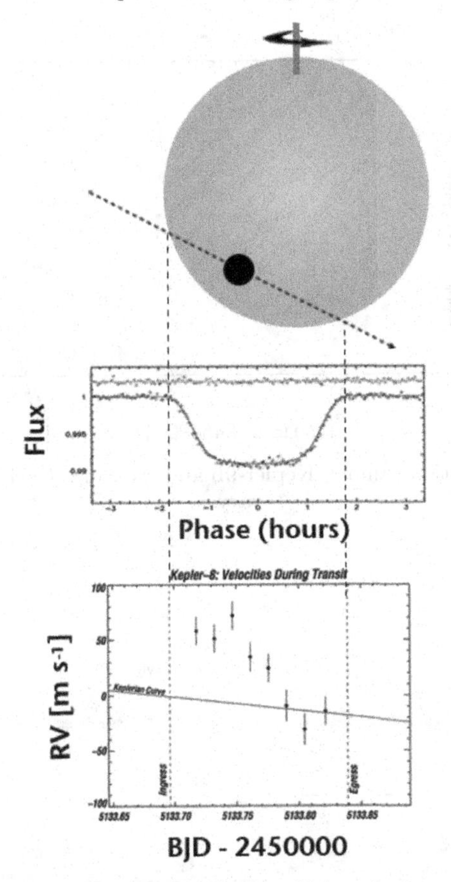

Figure 5. Rossiter-McLaughlin effect for Kepler-8b. Although the orbit plane is not perpendicular to the stellar spin axis, the transit shape is normal, but a strong asymmetrical shift in the spectral lines occurs.

planets with longer orbital period and cooler temperatures can be found and confirmed. The two Saturn-size planets (Kepler-9b and -9c) have orbital periods 19.2 days and 38.9 days and are substantially cooler. Figure 9 shows that four of the first five Kepler planets have densities less than that of a hydrogen-helium gas mixture of similar mass. Many of the planets discovered by ground-based observations (open circles) are consistent with the result. It is not known whether the calculation of the state properties of such mixtures must be revised or if it is the measurements that need revision.

Kepler−9*b* and −9*c* are the outer two planets and have temperatures lower than that of Mercury and Venus. The inner most planet Kepler−9*d* has a temperature well above the melting point of iron. It was validated by ruling out other explanations because the RV observations did not have sufficient precision to confirm this planet (Torres *et al.* 2011). It's density is undetermined and thus is not shown in figure 9.

8. Additional Highlights

The four star clusters shown in Figure 8 are in the Kepler FOV and span a very large range of ages, temperatures, and rotation periods. An investigation is underway to use the Kepler measurements to better define the surface relating these parameters. When the study is complete and the surface calibrated, it should be possible to determine

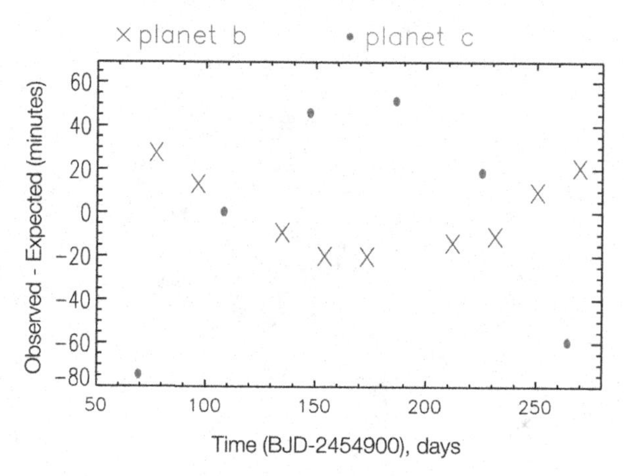

Figure 6. Mutual interactions among Kepler-9b and -9c cause the observed transit times to vary.

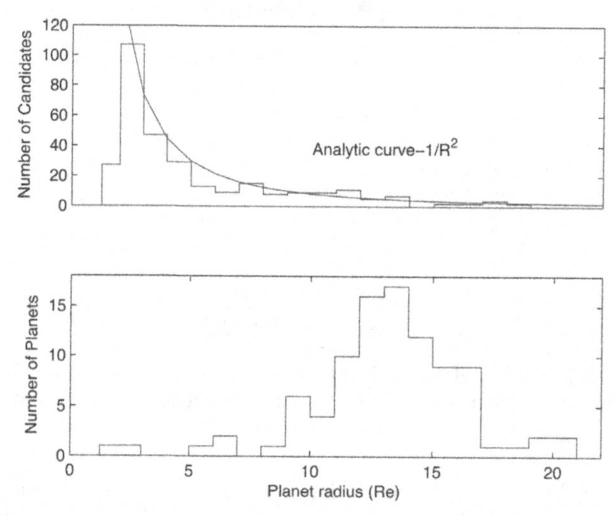

Figure 7. Size distribution. Upper panel: Kepler candidates. Lower Panel: Non-Kepler candidates planets discovered by the transit method and listed in the EPE as of 2010 December 7 (without Kepler planets).

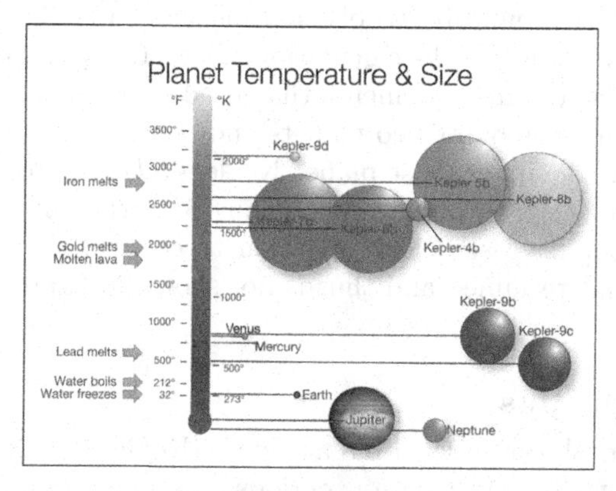

Figure 8. Relative sizes and temperatures of Kepler planets. Comparisons of candidate temperatures to melting points of common materials.

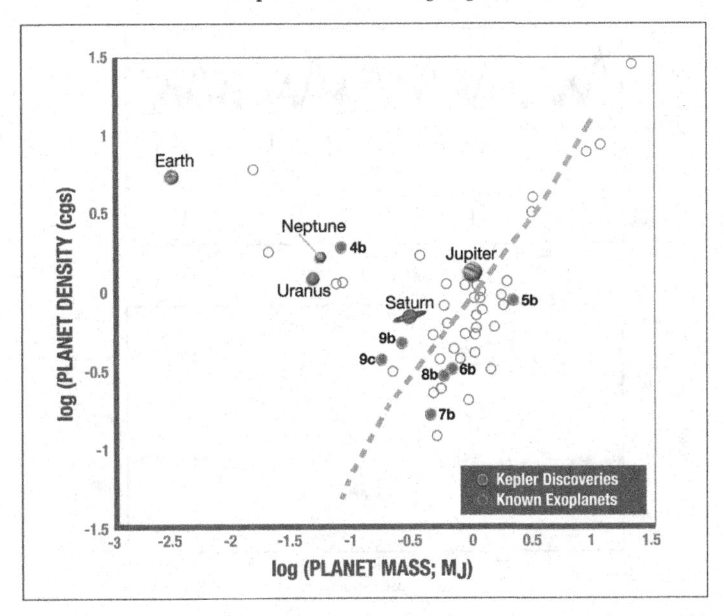

Figure 9. Comparison of confirmed planets with Solar system planets and expected density curve for gas giant planets. The dashed curve is a calculation of the density versus mass for a self gravitating hydrogen/helium sphere (Fortney 2008)

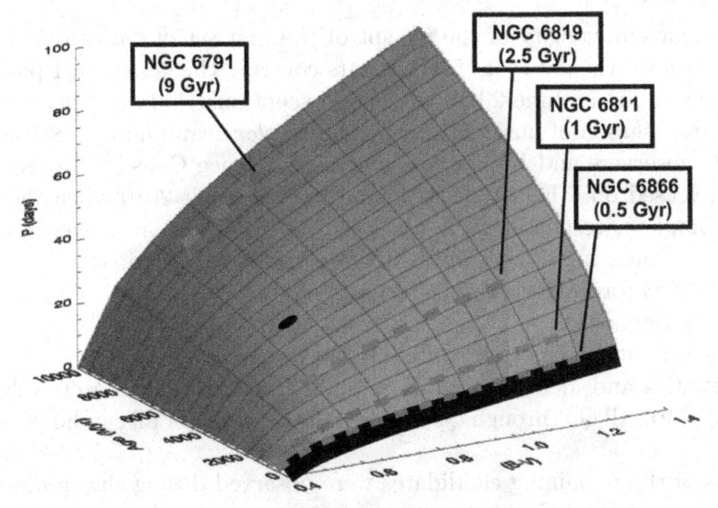

Figure 10. Expected relationship between stellar temperature, age, and rotation period. Black area marks the currently studied star clusters. Blue lines labels the four clusters being investigated using Kepler data.

stellar ages of field stars from measurements of their temperature and rotation period. The current relationship is tentative and based only on young clusters (marked in black) and the Sun's characteristics (Skumanich 1972; Barnes 2003).

The light curve shown in Figure 11 is one example of the many surprises that the Mission has produced. The observations indicate that the star temperature is 9400 K and the companion temperature is 12, 200 K. The size of the companion is 0.8 the radius of Jupiter. The light curves might be explained by an orbiting white dwarf, but white dwarfs are expected to be much smaller; approximately Earth-size. This example is one of two found in the first data set (Rowe *et al.* 2010).

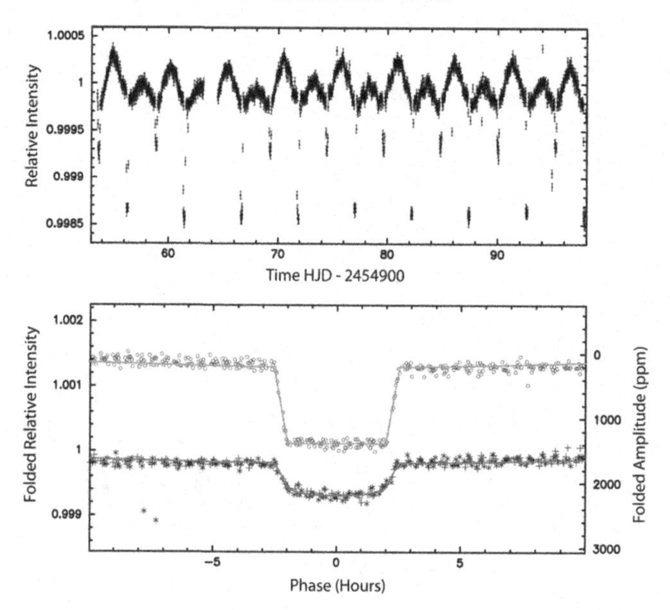

Figure 11. Light curve (upper panel) and folded light curves (bottom panel) showing 500 ppm transit and a 1300 ppm occultation.

9. Data Release

At the one-year anniversary of the receipt of the first set of data from the beginning of science operations, the data for 156, 097 stars covering the $Q0$ and $Q1$ periods became available to the public in June 2010, with two exceptions: 400 stars held back to allow completion of one season of observations by the *Kepler* team, and 2778 stars held back for the Guest Observers and Kepler Asteroseismic Science Consortium (KASC). These data will be released on 1 February 2011, and in November 2010 when the proprietary periods are complete, respectively. A total of 152, 919 stars are currently available at several levels of processing at the Multi-Mission Archive at the Space Telescope Science Institute (MAST †) for analysis by the community.

Summary of observations

- Over 700 stars with planetary candidates have been found.
- Characteristics and identification of 305 planetary candidates were released to the public in June 2010. All $Q0$ through $Q2$ data will be released for all candidates in February 2011.
- Hundreds of the remaining candidates were observed during the summer of 2010.
  - 800 high precision RV measurements of 45 stars
  - Active optics observations of 83 stars
  - Speckle observations of 72 stars
  - 750 reconnaissance spectra obtained for 550 stars
  - Centroid-movement plots for 129 stars.

A list of false positive events found in the released data is also available at the MAST. The identification of the false positives should help the community avoid wasting observation resources.

Characterization of the candidates is based on the stellar magnitude, effective temperature, and surface gravity of the star taken from the *KIC*. Also listed in the MAST archive are the orbital period, epoch, and an estimate of the size of each candidate.

† $http://archive.stsci.edu/kepler/kepler_fov/search.$

When only one transit is seen in the data, the epoch and period are calculated using data obtained subsequently. More information on the characteristics of each star can be obtained from the *KIC*. Several of the target stars show more than one series of planetary transit-like events and therefore have more than one planetary candidate. These candidate multi-planet systems are of particular interest to investigations of planetary dynamics.

10. Summary

The *Kepler* instrument is functioning well and many planetary candidates are being discovered. Data for over 300 candidates have been released so that the community can independently confirm the planets. Data for an additional several hundred candidates will be released in February 2011 when the analysis of this summer's follow up are available and when more of the false positive events have been identified and removed.

Acknowledgments

Funding for this Discovery mission is provided by the NASA Science Mission Directorate.

References

Barnes S. A. 2003, *ApJ*, 586, 464
Batalha, N. M, *et al.* 2010, *ApJL*, 713, L109
Borucki, W. J., *et al.* 2009, *Science*, 325, 709
Borucki, W. J., *et al.* 2010, *ApJL*, 713, L126
Caldwell, D. A., *et al.* 2010, *ApJL*, 713, L92
Chabrier, G., *et al.* 2009, *AIP Conf. Proc.*, 1094, 102
Dunham, E. W., *et al.* 2010, *ApJ*, 713, L136
Fortney, J. J. 2008 *ASP Conf. Ser.*, 398, 405
Gaudi, B. S., *et al.* 2005, *ApJ*, 628, L73
Gaudi, B. S., *et al.* 2007, *ApJ*, 655, 550
Gould, A., *et al.* 2003, *ApJL*, 591, L53
Gilliland, R. L., *et al.* 2010, *ApJL*, 713, L160
Holman, M. J., *et al.* 2010, *Science*, 330, 51
Jenkins, J. M., *et al.* 2010, *ApJL*, 713, L87
Jenkins, J. M., *et al.* 2010, *ApJ*, 724, 1108
Jenkins, J. M., *et al.* 2010, *ApJL*, 713, L120
Koch, D. G., *et al.* 2010, *ApJL*, 713, L79
Koch, D. G., *et al.* 2010, *ApJL*, 713, L131
Latham, D. W., *et al.* 2010, *ApJL*, 713, L140
Pal, A., *et al.* 2008, *ApJ*, 680,1450
Raymond, S. N., *et al.* 2006, *Science*, 313, 1413
Rowe, J. F., *et al.* 2010, *ApJL*, 713, L150
Skumanich, A. 1972, *ApJ*, 171, 565
Torres, G., *et al.* 2010, *ApJ*, 727, 24
Welsh, W. F. & Orosz, J. A. 2007, *ASP Conf Ser.*, 366, 176
Winn, J. N., *et al.* 2005, *ApJ*, 631, 1215

The Astrophysics of Planetary Systems: Formation, Structure, and Dynamical Evolution
Proceedings IAU Symposium No. 276, 2010
A. Sozzetti, M. G. Lattanzi & A. P. Boss, eds.

doi:10.1017/S1743921311019910

CoRoT mission highlights

Magali Deleuil[1], Pascal Bordé[2], Claire Moutou[1] and the *CoRoT* exoplanet science team

[1]Laboratoire d'Astrophysique de Marseille,
38 rue F. Joliot-Curie, 13388 Marseille Cedex 13, France
email: magali.deleuil@oamp.fr
email: claire.moutou@oamp.fr

[2]Institut d'Astrophysique Spatiale, Bât. 121, 91405 Orsay, France
email: pascal.borde@ias.u-psud.fr

Abstract. The *CoRoT* space telescope is detecting planets with the transit method for more than four years. Its tally includes hot jupiter planets with orbital periods up to 95 days but also the first Super-Earth, CoRoT-7b, whose density is similar to the Earth's one,as well as close-in brown dwarfs. We review the status of the CoRoT/Exoplanet program, including some elements of the multi step strategy of complementary observations. We then present some of the CoRoT exoplanetary systems and how they widen the range of properties of the close-in low mass population and contribute to our understanding of the properties of planets.

Keywords. planetary systems, techniques: photometric

1. The *CoRoT* mission

With the advent of space missions, the search of exoplanets has entered a new area. Launched in december 2006, *CoRoT* has opened the field, followed in march 2009 by Kepler. *CoRoT* is a modest size mission designed to explore the transiting exoplanet population at short orbital period. The second main goal of the mission is the study of stellar internal structure through asteroseismology. In addition, a number of fields in stellar physics benefit this extremely precise time-series photometry.

CoRoT flies on PROTEUS, a *CNES* recurrent plateform, and orbits the Earth on an inertial polar orbit at an altitude of 896 km. *CoRoT* has two continuous viewing zones (CVZ) which are two almost circular regions of 10° radius each, centered on the Galactic plane at right ascension 6^h 50^m and 18^h 50^m respectively. A small drift of the orbit has allowed to slightly extend these CVZ along the lifetime of the *CoRoT* mission. Observing any stellar field in one of these two directions minimizes the level of background light. These positions have been selected taking into account the stellar content of these regions, in terms of possible targets for both scientific programs. The observation strategy consists in observational periods ranging from $\simeq$ 25 days up to nearly six months. In order not to be blinded by the Sun, every six months, in April and October, the satellite is rotated by 180 deg with respect to the polar axis and fields in the opposite direction are pointed. These two CVZ allows to probe regions in the galactic plane with various stellar properties in terms of metallicity but also in terms of age, as thin and thick disk populations appear not mixed in exactly the same proportions in these low galactic fields (Gazzano *et al.* 2011). One might expect to get some insights into the physical conditions that prevail for planet formation mechanisms once a more complete planet statistics will be available.

Nominally scheduled for 3 years, the mission has been extended for three additional years, that is till March 2013. Up to September 2010, a total of 129 326 light curves has been collected. Their duration ranges from 21 days up to 152 days for the longest

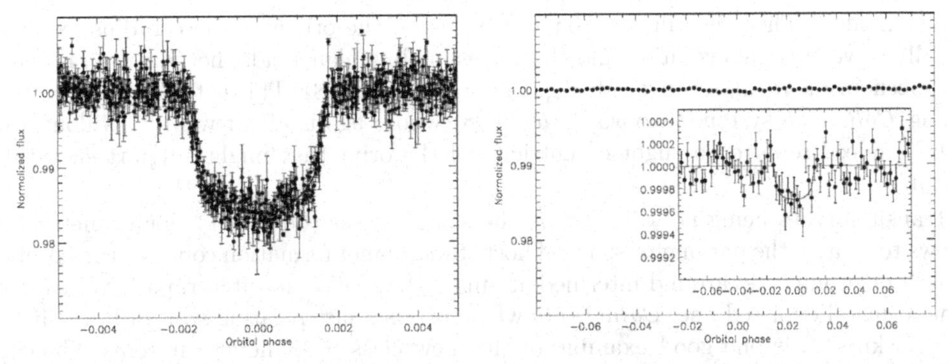

Figure 1. Two extreme cases of transit depth detected in the *CoRoT* light curves. Left : *CoRoT-9b* folded transit. Right : one of the smallest transit event plotted on the same scale, with a zoom on the transit.

run. The regular time sampling is 512 *s* which corresponds to the on-board co-addition of 16 exposures of 32 *s* each. This 32 *s* elementary integration time is preserved for one thousand of selected targets, referred to as over-sampled targets. During on-going *CoRoT* observations, the raw light curves are analyzed with the aim of detecting transits (Surace *et al.* 2008). According to this real-time detection called *Alarm mode*, the list of the over-sampled targets is updated, so that the possible forthcoming transit events will be accurately sampled at a cadence of 32 *s*. In addition, this allows to start follow-up observations for the most promising candidates during the same observation season. Indeed, according to the mission policy, the *CoRoT* CoIs community accesses the data six months after the end of the observation of a given field, when the complete set of light curves is fully reduced and calibrated.

Transits are detected in about 100 up to 300 light curves per run. It gives a total of about 3000 light curves in which transits have been found so far. The depth of these transits ranges from 0.02 % up to a few percents (Fig. 1). These transits are however predominantly eclipsing binaries. In that context, the long duration of the light curve is clearly an asset that allows to identify about 80% to 90% of the candidates as being stellar systems through secondary transits or light curve modulation. The remaining candidates are screened out by follow-up observations, using various techniques from photometric observations to radial velocity measurements. These follow-up observations play a major role in the *CoRoT* science and require a huge observational effort. They allow to assess the nature of the detected transiting body, stellar or planetary and, in the later case, to measure its mass. Bouchy *et al.* (2009) give a complete overview of the strategy for radial velocity observations and Deeg *et al.* (2009) present the ground-based photometric follow-up ones.

2. Family portrait

From 'Regular' hot Jupiters. Out of the 17 *CoRoT* planets secured at the time of the symposium, eleven belong to the giant planet population. As the regular members of this class, they exhibit a large spread in their orbital and physical characteristics. The planets density range from $\rho = 7.3\ \mathrm{g\,cm^{-3}}$ for extremely dense *CoRoT-14b* (Tingley *et al.* 2011) to $\rho = 0.22\ \mathrm{g\,cm^{-3}}$ in the case of *CoRoT-5b* whose host star is metal-depleted (Rauer *et al.* 2009). Three of them belong to the bloated population, that is planet whose radius is larger than can be accommodated by "standard" irradiated H-He planet models. There is still no clear consensus on the origin of these radius excesses. Tidally driven orbital evolution and the corresponding planet tidal heating even is one of the mechanisms regularly invoked (e.g. Jackson *et al.* 2008). This is the case of the very young *CoRoT-2b* system (Alonso *et al.* 2008) whose age is of a few Myr. Gillon *et al.* (2010) indeed reported a slight eccentricity of the orbit that might support such tidal origin.

Transit surveys being not affected by the same bias as the radial velocity method, it allows to enlarge the parameter space of host stars. Planet formation could be thus probed in new environments: around intermediate-mass stars that are often rapidly rotating or around metal depleted one. *CoRoT-11b* which orbits a fast rotating star with a $v \sin i =$ 40 ± 5 km s^{-1}, is one good example of this new class of planetary systems (Gandolfi *et al.* 2010).

Very few transiting planets present a marked eccentric orbit. With an eccentricity of 0.53 ± 0.04, *CoRoT-10b* is among the few exceptions. The planet whose orbital period is 13.2 days, experiences a 10.6-fold variation in insolation from the periastron to the apoastron Bonomo *et al.* (2010). At a much shorter orbital period of 5.35 days, *CoRoT-16b*, whose mass is 0.53 ± 0.8 M_{Jup}(Ollivier *et al.* 2011) is another possible case of eccentric orbit. The host star is faint with a r'-mag of 15.5 and further radial velocity measurements are affected by large uncertainties. According to MCMC modeling, even though the circular orbit couldn't be completely rule out, an orbital solution with a value of $e = 0.33 \pm 0.1$ is 30 times more probable. Kozai oscillations with a distant companion of stellar nature or by planet-planet scattering are the two mechanisms that might account for the high eccentricity. No hint of a non transiting companion from radial velocity monitoring or high angular resolution imaging has been so far reported for the two targets.

The long duration and high duty cycle of space-based observations allow to explore the transiting population over an extended range of orbital periods. Today *CoRoT* planets account for half of the transiting systems with an orbital period greater than 8 days. Among them, CoRoT-9b is of prime interest. First long orbital period transiting planet discovered by a transit survey, the planet orbits a solar like G3-type star in 95 days (Deeg *et al.* 2010). Its Jupiter's size and low eccentricity makes it a perfect representative of the largest known population of planets discovered by radial velocity surveys.

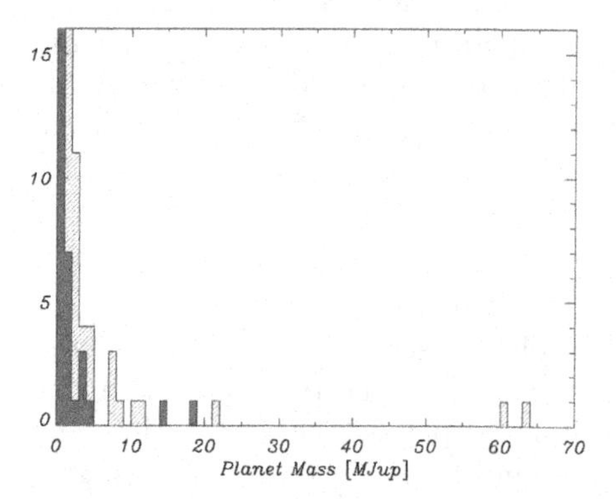

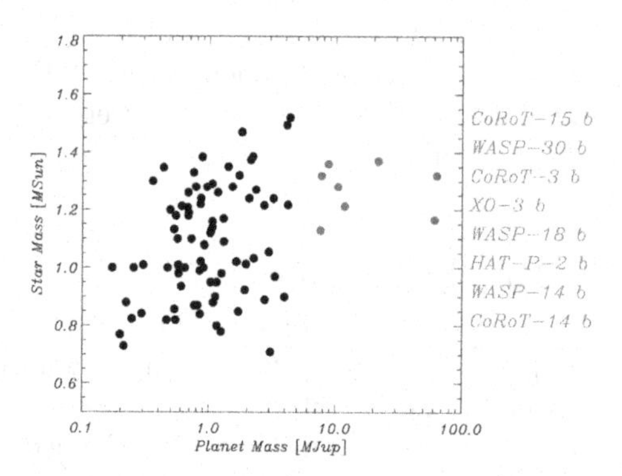

Figure 2. Left : Mass distribution of the exoplanets with orbital period less than 25 days; red: planets from radial velocity surveys, with a minimum mass only; green: transiting planets. Right : Mass of the host-star of transiting planets as a function of the mass of the planet.

To super-massive. Up to 2008, in mass range between 4 up to $\simeq$ 100 M_{Jup}, the mass distribution of the close-in population remained nearly un-populated. This paucity of massive sub-companions at short orbital period around solar-type stars was referred to as the brown-dwarf desert. As illustrated by the mass function distribution (Fig. 2), the transit surveys have recently started to populate this range of masses. First discovery of sub-stellar companion above the canonical value of 13M_{Jup} and an orbital period of about 4.5 days is *CoRoT-3b* (Deleuil *et al.* 2008). With a Jupiter's size but a mass of 21.6 M_{Jup}, it is located in the overlapping region between planets and brown dwarfs. According to planet formation models, core accretion or proto-planet collisions could result in planets with a mass up to 25M_{Jup}. The exact nature of *CoRoT-3b* thus remains an open question as it could belong to a new population of massive planets or to the extreme tail of the brown dwarfs mass distribution. By contrast, its mass of 63.5 M_{Jup} makes *CoRoT-15b* a bonefide transiting brown dwarf (Bouchy *et al.* 2010). These two massive close-in companions transit a F-type star and so does the new transiting brown dwarf *WASP-14b* (A. Collier-Cameron *et al.* this conference). The host-star's mass plotted in Figure 2 as a function of the companion's mass shows that, up to date, all close-in companions with a mass greater than 10M_{Jup} are hosted by slightly more massive stars than the Sun, F-type star typically. These stars have usually a large $v \sin i$. The lack of detection by radial velocity surveys would be in that case a simple selection bias.

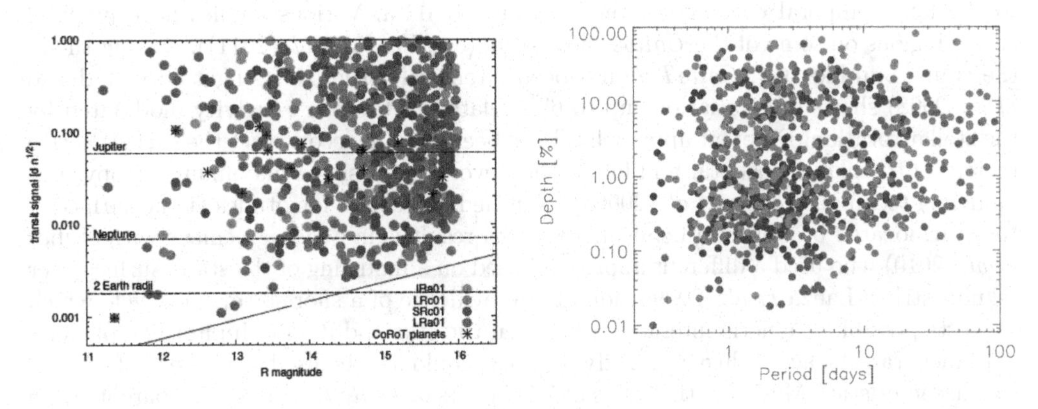

Figure 3. Left : Transit signal defined here as the product of the relative depth of the transit and the square root of the number of points in the transit, as a function of the *CoRoT* targets magnitude. Right: Depth of the transits as a function of the period of the event. The colored circles are all the candidates identified in the first runs according to the color codes in the bottom right of the left plot and the black stars the planets. (Adapted from Cabrera *et al.* (2009)).

And micro-planets. Space-based observations make the domain of small size planets around solar-like stars within reach. The detection capacity of *CoRoT* could be assessed using all the transit identified in the runs of the first year, whatever their nature, stellar or planetary, is. As illustrated in Figure 3, Neptune size planets can be detected whatever the magnitude of the star is, whereas super-Earth size planets down to 1 Earth radius can be discovered in the light curves of stars brighter than R $\simeq$ 14 only.

In the small-size regime, with a size of 0.57 R_{Jup}, *CoRoT-8b* has a density comparable to that of Neptune (Bordé *et al.* 2010), suggesting a massive core with a mass in the range 47 up to 64 $M_{\oplus}$ and a much smaller H - He envelope. In the mass - period diagram this planet lies in between the hot-Jupiter and the Super-Earth population, in a region where formation models predict a lack of planets. Its exact nature could be questioned as it might belong to the distribution tail of gaseous giants. Such massive planets with a high content of heavy elements are suggested to form preferentially around metal-rich stars, which is in agreement with the high metallicity of the host star.

With a depth of 0.35 *mmag CoRoT-7b* is the smallest planet discovered so far orbiting a late K-type star. As illustrated by Léger *et al.* (2009), in this mass domain the main difficulty lies in establishing the exact nature of the transiting companion. Number of follow-up observations were required to rule out the various blend scenarii. Even more challenging is the estimate of the mass of the planet. The CoRoT-7a's stellar activity gives indeed a stellar jitter that dominates the radial velocity signal (Queloz *et al.* 2009) making the planet's mass subject to controversies. According to the different studies carried out on the same radial velocity measurement set, the estimated mass of the planet varies from 2 $M_\oplus$ (Pont *et al.* 2011) up to 6.9 $M_\oplus$ (Hatzes *et al.* 2010) with an uncertainty as large as 20%. In addition, the exact number of planets in the system is also debated. There are indeed some evidences in the radial velocity signal for a second non-transiting planet (Queloz *et al.* 2009) and even a third one (Hatzes *et al.* 2010). But more data with an adapted observation strategy would be needed to clarify these issues.

3. Probing the star - planet interactions

Magnetic activity in the photosphere of the host stars could be also investigated thanks to the long temporal coverage of the photometric data. Various studies have mapped active regions on some of the *CoRoT* host stars (e.g. Lanza *et al.* (2011)). Among them, the young Sun-like star, *CoRoT-2b* has been intensively studied. Its light curve (Alonso *et al.* 2008) shows flux variations up to 6%, related to the star's activity modulated by the stellar rotation. Different approaches have been used to reconstruct the surface active regions of the planet host star and follow their evolution. Using a maximum entropy spot modeling technique Lanza *et al.* (2009a) identified two active longitudes at the surface of the star, located on opposite hemispheres. This result is in good agreement with Huber *et al.* (2010) who used a different approach based on a modeling of the stars' surface over regular strips. Lanza *et al.* (2009a) found also evidence of a short-term spot cycle which suggests possible magnetic interaction between the star and its hot Jupiter-like planet.

Planet transits, acting like a magnifying glass, could also be used to probe spots physical characteristics. Modeling the series of 77 transits of *CoRoT-2b* across its parent star's stellar disk, Silva-Valio *et al.* (2010) estimated a spot's size ranging from 0.2 to 0.7 planet radius, the spots temperature being between 3600 and 5000 K.

The stellar activity also impacts the planet's radius measurement by distorting the bottom of the transit. As a result, the transit depth is affected and could appear more shallower than in reality. In the case of *CoRoT-2b*, Silva-Valio *et al.* (2010) fitted the deepest transit of the series which yields a planet's radius $\simeq$ 3% larger than the value reported in the discovery paper which was calculated from the phase folded light curve.

Possible synchronization could also be searched for. The long duration light curves allows to derive the rotation period of the star. In the case of *CoRoT-4b*, Aigrain *et al.* (2008) established that it is a synchronized system, a result confirmed by Lanza *et al.* (2009b). Among the *CoRoT* planets, it remains however the only system for which the synchronization could be clearly detected.

4. Summary

At the time of the symposium, *CoRoT* has been in operation for 1383 days. With 17 planets fully characterized and a number of candidates still in the follow-up process, the instrument has fully demonstrated its ability to explore the close-in planet population over an extended range of sizes and properties from Super-Earth to temperate Jupiter-type planets. Transit detection being not limited to solar-type stars, *CoRoT* has also unexpectedly started to populate the brown dwarf desert. Whatever their nature, massive planets or brown dwarf, the formation of these massive companions appears taking place around F-type stars.

High precision photometric observations over long time span have also proved being valuable to probe the stellar activity and search for new hints of star - planet interactions such as synchronization. It has also the potential to bring new interesting information on the stars activity. This being one of the main limitation in planet detection in the low mass domain that might help our understanding of the stellar signatures and further their modeling. With the mission extension and further improvements of the detection from a better understanding of the *CoRoT* light curves, new discoveries are expected. *CoRoT* will contribute further to get a better picture of the planetary systems characteristics.

References

Aigrain, S., Collier-Cameron, A., Ollivier, M., *et al.* 2008, *A&A*, 488, L43
Alonso, R., Auvergne, M., Baglin, A., *et al.* 2008, *A&A*, 482, L21
Bonomo, A. S., Santerne, A., Alonso, R., *et al.* 2010, *A&A*, 520, 65
Bordé, P., Bouchy, F., Deleuil, M., *et al.* 2010, *A&A*, 520, 97
Bouchy, F., Moutou, C., & Queloz, D. 2009, *Proc. IAU Symp.* 253, 129
Bouchy, F., Deleuil, M., Guillot, T., *et al.* 2010, *A&A*, 525, 68
Cabrera, J., Fridlund, M., Ollivier, M., *et al.* 2009, *A&A* 506, 501
Deeg, H. J., Gillon, M., Shporer, A., *et al.* 2009, *A&A*, 506, 343
Deeg, H. J., Moutou, C., Erikson, A., *et al.* 2010, *Nature*, 464, 384
Deleuil, M., Deeg, H. J., Alonso, R., *et al.* 2008, *A&A*, 491, 889
Gandolfi, D., Hébrard, G., Alonso, R., *et al.* 2010, *A&A*, 524, 55
Gazzano, J.-C., de Laverny, P., Deleuil, M., *et al.* 2011, (*submitted to A&A*)
Gillon, M., Lanotte, A. A., Barman, T., *et al.* 2010, *A&A*, 511, 3
Hatzes, A. P., Dvorak, R., Wuchterl, G., *et al.* 2010, *A&A*, 520, 93
Huber, K. F., Czesla, S., Wolter, U., & Schmitt, J. H. M. M. 2010, *A&A*, 514, A39
Huber, K. F., Czesla, S., Wolter, U., & Schmitt, J. H. M. M. 2009, *A&A*, 508, 901
Jackson, B., Greenberg, R., & Barnes, R. 2008, *ApJ*, 678, 1396
Lanza, A. F., *et al.* 2009, *A&A*, 493, 193
Lanza, A. F., *et al.* 2009, *A&A*, 506, 255
Lanza, A. F., *et al.* 2011, *A&A*, 525, A14
Léger, A., *et al.* 2009, *A&A*, 506, 287
Ollivier, M., *et al.* 2011, *A&A, to be submitted*
Pont, F., Aigrain, S. & Zucker S. 2011, *MNRAS*, 411, 1953
Queloz, D., Bouchy, F., Moutou, C., *et al.* 2009, *A&A*, 506, 303
Rauer, H., Queloz, D., Csizmadia, Sz., *et al.* 2009, *A&A*, 506, 281
Silva-Valio, A., Lanza, A. F., Alonso, R., & Barge, P. 2010, textitA&A , 510, A25
Surace, C., Alonso, R., Barge, P. *et al.* 2008, *SPIE*, 7019, 111
Tingley, B., Endl , M., Gazzano J.-C., *et al.*, 2011, *A&A*, 528, A97
Wolter, U., Schmitt, J. H. M. M., Huber, K. F., Czesla, S., Müller, H. M., Guenther, E. W., & Hatzes, A. P. 2009, *A&A*, 504, 561

The Astrophysics of Planetary Systems: Formation, Structure, and Dynamical Evolution
Proceedings IAU Symposium No. 276, 2010
A. Sozzetti, M. G. Lattanzi & A. P. Boss, eds.

doi:10.1017/S1743921311019922

High-resolution spectroscopic view of planet formation sites

Zsolt Regály[1], Laszlo Kiss[1,2], Zsolt Sándor[3] and Cornelis P. Dullemond[3]

[1]Konkoly Observatory of the Hungarian Academy of Sciences, P.O. Box 67, H-1525 Budapest, Hungary
email: regaly@konkoly.hu

[2]Sydney Institute for Astronomy, School of Physics, University of Sydney, Australia

[3]Junior Research Group, Max-Planck-Institut für Astronomie, Königstuhl 17, D-69117 Heidelberg, Germany

Abstract. Theories of planet formation predict the birth of giant planets in the inner, dense, and gas-rich regions of the circumstellar disks around young stars. These are the regions from which strong CO emission is expected. Observations have so far been unable to confirm the presence of planets caught in formation. We have developed a novel method to detect a giant planet still embedded in a circumstellar disk by the distortions of the CO molecular line profiles emerging from the protoplanetary disk's surface. The method is based on the fact that a giant planet significantly perturbs the gas velocity flow in addition to distorting the disk surface density. We have calculated the emerging molecular line profiles by combining hydrodynamical models with semianalytic radiative transfer calculations. Our results have shown that a giant Jupiter-like planet can be detected using contemporary or future high-resolution near-IR spectrographs such as VLT/CRIRES or ELT/METIS. We have also studied the effects of binarity on disk perturbations. The most interesting results have been found for eccentric circumprimary disks in mid-separation binaries, for which the disk eccentricity - detectable from the asymmetric line profiles - arises from the gravitational effects of the companion star. Our detailed simulations shed new light on how to constrain the disk kinematical state as well as its eccentricity profile. Recent findings by independent groups have shown that core-accretion is severely affected by disk eccentricity, hence detection of an eccentric protoplanetary disk in a young binary system would further constrain planet formation theories.

Keywords. planetary systems: protoplanetary disks, hydrodynamics, line: profiles

---

1. Introduction

It is well known that circular Keplerian protoplanetary disks are expected to produce symmetric double-peaked molecular line profiles (Horne & Marsh 1986). Contrary to this simple symmetric disk assumption, asymmetric CO line profiles in the fundamental band have been observed in several cases. Grid-based numerical simulations of Kley & Dirksen (2006) have shown that local disk eccentricity might form in planet bearing disk near the gap. The theory of resonant excitation mechanisms in accretion disks of Lubow (1991) predicts that the circumstellar disks of close-separation young binaries might become fully eccentric due to the orbiting companion. Horne (1995) has shown that supersonic turbulence might cause observable line profile distortions. As the orbital velocity of gas parcels is highly supersonic in accretion disks, the disk eccentricity results in supersonic velocity distortions. Therefore distorted molecular line profiles are expected to form in giant planet bearing locally eccentric protoplanetary disks (Regály *et al.* 2010) and fully eccentric circumstellar disks of close-separation young binaries.

2. Spectral calculations combined with hydrodynamic simulations

In order to calculate the CO spectra, we need to know the temperature distribution in the disk. In our study the double layer flaring disk model of Chiang & Goldreich (1997) is assumed. In this model the disk is heated by the stellar irradiation and accretion processes. The incident stellar flux heats the disk atmosphere, which reprocesses the stellar light and irradiates the disk interior. The accretion processes heats the disk interior directly. In this way an optically thick interior and an optically thin atmosphere develop (Fig. 1). For low accretion rate, temperature inversion forms, resulting in emission spectra.

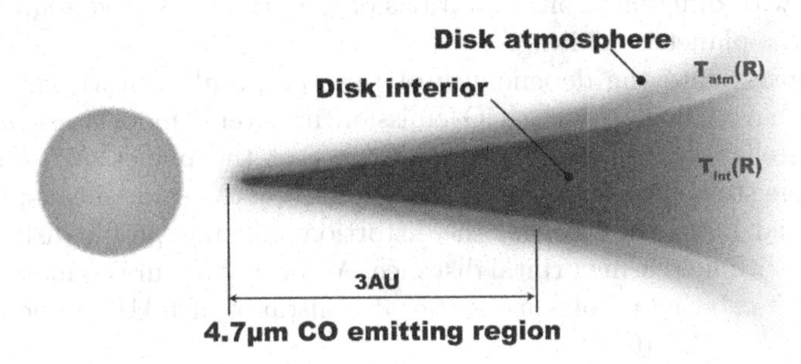

Figure 1. The double layer disk model in which the disk atmosphere is superheated.

The fundamental band CO emission is formed in the optically thin disk atmosphere above the optically thick cooler disk interior. In the optically thin approximation the $I(\nu, R, \phi)$ monochromatic intensity of the radiation at frequency ν from a given gas parcel at R, ϕ cylindrical coordinates is the sum of the dust continuum of the disk interior and the optically thin atmospheric CO emission

$$I(\nu, R, \phi, i) = B(\nu, T_d(R))e^{-\tau(\nu,R,\phi,i)} + B(\nu, T_g(R))(1 - e^{-\tau(\nu,R,\phi,i)}), \qquad (2.1)$$

where, $B(\nu, T)$ is the Planck function, $T_d(R)$ and $T_g(R)$ are the dust and gas temperatures in the disk interior and the disk atmosphere, respectively. Details of the spectral model can be found in Regály *et al.* (2010). For a disk perturbed by a companion (a planet or a secondary star) the velocity field is no longer circular Keplerian, in which case the local Doppler shift that affects the atmospheric optical depth ($\tau(\nu, R, \phi, i)$) can be given by

$$\begin{aligned} \Delta\nu(R, \phi, i) &= \frac{\nu_0}{c} \sin(i) \{[u_{\rm R}(R, \phi) \cos(\phi) - u_\phi(R, \phi) \sin(\phi)] \\ &\quad + [u_{\rm R}(R, \phi) \sin(\phi) + u_\phi(R, \phi) \cos(\phi)]\}, \end{aligned} \qquad (2.2)$$

where $u_{\rm R}(R, \phi)$ and $u_\phi(R, \phi)$ are the radial and the azimuthal velocity components of the orbiting gas parcels. In order to calculate the local Doppler shift in a perturbed disks with embedded planet or a secondary star, we solved the continuity and Navier-Stokes equations by the 2D grid-based hydrodynamic code FARGO (Masset 2000). In this way, the $u_{\rm R}(R, \phi)$ and $u_\phi(R, \phi)$ velocity components of gas parcels, were provided by the hydrodynamic simulations. For simplicity we used a locally isothermal equation of state for the gas. The α-type disk viscosity was applied (Shakura & Sunyaev 1987). The disk self-gravitation was neglected as the Toomre parameter is well above 1. For more details on the hydrodynamic simulations see Regály *et al.* (2010).

3. Planetary signal: local disk eccentricity due to a giant planet

According to our simulations the density distribution shows permanent elliptic shape near the gap (Fig. 2, left panel), confirming the results of Kley & Dirksen (2006). We found that the orbits of gas parcels are eccentric in the vicinity of the planet (Fig. 2, middle panel). As the magnitude of the velocity distortion of the gas parcels exceeds the local sound speed, distorted non-symmetric line profiles are expected to form.

We have calculated the fundamental band P10 non-blended CO line for 20, 40 and 60 degree of disk inclination in a perturbed disk assuming an $8M_{\rm Jup}$ mass planet orbiting $1\,M_\odot$ mass star at 1 AU (Fig. 2, right panel). The line shape is distorted and the relative strength of the distortions are increasing with increasing inclination angle. Calculating the line profiles with different orbital positions of the planet, we have found that the line shapes vary as the planet is orbiting.

In order to investigate the dependence of the line profile distortions on the model parameters, we have calculated the CO emission in several models assuming different planetary and stellar masses, and orbital distance of the planet. We found that the planetary and the stellar mass affect the line profiles by the same means: the larger the planetary or stellar mass, the stronger the distortion. The line profile distortion is found to be weakening with increasing orbital distance. According to our calculations, the signal of a Jovian planet orbiting a Solar mass star at a distance of 1 AU can be detected with CRIRES (Regály *et al.* 2010).

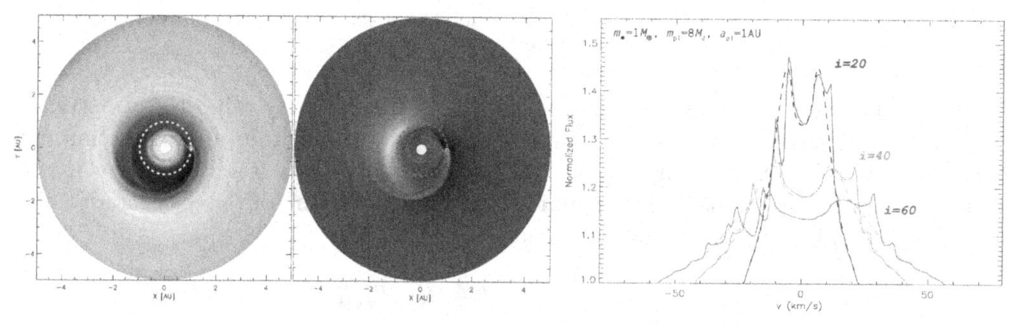

Figure 2. Surface density (*left panel*) and radial component of velocity (*middle panel*) distribution in a planet bearing protoplanetary disk. Distorted $V = 1 \rightarrow 0$ P10 CO lines (*right panel*) emerging from the gravitationally distorted disk, assuming 20°, 40° and 60° disk inclinations.

4. Asymmetric lines: fully eccentric disk due to a secondary star

Circumprimary disks in young binaries become fully eccentric via the gravitational perturbations of the secondary, in form of resonant excitation mechanisms (Lubow 1991; Kley *et al.* 2008). Thébault *et al.* (2006) and Paardekooper *et al.* (2008) have shown that the planetesimal accretion might be inhibited by the disk eccentricity in core-accretion models. Recently, Zsom *et al.* (2011) have shown that the dust coagulation process is also affected by the disk eccentricity in the core-accretion scenario.

According to our simulations, the average disk eccentricity is ~ 0.3 independently on the binary mass ratio and the magnitude of viscosity, assuming $h = 0.05$ aspect ratio for the disk, which is a reasonable assumption for a circumstellar disk around an average T Tauri star (Fig. 3, left panel). However, smaller disk eccentricity was found for thick ($h \geqslant 0.75$) and thin ($h \leqslant 0.03$) disks. We have found that the orbital eccentricity of the binary above 0.2 inhibits the development of disk eccentricity.

Calculating the $V = 1 \rightarrow 0$ P10 non-blended CO lines with 20° and 60° disk inclination angle, we have found a maximum $\sim 20\%$ red-blue peak asymmetry (Fig. 3, middle panel).

The profile asymmetry is due to the fact that the inner disk, where the fundamental band CO emission is formed, is fully eccentric. The peak asymmetry is inverted twice during a full disk precession, which takes about $\sim 6.5P_{\rm bin}$. Moreover, as the disk eccentricity slightly varies during one binary orbit, line profile variations might be observed in the line wings with $\sim P_{\rm bin}$.

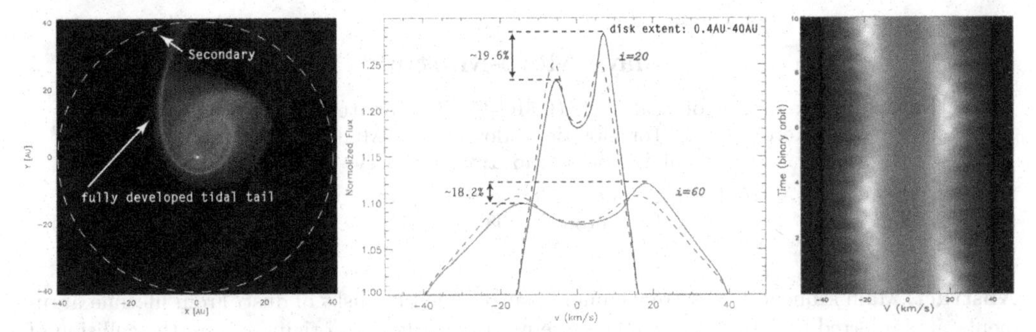

Figure 3. The circumprimary disk becomes eccentric due to the perturbations from the secondary. Asymmetric $V = 1 \rightarrow 0$ P10 CO line (*middle panel*), assuming 20° and 60° disk inclinations. Trailed spectra (*right panel*) covering a full disk precession shows the periodic variations in asymmetry with $\sim 6.5P_{\rm bin}$ and line wings with $\sim P_{\rm bin}$.

5. Summary

High-resolution 4.7μm CO molecular spectra provide us a tool to reveal disk eccentricity within the 3 AU regions, where terrestrial planets are expected to form. We have shown that Jovian planets still embedded in their protoplanetary disk can indirectly be detected with the 4.7 μm CO line profile distortions caused by the local disk eccentricity. We have also shown that significant CO line profile asymmetry and variations are expected from fully eccentric disks of close separation young binaries. By revealing disk eccentricity in close-separation young binaries the core-accretion model might be constrained further, since the planet formation seems to be strongly affected by the disk eccentricity.

Acknowledgements: This project has been supported by the "Lendület" program of the Hungarian Academy of Sciences and the DAAD-PPP mobility grant P-MÖB/841/.

References

Horne, K. & Marsh, T. R. 1986, *MNRAS*, 218 761
Chiang, E. I. & Goldreich, P. 1997, *ApJ*, 490, 368
Horne, K. 1995, *A&A*, 297, 273
Kley, W. & Dirksen, G. 2006, *A&A*, 447, 369
Kley, W., Papaloizou, J. C. B., & Ogilvie, G. I. 2008, *A&A*, 487, 671
Lubow, S. H. 1991, *ApJ*, 381, 259
Masset, F. 2000, *A&AS*, 141, 165
Paardekooper, S.-J., Thébault, P., & Mellema, G. 2008, *MNRAS*, 386, 973
Regály, Zs., Sándor, Zs., Dullemond, C. P., & van Boekel, R. 2010, *A&A*, 523, A69
Shakura, N. I. & Sunyaev, R. A. 1973, *A&A*, 24, 337
Thébault, P., Marzari, F., & Scholl, H. 2006, *Icarus*, 183, 193
Zsom, A., Dullemond, C. P., & Sándor, Zs. 2011, *A&A*, 527, A10

The Astrophysics of Planetary Systems: Formation, Structure, and Dynamical Evolution
Proceedings IAU Symposium No. 276, 2010
A. Sozzetti, M. G. Lattanzi & A. P. Boss, eds.

doi:10.1017/S1743921311019934

Characterizing planetesimal belts through the study of debris dust

Amaya Moro-Martín[1,2]

[1]Departamento de Astrofísica, CAB (CSIC-INTA), Instituto Nacional de Técnica Aeroespacial, Torrejón de Ardoz, 28850, Madrid, Spain
email: amaya@cab.inta-csic.es

[2]Department of Astrophysical Sciences, Princeton University, Peyton Hall, Ivy Lane, Princeton, NJ 08544, USA

Abstract. Main sequence stars are commonly surrounded by disks of dust. From lifetime arguments, it is inferred that the dust particles are not primordial but originate from the collision of planetesimals, similar to the asteroids, comets and KBOs in our Solar system. The presence of these debris disks around stars with a wide range of masses, luminosities, and metallicities, with and without binary companions, is evidence that planetesimal formation is a robust process that can take place under a wide range of conditions. Debris disks can help us learn about the formation, evolution and diversity of planetary systems.

Keywords. interplanetary medium, Kuiper Belt, circumstellar matter, stars: evolution, planetary systems

---

1. Why we care about debris disks

Circumstellar disks play a fundamental role in the formation of stars and planets. The accretion of mass onto the forming star is regulated by mass and angular momentum transfer mechanisms within the disk. With time, the mass reservoir of the cloud gets depleted and the gas-rich disk begins to dissipate in a time scale of about 6 Myr. The formation of gas giant planets needs to happen before this gas-rich disk dissipates, while the formation of terrestrial planets and massive planets beyond the ice line is not limited by the presence of gas in the disk and can continue for approximately 100 Myr; a critical step in this process is the formation of planetesimals. Observations with *Spitzer* show that there is evidence that *at least* 15% of mature stars (10 Myr–10 Gyr) of a wide range of masses (0.5–3 M_{Sun}) harbor planetesimal belts with sizes of 10s–100s AU. This evidence comes from the presence of an infrared emission in excess of that expected from the stellar photosphere, thought to arise from a circumstellar dust disk. The reason why these dust disks are evidence of the presence of planetesimals is because the lifetime of the dust grains is of the order of 0.01–1 Myr, much shorter than the age of the star (>10 Myr); therefore, the origin of these dust grains cannot be primordial, i.e. from the cloud of gas and dust where the star was born, but must be the result of on-going dust production generated by planetesimals, like the asteroids, comets and Kuiper Belt Objects (KBOs) in our Solar system; this is why we refer to these dust disks as *debris disks*. Debris disks are therefore evidence of the formation of planetesimals around other stars. The goal of this presentation is to describe how debris disks can shed light on the formation, evolution an diversity of planetary systems, helping us place our Solar system into context.

2. The Solar system debris disk

The Sun also harbors a disk of dust produced in the inner and outer Solar system by the asteroids, comets and KBOs. Scattered light and thermal emission observations of the inner part of the Solar system debris disk (known as the zodiacal cloud) can help determine the properties of the dust particles, revealing non-spherical, rapidly-rotating, irregular or fluffy aggregates, 10–100 μm in size, composed of a mixture of silicates and organic material with a low albedo; there is also a population of smaller 1 μm-sized grains made of crystalline olivine and hydrous silicate that accounts for a weak silicate emission feature at 10 μm (Reach *et al.* 2003). The thermal emission from these dust particles dominates the night sky between 5–500 μm, with a fractional luminosity of $f = L_{dust}/L_{\odot} \sim 10^{-8}$–$10^{-7}$ (Dermott *et al.* 2002), more than two orders of magnitude fainter than the extra-solar debris disks observed with *Spitzer*. Regarding the spatial distribution, the thermal emission from the zodiacal cloud shows long, narrow arcs that coincide with the perihelion passage of some short-period comets, and broader dust bands at low ecliptic latitudes, thought to originate from the disruption events that gave rise to the asteroidal families Themis, Koronis and Eos (Sykes & Greenberg 1986). The thermal emission also shows a ring-like structure that results from the trapping of dust particles in the exterior mean motion resonances (MMRs) with the Earth, with a 10% number density enhancement on the Earth's wake that results from the resonance geometry (Dermott *et al.* 1994). Secular perturbations with the planets are thought to account for the offset of the zodiacal cloud center with respect to the Sun, the inclination of the cloud with respect to the ecliptic and the cloud warp.

Nesvorny *et al.* (2010) argued that the splitting of Jupiter family comets accounts for 85% of the dust in the inner Solar system. The dust production rate of comets is difficult to estimate because the cometary activity is not steady: sublimation drives the cometary activity in the inner Solar system, however, there are also isolated flare-ups that can produce dust at large heliocentric distances. Cometary dust particles have been studied *in situ* in the case of comets Halley, Tempel 1 and Wild 2. Dust particles have also been detected *in situ* at different heliocentric distances by the spacecrafts *HEOS* (1 AU), *Hitten* (1 AU), *Helios* (0.3–1AU), *Galileo* (0.7–5 AU), *Pioneer* 8 and 9 (0.75–1.08 AU), *Ulysses* (1.3-2.3 AU), *Cassini* and *Pioneer* 10 and 11 (see review by Grün *et al.* 2001). The current dust production rate in the inner Solar system is of the order of 10^4 kg/s; the relative contribution of the different sources is still under debate and has likely changed with time. *Pioneer* 10 and 11 detected dust out to 18 AU and 13 AU, respectively (Humes 1980), and dynamical models indicate and that the Kuiper belt (KB) was likely the source of the dust detected beyond 10 AU (Landgraf *et al.* 2002). *Voyager* detected dust in the 30–60 AU Kuiper Belt region, with an estimated number density of $n \sim 2{\times}10^{-8}$ m^{-3} (Gurnett *et al.* 1997) that would correspond to a fractional luminosity of $f = L_{dust}/L_{star} \sim 4 \times 10^{-7}$ (Jewitt & Luu 2000). The dust production rate estimates in the outer Solar system are in the rage $(0.2$–$5){\times}10^4$ kg/s (from Voyager and Pioneer data, respectively; Jewitt & Luu 2000; Landgraf *et al.* 2002).

The dust production rate in the Solar system has changed significantly with time. It is thought that the Solar system was significantly more dusty in the past because the asteroid belt (AB) and the KB were more densely populated; with time, it became progressively less dusty as the planetesimal belts eroded away by mutual planetesimal collisions, leading to a $1/t$ decay in the dust thermal emission. It is expected that this decay was punctuated by large spikes that are due to large collisions happening stochastically (examples of stochastic events in the recent history of the Solar system are the fragmentation of the asteroids giving rise to the asteroidal families and the dust bands).

A major change in the dust production rate is expected to have occurred when the Solar system was $\sim$ 600 Myr old, at the time of the Late Heavy Bombardment (LHB), thought to be caused by the orbital migration of the giant planets that produced a resonance sweeping of the AB that made the asteroids orbits unstable, causing a large scale ejection of bodies into planet-crossing orbits (explaining the observed cratering record – Strom *et al.* 2005); the orbital migration of the planets also caused a major depletion of the KB as Neptune migrated outward (Gomes *et al.* 2005). After the LHB, there must have been a sharp decrease in the dust production rate due to the drastic depletion of planetesimal (Booth *et al.* 2009).

3. Extra-solar debris disks: lessons from Spitzer

Spitzer carried out extensive debris disks surveys. Taking advantage of the unprecedented sensitivity of the *Spitzer/MIPS* instrument, hundreds of debris disks were identified, allowing to characterize the frequency and properties of debris disks around stars of different spectral types, ages and environment. These are some of the main lessons learned.

3.1. *Debris disk frequency*

A survey of 328 solar-type FGK stars (30 Myr–3 Gyr) found that the frequency of 24 μm excess (tracing dust around 3–5 AU) is 14.7% at $<$ 300 Myr and 2% at $>$300 Myr, while at 70 μm (tracing dust around 28–75 AU), the excess rates are 6–10% and are fairly independent of age (Meyer *et al.* 2008; Hillenbrand *et al.* 2008; Carpenter *et al.* 2009). Debris disks are more common around A-type stars than around FGK stars: a survey of 160 A-type stars showed that 32% and $\geqslant$ 33% of stars show excess emission at the 3-σ confidence level at 24 μm at 70 μm, respectively (Su *et al.* 2006). On the other hand, debris disks are significantly less common around old M-type stars (Gautier *et al.* 2007), but this may be an observational bias because the peak emission of these colder disks would be at $\lambda > 70$ μm, i.e. beyond the wavelength where *Spitzer/MIPS* was most sensitive. On-going debris disks surveys with *Herschel* are doubling the number of disk detection rates made by *Spitzer* and, in fact, most of the new detected debris disks are found around cold late-type stars (Eiroa *et al.* in preparation, see proceeding in this volume). There is also evidence of the presence of planetesimals around white dwarfs: some of these evolved stars show infrared excesses and high levels of pollutants (elements other than the expected pure H and He), thought to arise from tidally disrupted planetesimals (Jura 2006). The presence of planetesimals around stars with a very wide range of spectral types, from M-type to the progenitors of white dwarfs – with several orders of magnitude difference in stellar luminosities – implies that planetesimal formation is a robust process that can take place under a wide range of conditions.

3.2. *Debris disk fractional luminosities*

Due to the limited sensitivity of the *Spitzer* debris disks surveys, the detected fractional luminosities are generally $f = L_{dust}/L_{*} \gtrsim 10^{-5}$; this is larger than those inferred for the Solar system's debris disk today : $f \sim 10^{-8} - 10^{-7}$ for the inner Solar system and the estimated $f \sim 10^{-7} - 10^{-6}$ for the outer Solar system. Assuming a gaussian distribution of debris disk luminosities and extrapolating from *Spitzer* observations (that show a steep increase in the number of detections with decreasing f), Bryden *et al.* (2006) concluded the fractional luminosity of an average debris disk around a solar-type stars could be between 0.1–10 $\times$ that of the Solar system debris disk. In other words, the observations are consistent with debris disks at the Solar system level being common (but too faint to be detected by *Spitzer*). On-going debris disks surveys with *Herschel*, sensitive to fainter and colder disks, are revisiting the frequency of disk detections.

3.3. *Debris disk evolution*

The study of the frequency and properties of debris disks around stars of different ages can shed light on the evolution of debris disks with time (see review by Wyatt 2008). Collisional models predict that the steady erosion of planetesimals will naturally lead to a decrease in the dust production rate; this slow decay will be punctuated by short-term episodes of increased activity triggered by large collisional events that can make the disk look an order of magnitude brighter. These models agree broadly with the observations derived from the *Spitzer* surveys. It is found that the frequency and fractional luminosities of debris disks around FGK stars with ages in the range 0.01–1 Gyr declines in a timescale of 100–400 Myr, but there is no clear evidence of a decline in the 1–10 Gyr age range (Trilling *et al.* 2008). This indicates that different physical processes might be dominating the evolution of the dust around the younger and the older systems. A possible scenario is that, at young ages, stochastic dust production due to individual collisions is more prominent, while at older ages, the steady grinding down of planetesimals dominates. The relative importance of these two processes is still under discussion. The *Spizer* surveys also showed that the evolution of dust around both A-type and FGK stars proceeds differently in the inner and outer regions, with the warmer dust (dominating the emission at 24 μm) declines faster than the colder dust (seen at 70 μm). This indicates that the clearing of the disk in the inner regions is more efficient, as would be expected from the shorter dynamical timescales. Regarding the issue of steady state vs. stochastic dust production, some systems show evidence that transient events dominate the dust production. This is the case of HD 69830, a star that shows no excess emission at 70 μm, but shows strong excess at 24 μm, with prominent spectral features in the *Spizter/IRS* wavelength range that are indicative of the presence of large quantities of small warm grains (Beichman *et al.* 2005). Because these small grains have very short lifetimes, it is inferred that the level of dust production is too high to be sustained for the age of the system (because the planetesimals would have not survived the inferred erosion rate), concluding that it is a transient event (Wyatt *et al.* 2007). HD 69830 is particularly interesting because it harbors three Neptune-like planets inside 1 AU (Lovis *et al.* 2006) and the best fit to the dust spectra is that of a disrupted, highly processed, low carbon P- or D-type asteroid (very common in the inner AB – Lisse *et al.* 2007), located near the 2:1 and 5:2 mean motion resonaces of the outermost planet; this raises the question whether the increased level of dust production is the result of gravitational perturbations by the planets.

As we mentioned above, there is evidence that the migration of the giant planets in the early Solar system had an important effect on the evolution of its debris disk: the drastic planetesimal clearing that resulted would have been associated with a sharp decreased in the dust production rate (Booth *et al.* 2009). Because the presence of hot Jupiters and multiple planets locked in resonances are evidence that planet migration has taken place in some planetary systems, a natural question to ask is whether debris disks observations show evidence of drastic planetesimal clearing events: a statistical study by Booth *et al.* (2009) concluded that less than 12% of solar-type star suffered LHB-type of events, but this is a preliminary result because *Spitzer* surveys were limited in sensitivity, so this issue needs to be revisited in the future with deeper surveys.

3.4. *Debris disk structure and inferred planetesimal location*

The few dozen debris disks that have been spatially resolved so far show a rich diversity of structural features, including narrow to wide belts, clumpy rings, sharp inner edges, brightness asymmetries, offsets of the dust disk center with respect to the central star, warps of the disk plane and spirals. Because some of these features have also been observed in the Solar system debris disk, where they are thought to be caused by the gravitational perturbations from the planets, a natural question to ask is whether the asymmetries observed in the extra-solar debris disks reveal the presence of unseen planetary companions: clumpy rings have been explained as dust and/or dust-producing planetesimals trapped in MMRs with a planet; warps can be the result of secular perturbations of a planet in an orbit inclined with respect to the planetesimal/dust disk; and spirals, offsets and brightness asymmetries might also be the result of secular perturbations, in this case of an eccentric planet that forces an eccentricity on the planetesimals and the dust. Because these structural features depend on the mass and orbit of the planet, this opens the possibility of using the study of the dust disk structure as a planet detection technique. The planets recently found in the Fomalhaut and β-Pic systems, predicted to exist to account for the structure observed in both disks, illustrate this idea (see proceedings by Kalas *et al.* and Lagrange *et al.* in this volume).

Even though most of the debris disks observed with *Spitzer* are spatially unresolved, limited information regarding the dust location can be extracted from the analysis of the disk spectral energy distribution (SED). A *Spitzer* survey of 328 FGK stars at 24 and 70 μm found that about 2/3 of the debris disks SEDs could be fitted with a single temperature blackbody ($T <$ 45–85 K) consistent with a ring-like configuration (Hillenbrand *et al.* 2008); detailed analysis of the excess spectra from 12–35 μm of 44 of these stars showed that the inner radii of these cavities are $\sim$ 40 AU for the disks with 70 μm excess, and $\sim$ 10 AU for the disks without 70 μm excess (Carpenter *et al.* 2009). Inner cavities are also common around more massive stars: a *Spitzer* survey of 52 A-type and and late B-type stars known to have debris disks showed that the majority of the disks (39/52) can be be fitted with a single-temperature blackbody with a median temperature of 190 K, corresponding to a characteristic distance of 10 AU (Morales *et al.* 2009). One caveat of the SED analysis is its degeneracy in the absence of spatially resolved observations and/or spectral features that are able to constrain the grain radius and composition. If the system is known to harbor planets, an additional constraint on the dust location can be obtained from dynamical simulations that study the effect of the planetary perturbations on the stability of the planetesimals' orbits, and that can identify the regions where the planetesimals could be stable and long-lived (Moro-Martín *et al.* 2007, 2010). To set tighter constraints on the planetesimal location, there is the need to obtain spatially resolved images and/or accurate photometric points from the mid-infrared to the submillimeter, so the inner and outer radius of the disk can be better determined. Observations with *Herschel*, *JWST* and *ALMA* will be very valuable for this purpose

4. Prospects for the future: Herschel and beyond

The on-going *Hershel* debris disks surveys DEBRIS and DUNES (Matthews *et al.* in preparation; Eiroa *et al.* in preparation – see proceeding in this volume) are designed to characterize debris disks around AFGKM stars at 70, 100 and 160 μm, with follow-up at 350, 450 and 500 μm; these observations are already doubling the *Spitzer* debris disk detection rate, increasing the number of spatially resolved disks, and allowing to characterize a new population of cold disks. *ALMA*'s unprecedented high spatial resolution will be able to advance in the study of debris disk structure, and to test the models of planet-disk interactions; its long wavelength observations will allow to better constrain the disks outer radii. Debris disks surveys with *JWST* in the near to mid-infrared will allow to characterize the warm dust component, setting constrains on the frequency of planetesimal formation in the terrestrial planet region, and identifying stars with low debris dust contamination that may be good targets for terrestrial planet detection. Deep debris disks surveys with *JWST* and *SPICA* (the approval of the latter is pending) will be able to characterize debris disks around stars of a wide range of spectral types, ages and environment, studying the debris disk evolution and the dust production rate as a function of stellar age (the latter will help identify systems undergoing LHB-type of events, and to assess whether the dynamical evolution of the Solar system was particularly benign). *SPICA/SAFARI* (if approved) will also be able to study the dust composition by carrying out a spectral survey of debris disks and, for nearby disks, spectral imaging (that will allow to trace the variation in the dust mineral content as a function of disk radius, that can be compared to the compositional gradient in the Solar system – also to be studied by *SPICA/SAFARI*). Regarding the Solar system, advancements need to be made in the study of its debris dust (e.g. with sample return missions to an asteroid and/or a comet, and dust detection experiments in the outer Solar system). Finally, there are programs to detect planets in debris disks stars using ground-based telescopes (e.g. *Subaru/HiCIAO*, *Gemini/GPI* and *VLT/SPHERE*) that will allow to study the planet-disk interaction. These are some of the future research lines in debris disk studies that will help us understand our Solar system in the context of the wide diversity of planetary systems.

References

Beichman, C. A., *et al.* 2005, *ApJ*, 626, 1061
Booth, M., *et al.* 2009, *MNRAS*, 399, 385
Bryden, G., *et al.* 2006, *ApJ*, 636, 1098
Carpenter, J. M., *et al.* 2009, *ApJS*, 181, 197
Dermott, S. F., *et al.* 1994, *Nature*, 369, 719
Dermott, S. F., *et al.* 2002, *Asteroids, Comets, and Meteors: ACM 2002*, 500, 319
Gautier, T., N. III, *et al.* 2007, *ApJ*, 667, 527
Gomes, R., Levison, H. F., Tsiganis, K., & Morbidelli, A. 2005, *Nature*, 435, 466
Gurnett, D. A., Ansher, J. A., Kurth, W. S., & Granroth, L. 1997, *Geoph. R. Lett.*, 24, 3125
Grün, B. A. S., *et al.* 2001 in: Grün *et al.* (eds.), *Interplanetary Dust* (Springer), 295
Hillenbrand, L. A., *et al.* 2008, *ApJ*, 677, 630
Humes, D. 1980, *J. Geophys. R.*, 85 (A/II), 5841
Jessberger, E. K. 2001, in: Grün *et al.* (eds.), *Interplanetary Dust* (Springer), 253
Jewitt, D. C. Luu, J. X. 2000, in: Mannings *et al.* (eds.), *Protostars and Planets IV*, 1201
Jura, M. 2006, *ApJ*, 653, 613
Landgraf, M., Liou, J.-C., Zook, H. A., & Grün, E. 2002, *AJ*, 123, 2857
Lisse, C. M., Beichman, C. A., Bryden, G., & Wyatt, M. C. 2007, *ApJ*, 658, 584
Lovis, C., Mayor, M., Pepe, F., Alibert, Y., Benz, W., *et al.* 2006, *Nature*, 441, 305
Meyer, M. R., *et al.* 2008, *ApJL*, 673, L181
Morales, F. Y., *et al.* 2009, *ApJ*, 699, 1067
Moro-Martín, A., *et al.* 2007, *ApJ, 668*, 1165
Moro-Martín, A., *et al.* 2010, *ApJ*, 717, 1123
Nesvorný, D., *et al.* 2010, *ApJ*, 713, 816
Reach, W. T., Morris, P., Boulanger, F., & Okumura, K. 2003, *Icarus*, 164, 384
Strom, R. G., Malhotra, R., Ito, T., Yoshida, F., & Kring, D. A. 2005, *Science*, 309, 1847
Su, K. Y. L., *et al.* 2006, *ApJ*, 653, 675
Sykes, M. V. & Greenberg, R. 1986, *Icarus*, 65, 51
Trilling, D. E., *et al.* 2008, *ApJ*, 674, 1086
Wyatt, M. C., *et al.* 2007, *ApJ*, 658, 569

The Astrophysics of Planetary Systems: Formation, Structure, and Dynamical Evolution
Proceedings IAU Symposium No. 276, 2010
A. Sozzetti, M. G. Lattanzi & A. P. Boss, eds.

doi:10.1017/S1743921311019946

The planet companion around β Pictoris

Anne-Marie Lagrange[1], Mickaël Bonnefoy[1], Gael Chauvin[1], Daniel Apai[2], David Ehrenreich[1], Anthony Boccaletti[3], Damien Gratadour[3], Daniel Rouan[3], David Mouillet[1], Sylvestre Lacour[3] and Markus Kasper[4]

[1] Laboratoire dAstrophysique de Grenoble, France
email: lagrange@obs.ujf-grenoble.fr, bonnefoy@obs.ujf-grenoble.fr, chauvin@obs.ujf-grenoble.fr, ehrenreich@obs.ujf-grenoble.fr, mouillet@obs.ujf-grenoble.fr,

[2] Space Telesope Sciente Institute, 3700 San Martin Dr. Baltimore, MD 21218, USA
email: apai@stsci.edu

[3] LESIA, Observatoire de Paris, place Jules Janssen, 92195 Meudon, France
email: anthony.boccaletti@obspm.fr, daniel.rouan@obspm.fr,damien.gratadour@obspm.fr, sylvestre.lacour@obspm.fr

[4] ESO, Karl Schwarzschild St, 2, 85748 Garching bei Muenchen, Germany
email: mkasper@eso.org

Abstract. The β Pic disk of dust and gas has been regarded as the prototype of young planetary systems since the 1980s and has revealed over the years an impressive amount of indirect signs pointing toward the presence of at least one giant planet. We present here the recently detected first giant planet around this star. We show how this planet could explain some very peculiar features of the star environment (disk, spectroscopic variability), and how it constrains the scenarios of planetary system formation (timescales, mechanisms).

Keywords. planetary systems: formation, techniques: high angular resolution

---

1. Planet formation processes

Understanding planetary systems formation and evolution has become one of the biggest challenges in astronomy, since the imaging of a debris disk around β Pictoris, in the eighties and the discovery of the first exoplanet around the solar-like star 51 Pegasi during the 90's. While $\simeq$ 500 close planets have been identified with radial velocity (RV) and transit technics, very few have been imaged and definitely confirmed (see Fig. 1 for the closest ones). These few cases already bring new insights in the way planets form. Indeed, while there are now strong pieces of evidence that short period ($\leqslant$ 5 AU) planets detected by RV have formed through core-accretion (CA), the origins of the directly imaged giant planets around stars/brown dwarfs remain debated. If the companions on very wide orbits (a few hundreds of AUs) formed in situ, the acting mechanism was probably stellar-like gaseous collapse, like binaries. This may be true also for 2M1207 b. In such cases, they would be more similar to brown dwarfs rather than planets. Based on various arguments (dynamical timescales, mass of planetesimals available at the current separations), an in-situ formation of HR 8799 bcd and A PsA b would rather involve instabilities (GI) within a circumstellar disk. Planetary-mass bodies might then form in different ways, depending on the initial conditions and in particular the mass of the parent star.

The physics of young giant planets is also still debated. Their brightness, predicted before by so-called "hot-start" models (Baraffe *et al.* 2003), has been recently questionned at young ages (Fortney *et al.* 2008). The latter predict planets $\simeq$ 5 mag. fainter at young ages than the "hot start" model. According to Fortney *et al.* differences in brightness are present until 50-100 Myr. The difference between both models is due to the modeling of the gas initial properties (temperature), depending on whether the accreting gas energy is radiated away or not. The predictions of planets luminosities is obviously a key issue for planet detection in imaging, especially for the forthcoming planet imagers (GPI, SPHERE) and direct detection and characterization of planets are of crucial importance to calibrate these models.

2. Discovery of β Pictoris b

In a new analysis of VLT/NACO high dynamics L'-band data of β Pictoris taken in Nov. 2003, we detected a faint, point-like signal at $\simeq$ 8 AU (proj. sep.) from the star, within the North-East side of the dust disk (Fig. 2; Lagrange *et al.* 2009a). We showed that this candidate companion was probably not a contaminant, but the data alone were not sufficient to clearly rule out this possibility. If bound, its $L' = 11.2 \pm 0.3$ magnitude would indicate a temperature of $\sim$ 1500 K and a mass of $\sim$ 9 M_{Jup} according to Lyon's group models. Follow-up observations in Jan. and Feb. 2009 (Lagrange *et al.* 2009b) did not detect the companion, in agreement with either the background hypothesis or the "planet" scenario.

In the fall of 2009, new images revealed a point-source in the SW side of the disk, 0.3" (i.e; less than 6 AU in proj. sep.) from the star, with a brightness consistent with that of the 2003 source (Fig. 2; Lagrange *et al.* 2010). These observations also definitely showed that the 2003 source was not a background source and is indeed orbiting β Pictoris. Its 2009 location furthermore implies a semi-major axis of about 8-15 U. It shows that the planet was observed after quadrature in 2003. While more precise orbital parameters are still to be determined by a careful orbital follow-up, β Pictoris b is now the closest planet ever imaged around a star. More recently, β Pictoris b was also detected at 4 microns with the recently available APP device on NACO (Quanz *et al.* 2010) as well as at Ks (Bonnefoy *et al.* 2011). In both cases, the constraints derived on the planet effective temperature are in agreement with those brought by the L' data.

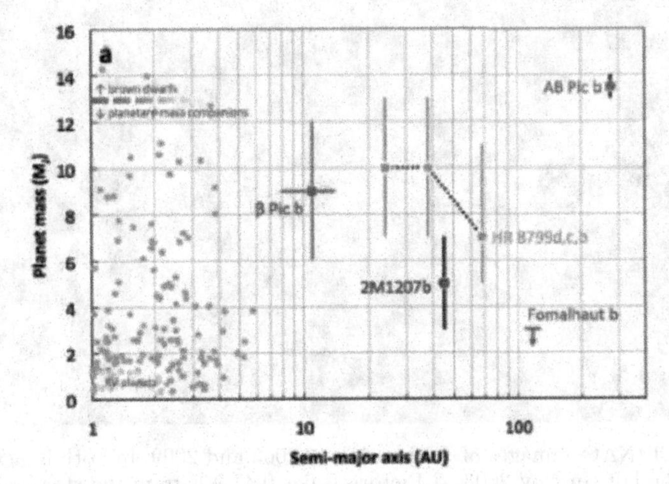

Figure 1. Planetary mass objects imaged close ($\leqslant$ 120 AUs) to stars (as of October 2010).

3. Constraints brought by β Pictoris b

Constraints on planet formation It is well accepted that giant planet formation should occur rapidly (less than a few Myr, as inferred from studies of dispersal timescales of young gaseous disks). Given the system young age (12^{+8}_{-4} Myr; Zuckerman *et al.* 2001), β Pictoris b now proves directly for the first time that giant planets can form in Myr time-scales in disks. Based on current models predictions according to its parent star's properties (age, star mass) developped by Kennedy and Kenyon (2008), and given its properties, β Pictoris b is probably the first planet imaged that could have formed via core accretion, like, our Solar System giant planets. It offers then the opportunity to constrain CA-related models in the future. We already note that the "cold start" evolutionary model proposed by Fortney *et al.*(2008) fails to reproduce its luminosity.

β Pictoris disk While about 20 debris disks –disks containing dust produced by collisions among larger rocky bodies – have been optically resolved today, β Pictoris remains the best-studied young system, with several indirect signs of the presence of planets on circular or low eccentric orbit (see a review in Lagrange *et al.* (2000), Freistetter *et al.* (2007)). Among the numerous peculiarities of the disk, is the so-called inner ($\simeq$ 60-80 AU) 5-6 degrees warp (Mouillet *et al.* 1997; Heap *et al.* 2000; Golimowski *et al.* 2004). More than 10 years ago, we interpreted and modeled this warp as the result of the gravitational perturbation of a massive body located on an inclined orbit, on a disk of planetesimals (Mouillet *et al.* 1997; Augereau *et al.* 2001). Such a configuration also explained the observed butterfly asymmetry (Kalas and Jewitt 1995). We derived in 1997 an analytical relation between the warp extension and the perturbing body properties. When using updated values of the star age and warp characteristics, it appears that the characteristics of β Pictoris b do verify this relation. However, within the present error bars, we cannot determine whether or not it is located in the inner warped disk or not. We note that if further observations show that the companion lies indeed in the warped disk, between 8 and 14 AU, then we will have a direct (independant from evolutionary models) constraint on its mass (to be less than 42 MJup).

Very interestingly, β Pictoris exhibits a very peculiar spectroscopic variability (see e.g. Vidal-Madjar *et al.* 1998; Lagrange *et al.* 2000) that has been since many years

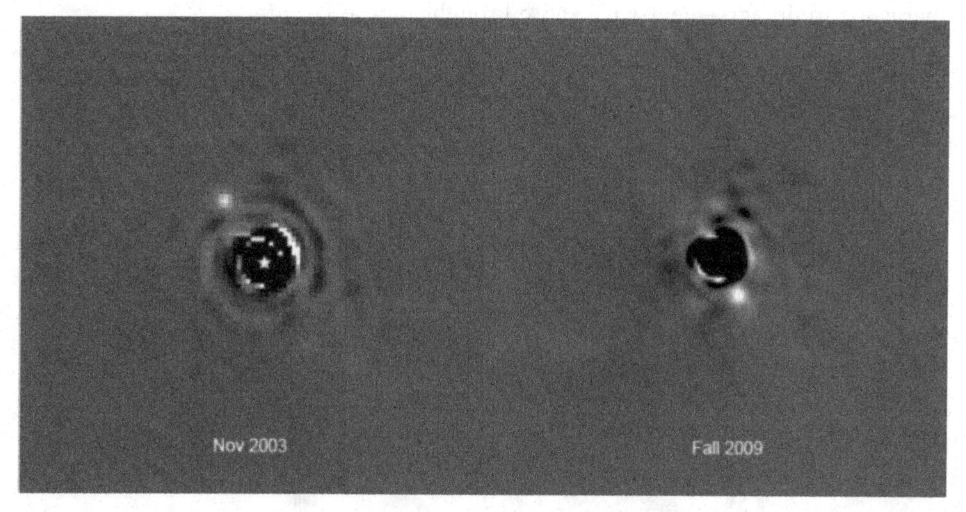

Figure 2. L' VLT/NACO images of β Pictoris b in 2003 and 2009. In both images, North is up and East is to the left. In Nov 2003, β Pictoris b lies 0.4" NE from the star, and in Fall 2009, 0.3" SW from the star.

attributed to the evaporation of star grazing comets (FEBs; Beust & Morbidelli 2005 and ref therein). β Pictoris b could be at the origin of the observed cometary infall.

Finally, Lecavelier des Etangs & Vidal-Madjar (2009) showed that the candidate companion observed in 2003 could be responsible for the peculiar photometric variation observed in 1981, provided its semi-major axis (circular case or assuming a low eccentricity) is in the range 7.6–8.7 AU, corresponding to periods in the range 16-20 years. The 2009 imaging data are still compatible with such a scenario. If confirmed, we could have access in a few years, to the transmission spectrum of a transiting giant planet located beyond the ice zone.

References

Augereau, J. C., Nelson, R., Lagrange, A. M., *et al.* 2001, *A&A* , 370, 447
Baraffe, I., Chabrier, G., Barman, T. S., *et al.* 2003, *A&A* , 402, 701
Beust, H. & Morbidelli, A. *Icarus*, 143, 170
Beuzit, J.-L., Boccaletti, A., Feldt, M., *et al.* 2010, *ASPC*, 430, 231
Bonnefoy, M., Lagrange, A. M., Chauvin, G. *et al.* 2011, *A&A* , 528, L15
Fortney, J. J., Marley, M. S., Saumon, D., & Lodders, K. 2008, *ApJ* , 683, 1104
Freistetter, F, Krivov, A. V. & Loehene, T. 2007, *A&A* , 466, 389
Golimovski, D. A., Henry, T. J., Krist, J. E., *et al.* 2004, *AJ* , 128, 1733
Heap, S., Lindler, D. J., Lanz, T. M., *et al.* 2000, *ApJ* , 539, 435
Kalas, P. & Jewitt, D. 1995, *AJ* , 110, 794
Kennedy, G. M. & Kenyon, S. J. 2008, *ApJ* 673, 502
Lagrange, A.-M., Backman, D., & Artymowicz, P. 2000, in *Protostars & Planets* IV, Univ. of Arizona Press, 639
Lagrange, A.-M., Gratadour, D., Chauvin, G., *et al.* 2009a, *A&A* , 493, L21
Lagrange, A.-M., Kasper, M., & Boccaletti, A., *et al.* 2009b, *A&A* , 506, L927
Lagrange, A.-M., Bonnefoy, M., Chauvin, G., *et al.* 2010, *Science* , 329, 57
Lecavelier des Etangs, A. & Vidal-Madjar, A. 2009, *A&A* , 497, 557
Quanz, S., Meyer, M. R., Kenworthy, M. A., *et al.* *ApJ* , 722, L49
Mouillet, D., Larwood, J. D., Papaloizou, J. C. B., & Lagrange A.-M. 1997, *MNRAS* , 292, 896
Vidal-Madjar, A., Lecavelier des Etangs, A., & Ferlet, R. 1998, *Planetary & Space Science*, 46, 629
Zuckerman, B., Song, I., Bessel, M. S., & Webb, R. A. 2001, *ApJ* , 562, L87

The Astrophysics of Planetary Systems: Formation, Structure, and Dynamical Evolution
Proceedings IAU Symposium No. 276, 2010
A. Sozzetti, M. G. Lattanzi & A. P. Boss, eds.

doi:10.1017/S1743921311019958

Theoretical predictions of mass, semimajor axis and eccentricity distributions of super-Earths

Shigeru Ida[1]

[1]Department of Earth & Planetary Science, Tokyo Institute of Technology, Ookayama 2-12-1 I2-10, Tokyo 152-8551, Japan
email: ida@geo.titech.ac.jp

Abstract. We discuss the effects of close scattering and merging between planets on distributions of mass, semimajor axis and orbital eccentricity, using population synthesis model of planet formation, focusing on the distributions of close-in super-Earths, which are being observed recently. We found that a group of compact embryos emerge interior to the ice line, grow, migrate, and congregate into closely-packed convoys which stall in the proximity of their host stars. After the disk-gas depletion, they undergo orbit crossing, close scattering, and giant impacts to form multiple rocky Earths or super-Earths in non-resonant orbits around ~ 0.1AU with moderate eccentricities of ~ 0.01–0.1. The formation of these planets does not depend on model parameters such as type I migration speed. The fraction of solar-type stars with these super-Earths is anti-correlated with the fraction of stars with gas giants. The newly predicted family of close-in super-Earths makes less clear "planet desert" at intermediate mass range than our previous prediction.

Keywords. planets and satellites: dynamical evolution and stability, planets and satellites: formation, planetary systems: protoplanetary disks

1. Introduction

A radial velocity survey suggests that 40–60% of solar-type stars bear super-Earths with mass up to $\sim 20M_\oplus$ and period up to 50 days and in many cases these super-Earths are members of multiple-planet systems (e.g., Mayor *et al.* 2009; Bouchy *et al.* 2009; Lo Curto *et al.* 2009). From a radial velocity survey for a controlled sample, Howard *et al.* (2010) derived a planetary mass function around solar-type stars. They found that about 12% and 7% of stars have close-in plantes in mass ranges of 3-10$M_\oplus$ and 10-30$M_\oplus$, respectively. The function monotonically decreases with planetary mass, not showing any deficit in intermediate masses ("planet desert") that theorectical simulations predicted (1-50$M_\oplus$ by Ida & Lin 2004a, 2008a; 1-10$M_\oplus$ by Mordasini *et al.* 2009b). Kepler transit survey also suggests non-existence of "planet desert" for close-in planets (Borucki *et al.* 2010).

Based on the conventional core accretion scenario, we constructed a population-synthesis planet formation model (Ida & Lin 2004a, 2004b, 2005, 2008a, 2008b). In the model, we derived prescriptions for coagulation of planetesimals to form rocky planetary embryos and icy/rocky cores, gas accretion onto the cores, and orbital migration of embryos and gas giants, from detailed simulation results. Mordasini *et al.* (2009a, b) constucted a similar model. All these models have neglected planet-planet interactions.

However, even with a modet amout of type I migration, embryos would migrate toward their host stars. Resonant trapping, close scattering, and collision would play an important role in the final configuration of close-in rocky planets. Gravitational perturbations by gas giants should also sculpture planetary systems as a whole. Here, we briefly summarize related N-body simulations, modeling of the N-body simulation results for population synthesis calculations, and the effects of planet-planet dynamical interactions on the predicted distributions of extrasolar planets, focusing on close-in super-Earths.

2. N-body simulations on formation of close-in super-Earths

Ogihara & Ida (2009) performed N-body simulation to study the accretion of planets from planetesimals near the disk inner edge. Inward type I migration of planets is halted either by truncation of gas at the disk edge or by resonant perturbation from an inner planet. They found that in the case of relatively slow type I migration, 20–30 planets are captured by mutual mean-motion resonances and extend from the disk inner edge to the regions well beyond 0.1AU. After disk gas depletion, these planets start orbit crossing and giant impacts, resulting in formation of several close-in super-Earths. The super-Earths thus formed are kicked out of resonances by strong scattering and collisions. This is in contrast to the fast migration case in which only several planets survive during the presence of the gas and they remain in stable resonant orbits even after disk gas is removed (Terquem & Papaloizou 2007; Ogihara & Ida 2009).

3. N-body simulations on formation of eccentric jupiters and close-in retrograde jupiters

Many of extrasolar gas giants so far discovered have large eccentricities (> 0.2). One of promising excitation mechanisms is orbital instability between gas giants called "jumping jupiter" process (e.g., Rasio & Ford 1996; Marzari & Weidenschilling 2000). After the disk gas depletion, secular perturbations among gas giants lead to their orbital crossing. Close scatterings usually result in ejection of one planet, leaving other giants in well-separated stable eccentric orbits. Secular perturbations from the giant planets in eccentric orbits may destabilize orbits of rocky and icy planets in board regions. The eccentricity distribution created by the scattering may be consistent with observed one (e.g., Chatterjee *et al.* 2008; Juric & Tremaine 2008).

The jumping jupiter process also forms close-in hot jupiters if it is combined with Kozai mechanism and tidal dissipation (Nagasawa *et al.* 2008). Nagasawa *et al.* (2008) found that in 30% of runs of N-body simulations the eccentricity of an inwardly scattered giant is further increased enough for tidal circularization by Kozai mechanism from outer giants that often have high inclinations acquired by mutual close scattering. The circularized probability of $\sim 30\%$ is one order of magnitude higher than that in the case of only two giants and that found only in final state of three giants case.

They also predicted that many of the circularized planets have retrograde orbits. When the eccentricity of the inner planet becomes close to unity, its orbital angular momentum is so low that small perturbations from the outer planets can easily makes the orbit retrograde. This prediction is consistent with recent Rossiter-MacLaughlin measurements.

These N-body simulations show that planet-planet interactions are important factors to configure orbital distributions of extrasolar gas giants as well as type II migration.

4. Modeling to planet-planet dynamical interactions

Ida & Lin (2010) constructed a prescription that approximates the process of eccentricity excitation and collisions (giant impacts) of rocky planetary embryos in gas free environment, which is briefly summarized as follows: 1) evaluate the timescale for embryo pairs to start orbit crossing ($\tau_{\rm cross}$) and identify the pair with the shortest $\tau_{\rm cross}$, 2) after such a time interval has elapsed, compute the expected statistical changes in their eccentricity and semimajor axis, and then identify all other embryos whose orbits would cross this pair, if these changes were implemented, 3) for this group of embryos, implement statistical changes in e and a due to repeated close scattering among themselves, preserving the conservation of total orbital energy, 4) identify pairs of impacting embryos based on their statistically weighted collisional probability, and 5) under the assumption that these events lead to cohesion, adjust both a and e of the merger product to satisfy the conservation of total Laplace-Runge-Lenz (LRL) vector. Although these comprehensive procedures are complicated to integrate, each step is based on well-studied celestial mechanics. Other than two empirical parameters, there is no need to introduce any arbitrary assumptions. The two parameters are also qualitatively inferred from celestial mechanics.

They found that the above prescription reproduces quantitative statistical properties of mass, semimajor axis and eccentricity distribution of final planets obtained in N-body simulations by Kokubo *et al.* (2006). Because this process itself includes chaotic features and the prescription includes Monte Carlo approach, comparison is meaningful only for statistical quantities such as mean values and their dispersion of physical quantities.

Ida & Lin (2010) found that a collision occurs only when an inner planet is near its apoastron and an outer one is near its periastron and such a collision results in small eccentricity of the merged body due to conservation of total LRL vector. As a result, eccentricities of final planets are usually much smaller than those during orbital crossing that are determined by surface escape velocity of the interacting planets. This demonstrates that this kind of modeling can reveal intrinsic physics that is not usually revealed by full simulations, such as N-body simulations.

They also showed that $\tau_{\rm cross}$ jumps up by orders of magnitude at every merging event, because $\tau_{\rm cross}$ increases with orbital spacing scaled by Hill radius and decrease in orbital eccentricity. Through repeated merging events on timescales of 10^7–10^8 years, the system eventually reaches a state with $\tau_{\rm cross}$ longer than main-sequence lifetime of solar-type stars.

Nagasawa & Ida (2011) derived a prescription to predict the final states of the jumping jupiter process. Since scatterings between gas giants have sufficient ability to eject planets from the systems, stable orbits are realized by ejection rather than merging. Since ejcetion dereases number of bodies, ejection also drastically increases $\tau_{\rm cross}$. In the case of three planets, only one ejection event is enough to raise $\tau_{\rm cross}$ over main-sequence lifetime of solar-type stars. Each step of this process is also based on well-studied celestial mechanics. Their results also quantitatively reproduce statistical features of N-body simulations.

5. Eccentricity trap

The coagulation of embryos to form close-in super-Earths sensitively depends on how migration of embryos is halted near the disk inner edge. In N-body simulation by Ogihara & Ida (2009), the innermost embryo is pinned to the disk inner edge (set at ~ 0.05AU) and 20–30 resonantly trapped embryos extend from the edge to the regions well beyond 0.1AU, in the slow migration case. Because the inner edge may correspond to the corotation radius with the host star's spin, tidal force from the host star does not decay these embryo orbits outside the disk inner edge. Consequently, the super-Earths which form through the giant impacts are distributed at $\sim$ 0.05–0.2AU in their simulations, which may be consistent with observed data.

If the innermost embryo were pushed inward into the inner cavity by torques from outer embryos, its orbit would be tidally decayed. Then, since many embryos are not retained and the embryo masses are usually smaller than the Earth mass, super-Earths are not formed from coagulation of these retained embryos in this case.

Because individual embryos are losing angular momentum through type I migration and the angular momentum is redistributed throughout the convoy with resonant interactions, large amount of angular momentum must be supplied to prevent the embryos from penetrating the disk edge. Ogihara *et al.* (2010) investigated this issue through orbital integration and analytical arguments. If the innermost planet has such large eccentricity that radial excursion is larger than width of inner disk edge, the planet suffers eccentricity damping due to disk-planet interaction only near the periastron. Then, the damping expands its orbit (increases semimajor axis). The eccentricity damping is fast enough to compensate for the angular momentum loss of all the outer embryos due to type-I migration. Because this mechanism requires continuous eccentricity excitation of the innermost planet, it works for comparable-mass embryos that are resonantly interacting with each other. Ogihara *et al.* (2010) called this mechanism as "eccentricity trap."

6. Formation of close-in planets

Using the prescription for giant impacts of rocky embryos and the "eccentricity trap" found by Ogihara *et al.* (2010), Ida & Lin (2010) systematically studied the formation process of non-resonant multiple close-in super-Earths found by Ogihara & Ida (2009).

Figure 1 shows an exmple of the integration of growth and migration of planets in a disk with a modest initial mass (2.5 times as massive as the minimum-mass solar nebula) and migration efficiency $C_1 = 0.1$, which is a scaling factor for type I migration defined by $\dot{a} = C_1 \dot{a}_{\rm Tanaka}$ where $\dot{a}_{\rm Tanaka}$ is migration rate derived by Tanaka *et al.* (2002). Seed

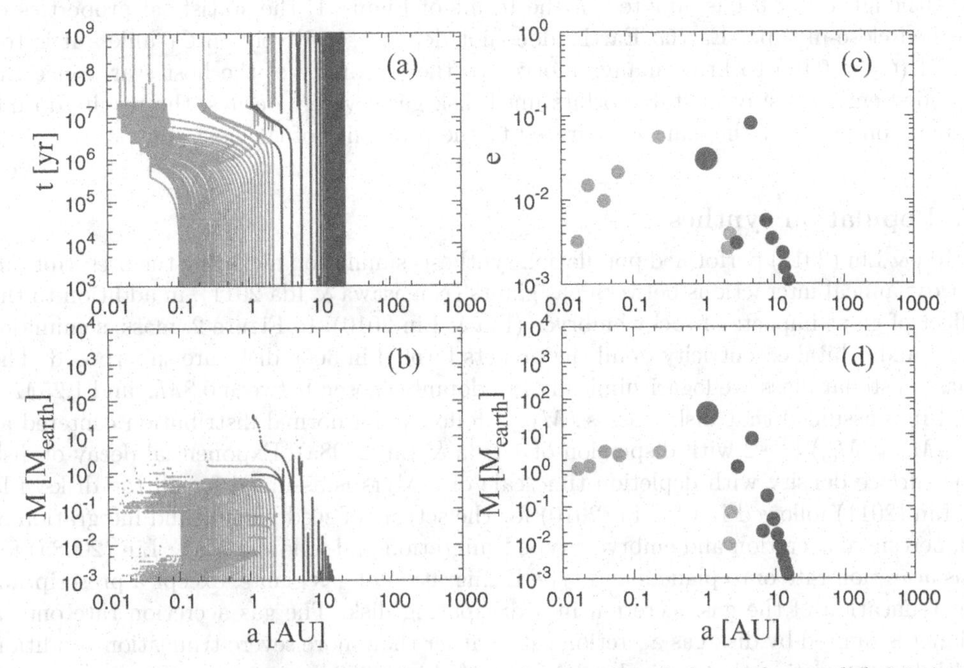

Figure 1. An example of integration of growth and migration of planets from a disk 2.5 times more massive than the minimum-mass solar nebula. (a) Time evolution, (b) mass evolution, (c) final eccentricities, and (d) final masses.

embryos are distributed at 0.1–30AU and are integrated including mutual dynamical interactions. Embryos' migration is stalled by "eccentricity trap" at the disk inner edge (at ~ 0.04AU).

In inner regions, embryo growth due to planetesimal accretion and their migration are so fast that they form a swarm of embryos trapped in mutual mean-motion resonances in the proximity of the host star (Figure 1a). Because type I migration is faster and embryo growth is slower for more massive bodies, type I migration is dominated for embryo masses larger than critical masses, $M_{\rm c,crit} \sim 0.1\text{–}1 M_\oplus$ (Figure 1b). Beyond an ice line, an icy core grows up to $\sim 10 M_\oplus$ and starts runaway gas accretion. In outermost regions, planetesimal growth is so slow that only small planets emerge (planets at < 5AU are ejected by the gas giant).

The mass, semimajor axis and eccentricity of final planets are plotted in Figure 1c and d. The innermost four planets have grown through giant impacts after gas depletion. Since a collision disspipates only a fraction of orbital energy, the resultant semimajor axis of the merger products is comparable to that of their progenitor embryos. In the absence of residual disk gas, these merger products do not undergo any further orbital decay and generally remain out of mean motion resonance with each other.

The velocity dispersion of the residual embryos is a fraction of their surface escape velocity. In the stellar proximity, it is much smaller than the local Keplerian velocity. Thus, the simulated eccentricities of close-in super-Earths are relatively small (~ 0.01–0.1).

The largest planet in the innermost region has mass more than $5 M_\oplus$. Since this planet acquired most of its mass after the gas is depleted, it cannot accrete a substantial gaseous envelope. The atmosphere of their progenitor embryos may also be ejected during giant impacts. This explains why some of the discovered super-Earths have masses larger than $10 M_\oplus$ without becoming gas giants.

Although $C_1 = 0.1$ is adopted in the result of Figure 1, the statistical properties of formed close-in super-Earths/Earths does not depend sensitively on C_1 unless it is too small ($C_1 < 0.01$) to bring many embryos to the proximity of the host star. Since the trapped embryos stay in stable orbits until disk gas severely decays, the results do not depend on how fast the emryos migrated to the proximity of the host star.

7. Population synthesis

Ida & Lin (2011) performed population synthesis simulation including the prescriptions for dynamical interactions between gas giants (Nagasawa & Ida 2011), in addition to the effect of giant impacts of rocky embryos (Ida & Lin 2010). In Figure 2, mass, semimajor axis, and orbital eccentricity of all the planets formed in 3000 disks are superposed. The mass of stellar mass are logarithimically ramdopmly chosen between $0.8 M_\odot$ and $1.25 M_\odot$, and it is assumed that disk masses ($M_{\rm disk}$) follow a log normal distribution centered at $\log(M_{\rm disk}/M_\odot) = -2$ with dispersion of 1 (Ida & Lin 2008a). Exponential decay of disk gas surface density with depletion timescale of 3 Myrs is assumed for all the disks. Ida & Lin (2011) followed Ida & Lin (2010) for the setting of seed planets and integration of planetesimal accretion and embryo's type I migration and followed Ida & Lin (2008a) for gas accretion rate onto planets and type II migration of gas giants, except a prescription for truncation of the gas accretion in a dissipating disk. The gas accretion rate onto a planet is limited by disk gas accretion rate, rather than more severe truncation condition with local depletion that was adopted in Ida & Lin (2008a).

Although eccentricity distributions of gas giant planets (Figures 2a and c) are important, we here focus on distributions of close-in super-Earths, which are based on the

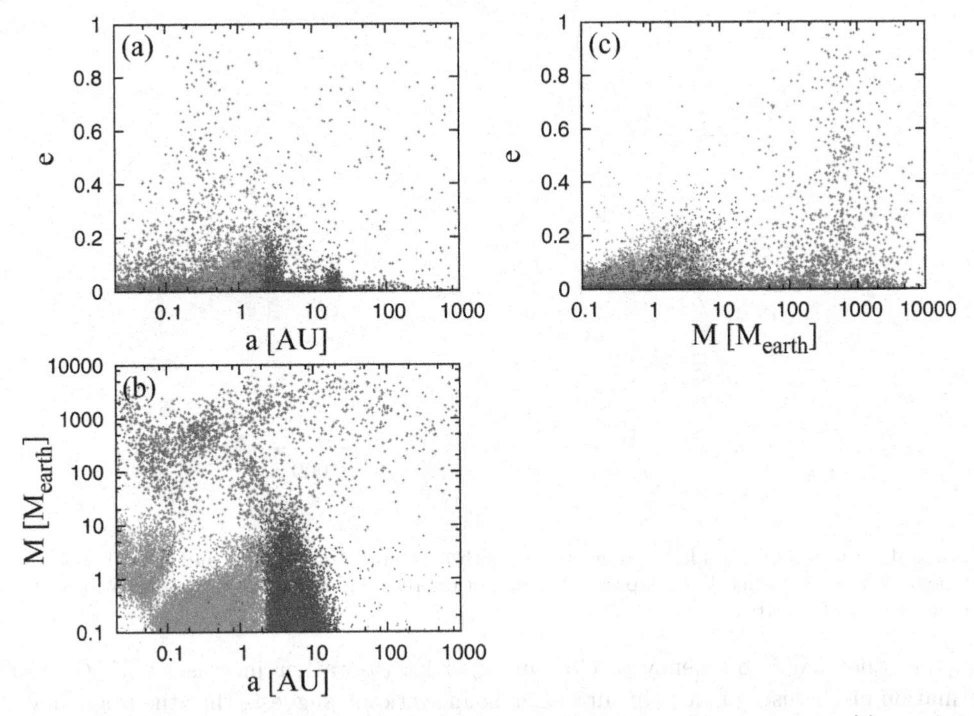

Figure 2. Distributions of simulated planets from 3000 disks on (a) eccentricity (e) - a plane, (b) semimajor axis (a) - mass (M) plan, and (c) e - M plane.

prescription in Ida & Lin (2010). In Figure 2b, a new pronounced population is found at $M \sim 1$–$10 M_\oplus$ and $a \sim 0.03$–0.15AU that did not exist at all in the previous simulations without dynamical interactions (e.g., Ida & Lin 2008). These close-in super-Earths are formed by the mechanism described in section 6. The eccentricities of these super-Earths are small (< 0.1).

Figure 3 shows dependence on C_1. Frequency of gas giants decreases sensitively with C_1, because for larger values of C_1, it is more difficult to form cores larger than a critical core mass of $>$ several $M_\oplus$ for the onset of runaway gas accretion (Ida & Lin 2008).

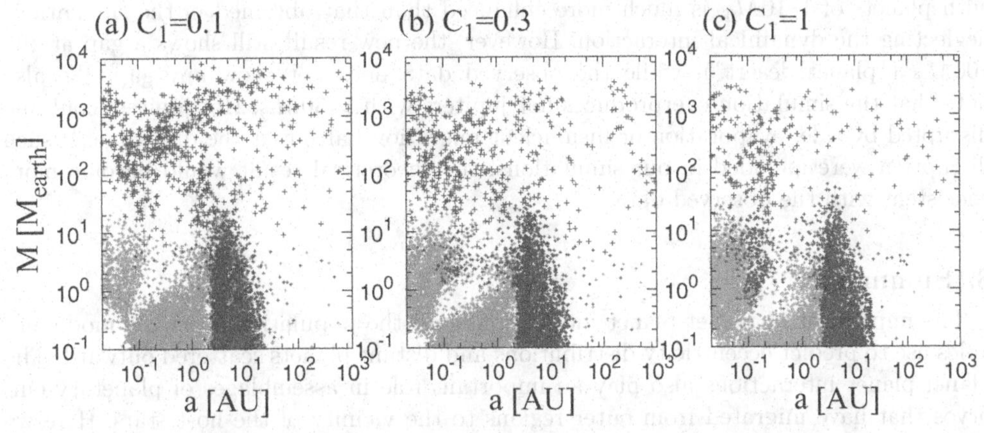

Figure 3. Distributions of simulated planets from 1000 disks on semimajor axis (a) - mass (M) plane for (a) $C_1 = 0.1$, (b) $C_1 = 0.3$, and (c) $C_1 = 1$.

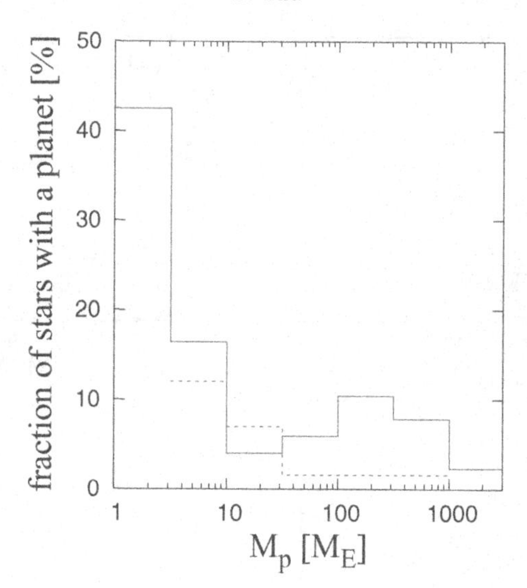

Figure 4. Fraction of stars harboring planets with corresponding masses and periods less than 50 days. Solid and dashed lines express our theoretical result in Figure 2 and observational data by Howard *et al.* (2010).

On the other hand, frequency of close-in super-Earths rather increases with C_1. The formation mechanism of close-in super-Earths in section 6 suggests that the frequency of these planets are almost independent of C_1. However, for smaller C_1, gas giants are more abundant and their perturbations destabilize orbits of super-Earths, so that surviving super-Earths are less abundant.

Since even in disks with not large initial dust-to-gas ratio, super-Earths can be formed from the accumulated embryos and there is no threshold mass like the critical core mass for runaway gas accretion, the dependence of frequency of super-Earths on host stars' metallicity would be much weaker than gas giants, which may be also consistent with observed data.

Figure 4 shows the fraction of stars harboring planets with corresponding masses and periods less than 50 days. Solid and dashed lines express our theoretical result in Figure 2 and observational data by Howard *et al.* (2010). The simulations including the dynamical interactions produce a new population of close-in super-Earths, so the fraction of stars with planets of 1–$10M_\oplus$ is much more enhanced than that obtained in the simulations neglecting the dynamical interaction. However, the new result still shows a gap at 10–$100M_\oplus$ ("planet desert"), while the observed data does not show any gap. We also note that the simulation overproduces hot jupiters with $> 30M_\oplus$. Hot jupiters could be disrupted by tide, evaporation or insufficient migration halting at the disk edge. If such disruption were included in our simulations, the theoretical result would become more consistent with the observed data.

8. Summary

The implement of planet-planet interactions into the population synthesis model enables use to predict eccentricity distributions and distant planets scattered outward. The planet-planet interactions also play an important role in assemblages of planetary embryos that have migrated from outer regions to the vicinity of the host stars. Here we described a scenario for formation of close-in multiple non-resonant super-Earths. Our model predicts ubiquity and weak dependence on stellar metallicity of these systems around solar-type stars and why they missed runaway gas accretion even if they have masses > several $M_{\oplus}$. We showed that the boundary condition at the disk edge plays an crucial role in the formation of these systems. For more reliable predictions, we need detailed investigation on evolution of the disk edge as well as the effects of tide and thermal evaporation.

References

Borucki, W., *et al.* 2010, *Science*, 327, 977
Bouchy, F., Mayor, M., Lovis, C., Udry, S., Benz, W., Bertaux, J. -L., Delfosse, X., Mordasini, C., Pepe, F., Queloz, D., & Segransan, D. 2009, *A&A*, 496, 527
Chatterjee, S., Ford, E. B., Matsumura, S., & Rasio, F. A. 2008, *ApJ*, 686, 580
Howard, A. W., Marcy, G. W., Johnson, J. A., Fischer, D. A., Wright, J. T., Isaacson, H., Valenti, J. A., Anderson, J., Lin, D. N. C., & Ida, S. 2010, *Science*, 330, 653
Ida, S. & Lin, D. N. C. 2004a, *ApJ*, 604, 388
Ida, S. & Lin, D. N. C. 2004b, *ApJ*, 616, 567
Ida, S. & Lin, D. N. C. 2005, *ApJ*, 626, 1045
Ida, S. & Lin, D. N. C. 2008a, *ApJ*, 673, 484
Ida, S. & Lin, D. N. C. 2008b, *ApJ*, 685, 584
Ida, S. & Lin, D. N. C. 2010, *ApJ*, 719, 810
Ida, S. & Lin, D. N. C. 2011, in preparation
Juric, M. & Tremaine, S. 2008, *ApJ*, 686, 603
Lo Curto, G., Mayor, M., Benz, W., Bouchy, F., Lovis, C., Moutou, C., Naef, D., Pepe, F., Queloz, D., Santos, N. C., Segransan, D., & Udry, S. 2010, *A&A*, 67, 4679
Marzari, F. & Weidenschilling, S. J. 2000, *Icarus*, 156, 570
Mayor, M., Udry, S., Lovis, C., Pepe, F., Queloz, D., Benz, W., Bertaux, J. -L., Bouchy, F., Mordasini, C., & Segransan, D. 2009, *A&A*, 493, 636
Mordasini, C., Alibert, Y., & Benz, W. 2009a, *A&A*, 501, 1139
Mordasini, C., Alibert, Y., Benz, W., & Naef, D. 2009b, *A&A*, 501, 1161
Nagasawa, M. & Ida, S. 2011, in preparation
Meyer, B. S., Clayton, D. D., & The, L.-S. 2000, *ApJ* (Letters), 540, L49
Ogihara, M., Duncan, M.., & Ida, S. 2010, *ApJ*, 721, 1184
Ogihara, M. & Ida, S. 2009, *ApJ*, 699, 824
Terquem, C. & Papaloizou, J. C. B. 2007, *ApJ*, 654, 1110
Rasio F. A., Ford, E. B. 1996, *Science*, 274, 954

The Astrophysics of Planetary Systems: Formation, Structure, and Dynamical Evolution
Proceedings IAU Symposium No. 276, 2010
A. Sozzetti, M. G. Lattanzi & A. P. Boss, eds.

doi:10.1017/S174392131101996X

Application of recent results on the orbital migration of low mass planets: convergence zones

Christoph Mordasini[1], Kai-Martin Dittkrist[1], Yann Alibert[2], Hubert Klahr[1], Willy Benz[2] and Thomas Henning[1]

[1]Max Planck Institute for Astronomy, Königstuhl 17, D-69117 Heidelberg, Germany
email: mordasini@mpia.de

[2]Physikalisches Institut, Sidlerstrasse 5, CH-3012 Bern, Switzerland

Abstract. Previous models of the combined growth and migration of protoplanets needed large ad hoc reduction factors for the type I migration rate as found in the isothermal approximation. In order to eliminate these factors, a simple semi-analytical model is presented that incorporates recent results on the migration of low mass planets in non-isothermal disks. It allows for outward migration. The model is used to conduct planetary populations synthesis calculations. Two points with zero torque are found in the disks. Planets migrate both in- and outward towards these convergence zones. They could be important for accelerating planetary growth by concentrating matter in one point. We also find that the updated type I migration models allow the formation of both close-in low mass planets, but also of giant planets at large semimajor axes. The problem of too rapid migration is significantly mitigated.

Keywords. planetary systems: formation, planetary systems: protoplanetary disks

1. Introduction

The timescales of orbital migration of low mass planets as found in linear, isothermal type I migration models are very short for typical protoplanetary disk conditions. They are in particular shorter than typical growth timescales (Tanaka *et al.* 2002). This means that most protoplanets would fall into the star before reaching a mass allowing giant planet formation ($\sim 10\,M_\oplus$). It is therefore not surprising that previous planetary population synthesis models have found that isothermal type I rates need to be reduced by large factors (10-1000) in order to reproduce the observed semimajor axis distribution of extrasolar planets (Ida & Lin 2008; Mordasini *et al.* 2009b; Schlaufman *et al.* 2009).

The necessity of such arbitrary reduction factors indicates a significant shortcoming in the understanding of the migration process. Additionally one finds that with universal reduction factors (independent of planet mass and distance), it is impossible to reproduce at the same time giant planets at several AU and close-in low mass planets (Mordasini *et al.* 2009b). This is inconsistent with observations (Howard *et al.* 2010).

Several mechanisms were proposed that could slow down type I migration. An approach that recently gained significant attention was a more realistic description of the thermodynamics in the protoplanetary disk, in order to drop the simplification of isothermality (e.g. Paardekooper & Mellema 2006; Kley & Crida 2008; Kley *et al.* 2009; Paardekooper *et al.* 2010; Baruteau & Lin 2010).

It was found that the migration rates (and even the direction of migration) in such more realistic models can be very different from the ones found in the isothermal limit. There are strong dependences on disk properties, like the temperature and the gas surface density gradient or the opacity, leading to different sub-regimes of type I migration.

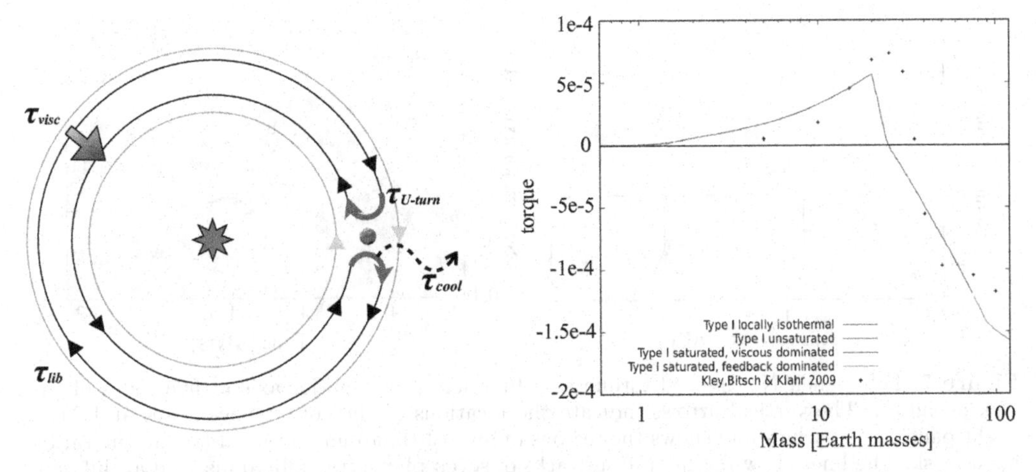

Figure 1. Left panel: Schematic representation of the timescales relevant to the migration regime of a low mass planet. Trajectories of different gas parcels are shown in a frame of reference rotating with the planet (big black circle) around the star (center). Right panel: Specific torque as a function of planetary mass from the analytical model (lines, indicating different regimes), compared to the radiation-hydrodynamic simulations of Kley *et al.* (2009) (crosses).

2. Updated type I migration model

We present results of incorporating these improved type I migration rates into the planet formation code of Alibert *et al.* (2005). The new migration model, which will be presented in details in Dittkrist *et al.* (in prep.), includes the following mechanisms:

Isothermal vs. adiabatic regime Gas parcels in the horseshoe region make a sharp U-turn close to the planet (Fig. 1, left). If during such a U-turn, the gas can cool quickly enough by radiation to equilibrate with the surrounding gas, the gas behaves in a locally isothermal way. On the other hand, if the U-turn timescale $\tau_{\mathrm{U-turn}}$ is short compared to the cooling timescale τ_{cool}, then the gas parcel keeps its entropy, and the process is approximately adiabatic. These two situations lead to different density distributions around the planet, which eventually translate into different torques. For both regimes we use the results of Paardekooper *et al.* (2010) for the migration rate.

Saturated vs. unsaturated regime The horseshoe region only contains a finite reservoir of angular momentum, which can cause the torque originating from this region to disappear after a finite time. In order to check if such a saturation of the horseshoe drag occurs, we compare the libration timescale τ_{lib} and the viscous timescale τ_{visc} across the corotation region (Fig. 1, left). If $\tau_{\mathrm{visc}} < \tau_{\mathrm{lib}}$ sustained outward migration is possible. In the other case, Lindblad torques (plus some residual horseshoe drag) are acting, driving usually inward migration. In some cases, the (outward) migration of the planet itself can keep the corotation region auto-unsaturated in a feedback effect.

Reduction of the gas surface density As the width of the horseshoe region increases with the planetary mass (e.g. Masset *et al.* 2006) and the viscous timescale increase with this width, saturation sets in when the planet reaches some mass, of order 10 $M_{\oplus}$. Such planets are particularly vulnerable to a rapid migration into the star, as the migration rate increases with planetary mass. On the other hand, starting gap formation reduces the gas surface density around the planet, reducing the torques. This is taken into account using the results of Crida & Morbidelli (2007).

Gap opening criterion The transition to type II migration is built on the results of Crida *et al.* (2006).

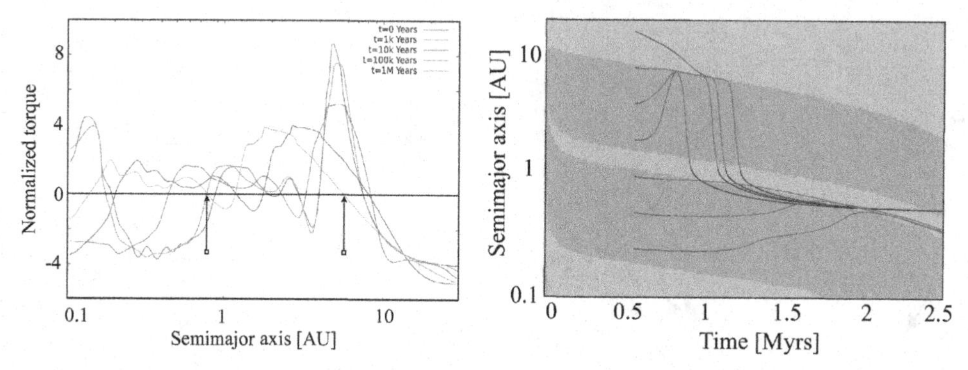

Figure 2. Left panel: Normalized torque as a function of semimajor axis at five times of the disk evolution. The vertical arrows indicate the locations of the convergence zones at 1 Myr. Right panel: The background shows the regions of inward (blue) and outward (green) migration in the disk. The lines show the migration tracks of seven planetary embryo inserted at different starting locations. Colors indicate migration regimes. Blue: locally isothermal. Green: adiabatic, unsaturated. Red: adiabatic, saturated. Brown: type II.

The right panel of Fig. 1 shows the specific torque (torque per mass unit) as a function of planetary mass. Positive torques correspond to outward migration. The line represents the result from our semi-analytical model, while crosses represent the 3D radiation-hydrodynamic simulations of Kley *et al.* (2009), indicating good agreement.

3. Convergence zones and migration tracks

With this model at hand, we have studied the migration of growing planets embedded in 1+1D standard alpha disk models (see Lyra *et al.* 2010).

The left panel of Fig. 2 shows the normalized torque in the adiabatic, unsaturated regime as a function of semimajor axis and time for a typical disk. The torque depends on the gradients of the temperature and gas surface density which change because of opacity transitions. There are special zones where the torque vanishes, and where the torque gradient is negative. This means that inward of these points, migration is directed outward, and outward of them, it is directed inward, so that these locations are migration traps onto which migrating planets converge.

The right panel shows the direction of migration as a function of time and semimajor axis. Green corresponds to outward, blue to inward migration. All protoplanets in the convergence zones migrate towards the stable zero torque locations. We find two convergence zones in agreement with Lyra *et al.* (2010), which seem to be a generic property. We note that the convergence zones are several AU wide, and therefore could concentrate a lot of matter in one point. This could have important implication for planetary growth, a process very recently studied by Sandor *et al.* (2011). We also see that the convergence zones themselves move inward on a viscous timescale (Paardekooper *et al.* 2010).

The lines show migration tracks of seven embryos (initial mass 0.6 $M_\oplus$) migrating and growing in the disk. Each planet was simulated separately. The inner three protoplanets migrate outward to the inner convergence zone and remain attached to it for the rest of the evolution. They remain low mass planets (3 to 6 $M_\oplus$). The outer four protoplanets migrate both inward and outward to the outer convergence zone. They stay there until the corotation torque saturates because of the mass growth and fast inward migration sets in. Around 1 AU however, the planetary cores become so massive that gas runaway accretion is triggered, and the migration changes into the slow (planet dominated) type II regime. The final masses of these planets are between 4 to 6 Jupiter masses.

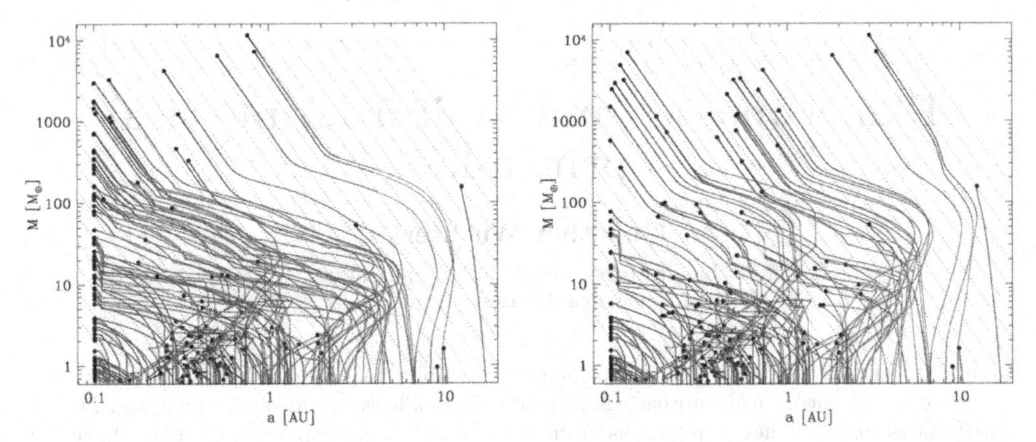

Figure 3. Concurrent growth and migration of protoplanets in the mass-distance plane. Planetary embryos start with an initial mass of 0.6 $M_{\oplus}$ at different initial semimajor axes. Final positions are indicated with large black dots. Colors indicate different migration regimes as in Fig. 2. The left panel shows the nominal model, while for the simulation on the right, saturation is assumed to set in at a four times larger mass.

4. Planetary population synthesis

Figure 3 shows planetary formation tracks in the mass-distance plane, illustrating how planetary embryos concurrently grow and migrate using the undated type I migration model. Other settings and probability distributions are similar as in Mordasini *et al.* (2009a). Planets were arbitrarily stopped when they migrate to 0.1 AU.

One sees that planets starting inside the inner convergence zone quickly migrate inward leading to close-in low mass planets. Further out, outward migration frequently occurs. Especially in the right panel, the imprint of the two convergence zones can be seen by two groups of giant planets. In the nominal model in the left panel, many embryos still migrate to 0.1 AU, but much less than with the full rate of Tanaka *et al.* (2002). It is clear that (giant) planet formation is no more suppressed by the loss of the embryos into the star. These results will be discussed in details in Dittkrist *et al.* (in prep.).

References

Alibert, Y., Mordasini, C., Benz, W., & Winisdoerffer, C. 2005, *A&A*, 434, 343
Baruteau, C. & Lin, D. N. C. 2010, *ApJ*, 709, 759
Crida, A., Morbidelli, A., & Masset, F. 2006, *Icarus*, 181, 587
Crida, A. & Morbidelli, A. 2007, *MNRAS*, 377, 1324
Dittkrist, K.-M., Mordasini, C., Alibert, Y., Klahr, H., Benz, W., & Henning, T. in prep.
Howard, A. W., *et al.* 2010, *Science*, 330, 653
Ida, S. & Lin, D. N. C. 2008, *ApJ*, 673, 487
Lyra, W., Paardekooper, S.-J., & Mac Low, M.-M. 2010, *ApJ*, 715, L68
Kley, W. & Crida, A. 2008, *A&A*, 487, L9
Kley, W., Bitsch, B., & Klahr, H. 2009, *A&A*, 506, 971
Masset, F. S., D'Angelo, G., & Kley, W. 2006, *ApJ*, 652, 730
Mordasini, C., Alibert, Y., & Benz, W. 2009a, *A&A*, 501, 1139
Mordasini, C., Alibert, Y., Benz, W., & Naef, D. 2009b, *A&A*, 501, 1161
Paardekooper, S.-J. & Mellema, G. 2006, *A&A*, 459, L17
Paardekooper, S.-J., Baruteau, C., Crida, A., & Kley, W. 2010, *MNRAS*, 401, 1950
Sandor, Z., Lyra, W., & Dullemond, C. 2011, *ApJ*, 728, L9
Schlaufman, K. C., Lin, D. N. C., & Ida, S. 2009, *ApJ*, 691, 1322
Tanaka, H., Takeuchi, T., & Ward, W. R. 2002, *ApJ*, 565, 1257

The Astrophysics of Planetary Systems: Formation, Structure, and Dynamical Evolution
Proceedings IAU Symposium No. 276, 2010
A. Sozzetti, M. G. Lattanzi & A. P. Boss, eds.

doi:10.1017/S1743921311019971

Planet masses and radii from physical principles

Guenther Wuchterl[1]

[1] Thüringer Landessternwarte, D-07778 Tautenburg, Germany
email: gwuchterl@tls-tautenburg.de

Abstract. Masses and radii are the primary observables to characterize exoplanets today. A self-consistent theoretical approach is presented that allows to calculate mass- and radius-distributions of exoplanet populations from basic physical principles and avoids the usual parametrisation of a multitude of processes.

Keywords. planetary systems: formation

1. Planet formation theory for the discovery era

An era of discovery in astronomy puts theory in a mixed situation: on the one hand new key objects are discovered that allow to adjust incompletely understood parts of the theories, on the other hand the nature of what is found may be severely determined by the selection bias that is ubiquitous at the frontier of astronomical endeavors. One approach in this situation is to try to work out as much as possible from the most basic principles, the ones that are the best tested. The results may be limited but they provide solid ground and help the discovery process itself. To prepare for the CoRoT mission Pečnik (2003); Pečnik (2005); Pečnik & Wuchterl (2005); Broeg (2006); Broeg (2009); Schönke (2005); Schönke (2007); and Wuchterl *et al.* (2007) worked out a theory that is based on similar principles and tools as the theory of stellar structure and evolution. It provides statistical information on the masses and radii of exoplanets.

2. Heuristics - in search of principles

The motivation and idea is twofold: (1) the stellar main-sequence provides understanding of the properties of most stars for most of the time without knowing the details of their formation process. The statistics of stars in the Hertzsprung-Russel diagram is dominated by evolutionary time-scales derived from stellar evolution physics, (2) detailed, fluid-dynamical calculations of planet formation showed convergence towards certain classes of planets (Wuchterl 1995; Wuchterl *et al.* 2000). Of course strong assumptions are necessary to work out a prediction from basic principles. For the stellar population and its main sequence these are the well known equilibria of stellar structure: hydrostatic and thermal equilibrium. While they are motivated by time-scale arguments their use is more a pragmatic assumption than a strict mathematical derivation. In all cases there is an implicit underlying assumption: the material manages to assemble into a star and the composition is the one that is observed. To construct an analogon, a *planetary main sequence* we first note that planets as stars are hydrostatic most of their time, a fact that is even part of the 2006 IAU definition. Secondly, while stars are in thermal equilibrium on the main sequence, planets do not have sufficient energy sources to sustain thermal equilibrium during most of their live. During formation, however they have significant energy resources, from contraction and planetesimal accretion. That leads to near equilibrium conditions in the protoplanets during most of their growth time (e.g. Wuchterl 1993 for a quantification of 'near' by radiation fluid-dynamics). Thus if we study the properties of the planetary equilibrium conditions during the formation era, we can estimate the *typical* properties of planets. After all, the planets have to grow through some of these equilibria and certainly end up near one of them, after dynamical events have settled down. It is interesting to note that planets, unlike stars are in or near thermal and hydrostatic equilibrium during a time when they are still embedded in the mass-reservoirs from which they are accumulated and are then able to exchange mass with the environment, i.e. the protoplanetary nebula. Thus there is some prospect to derive information on the typical mass of a planet from a study of planetary equilibria.

3. General theory of planet formation - assumptions and strategy

The goal is to build a theory for all detectable planets. Thus it should be, (a) general — for all stars and all orbital periods, (b) simple — with a direct relation to physics, (c) robust — with no a-priori selection of mechanisms, (d) predictive — contain no free parameters, (e) fast — make evaluation feasible to better than observed accuracy in the entire discovery space.

To calculate specific results in such a general approach we have to make strong assumptions to facilitate the task. We assume: (1) the diversity of nebulae: study planet formation for all gravitationally stable protoplanetary nebulae around stars of all masses; (2) the strong planetesimal hypothesis: there are always enough planetesimals; (3) the impossibility to chose a particular 'scenario' now and study all planets with cores of all masses including zero - we do not separate planets formed by nucleated instability from others, perhaps formed by a disk instability.

4. Isothermal planets and planetary mass

To understand the key physical properties of the planetary equilibria Pečnik (2003) first studied a simple model in detail - isothermal planets. He constructed the hydrostatic equilibria of an isothermal ideal gas around a core. The planetary envelope was embedded in the nebula inside the Hill-sphere at given distance from a star. Pečnik introduced a planetary 'phase diagram' and found that all planets fall in one of four classes in the core-mass, core-surface-density diagram: (I) mature terrestrial planets, (II) mature giant planets, (III) 'rock in the fog', (IV) protoplanets (Pečnik & Wuchterl 2005). The structure of the diagram being centred around a generalised critical core mass for planets. Schönke (2005) showed which of the isothermal planetary equilibria are stable and could find an analytical expression for the characteristic mass — close to but not identical with the critical one — and density, M_Ω, ϱ_Ω, resp., shaping the entire diagram, and thus defining a scale for planetary mass (Schönke 2007):

$$M_\Omega = \left(\frac{\frac{k_\mathrm{B} T}{G m_\mathrm{molecular}} \ln \left(\frac{4\pi a^3 \varrho_\mathrm{core}}{27 M_\mathrm{star}} \right)}{\left(\frac{4\pi \varrho_\mathrm{core}}{3} \right)^{\frac{1}{3}} - \frac{(3 M_\mathrm{star})^{\frac{1}{3}}}{a}} \right)^{\frac{3}{2}} ; \; \varrho_\Omega = \frac{\varrho_\mathrm{core}}{3} . \tag{4.1}$$

This expression gives a natural planetary mass derived from micro-physics - molecular mass, $m_\mathrm{molecular}$, primary mass, M_star, planetary core density, ϱ_core and orbital radius, a (G and k_B are the gravity and Boltzmann-constants, T, the temperature of planet and nebula gas). The planetary equilibria turned out to be governed by the self-gravity of the planetary envelopes in an interplay with the stellar tides forming the Hill-sphere (app. the Roche-Lobe in this regime). Planet formation can accordingly be understood as a compactification of a tenuous protoplanet to a dense planet driven by self-gravity.

5. Mass spectra for planets with energy transfer

The critical masses for isothermal planets are hundred times smaller than those of realistic planets because the inner parts of the envelopes become usually opaque and are heated significantly before the critical mass is reached. The temperature gradients due to planetesimal accretion require a treatment of energy transfer by radiation and convection as well as the use of realistic equations of state to account for the incompressible, liquid-like behaviour in the deep planetary interior. Broeg (2006), Broeg (2007b), Broeg (2009) constructed such planets for host stars of 0.4 to 2 solar masses and orbital periods of 1 to 64 days to estimate the planetary mass function in the discovery window of CoRoT. Results are available online for M to A stars and orbital periods up to 128 days: CoRoT - Mark 1: Dec. 2005; Mark 2: Sept. 2007; Mark 3: June 2008 (Broeg 2007a). These planetary mass spectra are typically bimodal with peaks near Jupiter's and Neptune's mass and they show a rich dependence on host star mass and orbital period, cf. Broeg (2007b), Broeg (2009).

6. The CoRoT pre-launch planetary mass function prediction

Two extra steps are necessary to derive epoch of observation mass functions that a planet search would find on the sky: (1) the stability of the planetary equilibria has to be investigated and (2) possible effects of mass loss during the planetary evolution into the present have to be estimated. On Dec. 26th, just before the launch of the CoRoT-satellite on Dec. 27th, 2006, Wuchterl *et al.* (2007), Lammer *et al.* (2007) made these estimates based on the above mass spectra, an isothermal stability analysis (Schönke 2005), and, for test-cases, detailed quasi-hydrostatic and non-linear, radiation fluid dynamical calculations. The main features of the resulting theoretical planetary mass functions for a stellar field, are a dominating population of Hot Neptunes and a double peaked mass function, with peaks near Jupiter's and Neptune's mass, cf. Fig. 1. The mass-distribution is probably robust against mass loss for periods starting at 4 d, cf. Lammer, this volume. For high resolution, extended period range and analysis of the underlying physics that sometimes leads to a triple-peak structure, see Broeg (2009).

7. Theoretical planet radii and transit signals

The above mass distributions do not contain any history, the are derived from the statistics of self-gravitating gas spheres that are embedded into protoplanetary nebulae with a finite but sometimes very small pressure. They are best viewed as *initial planetary mass functions* before and until the planets detach from the nebula, which acts as a mass- and thermal reservoir.

The evolution into the present, towards the epoch of observation, needs to be calculated to determine the corresponding planetary radii. That involves an assumption about how the nebula that corresponds to a given planet disappears and how the energy supply of the planet by planetesimal accretion decays. For simplicity and comparability with conventional planetary evolution calculations we assume that the nebula is dissipated, and the planetesimal accretion decays with a Ma time-scale and that the mass of a planet remains constant. Accordingly for every planet the nebula pressure at its

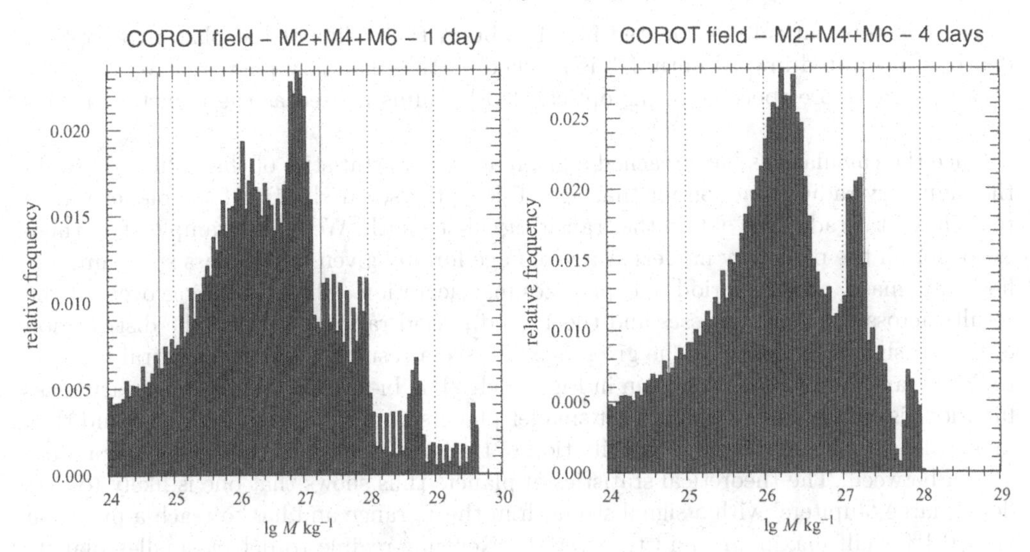

Figure 1. Pre CoRoT launch prediction (Dec. 2006) of the planetary mass function of a stellar population mix typical for a CoRoT field and planetary orbits of 1 (left) and 4 days. The relative frequency of planets is plotted vs. the decadic logarithm of planetary mass, thick (red) vertical lines mark Earth and Jupiter mass, resp., see Wuchterl *et al.* (2007).

Hill-sphere is exponentially lowered and the energy source due to planetesimal accretion is shut off. Thus every planet undergoes a decompression that is fast but on a thermal time-scale. Initially that is a test for secular (thermal) stability as the planet responds to being exposed to vacuum. As the contraction and cooling proceeds, the computation approaches a classical planetary evolution calculation with a plausible initial condition. Guaranteeing this convergence is the main motivation for keeping the mass constant. The key difference to conventional evolutionary calculations is that the initial conditions are diverse, with many protoplanets exhibiting a relatively high density trough-out their Hill-sphere. The broad spectrum of planets with its variety in core mass, nebula pressure and planetesimal accretion-rates assures that every physically possible evolution is calculated. Technically this is a pre-main sequence-like, quasi-hydrostatic calculation of

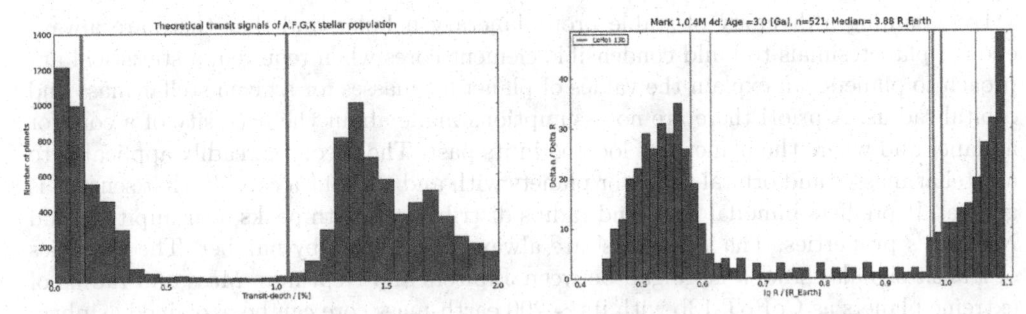

Figure 2. Left: Typical distribution of transit signals obtained from planetary evolutionary calculations for a sample planets orbiting A, F, G and K host stars with periods < 50 d. Right: Distribution of the planetary radii at an age of 3 Ga, calculated for the the CoRoT-13b orbital position. Dark (red) lines mark the radii consistent with the CoRoT-transit-light curve. Light (blue) lines mark the radii of solar system giant planets.

a radiative/convective gas shell (the planetary envelope) around a rigid core. The planet radiates into the radiation field of the star at its orbital radius. Thus irradiation by the star is automatically accounted for. To obtain the radius of the planet the optical depth is integrated inward until 2/3 is reached — the planetary photosphere. Thus the photospheric, more specifically the optical depth radius is used as the planetary radius here.

Once the calculations have reached a given age we do statistics of the radii and obtain the planetary radius function for that age. Taking the squared ratio of the planet radius to its host star radius we obtain the transit signal-strength. We built a sample of all these ratios for all the planetary models at a given age for any given initial mass spectrum, i.e. host star mass, orbital period and planetesimal accretion. Then, summing over all the results across the stellar masses and the 1–64 d period range we obtain the distribution of transit signals expected at the given age. We see a result for stars of spectral types K to A relevant for a CoRoT long run in Fig. 2, left. The bi-modality of the planetary mass functions is enhanced by the fact that smaller planets tend to have a larger core and thus higher density. This produces a distribution of transit signals with two peaks and a large gap in between. The theoretical statistics of planets thus shows that one is likely to first detect large 'Jupiters' with a signal strength in the %-range and has to reach a precision of $\sim 0.1\%$ (half maximum) and to enter the Neptune-regime to detect smaller planets with some probability. As a test of theory we confront it to CoRoT-13b (Cabrera *et al.* 2010), with its ~ 200 earth-mass heavy element core (Fig. 2, right). The extreme planet is located somewhat towards the lower end of the Jupiter-peak in the radius distribution but still predicted to occur relatively frequent.

8. Outlook - the statistical planetary mass-radius diagramm

The high resolution mass- and radius distributions allow to study the theoretical frequency of planets in the mass-radius diagram for a given age. A sample calculation for the planets of a solar mass star at 1 Ga with orbital periods of 1 and 64 days is shown in Fig. 3 together with known exoplanets (Oct. 2010).

9. Conclusion - results from the general theory of planet formation

By assuming a diversity of stable protoplanetary nebulae and that there are always enough planetesimals to build condensible element cores when required, a statistical approach to planets can explain the values of planetary masses for a given stellar mass and orbital radius. A-priori there are no assumptions made about the necessity of a core for a planet and where the planet was located in its past. The theory is readily applicable to all stellar masses and orbital radii, for planets with and without a core. In that sense it is general. It predicts bimodal mass and radius distributions with peaks near Jupiter's and Neptune's properties. The 'Neptunes' are always dominating by number. The statistics of transit signals shows a large gap between Jupiters and Neptunes. Mass and radius of extreme planets as CoRoT-13b with its ~ 200 earth-mass core can be explained as physically possible with a nebula origin and likely for the age of its host star and conditions at its orbit. The possibility to evaluate the mass- and radius distributions relatively fast allows a population-density approach to the planetary mass-radius diagram. Locations in the diagram that contain large numbers of physically possible objects with a nebula origin, and thus according to this theory likely planets of a given age, can be used much like isochrones in the conventional HRD or similar diagrams. The isochrones of stellar structure become two-dimensional distributions of planetary properties for given age.

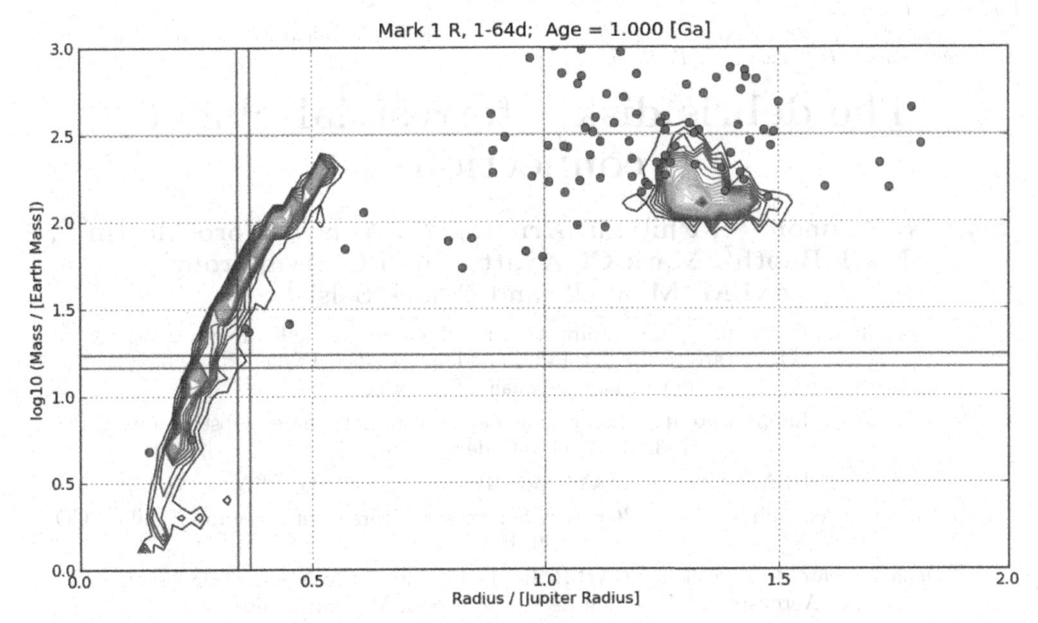

Figure 3. Theoretical population density (shaded contours) in the mass-radius diagram: initial mass functions from the statistics of core-envelope equilibria embedded in diverse protoplanetary nebula and subsequent planet-evolution calculations to determine the radii as a function of time. Shaded-contours point to many planets. Note the branch of core-dominated 'Neptunes' towards the left and the clump of 'Jupiters' in the upper right. Superimposed are exoplanets (dots) and Jupiter, Uranus and Neptune (lines).

References

Broeg, C. 2006, Ph.D. thesis, Friedrich-Schiller-Universität Jena, Germany. http://www.db-thueringen.de/servlets/DocumentServlet?id=6619

— 2007a, Mass spectra corot mark 2a. Http://www.space.unibe.ch/~broeg, http://www.space.unibe.ch/~broeg

— 2007b, *MNRAS*, 377, L44

Broeg, C. H. 2009, *Icarus*, 204, 15

Cabrera, J., Bruntt, H., Ollivier, M., *et al.* 2010, *A&A*, 522, A110+

Lammer, H., Penz, T., Wuchterl, G., *et al.* 2007, ArXiv Astrophysics e-prints. arXiv:astro-ph/0701565

Pečnik, B. 2005, Ph.D. thesis, Ludwig-Maximillians-Universität München, Germany. http://edoc.ub.uni-muenchen.de/archive/00005023/01/Pecnik_Bojan.pdf, http://www.znanost.org/ bonnie/work/Thesis.pdf

Pečnik, B. & Wuchterl, G. 2005, *A&A*, 440, 1183

Pečnik, B. 2003, Master's thesis, University of Zagreb, Croatia

Schönke, J. 2005, Diploma Thesis, AIU-Jena

Schönke, J. 2007, *Plan. & Sp. Sci.*, 55, 1299

Wuchterl, G. 1993, *Icarus*, 106, 323

— 1995, *Earth, Moon and Planets*, 67, 51

Wuchterl, G., Broeg, C., Krause, S., *et al.* 2007, arXiv:astro-ph/0701003

Wuchterl, G., Guillot, T., & Lissauer, J. J. 2000, *Protostars and Planets IV*, 1081

The Astrophysics of Planetary Systems: Formation, Structure, and Dynamical Evolution
Proceedings IAU Symposium No. 276, 2010
A. Sozzetti, M. G. Lattanzi, & A. P. Boss, eds.

doi:10.1017/S1743921311019983

The debris disk – terrestrial planet connection

Sean N. Raymond[1,2], Philip J. Armitage[3,4], Amaya Moro-Martín[5,6], Mark Booth[7], Mark C. Wyatt[7], John C. Armstrong[8], Avi M. Mandell[9] and Franck Selsis[1,2]

[1]Université de Bordeaux, Observatoire Aquitain des Sciences de l'Univers, 2 rue de l'Observatoire, BP 89, F-33271 Floirac Cedex, France
email: rayray.sean@gmail.com

[2]CNRS, UMR 5804, Laboratoire d'Astrophysique de Bordeaux, 2 rue de l'Observatoire, BP 89, F-33271 Floirac Cedex, France

[3]JILA, University of Colorado, Boulder CO 80309, USA

[4]Department of Astrophysical and Planetary Sciences, University of Colorado, Boulder CO 80309, USA

[5]Departamento de Astrofisica, CAB (CSIC-INTA), Instituto Nacional de Tecnica Aeroespacial, Torrejon de Ardoz, 28850, Madrid, Spain

[6]Department of Astrophysical Sciences, Princeton University, Peyton Hall, Ivy Lane, Princeton, NJ 08544, USA

[7]Institute of Astronomy, Cambridge University, Madingley Road, Cambridge, UK

[8]Department of Physics, Weber State University, Ogden, UT, USA

[9]NASA Goddard Space Flight Center, Code 693, Greenbelt, MD 20771, USA

Abstract. The eccentric orbits of the known extrasolar giant planets provide evidence that most planet-forming environments undergo violent dynamical instabilities. Here, we numerically simulate the impact of giant planet instabilities on planetary systems as a whole. We find that populations of inner rocky and outer icy bodies are both shaped by the giant planet dynamics and are naturally correlated. Strong instabilities – those with very eccentric surviving giant planets – completely clear out their inner and outer regions. In contrast, systems with stable or low-mass giant planets form terrestrial planets in their inner regions and outer icy bodies produce dust that is observable as debris disks at mid-infrared wavelengths. Fifteen to twenty percent of old stars are observed to have bright debris disks (at $\lambda \sim 70\mu m$) and we predict that these signpost dynamically calm environments that should contain terrestrial planets.

Keywords. planetary systems: formation, debris disks, methods: n-body simulations

---

1. Introduction

Circumstellar disks of gas and dust orbiting young stars are expected to produce three broad classes of planets in radially-segregated zones (e.g., Kokubo & Ida 2002). The inner disk forms terrestrial (rocky) planets because it contains too little solid mass to rapidly accrete giant planet cores. Terrestrial planets form in 10-100 million years (Myr) via collisional agglomeration of Moon- to Mars-sized planetary embryos and a swarm of 1-10^3 km sized planetesimals (Chambers 2004; Kenyon & Bromley 2006; Raymond *et al.* 2009).

From roughly a few to a few tens of Astronomical Units (AU), giant planet cores grow and accrete gaseous envelopes if the conditions are right. Despite their large masses, gas giants must form within the few million year lifetime of gaseous disks (Haisch *et al.*

2001) and be present during the late phases of terrestrial planet growth. The known extrasolar giant planets have a broad eccentricity distribution (Butler *et al.* 2006) that is quantitatively reproduced if dynamical instabilities occurred in 70-100% of all observed systems (Chatterjee *et al.* 2008; Juric & Tremaine 2008; Raymond *et al.* 2010). The onset of instability may be caused by the changing planet-planet stability criterion as the gas disk dissipates (Iwasaki *et al.* 2001), resonant migration (Adams & Laughlin 2003), or chaotic dynamics (Chambers *et al.* 2006), leading to a phase of planet-planet scattering and the removal of one or more planets from the system by collision or hyperbolic ejection (Rasio & Ford 1996; Weidenschilling & Marzari 1996).

Finally, in the outer regions of planetary systems the growth time scale exceeds the lifetime of the gas disk, and the end point of accretion is a belt of Pluto-sized (and smaller) bodies (Kenyon & Luu 1998). Debris disks trace these icy leftovers of planet formation. Debris disks consist of warm or cold dust observed around older stars, typically at infrared wavelengths ($\lambda \sim 10 - 100 \mu m$) and provide evidence for the existence of planetesimals because the lifetime of observed dust particles under the effects of collisions and radiation pressure is far shorter than the typical stellar age, implying a replenishment via collisional grinding of larger bodies (e.g., Wyatt 2008; Krivov 2010).

Here we perform a large ensemble of N-body simulations to model the interactions between the different radial components of planetary systems: formation and survival of terrestrial planets, dynamical evolution and scattering of giant planets, and dust production from collisional grinding. By matching the orbital distribution of the giant planets, we infer the characteristics of as-yet undetected terrestrial planets in those same systems. We calculate the spectral energy distribution of dust in the system by assuming that planetesimal particles represent a population of smaller objects in collisional equilibrium (as in Booth *et al.* 2009). We then correlate outcomes in the different radial zones and link to two key observational quantities: the orbital properties of giant planets and debris disks.

2. Methods

Our initial conditions include $9M_\oplus$ in planetary embryos and planetesimals from 0.5 to 4 AU, three giant planets on marginally stable orbits from Jupiter's orbital distance of 5.2 AU out to ~10 AU (depending on the masses), and an outer 10 AU-wide disk of planetesimals containing 50 $M_\oplus$. Giant planet masses were drawn randomly according to the observed exoplanet mass function (Butler *et al.* 2006) $\mathrm{d}N/\mathrm{d}M \propto M^{-1.1}$ for $M_{\mathrm{Sat}} < M < 3M_{\mathrm{Jup}}$. Each of our 160 fiducial simulations was integrated for 100-200 million years using the `Mercury` (Chambers 1999) hybrid integrator, paying careful attention to the energy conservation of the integrator. We post-process the simulations to compute the spectral energy distribution of dust by treating planetesimal particles as aggregates in collisional equilibrium (Dohnanyi 1969) to calculate the incident and re-emitted flux (Booth *et al.* 2009). Though certainly oversimplified, these initial conditions generically reproduce the predicted state of a planetary system at the time of the dissipation of the gaseous protoplanetary disk.

We do not include the effects of planetary migration for several reasons. First, current migration theories fail to reproduce most characteristics of the known exoplanets (Howard *et al.* 2010). Second, the effects of giant planet migration on terrestrial planet formation are thought to be weak(Raymond *et al.* 2006; Mandell *et al.* 2007; Fogg & Nelson 2007) compared with the effect of instabilities (Veras & Armitage 2006). Finally, dynamical instabilities are thought to occur when the disk is mostly or completely dissipated, *after* any migration has occurred such that the imprint of instabilities on the surviving planets should be far more pronounced.

3. Results

An unstable system evolves as follows (Figure 1). At early times accretion proceeds in the inner disk, the giant planets gravitationally interact, the outer disk is mostly passive, and the only significant interaction between the components is that the giant planets dynamically clear out nearby small bodies. The onset of instability causes a punctual increase in the giant planets' eccentricities and, depending on the details of the scattering event, the orbit of one or more giant planets intrudes into the inner and/or outer disk. Particles in the vicinity of a scattered giant planet are rapid destabilized to either be ejected from the system or collide with the central star, and this process continues until the instability concludes (usually with the ejection of one or two giant planets; e.g., Raymond *et al.* 2010). At this point, the surviving small bodies have been shaped by the dynamical instability but their continued evolution is governed by the new dynamical state of the surviving giant planets. The number and spacing of terrestrial planets that form depends on the eccentricities of the surviving embryos (if any), perturbed both during and after the instability. These perturbations span a continuous range but the outcome is quantized into a discrete number of terrestrial planets during the accretion process (Levison & Agnor 2003). If perturbations are weak – if the giant planets collide rather than scattering (or are dynamically stable or low-mass) – then embryos' eccentricities remain small and feeding zones narrow and several terrestrial planets form. For stronger giant planet perturbations, feeding zones widen and fewer terrestrial planets form, although the total mass in planets tends to decrease because stronger perturbations imply that the giant planets were scattered closer to the terrestrial planet region so more embryos end up on unstable orbits. In systems where embryos' radial excursions are comparable to the radial extent of the surviving disk only one planet forms, usually on an excited orbit. In the simulation from Figure 1, the lone terrestrial planet did not accrete from a disk of excited embryos but rather was the only planet to *survive* the instability. Perturbations during, not after, the instability determined the outcome. The outer disk's evolution is also governed by giant planet perturbations: icy planetesimals that survive the instability are subject to secular forcing that determines the collisional timescale and the rate of dust production.

The surviving giant planets in our fiducial simulations match the observed eccentricity distribution and we infer the properties of presumed terrestrial exoplanets (Figure 2). In 40-70% of *unstable* systems – depending on the initial giant planet masses – all terrestrial material is destroyed (Veras & Armitage 2006). However, matching the exoplanet orbital distribution requires a contribution of up to ~30-40% of stable systems(Juric & Tremaine 2008; Zakamska *et al.* 2011), which invariably form systems with two or more terrestrial planets. The orbits of surviving giant planets act as a measure of the strength of the instability, and as expected (Levison & Agnor 2003; Raymond *et al.* 2009) terrestrial planet formation is far less efficient for eccentric giant planets.

As a population, the surviving terrestrial planets have smaller eccentricities than the giant planets (Fig 2; median $e \approx 0.1$ for terrestrials, 0.25 for giants). Single terrestrial planets – formed in 10-20% of systems – have somewhat larger eccentricities than in multiple planet systems and undergo much larger oscillations in eccentricity and inclination

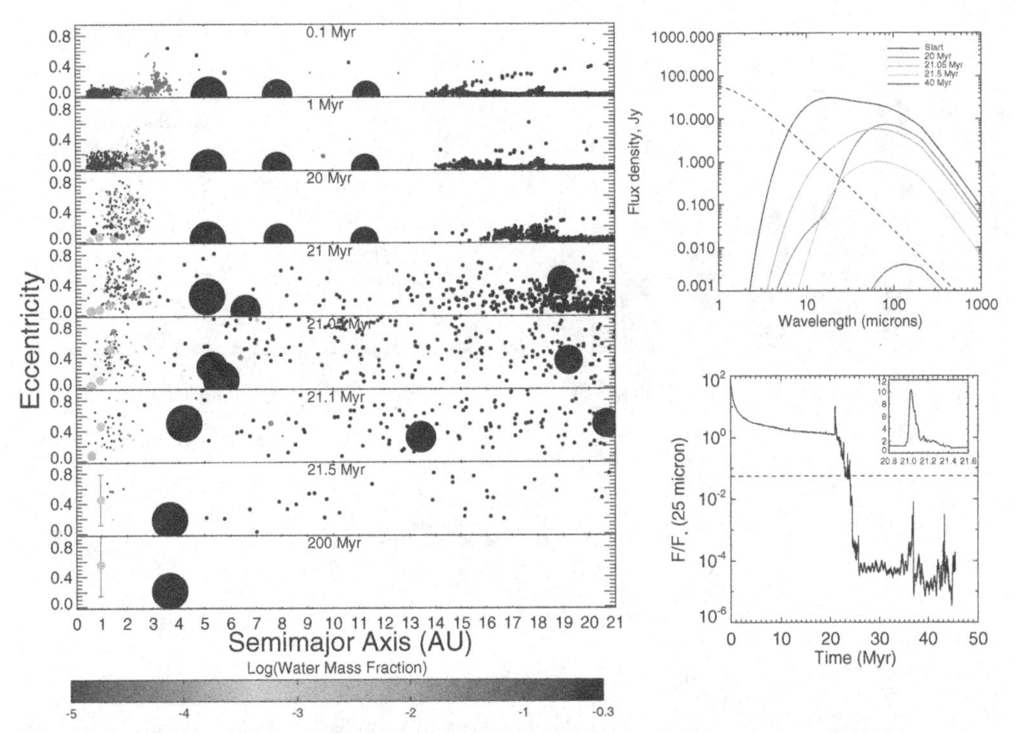

Figure 1. Evolution of a system with a relatively late (21 Myr) instability among the giant planets. **Left:** Snapshots in time of orbital eccentricity vs. semi-major axis for all particles; vertical bars denote $\sin(i)$ for terrestrial bodies with $M_p > 0.2\ M_\oplus$ and $i > 10°$. The particle size is proportional to the mass$^{1/3}$, but giant planets (large black circles) are not on this scale. Colors denote water content, assuming a Solar System-like initial distribution (Raymond *et al.* 2004). The surviving terrestrial planet has a mass of $0.72\ M_\oplus$, a stable orbit within the habitable zone, and a high eccentricity and inclination (and large oscillations in these quantities). **Top right:** The spectral energy distribution of the dust during five simulation snapshots, showing dramatic evolution during and immediately after the instability. The dashed line represents the stellar photosphere. **Bottom right:** The ratio of the dust-to-stellar flux at 25 microns as a function of time, including a zoom during the instability. The rough observational limit of the *MIPS* instrument on NASA's *Spitzer Space Telescope* is shown with the dashed line (Trilling *et al.* 2008). All planetesimal particles were destroyed as of 45 Myr via either collision or ejection. A movie of this simulation is available at http://www.obs.u-bordeaux1.fr/e3arths/raymond/Scattering_terrestrial_SED.mpg.

due to secular forcing from the giant planets.† If we scale our simulations to match the giant exoplanet semimajor axis-eccentricity distribution beyond 0.5-1 AU (appropriate given the dynamical regime), we can infer the radial distribution of terrestrial exoplanets in the known systems (Fig. 2). We predict a factor of a few higher abundance of terrestrial planets at a few tenths of an AU than at 1 AU because, given the typical giant-terrestrial planet spacing, planets at 1 AU require distant giant planets that are hard to detect by current methods. Planets within ~ 0.1 AU are sparsely populated because of the assumed inner edge of the embryo disk at 0.5 AU.

Icy planetesimals – whose collisional erosion creates debris disks – survive in dynamically calm environments where the giant planets were either stable, low-mass, or underwent a relatively weak instability. Indeed, there is a strong anti-correlation between the

† These oscillations certainly have important consequences for the planetary climate (Spiegel *et al.* 2010).

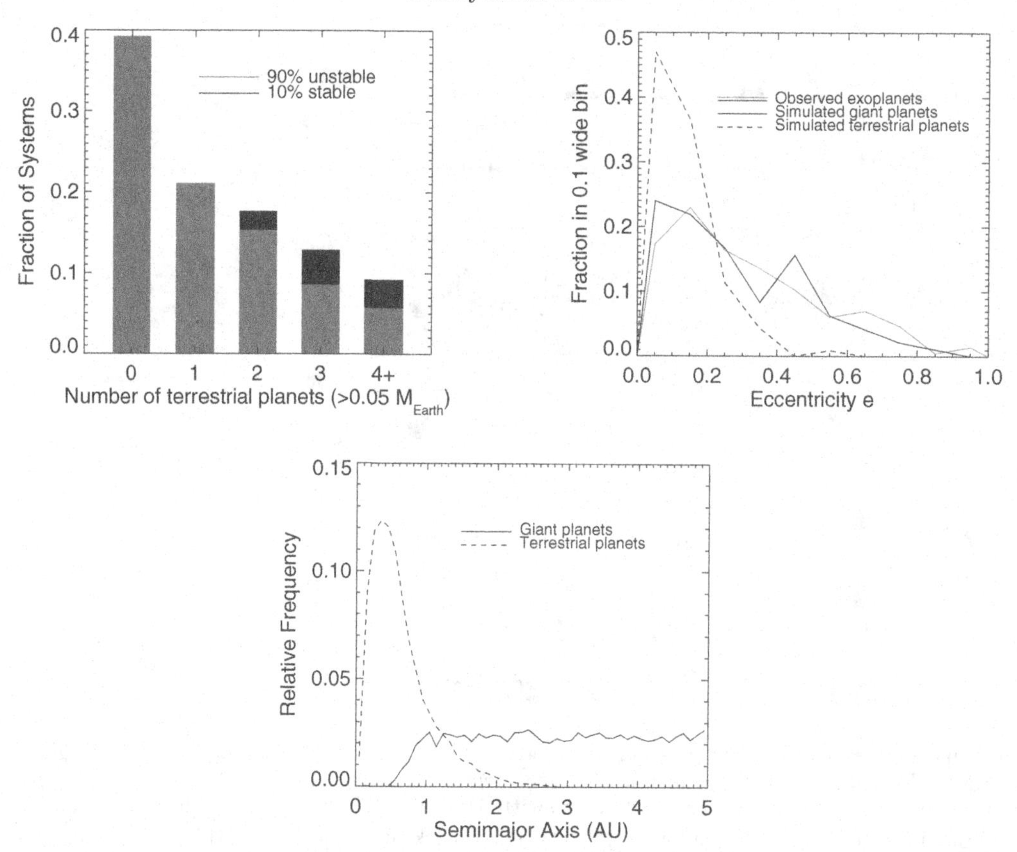

Figure 2. Properties of surviving terrestrial planets. **Top Left:** Distribution of the number of terrestrial planets ($M_p > 0.05\ M_\oplus$) in a sample weighted as 90% from unstable systems and 10% from stable systems. Note that the initial conditions for which the surviving giant planets match the observed exoplanet eccentricity distribution is degenerate between the giant planet masses in a given system and the minority contribution of systems with stable or low-mass giant planets (Raymond *et al.* 2010). **Top Right:** Eccentricity distribution of surviving giant and terrestrial planets, compared with the observed giant exoplanets beyond 0.2 AU. **Bottom:** Semimajor axis distribution of simulated terrestrial planets (dashed line), derived by scaling the innermost surviving giant planet in each simulation to match an assumed underlying distribution for relevant exoplanets that increases linearly from zero at 0.5 AU and is constant from 1-5 AU.

70 μm dust flux and giant planet eccentricity (Figure 3). Almost all lower-eccentricity ($e < 0.1 - 0.2$) giant planets are in systems with debris disks, but at higher eccentricities the fraction of dusty systems decreases as does the dust brightness itself. There exist some systems with high-eccentricity giant planets and bright dust emission, in agreement with the detected debris disks in known exoplanet systems (Moro-Martin *et al.* 2007). In these cases the dynamical instability tends to be asymmetric and confined to the inner planetary system and these are therefore not generally good candidates for terrestrial planets, which also agrees with the observed systems (Moro-Martin *et al.* 2010). The outcome of a given system depends critically on the details of the instability, which is determined by the giant planet masses (Raymond *et al.* 2010).

The total terrestrial planet mass also correlates with debris disk brightness (Fig. 3; the correlation holds for a range of stellar ages and wavelengths). This correlation exists because the inner and outer planetary system are both subject to the same dynamical environment: the violent instabilities that abort terrestrial planet formation also tend to

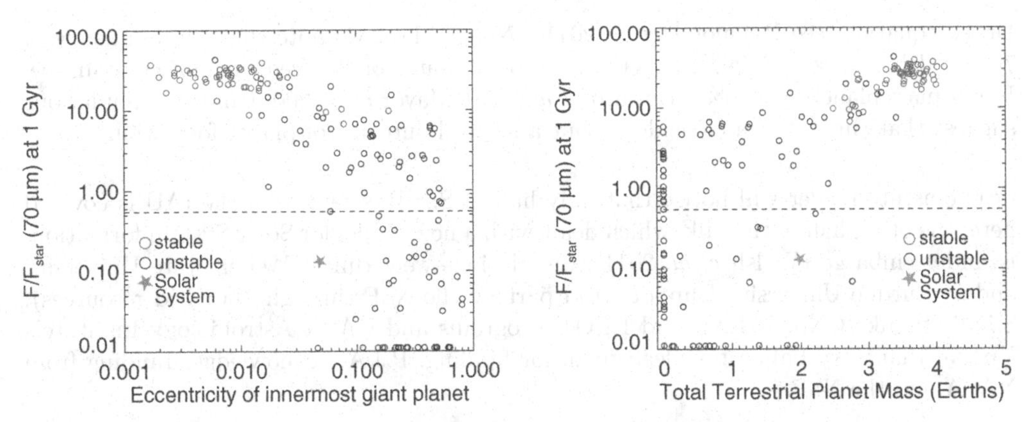

Figure 3. Correlations between debris disks and planets. **Left:** The dust-to-stellar flux ratio $F/F_\star$ at 70 μm after 1 Gyr of dynamical and collisional evolution as a function of the eccentricity of the innermost surviving giant planet. Unstable systems are in black, stable systems in red, and the dashed line represents an approximate threshold above which excess emission was detectable using *Spitzer* data (Trilling *et al.* 2008). The grey star is an estimate of the Solar System's debris disk flux roughly 900 Myr after the LHB (Booth *et al.* 2009). **Right:** $F/F_\star$ (70 μm) vs. the total mass in terrestrial planets. Debris disks, especially the brightest ones, trace systems that efficiently form terrestrial planets.

remove or erode their outer planetesimal disks. The correlation is not perfect as there exist "false positives" with bright dust emission and no terrestrial planets, corresponding to systems with asymmetric, inward-directed giant planet instabilities. Conversely, "false negatives" with terrestrial planets but little to no dust are systems that underwent asymmetric but outward-directed instabilities.

4. Discussion

The Solar System lies at the very edge of the debris disk correlations in Fig. 3 because of its combination of a rich terrestrial planet system, a low-eccentricity innermost giant planet, and a low dust flux. To a distant observer, the Solar System's faint debris disk would suggest a dust-clearing instability in the system's past. However, Jupiter's low-eccentricity orbit would imply that the instability was weak and that the system may in fact be suitable for terrestrial planets. This naive argument is remarkably consistent with our current picture of the LHB instability as an asymmetric, outward-directed instability that included a scattering event between Jupiter and an ice giant but not between Jupiter and Saturn (Morbidelli *et al.* 2010).

The observed statistics of debris disks show that 15-20% of solar-type stars have bright dust emission at 70 μm (Trilling *et al.* 2008; Carpenter *et al.* 2009). Our simulations show that debris disk systems generally represent dynamically calm environments that should have been conducive to efficient terrestrial accretion, and are therefore likely to contain systems of terrestrial planets. The candidate selection for terrestrial planet systems can be further improved by choosing systems with low-eccentricity giant planets and/or very bright dust emission (e.g., $F/F_{star} \geqslant 10$ at 70 μm). Given the existence of false positives and negatives, the significant uncertainties in our initial conditions, and the existence of other potentially important processes, a detailed statistical analysis is needed to use our simulations to estimate the frequency of debris disks into a robust estimate of the fraction of stars that will be found to harbor terrestrial planets (η_{Earth} from the famous

Drake equation; see Raymond *et al.* 2011). Nonetheless, we note that the frequency of 70 μm debris disks (15-20%) is very close to estimates of the frequency of close-in, few Earth-mass planets (10-30% – Howard *et al.* 2010; Mayor *et al.* 2009) and our simulations suggest that this is not a coincidence but a natural outcome of planet formation.

The astute reader will notice that only half of S.N.R.'s talk from the IAU is covered here; the other half of the talk, which dealt with a new model for Solar System formation, is under embargo (Walsh *et al.* 2011). Simulations were run at Weber State University and at Purdue University (supported in part by the NSF through TeraGrid resources). S.N.R. thanks CNRS's PNP and EPOV programs and NASA Astrobiology Institute's Virtual Planetary Laboratory lead team for funding. P.J.A. acknowledges funding from NASA and the NSF.

References

Adams, F. C. & Laughlin, G. 2003, *Icarus*, 163, 290
Booth, M., Wyatt, M. C., Morbidelli, A., Moro-Martín, A., & Levison, H. F. 2009, *MNRAS*, 399, 385
Butler, R. P., *et al.* 2006, *ApJ*, 646, 505
Carpenter, J. M., *et al.* 2009, *ApJS*, 181, 197
Chambers, J. E. 1999, *MNRAS*, 304, 793
Chambers, J. E., Wetherill, G. W., & Boss, A. P. 1996, *Icarus*, 119, 261
Chambers, J. E. 2004, *Earth and Planetary Science Letters*, 223, 241
Chatterjee, S., Ford, E. B., Matsumura, S., & Rasio, F. A. 2008, *ApJ*, 686, 580
Dohnanyi, J. S. 1969, *JGR*, 74, 2531
Fogg, M. J. & Nelson, R. P. 2007, *A&A*, 461, 1195
Haisch, K. E., Jr., Lada, E. A., & Lada, C. J. 2001, *ApJL*, 553, L153
Howard, A. W., *et al.* 2010, *Science*, 330, 653
Iwasaki, K., Tanaka, H., Nakazawa, K., & Hiroyuki, E. 2001, *PASJ*, 53, 321
Jurić, M. & Tremaine, S. 2008, *ApJ*, 686, 603
Kenyon, S. J. & Bromley, B. C. 2006, *AJ*, 131, 1837
Kenyon, S. J. & Luu, J. X. 1998, *AJ*, 115, 2136
Mandell, A. M., Raymond, S. N., & Sigurdsson, S. 2007, *ApJ*, 660, 823
Marzari, F. & Weidenschilling, S. J. 2002, *Icarus*, 156, 570
Mayor, M., *et al.* 2009, *A&A*, 507, 487
Morbidelli, A., Brasser, R., Gomes, R., Levison, H. F., & Tsiganis, K. 2010, *AJ*, 140, 1391
Moro-Martín, A., *et al.* 2007, *ApJ*, 658, 1312
Moro-Martín, A., Malhotra, R., Bryden, G., Rieke, G. H., Su, K. Y. L., Beichman, C. A., & Lawler, S. M. 2010, *ApJ*, 717, 1123
Rasio, F. A. & Ford, E. B. 1996, *Science*, 274, 954
Raymond, S. N., Armitage, P. J., & Gorelick, N. 2010, *ApJ*, 711, 772
Raymond, S. N., Mandell, A. M., & Sigurdsson, S. 2006, *Science*, 313, 1413
Raymond, S. N., O'Brien, D. P., Morbidelli, A., & Kaib, N. A. 2009, *Icarus*, 203, 644
Raymond, S. N., Quinn, T., & Lunine, J. I. 2004, *Icarus*, 168, 1
Spiegel, D. S., Raymond, S. N., Dressing, C. D., Scharf, C. A., & Mitchell, J. L. 2010, *ApJ*, 721, 1308
Trilling, D. E., et al. 2008, *ApJ*, 674, 1086
Veras, D. & Armitage, P. J. 2006, *ApJ*, 645, 1509
Weidenschilling, S. J. & Marzari, F. 1996, *Nature*, 384, 619
Wyatt, M. C. 2008, *ARA&A*, 46, 339
Zakamska, N. L., Pan, M., & Ford, E. B. 2011, *MNRAS*, 410, 1895

The Astrophysics of Planetary Systems: Formation, Structure, and Dynamical Evolution
Proceedings IAU Symposium No. 276, 2010
A. Sozzetti, M. G. Lattanzi & A. P. Boss, eds.

doi:10.1017/S1743921311019995

High-resolution simulations of planetesimal formation in turbulent protoplanetary discs

Anders Johansen[1], Hubert Klahr[2] and Thomas Henning[2]

[1]Lund Observatory, Lund University
Box 43, 221 00 Lund, Sweden
email: anders@astro.lu.se

[2]Max-Planck-Institut für Astronomie
Königstuhl 17, 69117 Heidelberg, Germany

Abstract. We present high resolution computer simulations of dust dynamics and planetesimal formation in turbulence triggered by the magnetorotational instability. Particles representing approximately meter-sized boulders clump in large scale overpressure regions in the simulation box. These overdensities readily contract due to the combined gravity of the particles to form gravitationally bound clusters with masses ranging from a few to several ten times the mass of the dwarf planet Ceres. Gravitationally bound clumps are observed to collide and merge at both moderate and high resolution. The collisional products form the top end of a distribution of planetesimal masses ranging from less than one Ceres mass to 35 Ceres masses. It remains uncertain whether collisions are driven by dynamical friction or underresolution of clumps.

Keywords. accretion, accretion disks, MHD, turbulence, planetary systems: formation, planetary systems: protoplanetary disks

---

1. Introduction

The formation of km-scale planetesimals from dust particles involves a complex interplay of physical processes, including most importantly collisional sticking (Weidenschilling 1997; Dullemond & Dominik 2005), the self-gravity of the sedimentary particle mid-plane layer (Safronov 1969; Goldreich & Ward 1973; Sekiya 1998; Youdin & Shu 2002; Johansen *et al.* 2007), and the motion and structure of the turbulent protoplanetary disc gas (Weidenschilling & Cuzzi 1993; Johansen *et al.* 2006; Cuzzi *et al.* 2008).

In the initial growth stages micrometer-sized silicate monomers readily stick to form larger dust aggregates (Blum & Wurm 2008). Further growth towards macroscopic sizes is hampered by collisional fragmentation and bouncing (Zsom *et al.* 2010), limiting the maximum particle size to a few cms or less. High speed collisions between a small impactor and a large target constitutes a path to net growth (Wurm *et al.* 2005), but the separation of small particles from large particles by turbulent diffusion limits the resulting growth rate (Johansen *et al.* 2008). Material properties are also important. Wada *et al.* (2009) demonstrated efficient sticking between ice aggregates consisting of 0.1 μm monomers at speeds up to 50 m/s.

Turbulence can play a positive role for growth by concentrating mm-sized particles on small scales (Cuzzi *et al.* 2008), near the dissipative scale of the turbulence, and m-sized particles on large scales in long-lived geostrophic pressure bumps surrounded by axisymmetric zonal flows (Johansen *et al.* 2009a). In the model presented in Johansen *et al.* (2007) approximately meter-sized particles settle to form a thin mid-plane layer in balance between sedimentation and turbulent stirring. Particles then concentrate in nearly axisymmetric gas high pressure regions, reaching local column densities up to ten times the average. The gravitational attraction between the particles in these overdense regions is high enough to initiate first a slow radial contraction, and as the local mass density becomes comparable to the Roche density, a full non-axisymmetric collapse to form gravitationally bound clumps with masses comparable to the dwarf planet Ceres.

Some of the open questions related to this gravoturbulent picture of planetesimal formation is to what degree the planetesimal size spectrum is numerically converged and the possible role of collisions and coagulation during the clumping and the gravitational collapse. Following a burst of planetesimal formation, the initial mass function can evolve rapidly by planetesimal mergers and accretion of unbound particles. Exploring the role of such physical effects requires both increased spatial resolution and increased simulation time.

In this paper we present high resolution and long time integration simulations of planetesimal formation in turbulence caused by the magnetorotational instability (MRI). We present the first evidence for collisions between newly formed planetesimals, observed at both moderate and high resolution, and indications that the initial mass function of planetesimals involve masses ranging from a few to several ten Ceres masses.

2. Simulation set up

We perform simulations solving the standard shearing box MHD/drag force/self-gravity equations for gas defined on a fixed grid and solid particles evolved as numerical superparticles. We use the Pencil Code, a sixth order spatial and third order temporal symmetric finite difference code†. The shearing box coordinate frame rotates at the Keplerian frequency Ω at an arbitrary distance r_0 from the central star. The axes are oriented such the x points radially away from the central gravity source, y points along the Keplerian flow, while z points vertically out of the plane.

Particle collisions become important inside dense particle clumps. In Johansen *et al.* (2007) the effect of particle collisions was included in a rather crude way by reducing the relative rms speed of particles inside a grid cell on the collisional time-scale, to mimic collisional cooling. Recently Rein *et al.* (2010) found that the inclusion of particle collisions changes the structure of small scale self-gravitating clumps that condense out of a turbulent flow.

Lithwick & Chiang (2007) presented a Monte Carlo method for superparticle collisions whereby the correct collision frequency can be obtained by letting nearby superparticle pairs collide on the average once per collisional time-scale. We have implemented this method in the Pencil Code and will present the algorithm and numerical tests in a paper in preparation (Johansen, Youdin, & Lithwick, in preparation). We let a particle interact with all other particles in the same grid cell. The collision time-scale $\tau_{\rm coll}$ between all particle pairs is calculated, and for each possible collision a random number P is chosen. If P is smaller than $\delta t/\tau_{\rm coll}$, where δt is the time-step set by the magnetohydrodynamics, then the two particles collide. The collision outcome is determined as if the two superparticles were actual particles with radii large enough to touch each other. By solving for momentum conservation and energy conservation, with the possibility for inelastic collisions to dissipate kinetic energy to heat and deformation, the two colliding particles acquire their new velocity vectors instantaneously.

All simulations include collisions with a coefficient of restitution of $\epsilon = 0.3$, meaning that each collision leads to the dissipation of approximately 90% of the relative kinetic energy to deformation and heating of the colliding boulders.

† See `http://code.google.com/p/pencil-code/`.

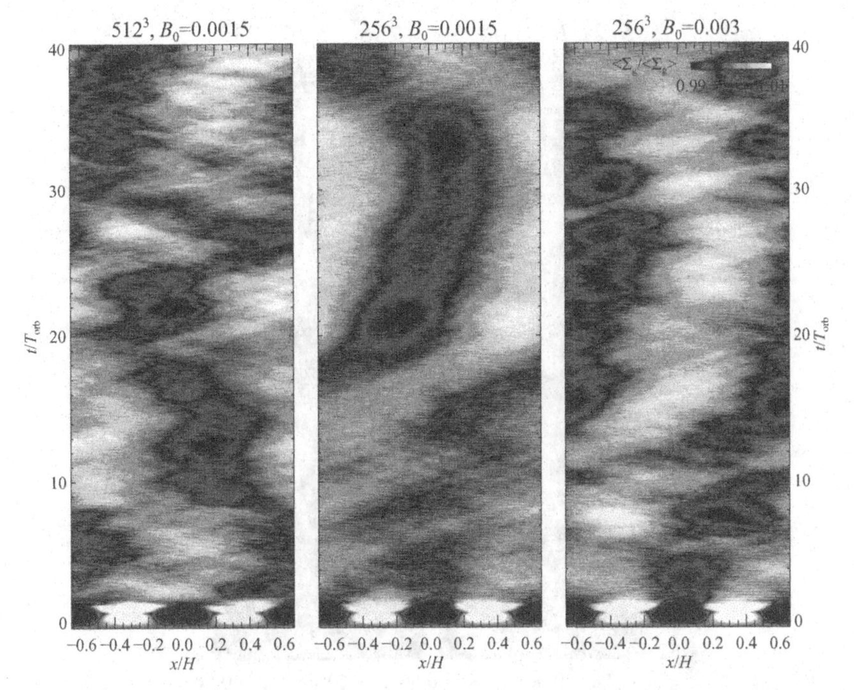

Figure 1. The gas column density averaged over the azimuthal direction, as a function of radial coordinate x and time t in orbits. Large scale pressure bumps appear with approximately 1% amplitude at both 256^3 and 512^3 resolution.

3. Results

An important feature of magnetorotational turbulence is the emergence of large scale slowly overturning pressure bumps (Fromang & Nelson 2005; Johansen *et al.* 2006). Large particles – pebbles, rocks, and boulders – are attracted to the center of pressure bumps, because of the drag force associated with the sub-Keplerian/super-Keplerian zonal flow envelope. In presence of a mean radial pressure gradient the trapping zone is slightly downstream of the pressure bump, where there is a local maximum in the combined pressure.

An efficient way to detect pressure bumps is to average the gas density field over the azimuthal and vertical directions. In Figure 1 we show the gas column density in the 256^3 and 512^3 simulations averaged over the y-direction, as a function of time. Large scale pressure bumps are clearly visible, with spatial correlation times of approximately 10-20 orbits. The pressure bump amplitude is around 1%, independent of both resolution and strength of the external field.

We release the particles at a time when the turbulence has saturated, but choose a time when there is no significant large scale pressure bump present. Thus we choose $t = 20T_{\rm orb}$ for the 256^3 simulation and $t = 32T_{\rm orb}$ for the 512^3 simulation (see Figure 1).

The particles immediately fall towards the mid-plane of the disc, before finding a balance between sedimentation and turbulent stirring. Figure 2 shows how the presence of gas pressure bumps has a dramatic influence on particle dynamics. The particles display column density concentrations of up to 4 times the average density just downstream of the pressure bumps. At this point the gas moves close to Keplerian, because the pressure gradient of the bump balances the radial pressure gradient there. The column density concentration is relatively independent of the resolution, as expected since the pressure bump amplitude is almost the same.

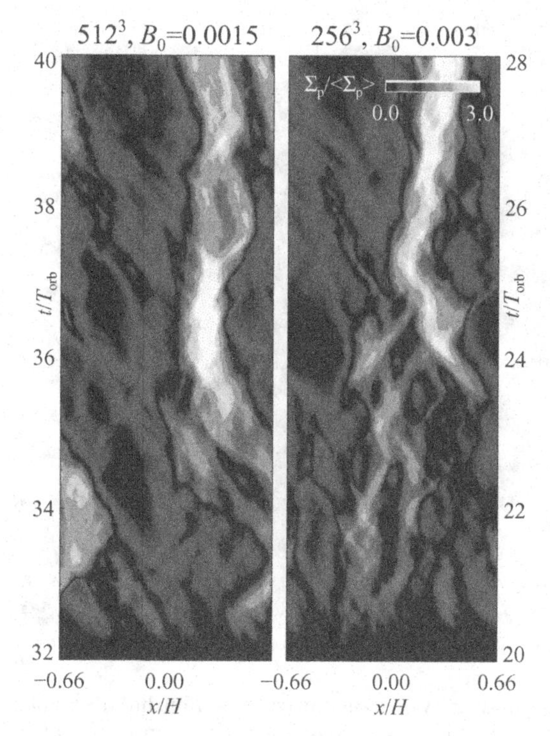

Figure 2. The particle column density averaged over the azimuthal direction, as a function of radial coordinate x and time t in orbits. The starting time was chosen to be slightly prior to the emergence of a pressure bump. The particles concentrate slightly downstream of the gas pressure bump, with a maximum column density between three and four times the mean particle column density.

A simulation including the self-gravity of the particles is presented in Figure 3. The Hill sphere of each bound clump is indicated in Figure 3, together with the mass of particles encompassed inside the Hill radius (in units of the mass of the dwarf planet Ceres)

4. Conclusions

At both moderate and high resolution we observe the close approach and merging of gravitationally bound clumps. Some concerns remain about whether these collisions are real, since our particle-mesh self-gravity algorithm prevents bound clumps from being smaller than a grid cell. Thus the collisional cross section is artifically large. Two observations nevertheless indicate that the collisions may be real: we observe planetesimal mergers at both moderate and high resolution and we see that the environment in which planetesimals merge is rich in unbound particles. Dynamical friction may thus play an important dissipative role in the dynamics and the merging (Goldreich *et al.* 2002).

The measured α-value of MRI turbulence at 512^3 is $\alpha \approx 0.003$. At a sound speed of $c_s = 500$ m/s, the expected collision speed of marginally coupled m-sized boulders is $\sqrt{\alpha} c_s \approx 25 - 30$ m/s. Johansen *et al.* (2007) showed that the actual collision speeds can be a factor of a few lower, because the particle layer damps MRI turbulence locally. In general boulders are expected to shatter when they collide at 10 m/s or higher (Benz 2000). Much larger km-sized bodies are equally prone to fragmentation as random gravitational torques exerted by the turbulent gas excite relative speeds higher than the gravitational escape speed (Ida *et al.* 2008; Leinhardt & Stewart 2009). A good environment for

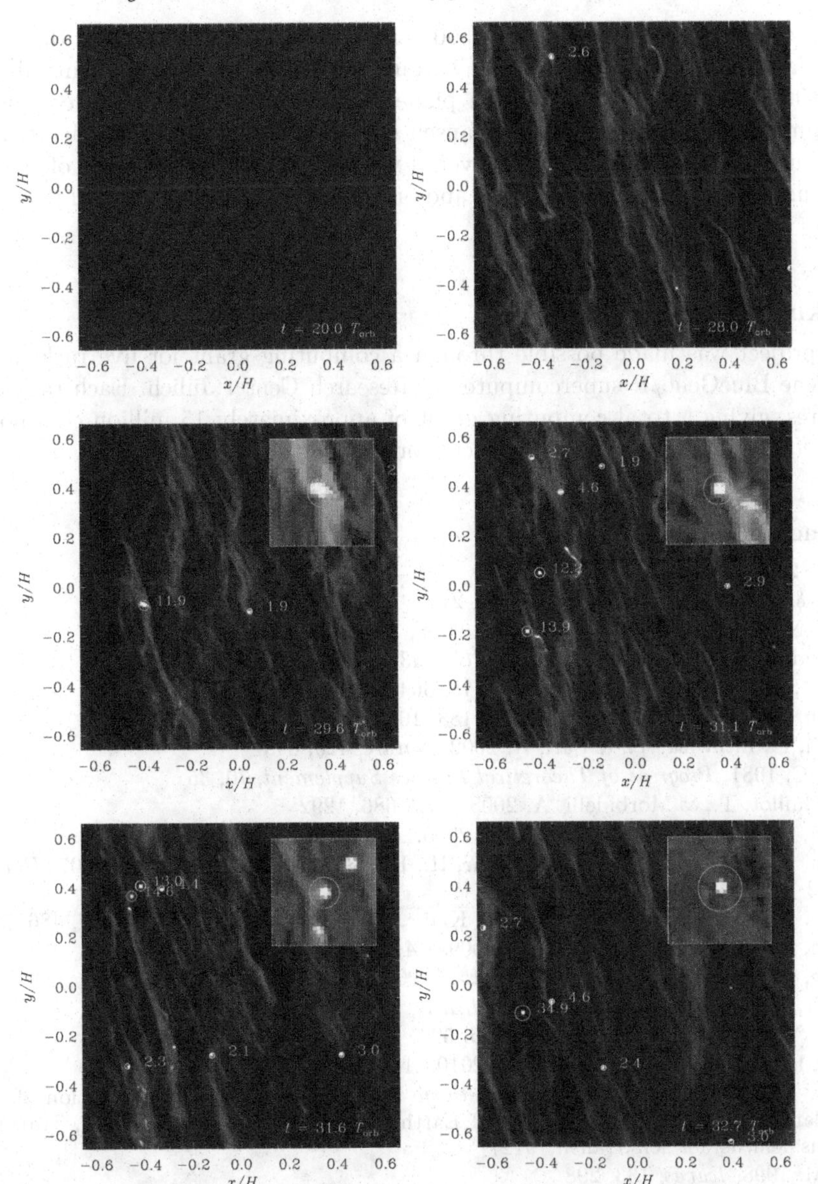

Figure 3. The particle column density as a function of time after self-gravity is turned on at $t = 20.0T_{\rm orb}$ for a simulation with 256^3 grid cells and 8×10^6 particles. The insert shows an enlargement of the region around the most massive bound clump. Each clump is labelled by its Hill mass in units of Ceres masses. The most massive clump at the end of the simulation contains a total particle mass of 34.9 Ceres masses, partially as the result of a collision between a 13.0 and a 14.6 Ceres mass planetesimal occuring at a time around $t = 31.6T_{\rm orb}$.

building planetesimals from boulders may require $\alpha < 0.001$. Johansen *et al.* (2009b) presented simulations with no MRI turbulence where turbulence and particle clumping is driven by the streaming instability (Youdin & Goodman 2005). They found typical collision speeds as low as a few meters per second.

A second reason to prefer weak turbulence is the amount of mass available in the disc. If we apply our results to $r = 5\,\mathrm{AU}$, then our dimensionless gravity parameter corresponds to a gas column density of $\Sigma_{\rm gas} \approx 1500\,{\rm g\,cm^{-2}}$, ten times higher than the Minimum Mass Solar Nebula (Hayashi 1981). Turbulence driven by streaming instability and Kelvin-Helmholtz instability can form planesimals for column densities comparable to the Minimum Mass Solar Nebula (Johansen *et al.* 2009b). Thus even moderate amounts of MRI turbulence seems to have an overall negative effect on the ability of the particle layer to undergo gravitational collapse and on the survival of boulders and planetesimals in face of shattering collisions.

5. Acknowledgements

This project was made possible through a computing grant for five rack months at the Jugene BlueGene/P supercomputer at Research Center Jülich. Each rack contains 4096 cores, giving a total computing grant of approximately 15 million core hours. We are grateful to Tristen Hayfield for discussions on particle load balancing.

References

Benz, W. 2000, *SSRv*, 92, 279
Blum, J. & Wurm, G. 2008, *ARA&A*, 46, 21
Cuzzi, J. N., Hogan, R. C., & Shariff, K., 2008, *ApJ*, 687, 1432
Dullemond, C. P. & Dominik, C. 2005, *A&A*, 434, 971
Fromang, S. & Nelson, R. P. 2005, *MNRAS*, 364, L81
Goldreich, P. & Ward, W. R. 1972, *ApJ*, 183, 1051
Goldreich, P., Lithwick, Y., & Sari, R. 2002, *Nature*, 420, 643
Hayashi, C. 1981, *Progress of Theoretical Physics Supplement*, 70, 35
Ida, S., Guillot, T., & Morbidelli, A. 2008, *ApJ*, 686, 1292
Johansen, A., Klahr, H., & Henning, Th. 2006, *ApJ*, 636, 1121
Johansen, A., Oishi, J. S., Low, M., Klahr, H., Henning, Th., & Youdin, A. 2007, *Nature*, 448, 1022
Johansen, A., Brauer, F., Dullemond, C., Klahr, H., & Henning, T. 2008, *A&A*, 486, 597
Johansen, A., Youdin, A., & Klahr, H. 2009a, *ApJ*, 697, 1269
Johansen, A., Youdin, A., & Mac Low, M.-M. 2009b, *ApJL*, 704, L75
Leinhardt, Z. M. & Stewart, S. T. 2009, *Icarus*, 199, 542
Lithwick, Y. & Chiang, E. 2007, *ApJ*, 656, 524
Rein, H., Lesur, G., & Leinhardt, Z. M. 2010, *A&A*, 511, A69
Safronov, V. S. 1969, *Evoliutsiia doplanetnogo oblaka* (English transl.: Evolution of the Protoplanetary Cloud and Formation of Earth and the Planets, NASA Tech. Transl. F-677, Jerusalem: Israel Sci. Transl. 1972)
Sekiya, M. 1998, *Icarus*, 133, 298
Youdin, A. N. & Shu, F. H. 2002, *ApJ*, 580, 494
Youdin, A. N. & Goodman, J. 2005, *ApJ*, 620, 459
Wada, K., Tanaka, H., Suyama, T., Kimura, H., & Yamamoto, T. 2009, *ApJ*, 702, 1490
Weidenschilling, S. J. & Cuzzi, J. N. 1993, in *Protostars and Planets III*, 1031
Weidenschilling, S. J. 1997, *Icarus*, 127, 290
Wurm, G., Paraskov, G., & Krauss, O. 2005, *Icarus*, 178, 253
Zsom, A., Ormel, C. W., Güttler, C., Blum, J., & Dullemond, C. P. 2010, *A&A*, 513, A57

The Astrophysics of Planetary Systems: Formation, Structure, and Dynamical Evolution
Proceedings IAU Symposium No. 276, 2010
A. Sozzetti, M. G. Lattanzi & A. P. Boss, eds.

doi:10.1017/S174392131102000X

Composition of massive giant planets

Ravit Helled[1], Peter Bodenheimer[2] and Jack J. Lissauer[3]

[1]Department of Earth and Space Sciences,
University of California, Los Angeles, CA 90095-1567, USA,
[2]University of California, Santa Cruz, CA 95064, USA
[3] NASA-Ames Research Center, Moffett Field, CA 94035, USA
email: rhelled@ucla.edu or r.helled@gmail.com

Abstract. The two current models for giant planet formation are core accretion and disk instability. We discuss the core masses and overall planetary enrichment in heavy elements predicted by the two formation models, and show that both models could lead to a large range of final compositions. For example, both can form giant planets with nearly stellar compositions. However, low-mass giant planets, enriched in heavy elements compared to their host stars, are more easily explained by the core accretion model. The final structure of the planets, i.e., the distribution of heavy elements, is not firmly constrained in either formation model.

Keywords. planets and satellites: formation, planetary systems: formation

1. Introduction

The topic of giant planet formation has been studied for decades. The discoveries of giant planets outside our solar system have given us the opportunity to test existing formation and interior models of gas giant planets. Transiting extrasolar giant planets provide important information about their bulk compositions, and a large range of compositions for these objects has been deduced (Guillot 2008). Observations of extrasolar gas planets have raised many challenges for giant planet formation theories, including the determination of planetary formation location, formation timescale, and planetary composition and structure. The interplay between theories and observations is expected to provide a clearer and more complete picture of giant planet formation.

The standard model for giant planet formation, *'core accretion'*, is based on the hypothesis that the formation of a giant planet begins with planetesimal coagulation and heavy-element core formation, followed by accretion of a gaseous envelope (Pollack *et al.* 1996; Lissauer *et al.* 2009). A second model for giant planet formation is *'gravitational (disk) instability'* in which gas giant planets form as a result of gravitational fragmentation in the disk surrounding the young star (Boss 1997; Durisen *et al.* 2007).

There are substantial differences between the two formation models, including formation timescale, the most favorable formation location, and the ideal disk properties for planetary formation. The different nature of the two formation mechanisms naturally leads to some expected differences in planetary composition. However, the final composition of a planet formed by each of these models depends upon the local disk properties in the region of planetary birth. As a result, no clear-cut criterion regarding planetary composition can be used to discriminate between these two formation mechanisms. Below we discuss the ranges of planetary compositions predicted by the two models, and their possible differences.

2. Core Accretion

In the core accretion model, first, dust grains accrete to form solid planetesimals that merge to form a solid core surrounded by a thin gaseous atmosphere. Solids, which typically consist of rocks, ices and organics, are accreted on a time scale of 1 to a few Myr, while the gas accretion rate initially falls well below the solid accretion rate. The solid accretion rate decreases significantly once the planetesimals in the planet's feeding zone are depleted, while the gas accretion rate increases steadily. Eventually, the gas accretion rate exceeds the accretion rate of solid planetesimals, and gas continues to be accreted at a nearly constant rate. Once the core mass and the mass of the gaseous envelope become about equal, a runaway gas accretion builds up the mass of the envelope rapidly while leaving the core at a nearly constant value. Gas accretion then stops either by dissipation of nebular gas or by gap opening (Lissauer *et al.* 2009). After that point the planet is practically isolated from the disk, and it contracts quasi-statically and cools on a time scale of 10^9 years (see Lissauer & Stevenson 2007 and references therein).

2.1. *Core Mass*

The core masses of giant planets formed by core accretion can vary from a few Earth masses ($M_\oplus$) up to even tens of $M_\oplus$. The growth of the core continues as long as planetesimals are present in the planet's feeding zone. The planet's feeding zone represents the region around the planet from which it can capture planetesimals, and its extent is normally taken to be about 4 Hill sphere radii inward and outward from the planet's orbit. The Hill sphere radius of the protoplanet is $R_H = a\left(\frac{M_p}{3M_\star}\right)^{1/3}$ where a is the planet's semimajor axis, M_p is the planet's mass, and M_* is the mass of the star. Once the feeding zone is depleted, the planetary object is nearly isolated and planetesimal accretion drops significantly. The mass at which isolation occurs is given by

$$M_{iso} \approx \frac{64}{\sqrt{3}}\pi^{3/2} M_\star^{-1/2} \sigma^{3/2} a^3 \tag{2.1}$$

where σ is the solid surface density (Lissauer 1993). Around a 1 $M_\odot$ star at 5.2 AU with $\sigma = 10$ g cm^{-2}, this mass is about 11.5 $M_\oplus$.

The final core mass $M_c \approx \sqrt{2}M_{iso}$ (Pollack *et al.* 1996) because the feeding zone for planetesimals expands during the slow gas accretion phase. Equation (2.1) relates the core mass and disk properties provided that neither the planet nor the planetesimals migrate substantially. First, the isolation mass increases with radial distance, so planets in wider orbits are predicted to have more massive cores. Second, the isolation mass is proportional to $\sigma^{3/2}$ (Lissauer 1987). The more solids available to the growing planet, the more massive the core can become. It is therefore clear that the core mass would increase with increasing σ, and that for a given disk mass (or metallicity), steeper density profiles would result in massive cores at smaller radial distances (see Helled & Schubert 2009 for details). Finally, M_{iso} is inversely proportional to the square root of stellar mass. Note that Movshovitz *et al.* (2010) find that core accretion can produce planets with cores of only a few earth masses.

2.2. *Planetary Composition*

The core accretion model does not predict a specific composition for the forming planet. The total heavy-element enrichment of the planet depends on (1) the planetary core mass, which, as just discussed, depends on disk properties, (2) on the amount of dust entrained within the accreted gaseous envelope, and (3) planetesimals (or even other planets) accreted subsequent to rapid gas accretion.

The total mass fraction of heavy elements in the planet can by given by $Z = (M_c + M_{Z_{env}})/M_p$, where $M_{Z_{env}}$ is the mass of heavy elements in the gaseous atmosphere. This simple relation demonstrates the dependence of the bulk composition on both the core mass and the abundance of the accreted gas. The relation implies that the final composition of a massive planet is strongly dependent on the accreted gas composition.

For relatively massive planets, the core contains only a very small fraction of the total mass. The predicted planetary Z depends on the accreted gas composition which could be that of the star, or dust-free or dust-enriched relative to the star. This point is illustrated in Fig. 1a, for different values of M_c, assuming that the composition of the gaseous envelope is solar. However, planets which are depleted in heavy elements compared to their host star could form as well. The gaseous envelope that is accreted could be depleted with solids

leading to a sub-stellar Z. Giant planets that are enriched with heavy elements compared to the star could form when cores are massive, when the accreted gas is enriched with solids, and when planetesimals are accreted at later stage of the planetary growth. It is therefore clear that the core accretion model could lead to the formation of planets with a wide range of final masses and compositions.

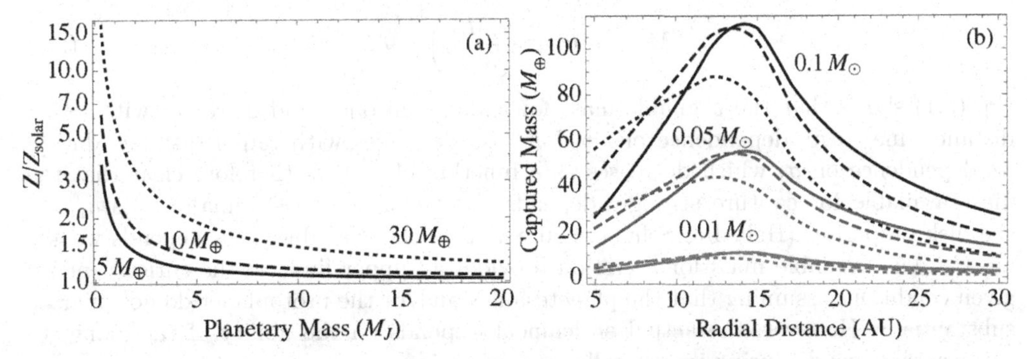

Figure 1. (**a**): metallicity of the planet divided by solar metallicity as a function of planetary mass for three different core masses: 5, 10, and 30 $M_\oplus$ (dotted, dashed, and solid curves, respectively) assuming a solar composition for the gas. (**b**): Captured solid mass by a planet formed by gravitational instability, as a function of the distance of the planet from the star. The black, blue and red curves are for disk masses of 0.1, 0.05, and 0.01 $M_\odot$, respectively. The solid, dashed, and dotted curves refer to density distributions proportional to $a^{-1/2}$, a^{-1} and $a^{-3/2}$, respectively (Helled & Schubert 2009).

3. Gravitational Instability

In the disk-instability model, planets are formed by fragmentation of the gas disk due to gravitational instabilities. Once a local instability occurs, a gravitationally bound sub-condensation region can be created. The fragment contracts, and eventually evolves to become a giant planet. Numerical investigations suggest that planets in wide orbits, such as those recently observed by direct imaging (Marois *et al.* 2008), could form by disk instability.

The evolution of planets formed by gravitational instability is often followed using standard stellar (planetary) evolution equations, under the approximation of spherical symmetry. The model typically takes the objects to be gravitationally bound, isolated, homogenous and static, with stellar composition (Bodenheimer *et al.* 1980; Helled *et al.* 2006).

The first evolutionary stage is the 'pre-collapse' stage in which the extended ($\approx$ 0.5 AU) and cold (internal temperatures $<$ 2000 K) protoplanet contracts quasi-statically. The duration of this stage is not well constrained but as a first approximation it can be taken to be inversely proportional to the square of the initial mass of the body (Helled & Bodenheimer 2011), so more massive protoplanets evolve faster. The pre-collapse stage ends when molecular hydrogen starts to dissociate at the center of the body and a hydrodynamic collapse ensues (Bodenheimer *et al.* 1980).

3.1. *Planetary Composition*

The initial composition of giant planets formed by gravitational instability is similar to the young disk's composition and is therefore stellar. However, the planet may subsequently be enriched in heavy elements by planetesimal capture (Helled *et al.* 2006; Helled & Schubert 2008).

The amount of solid mass available for capture depends on the solid surface density σ at the planetary location. As in the core accretion model, σ changes with the disk mass and its radial density profile, and with stellar metallicity. The total available mass of solids in the planet's feeding zone can be given by

$$M_{av} = 16\pi a^2 \left(\frac{M_p}{3M_\star} \right)^{1/3} \sigma. \tag{3.1}$$

Eq. (3.1) shows that the available mass for capture on one hand increases with radial distance due to its dependence on a^2, but also decreases with radial distance due to its dependence on σ, which decreases as a function of a. It is therefore clear that the mass available for capture strongly depends on the planetary semimajor axis a, and the disk properties (Helled & Schubert 2008). Also, the available mass depends weakly on M_p. The available mass for accretion provides an upper limit to the enrichment for given conditions assuming that the planetesimals and/or the protoplanet do not migrate substantially. However, the actual enrichment depends on the ability of the planet to capture these solids during its pre-collapse contraction.

The planetesimal accretion rate is given by Safronov (1969)

$$\frac{dm}{dt} = \pi R_{cap}^2 \sigma \Omega F_g \tag{3.2}$$

where R_{cap} is the protoplanet's capture radius, Ω is the protoplanet's orbital frequency, and F_g is the gravitational enhancement factor. Eq. (3.2) suggests that the accretion rate is significantly smaller at large radial distances due to its dependence on both σ and Ω. As a result, giant planets at wide orbits will have insufficient time for accretion, which leads to negligible enrichment of solids (Helled & Bodenheimer 2010). The protoplanet's capture radius depends on the planetary size and density, and is therefore larger with increasing planetary mass. Finally, the accreted mass depends on the available time for accretion. As discussed in Helled & Schubert (2008), enrichment is efficient as long as the protoplanet is extended and fills most of the area of its feeding zone, so planetesimals can be slowed down by gas drag, and be absorbed by the protoplanet. Therefore the longer the pre-collapse stage is, the longer the time available for accretion; thus low mass protoplanets have more time to accrete solids, although the mass available for accretion is limited because of its dependence on the total planetary mass.

The planetary enrichment can therefore change significantly from one system to another. Helled *et al.* (2006) have shown that a Jupiter-mass (M_J) clump formed at 5.2 AU could have accreted more than 40 $M_\oplus$ of heavy elements. Helled & Schubert (2009) have shown that a 1 M_J protoplanet formed between 5 and 30 AU could accrete 1-110 $M_\oplus$ of heavy elements, depending on disk properties, and concluded that in the disk instability model the final composition of a giant planet is strongly determined by its formation environment. Fig. 1b shows the heavy element enrichment found by Helled & Schubert (2009). The figure presents the captured mass in the first 10^5 years of the planetary evolution for all the cases considered.

Helled & Bodenheimer (2010) have found that protoplanets with masses between 3 and 7 M_J in wide orbits (24 to 68 AU) can accrete between tens and practically zero $M_\oplus$, with the negligible mass corresponding to the larger radial distances. In the disk instability model, it is predicted that the planetary enrichment with heavy elements strongly decreases with radial distance, and that giant planets in wide orbits would have nearly stellar composition.

3.2. *Core Mass*

In the disk instability model the newly formed protoplanets have no cores. However, cores could form by settling of solids towards the planetary center (DeCampli & Cameron 1979). Helled *et al.* (2008) presented a detailed analysis of coagulation and sedimentation of silicate grains in an evolving 1 M_J protoplanet including the presence of convection. It was found that during the initial contraction of the protoplanet (pre-collapse stage), which lasts several times 10^5 years, silicate grains can sediment to form a core both for convective and non-convective envelopes, although the sedimentation time is substantially longer if the envelope is convective. Grains made of ices and organics were found to dissolve in the planetary envelope, not contributing to core formation.

Helled & Schubert (2008) have investigated the topic of core formation for protoplanets with masses between 1 Saturn mass and 10 M_J. It was found that grain settling occurs in low-mass protoplanets which are cold enough and have a contraction time-scale long enough for the grains to grow and sediment to the center. The grain sedimentation process was found to be favorable for low-mass bodies, due to lower internal temperatures, lower convective velocities and longer contraction time-scales. In convective regions, the grains are carried by the convective eddies until they grow to sizes of 10 cm or larger (the exact value depends on the convection velocity which changes with depth, time and planetary mass). Then they are massive enough to decouple from the gas, sediment to the center and form a core. Protoplanets with masses of $\geqslant 5$ M_J were found to be too hot to allow core formation. In this mass range the grains evaporate in the planetary envelope, enriching it with refractory material (see Helled & Schubert, 2008 for details).

As a result, in the disk instability model, M_c is *not* simply proportional to the mass of the protoplanet. Low-mass clumps have the lowest convective velocities and longest evolution–properties that support core formation. The final core mass depends on the available solid mass within the planet's envelope. If a substantial amount of solids can be accreted before internal temperature are high enough to evaporate the grains, cores of several $M_\oplus$ could be formed (Helled *et al.* 2006; Helled & Schubert 2008).

4. Internal Structure

Neither model for giant planet formation can provide tight constraints on the planetary internal structure. In core accretion, one prediction would be that the core consists of rocks, organics, and ices, while in the disk instability model the cores would be composed of refractory material. Thus, if large pieces of ice and organic material enter the planetary envelope, they could settle to the core as well (Helled *et al.* 2006; Helled & Schubert 2008). In addition, core accretion simulations show that once the core mass reaches ~3 $M_\oplus$, volatiles in small solid bodies dissolve in the envelope and do not reach the core, while for simplification, the model typically assumes that the accreted planetesimals fall to the core. Also, it was found that a substantial amount of ices could stay in the envelope (Iaroslavitz & Podolak 2007).

Both of these formation models, although advanced in many ways, do not follow throughout the evolution the fate of the high-Z material. For example, accreted planetesimals could either fall to the center or evaporate in the envelope and mix with the gas, grains can slowly settle to the center, and high-Z material from the core could be mixed back up by core erosion. These processes (and more) suggest that the final internal structure of a giant planet is not well constrained, and cannot be directly related to the formation mechanism.

In the context of extrasolar giant planets, one must therefore think in terms of global or bulk composition, instead of planetary structure or atmospheric enrichment. Even for Jupiter and Saturn, for which the gravitational moments are measured, the internal structures are not well determined, and are often dependent on model assumptions, equations of state, etc. At present, the available information on the compositions of extrasolar giant planets is insufficient for discriminating between the two models for giant planet formation. Providing predictions of the internal structures, and atmospheric properties of giant planets is still very challenging for formation models.

References

Bodenheimer, P., Grossman, A. S., Decampli, W. M., Marcy, G., & Pollack, J. B. 1980, *Icarus*, 41, 293

Boss, A. P. 1997, *Science*, 276, 1836

Decampli, W. M. & Cameron, A. G. W. 1979, *Icarus*, 38, 367

Durisen, R. H., Boss, A. P., Mayer, L.;,Nelson, A. F., Quinn, T., & Rice, W. K. M. 2007, in *Protostars and Planets V*, B. Reipurth, D. Jewitt, and K. Keil (eds.), Univ. of Arizona Press, Tucson, 607

Guillot, T. 2008, in Nobel Symposium 135, *Physica Scripta*, 130, 014023

Helled, R. & Bodenheimer, P. 2010, *Icarus*, 207, 503

Helled, R. & Bodenheimer, P. 2011, *Icarus*, 211, 939

Helled, R., Podolak, M., & Kovetz, A. 2006, *Icarus*, 185, 64

Helled, R., Podolak, M., & Kovetz, A. 2008, *Icarus*, 195, 863

Helled, R. & Schubert, G. 2008, *Icarus*, 198, 156

Helled, R. & Schubert, G. 2009, *ApJ*, 697, 1256

Iaroslavitz, E. & Podolak, M. 2007, *Icarus*, 187, 600

Lissauer, J. J. 1987, *Icarus*, 69, 249

Lissauer, J. J. 1993, *ARA&A*, 31, 129

Lissauer, J. J. & Stevenson, D. J. 2007, in *Protostars and Planets V*, B. Reipurth, D. Jewitt, and K. Keil (eds.), University of Arizona Press, Tucson, 591

Lissauer, J. J., Hubickyj, O., D'Angelo, G., & Bodenheimer, P. 2009, *Icarus*, 199, 338

Marois, C., Macintosh, B., Barman, T., *et al.* 2008, *Science*, 322, 1348

Movshovitz, N., Bodenheimer, P., Podolak, M., & Lissauer, J. J. 2010, *Icarus*, 209, 616

Pollack, J. B., Hubickyj, O., Bodenheimer, P., Lissauer, J. J., Podolak, M., & Greenzweig, Y. 1996, *Icarus*, 124, 62

Safronov, V. S. 1969, *Evolution of the Protoplanetary Cloud and the Formation of the Earth and Planets* (Nauka, Moscow)

The Astrophysics of Planetary Systems: Formation, Structure, and Dynamical Evolution
Proceedings IAU Symposium No. 276, 2010
A. Sozzetti, M. G. Lattanzi & A. P. Boss, eds.

doi:10.1017/S1743921311020011

A new view on planet formation

Sergei Nayakshin[1]

[1]Department of Physics & Astronomy, University of Leicester, Leicester, LE1 7RH, UK
Sergei.Nayakshin@astro.le.ac.uk

Abstract. The standard picture of planet formation posits that giant gas planets are over-grown rocky planets massive enough to attract enormous gas atmospheres. It has been shown recently that the opposite point of view is physically plausible: the rocky terrestrial planets are former giant planet embryos dried of their gas "to the bone" by the influences of the parent star. Here we provide a brief overview of this "Tidal Downsizing" hypothesis in the context of the Solar System structure.

Keywords. planetary systems, planets and satellites: formation

---

1. Introduction

In the popular "core accretion" scenario (CA model hereafter; e.g., Safronov 1969; Wetherill 1990; Pollack *et al.* 1996), the terrestrial planet cores form first from much smaller solid constituents. A massive gas atmosphere builds up around the rocky core if it reaches a critical mass of about $10\,M_{\oplus}$ (e.g., Mizuno 1980). The CA model's main theoretical difficulty is in the very beginning of the growth: it is not clear how metre-sized rocks would stick together while colliding at high speeds, subject to high radial drifts into the parent star (Weidenschilling 1977, 1980), although gas-dust dynamical instabilities are suggested to help (e.g., Youdin & Goodman 2005; Johansen *et al.* 2007). Nevertheless, believed to be the only viable model for terrestrial planet formation, the model has enjoyed an almost universal support (e.g., Ida & Lin 2008).

This strongest asset of the theory – a "monopoly" on making terrestrial planets – is actually void. Recently, it has been proposed by (Boley *et al.* 2010; Nayakshin 2010a, 2011b, 2010b) that a modified version of the gravitational disc instability model for giant planet formation(Kuiper 1951; Boss 1998) may account for terrestrial planets as well, if gas clump migration (Goldreich & Tremaine 1980) and clump disruption due to tidal forces (McCrea & Williams 1965) are taken into account. This new scheme addresses (Nayakshin 2010b) all of the well known objections (Wetherill 1990; Rafikov 2005) to forming Jupiter in the Solar System via disc fragmentation.

The TD hypothesis is a new combination of earlier ideas and contains four important stages (Figure 1):

(1) Formation of gas clumps (which we also call giant planet embryos; GEs). As the protoplanetary disc cannot fragment inside $R \sim 50$ AU (Rafikov 2005; Boley *et al.* 2006), GEs are formed at somewhat larger radii. The mass of the clumps is estimated at $M_{\rm GE} \sim 10 M_J$ (10 Jupiter masses) (Boley *et al.* 2010; Nayakshin 2010a); they are intially fluffy and cool ($T \sim 100$ K), but contract with time and become much hotter (Nayakshin 2010a).

(2) Inward radial migration of the clumps due to gravitational interactions with the surrounding gas disc (Goldreich & Tremaine 1980; Vorobyov & Basu 2010; Boley *et al.* 2010; Cha & Nayakshin 2011).

(3) Grain growth and sedimentation inside the clumps (McCrea & Williams 1965; Boss 1998; Boss *et al.* 2002). If the clump temperature remains below 1400 – 2000K, massive terrestrial planet cores may form (Nayakshin 2011b), with masses up to the total high Z element content of the clump (e.g., ~ 60 Earth masses for a Solar metalicity clump of $10 M_J$).

(4) A disruption of GEs in the inner few AU due to tidal forces (McCrea 1960; McCrea & Williams 1965; Boley *et al.* 2010; Nayakshin 2010b) or due to irradiation from the star (Nayakshin 2010b) can result in (a) a smallish solid core and a complete gas envelope removal – a terrestrial planet; (b) a massive solid core, with most of the gas removed – a Uranus-like planet; (c) a partial envelope removal leaves a gas giant planet like Jupiter or Saturn. For (b), an internal energy release due to a massive core formation removes the envelope (Handbury & Williams 1975; Nayakshin 2011b).

It is interesting to note that it is the proper placement of step (1) into the outer reaches of the System and then the introduction of the radial migration (step 2) that makes this model physically viable. The theory based on elements (3,4) from an earlier 1960-ies scenario for terrestrial planet formation by McCrea (1960); McCrea & Williams (1965) were rejected by Donnison & Williams (1975) because step (1) is not possible in the inner Solar System. Similarly, the giant disc instability (Kuiper 1951; Boss 1998) cannot operate at $R \sim 5$ AU to make Jupiter (Rafikov 2005). It is therefore the proper placement of step (1) into the outer reaches of the System and then the introduction of the radial migration (step 2) that makes this model physically viable. The new hypothesis resolves (Nayakshin 2011a) an old mystery of the Solar System: the mainly coherent and prograde rotation of planets, which is unexpected if planets are built by randomly oriented impacts.

2. Solar System structure

The gross structure of the Solar System planets is naturally accounted for by the TD model. The innermost terrestrial planets are located within the tidal disruption radius of $r_t \sim 2 - 3$ AU (Nayakshin 2010b), so these are indeed expected to have no massive atmospheres. The asteroid belt in this scheme are the solids that grew inside the giant planet embryos but not made into the central core, and which were then left around the r_t. The gas giant planets are somewhat outside the tidal disruption radius, and thus have been only partially affected by tidal disruption/Solar irradiation.

The outer icy giant planets are too far from the Sun to have been affected strongly by it, so they are interesting cases of *self-disruption* in the TD model. In particular, 35 years ago, (Handbury & Williams 1975) suggested that the massive core formation in Uranus and Neptune evaporated most of their hydrogen envelopes. To appreciate the argument, compare the binding energy of the solid core with that of the GE. We expect the core of high-Z elements to have a density $\rho_c \sim$ a few g cm^{-3}. The radial size of the solid core, $R_{\rm core} \sim (3M_{\rm core}/4\pi\rho_c)^{1/3}$. The binding energy of the solid core is

$$E_{\rm bind,c} \sim \frac{3}{5}\frac{GM^2_{\rm core}}{R_{\rm core}} \approx 10^{41} \text{ erg} \left(\frac{M_{\rm c}}{10\,{\rm M}_\oplus}\right)^{5/3} . \tag{2.1}$$

The clump radius $R_{\rm GE} \approx 0.8$ AU at the age of $t = 10^4$ years, independently of its mass(Nayakshin 2010b), $M_{\rm GE}$. Thus, the GE binding energy at that age is

$$E_{\rm bind,GE} \sim \frac{3}{10}\frac{GM^2_{\rm GE}}{R_{\rm GE}} \approx 10^{41} \text{ erg} \left(\frac{M_{\rm GE}}{3M_J}\right)^{2} . \tag{2.2}$$

The two are comparable for $M_{\rm core} \sim 10\,{\rm M}_\oplus$. Radiation hydrodynamics simulations confirm such internal disruption events: the run labelled M0α3 in Nayakshin (2011b) made a $\sim 20\,{\rm M}_\oplus$ solid core that unbound all but $0.03\,{\rm M}_\oplus$ of the gaseous material of the original $10 M_J$ gas clump.

Future work on the TD hypothesis should address the outer Solar System structure (Kuiper belt; comet compositions, etc.). Detailed predictions for exo-planet observations are difficult as the model dependencies are non-linear (Nayakshin 2011b), but some predictions distinctively different from the CA scenario may be possible as planets loose rather than gain mass as they migrate inwards.

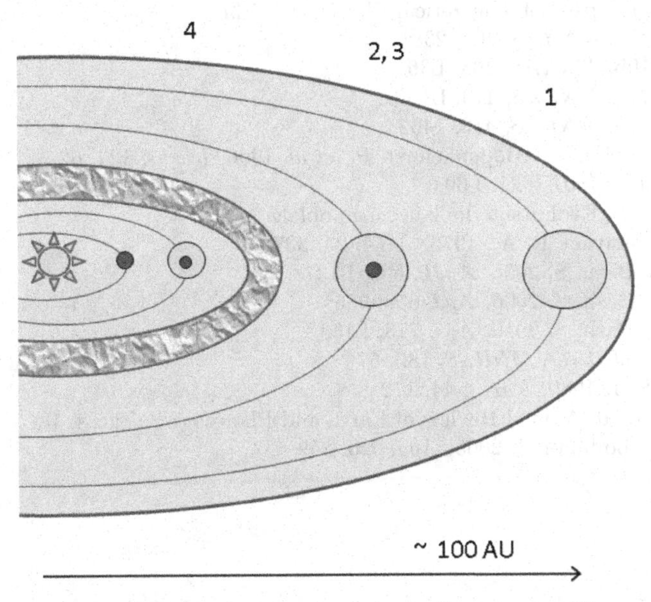

Figure 1. A cartoon of the Tidal Downsizing hypothesis. A protostar (the central Sun symbol) is surrounded by a massive $R \gtrsim 100$ gas disc (the larger grey oval). The four planet formation stages are schematically marked by numbers: (1) The formation of massive gas clumps (embryos) in the outer disc; (2) migration of the clumps closer in to the star, occurring simultaneously with (3) dust grains growth and (possibly) sedimentation into a massive solid core in the centre. The core is shown as a small brown sphere inside the larger gas embryo; (4) disruption of the embryo by tidal forces, irradiation or internal heat liberation. The brown pattern-filled donut-shaped area shows the solid debris ring left from an embryo disruption. The most inward orbit in the diagram shows a terrestrial-like planet, e.g., a solitary solid core whose gas envelope was completely removed. The planet on the next smallest orbit is a giant-like planet with a solid core that retained some of its gas envelope.

Acknowledgements

The author acknowledges the support of the STFC research council and the IAU travel grant to attend this exciting meeting.

References

Boley, A. C., Mejía, A. C., Durisen, R. H., Cai, K., Pickett, M. K., & D'Alessio, P. 2006, *ApJ*, 651, 517

Boley, A. C., Hayfield, T., Mayer, L., & Durisen, R. H., 2010, *Icarus*, 207, 509
Boss, A. P. 1998, *ApJ*, 503, 923
Boss, A. P., Wetherill, G. W., & Haghighipour, N. 2002, *Icarus*, 156, 291
Cha, S.-H. & Nayakshin, S. 2011, *MNRAS*, in press (arXiv:1010.1489)
Donnison, J. R. & Williams, I. P. 1975, *MNRAS*, 172, 257
Goldreich, P. & Tremaine, S. 1980, *ApJ*, 241, 425
Handbury, M. J. & Williams, I. P. 1975, AP&SS, 38, 29
Ida, S. & Lin, D. N. C. 2008, *ApJ*, 685, 584
Johansen, A., Oishi, J. S., Low, M., Klahr, H., Henning, T., & Youdin, A. 2007, *Nature*, 448, 1022
Kuiper, G. P. 1951, in 50th Anniversary of the Yerkes Observatory and Half a Century of Progress in Astrophysics, edited by J. A. Hynek, 357–+
McCrea, W. H. 1960, Royal Society of London Proceedings Series A, 256, 245
McCrea, W. H. & Williams I. P., 1965, Royal Society of London Proceedings Series A, 287, 143
Mizuno, H. 1980, Progress of Theoretical Physics, 64, 544
Nayakshin, S. 2010a, *MNRAS*, 408, 2381
Nayakshin, S. 2010b, *MNRAS*, 408, L36
Nayakshin, S. 2011a, *MNRAS*, 410, L1
Nayakshin, S. 2011b, *MNRAS*, 413, 1462
Pollack, J. B., Hubickyj, O., Bodenheimer, P., *et al.* 1996, *Icarus*, 124, 62
Rafikov, R. R. 2005, *ApJL*, 621, L69
Safronov, V. S. 1969, Evoliutsiia doplanetnogo oblaka.
Shakura, N. I. & Sunyaev R. A., 1973, *A&A*, 24, 337
Vorobyov, E. I. & Basu, S. 2005, *ApJL*, 633, L137
Vorobyov, E. I. & Basu, S. 2006, *ApJ*, 650, 956
Vorobyov, E. I. & Basu, S. 2010, *ApJ*, 713, L133
Weidenschilling, S. J. 1977, *MNRAS*, 180, 57
Weidenschilling, S. J. 1980, *Icarus*, 44, 172
Wetherill, G. W. 1990, Annual Review of Earth and Planetary Sciences, 18, 205
Youdin, A. N. & Goodman, J. 2005, *ApJ*, 620, 459

The Astrophysics of Planetary Systems: Formation, Structure, and Dynamical Evolution
Proceedings IAU Symposium No. 276, 2010
A. Sozzetti, M. G. Lattanzi & A. P. Boss, eds.

doi:10.1017/S1743921311020023

Formation of brown dwarfs and planets

Åke Nordlund[1]

[1]Centre for Star and Planet Formation and Niels Bohr International Academy,
University of Copenhagen, Blegdamsvej 17, 2100 Copenhagen, Denmark
email: aake@nbi.dk

Abstract. Brown dwarfs and massive planets have similar structures, and there is probably an overlap in mass between the most massive planets and the lowest mass brown dwarfs. This raises questions as to what extent the structures of the most massive planets and lowest mass brown dwarfs differ, and what similarities (or not) there might be between their formation mechanisms. Here I discuss these issues on the background of recent numerical simulations of star formation, new evidence from cosmochemistry about the conditions in the early solar system, and recently discovered mechanisms that can expedite planetesimal and possibly planet formation greatly.

Keywords. stars: formation, planetary systems: formation

1. Introduction

Due to the properties of gas at the low temperature and high densities that characterize the interiors of massive planets and low mass stars their typical radii are similar and do not vary much with mass (Chabrier *et al.* 2000; Baraffe *et al.* 2002; Chabrier *et al.* 2009), and in this respect their interior structures are quite similar. It is thus fair to ask what the difference is between massive planets and brown dwarfs. Attempts to answer this question based on for example the mass alone would be problematic, since there appears to be an overlap in mass (Baraffe *et al.* 2010).

A more fruitful line of discussion is to consider the abundance ratios of heavy elements to hydrogen and helium, especially in comparison to the abundances of the parent stars in the case of planets. Arguably the main physical distinction that can be made between massive planets and very low mass stars is that planets in general are underabundant in light elements, relative to the abundance of the host star. The main attention in this invited contribution is focussed on this difference, and on aspects of planet formation related to it.

It is indeed somewhat odd that, even though it is clear from first principles that the heavy elements residing in rocky planets and in the cores of gas and ice giants were once accompanied by the corresponding amounts of light elements, the issue of when and how the separation between light and heavy elements occurred is largely ignored in current planet formation paradigms. For reasons that are left essentially unexplained it is generally assumed that the protoplanetary disks that eventually built planets, at some convenient 'initial time' consisted of a suitable collection of planetesimals, with a gaseous component whose presence was needed mainly to supply the atmosphere of gas giants, and—when so called for—provide a migration mechanism that can be used to explain exoplanets with otherwise inexplicably small orbital radii.

Increased focus on the actual separation mechanism is not only prudent but may also provide an important background for new ideas about planet formation. Below I discuss some recent ideas that fall in this category and based on the discussion I propose a new planet formation paradigm to encompass these ideas.

2. Brown dwarf formation mechanisms

Star formation occurs (only) in dark and cold molecular clouds. These form by compression of the warm interstellar medium (ISM) due to turbulence, driven by energy input from supernovae (de Avillez & Breitschwerdt 2005, 2004, 2007) and other mechanisms such as large scale density waves (Shetty & Ostriker 2008; Kim *et al.* 2008, 2010). Once a region of the ISM has been compressed enough to become opaque (self-shielding) to the otherwise generally pervasive background UV-light from hot stars it cools very efficiently, until the temperature is stabilized at about 10-20 K, as a balance between heating by cosmic rays and cooling by infra-red radiation. The transition is fast and sharp; essentially a phase transition (McKee & Ostriker 2007).

In the warm medium the energy densities per unit volume of kinetic energy $\langle \rho u^2/2 \rangle$, thermal energy $\langle \rho e_{\rm internal} \rangle$, and magnetic energy $\langle B^2/2 \rangle$ are all similar. However, the sharp drop in temperature when going through the phase transition results in an increase in density (at some constant fiducial gas pressure) by about two orders of magnitude. Thus, while the thermal and magnetic energies after the phase transition initially remain similar, the kinetic energy per unit mass has been greatly increased. The situation is thus one with supersonic and super-Alfvénic turbulence. Indeed, the scenario explains *why* the situation arises; the turbulence becomes supersonic not because of 'driving', but because the sound speed drops while inertia maintains the velocity magnitude.

Note that, while the pressure of iso-thermal gas increases only linearly with compression magnetic pressure increases as the *square* of the compression. Because of this the compression ratio and fragment size is often determined by the magnetic pressure rather than by the gas pressure (Padoan & Nordlund 2002).

Turbulent fragments with masses larger than about a solar mass are essentially always dense enough to collapse by selfgravity, while fragments much smaller than a solar mass are dense enough in only a decreasing fraction (with fragment mass) of cases; this is the reason for the turn-over of the stellar initial mass function at the low mass end (Padoan & Nordlund 2002). But due to the random nature of turbulence even fragments with masses corresponding to low mass brown dwarfs occasionally reach high enough densities to collapse.

Brown dwarfs are indeed generally believed to form through the mechanism of gravitational instability although there are differences in opinion about details; i.e., whether brown dwarf formation happens in essentially the same manner as normal star formation, if premature abortion of accretion that would otherwise lead to the formation of a more massive ('normal') star is important (Reipurth & Clarke 2001), or if perhaps there could be a significant contribution due to fragmentation and gravitational collapse in disks around more massive proto-stars followed by ejection from the disk due to gravitational interactions (Stamatellos & Whitworth 2009). Whether or not such modified formation routes are significant, the main mechanism is essentially the same; gravitational collapse of a fraction of the pre-stellar core material, with unchanged ratios X:Y:Z captured in the brown dwarf, which thus abundance-wise is identical to larger mass stars formed in the same molecular cloud.

The fact the brown dwarf mass stars can form by turbulent fragmentation followed by gravitational collapse in molecular clouds with realistic values of supersonic and super-Alfvénic turbulence has been demonstrated by direct numerical simulations; e.g. by Padoan & Nordlund (2011). Some of the brown dwarfs created there have accretion histories that are cut short due to interactions and ejections from there natal cores, but that is the case also for normal stars. On the other hand, many brown dwarfs form simply because the fragment they form out of is compressed enough to become gravitationally unstable, even though the fragment mass lies in the brown dwarf regime. One concludes, therefore, that even though interactions and ejections play a role they are not *necessary* for the formation of brown dwarfs.

3. Planet formation mechanisms

While the formation of brown dwarfs is reasonably well understood and rests on a firm basis of numerical simulations and corresponding semi-analytical theory the same cannot be said about the theory of planet formation. In the context of massive planet formation there are two complementary scenarios, presumed to perhaps both be applicable, but in different regions of parameter space. The 'core accretion' scenario is widely accepted as being able to explain the formation of Jupiter and Saturn type gas giants, at least under some circumstances, while the 'gravitational instability' scenario is allowed to step in under circumstances where the core accretion scenario has little or no chance of working.

None of these scenarios is entirely convincing, for one thing precisely because none of them is general enough to work under all circumstances, but also for a number of specific reasons briefly mentioned below.

The core accretion scenario has problems creating even Jupiter and Saturn in a time short enough to also allow rocky planets to form without migrating into the Sun (Morishima *et al.* 2010), and Uranus and Neptune are required to initially form in close proximity to Jupiter and Saturn and to subsequently migrate to their present locations (Tsiganis *et al.* 2005). While this is in principle a possible work-around in the particular cases of Uranus and Neptune there are many other gas and ice giants discovered in exoplanetary systems, some at considerable distances from their host stars (sometimes of low mass), thus further aggravating the problems with time scales in the core accretion scenario.

Furthermore, in this scenario it is difficult to explain the abundance pattern of Jupiter, which has roughly a factor of three overabundance of most heavy elements, including noble gases (Owen *et al.* 1999). To achieve such an abundance pattern with core accretion one must rely on clathration of noble gases, a process that takes place only at extremely low temperatures. These ices must then be brought in to accrete onto Jupiter, in exactly the right proportions to reach the same overabundance as much more easily accreted elements. Unless one applies the popular 'Jupiter-is-there-so-it-must-have-happened' argument the explanation must be considered farfetched.

The core accretion scenario also inherits the weaknesses of the general paradigm for rocky planet formation; the meter-size drift barrier (Weidenschilling 1977), the problem of getting pebbles and rocks to grow by sticking rather than braking apart due to collisions, and the general problem of finalizing planet building in a time sufficiently short to prevent Type I migration from being a problem (Morishima *et al.* 2010).

The gravitational instability (GI) scenario, on the other hand, has a similar list of weaknesses, starting again with the need to invoke a complementary scenario in cases where the GI scenario cannot work. There are severe difficulties in explaining the composition of gas and (particularly) ice giants; their heavily modified abundance patterns can be achieved only be invoking extraordinary strong sources of external radiation to 'boil off' unwanted light element components of the original composition. Another weakness is that the expected dependence on metal abundance is the opposite of the one actually observed; it should be more difficult for the gravitational instability to operate at higher metallicity since larger opacities lead to longer cooling times (Cai *et al.* 2006). Ironically, the GI scenario seems to mainly survive to step in when the core accretion scenario cannot possibly work.

As discussed further below the recent discovery of a streaming instability (Youdin & Goodman 2005), its incorporation into a scenario for very rapid formation of planetesimals (Johansen & Youdin 2007; Johansen *et al.* 2007), and the recent extension to a scenario that involves 'pebble accretion' onto large asteroids (Johansen & Lacerda 2010) are very significant developments, which may turn out to be able to help resolve some of the main issues with the current planet formation scenarios; see also Boley & Durisen (2010) and Nayakshin (2010, this volume).

4. Evidence from cosmo-chemistry

Evidence from isotopic abundances can provide very important information about time scales and thermal histories of the early solar system. Live ^{26}Al and ^{60}Fe (half lives 0.72 Myr and 2.6 Myr, respectively) were present in the early solar system, which allows only a few million years to have passed since their production until the formation of the solar system. After being incorporated in solids, these short-lived radioactive isotopes function as potentially very accurate clocks, in that both parent and daughter nuclei are locked in minerals and differences in their ratios can be recovered today, using high precision mass spectrometers (e.g. Trinquier *et al.* 2008; Larsen *et al.* 2010). In addition ^{26}Al was present in sufficient quantities to melt the interior of even rather small asteroids (Baker *et al.* 2005), and more long lived radioactive isotopes are still important contributors to the internal heat balance of planets.

Combined evidence from several isotopes, particularly ^{54}Cr and ^{26}Mg, provide strong evidence for that these isotopes were initially homogeneously distributed, in that the abundance of heavy element enrichment products from different types of stars are strongly correlated in meteorite samples from different parts of the solar system (Thrane *et al.* 2008). Related evidence indicates that bulk portions of the solar system solids were subsequently subjected to strong thermal processing.

Recent improvements in the precision of ^{26}Al $\rightarrow$ ^{26}Mg analysis indicate, additionally, that the thermal processing caused significant heterogeneity of the ^{26}Al distribution (Larsen *et al.* 2010). This introduces significant uncertainties in previous age determinations based on the assumption that ^{26}Al was still homogeneously distributed at the time when the first solids formed in the solar system.

However, rather sensationally, the same analysis also shows an internal consistency of the measurements that puts an upper limit of only about 4000 years on the interval of time over which the (asteroid) parent body of the analyzed samples formed. This provides direct evidence for a very rapid planetesimal formation mechanism, consistent with recent astrophysical ideas and empirical evidence (Johansen & Lacerda 2010; Bottke *et al.* 2005; Morbidelli *et al.* 2009).

5. A new paradigm for planet formation

The 'streaming instability' mechanism (Johansen & Youdin 2007; Johansen *et al.* 2007), and related works (Boley & Durisen 2010; Nayakshin 2010) have opened up a new and very promising route, which could eventually lead to a much improved understanding of planet formation. One of the most important aspects of these new ideas is that they open up a way to not only avoid the meter size drift barrier (Weidenschilling 1977), but to actually turn the very same mechanism into an important part of planet formation. Johansen & Lacerda (2010) point out that accretion of cm-size particles ('pebbles') driven by the Weidenschilling (1977) mechanism leads to a natural explanation for a weak but clear tendency of prograde rotation of the largest asteroids. Indeed, one may ask "Why stop at planetesimals?"; there seems to be no reason why the rapid growth of planetesimals due to 'pebble accretion' should not continue to build larger bodies, since growing mass leads to a corresponding growth of the Hill sphere, and thus presumably continued rapid accretion. Indeed such a mechanism could also be a natural explanation for the prograde and rapid rotation of a majority of the planets (Nayakshin this volume).

The cosmo-chemical evidence for very short times scales, and for bulk thermal processing at high temperatures in the early solar system also calls for shifting the perspective and focusing on processes involving both dynamical and thermal interaction of gas and solids. Even the structure of the early Sun itself may depend critically on details of the accretion history (Vorobyov & Basu 2005; Baraffe & Chabrier 2010; Inutsuka *et al.* 2010).

These new pieces of evidence and ideas related to the interaction of gas and solids in the early solar system and other proto-planetary systems are components of what could indeed develop into a new paradigm for planet formation. Major issues to be focussed upon in this paradigm are "When and how were heavy elements (Z) separated from the light element components (X+Y) of the gas?", "What were the main mechanisms responsible for thermal processing of gas and solids?", and "How did the strong mass fractionation that is so evident in the atmospheres of rocky planets happen?".

Whether one thinks the outcome of research along these lines of questioning may change or current picture of planet formation drastically or not, it is certainly important to try to answer these and other questions that may be inspired by such a new paradigm.

6. Atmospheres of Planets

As an example where taking the point of view of the new paradigm may turn out to be relevant let's explore the consequences of assuming that 'pebble accretion' is sufficiently efficient to rapidly form rocky cores with planetary masses, embedded in a dense gaseous disk, and that the rocky cores will attract massive gaseous envelopes with temperatures of the order 1500–2000 K.

Such initial states have the interesting property that they evolve, rather independently of details of the initial states, towards rocky planets with atmospheres broadly consistent with the atmospheres that our solar system planets have (or can be surmised to have had). To see this, assume at first that the gaseous atmospheres are so massive that they dominate the total mass inside the systems Hill sphere.

The total optical depth of such an atmosphere is of the order

$$\tau \sim \kappa M R_{\rm Hill}^{-2}, \tag{6.1}$$

where κ is the opacity (of the order of 1 cm^2 g^{-1} or larger for gas+dust at these temperatures). For $M = M_{\rm Earth}$ the estimated optical depth is of the order of several times 10^5. Such an atmosphere is thus extremely optically thick and evolves, at least on short time scales, nearly adiabatically.

The escape speed becomes formally zero at the Hill sphere, but one can estimate its order of magnitude at radii somewhat smaller than the Hill radius from

$$v_{\rm escape}^2 = \frac{GM}{R_{\rm Hill}}. \tag{6.2}$$

Even for masses as large as $M = 10 M_{\rm Earth}$ the escape speed estimate is less that a thousand meters per second. At T=1500 K the thermal speeds of hydrogen molecules, helium atoms, and nitrogen molecules on the other hand are $\sim$ 2500, 1800, and 700 m s^{-1}, respectively. Hydrogen molecules and helium atoms thus escape easily, via the 'hydrodynamic escape' mechanism, while nitrogen molecules and other heavy molecules and atoms escape less easily. This would lead to strong mass fractionation, as is indeed observed to have occurred in the Earth's and Mars' atmospheres (Pepin 1991).

The hydrodynamic escape of hydrogen and helium, which in this phase are the main constituents of the atmosphere, leads to a decrease of mass, which in turn leads to an even more rapid escape of the light elements, since the reduction of the Hill radius with mass is more than compensated for by the reduction of the force of gravity. (The mass loss takes place at approximately constant internal entropy, which makes it possible to predict and include also the change in temperature when evaluating the evolution of the balance between gravitational and gas pressure forces.)

The hydrodynamic escape thus continues first of all to the point when the mass of the gaseous envelope becomes smaller than the mass of the rocky core. From that point on the Hill radius remains approximately unchanged, while the atmospheric density continues to decrease by hydrodynamic escape. The decreasing mass density implies a corresponding tendency for an adiabatic temperature decrease, but this is likely to be counteracted by heating from the hot surface of the rocky planet core.

The hydrodynamic escape in the neighborhood of the Hill radius is thus bound to continue until the atmosphere becomes sufficiently transparent to cool efficiently. One can estimate the total mass of the remaining atmosphere by requiring that it has an optical depth of the order of unity if it fills the Hill sphere. From Eq. 6.1 above one concludes that this does not happen until the remaining atmosphere weighs only a fraction of the order of 10^{-6} to 10^{-5} of the mass of an Earth size rocky planet core. But once the optical depth of the atmosphere becomes small enough the temperature drops and the remaining atmosphere shrinks to a size compatible with the lower temperature, thus effectively shutting down the hydrodynamic escape mechanism.

Mass fractions of 10^{-6} to 10^{-5} are consistent with the current atmospheres of Earth and Venus (note that the large difference between the atmospheric pressures on Earth and Venus is mainly due to pressure from CO_2 while the partial pressures of N_2 for example do not differ by more than a factor of a few). The estimate is also consistent with a significantly higher initial atmospheric pressure on Mars than the current one, as required by the evidence for earlier abundant presence of liquid water. However, the escape velocity at the top of Mars atmosphere is so low that even nitrogen and oxygen molecules have thermal speeds where molecules in the tail of the Maxwell distribution can escape, and as evidenced by the very low pressure of the current Martian atmosphere the effect has been sufficient to remove a large fraction of the atmosphere that must once have been present.

The same fate would even more rapidly and efficiently result from any attempt to endow Mercury with an atmosphere, so even if it would have attracted one during the period with a high pressure gaseous proto-planetary disk the current state, with practically no atmosphere, is also consistent with the assumptions made here.

Jupiter, on the other hand, was heavy enough that even if the inial mass was consistent with overall solar abundances (and thus with an initial Jupiter mass about three times larger than now) the estimates of escape speeds and thermal speeds show that at least the outermost layers of an initially hot atmosphere would have been lost. However, already at a radius of about half the Hill radius the escape speed exceeds the likely thermal speeds of hydrogen molecules and helium atoms. This is fully consistent with the fact that Jupiter seems to have retained about 1/3 of an atmosphere with initially solar abundance (a much simpler explanation for Jupiters pattern of abundances of heavy elements than accretion of planetesimals after creation).

Thus, estimates of the masses of planetary atmospheres based on the assumptions above appear to be fully consistent with current planetary atmospheres (and lack thereof).

7. Summary and conclusions

The similarity of the internal structure of giant planets and brown dwarfs and their overlapping mass intervals makes it difficult to distinguish between massive planets and very low mass brown dwarfs. However, in general planets are underabundant in the lightest elements (hydrogen and helium) relative to the primordial composition of the gas that formed their host stars, a fact that must reflect their different mode of formation.

A mechanisms that can separate the gaseous and solid components was discussed by Weidenschilling (1977), but in a context where its consequences actually constituted a major problem for planet formation. Recent developments (Johansen & Lacerda 2010; Boley & Durisen 2010; Nayakshin 2010) have made it clear that the mechanism can be turned around into being instead beneficial for planet formation, and the mechanism has already been proven to allow very rapid formation of asteroid-size bodies. Possibly this 'pebble accretion' mechanism could turn out to also be the main mechanism responsible for the formation of rocky planets and giant planet cores.

In the mean time it is already clear that if planet sized rocky cores form in a gaseous proto-planetary disk then Earth sized planets will naturally acquire atmospheres that are essentially devoid of the light elements and that have total masses compatible with the current atmospheres of Earth and Venus. The atmospheres of planets built around smaller rocky cores will suffer heavy atmospheric losses to space, while much more massive cores will be able to hold on to atmospheres compatible with the ones belonging to gas and ice giants.

With such a paradigm water would be a natural and initial part of planetary atmospheres, rather than something that would have to be added to it afterwards via comets or water bearing planetesimals.

In the new paradigm the abundance pattern of Jupiter may be understood as simply reflecting the loss of about 2/3 of the initial amount of light elements, rather than being caused by 'enrichment' of a gaseous atmosphere bombarded by a carefully instrumented collection of rocks and clathrated ices.

The new paradigm also supplies a natural explanation for the heavy mass fractionation of the atmospheres of Venus, Earth and Mars, something that has otherwise been long forgotten and neglected (Pepin 1991).

In conclusion, a renewed focus on the mechanisms that separated solids and gases in the early solar system is of interest by itself, and may in the end turn out to fundamentally change our understanding of planet formation.

References

Baker, J., Bizzarro, M., Wittig, N., Connelly, J., & Haack, H. 2005, *Nature*, 436, 1127

Baraffe, I. & Chabrier, G. 2010, *A&A*, 521, A44+

Baraffe, I., Chabrier, G., Allard, F., & Hauschildt, P. H. 2002, *A&A*, 382, 563

Baraffe, I., Chabrier, G., & Barman, T. 2010, Reports on Progress in Physics, 73, 016901

Boley, A. C. & Durisen, R. H. 2010, *ApJ*, 724, 618

Bottke, W. F., Durda, D. D., Nesvorný, D., Jedicke, R., Morbidelli, A., Vokrouhlický, D., & Levison, H. 2005, *Icarus*, 175, 111

Cai, K., Durisen, R. H., Michael, S., Boley, A. C., Mejía, A. C., Pickett, M. K., & D'Alessio, P. 2006, *ApJ* (Letters), 636, L149

Chabrier, G., Baraffe, I., Allard, F., & Hauschildt, P. 2000, *ApJ*, 542, 464

Chabrier, G., Baraffe, I., Leconte, J., Gallardo, J., & Barman, T. 2009, in American Institute of Physics Conference Series, Vol. 1094, American Institute of Physics Conference Series, ed. E. Stempels, 102–111

de Avillez, M. A. & Breitschwerdt, D. 2004, *A&A*, 425, 899

—. 2005, *A&A*, 436, 585

—. 2007, *ApJ* (Letters), 665, L35

Inutsuka, S., Machida, M. N., & Matsumoto, T. 2010, *ApJ* (Letters), 718, L58

Johansen, A. & Lacerda, P. 2010, *MNRAS*, 404, 475

Johansen, A., Oishi, J. S., Mac Low, M., Klahr, H., Henning, T., & Youdin, A. 2007, *Nature*, 448, 1022

Johansen, A. & Youdin, A. 2007, *ApJ*, 662, 627

Kim, C., Kim, W., & Ostriker, E. C. 2008, *ApJ*, 681, 1148

—. 2010, *ApJ*, 720, 1454

Larsen, K., Trinquier, A., Paton, C., Ivanova, M., Nordlund, Å., Krot, A. N., & Bizzarro, M. 2010, *Meteoritics and Planetary Science Supplement*, 73, 5202

McKee, C. F. & Ostriker, E. C. 2007, *ARA&A*, 45, 565

Morbidelli, A., Bottke, W. F., Nesvorný, D., & Levison, H. F. 2009, *Icarus*, 204, 558

Morishima, R., Stadel, J., & Moore, B. 2010, *Icarus*, 207, 517

Nayakshin, S. 2010, *MNRAS*, 408, L36

Nayakshin, S. 2011, this volume

Owen, T., Mahaffy, P., Niemann, H. B., Atreya, S., Donahue, T., Bar-Nun, A., & de Pater, I. 1999, *Nature*, 402, 269

Padoan, P. & Nordlund, Å. 2002, *ApJ*, 576, 870

Padoan, P. & Nordlund, A. 2011, *ApJ*, 730, id.40

Pepin, R. O. 1991, *Icarus*, 92, 2

Reipurth, B. & Clarke, C. 2001, *AJ*, 122, 432

Shetty, R. & Ostriker, E. C. 2008, *ApJ*, 684, 978

Stamatellos, D. & Whitworth, A. P. 2009, *MNRAS*, 400, 1563

Thrane, K., Nagashima, K., Krot, A. N., & Bizzarro, M. 2008, *ApJ* (Letters), 680, L141

Trinquier, A., Bizzarro, M., Ulfbeck, D., Elliott, T., Coath, C. D., Mendybaev, R. A., Richter, F. M., & Krot, A. N. 2008, *Geochimica et Cosmochimica Acta Supplement*, 72, 956

Tsiganis, K., Gomes, R., Morbidelli, A., & Levison, H. F. 2005, *Nature*, 435, 459

Vorobyov, E. I. & Basu, S. 2005, *ApJ* (Letters), 633, L137

Weidenschilling, S. J. 1977, *MNRAS*, 180, 57

Youdin, A. N. & Goodman, J. 2005, *ApJ*, 620, 459

The Astrophysics of Planetary Systems: Formation, Structure, and Dynamical Evolution
Proceedings IAU Symposium No. 276, 2010
A. Sozzetti, M. G. Lattanzi & A. P. Boss, eds.

doi:10.1017/S1743921311020035

Direct imaging and spectroscopy of planets and brown dwarfs in wide orbits†

Mariangela Bonavita[1], Ray Jayawardhana[1], Markus Janson[1] and David Lafrenière[2]

[1]Department of Astronomy and Astrophysics, University of Toronto, 50 St. George Street M5S 3H4 Toronto ON Canada
email: bonavita@lepus.astro.utoronto.ca

[2]Département de Physique, Université de Montréal, C.P. 6128 Succ. Centre-Ville, Montréal, QC H3C 3J7, Canada

Abstract. Recent direct imaging discoveries of exoplanets have raised new questions about the formation of very low-mass objects in very wide orbits. Several explanations have been proposed, but all of them run into some difficulties, trying to explain all the properties of these objects at once. Here we present the results of a deep adaptive optics imaging survey of 85 stars in the Upper Scorpius young association with Gemini, reaching contrasts of up to 10 magnitudes. In addition to identifying numerous stellar binaries and a few triples, we also found several interesting sub-stellar companions. We discuss the implications of these discoveries, including the possibility of a second pathway to giant planet formation.

Keywords. brown dwarfs, planetary systems, stars: low-mass, stars: pre-main-sequence

1. Introduction

The detection and characterization of low-mass sub-stellar companions on large orbits (over several tens of AU) around stars is of great importance for our understanding of planet, brown dwarf, and star formation, as well as for our understanding of the dynamical evolution of such companions in multiple systems or in circumstellar disks. Currently, most formation models of planets or low-mass brown dwarf companions - core accretion (e.g., Pollack *et al.* 1996), gravitational instability (e.g., Boss 1997), and fragmentation of a pre-stellar core (e.g., Padoan & Nordlund 2002; Bate *et al.* 2003) - cannot easily explain the existence of companions at large orbital separations, like those recently discovered via direct imaging (see e.g. Marois *et al.* 2008; Lafrenière 2008, 2010). Those companions have separations of tens to several hundreds of AU and mass ratios of the order of 0.01 relative to their primaries. In this context a good determination of the semi-major axis, mass, and mass ratio distributions of far out giant planets will provide a powerful tool to understand the formation mechanisms at play and distinguish between star-like and planet-like origins. In addition to this, wide sub-stellar companions are good benchmark objects for the study of brown dwarf/planetary atmospheres and evolution models. They have in general the same age and metallicity of the host star, their distance can often be easily inferred, and their high separation from their star in the sky makes them the easiest companions to study.

† Talk was delivered by R. Jayawardhana (rayjay@astro.utoronto.ca)

2. AO imaging survey in Upper Scorpius

Here we present the outcomes of our survey of 85 young stars belonging to the 5 Myr old Upper Sco association. The observations were taken using the NIRI camera (Hodapp *et al.* 2003) and the ALTAIR adaptive optic system at the Gemini North Telescope, the sensitivity being such to allow us to reach the planetary domain at separations higher than $\sim$ 60 AU (See Figs. 1), leading to the discovery/confirmation of 4 new sub-stellar companions. The characteristics of the companions are summarized in Table 1, whether Figs. 2 and 3 show the AO images of each target.

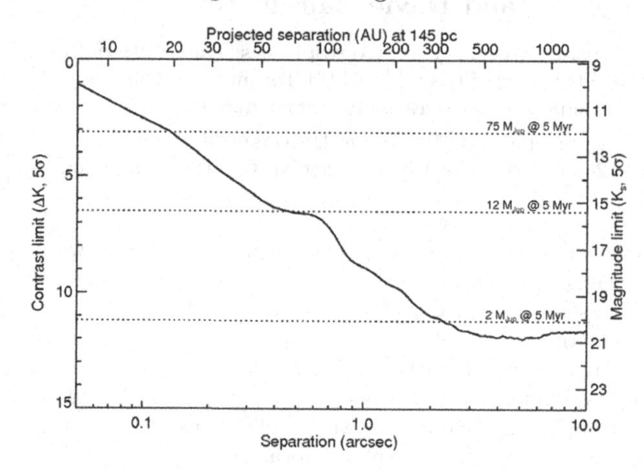

Figure 1. Detection limits of our observations, as a function of the separation. The dotted lines indicate the expected contrast for various planetary masses.

Table 1. Principal characteristics of the new companions and their host stars

ID	M_{Star} ($M_{\odot}$)	M_{Comp} (M_{Jup})	Separation (AU)	Reference
1RXS 1609	0.85	~ 8	330	Lafrenière *et al.* (2008, 2010)
HIP 78530	2.50	20	710	Lafrenière *et al.* (submitted)
1RSX 1541-2656	0.70	30-40	850	Lafrenière *et al.* (in prep.)
UpSco 1610-1913	1.80	70	915	Lafrenière *et al.* (in prep.)

The smallest of these companions, first announced in September 2008 and now confirmed with common proper motion (see Lafrenière *et al.* 2010), is a 8 M_{Jup} planet around the roughly solar-mass star 1RXS 1609 (Fig. 2(a)). We have obtained multi-band infrared imaging and spectroscopy of it, which reveal evidence of its youth, temperature (1800K) and mass as well as absorption features due to CO, H2, and H2O. The planet's projected separation of 330 AU is quite surprising for such a low-mass object.

Two of the other sub-stellar companions in our Upper Sco sample have even larger projected separations:

- A 30-40 M_{Jup} companion at $\sim$850 AU from UpSco 1610-1913 (Fig. 3(b)). The primary has a tight companion (0.14", 20 AU see Kraus 2008), so this forms a triple system. Interestingly, the tight companion could be at the star/BD boundary, 80 M_{Jup}.
- A 70-80 M_{Jup} companion at 6.3" (915 AU) from 1RXS 1541-2526 (Fig. 3(a)).

In addition to those, a 20 M_{Jup} companion has been imaged at 710 AU from HIP 78530 (Fig. 2(b)). The latter has also been confirmed to be co-moving, via proper motion measurements (Lafrenière *et al.* submitted).

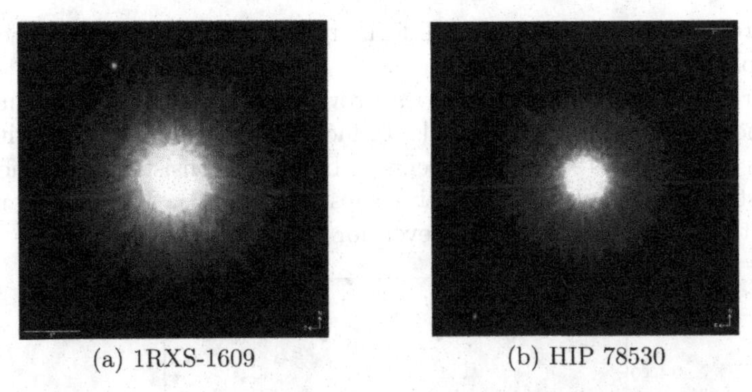
(a) 1RXS-1609 (b) HIP 78530

Figure 2. Composite of J-, H- and K-band images of 1RXS-1609 (a) and HIP 78530 (b) and their companions, respectively in the upper left and lower left corners.

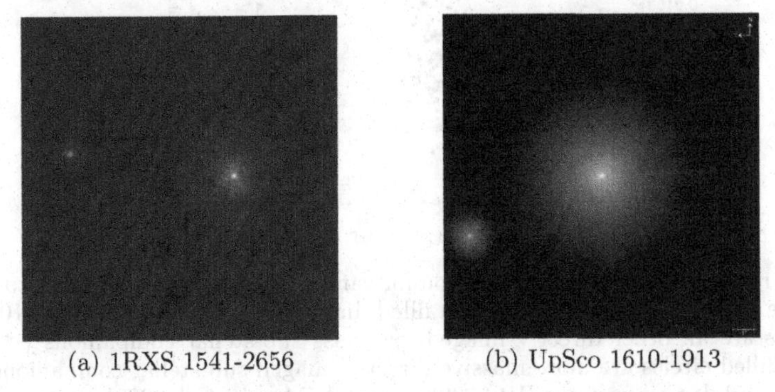
(a) 1RXS 1541-2656 (b) UpSco 1610-1913

Figure 3. K-band images of 1RXS 1541-2526 (a) and UpSco 1610-1913 (b) and their companions, respectively in the upper left and lower left corners.

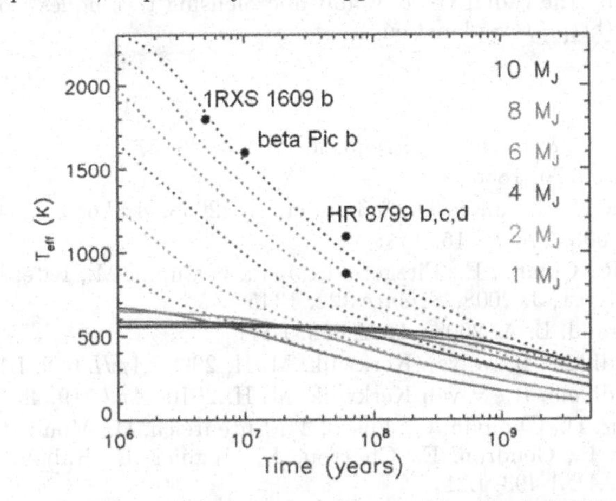

Figure 4. Comparison of hot-start (dotted lines) and cold-start (solid lines) evolutionary models (see Fortney *et al.* (2008) for the original plot) with the values corresponding to β Pic b (Quanz *et al.* 2010), HR 8799 b, c, d (Marois *et al.* 2008) and 1RXS 1609b (Lafrenière *et al.* 2008, 2010)

3. Discussion and Conclusions

The separation of the companions we found in our survey places them among the widest known sub-stellar companions to stars. Moreover, the extreme mass-ratio of HIP 78530 and 1RXS 1609 are comparable to those of directly imaged planets, even though the mass of HIP 78530b is above the deuterium-burning limit. Thus, the existence of these objects presents new challenges to all formation scenarios, blurring the distinction between giant planets and brown dwarfs.

On the other hand, these new discoveries provide observational constraints for the planet atmosphere and evolutionary models. In fact 1RXS-1609b, together with β Pic b and the planets in the HR 8799 system, seems to be better consistent with the hot-start than the cold-start models at the respective ages (see Fig. 4). The cold-start models, never reaching temperature above 700 K, even for the youngest ages.

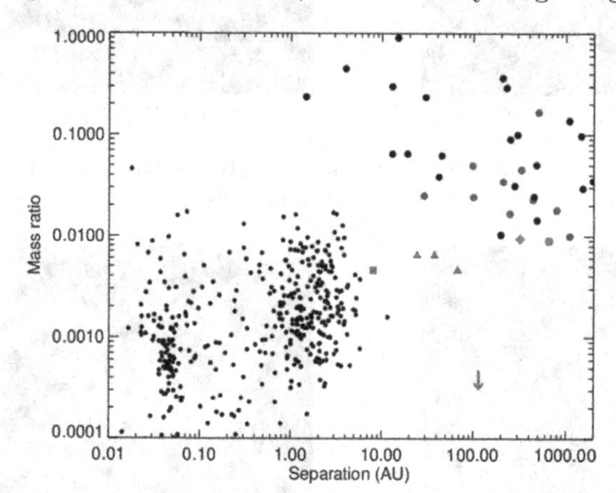

Figure 5. Mass ratio as a function of separation for various sub-stellar companions to stars. The green filled circle is HIP 78530b and the green filled diamond is 1RXS J160929.1- 210524b. The blue filled circles are all other directly imaged low-mass sub-stellar companions ($< 25 M_{Jup}$). The large black filled circles are more massive directly imaged sub-stellar companions. The red triangles, square and down arrow are HR 8799b, c and d (Marois*et al.* 2008), β Pic b (Lagrange *et al.* 2009, 2010) and Fomalhaut b (Kalas *et al.* 2008), respectively. The small black circles indicate planets found by the radial velocity and microlensing techniques from the Extrasolar Planets Encyclopaedia (http://exoplanet.eu/).

References

Bate, M. R., Bonnell, I. A., & Bromm, V. 2003, *MNRAS*, 339, 577

Boss, A. P. 1997, *Science*, 276, 1836

Fortney, J. J., Marley, M. S., Saumon, D., & Lodders, K. 2008, *ApJ* 683, 1104

Hodapp, K. W., *et al.* 2003, *PASP* 115, 1388

Kalas, P., Graham, J. R., Chiang, E., Fitzgerald, M. P., Clampin, M., Kite, E. S., Stapelfeldt, K., Marois, C., & Krist, J. 2008, *Science* 322, 1345

Kraus, A. L. & Hillenbrand, L. A. 2008, *ApJL* 686, L111

Lafrenière, D., Jayawardhana, R., & van Kerkwijk, M. H. 2008, *ApJL* 689, L153

Lafrenière, D., Jayawardhana, R., & van Kerkwijk, M. H. 2010, *ApJ* 719, 497

Lagrange, A., Gratadour, D., Chauvin, G., Fusco, T., Ehrenreich, D., Mouillet, D., Rousset, G., Rouan, D., Allard, F., Gendron, E., Charton, J., Mugnier, L., Rabou, P., Montri, J., & Lacombe, F. 2009, *A&A* 493, L21

Lagrange, A., Bonnefoy, M., Chauvin, G., Apai, D., Ehrenreich, Boccaletti, A., Gratadour, D., Rouan, D., Mouillet, D., & Lacour, S., Kasper M. 2010, *Science* 329, 57

Marois, C., Macintosh, B., Barman, T., Zuckerman, B., Song, I., Patience, J., Lafrenière, D., & Doyon, R. 2008, *Science* 322, 1348

Quanz, S. P., Meyer, M. R., Kenworthy, M. A., Girard, J. H. V., Kasper, M., Lagrange, A. M., Apai, D., Boccaletti, A., Bonnefoy, M., Chauvin, G., Hinz, P. M. & Lenzen, R. *ApJL* 722, L49

Padoan, P. & Nordlund, A. 2002, *ApJ*, 576, 870

Pollack, J. B., Hubickyj, O., Bodenheimer, P., Lissauer, J. J., Podolak, M., & Greenzweig, Y. 1996, *Icarus*, 124, 62

The Astrophysics of Planetary Systems: Formation, Structure, and Dynamical Evolution
Proceedings IAU Symposium No. 276, 2010
A. Sozzetti, M. G. Lattanzi & A. P. Boss, eds.

doi:10.1017/S1743921311020047

A possible dividing line between massive planets and brown-dwarf companions

Johannes Sahlmann[1], Damien Ségransan[1], Didier Queloz[1] and Stéphane Udry[1]

[1]Observatoire de Genève, Université de Genève, 51 Chemin Des Maillettes, 1290 Sauverny, Switzerland. email: johannes.sahlmann@unige.ch

Abstract. Brown dwarfs are intermediate objects between planets and stars. The lower end of the brown-dwarf mass range overlaps with the one of massive planets and therefore the distinction between planets and brown-dwarf companions may require to trace the individual formation process. We present results on new potential brown-dwarf companions of Sun-like stars, which were discovered using CORALIE radial-velocity measurements. By combining the spectroscopic orbits and Hipparcos astrometric measurements, we have determined the orbit inclinations and therefore the companion masses for many of these systems. This has revealed a mass range between 25 and 45 Jupiter masses almost void of objects, suggesting a possible dividing line between massive planets and sub-stellar companions.

Keywords. brown dwarfs, stars: low-mass, binaries: spectroscopic, stars: statistics, techniques: radial velocities, astrometry

---

1. Introduction

Radial-velocity (RV) studies show that close-in (< 10 AU) brown-dwarf companions to Sun-like stars are rare compared to planets and stars (Marcy & Butler 2000; Halbwachs *et al.* 2003; Grether & Lineweaver 2006). Precision-RV surveys have found ~50 of these objects with $M_2 \sin i = 13 - 80\, M_J$, which we adopt as working-definition of the brown-dwarf mass range, regardless of the object's formation history, composition, and membership in a multiple system (Sozzetti & Desidera 2010; Sahlmann *et al.* 2010, henceforth SA10). Brown-dwarf (BD) companions have also been found in close orbit around M-dwarfs and in wide orbits (> 10 AU) around stars, but these are not discussed in this contribution. There are several proposed scenarios for the formation of brown-dwarf companions, but for an individual object we can assume that it either formed like a planet or like a high mass-ratio binary (Leconte *et al.* 2009). In particular, the discovery of transiting candidates (CoRoT-3b, CoRoT-15b, WASP-30b) in very close orbit around their F-type host stars challenge the current planet taxonomy.

2. Search for brown-dwarf companions in the CORALIE survey

On the basis of the well-defined and uniform CORALIE planet-search sample (Udry *et al.* 2000), we have conducted a search for close BD companions of Sun-like stars (SA10). In combination with the astrometric measurements of Hipparcos, we were able to determine the orbit inclinations for many of these systems, which always rendered M-dwarf companions. All 21 BD companion candidates in the CORALIE survey are listed in Table 1. The orbital parameters of the potential BD companions, as derived from the RV solution of the respective reference papers, are shown in Fig. 1. In relation to the sample

Table 1. BD candidates in the CORALIE survey

Object	Ref.	$M_2 \sin i$ (M_J)	M_2 (M_J)	BD?	Object	Ref.	$M_2 \sin i$ (M_J)	M_2 (M_J)	BD?
HD3277	(1)	64.7	344.2	No	HD112758	(2)	34.0	210.0	No
HD4747	(1)	46.1	...		HD154697	(1)	71.1	151.9	No
HD17289	(1)	48.9	547.4	No	HD162020	(4)	14.4	...	
HD18445	(2)	45.0	175.0	No	HD164427A	(1)	48.0	269.9	No
HD30501	(1)	62.3	89.6	No	HD167665	(1)	50.6	...	
HD38529	(3)	13.4	17.6	Yes	HD168443	(5)	18.1	...	
HD43848	(1)	24.5	101.8	No	HD189310	(1)	25.6	...	
HD52756	(1)	59.3	...		HD202206	(6)	17.4	...	
HD53680	(1)	54.7	226.7	No	HD211847	(1)	19.2	...	
HD74014	(1)	49.0	...		HD217580	(2)	68.0	170.0	No
HD89707	(1)	53.6	...						

References: (1) Sahlmann *et al.* (2011); (2) Zucker & Mazeh (2001); (3) Benedict *et al.* (2010); (4) Udry *et al.* (2002); (5) Wright *et al.* (2009); (6) Correia *et al.* (2005).

size of 1600 stars, this yielded a frequency of 1.3 % of candidate BD companions to Sun-like stars, assuming that all candidates have been discovered. Of these 21 candidates, ten are stellar companions with masses $M_2 > 80\,M_J$. Thus, less than 0.6 % of Sun-like stars have close-in BD companions. One candidate was confirmed to a brown dwarf with HST astrometry (Benedict *et al.* 2010).

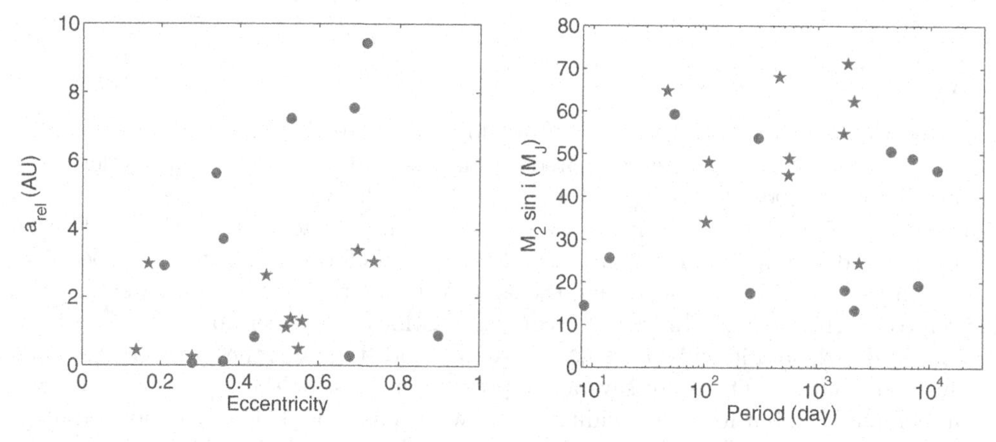

Figure 1. Orbital parameters and $M_2 \sin i$ of the 21 potential brown-dwarf companions characterised with CORALIE. Blue asterisks show the M-dwarf companions and red dots indicate the remaining candidates and HD 38527. For display clarity, error bars are not shown. They can be found in the respective references given in Table 1.

Figure 2 shows the distribution of $M_2 \sin i$ for 85 sub-stellar companions characterised with CORALIE†. Below $4\,M_J$, the sensitivity of the CORALIE survey is limited by the RV-measurment accuracy and timespan. At minimum masses higher than $4\,M_J$, the distribution shows approximately a linear decrease (in $\partial N / \partial \log M_2$) in the number of companions and extends into the brown-dwarf mass range. Above $45\,M_J$, a large population of companions appears, although it is drastically reduced by the astrometric

† The list of objects is taken from the series of CORALIE papers I-XVI (see Ségransan *et al.* 2010), from SA10, and from Marmier *et al.* (*in preparation*).

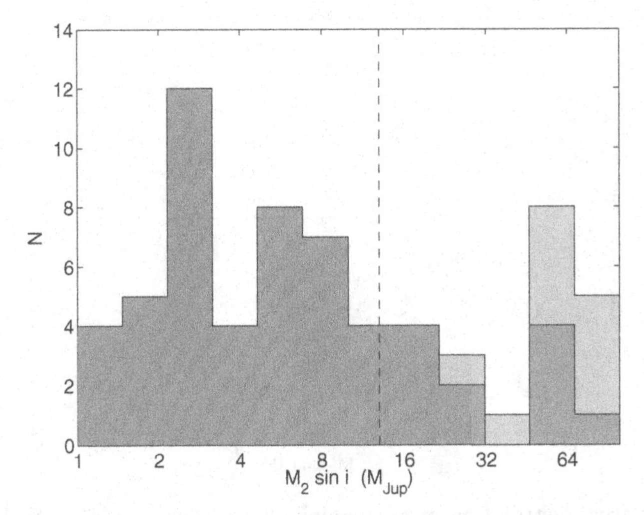

Figure 2. Minimum mass histogram of sub-stellar companions characterised with CORALIE. The vertical dashed line indicates the $13\,M_J$ boundary. Light-grey areas show the companions for which SA10 found a mass higher than $80\,M_J$.

analysis of SA10. The $M_2 \sin i$ distribution of M-dwarf companions is shown in the light-grey histogram of Fig. 2.

3. Companion mass function

We show the cumulative distribution of minimum masses in Fig. 3. The initial curve, including all 21 candidates, shows a steady increase of objects over the brown-dwarf mass range (dashed blue curve), though more than half of the objects have $M_2 \sin i > 45\,M_J$. After removal of the 10 stellar companions (solid black curve), the cumulative distribution exhibits a particular shape: it shows a steep rise in the $\sim 13 - 25\,M_J$ region followed by a flat region spanning $\sim 25 - 45\,M_J$ void of companions. At masses higher than $\sim 45\,M_J$, the distribution rises again up to $\sim 60\,M_J$. Above $\sim 60\,M_J$, no companion is left.

The distribution function's bimodal shape is incompatible with any monotonic companion mass function in the brown-dwarf and low-mass-star domain (approximately $13\,M_J - 0.6\,M_\odot$). Instead, the observed distribution can be explained by the detection of the high-mass tail of the planetary companions, which reaches into the brown-dwarf domain and contributes to the companions with $M_2 \sin i < 25\,M_J$, and the low-companion-mass tail of the binary star distribution with with $M_2 \sin i > 45\,M_J$. The interjacent mass-range is void of objects and represents a possible dividing line between massive planets and brown-dwarf companions obtained solely from observations.

4. Discussion

The mass range of $25 - 45\,M_J$, which is void of companions, comprises the minimum of the companion mass function at $43^{+14}_{-23}\,M_J$ derived by Grether & Lineweaver (2006) for the 50 pc sample, which corresponds to the volume limit of the CORALIE survey. Hence, our result is in agreement with the conclusions of Grether & Lineweaver (2006), which were based on extrapolation of the planet and binary distribution function into the brown-dwarf mass range. Still, the number of stars with brown-dwarf companions is limited, but both planet-search programmes (Diaz *et al.* 2011; Santos *et al.* 2011)

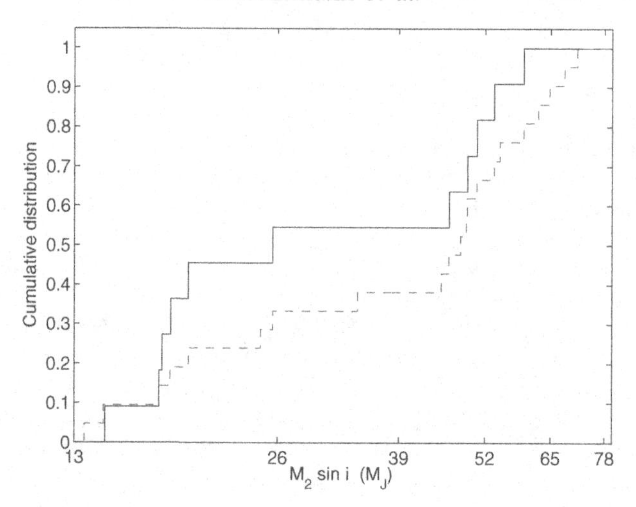

Figure 3. Cumulative distribution of companions with $M_2 \sin i = 13 - 80\,M_J$ characterised with CORALIE (blue dashed line). The black solid line shows the distribution after removal of the M-dwarf companions (SA10).

and more dedicated surveys (Lee *et al.* 2011) will provide us with many more candidates and will allow us to better trace the shape of the companion mass function. The GAIA astrometry mission (e.g. Lindegren 2010) will detect and characterise a wealth of brown-dwarf companions and, not being affected by the inclination ambiguity, eventually allow us to identify their mass function.

References

Benedict, G. F., McArthur, B. E., Bean, J. L., Barnes, R., Harrison, T. E., Hatzes, A., Martioli, E., & Nelan, E. P. 2010, *AJ*, 139, 1844

Correia, A. C. M., Udry, S., Mayor, M., Laskar, J., Naef, D., Pepe, F., Queloz, D., & Santos, N. C. 2005, *A&A*, 440, 751

Diaz, R., The SOPHIE Consortium 2011, *EPJ Web of Conferences*, 11, 02006

Grether, D. & Lineweaver, C. H. 2006, *ApJ*, 640, 1051

Halbwachs, J. L., Mayor, M., Udry, S., & Arenou, F. 2003, *A&A*, 397, 159

Leconte, J., Baraffe, I., Chabrier, G., Barman, T., & Levrard, B. 2009, *A&A*, 506, 385

Lee, B. L., Ge, J., Fleming, S. W., *et al.* 2011, *ApJ*, 728, i.d32

Lindegren, L. 2010, *Proceedings IAU Symposium*, 261, 296

Marcy, G. W. & Butler, R. P. 2000, *PASP*, 112, 137

Sahlmann, J., Segransan, D., Queloz, D., Udry, S., Santos, N. C., Marmier, M., Mayor, M., Naef, D., Pepe, F., & Zucker, S. 2011, *A&A*, 525, A95

Santos, N. C., Mayor, M., Bonfils, X., Dumusque, X., Bouchy, F., Figueira, P., Lovis, C., Melo, C., Pepe, F., Queloz, D., Ségransan, D., Sousa, S. G., & Udry, S. 2011, *A&A*, 526, A112

Ségransan, D., Udry, S., Mayor, M., Naef, D., Pepe, F., Queloz, D., Santos, N. C., Demory, B.-O., Figueira, P., Gillon, M., Marmier, M., Mégevand, D., Sosnowska, D., Tamuz, O., & Triaud, A. H. M. J. 2010, *A&A*, 511, A45+

Sozzetti, A. & Desidera, S. 2010, *A&A*, 509, A26+

Udry, S., Mayor, M., Naef, D., Pepe, F., Queloz, D., Santos, N. C., Burnet, M., Confino, B., & Melo, C. 2000, *A&A*, 356, 590

Udry, S., Mayor, M., Naef, D., Pepe, F., Queloz, D., Santos, N. C., & Burnet, M. 2002, *A&A*, 390, 267

Wright, J. T., Upadhyay, S., Marcy, G. W., Fischer, D. A., Ford, E. B., & Johnson, J. A. 2009, *ApJ*, 693, 1084

Zucker, S. & Mazeh, T. 2001, *ApJ*, 562, 549

The Astrophysics of Planetary Systems: Formation, Structure, and Dynamical Evolution
Proceedings IAU Symposium No. 276, 2010
A. Sozzetti, M. G. Lattanzi & A. P. Boss, eds.

doi:10.1017/S1743921311020059

The visitor from an ancient galaxy: A planetary companion around an old, metal-poor red horizontal branch star

Rainer J. Klement[1], Johny Setiawan[1], Thomas Henning[1], Hans-Walter Rix[1], Boyke Rochau[1], Jens Rodmann[2] and Tim Schulze-Hartung[1]

[1]Max-Planck-Institute for Astronomy
Königstuhl 17, D-69121 Heidelberg, Germany
email: klement@mpia.de
[2]European Space Agency, Space Environment and Effects Section, ESTEC

Abstract. We report the detection of a planetary companion around HIP 13044, a metal-poor red horizontal branch star belonging to a stellar halo stream that results from the disruption of an ancient Milky Way satellite galaxy. The detection is based on radial velocity observations with FEROS at the 2.2-m MPG/ESO telescope. The periodic radial velocity variation of $P = 16.2$ days can be distinguished from the periods of the stellar activity indicators. We computed a minimum planetary mass of 1.25 M_{jup} and an orbital semimajor axis of 0.116 AU for the planet. This discovery is unique in three aspects: First, it is the first planet detection around a star with a metallicity much lower than few percent of the solar value; second, the planet host star resides in a stellar evolutionary stage that is still unexplored in the exoplanet surveys; third, the planetary system HIP 13044 most likely has an extragalactic origin in a disrupted former satellite of the Milky Way.

Keywords. galaxy: solar neighborhood, planetary systems: formation, stars: horizontal-branch, stars: individual (HIP 13044), techniques: radial velocities

1. Introduction

Do planets exist in external galaxies? There is no reason that they should not. However, directly observing stars in one of the Milky Way's satellites in order to obtain precise enough radial velocity or transit photometry measurements to detect planetary signals is not possible because the galaxies are too far away.† Although microlensing would in principal allow for extra-galactic planet detections, this phenomenon is intrinsically non-repeatable, and no microlensing planets in other galaxies have been confirmed yet (see, however, Ingrosso *et al.* 2009). Nevertheless, we know that certain stars in our solar neighborhood have been accreted from satallite galaxies in the past and in fact, examples of such accretion events are still observable today (e.g. Belokurov *et al.* 2006). Accreted stars therefore provide an opportunity to indirectly search for planets that originated in external galaxies and travelled into our Galaxy along with their host stars. In particular, stars belonging to nearby stellar halo streams are ideal targets to search for extra-galactic planets. In the inner halo, where the orbital frequencies are high, such streams are no

† With a distance modulus of ~14.5, the Canis Major dwarf galaxy could pose an exception, allowing at least principally for planet searches around giant stars; but it is located behind the Milky Way's disk ($l = 240°$, $b = -8°$) and therefore heavily obscured by dust and foreground light (Martin *et al.* 2004; Bellazzini *et al.* 2004).

Table 1: Important stellar parameters of HIP 13044.

Parameter	Value	Unit	Reference
Spectral type	F2		SIMBAD
m_V	9.94	mag	*Hipparcos*
$T_{\rm eff}$	6025±63	K	Carney *et al.* (2008a); Roederer *et al.* (2010)
R_*	6.7±0.3	$R_\odot$	Carney *et al.* (2008a)
[Fe/H]	-2.09 ± 0.26		Beers *et al.* (1990); Chiba & Beers (2000); Carney *et al.* (2008b); Roederer *et al.* (2010)
$v \sin i$	8.8±0.8	$\mathrm{km\,s^{-1}}$	Carney *et al.* (2008a)
	11.7±1.0	$\mathrm{km\,s^{-1}}$	own measurement

longer spatially coherent, but they still share similar chemical and dynamical properties that depend on the composition and orbit of their progenitor system and allow for identifying individual stream members (see e.g. the review by Klement 2010).

The most significant stellar halo stream in the solar neighborhood is the one discovered by Helmi *et al.* (1999). Its members have kinematics that clearly separate them from the bulk of other halo stars. The stream's progenitor system possibly resembled a galaxy similar to Fornax or the Sagittarius dwarf, and has been disrupted about 6–9 Gyr ago (Helmi *et al.* 1999; Kepley *et al.* 2007). The Red Horizontal Branch (RHB) star HIP 13044 is a confirmed member of the Helmi stream (Helmi *et al.* 1999; Re Fiorentin *et al.* 2005; Kepley *et al.* 2007; Roederer *et al.* 2010). Its stellar parameters, in particular the low metallicity and relatively large projected rotational velocity (Table 1), mark this star as an interesting target for a planet search.

2. Data and Analysis

We have obtained 36 radial velocity (RV) measurements of HIP 13044 between September 2009 and July 2010 with FEROS, a high-resolution spectrograph (R = 48,000) attached at the 2.2 meter Max-Planck Gesellschaft/European Southern Observatory telescope, located at the La Silla observatory in Chile (Kaufer *et al.* 2000). RV values have been obtained through cross-correlating the stellar spectrum with a numerical template designed for stars of the spectral type F0 and containing 550 selected spectral lines. Typical uncertainties of the RV values are ~ 50 m s^{-1}.

The variation of the RV between our observations at different epochs has a semi-amplitude of 120 m s^{-1} (Fig. 2). The Generalized Lomb Scargle (GLS) periodogram (Zechmeister & Kürster 2009) reveals a significant RV periodicity at P = 16.2 days with a False Alarm Probability (FAP) of 5.5×10^{-6}. Additional analysis by a Bayesian algorithm (Gregory 2005) confirms this period. To find out whether stellar activity (moving/rotating surface inhomogeneities or stellar pulsations) is responsible for producing the observed RV variation, we investigated different stellar activity indicators.

We analyzed the variation of two independent spectroscopic activity indicators: the bisector velocity spans (Hatzes 1996) of all 550 lines and the equivalent width (EW) of the infrared CaII line at λ = 849.8 nm. The GLS periodogram analyses for both the BVS and the Ca II line EW variations revealed significant peaks at similar periodicities ($P_{\rm BVS}$ = 5.02 d, FAP$_{\rm BVS}$ = 1.4×10^{-5} and $P_{\rm EW}$ = 6.31 d, FAP$_{\rm EW}$ = 4.4×10^{-6}). We attribute these variations in the spectral line indicators to rotational modulation caused by star spots or large granulation cells. We point out that we have found no hints for a period around 16 days in the BSV and EW variations.

Photometric observations of HIP 13044 have been made by *Super-WASP*. There exist 3620 high-precision measurements of HIP 13044 in the public *Super-WASP* archive. After removing 10 data points with high error bars, the Lomb-Scargle periodogram shows that there is no signal for a $\sim$ 16 d period (Fig. 1b). Instead, there are several marginally significant peaks between a few hours and a few days (FAP $< 1\%$), with the ones at 1.4 d and 3.5 d beeing the most significant. These two peaks are, however, most likely harmonic to each other: $1.4^{-1} + 3.5^{-1} = 1$. It is expected that HIP 13044 oscillates only at pulsationally unstable overtones of high order (Xiong *et al.* 1998). Observations of one RHB star in the metal-poor globular cluster NGC 6397 (Stello & Gilliland 2009) as well as theoretical predictions (Xiong *et al.* 1998) set these periods in the range of a few hours to a day or so. We caution, however, that no clear theoretical predictions for a star with parameters similar to HIP 13044 exist, and it could be possible that some high-order oscillations are able to explain the 1.4 or 3.5 day signal. More important, however, is that we found no signal of a period around 16.2 d in the photometric data.

3. Discussion and Conclusions

From the lack of a $\sim$ 16 d period in both photometric and spectroscopic activity indicators, we conclude that a (sub-stellar) companion remains the most likely hypothesis for the observed 16.2 d RV variation. Table 2 lists the orbital solution for HIP 13044 b that we have computed from the stellar reflex motion. While the semi-major axis is not unusual for a giant extra-solar planet, the non-circular orbit ($e = 0.25$) for such a close-in companion is. One has to keep in mind, however, that the star's red giant branch (RGB) phase, which preceeds the RHB phase, probably has changed the orbital properties of the planet. From the lack of close-in ($\lesssim 0.5$ AU, e.g. Sato *et al.* 2008) giant planets around RGB stars, it seems plausible that such planets, if they existed, have been engulfed during the expansion of the stellar atmosphere (although other explanations exist, e.g. Currie 2009). Such planet-swallowing would spin up the star and increase the mass loss (Soker 1998; Carlberg *et al.* 2009), and indeed enhanced rotation among RGB and RHB stars has been found (Carney *et al.* 2003, 2008a). Interestingly, HIP 13044 is such

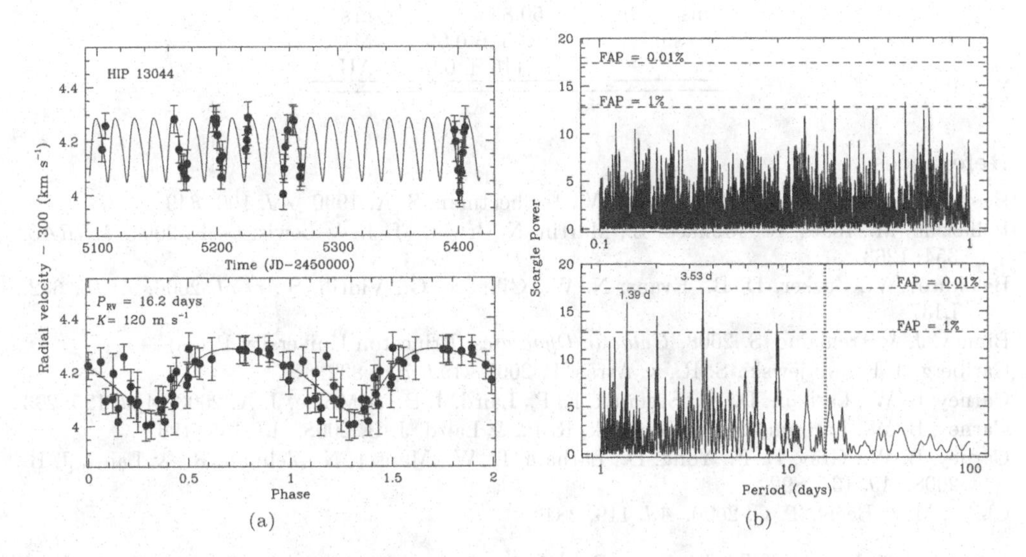

Figure 1: (a) RV variation of HIP 13044. (b) LS-periodogram of the *Super-WASP* data.

a fast rotator † too, so possibly HIP 13044 b was part of a multiple planetary system of which some planets have not survived the RGB phase. The survival of HIP 13044 b during that phase is theoretically possible under certain circumstances (Soker 1998). It is also possible that the planet's orbit decayed through tidal interaction with the stellar envelope. However, a prerequisite to survival is then that the mass loss of the star stops before the planet would have been evaporated or accreted. Assuming asymmetric mass loss, velocity kicks could have increased the eccentricity of HIP 13044 b to its current, somewhat high value (Heyl 2007). The same could be achieved by interaction with a third body in the system. The formation of the planet(s) around such a metal-poor star is possible via gravitational instability. Even if we assume that the star's metallicty was higher in the past, the fact that it belongs to the chemically homogenous Helmi stream ($\langle[\mathrm{Fe/H}]\rangle = -1.8 \pm 0.4$ for 33 stream members from Kepley *et al.* 2007; Klement *et al.* 2009) still implies the lowest metallicty of a planet host star found to date. The reason why most planet searches around metal-poor stars failed could be twofold: first, most planets around such stars should reside at large orbital distances (e.g. Marchi 2007; Santos *et al.* 2010); we could have been lucky that the RGB phase lead to an orbital decay of HIP 13044 b. Second, most planet searches tried to observed a large sample simultaneously (e.g. Sozzetti *et al.* 2009), in this way eventually not obtaining enough data points for a significant RV signal.

We can be pretty sure that the planet was not captured from another star inside the Milky Way, because the time on which stellar encounters play a role exceeds the Hubble time for galaxies like the Milky Way (Binney & Tremaine 2008). The stream membership of HIP 13044 therefore implies an extra-Galactic origin of HIP 13044 b.

Table 2: Orbital parameters of HIP 13044 b

P	16.2 ± 0.3	days
K_1	119.9 ± 9.8	$\mathrm{m\,s^{-1}}$
e	0.25 ± 0.05	
ω	219.8 ± 1.8	deg
$JD_0 - 2450000$	5109.78 ± 0.02	days
χ^2	32.35	$\mathrm{m\,s^{-1}}$
rms	50.86	$\mathrm{m\,s^{-1}}$
$m_2 \sin i$	1.25 ± 0.05	$\mathrm{M_{Jup}}$
a	0.116 ± 0.01	AU

References

Beers, T. C., Kage, J. A., Preston, G. W., & Shectman, S. A. 1990, *AJ*, 100, 849

Bellazzini, M., Ibata, R., Monaco, L., Martin, N., Irwin, M. J., & Lewis, G. F. 2004, *MNRAS*, 354, 1263

Belokurov, V., Zucker, D. B., Evans, N. W., Gilmore, G., Vidrih, S., *et al.* 2006, *ApJL*, 642, L137

Binney, J. & Tremaine, S. 2008, *Galactic Dynamics* (Princeton University Press)

Carlberg, J. K., Majewski, S. R., & Arras, P. 2009, *ApJ*, 700, 832

Carney, B. W., Latham, D. W., Stefanik, R. P., Laird, J. B., & Morse, J. A. 2003, *ApJ*, 125, 293

Carney, B. W., Latham, D. W., Stefanik, R. P., & Laird, J. B. 2008, *AJ*, 135, 196

Carney, B. W., Gray, D. F., Yong, D., Latham, D. W., Manset, N., Zelman, R., & Laird, J. B. 2008, *AJ*, 135, 892

Chiba, M. & Beers, T. C. 2000, *AJ*, 119, 2843

† $v_{\mathrm{rot}} \approx 60$ km s^{-1}, following from $P_{\mathrm{rot}}/\sin i = 2\pi R_*/(v_{\mathrm{rot}} \sin i)$, with R_* from Table 1 and $P_{\mathrm{rot}} \approx 5.5$ d from the BVS and EW variations.

Currie, T. 2009, *ApJL*, 694, L171

Gregory, P. C. 2005, *Bayesian Logical Data Analysis for the Physical Sciences: A Comparative Approach with 'Mathematica' Support* (Cambridge University Press)

Hatzes, A. P. 1996, *PASP*, 108, 839

Helmi, A., White, S. D. M., de Zeeuw, P. T., & Zhao, H. 1999, *Nature*, 402, 53

Heyl, J. 2007, *MNRAS*, 382, 915

Ingrosso, G., Calchi Novati, S., De Paolis, F., Jetzer, Ph., Nucita, A. A., & Zakharov, A. F. 2009, *MNRAS*, 399, 219

Kaufer, A., Stahl, O., Tubbesing, S., Norregaard, P., Avila, G., *et al.* 2000, in: M. Iye & A. F. Moorwood (eds.), *Proc. SPIE*, 4008, 459

Kepley, A. A., Morrison, H. L., Helmi, A., Kinman, T. D., Van Duyne, J., *et al.* 2007, *AJ*, 134, 1579

Klement, R., Rix, H.-W., Flynn, C., Fuchs, B., Beers, T. C., *et al.* 2009, *ApJ*, 698, 865

Klement, R. J. 2010, *A&AR*, 18, 567

Marchi, S. 2007, *ApJ*, 666, 475

Martin, N. F., Ibata, R. A., Bellazzini, M., Irwin, M. J., Lewis, G. F., & Dehnen, W. 2004, *MNRAS*, 348, 12

Re Fiorentin, P., Helmi, A., Lattanzi, M. G., & Spagna, A. 2005, *A&A*, 439, 551

Roederer, I. U., Sneden, C., Thompson, I. B., Preston, G. W., & Shectman, S. A. 2010, *ApJ*, 711, 573

Santos, N. C., Mayor, M., Benz, W., Bouchy, F., Figueira, P., *et al.* 2010, *A&A*, 512, A47

Sato, B., Izumiura, H., Toyota, E., *et al.* 2008, *PASJ*, 60, 539

Soker, N. 1998, *AJ*, 116, 1308

Sozzetti, A., Torres, G., Latham, D. W., *et al.* 2009, *ApJ*, 697, 544

Stello, D. & Gilliland, R. L. 2009, *AJ*, 700, 949

Xiong, D. R., Cheng, Q. L., & Deng, L. 1998, *ApJ*, 500, 449

Zechmeister, M. & Kürster, M. 2009, *A&A*, 496, 577

PART 2:

STRUCTURE AND ATMOSPHERES

The Astrophysics of Planetary Systems: Formation, Structure, and Dynamical Evolution
Proceedings IAU Symposium No. 276, 2010
A. Sozzetti, M. G. Lattanzi & A. P. Boss, eds.

doi:10.1017/S1743921311020060

Statistical patterns in ground-based transit surveys

Andrew Collier Cameron[1]

[1]SUPA, School of Physics & Astronomy, University of St Andrews,
North Haugh, St Andrews KY16 9SS, UK
email: andrew.cameron@st-andrews.ac.uk

Abstract. As the number of known transiting planets from ground-based surveys passes the 100 mark, it is becoming possible to perform meaningful statistical analyses on their physical properties. Caution is needed in their interpretation, because subtle differences in survey strategy can lead to surprising selection effects affecting the distributions of planetary orbital periods and radii, and of host-star metallicity. Despite these difficulties, the planetary mass-radius relation appears to conform more or less to theoretical expectations in the mass range from Saturns to super-Jupiters. The inflated radii of many hot Jupiters indicate that environmental factors can have a dramatic effect on planetary structure, and may even lead to catastrophic loss of the planetary envelope under extreme irradiation. High-precision radial velocities and secondary-eclipse timing are yielding eccentricity measurements of exquisite precision. They show some hot Jupiters to be in almost perfectly circular orbits, while others remain slightly but significantly eccentric.

Keywords. planetary systems, techniques: radial velocities, techniques: photometric

1. Introduction

Among the 110 or so transiting planets discovered up to the end of 2010, 79 have been accounted for by five major ground-based wide-field photometric surveys: OGLE-III (Udalski *et al.* 2002), the Transatlantic Exoplanet Survey (TrES; Alonso *et al.* 2004), the Hungarian Automated Transit Network (HATNet; Bakos *et al.* 2004), the Wide-Angle Search for Planets (WASP; Pollacco *et al.* 2006) and the XO Project (McCullough *et al.* 2005). The last four of these employ commercial camera lenses of 11 cm aperture and 200mm focal length, backed by large-format CCD detectors giving fields of view of order 8 degrees square per camera. They achieve differential photometric precisions of order 0.01 magnitude, sufficient to detect transits of Jupiter-sized planets, on stars with $V = 12$. The host stars are thus bright enough that the essential radial-velocity follow-up can be carried out using high-precision radial-velocity spectrometers on telescopes of moderate size. The SOPHIE spectrograph on the 1.9-m telescope at Haute-Provence, the CORALIE spectrometer on the 1.2-m Swiss Euler telescope at La Silla, the FIES spectrograph on the 2.5-m Nordic optical Telescope on La Palma, and the TRES instrument on the 1.5-m Tillinghast reflector telescope on Mount Hopkins have provided the large amounts of radial-velocity followup time needed to eliminate astrophysical false positives and determine planetary masses.

2. Survey strategies and system properties

A number of important statistical trends are beginning to emerge from these surveys. Their interpretation must, however, be tempered by an understanding of the systematic errors inherent in such surveys. For example, Charbonneau (this volume) notes that the distributions of host-star metallicity, planet radius and orbital period differ significantly between the two most prolific surveys, HATNet and WASP.

HATNet operates sites at widely-distributed longitudes, whereas WASP operates from a single site in each hemisphere. This enables HATNet to detect long-period planets with greater efficiency than WASP, whose ability to amass sufficient transits to secure a detection in a single season is hampered by a relatively low day/night duty cycle. WASP employs a raster observing pattern, sweeping the 8 cameras on each mount across 6 or 7 hours in RA with a cadence of about 10 minutes. The very wide area coverage of this strategy enables WASP to detect rare objects with relatively deep transits despite its low cadence and sparse duty cycle. HATNet's more intensive observing strategy and multi-longitude duty cycle is better suited to the detection of shallow transits and hence smaller planets.

A more puzzling difference between the two surveys is that the host stars of the WASP planets have a distribution of metallicities whose median [Fe/H] is close to solar, whereas the HAT planet hosts have a median [Fe/H] that is ~ 0.2 dex more metal rich. The HATNet cameras employ Cousins I filters, whereas the WASP cameras use only a red blocking filter with a cutoff at 750 nm. The median V magnitude of the HAT planet-host stars is nearly a magnitude brighter than the WASP median. Experiments with the Besançon galactic model (Robin *et al.* 2003) show that, at the intermediate galactic latitudes where these surveys are conducted, stars of luminosity class V show a steady decrease in median metallicity with increasing V magnitude. Although this trend is not in itself sufficient to explain the difference in the two samples, metal-poor stars of a given I magnitude also appear brighter in the WASP cameras owing to their lower line blanketing. Although further investigation is needed, a combination of these two effects provides a plausible explanation of the difference in metallicity between the HAT and WASP host-star populations.

3. Mass-radius relation for transiting planets

The mass-radius relation for transiting planets provides the most basic test of our understanding of their interior structure and composition. The fundamental theory was developed by Zapolsky & Salpeter (1969): at low masses, cold spheres of a given composition follow a mass-radius relation governed by electrostatic forces, yielding densities almost independent of mass such that $R \propto M^{1/3}$. For planets of roughly Neptune mass or less, the spread in the mass-radius relation is effectively a composition sequence, and the density of a planetary body is a fairly reliable guide to its likely composition (Seager *et al.* 2007). At masses comparable to that of Jupiter, the dependence on mass flattens off and eventually turns over to follow a relation of the form $R \propto M^{-1/3}$, in which non-relativistic electron degeneracy pressure is balanced by gravitation. Most of the planets found in ground-based transit surveys (Fig. 1) are in the Jovian mass regime, where radius is more or less independent of mass. The few objects found in the brown-dwarf desert, such as the 60 M_{Jup} WASP-30b, have systematically smaller radii than the Jupiter-mass planets, as do the Saturn-mass and Neptune-mass planets. This pattern is in broad agreement with theoretical expectations. Within each mass range, however, there is a substantial spread in planet radii.

Fressin *et al.* (2007) modelled the radius anomalies of hot Jupiters in terms of their rock/ice core masses and equilibrium temperatures. The equilibrium temperature of a planet is obtained by balancing the irradiating power against re-radiated power assuming a black-body spectrum and re-radiation from a specified fraction of the planetary surface.

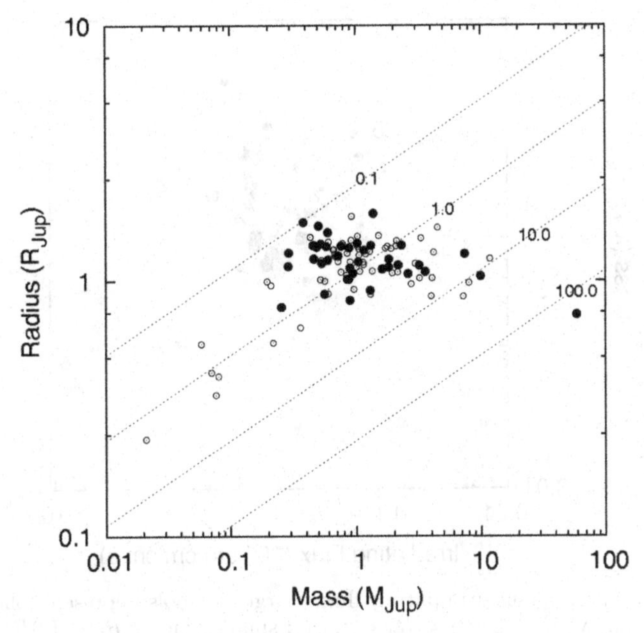

Figure 1. Planet radius versus planet mass. Dashed lines denote contours of constant density labelled in Jovian units. Filled symbols denote WASP planets. While there is a substantial spread in radii at each mass, the lower envelope attains its greatest radius at masses close to that of Jupiter.

Fressin *et al.* assumed that 0.5% of the incoming power was dissipated in the planetary core. In these models, the resulting inflation of the planetary radius increases dramatically with decreasing mass. Enoch *et al.* (2010) demonstrated that a strong correlation between planet radius and irradiating flux is indeed present among transiting planets in the mass range between 0.4 and 0.7 M_{Jup}.

As Fig. 2 shows, the most inflated gas-giant planets at a given mass tend also to be the most strongly irradiated. There appears to be a sharp upper boundary to the irradiating flux, beyond which no planets are found. The critical irradiating flux at the boundary is mass-dependent, with more massive planets apparently being able to sustain greater irradiating fluxes. Baraffe *et al.* (2004) predicted that the response of a gas-giant planet to mass loss driven by strong irradiation could lead to catastrophic evaporation of the gaseous planetary envelope. The runaway occurs when the evaporation timescale $m/\dot{M}$ becomes significantly shorter than the Kelvin-Helmholtz timescale of the envelope.

4. Orbital eccentricities

Among exoplanets in general, high orbital eccentricities tend to be the rule rather than the exception. At the small orbital separations less than about 0.07 AU that favour the discovery of transiting planets, however, low orbital eccentricities are found to be the norm. Planets found in radial-velocity surveys at these small separations also include a significant number of objects with orbital eccentricities indistinguishable from zero.

Among these close-orbiting planets, however, a substantial minority are reported to show significant orbital eccentricities. Some caution must be exercised in interpreting the apparent significance of eccentricities derived from radial-velocity curves. When modelling RV curves it is common practice to use $h = e\cos\omega$ and $k = e\sin\omega$ as fitting parameters, since these two parameters are far less strongly correlated than e and ω

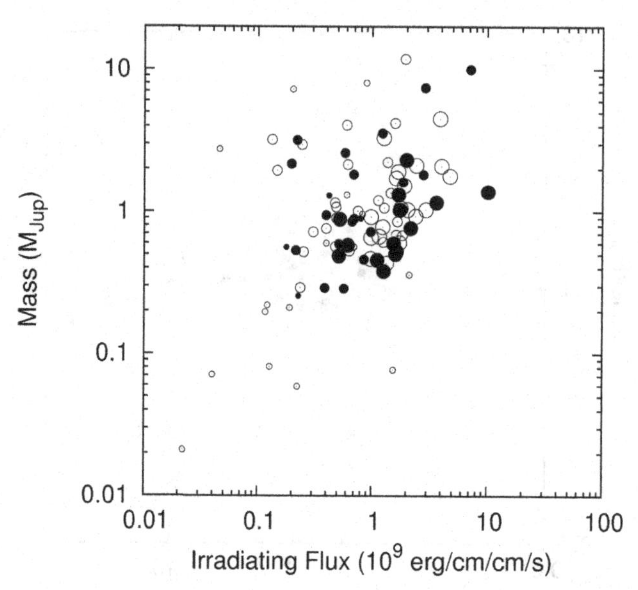

Figure 2. Planet mass versus irradiating flux. Large symbols represent planets with radii $R_p > 1.3R_J$; medium symbols $1.0R_J < R_p < 1.3R_J$; small symbols $R_p < 1.0R_J$. Filled symbols denote WASP planets. The most inflated gas-giant planets cluster against a sharp upper limit to the irradiating flux, beyond which no gas-giant planets are found at a given mass.

(Ford 2006). As Ford points out in the same paper, however, this convenience carries a price. If the mass of the planet is low, the amplitude K of the stellar reflex orbit may be little greater than the uncertainty on a single RV observation. Even if the reality of the orbit is established beyond question, the eccentricity may be poorly determined. Any asymmetry in the distribution of the RV observations around the orbit can then produce a spurious eccentricity. This effect is seen clearly in the right-hand panel of Fig. 3: both the eccentricity and its uncertainty increase markedly with decreasing radial-velocity amplitude. Ford notes that when the eccentricity is poorly constrained, the use of h and k as fitting parameters in a Markov-chain Monte-Carlo fitting algorithm implicitly imposes a prior on e that is linearly proportional to e. This is readily understood by considering the areas of annuli of radius e and width de in (h, k) space where the algorithm is executing a random walk.

Binary-star orbit modelling suffers from the same problem, as was discussed nearly forty years ago by Lucy & Sweeney (1971). They devised a simple F-test approach to determining the "false-alarm" probability that the fitted eccentricity would exceed a given value by chance if the true orbit is circular. As a rule of thumb, Lucy & Sweeney calculate that a circular orbit will yield a spurious detection of 2.45σ or more, with a probability of 5 percent. This is a best-case scenario, assuming the observations to be uniformly distributed in phase. To be on the safe side it is always advisable to examine carefully the improvement in the fit that results from fitting an eccentric as opposed to a circular orbit. Several simple but powerful statistical tools are available to achieve this. The Lucy-Sweeney test and the Bayesian Information Criterion e.g. Liddle 2007) are both effective at determining the true level of significance of an eccentricity detection arising from a given data set, in terms of the number of data points and the improvement in χ^2 produced by the introduction of two additional free parameters in the model.

So which eccentricity determinations is it safe to believe? Among the 30 WASP planets currently listed at exoplanet.eu, only WASP-8b, WASP-14 and HAT-P-14b(=WASP-27b)

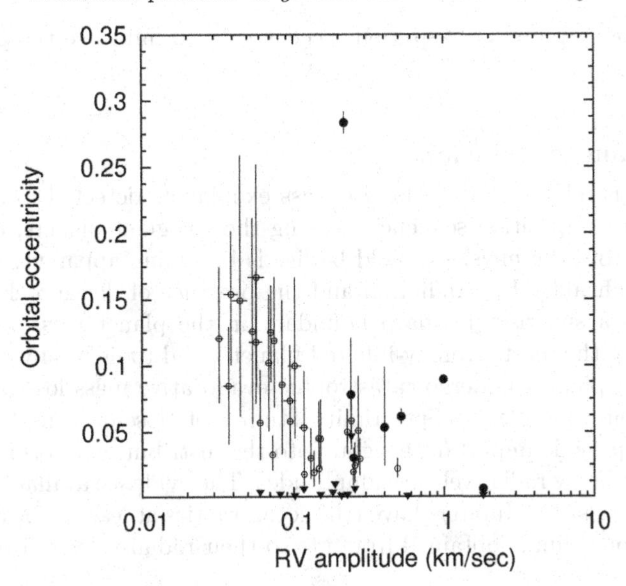

Figure 3. Fitted orbital eccentricity versus orbital velocity semi-amplitude K. Filled dots denote fits for which the Lucy-Sweeney test yields a probability less than 5% that the improvement in the fit for an eccentric orbit could have arisen by chance from an underlying circular orbit. Filled inverted triangles denote upper limits on $e\cos\omega$ derived from secondary-eclipse timing.

have orbital eccentricities that have a greater than 97% chance of being significantly eccentric on the basis of the original discovery data. Even the massive, close-orbiting WASP-18b (Hellier 2009) has an eccentricity which, despite appearing significant at the 3.3-σ level in the discovery data, has a 17% chance of being spurious according to the Lucy-Sweeney test. Subsequent RV observations by Triaud *et al.* (2010), however, reduce the Lucy-Sweeney false-alarm probability to 2.7%, showing WASP-18b to have a small but genuinely significant $e\sin\omega$.

In other cases where eccentricities of marginal significance have been detected, *SPITZER* and ground-based secondary-eclipse mid-times and durations place tight constraints on $e\cos\omega$ and to a lesser extent $e\sin\omega$. Nymeyer *et al.* (2010), for example, find $e\cos\omega = 0.0002\pm0.0005$ for WASP-18b. WASP-1b, WASP-2b have $e\cos\omega$ less than 0.003 and 0.004 respectively Wheatley *et al.* (2010). Campo *et al.* (2011) obtain $e\cos\omega = 0.0016\pm0.0007$ for WASP-12b, while admitting $e\sin\omega = 0.063\pm0.014$ constrained mainly by spectroscopic radial velocities. The Lucy-Sweeney test indicates a 17% chance that this component too could be spurious. Other published secondary-eclipse timings from *SPITZER* reveal insignificant displacements of secondary eclipse from phase 0.5 in HD 209458b (Deming *et al.* 2005), HAT-P-1b (Todorov *et al.* 2010) and TrES-2b (O'Donovan *et al.* 2010), a small but marginally significant displacement in CoRoT-2b (Gillon *et al.* 2010) and a clear confirmation of the orbital eccentricity of GJ436b (Deming *et al.* 2007) .

Orbital eccentricity has the potential to bias our estimates of stellar masses and radii. The total duration, depth and shape of the transit profile determine the planet's radius from the directly-measured stellar density and an estimate of the stellar mass (Seager & Mallén-Ornelas 2003). The estimate of the scaled stellar radius R_*/a scales as $1 + e\sin\omega$ to first order. If transit occurs near periastron, when the planet is travelling fastest, the stellar and planet radii will be underestimated by this factor. The secondary-eclipse timings obtained so far thus suggest that it is generally safer to assume a circular orbit than to allow a potentially-spurious eccentricity to influence the planetary radius estimate.

5. Summary and conclusions

The mass-radius relation for the lowest-mass exoplanets detected from the ground is consistent with a composition sequence. Among the gas-giant planets, the turn-over in radius expected from the physics of cold bodies is becoming apparent, but the relation is broadened significantly by irradiation and the presence of dense rocky cores in some planets. There is a surprisingly sharp boundary in the planet mass - irradiating flux plane, along which the most strongly-inflated planets tend to lie at any given mass. This may indicate that planets undergo catastrophic evaporative mass loss beyond a critical level of irradiation. Secondary-eclipse timing studies of close-orbiting low-mass planets are beginning to provide important insights into the distribution of orbital eccentricities among planets with low radial-velocity amplitudes. The evidence available so far suggests that the orbits of most hot Jupiters have true eccentricities at least an order of magnitude lower than the upper limits obtained from fits to their radial-velocity orbits.

References

Alonso, R., *et al.* 2004, *ApJ*, 613, L153

Bakos, G., Noyes, R. W., Kovács, G., Stanek, K. Z., Sasselov, D. D., & Domsa, I. 2004, *PASP*, 116, 266

Baraffe, I., Selsis, F., Chabrier, G., Barman, T. S., Allard, F., Hauschildt, P. H., & Lammer, H. 2004, *A&A*, 419, L13

Campo, C. J., *et al.* 2011, *ApJ*, 727, 125

Deming, D., Seager, S., Richardson, L. J., & Harrington, J. 2005, *Nature*, 434, 740

Deming, D., Harrington, J., Laughlin, G., Seager, S., Navarro, S. B., Bowman, W. C., & Horning, K. 2007, *ApJ*, 667, L199

Enoch, B., *et al.* 2011, *MNRAS*, 410, 1631

Ford, E. B. 2006, *ApJ*, 642, 505

Fressin, F., Guillot, T., Morello, V., & Pont, F. 2007, *A&A*, 475, 729

Gillon, M., *et al.* 2010, *A&A*, 511, A3

Hellier, C., *et al.* 2009, *Nature*, 460, 1098

Liddle, A. R. 2007, *MNRAS*, 377, L74

Lucy, L. B. & Sweeney, M. A. 1971, *AJ*, 76, 544

McCullough, P. R., Stys, J. E., Valenti, J. A., Fleming, S. W., Janes, K. A., & Heasley, J. N. 2005, *PASP*, 117, 783

Nymeyer, S., *et al.* 2010, *arXiv:1005.1017*

O'Donovan, F. T., Charbonneau, D., Harrington, J., Madhusudhan, N., Seager, S., Deming, D., & Knutson, H. A. 2010, *ApJ*, 710, 1551

Pollacco, D. L., *et al.* 2006, *PASP*, 118, 1407

Robin, A. C., Reylé, C., Derrière, S., & Picaud, S. 2003, *A&A*, 409, 523

Seager, S. & Mallén-Ornelas, G. 2003, *ApJ*, 585, 1038

Seager, S., Kuchner, M., Hier-Majumder, C. A., & Militzer, B. 2007, *ApJ*, 669, 1279

Todorov, K., Deming, D., Harrington, J., Stevenson, K. B., Bowman, W. C., Nymeyer, S., Fortney, J. J., & Bakos, G. A. 2010, *ApJ*, 708, 498

Triaud, A. H. M. J., *et al.* 2010, *A&A*, 524, A25

Udalski, A., *et al.* 2002, *AcA*, 52, 1

Wheatley, P. J., *et al.* 2010, *arXiv:1004.0836*

Zapolsky, H. S. & Salpeter, E. E. 1969, *ApJ*, 158, 809

The Astrophysics of Planetary Systems: Formation, Structure, and Dynamical Evolution
Proceedings IAU Symposium No. 276, 2010
A. Sozzetti, M. G. Lattanzi & A. P. Boss, eds.

doi:10.1017/S1743921311020072

The spectra of low-temperature atmospheres: Lessons learned from brown dwarfs

Adam J. Burgasser [1]

[1]University of California, San Diego 9500 Gilman Drive La Jolla, CA 92093, USA
email: aburgasser@ucsd.edu

Abstract. Indirect and direct spectroscopic studies of exoplanets are beginning to probe the most prominent chemical constituents and processes in their atmospheres. However, studies of equivalently low-temperature brown dwarfs have been taking place for over a decade. In this review, I summarize some of the results of detailed spectroscopic, photometric and polarimetric studies of brown dwarfs of various effective temperatures, surface gravities and metallicities, highlighting the insight gained into the chemistry and cloud formation of planetary-like atmospheres. Nonequilibrium chemistry and variations in cloud properties are singled out as critical ingredients for interpreting exoplanet spectra. I also discuss recent direct spectroscopic studies of exoplanet atmospheres, both close to and widely-separated from their host star, and propose that the latter are better analogs to isolated brown dwarfs.

Keywords. astrochemistry, stars: atmospheres, stars: low-mass, brown dwarfs, planetary systems

1. Introduction

Our understanding of exoplanet atmospheres is proceeding rapidly, thanks to the fairly recent development of indirect and direct photometric and spectroscopic observational studies of these objects.† Reflectance spectroscopy of transiting planets has revealed key molecular constituents, temperature inversions, and the presence of equatorial jets on the dayside hemispheres of transiting planets (e.g., Charbonneau *et al.* 2005; Deming *et al.* 2005; Burrows *et al.* 2007; Grillmair *et al.* 2007; Knutson *et al.* 2007; Swain *et al.* 2009). Transmission spectroscopy has detected atomic and molecular constituents, hazes, and evaporating exospheres along the termini of transiting exoplanets (e.g., Charbonneau *et al.* 2002; Vidal-Madjar *et al.* 2003; Tinetti *et al.* 2007; Swain *et al.* 2008). Most recently, the detection of young exoplanets through high-contrast imaging has opened up opportunities to directly measure emergent spectral energy distributions, atmospheric compositions and temperatures (e.g., Bowler *et al.* 2010; Hinz *et al.* 2010). The novelty of these investigations have been their ability to overcome the extreme contrast ratio between planet and host star, although results are not without controversy (e.g., Gibson *et al.* 2010).

More easily detectable analogs of warm exoplanets are the brown dwarfs. These "failed stars", with masses insufficient to sustain core hydrogen fusion ($M \lesssim 0.072\ M_{\odot}$; Kumar 1962; Hayashi & Nakano 1963), cool to the point at which their atmospheres have similar temperatures as those of exoplanets. Since their discovery in 1995‡, well over 500 brown

† For recent reviews, see Tinetti & Beaulieu (2009) and Lopez-Morales (2010); also see contributions from M. Swain, G. Tinetti, B. Croll, A. M. Mandell in these proceedings.

‡ On a historical note, the first examples of a hot Jovian planet, 51 Peg, and a cool brown dwarf, Gliese 229B, were reported at the same conference, the 9th Cambridge Workshop on Cool Stars, Stellar Systems and the Sun in Florence, Italy.

dwarfs have been uncovered, mostly isolated sources in the vicinity of the Sun. These objects have temperatures as cool as ≈500 K (e.g., Lucas *et al.* 2010), ages ranging from ~1 Myr newborns to ~10 Gyr halo objects, and masses extending from the hydrogen burning limit to below the deuterium burning limit (~13 M_{Jup}; Chabrier & Baraffe 2000). Their varied and complex spectral energy distributions have given rise to three new spectral classes: L dwarfs, T dwarfs and the still putative Y dwarfs (see Kirkpatrick 2005 for a recent review). Most relevant to this review, the inferred photospheric parameters of known brown dwarfs—pressure, temperature, and to limited degree composition—are equivalent to those for the majority of known exoplanets. As a result, the spectra of exoplanets and brown dwarfs, and the underlying chemistry and atmospheric processes that shape these spectra, are expected to have important similarities.

2. Brown Dwarf Photospheric Gas Chemistry

The spectra of brown dwarfs are characterized by the forest of atomic and molecular absorption features that reflect their chemistry-rich atmospheres. The evolution of a cooling brown dwarf is matched by the evolution of photospheric gas species, from the strong metal oxide bands (TiO, VO, CO) present in young M dwarfs, to the emergence of metal hydrides (FeH, CrH, MgH, and CaH) and neutral alkali lines in the L dwarfs, to the substantial shift to planetary-like H_2O, CH_4 and NH_3 bands in the T dwarfs; collision-induced H_2 absorption also plays a prominent role in shaping L and T dwarf spectra. Yamamura *et al.* (2010) has recently added CO_2 to this list, detected in the 4.0–4.5 μm spectroscopy of T dwarfs with AKARI. The neutral alkalis play a surprisingly prominent role in shaping cool brown dwarf spectra, through the heavily pressure-broadened Na I and K I doublet lines whose 1000 Å-wide wings effectively suppress all optical emission from these sources (e.g., Burrows & Volobuyev 2003). Hydrogen sulfide (H_2S), phospine (PH_3) and more complex organic molecules are also expected to be present in brown dwarf atmospheres but have not yet been detected. Heavier metals such as Mg, Al, SI, and Fe end up in condensate grain particles in the L dwarfs, and eventually settle below the visible photosphere (Visscher *et al.* 2010, see next section)..

These species are predicted from equilibrium chemistry calculations, but in a few cases the observed abundances are skewed by non-equilibrium chemistry (Griffith & Yelle 1999; Saumon *et al.* 2003; Hubeny & Burrows 2007). Chemical conversions between molecular species have finite rates that are both temperature and pressure dependent. Reactions with long chemical timescales can be slow to adjust to changes in temperature and pressure that accompany vertical mixing in a turbulent photosphere, resulting in locally non-equilibrium abundances. Two key reactions studied in brown dwarf atmospheres are

$$CO + 3H_2 \longleftrightarrow CH_4 + H_2O \tag{2.1}$$

$$N_2 + 3H_2 \longleftrightarrow 2NH_3. \tag{2.2}$$

In both reactions, the products are favored at lower temperatures, but only after breaking double and triple bonds in the reactants. As a result, deep hot gas mixed upward into the cooler photosphere retains enhanced CO and reduced CH_4, NH_3 and H_2O abundances with respect to thermodynamic equilibrium (the symmetric diatom N_2 does not produce a distinct band), with measurable consequences (Noll *et al.* 1997; Saumon *et al.* 2007; Figure 1). Nonequilibrium chemistry appears to be universal in L and T dwarf atmospheres, and hence important for exoplanet spectra (e.g., Fortney *et al.* 2008b).

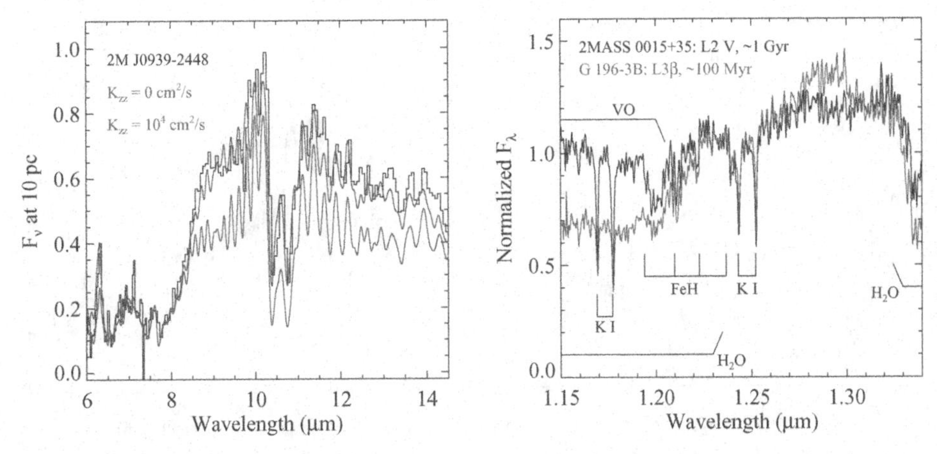

Figure 1. (Left): The mid-infrared spectrum of the T dwarf 2MASS J0939-2448 (black line), compared to two T_{eff} = 700 K, $\log g$ = 5.0 (cgs) from Saumon & Marley (2008), one assuming equilibrium chemistry (red line) and one assuming non-equilibrium chemistry (blue line, with vertical circulation parameterized by a diffusion coefficient K_{zz}). The latter produces weaker NH_3 absorption and an overall better fit to the observational data. (Rght): *J*-band spectra of two early-type L dwarfs, a ~1 Gyr field source and a ~100 Myr companion to a nearby young star (G 196-3B; see contribution by M. R. Zapatero Osorio). The spectral deviations seen in the younger source–enhanced VO absorption, weakened alkalis–are a subset of the gravity-sensitive features related to gas pressure effects in brown dwarf photospheres.

Chemical abundances are also modulated by gas pressure and metallicity. Photospheric gas pressure is a proxy for surface gravity, which in turn depends on the mass and age of a brown dwarf. As such, young, low-mass brown dwarfs with low surface gravities exhibit distinct spectral peculiarities, including redder near-infrared colors arising from reduced collision-induced H_2 absorption (e.g., Kirkpatrick *et al.* 2006; Allers *et al.* 2007), weakened alkali lines and narrower line wings (e.g., Martín *et al.* 1999; Cruz *et al.* 2009), and modified chemistry arising from pressure effects in conversion reactions†. Metallicity effects can both mimic photospheric gas pressure effects, since $P_{phot} \propto g/\kappa_R$, where g is the surface gravity and κ_R the metallicity-dependent Rosseland mean opacity; and change the overall chemistry through compositional variations. These effects have been studied in a handful of metal-poor ([M/H] = -1...-2) halo L dwarfs (Burgasser *et al.* 2005), which exhibit increased absorption from pressure-sensitive species (H_2, alkalis), higher ratios of metal hydride to metal oxide absorption, and decreased condensate cloud production (Burgasser *et al.* 2003, 2007; Gizis & Harvin 2006; Reiners & Basri 2006). Modestly metal-rich brown dwarfs have been identified as companions to metal-rich stars, and exhibit subtle spectral discrepancies, including a possible enhancement in condensate opacity (e.g., Looper *et al.* 2008).

3. Brown Dwarf Clouds

Equilibrium chemistry at low temperatures includes the formation of condensate (solid and liquid) molecular species. In brown dwarfs, these species appear to be constrained to discrete "cloud" layers by competing processes of vertical upwelling, grain growth and gravitational settling (Ackerman & Marley 2001; Cooper *et al.* 2003; Woitke & Helling

† Both reactions listed above are pressure sensitive, with low pressures favoring the reactants. As a result, the onset of CH_4 absorption, signifying the start of the T dwarf sequence, occurs at lower temperatures for lower surface gravity brown dwarfs—and exoplanets.

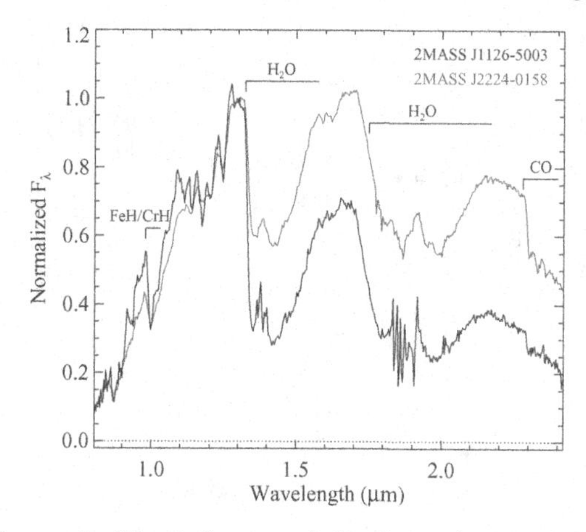

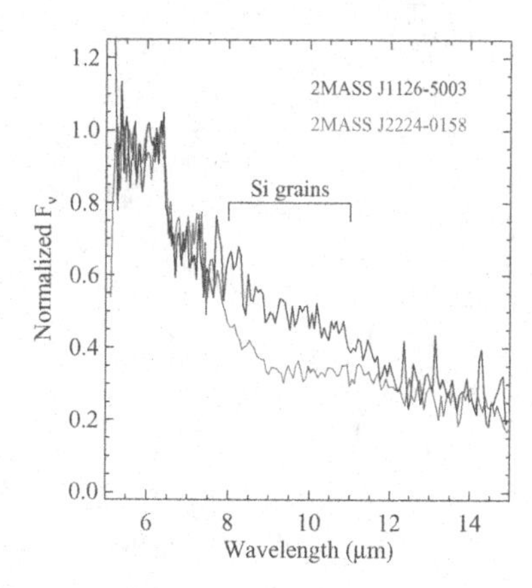

Figure 2. The influence of clouds on brown dwarf spectra. (Left): Near-infrared spectra of two equivalently (optically) classified L dwarfs, illustrating the reddening effects of condensate opacity. The spectra are normalized at 1.25 μm, with the "cloudier" object (2MASS J2224-0158) indicated in red. (Right): Mid-infrared spectra of the same two L dwarfs, highlighting the 8-11 μm silicate grain feature in the cloudier source. These spectra are normalized at 6 μm.

2004; Tsuji 2005). There is abundant evidence of for the presence of condensates in brown dwarf atmospheres, including the red near-infrared colors of late-type M and L dwarfs (e.g., Tsuji *et al.* 1996; Marley *et al.* 2002), the disappearance of gas precursors of condensate species (e.g., Kirkpatrick *et al.* 1999; Lodders 2002), the retention of alkali species (e.g., Lodders 1999; Burrows & Sharp 1999), "muted" near-infrared H_2O bands resulting from competing condensate grain opacity (e.g., Jones & Tsuji 1997) the direct detection of the 8-11 μm silicate grain feature (Cushing *et al.* 2006), and spectral model fits indicating the presence of condensate opacity (e.g., Cushing *et al.* 2008; Stephens *et al.* 2009; see Figure 2).

The opacity effects of condensates are most prominent in the infrared spectra of L dwarfs, where they cap the JHK flux peaks, minima in gas opacity (Ackerman & Marley 2001). However, the properties of L dwarf clouds are not universal. Significant source-to-source variations in near-infrared colors (by up to 1.5 mag) and spectral morphology among equivalently classified L dwarfs have been attributed in part to differences in cloud structure and/or particle sizes (Knapp *et al.* 2004; Burgasser *et al.* 2008; Marley *et al.* 2010, Figure 2). Temporal variability observed from magnetically inactive L dwarfs has been attributed to clouds, specifically rotational modulation of cloud structures (e.g., Goldman 2005; Artigau *et al.* 2009; see contribution by J. Radigan). The oblate cloudy atmospheres of rapid rotating L dwarfs can also explain observed linear polarizations of up to 2.5% (Sengupta & Krishan 2001; Ménard *et al.* 2002; Zapatero Osorio *et al.* 2005).

Cloud properties also evolve as a brown dwarf cools into a T dwarf. In general, T dwarf spectra are well-matched to "condensate-free" models, in accord with models of condensate clouds that sink below the visible photosphere (e.g., Marley *et al.* 2002; Burrows *et al.* 2006). What is surprising, however, is that the condensate opacity is observed to disappear more rapidly than models predict, as indicated by a remarkable brightening in 1 μm fluxes across the L/T transition (Dahn *et al.* 2002; Tinney *et al.* 2003; Vrba *et al.* 2004). This brightening has been observed in coeval pairs of L and T dwarfs, ruling out sample heterogeneity arguments (Burgasser *et al.* 2006; Liu *et al.* 2006). The driving mechanism for the sudden loss of photospheric condensates, whether by an onset of patchiness in the global cloud layer (Burgasser *et al.* 2002; Marley *et al.* 2010) or a global

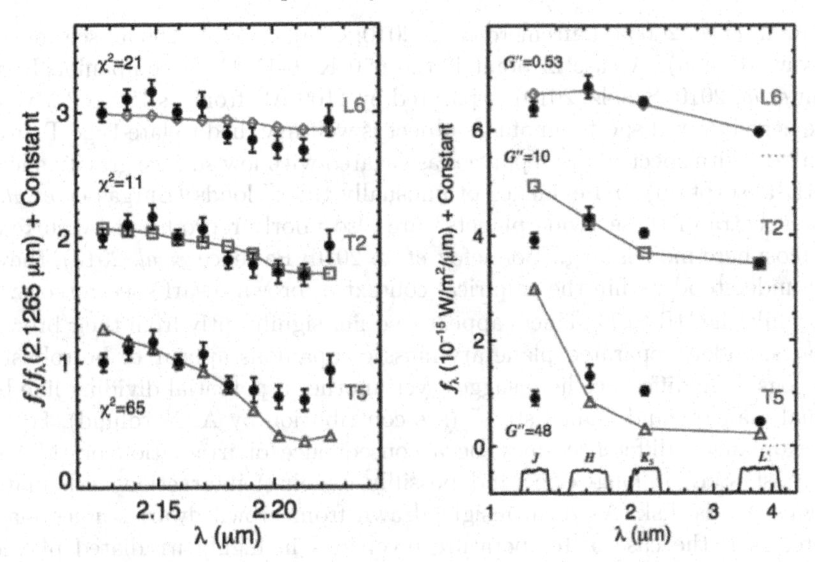

Figure 3. Near-infrared spectrum of the exoplanet HR 8799b, from Bowler *et al.* (2010). (Left): K-band spectrum (circles with error bars) compared to three spectral templates of types L6, T2 and T5. The T2 template provides a reasonable fit. (Right): Broad-band *JHKL* photometry compared to the same templates, now showing a pronounced mismatch with the T2 template.

phase change in grain sedimentation (Knapp *et al.* 2004), remains an open question. Ongoing theoretical and observational studies are also exploring poorly-constrained grain size distributions and compositions, grain surface chemistry, cloud structure and atmospheric dynamics, and non-equilibrium effects in condensate formation (Helling *et al.* 2008b; Witte *et al.* 2009; Freytag *et al.* 2010).

4. Observed Spectra of Exoplanets: Does Distance Matter?

Much of our understanding of brown dwarf atmospheres comes from the comparison of detailed spectroscopic measurements to theoretical atmosphere models, and the ongoing development of those models (e.g., Burrows *et al.* 2002; Saumon & Marley 2008). Examples of both excellent and very poor model fits to observational data can be found in the literature (e.g., Stephens *et al.* 2009), with problems driven primarily by incomplete molecular opacities, inaccurate treatment of cloud physics, and uncertainties in elemental abundances (see Helling *et al.* 2008a and contribution by A. Becker). The recent direct detection of exoplanets around Fomalhaut (Kalas *et al.* 2008), HR 8799 (Marois *et al.* 2008), and β Pictoris (Lagrange *et al.* 2009, 2010, see contribution by A.-M. Lagrange) has opened to door for direct spectroscopic investigations of these objects as well. Two studies have explored the spectroscopic properties of planets in the HR 8799 system: the mid-infrared spectrum of HR 8799c (Janson *et al.* 2010) and the near-infrared spectrum of HR 8799b (Bowler *et al.* 2010)†. Both studies conclude that current models have difficulty reproducing exoplanet spectra, but in the case of HR 8799b there also appears to be major discrepancies with brown dwarf empirical templates (Figure 3). Does this mean that brown dwarfs are *not* useful analogs for exoplanet atmospheres?

There are in fact several widely-separated ($\gtrsim$100 AU), planetary-mass companions to young stars that show excellent agreement with their brown dwarf counterparts

† There are in addition to several multi-band photometric studies of these and other directly imaged planets.

(e.g., Chauvin *et al.* 2005a; Lafrenière *et al.* 2010; Chauvin *et al.* 2005b; see contribution by R. Jayawardhana). A case in point is the 650 K, 6-11 M_{Jup} companion Ross 458C (Goldman *et al.* 2010; Scholz 2010), separated by 1100 AU from its 150-800 Myr host binary. The near-infrared spectrum of this object is well-matched to late-type T dwarf field analogs, albeit with spectral discrepancies associated with low surface gravity effects (i.e., reduced H_2 absorption) and evidence of unusually thick clouds (Burgasser *et al.* 2010). While the spectra of these "wide planets" are also poorly reproduced by current theoretical atmosphere models (e.g., Bonnefoy *et al.* 2010; Patience *et al.* 2010), they are at least well-understood within the empirical context of brown dwarf spectroscopy.

Hence, while the HR 8799 planets appear to differ significantly from their brown dwarf counterparts, widely-separated planetary-mass companions appear to be robust brown dwarf analogs. The difference here suggest yet another a potential dividing line between "traditional planets" and "failed stars" (see contribution by A. Nordlund). For closely-orbiting exoplanets, differences may be a consequence of irradiation or tidal heating from the host star, or long-term (and possibly ongoing) interaction with material in the system's debris disk. As such, insight drawn from brown dwarf studies may prove incomplete, as is the case of temperature inversions in highly irradiated planets (e.g. Fortney *et al.* 2008a). Nevertheless, brown dwarf spectroscopy provides a "zeroth-order" empirical calibration for models to which additional physics can be added. Given the rich diversity of brown dwarf atmospheres already observed, it is clear that there is even more to come as exoplanet spectroscopy becomes a more widely employed technique.

References

Ackerman, A. S. & Marley, M. S. 2001, *ApJ*, 556, 872

Allers, K. N., Jaffe, D. T., Luhman, K. L., Liu, M. C., Wilson, J. C., Skrutskie, M. F., Nelson, M., Peterson, D. E., *et al.* 2007, *ApJ*, 657, 511

Artigau, É., Bouchard, S., Doyon, R., & Lafrenière, D. 2009, *ApJ*, 701, 1534

Bonnefoy, M., Chauvin, G., Rojo, P., Allard, F., Lagrange, A., Homeier, D., Dumas, C., & Beuzit, J. 2010, *A&A*, 512, A52+

Bowler, B. P., Liu, M. C., Dupuy, T. J., & Cushing, M. C. 2010, *ApJ*, 723, 850

Burgasser, A. J., Cruz, K. L., Kirkpatrick, J. D. 2007, *ApJ*, 657, 494

Burgasser, A. J., Kirkpatrick, J. D., Burrows, A., Liebert, J., Reid, I. N., Gizis, J. E., McGovern, M. R., Prato, L., *et al.* 2003, *ApJ*, 592, 1186

Burgasser, A. J., Kirkpatrick, J. D., Cruz, K. L., Reid, I. N., Leggett, S. K., Liebert, J., Burrows, A., & Brown, M. E. 2006, *ApJS*, 166, 585

Burgasser, A. J., Kirkpatrick, J. D., & Lépine, S. 2005, in ESA Special Publication, Vol. 560, 13th Cambridge Workshop on Cool Stars, Stellar Systems and the Sun, ed. F. Favata & *et al.*, 237–+

Burgasser, A. J., Looper, D. L., Kirkpatrick, J. D., Cruz, K. L., & Swift, B. J. 2008, *ApJ*, 674, 451

Burgasser, A. J., Marley, M. S., Ackerman, A. S., Saumon, D., Lodders, K., Dahn, C. C., Harris, H. C., *et al.* 2002, *ApJL*, 571, L151

Burgasser, A. J., Simcoe, R. A., Bochanski, J. J., Saumon, D., Mamajek, E. E., Cushing, M. C., Marley, M. S., McMurtry, C., *et al.* 2010, *ApJ*, 725, 1405

Burrows, A., Burgasser, A. J., Kirkpatrick, J. D., Liebert, J., Milsom, J. A., Sudarsky, D., & Hubeny, I. 2002, *ApJ*, 573, 394

Burrows, A., Hubeny, I., Budaj, J., Knutson, H. A., & Charbonneau, D. 2007, *ApJL*, 668, L171

Burrows, A. & Sharp, C. M. 1999, *ApJ*, 512, 843

Burrows, A., Sudarsky, D., & Hubeny, I. 2006, *ApJ*, 640, 1063

Burrows, A. & Volobuyev, M. 2003, *ApJ*, 583, 985

Chabrier, G. & Baraffe, I. 2000, *ARA&A*, 38, 337

Charbonneau, D., Allen, L. E., Megeath, S. T., Torres, G., Alonso, R., Brown, T. M., Gilliland, R. L., Latham, D. W., *et al.* 2005, *ApJ*, 626, 523
Charbonneau, D., Brown, T. M., Noyes, R. W., & Gilliland, R. L. 2002, *ApJ*, 568, 377
Chauvin, G., Lagrange, A.-M., Dumas, C., Zuckerman, B., Mouillet, D., Song, I., Beuzit, J.-L., & Lowrance, P. 2005a, *A&A*, 438, L25
Chauvin, G., Lagrange, A.-M., Zuckerman, B., Dumas, C., Mouillet, D., Song, I., Beuzit, J.-L., Lowrance, P., *et al.* 2005b, *A&A*, 438, L29
Cooper, C. S., Sudarsky, D., Milsom, J. A., Lunine, J. I., & Burrows, A. 2003, *ApJ*, 586, 1320
Cruz, K. L., Kirkpatrick, J. D., & Burgasser, A. J. 2009, *AJ*, 137, 3345
Cushing, M. C., Marley, M. S., Saumon, D., Kelly, B. C., Vacca, W. D., Rayner, J. T., Freedman, R. S., Lodders, K., *et al.* 2008, *ApJ*, 678, 1372
Cushing, M. C., Roellig, T. L., Marley, M. S., Saumon, D., Leggett, S. K., Kirkpatrick, J. D., Wilson, J. C., Sloan, G. C., *et al.* 2006, *ApJ*, 648, 614
Dahn, C. C., Harris, H. C., Vrba, F. J., Guetter, H. H., Canzian, B., Henden, A. A., Levine, S. E., Luginbuhl, C. B., *et al.* 2002, *AJ*, 124, 1170
Deming, D., Seager, S., Richardson, L. J., & Harrington, J. 2005, *Nature*, 434, 740
Fortney, J. J., Lodders, K., Marley, M. S., & Freedman, R. S. 2008a, *ApJ*, 678, 1419
Fortney, J. J., Marley, M. S., Saumon, D., & Lodders, K. 2008b, *ApJ*, 683, 1104
Freytag, B., Allard, F., Ludwig, H., Homeier, D., & Steffen, M. 2010, *A&A*, 513, A19+
Gibson, N. P., Pont, F., & Aigrain, S. 2010, ArXiv e-prints
Gizis, J. E. & Harvin, J. 2006, *AJ*, 132, 2372
Goldman, B. 2005, Astronomische Nachrichten, 326, 1059
Goldman, B., Marsat, S., Henning, T., Clemens, C., & Greiner, J. 2010, *MNRAS*, 405, 1140
Griffith, C. A. & Yelle, R. V. 1999, *ApJL*, 519, L85
Grillmair, C. J., Charbonneau, D., Burrows, A., Armus, L., Stauffer, J., Meadows, V., Van Cleve, J., & Levine, D. 2007, *ApJL*, 658, L115
Hayashi, C. & Nakano, T. 1963, Progress of Theoretical Physics, 30, 460
Helling, C., Ackerman, A., Allard, F., Dehn, M., Hauschildt, P., Homeier, D., Lodders, K., Marley, M., *et al.* 2008a, *MNRAS*, 391, 1854
Helling, C., Dehn, M., Woitke, P., & Hauschildt, P. H. 2008b, *ApJL*, 675, L105
Hinz, P. M., Rodigas, T. J., Kenworthy, M. A., Sivanandam, S., Heinze, A. N., Mamajek, E. E., & Meyer, M. R. 2010, *ApJ*, 716, 417
Hubeny, I. & Burrows, A. 2007, *ApJ*, 669, 1248
Janson, M., Bergfors, C., Goto, M., Brandner, W., & Lafrenière, D. 2010, *ApJL*, 710, L35
Jones, H. R. A. & Tsuji, T. 1997, *ApJL*, 480, L39+
Kalas, P., Graham, J. R., Chiang, E., Fitzgerald, M. P., Clampin, M., Kite, E. S., Stapelfeldt, K., Marois, C., *et al.* 2008, Science, 322, 1345
Kirkpatrick, J. D. 2005, *ARA&A*, 43, 195
Kirkpatrick, J. D., Barman, T. S., Burgasser, A. J., McGovern, M. R., McLean, I. S., Tinney, C. G., & Lowrance, P. J. 2006, *ApJ*, 639, 1120
Kirkpatrick, J. D., Reid, I. N., Liebert, J., Cutri, R. M., Nelson, B., Beichman, C. A., Dahn, C. C., Monet, D. G., *et al.* 1999, *ApJ*, 519, 802
Knapp, G. R., Leggett, S. K., Fan, X., Marley, M. S., Geballe, T. R., Golimowski, D. A., Finkbeiner, D., Gunn, J. E., *et al.* 2004, *AJ*, 127, 3553
Knutson, H. A., Charbonneau, D., Allen, L. E., Fortney, J. J., Agol, E., Cowan, N. B., Showman, A. P., Cooper, C. S., *et al.* 2007, *Nature*, 447, 183
Kumar, S. S. 1962, *AJ*, 67, 579
Lafrenière, D., Jayawardhana, R., & van Kerkwijk, M. H. 2010, *ApJ*, 719, 497
Lagrange, A., Bonnefoy, M., Chauvin, G., Apai, D., Ehrenreich, D., Boccaletti, A., Gratadour, D., Rouan, D., *et al.* 2010, Science, 329, 57
Lagrange, A., Gratadour, D., Chauvin, G., Fusco, T., Ehrenreich, D., Mouillet, D., Rousset, G., Rouan, D., *et al.* 2009, *A&A*, 493, L21
Liu, M. C., Leggett, S. K., Golimowski, D. A., Chiu, K., Fan, X., Geballe, T. R., Schneider, D. P., & Brinkmann, J. 2006, *ApJ*, 647, 1393
Lodders, K. 1999, *ApJ*, 519, 793

—. 2002, *ApJ*, 577, 974

Looper, D. L., Kirkpatrick, J. D., Cutri, R. M., Barman, T., Burgasser, A. J., Cushing, M. C., Roellig, T., McGovern, M. R., *et al.* 2008, *ApJ*, 686, 528

Lopez-Morales, M. 2010, ArXiv e-prints

Lucas, P. W., Tinney, C. G., Burningham, B., Leggett, S. K., Pinfield, D. J., Smart, R., Jones, H. R. A., Marocco, F., *et al.* 2010, *MNRAS*, 408, L56

Marley, M. S., Saumon, D., & Goldblatt, C. 2010, *ApJL*, 723, L117

Marley, M. S., Seager, S., Saumon, D., Lodders, K., Ackerman, A. S., Freedman, R. S., & Fan, X. 2002, *ApJ*, 568, 335

Marois, C., Macintosh, B., Barman, T., Zuckerman, B., Song, I., Patience, J., Lafrenière, D., & Doyon, R. 2008, Science, 322, 1348

Martín, E. L., Delfosse, X., Basri, G., Goldman, B., Forveille, T., & Zapatero Osorio, M. R. 1999, *AJ*, 118, 2466

Ménard, F., Delfosse, X., & Monin, J.-L. 2002, *A&A*, 396, L35

Noll, K. S., Geballe, T. R., & Marley, M. S. 1997, *ApJL*, 489, L87+

Patience, J., King, R. R., de Rosa, R. J., & Marois, C. 2010, *A&A*, 517, A76+

Reiners, A. & Basri, G. 2006, *AJ*, 131, 1806

Saumon, D. & Marley, M. S. 2008, *ApJ*, 689, 1327

Saumon, D., Marley, M. S., Leggett, S. K., Geballe, T. R., Stephens, D., Golimowski, D. A., Cushing, M. C., Fan, X., *et al.* 2007, *ApJ*, 656, 1136

Saumon, D., Marley, M. S., Lodders, K., & Freedman, R. S. 2003, in IAU Symposium, Vol. 211, Brown Dwarfs, ed. E. Martín, 345–+

Scholz, R. 2010, *A&A*, 515, A92+

Sengupta, S. & Krishan, V. 2001, *ApJL*, 561, L123

Stephens, D. C., Leggett, S. K., Cushing, M. C., Marley, M. S., Saumon, D., Geballe, T. R., Golimowski, D. A., Fan, X., *et al.* 2009, *ApJ*, 702, 154

Swain, M. R., Tinetti, G., Vasisht, G., Deroo, P., Griffith, C., Bouwman, J., Chen, P., Yung, Y., *et al.* 2009, *ApJ*, 704, 1616

Swain, M. R., Vasisht, G., & Tinetti, G. 2008, *Nature*, 452, 329

Tinetti, G. & Beaulieu, J. 2009, in IAU Symposium, Vol. 253, IAU Symposium, 231–237

Tinetti, G., Vidal-Madjar, A., Liang, M., Beaulieu, J., Yung, Y., Carey, S., Barber, R. J., Tennyson, *et al.* 2007, *Nature*, 448, 169

Tinney, C. G., Burgasser, A. J., & Kirkpatrick, J. D. 2003, *AJ*, 126, 975

Tsuji, T. 2005, *ApJ*, 621, 1033

Tsuji, T., Ohnaka, K., & Aoki, W. 1996, *A&A*, 305, L1+

Vidal-Madjar, A., Lecavelier des Etangs, A., Désert, J., Ballester, G. E., Ferlet, R., Hébrard, G., & Mayor, M. 2003, *Nature*, 422, 143

Visscher, C., Lodders, K., & Fegley, B. 2010, *ApJ*, 716, 1060

Vrba, F. J., Henden, A. A., Luginbuhl, C. B., Guetter, H. H., Munn, J. A., Canzian, B., Burgasser, A. J., Kirkpatrick, J. D., *et al.* 2004, *AJ*, 127, 2948

Witte, S., Helling, C., & Hauschildt, P. H. 2009, *A&A*, 506, 1367

Woitke, P. & Helling, C. 2004, *A&A*, 414, 335

Yamamura, I., Tsuji, T., & Tanabé, T. 2010, *ApJ*, 722, 682

Zapatero Osorio, M. R., Caballero, J. A., & Béjar, V. J. S. 2005, *ApJ*, 621, 445

The Astrophysics of Planetary Systems: Formation, Structure, and Dynamical Evolution
Proceedings IAU Symposium No. 276, 2010
A. Sozzetti, M. G. Lattanzi & A. P. Boss, eds.

doi:10.1017/S1743921311020084

New transiting exoplanets from the SuperWASP-North survey

Francesca Faedi[1], Susana C. C. Barros[1], Don Pollacco[1], Elaine K. Simpson[1], James McCormac[1], Victoria Moulds[1], Chris Watson[1], Ian Todd[1], F. Keenan[1], Alan Fitzsimmons[1], Yilen Gómez Maqueo Chew[1] and the WASP Consortium

[1]Astrophysics Research Centre, School of Mathematics & Physics, Queen's University Belfast, University Road, Belfast, BT7 1NN, UK
email: f.faedi@qub.ac.uk

Abstract. The Wide Angle Search for Planet (WASP) project is one of the leading projects in the discovery of transiting exoplanets. We present 1) the current status of the WASP-North survey, 2) our recent exoplanet discoveries, and 3) we exemplify how these results fit into our understanding of transiting exoplanet properties and how they can help to understand exoplanet diversity.

Keywords. planetary systems, techniques: photometric, techniques: radial velocities

---

1. The SuperWASP-North survey

The SuperWASP-North observatory in La Palma consists of 8 cameras each with a Canon 200-mm f/1.8 lens coupled to an Andor e2v 2048×2048 pixel back-illuminated CCD (Pollacco *et al.* 2006). This configuration gives a pixel scale of 13.7″/pixel which corresponds to a field of view of 7.8×7.8 square degrees per camera.

In October 2008, we introduced an electronic focus control and we also started stabilisation of the temperature of the SuperWASP-North camera lenses. Prior to this upgrade, night-time temperature variations affected the focal length of the WASP lenses altering the FWHM of stars. This introduced trends in the data that mimic partial transits, especially at the beginning and end of the night when the temperature variation is more extreme. These effects are not corrected by the SYSREM (Tamuz *et al.* 2005; Cameron *et al.* 2006) and FTA (Kovacs *et al.* 2005) detrending algorithms because they are position-dependent and do not affect all stars in the same manner. To reduce this source of systematic noise, heating strips were placed around each lens so that their temperature is maintained above ambient at 21 degrees. In addition, we significantly improved the focus of each of the lenses, and now can be done remotely. This upgrade has been crucial to improve the signal-to-noise of WASP light-curves, allowing the detection of planets, previously hidden in the RMS scatter of the data (e.g. WASP-38b; Barros *et al.* 2011). Moreover, from January 2008 both instruments, WASP-South and SuperWASP-North, have been monitoring an equatorial region of sky (-20⩽Dec⩽+20) significantly increasing the amount of data collected on each planet candidate. This has been a key element, together with our improvents to the SuperWASP-North system, for the detetion of the most recent WASP planets. The WASP planet detection rate is better than 1 planet for every 6–7 candidates.

1.1. *Follow-up campaign*

We perform our spectroscopic follow-up campaign using the higly pressure-stabilised Echelle spectrographs SOPHIE mounted on the 1.93m telescope of the Observatoire de Haute Provence (Perruchot *et al.* 2008; Bouchy *et al.* 2009), and CORALIE on the 1.2m Swiss Euler telescope at ESO La Silla Observatory in Chile (Baranne *et al.* 1996; Queloz *et al.* 2000; Pepe *et al.* 2002). In addition, we also use the Fiber-Fed Echelle spectrograph (FIES) mounted on the 2.56m Nordic Optical Telescope in La Palma.

We obtain high precision high signal-to-noise transit light-curves of WASP planets using the LCOGT 2.0m Faulkes-North and South telescopes situated on Haleakala, Hawaii, and Siding Springs, Australia, respectively; the robotic 2.0-m Liverpool Telescope (LT) equipped with the RISE frame transfer CCD, located on La Palma (Gibson *et al.* 2008; Steele *et al.* 2008), and finally the Euler-Swiss telescope in La Silla.

2. New SuperWASP discoveries

Here we present an overview of the newly discovered WASP planets.

2.1. *WASP-37b*

WASP-37b is a hot Jupiter planet with a radius of $R_{pl} = 1.16^{+0.07}_{-0.06}$ R_{Jup}, a mass of $M_{pl} = 1.80 \pm 0.17 M_{Jup}$ and an orbital period of P=3.6 days (Simpson *et al.* 2011). It has a high surface gravity compared to other planets with similar orbital periods ($\log g_{pl} = 3.48^{+0.03}_{-0.04}$) see also Southworth (2010). WASP-37b is orbiting a very old (age$\sim$11 Gyr) metal poor ([Fe/H]=-0.4$\pm$0.12 dex) star, firmly in the tail of the metallicity distribution for exoplanet host stars, which is probably a member of the thick disc population. Even if theoretical models of Fortney *et al.* (2007) and Baraffe *et al.* (2008) do not cover the age range of the system, because planetary radii are thought to decrease with age, we find that WASP-37b has an inflated radius and probably no core. This is consistent with the correlation between core mass and metallicity (Guillot *et al.* 2006; Burrows *et al.* 2007). More details on the system, follow-up spectroscopy and photometry are presented in Simpson *et al.* (2011).

2.2. *WASP-38b*

WASP-38b (Barros *et al.* 2011) is a long period (P= 6.87d), massive (2.69$\pm$0.06M_{Jup}) planet in an eccentric orbit (e = 0.031). WASP-38b does not suffer from the radius anomaly mentioned above ($R_{pl} = 1.09 \pm 0.03 R_{Jup}$) and according to Fortney *et al.* (2008) belongs to the 'pL' class of planets, which show no temperature inversion in their atmosphere. WASP-38b is the forth exoplanet with P$>$ 6 days discovered in a ground-based transit survey (the remaining three are WASP-8b, Queloz *et al.* 2010, HAT-P-15b Kovacs *et al.* 2010, and HAT-P-17b Howard *et al.* 2010). The smaller transit probability, the longer duty cycle and transit duration of these systems, coupled with the restricted observing time from a single site make long period systems more challenging to detect. In the case of WASP-38b, it has been crucial to reduce the systematic noise in the WASP light-curve which ultimately allowed the discovery of the planet. However, we note that from radial velocity surveys there appears to be a depletion of planets between $\sim$0.1-1AU (Udry *et al.* 2003). Details of the system, as well as, follow-up spectroscopy, and photometry are presented in Barros *et al.* (2011).

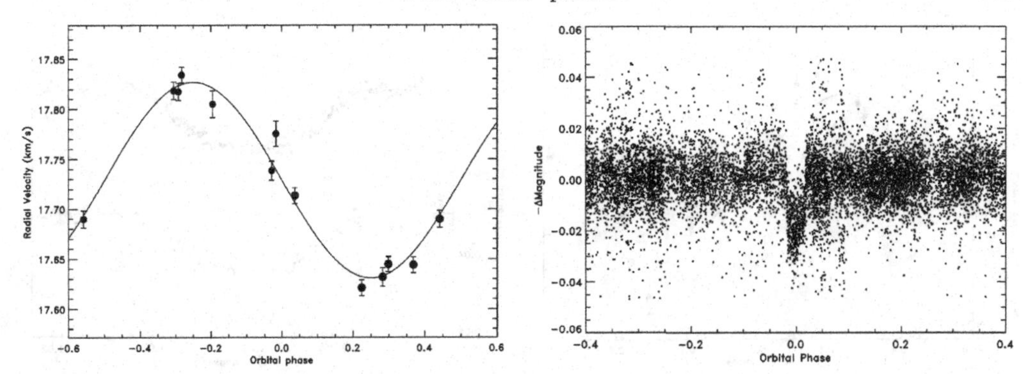

Figure 1. Left-panel, WASP-35b radial velocity curve FIES and CORALIE data. Right-panel WASP discovery light-curve. Follow-up photometry is scheduled for WASP-35b.

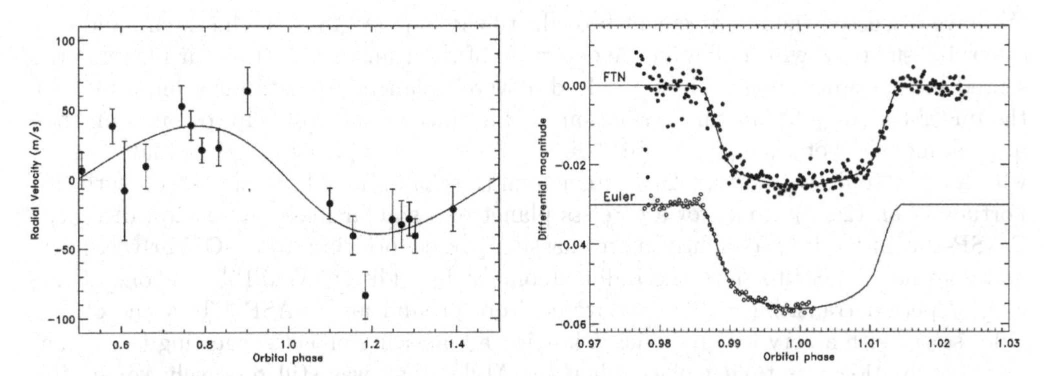

Figure 2. Left-panel, WASP-39b radial velocity curve SOPHIE and CORALIE data. Right-panel, FTN and Euler light-curves.

2.3. *WASP-35b, WASP-39b and WASP-40b*

WASP-35b, WASP-39b and WASP40b are all sub-Jupiter mass planets found in the joint equatorial region of sky observed simultaneously by both WASP telescopes. Figure 1 shows the radial velocity follow-up curve and the WASP discovery photometry of WASP-35b. This is the first WASP planet identified in the joint equatorial region and it is a sub-Jupiter mass planet with an estimated mass of $\sim$0.7M$_{Jup}$ and radius $\sim$1.3R$_{Jup}$ orbiting a slightly metal poor star every 3 days (Enoch *et al.* in prep.). We present our discovery of WASP-39b in Figure 2. This is a Saturn mass planet, orbiting a late G type star every $\sim$4 days. From the data presented above we obtained a preliminary estimate of the planetary mass of $\sim$0.3M$_{Jup}$ and radius of $\sim$1.3R$_{Jup}$. This suggest that WASP-39b is a very low density planet with an inflated radius (Faedi *et al.* in prep).

Finally, in Figure 3 we present the radial velocity and photometry follow-up data of WASP-40b. WASP-40b is orbiting a late G/early K dwarf star with an orbital period of $\sim$3 days, and has an estimated mass of $\sim$0.6M$_{Jup}$ and radius of $\sim$1R$_{Jup}$. The radial velocity residuals of WASP-40b, after subtracting the fitted model, also suggest the presence of an additional signal. However more data are needed before any conclusion can be drawn (West *et al.* in prep).

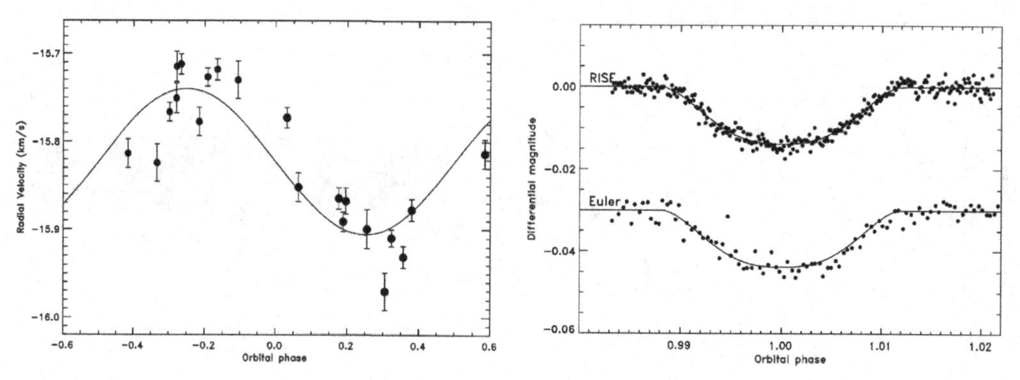

Figure 3. Left-panel, WASP-40b radial velocity curve SOPHIE and CORALIE data. Right-panel, follow-up photometry with RISE and Euler.

3. Conclusion

We have discussed the improvement brought to the SuperWASP-North system, and our observing strategy, which allowed the detection of the transiting extra-solar planets presented in this proceedings. These newly discovered planets are extremely important to the understanding of planetary formation, and evolution, and will help to constrain theoretical models. For example, WASP-39b is a Saturn-mass planet orbiting a late G star with a highly inflated radius, >20% larger than the R_{pl} obtained by comparison with the Fortney *et al.* (2007) model for a coreless planet of a similar mass and orbital distance. WASP-40b instead, is a slightly more massive planet, orbiting a late G/Early K star, and does not appear to show the radius anomaly. In addition, WASP-38b is one of the longest period transiting planet discovered from ground and WASP-37b is one of the older stars, with a very low metallicity, hosting a transiting planet suggesting that giant planet formation was taking place when the Milky Way was still relatively young. To date in the WASP archive there are 150-200 planet candidates which will be followed up in the forecoming seasons. We thus expect, more interesting systems to be dicovered by SuperWASP in the near future.

References

Anderson, D. R., *et al.* 2010, *ApJ*, 709, 159
Baraffe, I., Chabrier, G., & Barman, T. 2008, *A&A*, 482, 315
Baranne, A., Queloz, D., Mayor, M., *et al.* 1996, *A&ASS*, 119, 373
Barros, S. C. C., *et al.* 2011, *A&A*, 525, A54
Bouchy, F., *et al.* 2009 *A&A*, 505, 853
Burrows, A., Hubeny, I., Budaj, J., & Hubbard, W. B. 2007, *ApJ*, 661, 502
Charbonneau, D., Brown, T. M., Noyes, R. W., & Gilliland, R. L. 2002, *ApJ*, 568, 377
Charbonneau, D., Knutson, H. A., Barman, T., Allen, L. E., Mayor, M., Megeath, S. T., Queloz, D., & Udry, S. 2008, *ApJ*, 686, 1341
Collier Cameron, A., *et al.* 2006, *MNRAS*, 373, 799
Fortney, J. J., Marley, M. S., & Barnes, J. W. 2007, *ApJ*, 659, 1661
Fossey, S. J., Waldmann, I. P., & Kipping, D. M. 2009, *MNRAS*, 396, L16
Hebb, L., *et al.* 2010, *ApJ*, 708, 224
Howard, A. W., *et al.* 2010, *arXiv:1008.3898*
Kovács, G., Bakos, G., & Noyes, R. W. 2005, *MNRAS*, 356, 557
Kovács, G., *et al.* 2010, *ApJ*, 724, 866
Garcia-Melendo, E. & McCullough, P. R. 2009, *ApJ*, 698, 558
Gibson, N. P., *et al.* 2008, *A&A*, 492, 603

Guillot, T. 2005, *Annual Review of Earth and Planetary Sciences*, 33, 493
Guillot, T., Santos, N. C., Pont, F., Iro, N., Melo, C., & Ribas, I. 2006, *A&A*, 453, L21
Latham, D. W., *et al.* 2010, *ApJL*, 713, L140
Laughlin, G., Deming, D., Langton, J., Kasen, D., Vogt, S., Butler, P., Rivera, E., & Meschiari, S. 2009, *Nature*, 457, 562
Moutou, C., *et al.* 2009, *A&A*, 498, L5
Naef, D., *et al.* 2001, *A&A*, 375, L27
Pepe, F., Mayor, M., Galland, F., Naef, D., Queloz, D., Santos, N. C., Udry, S., & Burnet, M. 2002, *A&A*, 388, 632
Perruchot, S., *et al.* 2008, *in Society of Photo-Optical Instrumentation Engineers (SPIE) Conference Series*, 7014,
Pollacco, D., *et al.* 2006, *PASP*, 118, 1407
Queloz, D., *et al.* 2000, *A&A*, 354, 99
Queloz, D., *et al.* 2010, *A&A*, 517, L1
Simpson, E. K., *et al.* 2011, *AJ*, 141, id.8
Southworth, J. 2010, *MNRAS*, 408, 1689
Steele, I. A., Bates, S. D., Gibson, N., Keenan, F., Meaburn, J., Mottram, C. J., Pollacco, D., & Todd, I. 2008, *in Society of Photo-Optical Instrumentation Engineers (SPIE) Conference Series*, 7014,
Swain, M. R., Vasisht, G., Tinetti, G., Bouwman, J., Chen, P., Yung, Y., Deming, D., & Deroo, P. 2009, *ApJL*, 690, L114
Tamuz, O., Mazeh, T., & Zucker, S. 2005, *MNRAS*, 356, 1466
Udry, S., Mayor, M., & Santos, N. C. 2003, *A&A*, 407, 369
Vidal-Madjar, A., Lecavelier des Etangs, A., Désert, J.-M., Ballester, G. E., Ferlet, R., Hébrard, G., & Mayor, M. 2003, *Nature*, 422, 143

The Astrophysics of Planetary Systems: Formation, Structure, and Dynamical Evolution
Proceedings IAU Symposium No. 276, 2010
A. Sozzetti, M. G. Lattanzi & A. P. Boss, eds.

doi:10.1017/S1743921311020096

NICMOS spectroscopy of HD 189733b

Mark R. Swain[1], Pieter Deroo[1] and Gautam Vasisht[1]

[1]Jet Propulsion Laboratory, California Institute of Technology, 4800 Oak Grove, Pasadena, California, 91109, United States
email: Mark.R.Swain@jpl.nasa.gov

Abstract. Spectral features corresponding to methane and water opacity were reported based on transmission spectroscopy of HD 189733b with Hubble/NICMOS. Recently, these data, and a similar data set for XO-1b, have been reexamined in Gibson *et al.* (2010), who claim they cannot reliably reproduce prior results. We examine the methods used by the Gibson team and identify two specific issues that could act to increase the formal uncertainties and to create instability in the minimization process. This would also be consistent with the GPA10 finding that they could not identify a way to select among the several instrument models they constructed. In the case of XO-1b, the Gibson team significantly changed the way in which the instrument model is defined (both with respect to the three approaches they used for HD 189733b, and the approach used by previous authors); this change, which omits the effect of the spectrum position on the detector, makes direct intercomparison of results difficult. In the experience of our group, the position of the spectrum on the detector is an important element of the instrument model because of the significant residual structure in the NICMOS spectral flat field. The approach of changing instrument models significantly complicates understanding the data reduction process and interpreting the results. Our team favors establishing a consistent method of handling NICMOS instrument systematic errors and applying it uniformly to data sets.

Keywords. techniques: spectroscopic, infrared: stars, planetary systems

---

1. Introduction

Measurements with the NICMOS instrument on the Hubble space telescope demonstrated that molecular spectroscopy of exoplanet atmospheres was possible and provided the first detection of methane in a planet orbiting another star (Swain *et al.* 2008). From this initial result, molecular spectroscopy of exoplanet atmospheres has grown rapidly, using Spitzer (Grillmair *et al.* 2008), Hubble (Swain *et al.* 2009a, 2009b), and ground-based measurements (Swain *et al.* 2010; Thatte *et al.* 2010; Snellen *et al.* 2010). The alkali element sodium has also been detected in exoplanet atmospheres using Hubble, (Charbonneau *et al.* 2002) and ground-based measurements have also been made (Redfield *et al.* 2008; Snellen *et al.* 2008). Spectroscopic measurements reporting an absence of atomic or molecular features have also been reported (Pont *et al.* 2008, 2009), with useful constraints being placed by the measurements. At the present time, numerous teams have exoplanet spectroscopy programs in place, and at least one purpose-built instrument is being constructed (Jurgenson *et al.* 2010). In short, the era of exoplanet characterization via spectroscopy is at hand.

Although the exoplanet spectroscopy field is broadening rapidly, the role of measurements with Hubble/NICMOS is currently unique. Operating in the near-IR, HST/NICMOS has produced published observations of four exoplanet systems, and molecules were detected in three; transmission and emission spectra were obtained, and the measurements show that water, methane, and carbon dioxide are routinely present in hot-Jupiter atmospheres. Simply put, HST/NICMOS has made an enormous contribution to the area of exoplanet spectroscopy; the fact that the NICMOS instrument is currently not available, with no specific plans to return it to operation, is certainly a setback for the exoplanet community. Given the large and unique impact of the NICMOS instrument, independent confirmation of results is a high priority.

In the recent paper by Gibson, Pont & Aigrain (2010; hereafter GPA10), the authors reanalyzed transmission spectroscopic measurements of HD 189733b (Swain, Vasisht & Tinetti 2008; hereafter SVT08), and XO-1b (Tinetti *et al.* 2010) and find that they cannot reliably reproduce the previous results. Here we explore the differences in method that lead the two teams to such different conclusions.

2. Observations & Methods

The observational scheme for NICMOS exoplanet spectroscopy consists of a series of spectrophotometric observations, $F(t)$, centered on the exoplanet eclipse. For a transmission spectrum, a transit of the planet is observed, and by measuring the difference between in and out-of-eclipse (through modeling with a theoretical light-curve), the exoplanet spectrum is measured. For HD 189733b, two Hubble orbits prior to the transit and two orbits after the transit are observed to establish the out-of-eclipse baseline. For XO-1b, two orbits before and one after transit are observed.

The Hubble/NICMOS instrument introduces systematic effects and the most significant of these are non-random changes in the location of the spectrum with respect to the focal plane array (FPA) pixel grid; the changes, caused by both displacement and

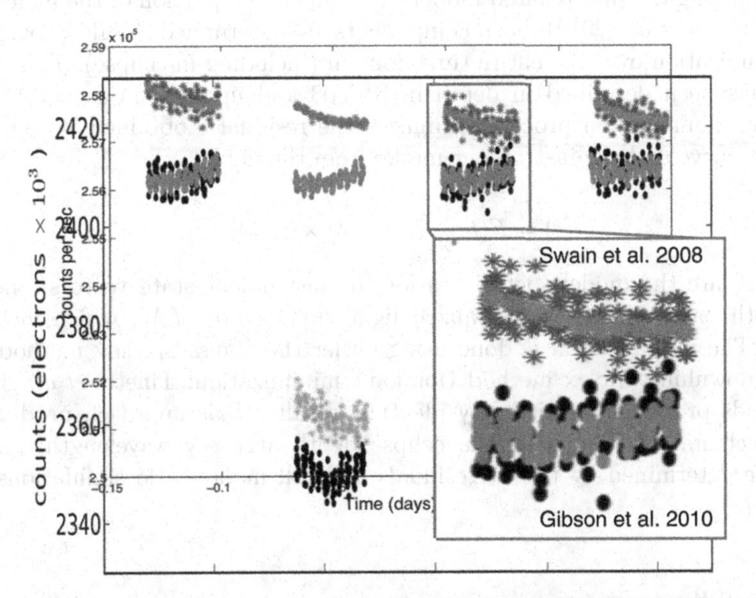

Figure 1. Comparison of the instrument model in SVT08 and GPA10. This figure is the combination of Figure 7 in GPA10 and Figure 7 in the supplementary index of SVT08 over-plotted and rescaled to the same range (the scaling is such that the depth of the primary eclipse is the same). In the inset, we zoom to compare the instrument model of SVT08 and GPA10, which is shown in green in both publications. While the photometry in GPA10 and SVT08 both show similar scatter, the model for the instrument systematic effect in GPA10 is > 3 times larger scatter than the model of SVT08. The large scatter may indicate a poorly determined instrument model that may be factor in producing both the larger uncertainty and decorrelation instability reported by GPA10.

change in the shape of the point spread function, interact with the residual uncertainty ($\sim 5\,\%$) in the NICMOS FPA flat field. If uncorrected, these systematic errors limit NICMOS spectro-photometry to about 1000 ppm per spectral channel (in the defocused mode used for exoplanet observations). These systematic errors can be corrected using an instrument model to separate instrument effects from astrophysical modulation of the spectrophotometric time series. Our approach is to use linear models because squared terms amplifiy the noise and reduce the dynamic range of a measurement; if a measurement m is composed of $m = a + b$ where a is the real signal and b is a noise term, $[a + b]^2 \neq a^2 + b^2$. To insure a linear approximation is a valid model, we restrict our analysis to data that have relatively small changes ($\leqslant 2\,\%$ of the size of the box used to capture the PSF) in the shift in the position of the spectrum, and we have tested this validity domain both by simulations (synthetic illumination of an actual NICMOS flat field) and real data.

The instrument model we use is constructed from seven instrument state vectors, collectively identified as Ψ_i, where the index i identifies the individual state vectors, which are x position, y position, the angle of the spectrum on the detector, the full width at half maximum of the PSF, the Hubble telescope orbital phase, the orbital phase squared (the error in the quantity is minimal so we admit it as a squared term), and the FPA temperature. With these state vectors, the observed spectrophometry is modeled using (1) an instrument model constructed using a linear combination of the state vectors, and (2) a light curve model with the radius ratio between planet and star as a free parameter. In SVT08, the two components of the spectrophotometric model are determined separately by establishing the instrument model using the out-of-eclipse data and then applying the interpolated model to the in-eclipse portion of the light curve. For XO-1b in Tinetti *et al.* (2010), both components are determined simultaneously through a joint minimization over the entire time-domain (including in-eclipse). This calibration approach has been described in detail in SVT08 and updated in Swain *et al.* (2009, 2009b). The minimization process minimizes the residual ϵ obtained by removing the eclipse light curve and the instrument model from the data:

$$\epsilon(t) = F(t) - LM(d,t) \times \Sigma_i \beta_i \dot{\Psi}_i(t) \tag{2.1}$$

where the β_i are the model coefficients for the instrument state vectors, and d is the eclipse depth parameter for an idealized light curve model LM, which includes limb darkening. The minimization is done using either the Gauss-Markov method (SVT08) or using a downhill simplex method (for joint minimization; Tinetti *et al.* 2010), with both methods producing statistically identical results. This approach produces an exoplanet spectrum by estimating the eclipse depth at every wavelength, and the error bars are determined by the fit residuals and full monte-carlo simulations for error propagation.

3. Discussion

The GPA10 approach differs significantly from the SVT08 approach, and this complicates comparing the methods in detail. The first major difference is the use by GPA10 of multiple instrument models; state vectors are sometimes included, sometimes excluded, and sometimes included as squared terms. The second major difference is the use of a residual permutation algorithm (RPA), which GPA10 uses to feed back permuted, post-decorrelated residuals into the predecorrelated light curve. Use of the RPA in this manner (the feedback step), prior to application of all systematic error removal steps, will inject noise power from an otherwise correctable systematic error into the light-curve parameter determination. It appears that GPA10 did apply the RPA approach prior to correcting the 'channel-to-channel' variations (SVT08), which have also been termed 'gain-like' variations Burke *et al.* 2010. This could contribute to the larger uncertainties in the GPA10 analysis.

There is one aspect of the GPA10 and SVT08 methods that appears to be directly comparable; this is the comparison of one particular instantiation of the GPA10 instrument models and the instrument model from SVT08. For clarity, we will term this model GPA_189_m1 as it differs substantially from two other instrument models, GPA_189_m2 and GPA_189_m3, that GPA10 also discuss. GPA_189_m1 is defined using the same state vectors as SVT08, and Fig. 1 shows a direct comparison of the SVT08 and the GPA_189_m1 instrument models. The comparison shows that the GPA_189_m1 insrument model has an internal scatter that is $\sim$3 times larger than the internal scatter of the SVT08 instrument model. This cannot be the result of worse photon noise in GPA10; based on the number of electrons and taking into account the analog-digital conversion gain of 6.5 (electrons/DN), the photon noise of the GPA10 photometry should be 20 % better than the photon noise % in the SVT08 model. A possible explanation for the increased scatter in the GPA10_189_m1 instrument model could be a poorly determined state vector; reprocessing the original SVT08 data set with our group's current pipeline (the original scripts are missing) suggests that the angle vector determination might be a contributor to the difference between the GPA10_189_m1 instrument model and the SVT08 instrument model. However, this is only a suggestion, and a detailed comparison of both the instrument state vectors and the model coefficients would be necessary to establish a clear causal connection. Regardless of the origin, the larger scatter in the GPA10_189_m1 instrument model is a possible contributor to larger formal uncertainties in the final spectrum, and we note that the spectrum derived using the GPA10_189_m1 model is reasonably consistent (albeit with larger errors) with the SVT08 result.

An interesting question to consider is the possible effects of larger uncertainties in model selection and in decorrelation stability. As discussed above, both (1) the details of the RPA method implementation, and (2) the way in which the instrument models were constructed could act to increase the formal uncertainties in the GPA10 analysis. It is possible that this would have made the decorrelation process sufficiently marginal, by effectively reducing the signal to noise of the data, that it made it difficult for GPA10 to identify a valid instrument model. One indication consistent with this hypothesis is that GPA10 found decorrelation results that were unstable in that the result changed when portions of the data were omitted; this contrasts strongly with the situation SVT08 found, where portions of the data were omitted and the decorrelation results were stable (see SVT08 Supplemental Information).

A detailed comparison of Tinetti *et al.* (2010) results with GPA10 is not really possible because of the radically different instrument model favored by GPA10, which omits the x, y, and angular position information of the spectrum on the detector. We found this change in the definition of the instrument model surprising. In the experience of our group, the position of the spectrum on the detector is an important element of the instrument model because of the significant residual structure in the NICMOS spectral flat field. The instrument model used by GPA10 for XO-1b also differs from the instrument models they constructed for the HD 189733b observations. This approach of constantly changing instrument models, together with the uncertainties in the implementation of the RPA method and in the model determination, makes it very difficult to understand what is going on.

The view in our group is that we employ a consistent instrument model based on the definition in section 2. Using this method, we have published four exoplanet spectra with Hubble/NICMOS data (two transmission spectra and two emission spectra). The method does not work on all Hubble/NICMOS data sets; sometimes there are jumps in the position of the spectrum which are simply too large, and we do not attempt to recover a spectrum in these cases. Our objective for the Hubble/NICMOS measurement has been to enable the process of comparative planetology with near-IR spectra based on a consistent measurement method. In all of the cases where we published Hubble/NICMOS spectra, we tested the stability of the decorrelation by exclusion of some data; this is standard process for our data analysis. In none of these (published) cases did we find a problem with stability of the results. An ideal test of the Hubble/NICMOS results would be confirming observations with a different instrument. Recently reported, ground-based, near-IR emission spectra of HD 189733b (Swain *et al.* 2010; Thatte *et al.* 2010) are consistent with the previous Hubble/NICMOS measurements (Swain *et al.* 2009a); this detection of the same spectral shape by two completely different instruments, with two completely different sets of systematics, is a strong indication that both instruments are measuring the exoplanet emission spectrum.

4. Conclusions

There are several important differences between the approach used by our team and the approach used by GPA10. We have identified two potential ways in which additional noise may be incorporated in the GPA10 calibration process. This additional noise may, in turn, impact the GPA10 decorrelation stability tests and thus some of their conclusions. However, the most significant difference in the approach used by GPA10 and our team may reflect a philosophical difference. GPA10 favor a fundamentally different instrument model for each of the three objects they consider; in contrast, our team has used a single, consistent instrument model for the four reported Hubble/NICMOS spectra. We strongly favor a consistent calibration approach to facilitate intercomparion of exoplanet spectra.

5. Acknowledgements

The research was carried out at the Jet Propulsion Laboratory, California Institute of Technology, under contract with the National Aeronautics and Space Administration.

References

Burke, C. *et al.* 2010, *ApJ*, 719, 1796

Charbonneau, D., Brown, T. M., & Noyes, R. W., Gilliland R. L. 2002, *ApJ*, 568, 377

Gibson, N. P., Pont, F., & Aigrain, S. 2011, *MNRAS*, 411, 2199

Grillmair, C. J., *et al.* 2008, *Nature*, 456, 767

Jurgenson, C., *et al.* 2010, *SPIE*, 7735, 43J

Pont, F., Knutson, H., Gilliland, R. L., Moutou, C., & Charbonneau, D. 2008, *MNRAS*, 385, 109

Pont, F., Gilliland, R. L., Knutson, H., Holman, M., & Charbonneau, D. 2009, *MNRAS*, 393, 6

Redfield, S., Endl M., Cochran, W. D., & Koesterke, L. 2008, *ApJ* (Letters), 673, 87

Swain, M. R., Vasisht, G., & Tinetti, G.. 2008, *Nature*, 452, 329

Swain, M. R., Vasisht, G., Tinetti, G., Bouwman, J., Chen, P., Yung, Y., Deming, D., & Deroo, P. 2009a, *ApJ* (Letters), 690, 114

Swain, M. R., *et al.* 2009b, *ApJ* (Letters), 704, 1616

Swain, M. R., *et al.* 2010, *Nature*, 463, 637
Snellen, I. A. G.., Albrecht, S., De Mooij, E. J. W.., & Le Poole, R. S. 2008, *A&A*, 483, 375
Snellen, I. A. G.., de Kok, R. J., De Mooij, E. J. W.., & Albrecht, S. 2010, *Nature*, 465, 1049
Thatte, A., Deroo, P., & Swain, M. R. 2010, *A&A*, 523, 35
Tinetti, G., Deroo, P., Swain, M. R., Griffith, C. A., Vasisht, G., Brown, L. R., Burke, C., & McCullough, P. 2010, *ApJ* (Letters), 712, 139

The Astrophysics of Planetary Systems: Formation, Structure, and Dynamical Evolution
Proceedings IAU Symposium No. 276, 2010
A. Sozzetti, M. G. Lattanzi & A. P. Boss, eds.

doi:10.1017/S1743921311020102

Lessons from detections of the near-infrared thermal emission of hot Jupiters

Bryce Croll[1]

[1]Department of Astronomy and Astrophysics, University of Toronto, 50 St. George Street, Toronto ON M5S 3H4 email: croll@astro.utoronto.ca

Abstract. There have recently been a flood of ground-based detections of the near-infrared thermal emission of a number of hot Jupiters. Although these near-infrared detections have revealed a great deal about the atmospheric characteristics of individual hot Jupiters, the question is: what information does this ensemble of near-infrared detections reveal about the atmospheric dynamics and reradiation of all hot Jupiters? I explore whether there is any correlation between how brightly these planets shine in the near-infrared compared to their incident stellar flux, as was theoretically predicted to be the case. Secondly, I look for whether there is any correlation between the host star's activity and the planet's near-infrared emission, like there is in the mid-infrared, where Spitzer observations have revealed a correlation between the host star activity with the presence, or lack thereof, of a temperature inversion and a hot stratosphere.

Keywords. eclipses, infrared: planetary systems, planetary systems, techniques: photometric

1. Introduction

Near-infrared observations of the thermal emission of hot Jupiters are crucial to our understanding of these exotic worlds. The blackbodies of the hottest of the hot Jupiters peak in the near-infrared, and it is thus in this wavelength regime that we will obtain the best constraints on the properties of their atmospheres. Near-infrared detections will help to constrain their bolometric luminosities, their temperature-pressure profiles at depth, and the efficiency of day/night redistribution of heat deep in their atmospheres. The desire to achieve such goals led to the program that I have been involved in using the Wide-field Infrared Camera on the Canada-France-Hawaii Telescope to detect the thermal emission of several hot Jupiters in the near-infrared (Croll *et al.* 2010a; Croll *et al.* 2010b; Croll *et al.* 2011). Truly understanding even the reradiation of hot Jupiters, however, will require multiple-band detections in the near- and mid-infrared of the thermal emission of a great number of hot Jupiters. There have been a series of multiple band detections of the thermal emission of hot Jupiters to date, mostly in the mid-infrared from observations obtained with the Spitzer Space Telescope. These multiple-band detections have revealed a wealth of information about these worlds, including that the most highly irradiated hot Jupiters orbiting inactive hosts seem to harbor hot stratospheres and temperature inversions (e.g. Knutson *et al.* 2008a; Charbonneau *et al.* 2008; Machalek *et al.* 2008; Knutson *et al.* 2008b; Knutson *et al.* 2010). One could imagine that near-infrared constraints, where the majority of the flux of these planets emerges, would be equally informative - especially if we are able to achieve multiple-band near-infrared detections. Here we attempt to determine what lessons, if any, can be taken away from all the ground-based near-infrared detections of the thermal emission of hot Jupiters that have been achieved to date, including two multiple-band detections.

2. Comparison of near-infrared detection of the thermal emission of hot Jupiters to date

There have been a wealth of recent detections from the ground of the thermal emission of hot Jupiters in the near-infrared. Detections with greater than 3σ confidence include†: TrES-3b in Ks-band (de Mooij & Snellen 2009; Croll *et al.* 2010b), OGLE-TR-56b in z'-band (Sing & Lopez-Morales 2009), WASP-12b in z'-band (Lopez-Morales *et al.* 2010) and in J, H & Ks (Croll *et al.* 2011), CoRoT-1b at ~2.1 μm (Gillon *et al.* 2009) and in Ks (Rogers *et al.* 2009), TrES-2 in Ks (Croll *et al.* 2010a), WASP-19 in H (Anderson *et al.* 2010) and Ks (Gibson *et al.* 2010), HAT-P-1 in Ks (de Mooij *et al.* 2010), as well as our own WASP-3 detection in Ks-band (Croll *et al.* in prep.). Given this recent flurry of detections it is high-time to search for possible trends revealed by these observations.

One such possible correlation is between the incident stellar flux and the efficiency of redistribution of heat from the day to the nightside, which we parameterize with a joint constraint on the Bond albedo and the reradiation factor, $f \times (1 - A_B)$. As, the geometric and Bond albedoes of many hot Jupiters have been measured (Rowe *et al.* 2008; Snellen *et al.* 2009; Alonso *et al.* 2009; Alonso *et al.* 2009b; Snellen *et al.* 2010) or inferred to be close to zero (Cowan & Agol 2011), this joint constraint should be in most cases largely identical to a constraint on the reradiation factor, f. Fortney *et al.* 2008 predicted that more highly irradiated hot Jupiters would have temperature inversions and feature more inefficient redistribution of heat to their nightsides than their less highly irradiated cousins, which were predicted to lack such inversions; although, Spitzer observations have not universally supported this theory (i.e. Machalek *et al.* 2008; Fressin *et al.* 2010), it is worth investigating if at least the redistribution of heat is supported at near-infrared wavelengths. We thus plot the correlation between the inferred equilibrium temperatures of these planets, a proxy for incident stellar flux, and their observed near-infrared brightness temperatures, and joint constraints on $f \times (1 - A_B)$ (Figure 1, top and middle panel). HAT-P-1b's Ks-band thermal emission is an obvious outlier; we note that de Mooij *et al.* (2011) highlighted the possibility that unaccounted for systematic errors may have affected their measurement of the Ks-band thermal emission of this planet. Even ignoring the HAT-P-1b measurement, the efficiency of day-to-nightside redistribution in the atmospheric layers probed by the near-infrared does not decrease monotonically with increasing effective temperature. With still so few near-infrared measurements it is unclear whether there is, or is not, a trend with increasing equilibrium temperature.

Recently, Knutson *et al.* 2010 suggested that there was a trend between the activity of the hot Jupiter host stars (as measured by the Ca II H & K activity index, $log[R'_{HK}]$), and the depths of the Spitzer/IRAC secondary eclipses. Specifically, Knutson *et al.* 2010 suggested that hot Jupiters orbiting active stars (higher $log[R'_{HK}]$ values) lacked temperature inversions, while those orbiting less active stars (lower $log[R'_{HK}]$ values) displayed signs of temperature inversions. One possible explanation for this phenomenon is that active stars should have increased UV flux, which may destroy the high altitude absorber that would otherwise cause the hot stratosphere and the temperature inversion. It is predicted that planets with and without temperature inversions, should have

† We note that we do not include detections from space using the NICMOS instrument on the Hubble Space Telescope (of HD 189733b [Swain *et al.* 2009a], and HD 209458b [Swain *et al.* 2009b]), as the analysis of this data has recently been called into question (Gibson *et al.* 2011). Also, we do not include the spectroscopic detection of the dayside spectrum of HD 189733b in the near-infrared using the SPEX instrument on the NASA Infrared Telescope Facility (Swain *et al.* 2010), as some of the features in this spectrum have likely been ruled out with another data-set with high confidence (Mandell *et al.* 2011).

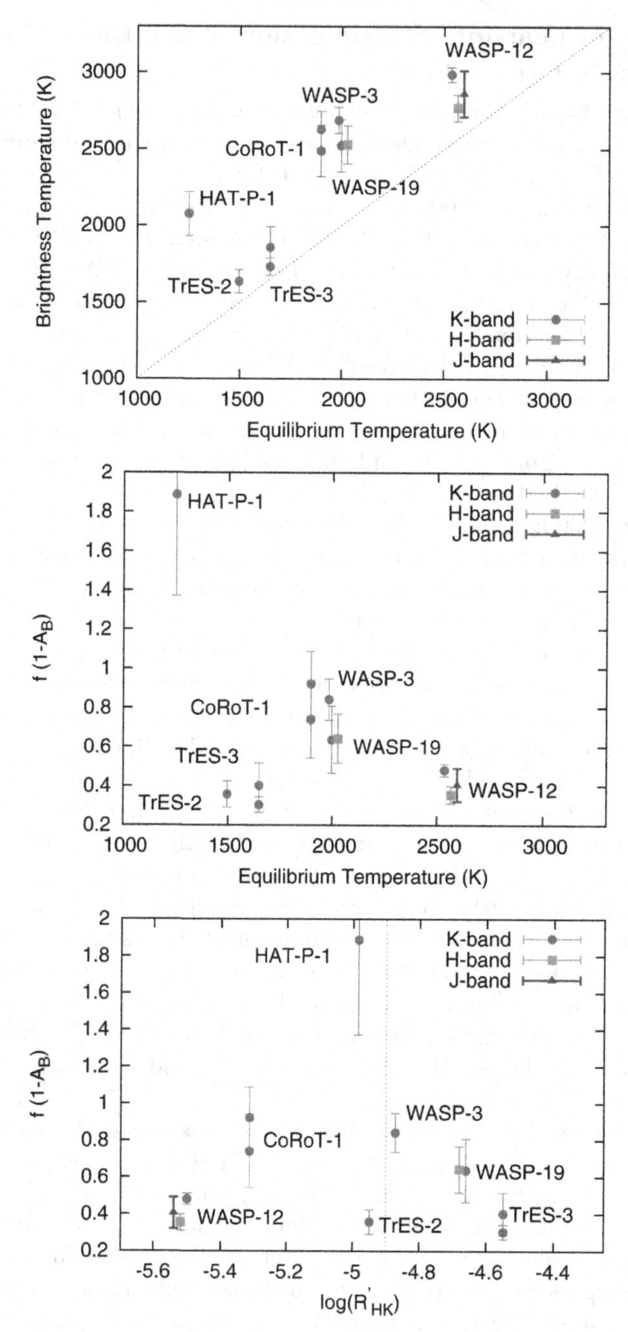

Figure 1. Various thermal emission detections from broadband, ground-based photometry in the JHK-bands (or at wavelengths that overlap with those filters). Top: The observed brightness temperatures compared to the equilibrium temperatures of the planets. The dashed line indicates a one-to-one correlation between brightness and equilibrium temperatures. Middle: The joint Bond albedo reradiation factors, $f \times (1 - A_B)$, compared to the equilibrium temperatures. Bottom: $f \times (1 - A_B)$ compared to the Ca II H & K activity index obtained from Knutson *et al.* 2010. The dotted vertical line indicates the dichotomy between hot Jupiters that Knutson *et al.* 2010 (and references therein) observed to have temperature inversions ($log(R'_{HK})$ <-4.9; to the left of the line) and those that did not ($log(R'_{HK})$ >-4.9; to the right of the line). In all plots we marginally offset the thermal emission measurements of the same planet in different near-infrared bands (JHK) along the x-axis for clarity.

observable differences in the near-infrared as well; as already mentioned, planets with a hot stratosphere, should have increased thermal emission in the mid-infrared, and are therefore expected to reradiate less at shorter wavelengths in the near-infrared (Hubeny *et al.* 2003). We note that all those planets orbiting less active stars with ground-based near-infrared detections have been reported to have temperature inversions: CoRoT-1b (Deming *et al.* 2011), HAT-P-1b (Todorov *et al.* 2010 report a modest temperature inversion), WASP-12b (Madhusudhan *et al.* 2010 report a weak inversion) and TrES-2b (O'Donovan *et al.* 2010, although see Madhusudhan & Seager 2010). Of the sample orbiting active stars with near-infrared thermal emission detections, TrES-3b (Fressin *et al.* 2010) has been reported to lack an inversion as expected, while the Spitzer/IRAC eclipse depths of WASP-19b and WASP-3b have yet to be reported. However, as Figure 1 (bottom panel) shows there is no clear trend between more active hot Jupiter hosts and increased near-infrared thermal emission.

A greater number of near-infrared thermal emission detections of hot Jupiters, along with confirmation of the current detections, will be required to elucidate whether any of the predicted trends have been observed or can be confidently ruled out.

References

Anderson, D. R., *et al.* 2010, *A&A*, 513, L3
Alonso, R., *et al.* 2009a, *A&A*, 501, L23
Alonso, R., *et al.* 2009b, *A&A*, 506, 353
Cowan, N. B. & Agol, E. 2011, *ApJ*, 729, 54
Charbonneau, D., *et al.* 2008, *ApJ*, 686, 1341
Croll, B., *et al.* 2010a, *ApJ*, 717, 1084
Croll, B., *et al.* 2010b, *ApJ*, 718, 920
Croll, B., *et al.* 2011, *AJ*, 141, 30
de Mooij, E. J. W.. & Snellen, I. A. G.. 2009, *A&A*, 493, L35
de Mooij, E. J. W.., *et al.* 2011, *A&A*, 528, A49
Deming, D., *et al.* 2011, *ApJ*, 726, 95
Fressin, F., *et al.* 2010, *ApJ*, 711, 374
Gibson, N. P., *et al.* 2010, *MNRAS*, 404, L114
Gibson, N. P., *et al.* 2011, *MNRAS*, 411, 2199
Gillon, M., *et al.* 2009, *A&A*, 506, 359
Hubeny, I., *et al.* 2003, *ApJ*, 594, 1011
Knutson, H., *et al.* 2008a, *ApJ*, 673, 526
Knutson, H., *et al.* 2008b, *ApJ*, 691, 866
Knutson, H., *et al.* 2010, *ApJ*, 720, 1569
Lopez-Morales, M., *et al.* 2010, *ApJ*, 716, L36
Machalek, P., *et al.* 2008, *ApJ*, 684, 1427
Madhusudhan, N., *et al.* 2010, *Nature*, 469, 64
Madhusudhan, N. & Seager, S. 2010, *ApJ*, 725, 261
Mandell, A. M., *et al.* 2011, *ApJ*, 728, 18
O'Donovan, F. T., *et al.* 2010, *ApJ*, 710, 1551
Rogers, J. C., *et al.* 2009, *ApJ*, 707, 1707
Rowe, J. F., *et al.* 2008, *ApJ*, 689, 1345
Sing, D. K. & Lopez-Morales, M. 2009, *A&A*, 493, L31
Snellen, I. A. G., *et al.* 2009, *Nature*, 459, 543
Snellen, I. A. G., *et al.* 2010, *A&A*, 513, 76
Swain, M. R., *et al.* 2009a, *ApJ*, 690, L114
Swain, M. R., *et al.* 2009b, *ApJ*, 704, 1616
Swain, M. R., *et al.* 2010, *Nature*, 463, 637
Todorov, K., *et al.* 2010, *ApJ*, 708, 498

The Astrophysics of Planetary Systems: Formation, Structure, and Dynamical Evolution
Proceedings IAU Symposium No. 276, 2010
A. Sozzetti, M. G. Lattanzi & A. P. Boss, eds.

doi:10.1017/S1743921311020114

A NIR spectrum of a hot Jupiter from the ground: Preliminary results

Avi M. Mandell[1], L. Drake Deming[1], Geoffrey A. Blake[2], Heather A. Knutson[3], Michael J. Mumma[1], Geronimo L. Villanueva[1,4] and Colette Salyk[5]

[1]NASA GSFC, [2]Caltech, [3]UC Berkeley, [4]Catholic University, [5]UT Austin
email: avi.mandell@nasa.gov

Abstract. High resolution NIR spectroscopy offers an excellent complement to the expanding dataset of transit and secondary eclipse observations of exo-planets with Spitzer that have provided the bulk of our understanding of the atmospheres and internal structure of these objects. High-resolution data can quantify the vertical temperature structure by isolating specific spectral lines formed at various depths. The presence of an opaque absorbing layer can also be inferred - and its pressure level determined quantitatively - via its effect on spectral line intensities.

We have analyzed data for a single secondary eclipse of the bright transiting exo-planet host star HD189733 at L-band wavelengths ($3-4\,\mu$m) using the NIRSPEC instrument on Keck-II. We utilize a sophisticated first-order telluric absorption modeling technique that, combined with a calibration star, has already been proven to remove the effects of varying atmospheric transmittance and allow us to reach unprecedented S/N. We are conducting validation of the final data reduction products and developing high-resolution atmospheric models for comparison, but we have already been able to rule out emission from methane as reported by Swain *et al.* (2010). We present preliminary results and discuss future plans for analysis and observations.

Keywords. infrared: planetary systems, radiative transfer, techniques: spectroscopic

1. Introduction

Ground-based observations are now pushing the boundaries of our knowledge of exo-planet atmospheres. Near-infrared photometry of the hottest exoplanets is now possible using ground-based observatories (Gillon *et al.* 2009; Alonso *et al.* 2010; de Mooij & Snellen 2009), and measurement of atmospheric absorption in strong atomic lines has also been successful from the ground (Redfield *et al.* 2008; Snellen *et al.* 2008). However, ground-based spectroscopic analysis of exoplanet atmospheres is in its infancy. Even at infrared wavelengths, short-period exoplanets are very dim compared with their parent star (at $5\,\mu$m the planet-star contrast is $0.002-0.003$, while at $1\,\mu$m the contrast drops to 0.0005), and many previous attempts to reach these levels of precision have been frustrated by the fundamental difficulties of observing in the infrared: emission and absorption by our own atmosphere. There have been several recent announcements of breakthroughs in ground-based spectroscopy: Snellen *et al.* (2010) announced a detection of wind motions in the atmosphere of HD209458 using high-resolution data of CO at $2\,\mu$m from the VLT/CRIRES spectrograph, and a startling ground-based detection of a possible bright emission feature at $3.5\,\mu$m from the atmosphere of the exoplanet HD 189733b was announced by Swain *et al.* (2010). While these early detections may eventually prove to be spurious, they emphasize the impact that ground-based spectroscopy will have on the field of exoplanet characterization.

With a new generation of extremely large ground-based telescopes now being planned (Hook 2009), achieving successful ground-based spectroscopy of molecular features in exoplanet atmospheres becomes a major priority. Ground-based facilities not only provide a huge increase in collecting power and available observing time, but the resolving power of available instruments covers both low-resolution broad-spectrum instruments as well as high-resolution echelle spectrometers. Increasing spectral resolving power provides increasing diagnostic power with respect to both the chemical and thermal structure of the atmosphere. If ground-based capabilities can be successfully utilized for exoplanet spectroscopy, they can quickly revolutionize our understanding of the atmospheric structure of known exoplanets and eventually prepare us to explore the complex atmospheres of rocky super-Earths in the Habitable Zone.

2. Towards A Measurement of The Day-Side Spectrum for HD 189733b at High Resolution

In this article we describe a new strategy for removing telluric contamination and instrumental systematics to a very high degree, which has enabled us to push to the precision required to measure an exoplanet spectrum (S/N > 1000) for observations over the full range of near-IR wavelengths (1-5 μm). We have developed custom data reduction algorithms that enable us to achieve fractional accuracy of less than a part in a thousand at wavelengths with significant telluric absorption at high resolution (see Figure 1). These analysis techniques were previously used to detect new molecular emission features from warm gas in circumstellar disks (Mandell *et al.* 2008), reaching line-continuum contrast sensitivities of better than 1/2000 at L-band wavelengths. To correct for changing atmospheric conditions and airmass, we utilize terrestrial spectral transmittance models synthesized with the LBLRTM atmospheric modeling code (Clough *et al.* 2005) based on line parameters from the HITRAN 2008 molecular database updated with the latest spectroscopic parameters (Rothman *et al.* 2009). We fit models for every AB image set, enabling us to compensate for changes over the night and between stellar targets, and we perform the same telluric removal routine on both the science target and the calibration star. The post-telluric removal residuals of the science and calibration stars are then differenced to remove remnant fringes due to internal reflections in the detector and other instrumental artifacts, as well as minor errors in the telluric model such as improper pressure broadening and isotopic ratios.

Stellar absorption features are removed by creating a stellar template from the data, using either a transit/eclipse method or averaging spectra that cover a wide range of radial velocity changes in the planet spectrum due to the rapid motion of hot exoplanets around their parent star. We then subtract this stellar template from each AB set in the data after shifting the template to the correct planetary velocity shift, leaving final residuals for each AB set with only the signal from the planet. We then combine all the residuals, shifting all the residual spectra to compensate for the planet's orbital motion. This relies on an accurate ephemeris for the orbit of the planet; however, this requirement can also allow us to solve for the planetary velocity exactly by fitting a model to the data using different orbital solutions. This process achieves results corresponding to an rms noise only slightly larger (a factor of 1.5) than that expected from the photon statistics.

The first exoplanet we have targeted is HD 189733b, due to its close proximity to the Sun (19 parsecs) and the favorable planet-star contrast expected based on Spitzer observations. We observed the star on July 13, 2009 with NIRSPEC on Keck at a resolving power of R= 25,000, acquiring a total of 200 images before, during, and after the secondary eclipse. HD 189733 is a main-sequence K-type star, and the 1.1 M_J planet

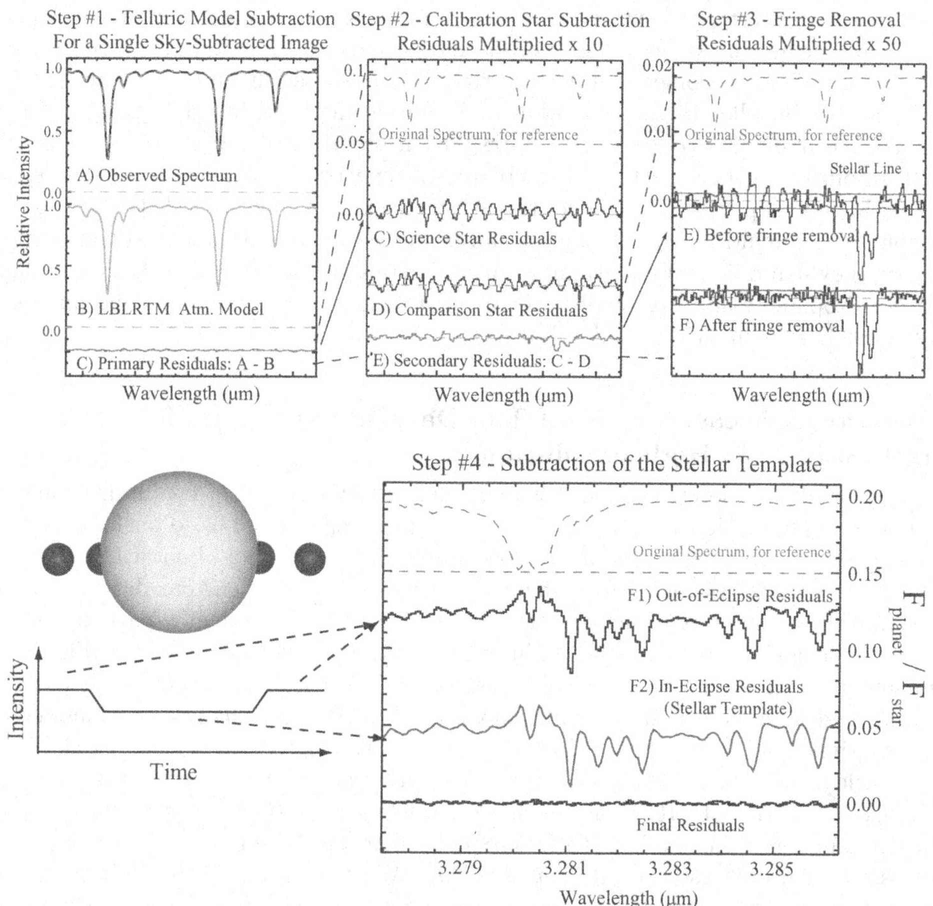

Figure 1. Demonstration of our data analysis procedure. In Step 1, synthetic terrestrial atmospheric models (B) are fitted separately to both the science and comparison stars (A). In Step 2, the two sets of residuals (C and D) are then differenced to remove second-order instrumental or atmospheric features, leaving a secondary residual (E). In Step 3, remaining fringing is removed using a Fourier filter (F). This process is repeated for each AB set. In Step 4, stellar features are removed by subtracting the residuals from data during the eclipse of the planet (F2) from the combined residuals of the out-of-eclipse data (F1). The fractional accuracy on the stellar continuum in the final residuals is greater than 1 part in 1000.

orbits in a circular orbit at 0.03 AU. We chose to observe a secondary eclipse of the planet at L-band wavelengths ($3 - 4\,\mu$m); our current wavelength coverage spans 4 spectral orders between $3.25\,\mu$m and $3.8\,\mu$m, each covering $0.05\,\mu$m. For observations of the day-side of hot exoplanets, longer wavelengths provide the best planet-star contrast, and sky background levels are still relatively low at these wavelengths. The second important advantage of the L-band spectral region is the large number of strong transitions of molecular species such as H_2O, CH_4, and other simple molecules present. We primarily target spectroscopic features of water because it is abundant in hot Jupiters, and has many hundreds of well-determined "hot" transitions (Barber *et al.* 2006) that we expect to be prominent in the exoplanet spectrum, but not excited in the telluric spectrum. The spectral coverage will also include methane near $3.3\,\mu$m, as well as additional trace species such as OH ($3.28\,\mu$m) and H_3^+ ($3.67\ \mu$m), both of which are observed in Solar

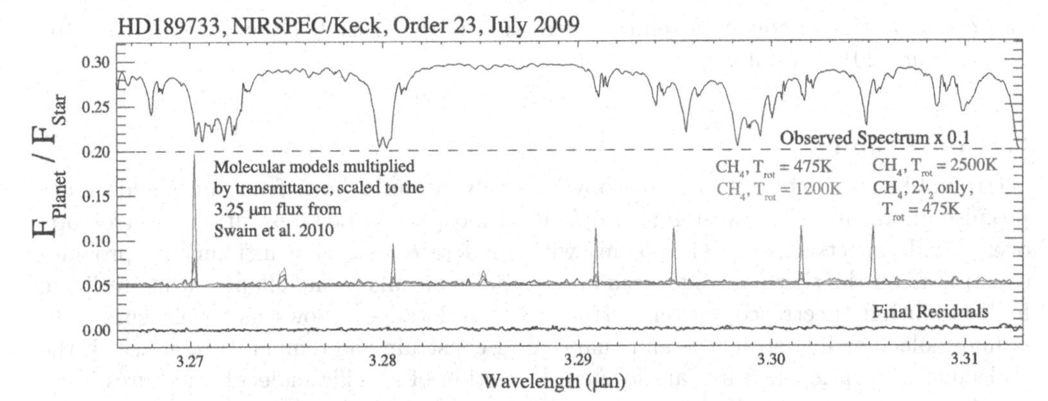

Figure 2. Results from our data reduction, with candidate molecular models for methane using the CH4@Titan line lists overplotted. The upper black trace is the original data, and the bottom black trace are our residuals after removing telluric and stellar features, shifting the data to the correct radial velocity and combining it, and then removing large-scale gradients in the continuum with a high-pass filter. We detect no emission at any of the expected positions, with upper limits more than 10 times below the total intensity of the expected emission for each model.

System objects and which may be excited by radiative pumping in the upper atmosphere of the exoplanet (cf. Shkolnik *et al.* (2006)).

2.1. *Non-Detection of Expected Bright L-Band Emission*

Our data has a theoretical contrast limit of 0.0005, and our current analysis reaches an rms of 0.0011. This is not sufficient to detect an unambiguous spectrum of the planet based on the expect line-to-continuum contrast from models, but our current level of precision allows us to easily search for evidence of the bright non-LTE emission from the exoplanet HD 189733b at 3.25 μm recently reported by Swain *et al.* (2010) based on observations at low spectral resolving power ($\lambda/\delta\lambda \approx 30$). Non-LTE emission lines from gas in an exoplanet atmosphere will not be significantly broadened by collisions, so the measured emission intensity per resolution element must be substantially brighter when observed at high spectral resolving power, and we should be able to easily detect and characterize any planetary emission. We modeled the expected signal using ro-vibrational emission models of molecular species with transitions in the relevant wavelength region using a wide range of rotational excitation temperatures and vibrational level intensities, scaled to the results presented by Swain *et al.* (2010), in order to predict the line flux required at high resolution. We then compared it to our data; no flux was present at any of the expected transition frequencies between 3.27 and 3.31 μm, with upper limits 30 times smaller than the expected line fluxes, and similar limits were set for the other spectral orders we analyzed (Mandell *et al.* 2011).

The conditions that would lead to broad emission features beyond our detection limits are extremely difficult to reconcile with realistic models and previous observations. Additionally, our analysis indicates that the emission, if real, is too bright to be produced by fluoresence. Our wavelength region, even including all four spectral orders from our data, only covers a small section of the spectrum published by S10, and we cannot rule out an emission mechanism that produces flux outside our band passes. Additionally, we cannot rule out an exotic highly time-variable stellar emission process such as charged-particle excitation due to flares. However, we regard these explanations as improbable, and conclude that inadequate telluric correction is the most likely explanation for the Swain *et al.* (2010) results.

3. Conclusion

Ground-based exoplanet spectroscopy has only recently born fruit, but holds great promise in our quest to investigate exoplanet atmospheres. Low-resolution spectroscopy over a wide spectral range ($1 - 5\,\mu$m) with moderate-resolution instruments provides constraints on the thermal continuum and overall molecular band shapes, while utilizing high-resolution spectroscopy over narrower spectral ranges allows us to measure individual molecular line strengths and shapes and constrain the temperature at which the molecular absorption features are formed. Detection of specific molecular features in an exoplanet spectrum will allow us to quantify temperature versus height by isolating combinations of lines formed over a range of height and comparing their intensities.These analysis techniques will allow us to probe the chemical and thermal structure of exoplanet atmospheres by measuring the strengths and line shapes for individual molecular transitions of important molecular constituents such as H_2O, CH_4, and CO at moderate and high resolution, removing degeneracies in the fitting of molecular bands and directly probing the vertical temperature structure of the atmosphere and allowing us to potentially resolve much of the current ambiguity in the modeling of exoplanet atmospheres (Madhusudhan & Seager 2009). Longer wavelengths (3-5 μm) can provide much better sensitivity for secondary eclipse measurements (which measures the thermal emission from the day side of the planet), while shorter wavelengths favor transit spectroscopy for planets with large atmospheric scale-heights. By combining multiple regimes in spectral resolving power and wavelength range, these observations hold the potential to revolutionize our ability to characterize the atmospheric structure of exoplanets in a variety of stellar and orbital configurations, and will help to prepare us for future observations with JWST.

Acknowledgements

A.M.M. was supported by the Goddard Center for Astrobiology and the NASA Postdoctoral Fellowship Program. H.A.K. is supported by a fellowship from the Miller Institute for Basic Research in Science.

References

Alonso, R., Deeg, H. J., Kabath, P., & Rabus, M. 2010, *AJ*, 139, 1481
Barber, R. J., Tennyson, J., Harris, G. J., & Tolchenov, R. N. 2006, *MNRAS*, 368, 1087
Clough, S. A., Shephard, M. W., Mlawer, E. J., *et al.* 2005, *JQSRT*, 91, 233
de Mooij, E. J. W. & Snellen, I. A. G. 2009, *A&A*, 493, L35
Gillon, M., Demory, B.-O., Triaud, A. H. M. J., *et al.* 2009, *A&A*, 506, 359
Hook, I. 2009, in: *Science with the VLT in the ELT Era*, 225
Madhusudhan, N. & Seager, S. 2009, *ApJ*, 707, 24
Mandell, A. M., Mumma, M. J., Blake, G. A., *et al.* 2008, *ApJ*, 681, L25
Mandell, A. M., Deming, L. D., Blake, G. A., *et al.* 2011, *ApJ*, 728, id.18
Redfield, S., Endl, M., Cochran, W. D., & Koesterke, L. 2008, *ApJ*, 673, L87
Rothman, L. S., Gordon, I. E., Barbe, A., *et al.* 2009, *JQSRT*, 110, 533
Shkolnik, E., Gaidos, E., & Moskovitz, N. 2006, *AJ*, 132, 1267
Snellen, I. A. G., Albrecht, S., de Mooij, E. J. W., & Poole, R. S. L. 2008, *A&A*, 487, 357
Snellen, I. A. G., de Kok, R. J., de Mooij, E. J. W., & Albrecht, S. 2010, *Nature*, 465, 1049
Swain, M. R., Deroo, P., Griffith, *et al.* 2010, *Nature*, 463, 637

The Astrophysics of Planetary Systems: Formation, Structure, and Dynamical Evolution
Proceedings IAU Symposium No. 276, 2010
A. Sozzetti, M. G. Lattanzi & A. P. Boss, eds.

doi:10.1017/S1743921311020126

A multi-wavelength analysis of the WASP-12 planetary system

Luca Fossati[1], Carole A. Haswell[1] and Cynthia S. Froning[2]

[1]Department of Physics and Astronomy, Open University,
Walton Hall, Milton Keynes MK7 6AA, UK
email: l.fossati@open.ac.uk, C.A.Haswell@open.ac.uk

[2]Center for Astrophysics and Space Astronomy, University of Colorado,
593 UCB, Boulder, CO 80309-0593, USA
email: cynthia.froning@colorado.edu

Abstract. WASP-12 is a 2 Gyr old solar type star, hosting WASP-12b, one of the most irradiated transiting planets currently known. We observed WASP-12 in the UV with the Cosmic Origins Spectrograph (COS) on HST. The light curves we obtained in the three covered UV wavelength ranges, all of which contain many photospheric absorption lines, imply effective radii of $2.69 \pm 0.24\,R_J$, $2.18 \pm 0.18\,R_J$, and $2.66 \pm 0.22\,R_J$, suggesting that the planet is surrounded by an absorbing cloud which overfills the Roche lobe. We clearly detected enhanced transit depths at the wavelengths of the MgII h&k resonance lines. Spectropolarimetric analysis of the host star was also performed. We found no global magnetic field, but there were hints of atmospheric pollution, which might be connected to the very unusual activity of the host star.

Keywords. ultraviolet: stars, stars: magnetic fields, stars: individual (WASP-12), stars: activity, stars: abundances

1. Introduction

Transiting exoplanets provide a huge amount of information on both properties and evolution of planets outside our solar system. In particular close-in exoplanets are good 'laboratories' to test and constraint models of exoplanet atmospheres and dynamical evolution. These objects undergo a very strong star-planet interaction to the extent that the stellar activity is influenced by the presence of the planet (e.g. Shkolnik *et al.* 2005) and the star controls the planet's life-time (Li *et al.* 2010; Fossati *et al.* 2010a).

WASP-12 is a solar-type star (F9V), hosting one of the hottest, most bloated known exoplanets. It has one of the shortest known orbital periods, and is extremely close to the host star (Hebb *et al.* 2009). WASP-12b provides a key test of theories of planet evaporation and star-planet interactions.

2. The HST observations

UV observations of HD 209458 b, the prototype transiting hot Jupiter, were of the far-UV Ly α emission line. The hydrogen abundance makes this an attractive line to observe, but the temporal and spatial variability of stellar Ly α emission is a highly undesirable complicating factor. For this reason, and to obtain better signal to noise, we observed WASP-12 in the near-UV where there are many other resonance lines including the very strong MgII h&k lines. Observations were performed with COS (see Osterman *et al.* 2011) on HST for five consecutive HST orbits on 2009 September 24th (see Fossati *et al.* (2010a) for more details on the data reduction).

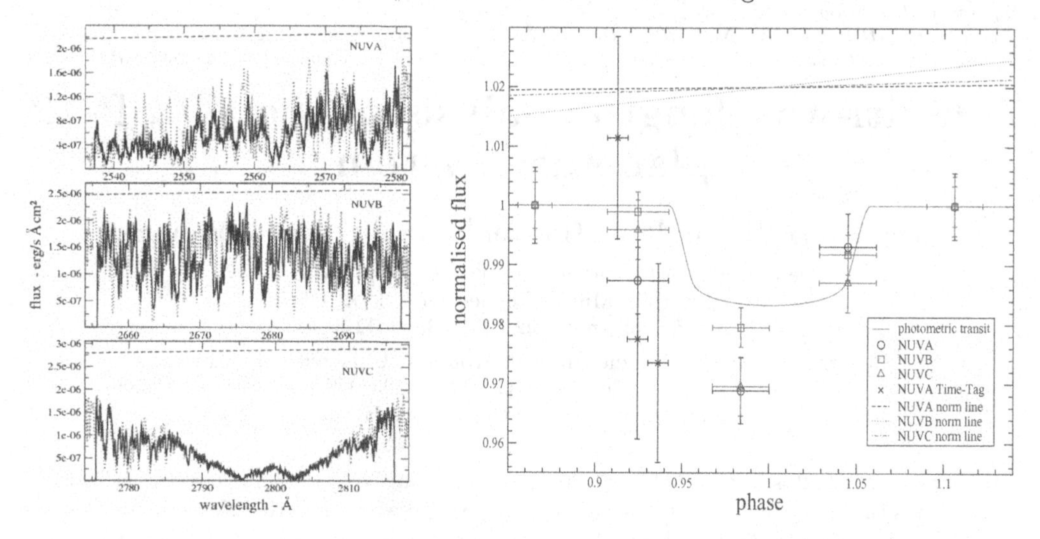

Figure 1. Left panel: Comparison between the observed fluxes of WASP-12 (full line) and LLMODELS synthetic fluxes (dotted line). The dashed line shows the modeled level of the stellar continuum flux. The three observed spectral ranges are defined as NUVA, NUVB and NUVC, from top to bottom. **Right panel:** Light curve obtained for each observed wavelength range (NUVA: open circles - NUVB: open squares - NUVC: open triangles). The horizontal error bar defines the orbital phase range covered by each observation. Vertical uncertainties come from a Poissonian treatment of the error bars. The full line shows the photometric transit light curve by Hebb *et al.* (2009). The three light curves were normalised to the line passing through the out-of-transit photometric points. The earlier ingress is shown by the crosses (from Fossati *et al.* 2010a).

The left panel of Fig. 1 shows the total summed spectrum in comparison with synthetic fluxes from the LLMODELS stellar model atmosphere code (Shulyak *et al.* 2006), assuming fundamental parameters and abundances by Fossati *et al.* (2010b).

The right panel of Fig. 1 compares the transit light curves obtained in each observed wavelength range with a fit to optical photometry (Hebb *et al.* 2009). As shown on the left panel of Fig. 1, the NUVB wavelength range is the closest to the stellar continuum and shows a transit depth that matches, at $\sim 1\sigma$, the transit light curve derived from optical photometry. In the NUVA and NUVC wavelength ranges we obtained a deeper transit at $\sim 2.5\sigma$ level. The NUVC spectral region is clearly dominated by the Mg II resonance lines, which are probably responsible for the detected extra depth in the transit light curve. On the other hand, the NUVA spectral region does not include dominant lines, but shows the presence of several resonance lines of various metals, in addition to Mg I and Fe I lines from low energy levels, which dominate the stellar spectrum. These spectral features, formed mostly by neutral atoms, are likely to cause the observed deeper transit.

From the transit depth in each of the three observed wavelength ranges we obtained three values of the effective planet radius (NUVA: $2.69 \pm 0.24\,R_J$, NUVB: $2.18 \pm 0.18\,R_J$, and NUVC: $2.66 \pm 0.22\,R_J$), concluding that the radii obtained from the NUVA and NUVC wavelength regions exceed the planet Roche lobe ($2.36\,R_J$), implying that WASP-12 b is currently losing mass.

3. WASP-12 accretion disk and possible atmospheric pollution

Fig. 1 shows that the NUVA flux during the second exposure lies below the out-of-transit level by $\sim 2\sigma$. We divided this particular exposure into three equal sub-exposures,

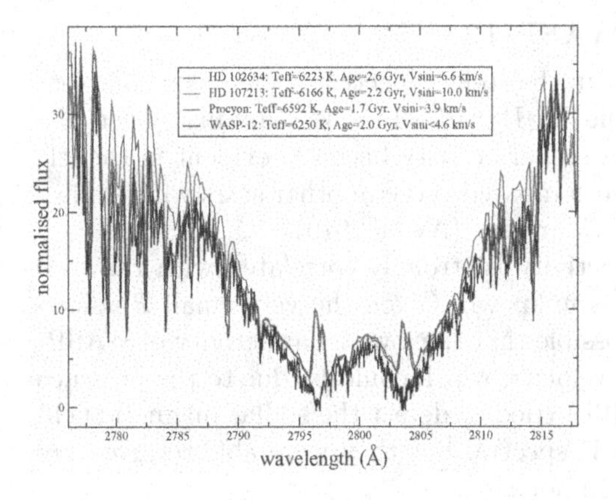

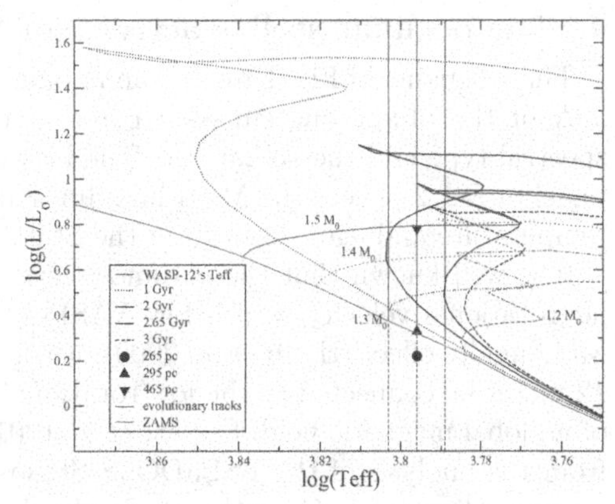

Figure 2. Left panel: Comparison of the Mg II h&k resonance line profiles between WASP-12, observed with COS, and three comparison stars, observed with STIS. Two of the comparison stars have effective temperature and age similar to the one of WASP-12. The third comparison star is Procyon, which is known to have a particularly low activity level. **Right panel:** Position of WASP-12 on the HR diagram assuming three different stellar distances: 265 pc (circle), 295 pc (triangle), and 465 pc (inverted triangle). The dotted, thin full and dashed lines show isochrones from Marigo *et al.* (2008) corresponding to ages of 1 Gyr, 2 Gyr and 3 Gyr, respectively, encompassing the possible age range of WASP-12. The thick full line is the 2.65 Gyr isochrone we argue being the maximum possible age for WASP-12. Evolutionary tracks from Girardi *et al.* (2000) for 1.5 $M/M_{\odot}$, 1.4 $M/M_{\odot}$, 1.3 $M/M_{\odot}$and 1.2 $M/M_{\odot}$, from top to bottom, are also shown. Both isochrones and evolutionary tracks assume a metallicity Z of 0.03. The vertical lines show the WASP-12's temperature range; these lines change from full to dashed below the ZAMS.

plotted as crosses in Fig. 1 (right panel). These suggest an early ingress in the NUVA spectral region.

Lai *et al.* (2010) and Vidotto *et al.* (2010) suggested that the early ingress is caused by a bow shock ahead of the planetary orbital motion, formed by the collision of the planet material with the stellar wind.

Li *et al.* (2010) predicted a loss of material from WASP-12 b, suggesting that this material forms a disk before falling onto the stellar surface, polluting the photosphere. Since we expect the planet metallicity to be higher than that of the star (as in the solar system), this effect would lead to an increased metallicity of the stellar photosphere.

To explore this possibility, we observed WASP-12 using the ESPaDOnS spectropolarimeter at the Canada-France-Hawaii Telescope (CFHT) on the 3rd and 5th of January 2010, in 'polarimetric' mode (see Fossati *et al.* (2010b) for more details).

Making use of the ESPaDOnS spectra, we performed a detailed parameter determination and abundance analysis, which lead to a set of high precision stellar parameters ($T_{\rm eff} = 6250 \pm 100$ K, $\log g = 4.2 \pm 0.2$, $\upsilon_{\rm mic} = 1.2 \pm 0.3$ km s^{-1}, [Fe/H] $= 0.32 \pm 0.12$ dex, $Z = 0.021 \pm 0.002$, $\upsilon \sin i < 4.6 \pm 0.5$ km s^{-1}, $\upsilon_{\rm macro} = 4.75$–$7.0 \pm 0.6$ km s^{-1}) and abundances for 39 ions (see Fossati *et al.* 2010b). We examined the correlation between the relative abundance of various elements and their condensation temperature, but no clear trend was found. We also compared the WASP-12 abundances with those obtained by other authors on a large number of stars with a similar $T_{\rm eff}$, concentrating on stars that are not known planet hosts, in the temperature range 6000–6500 K, and with a [Fe/H]>0.0 dex. The information gathered from this comparison lead to the conclusion that there is the possibility that the WASP-12 abundances of Na, Si, and Ni do not really match the abundance pattern that the WASP-12 Fe abundance would suggest (see Fossati *et al.* (2010b) for more details).

4. The peculiar stellar activity of WASP-12

The left panel of Fig. 1 reveals an anomaly in the stellar spectrum: there is a complete lack of the ubiquitous emission cores in the Mg II h&k. This is stunning, given the spectral type and the stellar age. The lack of stellar activity becomes evident in the left panel of Fig. 2, where the Mg II h&k lines are compared those of other stars with similar temperature and age, taken from the StarCAT archive (Ayres 2010).

It is well known that the chromospheric activity is strongly correlated with the stellar rotational velocity, which for WASP-12 is unknown. Given the very small Rossiter-McLaughlin effect (Hebb *et al.* 2009), it is possible that the low stellar activity of WASP-12 might be connected to the low rotational velocity, which could be due to the presence of a global magnetic field. Fossati *et al.* (2010b) tried to detect the stellar magnetic field from the analysis of the ESPaDOnS Stokes V spectra, but they were able to give only an upper limit of 10 G on the longitudinal component.

It is also possible that the star is passing through a period of low activity; future observations can confirm or reject this hypothesis. The right panel of Fig. 2 excludes stellar ages >2.65 Gyr, given the well established WASP-12 $T_{\rm eff}$ and metallicity. About half of the main-sequence lifetime remains (see Fossati *et al.* (2010b) for more details).

Other possibilities to be explored include the lack of the typical signs of stellar activity, such as emission in the cores of the MgII resonance lines which might be absorbed by the material lost by the planet and falling onto the star.

References

Ayres, T. R. 2010, *ApJS*, 187, 149

Fossati, L., Haswell, C. A., Froning, C. S., Hebb, L., Holmes, S., Kolb, U., Helling, C., Carter, A., Wheatley, P., Cameron, A. C., Loeillet, B., Pollacco, D., Street, R., Stempels, H. C., Simpson, E., Udry, S., Joshi, Y. C., West, R. G., Skillen, I., & Wilson, D. 2010, *ApJL*, 714, 222

Fossati, L., Bagnulo, S., Elmasli, A., Haswell, C. A., Holmes, S., Kochukhov, O., Shkolnik, E. L., Shulyak, D. V., Bohlender, D., Albayrak, B., Froning, C., & Hebb, L. 2010, *ApJ*, 720, 872

Girardi, L., Bressan, A., Bertelli, G., & Chiosi, C. 2000, *A&AS*, 141, 371

Hebb, L., Cameron, A. C., Loeillet, B., Pollacco, D., Hébrard, G., Street, R. A., Bouchy, F., Stempels, H. C., Moutou, C., Simpson, E., Udry, S., Joshi, Y. C., West, R. G., Skillen, I., Wilson, D. M., McDonald, I., Gibson, N. P., Aigrain, S., Anderson, D. R., Benn, C. R., Christian, D. J., Enoch, B., Haswell, C. A., Hellier, C., Horne, K., Irwin, J., Lister, T. A., Maxted, P., Mayor, M., Norton, A. J., Parley, N., Pont, F., Queloz, D., Smalley, B., & Wheatley, P. J. 2009, *ApJ*, 693, 1920

Lai, D., Helling, C., & van den Heuvel, E. P. J. 2010, *ApJ*, 721, 923

Li, S., Miller, N., Lin, D. N. C., & Fortney, J. J. 2010, *Nature*, 463, 1054

Marigo, P., Girardi, L., Bressan, A., Groenewegen, M. A. T., Silva, L., & Granato, G. L. 2008, *A&A*, 482, 883

Osterman, S., *et al.* 2011, *Ap&SS*, in press

Shkolnik, E., Walker, G. A. H., Bohlender, D. A., Gu, P.-G., & Kürster, M. 2005, *ApJ*, 622, 1075

Shulyak, D., Tsymbal, V., Ryabchikova, T., Stütz, Ch., & Weiss, W. W. 2004, *A&A*, 428, 993

Vidotto, A. A., Jardine, M., & Helling, C. 2010, *ApJL*, 722, L168

The Astrophysics of Planetary Systems: Formation, Structure and Dynamical Evolution
Proceedings IAU Symposium No. 276, 2010
A. Sozzeti, M. G. Lattanzi & A. P. Boss, eds.

doi:10.1017/S1743921311020138

The *Spitzer* search for the transits of HARPS low-mass planets

Michaël Gillon[1], Brice-Olivier Demory[2], Drake Deming[3], Sara Seager[2], Christophe Lovis[4] and the HARPS team

[1]Institut d'Astrophysique et de Géophysique, Université de Liège, 17 Allée du 6 Août, Bat. B5C, 4000 Liège, Belgium
email: michael.gillon@ulg.ac.be

[2]Department of Earth, Atmospheric and Planetary Sciences, Massachusetts Institute of Technology, 77 Massachusetts Ave., Cambridge, MA 02139, USA

[3]Astronomy Department, University of Maryland, College Park, MD 20742-2421, Maryland, USA

[4]Observatoire de Genève, Université de Genève, 51 Chemin des Maillettes, 1290 Sauverny, Switzerland

Abstract. Radial velocity, microlensing and transit surveys have revealed the existence of a large population of low-mass planets in our Galaxy, the so-called 'Super-Earths' and 'Neptunes'. The understanding of these objects would greatly benefit from the detection of a few of them transiting bright nearby stars, making possible their thorough characterization with high signal-to-noise follow-up measurements. Our HARPS Doppler survey has now detected dozens of low-mass planets in close orbit around bright nearby stars, and it is highly probable that a few of them do transit their host star. In this context, we have set up an ambitious *Spitzer* program devoted to the search for the transits of the short period low-mass planets detected by HARPS. We present here this program and some of its first results.

Keywords. planetary systems, eclipses, techniques: photometric, techniques: radial velocities

1. Introduction

Transiting exoplanets are definitely key objects for our study of planetary systems. Except for the planets of our own solar system, they are the only planets we can accurately estimate for mass, radius, and, by inference, constrain the internal composition. Furthermore, the geometry of their orbit relative to Earth makes possible the study of their atmospheric properties without having to resolve them from their host star (see contribution by Charbonneau, this volume). About 100 transiting planets have been detected so far, most of them by dedicated photometric surveys. Among them, the 'hot Jupiters' HD 209458b and HD 189733b are certainly the ones that have been best characterized so far (see contribution by Tinetti, this volume), thanks to the brightness of their host stars (K=6.3 and 5.5). The galore of results obtained for these two planets and a few others have opened a new field of astronomy: exoplanetary science.

Radial velocity (RV) and microlensing surveys have revealed the existence of a large population of planets with a mass of a few to ~20 Earth masses (Lovis *et al.* 2009; Sumi *et al.* 2009). Based on their mass (or minimal mass for RV planets), these planets are loosely classified as 'super-Earths' ($M_p \leqslant 10\ M_\oplus$) and 'Neptunes' ($M_p > 10\ M_\oplus$). This classification is based on the theoretical limit for gravitational capture of H/He, $\sim 10\ M_\oplus$ (Rafikov 2006), and thus implicitly assumes that 'Neptunes' are predominantly ice giants with a significant H/He envelope, and that most 'super-Earths' are massive terrestrial planets. In fact, the exact nature of these low-mass planets remains mysterious. A few of them transit their parent star, but only the 'hot Neptune' GJ 436b (Gillon *et al.* 2007) and the 'mini-Neptune' GJ 1214b (Charbonneau *et al.* 2009) are good targets for a thorough characterization with existing or near-to-come instruments, thanks to the small size and infrared brightness of their parent M-dwarf stars. It is now desirable to extend our understanding of low-mass planets towards solar-type host stars, and it requires the detection of a few 'hot Neptunes' and 'super-Earths' transiting bright nearby FGK-dwarf stars. Doppler surveys target such bright solar-type stars, and searching for the transits of the low-mass planets detected by RV measurements is thus an obvious method of detecting transiting low-mass planets suitable for a thorough characterization.

Our HARPS Doppler survey has now detected a few dozens short-period low-mass planets, enough to make highly probable that a few of them do transit. Detecting these shallow transits could not be done from the ground, and requires a space-based instrument that not only can reach extremely high photometric precisions, but also that can monitor the same star for a few dozens of hours (see discussion in Gillon *et al.* 2010). Indeed, the small amplitude of the RV signals of low-mass planets, their tendency to be found in multiple-planet systems (Lovis *et al.* 2009), and also the RV low-frequency noise of the host star itself, make extremely difficult a precise estimation of the transit timings, and rather large time windows have to be probed. In this context, we have concluded that the best instrumental choice for our transit search was the *Spitzer Space Telescope*. Due to its heliocentric orbit, it can monitor most of the stars for weeks to months during their visibility window, and it has demonstrated on many instances its capacity to detect eclipses with an amplitude of a few hundreds of ppm (e.g. Stevenson *et al.* 2010). We have thus set up a *Spitzer* program devoted to the search for the transits of HARPS low-mass planets. This program has been presently divided in a cycle 5 (cryogenic) *Spitzer* program that targeted the 'super-Earth' HD 40307b (Mayor *et al.* 2009), and a cycle 6 100-hr *Warm Spitzer* program targetting nine other low-mass planets. The results of our HD 40307b program were presented in Gillon *et al.* (2010), and we present here preliminary results for our *Warm Spitzer* program.

2. Overview

Our nine targets were (or will be) observed with the IRAC instrument, at 3.6 or 4.5 μm, the two only channels that have remained operational after the depletion of *Spitzer*'s cryogen in 2009. Table 1 presents these targets (including HD 40307b). For each of them, Table 1 gives the K-magnitude of the host star, the minimal planetary mass measured from RVs, a minimal value for the expected transit depth assuming a pure iron composition (Seager *et al.* 2007), the geometric transit probability, the used IRAC channel and a reference to the RV discovery paper, if any. Including HD 40307b, the formal *a priori* probability that at least one of these planets transits is $\sim$60%. For each of them, our strategy is to deduce from the RVs the 2-σ transit window, possibly after getting new RVs at well chosen phases to improve the transit ephemeris, then to monitor continuously the transit window with *Spitzer*. Once *Spitzer* data are gathered, they are analyzed globally with the RVs for the system, using the Bayesian MCMC method described in Gillon *et al.* (2010). In this analysis, a prior on the planet size is used to avoid detecting spurious very low-amplitude transits. The main final output of the analysis is a posterior transit probability for the planet.

The *Warm Spitzer* photometric time-series are affected by systematic effects, and getting a photometric precision high enough to detect a transit of a few hundreds of ppm requires to take them properly into account. The IRAC 3.6 and 4.5 μm InSb detectors

Table 1. Targets observed in our HARPS-*Spitzer* program.

Target	K	$M \sin i$ $[M_\oplus]$	$Min\ (R_p/R_*)^2$ [ppm]	P_{tr} [%]	IRAC channel [μm]	Obs. duration [hrs]	Reference
HD 10180 c	5.9	13.1 ± 0.5	300	8.5	4.5	10.3	Lovis *et al.* 2011
HD 40307 b	4.8	4.3 ± 0.3	220	7	8	26.8	Mayor *et al.* 2009
HD 47186 b	6.0	23.0 ± 0.5	450	10	3.6	5	Bouchy *et al.* 2009
HD 115617 b	3.0	5.1 ± 0.5	130	9	4.5	12.5	Vogt *et al.* 2010
HD 125612 c	6.8	18 ± 0.8	400	9	3.6 & 4.5	30.3	Lo Curto *et al.* 2010
HARPS-19 b	4.8	11.2 ± 0.9	-[1]	21[1]	3.6	5.9	Lovis *et al.* , in prep.
HD 215497 b	6.8	5.5 ± 0.6	200	8	4.5	13.4	Lo Curto *et al.* 2010
HD 219828 b	6.5	19.1 ± 1.2	95	14	4.5	11	Melo *et al.* 2007
GJ 3634 b	7.5	6.6 ± 1.1	750	6.5	4.5	6.6	Bonfils *et al.* 2011
55 *Cnc* e	4.0	9.5 ± 0.9	210	11	4.5	5	Dawson & Fabrycky 2010

Notes:
[1] Because of its high orbital eccentricity ($e \sim 0.45$), the occultation probability for HARPS-19 b is much larger than its transit probability, so we monitored the *occultation* window with *Spitzer*.

show a large intra-pixel variability that, combined to the jitter of the telescope, leads a strong correlation of the measured fluxes with the stellar position on the chip (Knutson *et al.* 2008). Furthermore, we have noticed that some of our 3.6 μm data show a low-frequency *evolution* of the correlation between fluxes and y-positions, and also a sharp increase of the effective gain during the first hour of observation, similar to the well-documented 'ramp' effect of 8 μm cryogenic data. We model the combination of all these effects by polynomial functions of the time (and/or its logarithm) and the stellar position on the detector. These models for systematics are part of our global model in the MCMC analysis. We also use the calibration method presented by Ballard *et al.* (2010) to cross-check our results. Fig. 1 shows our *Spitzer* photometric time-series for HD 47186, before and after correction for the systematics.

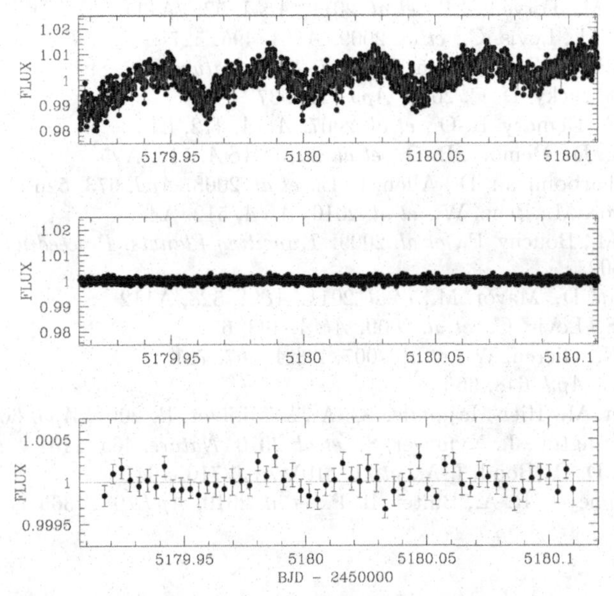

Figure 1. HD 47186 3.6 μm light curve, before (*top*) and after (*middle*) correction for the systematics. *Bottom*: calibrated light curve binned per interval of 5 minutes. Its *rms* is 130 ppm.

3. First results and perspectives

HD 47186 b. This 'hot Neptune' was detected by HARPS in 2009 (Bouchy *et al.*). Thanks to the large amplitude of its orbital RV signal ($K \sim 9$ m.s^{-1}) and to the high precision of the HARPS measurements, its 2-σ transit window was only 5hr. We observed it with *Spitzer*-IRAC at 3.6 μm. Fig. 1 shows the resulting photometry before and after correction for the systematics. No transit is detected. Our global analysis of the *Spitzer* and HARPS data leads to a posterior transit probability of only 0.4%.

HD 10180 c. This 'hot Neptune' was detected recently with six other planets around the nearby solar-type star HD 10180 (Lovis *et al.* 2011). Our *Warm Spitzer* photometry allows us to reject the transiting nature of this planet to a high level of confidence, its posterior transit probability being ~0.7%.

GJ 3634 b. GJ 3634 is a M2.5V star at 17.5 parsec from the solar system. This red dwarf is one of the ~400 targets of our HARPS M-dwarf program. We recently detected a $6.6 \pm 1.1 M_{\oplus}$ 'super-Earth' in very short period orbit around it ($P = 2.64561 \pm 0.0007$ days, Bonfils *et al.* 2011). Our 6-hour long *Warm Spitzer* light curve excludes that a transit occurs within the 2-σ transit window, decreasing the probability that GJ 3634 b undergoes transit down to 0.5%.

The data obtained for HARPS-19, HD 115617, HD 125612, HD 215497 and HD 219828 are still under analysis, but unfortunately our preliminary light curves do not show any obvious transit. Our remaining target for our cycle 6 program, 55 *Cnc*, has still to be observed by *Spitzer*. The prior probability that none of our 10 targets transit was ~ 40%. Still, if we consider all the short-period low-mass planets (~ 35 planets) detected by HARPS so far, the prior probability that at least one of them transits is better than 95%, giving us a strong motivation to pursue this program in the future *Spitzer* cycle 7.

References

Ballard, S., Charbonneau, D., Deming, D., *et al.* 2010, *PASP*, 122, 1341
Bonfils, X., Gillon, M., Forveille, T., *et al.* 2011, *A&A*, 528, A111
Bouchy, F., Mayor, M., Lovis, C., *et al.* 2009, *A&A*, 496, 527
Charbonneau, D., Berta, Z. K., Irwin, J., *et al.* 2009, *Nature*, 462, 891
Dawson, R. I. & Fabrycky, D. C. 2010, *ApJ* 722, 937
Gillon, M., Pont, F., Demory, B.-O., *et al.* 2007, *A&A*, 472, L13
Gillon, M., Deming, D., Demory, B.-O., *et al.* 2010, *A&A*, 518, A25
Knutson, H. A., Charbonneau, D., Allen, L. E., *et al.* 2008, *ApJ*, 673, 526
Lo Curto, G., Mayor, M., Benz, W., *et al.* 2010, *A&A*, 512, A48
Lovis, C., Mayor, M., Bouchy, F., *et al.* 2009, *Transiting Planets, Proceedings of the IAU Symposium 253*, 502
Lovis, C., Ségransan, D., Mayor, M., *et al.* 2011, *A&A*, 528, A112
Mayor, M., Udry, S., Lovis, C., *et al.* 2009, *A&A*, 493, 639
Melo, C., Santos, N., Gieren, W., *et al.* 2007, *A&A*, 467, 721
Rafikov, R. R. 2006, *ApJ*, 648, 666
Seager, S., Kuchner, M., Hier-Majumder, C. A., & Militzer, B. 2007, *ApJ*, 669, 1279
Stevenson, K., Harrington, J., Nymeyer, S., *et al.* 2010, *Nature*, 464, 1161
Sumi, T., Bennett, D. P., Bond, I. A., *et al.* 2010, *ApJ*, 710, 1641
Vogt, S. S., Wittenmeyer, R. A., Bulter, R. P., *et al.* 2010, *ApJ* 708, 1366

The Astrophysics of Planetary Systems: Formation, Structure, and Dynamical Evolution
Proceedings IAU Symposium No. 276, 2010
A. Sozzetti, M. G. Lattanzi & A. P. Boss, eds.

doi:10.1017/S174392131102014X

Understanding exoplanet formation, structure and evolution in 2010

Gilles Chabrier[1,2], Jérémy Leconte[1] and Isabelle Baraffe[2,1]

[1]École Normale Supérieure de Lyon, CRAL (CNRS, UMR 5574), F-69364 Lyon cedex 07, France

[2]Physics & Astronomy, University of Exeter, Exeter EX4 4PE, UK
(chabrier, jeremy.leconte, ibaraffe@ens-lyon.fr)

Abstract. In this short review, we summarize our present understanding (and non-understanding) of exoplanet formation, structure and evolution, in the light of the most recent discoveries. Recent observations of transiting massive brown dwarfs seem to remarkably confirm the predicted theoretical mass-radius relationship in this domain. This mass-radius relationship provides, in some cases, a powerful diagnostic to distinguish planets from brown dwarfs of same mass, as for instance for Hat-P-20b. If confirmed, this latter observation shows that planet formation takes place up to at least 8 Jupiter masses. Conversely, observations of brown dwarfs down to a few Jupiter masses in young, low-extinction clusters strongly suggests an overlapping mass domain between (massive) planets and (low-mass) brown dwarfs, i.e. no mass edge between these two distinct (in terms of formation mechanism) populations. At last, the large fraction of heavy material inferred for many of the transiting planets confirms the core-accretion scenario as been the dominant one for planet formation.

Keywords. planets and satellites: general

---

1. Planet internal structure and evolution

1.1. *General overview*

The realm of extrasolar planet discoveries now extends from gaseous giants of several Jupiter masses down to objects of a few Earth masses. Detailed models of planet structure and evolution have been computed by different groups (Fortney *et al.* 2007, Baraffe *et al.* 2008, Burrows *et al.* 2007, Leconte *et al.* 2009; see Baraffe *et al.* 2010 for a recent review). These calculations include various internal compositions, based on presently available high-pressure equations of state (EOS) for materials typical of planetary interiors. A detailed discussion and a comparison of these models can be found in Baraffe *et al.* (2008)†. This latter paper also explores the effect of the location of the heavy element material in the planet, either all gathered at depth as a central core or distributed throughout the gaseous H/He envelope, on the planet's radius evolution. These different possible distributions of heavy elements can in some cases have an important impact on the planet's contraction. This paper also shows that the presence of even a modest gaseous (H/He) atmosphere hampers an accurate determination of the planet's internal composition, as the highly compressible gas contains most of the entropy of the planet and thus governs its cooling and contraction rate. In such cases, only the average internal composition of the planet can be inferred from a comparison of the models with the observed mass and radius determinations, for transiting objects.

† Models are available at http://perso.ens-lyon.fr/isabelle.baraffe/PLANET08/

Objects below about 10 Earth-masses, globally denominated Super-Earth or Earth-like planets, on the other hand, are not massive enough to retain a significant gaseous atmosphere by gravitational instability (Mizuno 1980, Stevenson 1982, Rafikov 2006). For these objects, the lack of a substantial gaseous atmosphere allows a more detailed exploration of the planet's internal composition than for the gaseous planets (Valencia *et al.* 2007, Seager *et al.* 2007, Sotin *et al.* 2007). It should be kept in mind, however, that, even for these Super-Earth or Earth-like planets, present uncertainties in the EOS of the various heavy elements (e.g. H_2O, Fe) under relevant P-T interior conditions prevent an accurate determination of their internal composition (see e.g. Fig. 2 of Baraffe *et al.* 2008). The melting lines of water or Iron are not even known under such conditions, so the exact thermodynamics state of these elements is unknown. The situation should improve in the coming years with the advent of high-pressure experiments conducted with the high-power laser facilities developed in the US and in France.

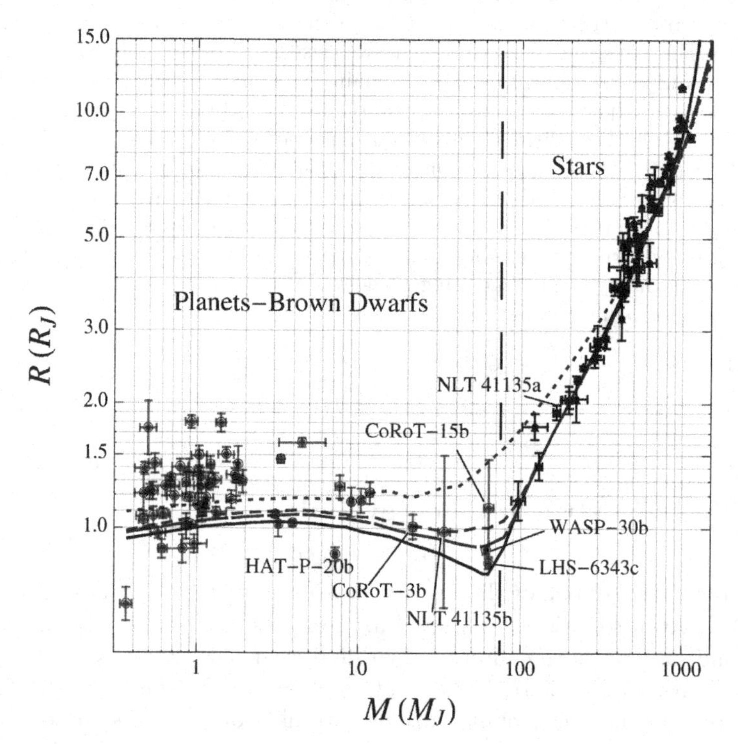

Figure 1. Mass-radius relationship from the stellar to the planetary regime, from one solar mass to one Saturn mass. The four curves display four isochrones, namely, from top to bottom, 10^8 (dot), 5×10^8 (short-dash), 10^9 (long-dash) and 5×10^9 (solid) yr. Some objects are identified on the figure, including the recent field M-dwarf/BD system NLTT 41135a,b (Irwin *et al.* 2010). The group of top 4 objects at $\sim 10 M_J$ includes Wasp 14b, Hat-P-2b, Wasp 18b and XO 3b.

Figure 1 displays the overall mass-radius relationship in the stellar and substellar domains, from a solar mass down to a Saturn mass. The lines denote the low-mass star, brown dwarf and planet models of the Lyon group for 4 isochrones. The vertical dash-line corresponds to the mass limit to reach thermal equilibrium, i.e. balance between nuclear H-burning energy and gravitational contraction energy, $M_{\rm HBMM} = 0.075\,M_\odot$ (Chabrier & Baraffe 1997). This defines the limit between the stellar and brown dwarf domains. The general behaviour of this m-R relationship is discussed in detail in Chabrier & Baraffe (2000) and Chabrier *et al.* (2009) and will not be repeated here, where we will focus on the brown dwarf and planetary part of the domain. It is important however, to point out the recent observations of the 61 M_J and 63 M_J transiting brown dwarfs WASP-30b (Anderson *et al.* 2010) and LHS 6343 C (Johnson *et al.* 2010), respectively, which remarkably confirm the predicted theoretical mass-radius relation in the brown dwarf domain (Chabrier & Baraffe 2000).

As seen in the figure and explored in detail in Baraffe *et al.* (2008) and Leconte *et al.* (2009), for several of these transiting planets, the observed mass-radius relation can be adequately explained by the planet "standard" evolution models mentioned in §1.1. A typical case, for instance, is CoRoT-Exo-4b, a $0.72 M_J$ planet whose 1.17 R_J radius is reproduced at the 1σ level by a model including a $10\, M_\oplus$ water core surrounded by a gaseous H/He envelope, i.e. a global $\sim$ 5% mass fraction of heavy material, more than twice the solar value (Leconte *et al.* 2009)†. Choosing rock or a mixture of water and rock instead of water as the main component of the core only slightly changes this value. Several other transiting planet mass-radius signatures are well explained by standard models with moderate to high (up to $\sim$ 95% for Neptune-mass planets) heavy element enrichment, as expected from the standard "core accretion" scenario for planet formation (see §4).

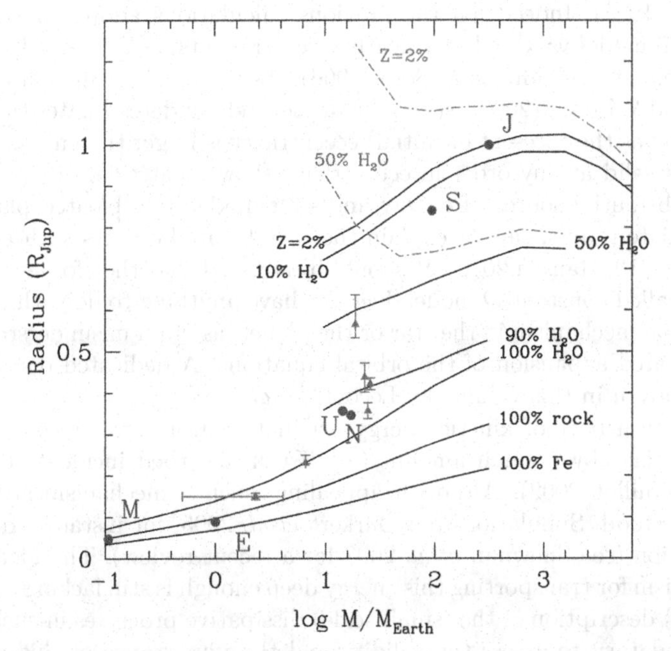

Figure 2. Planetary radii at 4.5 Gyr as a function of mass, from 0.1 $M_\oplus$ to 20 M_J. Models with solar metallicity (Z = 2%) and with different amounts of heavy material (water, "rock" (i.e. olivine or dunite), or iron) are shown (Baraffe *et al.* 2008, Fortney *et al.* 2007). Solid curves are for non-irradiated models while dash-dotted curves correspond to irradiated models at 0.045 AU from a Sun. The positions of Mars, the Earth, Uranus, Neptune, Saturn and Jupiter are indicated by solid points, while the most recent transiting Earth-like (Corot-Exo-7b, GJ1214b) and Neptune-mass (Hat-P-26b, GJ436b, Kepler-4b, Hat-P-11b) planets are indicated by solid triangles.

Figure 2 focuses on the lowest mass part of the planetary domain, from 20 Jupiter masses down to Mars, i.e. going from gaseous giants to nearly incompressible matter.

† Models are available at:
http://perso.ens-lyon.fr/jeremy.leconte/JLSite/JLsite/Exoplanets_Simulations.html

The figure displays the behaviour of the mass-radius relationship in this domain for various internal compositions, and highlights also the impact of stellar irradiation for a typical HD209458b-like system on the radius of a gas-dominated planet. Also indicated on the figure are the locations of the Solar System planets and of the recently discovered Earth-like‡, Super Earth or Neptune-like transiting objects.

1.2. *The planet radius anomaly*

On the other hand, as seen in Figure 1, a large number of transiting planets exhibit a radius significantly larger than predicted by the theory, even when including irradiation effects from the parent star. Denoting $R_{\rm irrad}$ such a theoretical radius, the *radius anomaly* of "Hot Jupiters" is thus defined as $(R_{\rm obs} - R_{\rm irrad})/R_{\rm irrad}$ (see e.g. Leconte *et al.* 2009, 2011b or Fig. 10 of Baraffe *et al.* 2010). Several physical mechanisms have been suggested to explain this radius anomaly. The most promising ones are discussed in details in Baraffe *et al.* (2010) and are quickly summarized below:

- tidal heating due to circularization of the orbit, as originally suggested by Bodenheimer *et al.* (2001). This suggestion has been revisited recently by Leconte *et al.* (2010) using orbital equations which are valid at any order in eccentricity (Hut 1981, see also Eggleton *et al.* 1998). Indeed, all the previous calculations addressing this issue were based on a tidal model valid only for nearly circular orbits, as developed initially for our Solar System planets (Goldreich & Soter 1966). As rigorously demonstrated in Leconte *et al.* (2010) and Wisdom (2008), such a model severely underestimates the tidal dissipation rate as soon as the (present or initial) eccentricity is larger than about 0.2-0.3. Using tidal equations valid at any order in eccentricity shows that tidal dissipation, although providing a substantial source of energy and - for moderately bloated planets - leading to the appropriate radius, cannot explain the very bloated objects such as HD 209458b (Leconte *et al.* 2010, Hansen 2010). It should be stressed that the aforementioned limitation of the so-called constant-Q model does not have anything to do with the description of the dissipation mechanism in the star or the planet, as often misunderstood, but stems from the truncated expansion of the orbital equations. A dedicated discussion of these tidal effects is given in this volume by Leconte *et al.*

- downward transport of kinetic energy originating from strong winds generated at the planet's surface by a small amount ($\sim 1\%$) of absorbed incident stellar radiation (Showman & Guillot 2002). Although appealing, such a mechanism still needs to be correctly understood. Simulations par Burkert *et al.* 2005, for instance, do not produce such a dissipation (see Shownan *et al.* 2008 for a recent review). The identification of a robust mechanism for transporting this energy deep enough is still lacking and an accurate (so far missing) description of the (small-scale) dissipative processes in such natural heat engines is mandatory to assess the validity and the importance of this mechanism for hot-Jupiters (see e.g. Goodman 2009).

- ohmic dissipation in the ionized atmosphere of hot-Jupiters (Batygin & Stevenson 2010). This scenario has received some support from recent 3D resistive MHD atmospheric circulation simulations of HD 209458b's weakly ionized atmosphere (Perna *et al.* 2010). According to these simulations, for magnetic field strengths $B \gtrsim 10$ G, enough ohmic dissipation occurs at deep enough levels (from a few bars to several tens of bars) to affect the internal adiabat and to slow down enough the planet's contraction to yield a significantly inflated radius. These results have to be confirmed by further studies, as quantifying the impact of non-ideal MHD terms and induced currents in numerical simulations is a challenging task.

‡ Note that Fig. 2 displays the revised 1σ mass determination of Corot-7b (Pont *et al.* 2011).

- enhanced opacities ($\sim 10\times$ the solar mixture) in hot-Jupiter atmospheres, stalling the planet's cooling and contraction (Burrows *et al.* 2007). It should be stressed, however, that if the planet H/He envelope's global *metallicity* is enhanced at this level, the increased molecular weight will cancel or even dominate the opacity effect and will lead to a similar or smaller radius than the one obtained with solar metallicity (Guillot 2008). This scenario is thus so far an ad-hoc procedure and enhanced sources of opacities for a global solar-like metallicity must be identified, both theoretically and observationally.

- inefficient (layered or oscillatory) convection in the planet's interior, due to a gradient of heavy elements either inherited from the formation stages or due to core erosion during the planet's evolution (Chabrier & Baraffe 2007). Although layered convection is observed in many situations in Earth lakes or oceans, due to the presence of salt concentrations (the so-called thermohaline convection), it remains unclear, however, whether this process can occur and persist under giant planet interior conditions.

In contrast to the first three scenarios, the last two ones (i) do not invoke an extra source of heating in the planet but rather an hampered output flux during the evolution, leading to a slower contraction rate, (ii) do not necessarily apply to short-period, irradiated planets only but could possibly also occur in planets at large orbital distances.

None of these mechanisms has either been confirmed or ruled out so far. Note that they are not exclusive from each other and it might be possible that they all contribute, at some level, to the puzzling anomalously large radius problem.

2. Departure of short-period planets from sphericity. Effect on the transit light curve and radius determination.

Recent observations have shown that information about the departure of the planet from sphericity, due to rotationally or tidally induced forces, can be obtained from the analysis of planet transit light curves (Welsh *et al.* 2010, Carter & Winn 2010a,b). Because of the tidal and/or rotational deformation (of both the planet and the star) the observed transit cross-section is smaller than the one corresponding to the genuine equilibrium radius of the planet (the one given by definition by 1D structure and evolution models). Recently, Leconte *et al.* (2011a) have investigated such a deformation of short-period planets and have shown that this deformation can have a non negligible impact on (i) the depth of the light curve itself, (ii) the radius of the planet inferred from this light curve. The impact on the depth of the transit is found to be of the order of a few percents for planets orbiting within about 0.04 AU from their host star, and can reach almost 10% for the least massive short-period planets, such as e.g. WASP-19b or WASP-12b, leading to a $\sim 5\%$ effect on the planet's radius determination. These effects must be correctly taken into account when determining the proper equilibrium radius of the planet from the transit observations, to be compared with the 1D theoretical models. As mentioned above, and demonstrated in Leconte *et al.* (2011a), the radius correction on the planet will always lead to a *larger* radius determination than the one obtained when ignoring aspherical deformation, therefore increasing the radius anomaly mentioned in §1.2. Leconte *et al.* (2011a) derive analytical expressions to take these deformation effects into account and to calculate the planet's proper triaxial shape (and thus proper equilibrium radius), for various relevant transiting planetary system conditions (see §5 of Leconte *et al.* 2011a). Using these analytical expressions, one can straightforwardly derive the correct transit depth and planet's radius from the observed (distorted) transiting object.

3. The Brown Dwarf/Planet overlapping mass regime

The distinction between BDs and giant planets has become these days a topic of intense debate. In 2003, the IAU has adopted the deuterium-burning minimum mass, $\sim 10\,M_{\rm J}$, as the official distinction between the two types of objects. We have already discussed the inadequacy of this limit in previous reviews (see e.g. Chabrier *et al.* 2007). As discussed in 4 below, brown dwarfs and planets, although issued from two different formation mechanisms, probably overlap in mass, so that there is no common mass limit between these two populations. Therefore, the recent transit detection of massive companions in the substellar regime ($5\,M_{\rm J} \lesssim M_p \lesssim M_{\rm HBMM}$) raises the questions about their very nature: planet or brown dwarf ?

As mentioned in §1, an internal heavy material enrichment yields a smaller radius, for a given mass, than a solar composition body. Therefore, the m-R relationship provides in principle a powerful diagnostic to distinguish planets from BDs in their overlapping mass domain. In practice, this diagnostic cannot always be obtained. As shown in Leconte *et al.* (2009), for objects such as CoRoT-3 b (Deleuil *et al.* 2008) or HAT-P-2 b, with the revised radius determination (Pál *et al.* 2010), the situation remains ambiguous. On one hand, the observations are consistent with these objects being irradiated solar-metallicity brown dwarfs. On the other hand, given the impossibility so far to assess the nature and, more importantly, the impact of the missing mechanism responsible for the anomalously large radius observed in some short-period planets (see §1.2), these objects can also be strongly inflated irradiated planets, with a substantial metal enrichment. As seen in Figure 1, several substellar objects in the mass range $\sim$ 3-20 $M_{\rm J}$ belong to this category, i.e. have a radius consistent with the object being either an irradiated brown dwarf or planet. As discussed in Leconte *et al.* (2009), this ambiguity can be resolved only in the case where the observed radius is significantly *smaller* than predicted for solar or nearly-solar metallicity (irradiated) objects. As mentioned above, this indeed reveals the presence of a significant global amount of heavy material in the transiting object's interior, as expected from planets formed by core accretion. This is, for instance, the case of Hat-P-20b (see Figure 1, and Fig. 2 of Leconte *et al.* 2011b), which is too dense to be a brown dwarf. According to the calculations of Leconte *et al.* (2011b), this object's radius determination implies more than 340 $M_\oplus$ of heavy material in the planet, i.e. a $Z \gtrsim 15\%$ global mass fraction. Such a heavy material mass fraction is compatible with, although at the upper end of, planet formation efficiency in protoplanetary disks (see eqn.(1) of Leconte *et al.* 2009), according to models of planet formation by core accretion (Mordasini *et al.* 2009). If the mass and radius of Hat-P-20b are confirmed, this object proves that planets can form up to at least $8M_{\rm J}$.

On the other hand, the brown dwarf status of objects such as CoRoT-15 b (Bouchy *et al.* 2011), WASP-30 b (Anderson *et al.* 2011) or LHS 6343 C (Johnson *et al.* 2011) can not be questioned given their mass. Such masses can not be produced by the core accretion mechanism for planet formation, nor by gravitational instability in a disk at this orbit. As mentioned in §1, the radius determination of these objects (at least the two last ones, given the large error bar for CoRoT 15 b) confirms remarkably well the predicted m-R relationship in the BD domain. Comparison with this theoretical relation also shows that these objects are not significantly inflated, a consequence of the smaller incident flux contribution with respect to these object intrinsic internal energy compared with smaller objects.

4. Constraints on planet formation mechanisms

The observation of free floating objects with masses of the order of a few Jupiter masses in (low extinction) young clusters (Caballero *et al.* 2007) shows that star and BD formation extends down to Jupiter-like masses, with a limit set up most likely by the opacity-limited fragmentation, around a few Jupiter-masses (Boyd & Whitworth 2005). Observations show that young brown dwarfs and stars share the same properties and are consistent with BDs and stars sharing the same formation mechanism (Andersen *et al.* 2008, Joergens 2008; see Luhman *et al.* 2007 for a review), as supported by analytical theories of gravo-turbulent collapse of molecular clouds (Padoan & Nordlund 2004, Hennebelle & Chabrier 2008). On the other hand, the fundamentally different mass distributions of exoplanets detected by radial velocity surveys (Udry & Santos 2007) clearly suggests a different formation mechanism. The detection of transiting planets whose radius implies a large enrichment in heavy material, as mentioned in the previous sections, strongly supports the so-called core accretion scenario for planet formation (Pollack *et al.* 1996, Alibert *et al.* 2005, Mordasini *et al.* 2009). Conversely, this large heavy material enrichment clearly excludes the gravitational instability scenario (Boss 1997). The only remaining, although uncertain possibility for this latter is the formation of planets at very large distances ($\gtrsim$100 AU), for the disk, assuming it is massive enough, to be cold enough to violate the Toomre stability condition (Rafikov 2005, Whitworth & Stamatellos 2006; see Dullemond *et al.* 2009 for a recent review on this issue).

According to these two different *dominant* formation mechanisms for stars/BDs and planets, these latter are supposed to have a substantial enrichment in heavy elements compared with their parent star, as observed for our own solar giant planets, whereas BDs of the same mass, issued dominantly from the gravoturbulent collapse of a cloud, should have the same composition as their parent cloud, ie a $Z \sim 2\%$ heavy element mass fraction for a solar-like environment. Furthermore, the aforementioned brown dwarf detections down to a few (~ 5) Jupiter masses, below the deuterium-burning limit, and the planetary nature of Hat-P-20b (if confirmed by further observations) are evidences that there is probably *no mass edge between planets and brown dwarfs but instead that these two populations of astrophysical bodies overlap.*

5. Conclusion and perspectives

In this review, we have examined our present understanding and non-understanding of exoplanet formation, structure and evolution. The results can be summarized as follows:

• the theoretical mass-radius relationship in the brown dwarf and planetary regime seems to be confirmed by recent radius determinations of transiting massive brown dwarfs. When the object's radius is smaller than the one predicted for a gaseous body with solar composition, this m-R relationship enables us to distinguish planets from brown dwarfs in their overlapping mass domain and thus provides a key diagnostic to identify these two distinct populations. In other cases, the diagnostic remains ambiguous and the very nature of the transiting object can not be determined.

• Present models of planet interior structure and evolution stand on relatively robust grounds and enable us to infer with reasonable confidence the gross internal structure and composition of these objects. Uncertainties in the EOS of various elements under the relevant conditions, however, prevent a detailed determination of this composition.

• a large fraction of *gas dominated* transiting planets still exhibit a radius significantly larger than predicted by the models. Several physical mechanisms have been proposed to solve this "radius anomaly" problem but, so far, no firm conclusion about which one, if any, of these mechanisms is the correct one has been reached.

- tidal energy dissipation due to circularization of the orbit, in the planet's interior, although providing a significant extra source of energy to the planet, has been shown not to be sufficient to explain the radius of the most bloated planets, including HD 209458-b. Indeed, when properly calculating the orbital evolution equations in case of a finite present or initial eccentricity, tidal dissipation is shown to occur too quickly during the planet's evolution to explain its present radius. Although a proper treatment of the contribution of dynamical tides is presently lacking, the equilibrium tide contribution calculated with the complete tidal equations still provides a lower limit for tidal dissipation and must be correctly calculated. Interestingly enough, recent observations of spin-orbit misalignment for planets orbiting F stars seem to point to a tidal dissipation in the star, and thus to a dynamical evolution of the system, which depends on the stellar mass, more precisely on the size of the stellar outer convection zone (Winn *et al.* 2010).
- an update of the presently discovered transiting systems confirms the previous analysis of Levrard *et al.* (2009). Only a handful of these systems have enough total angular momentum to reach an orbital equilibrium state. For the vast majority of these systems, the planet experiences ongoing orbital decay and will eventually merge with the star, with the dynamical evolution timescale for the orbit semimajor axis and the stellar spin and obliquity being essentially the lifetime of the system itself (Levrard *et al.* 2009, Matsumura *et al.* 2010).
- departure of both the parent star and the transiting planet from sphericity, because of either rotational or tidal forces, affects both the depth of the transit light curve and the planet's radius determination, and leads to an *underestimate* of this latter. This bias must be corrected to get a proper determination of the planet's genuine equilibrium radius, the one calculated with 1D structure models.
- observations of Hat-P-20-b, if confirmed, show that planets form up to at leat about 8 M_J and thus the brown dwarf and planet mass regimes very likely overlap. Therefore, there is no common mass limit between these two populations of astrophysical bodies, stressing again the inadequacy of the definition put forward by the IAU.
- the large number of transiting planets whose radius implies a substantial fraction of heavy material strongly supports the core accretion scenario formation for planets. In this scenario, the planet embryo originates from accretion of solids onto a core in the protoplanetary disk, leading eventually to dynamical accretion of gas dominated material above about 10 $M_{\oplus}$. Conversely, this same large metal enrichment excludes the gravitational instability scenario as the dominant formation mechanism for planets.

Acknowledgements

The research leading to these results has received funding from the European Research Council under the European Community's 7th Framework Programme (FP7/2007-2013 Grant Agreement no. 247060).

References

Alibert Y., Mordasini C., Benz W., & Winisdoerffer, C. 2005, *A&A*, 434, 343
Anderson, D. *et al.* 2010, *ApJ*, 726, L19
Andersen, M., Meyer, M., Greissl, J., & Aversa, A. 2008, *ApJ*, 683, L183

Boss, A. 1997, *Science*, 276, 1836
Baraffe, I., Chabrier, G., & Barman, T. 2008, *A&A*, 482, 315
Baraffe, I., Chabrier, G., & Barman, T. 2010, *Reports on Progress in Physics*, 73, 016901
Batygin, K. & Stevenson, D. J. 2010, *ApJ*, 714, L238
Bodenheimer, P., Lin, D. N. C., & Mardling, R. A. 2001, *ApJ*, 548, 466
Bouchy, F., *et al.* 2011, *A&A*, 525, A68
Boyd, D. & Whitworth, A. 2005, *A&A*, 430, 1059
Burkert, A., Lin, D., Bodenheimer, P., Jones, C., & Yorke, H. 2005, *ApJ*, 618, 512
Burrows, A., Hubeny, I., Budaj, J., & Hubbard, W. B. 2007, *ApJ*, 661, 502
Caballero, J., *et al.* 2007, *A&A*, 470, 903
Carter, J. & Winn, J. 2010a, *ApJ*, 709, 1219
Carter, J. & Winn, J. 2010b, *ApJ*, 716, 850
Chabrier, G. & Baraffe, I. 1997, *A&A*, 327, 1039
Chabrier, G. & Baraffe, I. 2000, *ARA&A*, 38, 337
Chabrier, G. & Baraffe, I. 2007, *ApJ*, 661, L81
Chabrier, G., *et al.* 2007, *Protostars and Planets V*, B. Reipurth, D. Jewitt, & K. Keil (eds.), University of Arizona Press, 951, 623
Chabrier, G., Baraffe, I., Leconte, J., Gallardo, J., & Barman, T. 2009, *AIPC*, 1094, 102
Deleuil, M., *et al.* 2008, *A&A*, 491, 889
Dullemond, C., Durisen, R., & Papaloizou, J. 2009, *Structure Formation in Astrophysics*, Ed. G. Chabrier, Cambridge U. Press
Eggleton P., Kiseleva, L. & Hut, P. 1998, *ApJ*, 499, 853
Fortney, J. J., Marley, M. S., & Barnes, J. W. 2007, *ApJ*, 659, 1661
Goldreich, P. & Soter, S. 1966, *Icarus*, 5, 375
Goodman, J. 2009, *ApJ*, 693, 1645
Guillot, T. 2008, *Physica Scripta* 130, 014023
Hansen, B. 2010, *ApJ*, 723, 285
Hennebelle, P. & Chabrier, G. 2008, *ApJ*, 684, 395
Hut, P. 1981, *A&A*, 99, 126
Irwin, J., *et al.* 2010, *ApJ*, 718, 1353
Joergens, V. 2008, *A&A*, 492, 545
Johnson, J., *et al.* 2011, *ApJ*, 730, id.79
Leconte, J., Baraffe, I., Chabrier, G., Barman, T., & Levrard, B. 2009, *A&A*, 506, 385
Leconte, J., Chabrier, G., Baraffe, I., & Levrard, B. 2010, *A&A*, 516, A64
Leconte, J., Lai, D., & Chabrier, G. 2011a, *A&A*, 528, A41
Leconte, J., Chabrier, G., Baraffe, I., & Levrard, B. 2011b, *EPJ Web of Conferences*, 11, id.03004
Levrard, B., Winisdoerffer, C., & Chabrier, G. 2009, *ApJ*, 692, L9
Luhman, K. L., *et al.* 2007, *Protostars and Planets V*, B. Reipurth, D. Jewitt, and K. Keil (eds.), University of Arizona Press, 951, 443
Matsumura, S., Peale, S., & Rasio, F. 2010, *ApJ*, 725, 1995
Mizuno, Z. 1980, *Prog. Th. Phys.*, 64, 544
Mordasini, C., Alibert, Y., & Benz, W. 2009, *A&A*, 501, 1139
Padoan, P. & Nordlund, A. 2004, *ApJ*, 617, 559
Pál, A., *et al.* 2010, *MNRAS*, 401, 2665
Perna R., Menou, K. & Rauscher, E. 2010, *ApJ*, 724, 313
Pollack, J., *et al.* 1996, *Icarus*,124, 62
Pont, F., Aigrain, S., & Zucker, S. 2011, *MNRAS*, 411, 1953
Rafikov, R. 2005, *ApJ*, 621, L69
Rafikov, R. 2006, *ApJ*, 648, 666
Seager, S., Kuchner, M., Hier-Majumder, C. A., & Militzer, B. 2007, *ApJ*, 669, 1279
Showman, A. & Guillot, T. 2002, *A&A*, 385, 166
Showman A., Menou K., & Cho J. 2008, *ASP Conf. Ser.*, 398, 419
Sotin, C., Grasset, O., & Mocquet, A. 2007, *Icarus*, 191, 337
Stevenson, D. J. 1982, *Ann. Rev. of earth and planetary sc.*, 10, 257

Udry, S. & Santos, N. 2007, *ARA&A*, 45, 397
Valencia, D., Sasselov, D. D., & O'Connell, R. J. 2007, *ApJ*, 665, 1413
Welsh, W., *et al.* 2010, *ApJ*, 713, L145
Whitworth, A. & Stamatellos, D. 2006, *A&A*, 458, 817
Winn, J., Fabrycky, D., Albrecht, S., & Johnson, J. 2010, *ApJ*, 718, L145
Wisdom, J. 2008, *Icarus*, 193, 637

The Astrophysics of Planetary Systems: Formation, Structure, and Dynamical Evolution
Proceedings IAU Symposium No. 276, 2010
A. Sozzetti, M. G. Lattanzi & A. P. Boss, eds.

doi:10.1017/S1743921311020151

Composition of transiting and transiting-only super-Earths

Diana Valencia[1*]

[1]Université de Nice-Sophia Antipolis, CNRS UMR 6202,
Observatoire de la Côte d'Azur, B.P. 4229, 06304 Nice Cedex 4, France
email: dianav@mit.edu

*Now at 54-1710, Earth, Atmospheric and Planetary Sciences Department,
Massachusetts Institute of Technology, 77 Massachusetts Ave, Cambridge, MA, 02139

Abstract. The relatively recent detections of the first three transiting super-Earths mark the beginning of a subfield within exoplanets that is both fruitful and challenging. The first step into characterizing these objects is to infer their composition given the degenerate character of the problem. The calculations show that Kepler-10b has a composition between an Earth-like and a Mercury-like (enriched in iron) composition. In contrast, GJ 1214b is too large to be solid, and has to have a volatile envelope. Lastly, while three of the four reported mass estimates of CoRoT-7b allow for a rocky composition, one forbids it and can only be reconciled with significant amounts of water vapor. In addition to these three transiting low-mass planets, there are now more than one thousand Kepler planets with only measured radius. Even without a mass measurement ("transiting-only") it is still possible to place constraints on the amount of volatile content of the highly-irradiated planets, as their envelopes, if present, are flared. Using Kepler-9d as an example, we estimate its water vapor, or hydrogen and helium content to be less than 50% or 0.1% by mass respectively.

Keywords. stars: individual (CoRoT-7, GJ 1214, Kepler-9, Kepler-10), planets and satellites: physical evolution, planets and satellites: interiors

---

1. Introduction

In the quest for finding habitable worlds, the efforts to detect small planets are starting to pay off with the discovery of the first three transiting super-Earths: CoRoT-7b, GJ1214b and Kepler-10b. They exemplify the fruitful results of three different missions that have the potential to detect low-mass planets: the french-led CoRoT mission (Bordé *et al.* 2003), the MEarth ground mission that surveys the nearest M dwarfs stars (Irwin *et al.* 2009) and the space mission Kepler, that can detect planets as small as Earth thanks to its unprecedented precision (Borucki *et al.* 2003). In the next few years, the count for super-Earths is expected to grow, especially from objects being measured by Kepler. However, many of them will not have measured masses as Kepler's targets are a challenge for the radial velocity telescopes observing the same field of view. We will have to wait for HARPS-NEF (Latham 2007) to be built before measuring their masses.

Armed with masses and radii, and an appropriate internal structure model we can infer the composition of super-Earths. Owing to the degenerate character of the problem, there is no unique solution, and what we can infer is bounds to the composition. However, the very short period planets have an additional constraint, as their insolation values are very high making their atmospheres susceptible to escape. This is the case for the first three transiting planets, especially CoRoT-7b and Kepler-10b that have orbital periods of less than one day! In addition, there is one hot transiting-only planet, Kepler-9d, for which only the radius is known. Despite the lack of information on its mass, the planet is also so irradiated, that it is possible to place some constraints on its composition.

In these proceedings I will describe what we have learned about the composition of each of the transiting low-mass planets, and use Kepler-9d as an example to show what can be inferred for transiting-only hot super-Earths.

2. Model

In contrast to the structure of the gaseous planets, which have very small cores compared to their fluid H/He envelopes, super-Earths and to some extent mini-Neptunes are mostly dominated by their rocky/icy cores, and their atmospheres play only a small role in terms of their bulk composition. This is why an internal structure model that takes into account the complexity of rock and ice structure, as seen in the terrestrial planets and icy satellites in our Solar System, is the appropriate one for super-Earths. Despite having non-massive gaseous envelopes (up to several earth-masses), mini-Neptunes call for a model that calculates correctly the temperature structure of the envelope, as it has a substantial effect on the density of water vapour and H/He.

The results presented here were obtained by combining the internal structure model by Valencia *et al.*(Valencia *et al.* 2006, 2007) for the rocky/icy interiors with the internal structure model CEPAM (Guillot & Morel 1995; Guillot 2005) for the gaseous envelope composed of H_2O and/or H/He. While the former model has been applied and tested for super-Earths and the solid planets in the solar system, the latter has been used successfully to model the giant planets in the solar system and in extra solar systems. The boundary condition at the solid-gaseous interface satisfies continuity in pressure and mass. Valencia *et al.* 2010 discuss details of the combined model.

3. Results

3.1. *CoRoT-7b and Kepler-10b*

CoRoT-7b and Kepler-10b are planets that share similar features: their radii, their period, the type of host star and perhaps their mass. Their major differences are their age, and the uncertainty on their masses. CoRoT-7 is a very active star, which makes the radial velocity data very noisy. Several studies have suggested different values for the mass, with a combined uncertainty of $1 - 10\,M_E$ (see Table 1). However, three of the four studies suggest that the mass is compatible with a rocky composition.

Kepler-10b, on the other hand, has a well determined mass and an exquisitely well determined radius $R = 1.416^{+0.033}_{-0.036} R_E$ (Batalha *et al.* 2011), thanks to the tight constraints on the star allowed by asteroseismology. The object is compact enough that a rocky composition is inferred.

The range of possible rocky compositions vary mostly due to the amount of total iron the planet has, which is distributed mostly in the core and some in the mantle, unless the planet is undifferentiated. To span the range of rocky compositions we consider two unlikely extremes: a pure iron planet, and a planet devoid of any iron (close to a Moon-like composition); and two compositions present in the solar system: Earth-like (33% iron core, 67% silicate mantle, with an iron to silicate ratio of 2), and a composition enhanced in iron similar to that of Mercury (63% iron core, 37% silicate mantle).

CoRoT-7b's first mass $M = 4.8 \pm 0.8\,M_E$ (Queloz *et al.* 2009) and radius $R = 1.68 \pm 0.09\,R_E$ estimations (Léger *et al.* 2009) made it a planet lighter than Earth (see Figure 1). With improved stellar parameters, the radius was revised to a smaller value of

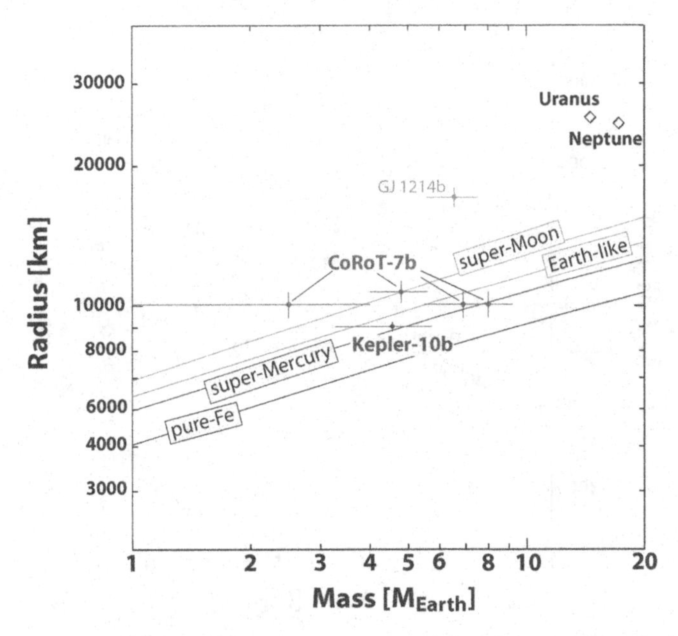

Figure 1. CoRot-7b and Kepler-10b's composition as rocky planets. The relationships for four rocky compositions: a super-moon (no iron), Earth-like (67% silicate mantle with 10% iron by mol + 33% iron core), super-Mercury (in this study as 37% silicate mantle with no iron by mol + 63% iron core), and pure iron are shown. Data for Kepler-10b, and CoRoT-7b (with its corresponding four mass estimates, and revised radius value – see text for references) are shown. Uranus, Neptune and GJ 1214b are shown for reference

$R = 1.58 \pm 0.10\, R_E$ (Bruntt *et al.* 2010). This more compact scenario made the planet more 'Earth-like'. Subsequently, the mass of the planet has been intensely revised. While Hatzes *et al.* (2010) and Ferraz-Mello *et al.* (2011) both obtain larger values of $M = 6.9 \pm 1.4\, M_E$ and $M = 8.0 \pm 1.2\, M_E$ respectively, making the planet denser and compatible with a composition between Earth-like and Mercury-like, Pont *et al.* (2011) suggests a mass of only $M = 1 - 4\, M_E$ which can not be reconciled with a rocky composition. This low value can only be satisfied with significant amounts of volatiles.

On the other hand, the mass and radius of Kepler-10b is concordant with a composition between an Earth-like and a Mercury-like planet. Thus, if we disregard the Pont *et al.* (2011) suggested mass value for CoRoT-7b, both planets appear to be almost identical in composition. This stresses the importance of establishing the reliability of Pont *et al.* (2011) treatment to infer the mass of CoRoT-7b.

While we might be tempted at first to classify these planets as rocky based on their bulk densities and proximity to their host stars, only through a rigorous analysis on the likelihood of other compositions can we be sure that they are indeed telluric planets. To this effect, we considered planets with different amounts of water vapor or a hydrogen and helium mixture above an Earth-like nucleus, combined with a simple analysis of atmospheric escape to determine if the timescales of evaporation of these envelopes are consistent with the ages of the planets (Valencia *et al.* 2010). Figure 2 shows the results. Based on a simple hydrodynamic espace treatment the timescale of evaporation of water vapor is around a gigayear, while for H-He it is only a few million years. Given the compact size of both planets, and the short timescale of evaporation, we can infer that there is no H-He in both Kepler-10b and CoRoT-7b, even with the low mass value suggested by Pont *et al.* (2011). It is also clear that Kepler-10b is too compact to allow for any

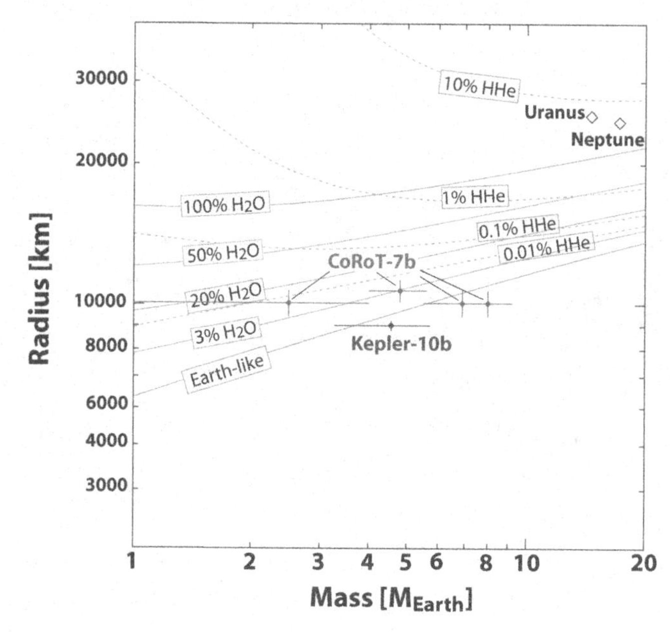

Figure 2. Volatile content of CoRot-7b and Kepler-10b's. Data for CoRoT-7b and Kepler-10b is shown. Two families of volatile content are shown: (in pink) 0.01, 0.1, 1, and 10% by mass of H-He and (in blue) 3, 20, 50, 100% by mass of H_2O above an Earth-like nucleus.

significant water vapor, plus the fact that it is a much older planet (at least 8 billion years old) suggests the evaporation would have taken place for longer. Therefore, it is safe to assume it is a rocky planet.

For CoRoT-7b, the timescale of evaporation of water vapor is of the same order of magnitude as the age of the system. Thus, it is possible that water vapor has not evaporated completely from the planet. According to Pont *et al.* (2011)'s estimate, CoRoT-7b has at most a few tens of percents (less than 40%) of water vapor (see Fig. 2). For the larger mass values suggested by Queloz *et al.* (2009), and Hatzes *et al.* (2010) , there is less than a few percents of water vapor. And for the mass values suggested by Ferraz-Mello *et al.* (2011) to have some water vapor, the solid nucleus beneath has to be significantly enhanced in iron with respect to an Earth-like composition.

3.2. *GJ 1214b*

In contrast to CoRoT-7b and Kepler-10b, there is no ambiguity that GJ 1214b has a volatile envelope. This planet has a radius that is about 1000 km larger than if it was made of pure ice, the lightest composition for a solid planet. The important question is the nature of this envelope. As a starting point, we consider planets that have different amounts of water vapor above an Earth-like nucleus, up to a composition that is of pure H_2O (see Fig. 3). The boundary condition we used is based on the calculations by Miller-Ricci & Fortney (2010) for the pressure-temperature of the atmosphere of this planet. For different compositions, they obtain a value close to 1000 K at 10 bars. For this boundary condition, we find that GJ 1214b can be made of 100% water vapor. However, this composition is unlikely. Before water can condense out of the solar nebula, refractory material has to condense out. An upper limit to the ratio of water to refractory material can be estimated from the composition of comets, with a dust to gas ratio of 1-2. Therefore, given the fact that we expect some rocky material to be present in the composition of this planet, we also expect a component ligher than water. We propose this

Table 1. Data for transiting super-Earths and Kepler-9d

	Mass (M_E)	Radius (R_E)	Period (days)	Teff (K)	Age (Gy)	Ref
CoRoT-7b		1.68±0.09	0.854	1800-2550[a]	1.2-2.3	Léger *et al.* (2009)
	4.8±0.8					Queloz *et al.* (2009)
		1.58±0.10				Bruntt *et al.* (2010)
	6.9±1.4					Hatzes *et al.* (2010)
	8.0±1.2					Ferraz-Mello *et al.* (2011)
	1-4					Pont *et al.* (2011)
Kepler-10b	$4.56^{+1.17}_{-1.29}$	$1.416^{+0.033}_{-0.036}$	0.837	2150-3050[a]	11.9±4.5	Batalha *et al.* (2011)
GJ 1214b	6.55±0.98	2.678±0.13	1.58	500-700 [b]	3-10	Charbonneau *et al.* (2009)
Kepler-9d	-		1.59	1620-2300[a]	2-4	Holman *et al.* (2010)
	-	$1.64^{+0.19}_{-0.14}$		or 1800-2500 [b]		Torres *et al.* (2011)

Notes:
[a] Effective temperatures are calculated with an albedo = 0 assuming a rocky composition, assuming no and full redistribution over the planetary surface.
[b] Effective temperature calculated with an albedo = 0.3 assuming a water composition, assuming no and full redistribution over the planetary surface.

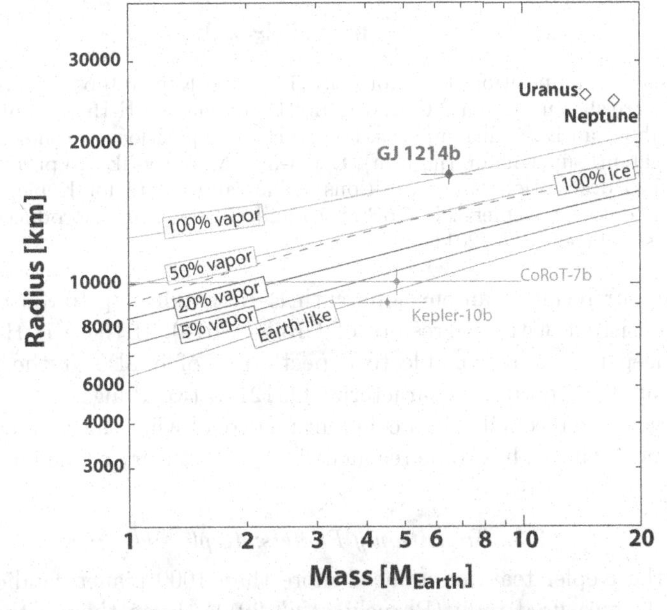

Figure 3. GJ 1214b volatile composition. Data for CoRoT-7b and Kepler-10b is shown. Mass-radius relationships for a composition of 5, 20, 50, 100% by mass of H_2O above an Earth-like nucleus. The pure-ice line shows that GJ 1214b is a vapor planet.

to be a hydrogen and helium mixture (see Rogers & Seager (2010) for other suggestions). We calculate the maximum amount of H-He to be 8% by mass by adding varying amounts of refractory material in the form of an Earth-like composition. An important note to make is that this result hinges on the boundary condition for the atmosphere that we chose. At this point, we are testing the sensitivity of this result.

One advantage of characterizing GJ1214b, is that the system is amenable to further for observations. In fact, two different groups have obtained transmission spectra to constrain the composition of the atmosphere at the millibar lever. While Croll (2011) suggest that the composition is H-He, Bean *et al.* (2010) report a non-detection. To reconcile their results, the latter authors argue that they might be seeing hazes.

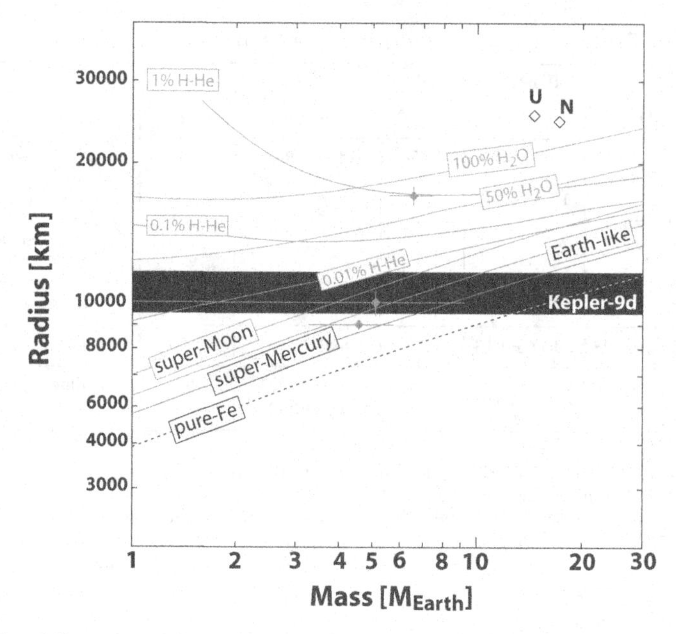

Figure 4. Kepler-9d's composition. Data for CoRoT-7b and Kepler-10b is shown. Two families of volatile content are shown: (in pink) 0.01, 0.1, and 1% by mass of H-He and (in blue) 50, 100% by mass of H_2O above an Earth-like nucleus. Four rocky compositions are shown varying in the amount of iron content: super-moon (no iron), Earth-like, Mercury-like, to pure-iron planet (see figure 1 for detailed description of compositions). The radius data for Kepler-9d is shown in black. In grey, the data for Kepler-10b, CoRoT-7b and GJ1214b are reproduced for reference. For related work see Havel *et al.* (2011).

At first value, our result of an envelope with H-He making up to 8% by mass of the planet seems consistent with the results of Croll (2011). If there is H-He at the 10 bar level and deeper, it is reasonable to expect some of it also at the millibar level, more so than water. To properly characterize GJ1214b two things have to happen: the observations have to be reconciled in a coherent picture of what the composition is at the millibar lever, and then this has to be reconciled with the theory of the internal structure of the planet.

3.3. *Transit-only Planets: Kepler-9d*

Early in 2011, the Kepler team announced more than 1000 planet candidates, most of them having only measured radii (Borucki *et al.* 2011). Even though there cannot be any absolute inferences on the composition of a planet without a mass measurement, it is still possible to put limits on the amount of volatiles of the short-period planets that only have a radius measurement. An example is Kepler-9d with a measured radius of $1.64^{+0.19}_{-0.14}\ R_E$ (Torres *et al.* 2011). This is the smallest object orbiting Kepler-9 at 1.59 days, in a system with two saturn-like planets in close resonance at 19.2 and 38.9 day periods (Holman *et al.* 2010). Even though there is no confirmation from radial velocity observations, (Torres *et al.* 2011) have done a careful analysis to assess the likelihood of false positives from which they conclude that Kepler-9d is most likely a planet. A maximum mass of 15-20 M_E is estimated from a non-detection in the radial velocity measurements (private communication with M. Holman).

Figure 4 shows the radius range for Kepler-9d and different mass-radius relationships. One advantage of the short-period planets is that if they have envelopes they are very hot and expanded, limiting the amount of volatiles present in the composition for a given radius constraint. This is exemplified by the flaring of the mass-radius curves for H-He compositions for planets with masses below 5 M_E and to some extent of the H_2O compositions for planets with masses below 2 M_E. These planets do not have enough gravity to keep tightly bounded their very hot and expansive envelopes. This means that for Kepler-9d we can rule out compositions with more than a few tens in 10000 parts of H-He by mass, and more than 50% water vapor content.

On the other hand, for this planet to be rocky, its mass would have to be more than 3 M_E. Larger masses would call for more iron content, with an Earth-like composition corresponding to a mass range between 4-10 M_E, a Mercury-like composition to the range 7-15 M_E, and a pure iron composition to the range 13-30 M_E. The latter would be an unlikely composition, perhaps possible only after complete evaporation of a silicate mantle. Thus, based on physical grounds, the absolute maximum mass for Kepler-9d is 30 M_E, although a more realistic upper limit is $\sim$15 M_E, when using Mercury's composition as a proxy.

4. Conclusions

Now that we have formally entered the era of super-Earths with the first transiting such planets having measured masses and radii, we can start placing constraints on their composition. Given the degenerate character of the problem, the first challenge is to differentiate the planets that are mainly solid (rocky or icy) from the ones that have non-negligible volatile envelopes (more than a few percent by mass). This task plus the bias towards detecting short-period planets brings to light the importance of better understanding and modelling atmospheric escape, as these planets are highly irradiated by their host stars. Any composition that allows for a volatile envelope has to be reconciled with the planet's history of mass loss.

From this initial small sample of low-mass planets, there are two robust although simple conclusions: Kepler-10b is a rocky planet, while GJ 1214b has a significant amount of volatiles (better termed a mini-Neptune). Kepler-10b's composition ranges in iron content from an Earth-like (67% silicate mantle + 33% iron core) to a Mercury-like (63% silicate mantle + 37% iron core) composition.

Although it is clear that GJ 1214b has a volatile envelope, it is unclear what the nature of of this envelope is, and especially if it is water vapor dominated or H-He dominated. From internal structure models, the maximum amount of H-He possible is 8% by mass. In addition, the two reported observations on the scale height, and therefore the composition of the atmosphere (at the millibar level) seem contradicting. We stand to learn a lot about the mini-Neptunes through a careful characterization of GJ 1214b.

On the other hand, the nature of CoRoT-7b is controversial, given the large uncertainty on its mass. Three of the four studies point towards a plausible rocky composition, and one forbids it. Determining if this planet has a mass less than 4 M_E is key, to classify it as a super-Earth or a mini-Neptune, with implications about its origin and evolution. In any case, we can rule H-He in the envelope given the relatively small size of the planet, and the short timescale of evaporation of such an envelope compared to the age of the system.

Lastly, even though it is impossible to characterize transit-only planets, we can still estimate the maximum mass of these planets and put constraints on their amount of volatile content (especially for the hot ones). One example is Kepler-9d, for which we find that a reasonable maximum mass based on physical grounds is 15 M_E (consistent with a non-detection given the precision of the radial velocity analysis), a maximum amount of water vapor at the 50% by mass level, and less than 0.1% of H-He.

References

Batalha, N. M., Borucki, W. J., Bryson, S. T., Buchhave, L. A., Caldwell, D. A., *et al.* 2011, *ApJ*, 729, 27

Bean, J. L., Kempton, E. M.-R., & Homeier, D. 2010, *Nature*, 468, 669

Bordé, P., Rouan, D., & Legér, A. 2003, *A&A*, 405, 1137

Borucki, W. J., Koch, D., Basri, G., Brown, T., Caldwell, D., *et al.* 2003, in Proceedings of the Conference on Towards Other Earths: DARWIN/TPF and the Search for Extrasolar Terrestrial Planets, *ESA SP-539*, 69

Borucki, W. J., Koch, D. G., Basri, G., Batalha, N., Brown, T. M., *et al.* 2011, *ApJ*, 736, A19

Bruntt, H., Deleuil, M., Fridlund, M., Alonso, R., Bouchy, F., Hatzes, A., Mayor, M., Moutou, C., & Queloz, D. 2010, *A&A*, 519, A51

Charbonneau, D., Berta, Z. K., Irwin, J., Burke, C. J.,Nutzman, P., *et al.* 2009, *Nature*, 462, 891

Croll, B. 2011, *This Volume*

Ferraz-Mello, S., Tadeu dos Santos, M., Beauge, C., Michtchenko, T. A., & Rodriguez, A. 2011, *A&A*, 531, A161

Guillot, T. 2005, *Annual Review of Earth and Planetary Sciences*, 33, 493

Guillot, T. & Morel, P. 1995, *A&A*, 109, 109

Hatzes, A. P., Dvorak, R., Wuchterl, G., Guterman, P., Hartmann, M., Fridlund, M., Gandolfi, D., Guenther, E., & Pätzold, M. 2010, *A&A*, 520, A93

Havel, M., Guillot, T., Valencia, D., & Crida, A. 2011, *A&A*, 531, A3

Holman, M. J., Fabrycky, D. C., Ragozzine, D., Ford, E. B., *et al.* Steffen, *et al.* 2010, *Science*, 330, 51.

Irwin, J., Charbonneau, D., Nutzman, P., & Falco, E. 2009, *Proc. IAU Symp.*, 253, 37

Latham, D. W. 2007, *BAAS*, 38, 234

Léger, A., Rouan, D., Schneider, J., Barge, P., Fridlund, M., *et al.* 2009 *A&A*, 506, 287

Miller-Ricci, E. & Fortney, J. J. 2010, *ApJ*, 716, L74

Pont, F., Aigrain, S., & Zucker, S. 2011, *MNRAS*, 411, 1953

Queloz, D., Bouchy, F., Moutou, C., Hatzes, A., Hébrard, G., *et al.* 2009, *A&A*, 506, 303

Rogers, L. A. & Seager, S. 2010, *ApJ*, 716, 1208

Torres, G., Fressin, F., Batalha, N. M., Borucki, W. J., Brown, T. M., *et al.* 2011, *ApJ*, 727, 24

Valencia, D., O'Connell, R. J., & Sasselov, D. D. 2006, *Icarus*, 181, 545

Valencia, D., Sasselov, D. D., & O'Connell, R. J. 2007, *ApJ*, 656, 54

Valencia, D., Ikoma, M., Guillot, T., & Nettelmann, N. 2010, *A&A*, 516, A20+

The Astrophysics of Planetary Systems: Formation, Structure, and Dynamical Evolution
Proceedings IAU Symposium No. 276, 2010
A. Sozzetti, M. G. Lattanzi & A. P. Boss, eds.

doi:10.1017/S1743921311020163

GJ 1214b and the prospects for liquid water on super Earths

Leslie A. Rogers[1] **and Sara Seager**[2]

[1]Department of Physics, Massachusetts Institute of Technology
37-602, 77 Massachusetts Ave., Cambridge, MA 02139, USA
email: larogers@mit.edu

[2]Department of Earth, Atmospheric, and Planetary Sciences, Department of Physics
Massachusetts Institute of Technology
54-1626, 77 Massachusetts Ave., Cambridge, MA 02139, USA
email: seager@mit.edu

Abstract. GJ 1214b is one of the first discovered transiting planets having mass (6.55 $M_\oplus$) and radius (2.678 $R_\oplus$) smaller than Neptune. To account for its low average density (1870 $\mathrm{kg\,m^{-3}}$), GJ 1214b must have a significant gas component. We use interior structure models to constrain GJ 1214b's gas envelope mass, and to explore the conditions needed to achieve within the planet pressures and temperatures conducive to liquid water. We consider three possible origins for the gas layer: direct accretion of gas from the protoplanetary nebula, sublimation of ices, and outgassing from rocky material. Despite having an equilibrium temperature below 647 K (the critical temperature of water) GJ 1214b does not have liquid water under most conditions we consider. Even if the outer envelope is predominantly sublimated water ice, in our model a low intrinsic planet luminosity (less than 2 TW) is needed for the water envelope to pass through the liquid phase; at higher interior luminosities the outer envelope transitions from a vapor to a super-fluid then to a plasma at successively greater depths.

Keywords. Stars: individual (GJ1214), planetary systems

---

1. Introduction

The MEarth transiting planet, GJ 1214b (Charbonneau *et al.* 2009), is exciting for many reasons. Firstly, it lies in a mass and density regime for which there are no solar system analogs; with $M_p = 6.55 \pm 0.98\ M_\oplus$ and $R_p = 2.678 \pm 0.13\ R_\oplus$ GJ 1214b is smaller than the ice giants Neptune and Uranus, while larger than the terrestrial Earth, Venus, and Mars. Secondly, GJ 1214b has a low enough density ($\rho_p = 1870 \pm 400\ \mathrm{kg\,m^{-3}}$) that it cannot be composed of rocky and iron material alone, and almost certainly contains a gas component. Finally, orbiting with a 1.5803952 ± 0.0000137 day period around an $L_* = 0.00328 \pm 0.00045\ L_\odot$ M dwarf, GJ 1214b has an equilibrium temperature below 647 K, the critical point of water ($T_{eq} = 555$ K for a Bond albedo of 0).

We use planet interior structure models to constrain the bulk composition of GJ 1214b. With our current knowledge of GJ1214b (its mass, radius and orbit) we can not pinpoint GJ 1214b's unique true interior composition (see, e.g. Valencia *et al.* 2007; Adams *et al.* 2008), but we can quantify the range of possibilities that are consistent with the transit and radial velocity observations. In this proceedings we focus on bounding the mass of GJ 1214b's gas envelope and on studying GJ 1214b's prospects for harboring liquid water. We examine three possible sources for the GJ 1214b gas layer (direct accretion of gas from the protoplanetary nebula, sublimation of ices, and outgassing from rocky material) and consider end-member scenarios in which only one gas-layer source was important in forming the current gas layer on GJ 1214b.

2. Model

We model GJ 1214b as a spherically symmetric differentiated planet in hydrostatic equilibrium. We consider up to four layers within the planet: an iron core, silicate mantle, ice layer, and gas envelope. The gas envelope is assumed to be in radiative-convective equilibrium, with a thin outer radiative zone surrounding a convective zone at greater depths. Within the outer radiative zone we adopt the temperature profile from Equation (45) of Hansen (2008), an analytic "two-stream" solution to the gray equations of radiative transfer for a plane-parallel irradiated atmosphere. The onset of convective instabilities ($0 < (\partial \rho / \partial s)_P \, ds/dm$) determines depth of the transition to the convective layer of the gas envelope. In the convective regime, we adopt an adiabatic temperature profile.We neglect thermal effects within the core, mantle and ice layer since at the high pressures found in these interior layers, thermal corrections have only a small effect on the mass-density (Seager *et al.* 2007). A more detailed description of our model (including the equations of state employed) can be found in Rogers & Seager (2010a) and Rogers & Seager (2010b).

3. Results

3.1. *Direct Accretion Scenario*

We first consider the possibility that GJ 1214b's gas layer originated from direct accretion from the protoplanetary disk and comprises a roughly solar mixture of H/He. In this case, we take a Neptune-inspired template with four chemically distinct layers in the planet interior: an iron core, silicate mantle, water-ice layer, and H/He envelope with $Y = 0.28$. Figure 1 shows how the planet mass may be distributed among the four layers and still be consistent with GJ 1214b's measured mass and radius.

In this scenario, GJ 1214b requires between 10^{-4} and 6.8% H/He by mass to account for the measured planet mass and radius within 1 σ. This range of possible gas layer masses not only includes observational uncertainties, but also uncertainties in the planet's atmospheric pressure-temperature (PT) profile and the iron:silicate:ice mass ratio of the inner three layers. In proportion to the planet mass, the H/He envelope on GJ1214b would be more massive than Venus gas layer ($\sim 10^{-4} M_p$) but less massive than Neptunes (0.05 M_p to 0.15 M_p).

3.2. *Sublimation of Ices Scenario*

We next explore the scenario in which a layer of vapors from sublimated ices accounts for GJ 1214b's transit radius. This could be the case if GJ 1214b did not manage to accrete or retain any H/He from the protoplanetary disk, yet still formed from ice-rich material beyond the snow line. For this scenario, we model the planet interior with an iron core, silicate mantle, and water envelope and show in Figure 2 how GJ 1214b's mass may be distributed between these three layers. A sublimated vapor dominated envelope on GJ 1214b is possible only if water accounts for a large fraction of planet mass. To account for the observed planet mass and radius within their 1, 2, and 3σ observational uncertainties, at least 47%, 24%, and 6% water by mass are required, respectively.

GJ 1214b does not contain liquid water in any of our model interiors displayed in Figure 2. Instead, the putative GJ 1214b water envelopes begin in the vapor phase at low pressures, then continuously transition to a super-fluid at $P = 22.1$ MPa (the critical pressure of water), before eventually becoming an electronically conductive dense fluid plasma at greater depths. To obtain liquid water in our model interior, we must decrease GJ 1214b's assumed intrinsic energy flux below 5×10^{-4} W m^{-2}. For comparison, Earth's

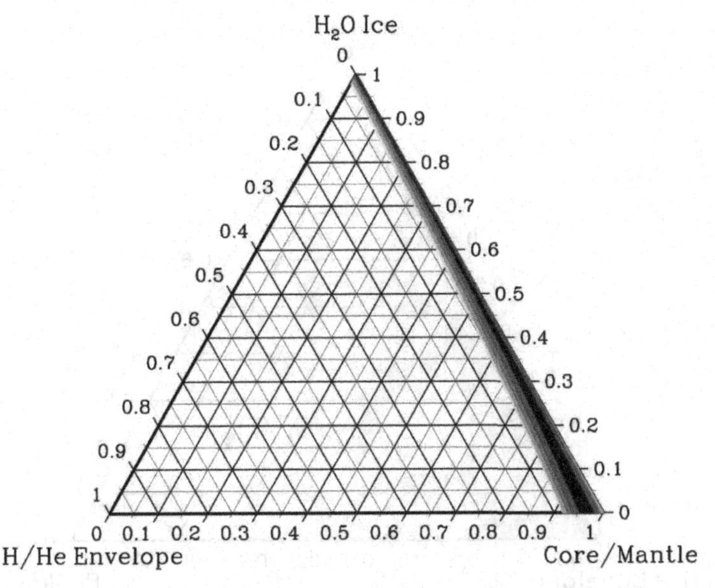

Figure 1. Ternary diagram for the scenario where GJ 1214b's gas layer formed through direct accretion of H/He. The relative contributions of the iron core, $Mg_{0.9}Fe_{0.1}SiO_3$ silicate mantle, H_2O ices, and H/He envelope to the mass of the planet are plotted. The core and mantle are combined together on a single axis, with the perpendicular distance from the lower right vertex determined by the fraction of the planet's mass in the two innermost planet layers. The black shaded region denotes the interior compositions that are consistent with the nominal planet mass and radius ($M_p = 6.55\ M_\oplus$, $R_p = 2.678\ R_\oplus$) for our fiducial atmospheric PT profile. The H/He mass fraction has a spread in this case due to the range of possible core-to-mantle mass ratios. The span of plausible interior compositions widens to the darkest grey (blue online) shaded area when we consider the uncertainties on the planet's albedo and intrinsic luminosity. Successively lighter shades of grey (colored red, green, and yellow online) denote compositions that are consistent with M_p and R_p to within 1, 2, and 3σ of their observational uncertainties, respectively, when uncertainties in the atmospheric parameters are also included. For an explanation of how to read ternary diagrams, see e.g., Valencia *et al.* (2007).

internal heat flux is 0.087 W m^{-2} (Turcotte & Schubert 2002). We therefore predict that, in this scenario, GJ 1214b would need a cold interior in order to harbor liquid water.

3.3. *Outgassing Scenario*

We now explore the possibility that GJ 1214b is a rocky planet with an outgassed atmosphere contributing to its transit radius. In this scenario GJ 1214b formed from rocky planetesimals (without retaining significant amounts of primordial gas or icy material) that when heated during planet formation released volatiles to form the planet's gas envelope.

We find that if GJ 1214b is a rocky super Earth with an outgassed atmosphere, its gas layer would need to be Hydrogen rich. In principle, outgassing can produce a wide range of gas layer compositions depending on the composition of the primordial rocky material (Elkins-Tanton & Seager 2008; Schaefer & Fegley 2009). However, there is also a limit on how much gas rocky material can outgas, determined by its initial volatile content and oxidation state. In order for the outgassing scenario to be feasible GJ 1214b's gas envelope must have both a mass low enough to be produced by outgassing and a volume large enough to account for the transit radius. H_2 is the only species that both is light enough and can plausibly be outgassed in sufficient quantities (e.g. by ordinary H, L, LL and high iron enstatite EH chondritic material) to satisfy this criterion.

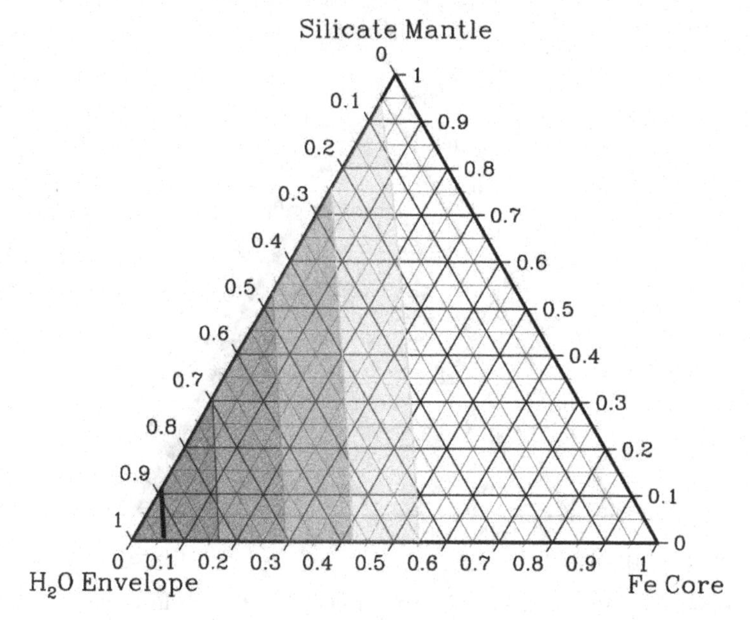

Figure 2. Ternary diagram for the scenario in which GJ 1214b's envelope is dominated by vapor from sublimated water ice. The fractions of the planet's mass in the Fe core, the $Mg_{0.9}Fe_{0.1}SiO_3$ silicate mantle, and the water envelope are plotted on the three axes. The solid black curve represents the locus of interior compositions that are consistent with the nominal measured planetary mass and radius for our fiducial PT profile. The colors in this figure have the same designations as in Figure 1.

4. Summary

We consider three scenarios for the origin of GJ 1214b's gas layer: direct accretion of nebular gas, sublimation of ices, and outgassing. Based on the measured planet mass and radius alone, none of the scenarios can be ruled out. Future measurements of GJ 1214b's transmission spectrum may help to distinguish between the vapor dominated and hydrogen-dominated cases (Miller-Ricci *et al.* 2009). We also find that, under most conditions we consider, GJ 1214b would not have liquid water. Its PT profile is simply too hot to go through the liquid water phase.

References

Adams, E. R., Seager, S., & Elkins-Tanton, L. 2008, *ApJ*, 673, 1160

Charbonneau, D., Berta, Z. K., Irwin, J., Burke, C. J., Nutzman, P., Buchhave, L. A., Lovis, C., Bonfils, X., Latham, D. W., Udry, S., Murray-Clay, R. A., Holman, M. J., Falco, E. E., Winn, J. N., Queloz, D., Pepe, F., Mayor, M., Delfosse, X., & Forveille, T. 2009, *Nature*, 462, 891

Elkins-Tanton, L. T. & Seager, S. 2008, *ApJ*, 685, 1237

Hansen, B. M. S. 2008, *ApJS*, 179, 484

Miller-Ricci, E., Seager, S., & Sasselov, D. 2009, *ApJ*, 690, 1056

Rogers, L. A. & Seager, S. 2010a, *ApJ*, 712, 974

—. 2010b, *ApJ*, 716, 1208

Schaefer, L. & Fegley, Jr., B. 2009, *ApJ*, 703, L113

Seager, S., Kuchner, M., Hier-Majumder, C. A., & Militzer, B. 2007, *ApJ*, 669, 1279

Turcotte, D. L. & Schubert, G. 2002, Geodynamics (Cambridge, UK: Cambridge University Press)

Valencia, D., Sasselov, D. D., & O'Connell, R. J. 2007, *ApJ*, 665, 1413

The Astrophysics of Planetary Systems: Formation, Structure, and Dynamical Evolution
Proceedings IAU Symposium No. 276, 2010
A. Sozzetti, M. G. Lattanzi & A. P. Boss, eds.

doi:10.1017/S1743921311020175

Physical state of the deep interior of the CoRoT-7b exoplanet

Frank W. Wagner[1], Frank Sohl[1], Tina Rückriemen[1] and Heike Rauer[1,2]

[1]Institute of Planetary Research, German Aerospace Center (DLR), Berlin, Germany
email: Frank.Wagner@dlr.de

[2]Center of Astronomy and Astrophysics, Berlin Institute of Technology, Berlin, Germany

Abstract. The present study takes the CoRoT-7b exoplanet as an analogue for massive terrestrial planets to investigate conditions, under which intrinsic magnetic fields could be sustained in liquid cores. We examine the effect of depth-dependent transport parameters (e.g., activation volume of mantle rock) on a planet's thermal structure and the related heat flux across the core mantle boundary. For terrestrial planets more massive than the Earth, our calculations suggest that a substantial part of the lowermost mantle is in a sluggish convective regime, primarily due to pressure effects on viscosity. Hence, we find substantially higher core temperatures than previously reported from parameterized convection models. We also discuss the effect of melting point depression in the presence of impurities (e.g., sulfur) in iron-rich cores and compare corresponding melting relations to the calculated thermal structure. Since impurity effects become less important at the elevated pressure and temperature conditions prevalent in the deep interior of CoRoT-7b, iron-rich cores are likely solid, implying that a self-sustained magnetic field would be absent.

Keywords. planets and satellites: individual (CoRoT-7b), planets and satellites: formation, equation of state, methods: numerical

---

1. Introduction

CoRoT-7b is probably the most prominent discovery of the CoRoT space mission as it is the first extrasolar planet below ten Earth masses with firm observational constraints on planetary mass *and* radius. Although the planet's exact mass is still subject to debate (e.g, Pont *et al.* 2011; Hatzes *et al.* 2010), a total mass of about five times that of the Earth seems to be most favored (Bruntt *et al.* 2010; Queloz *et al.* 2009). Combined with the measured radius of (1.58 ± 0.10) $M_\oplus$ (Bruntt *et al.* 2010), this suggests an average compressed density of (7.2 ± 1.8) Mg m^{-3} for CoRoT-7b, being consistent with a terrestrial bulk composition (Fig. 1). In the following, we take CoRoT-7b as a type example for a terrestrial exoplanet to investigate the physical state and thermal structure of planetary interiors. The purpose of this study is to understand the necessary conditions, under which an intrinsic magnetic field could be sustained in liquid cores of massive terrestrial planets.

2. Method

We use a one-dimensional, four-layer structural model that is similar to the model approach disclosed on the occasion of the XXVIIth IAU General Assembly held in Rio de Janeiro, Brazil (Wagner *et al.* 2010). The interior structure of CoRoT-7b is obtained by solving the mass and energy balance equations in conjunction with an equation of state (EoS) for the internal density distribution. The Birch-Murnaghan EoS (Birch 1947) and

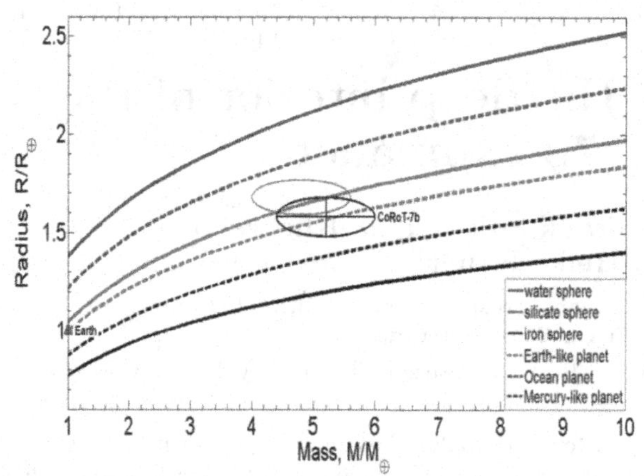

Figure 1. Mass-radius relations for differentiated exoplanets (dashed lines) and homogeneous, self-compressible spheres of water, silicate, and iron (solid lines) ranging from 1 to 10 $M_\oplus$. The relative position of the Earth is indicated by the triangle. Whereas the gray ellipse denotes the previously reported range for total mass M_p (Queloz *et al.* 2009) and planetary radius R_p (Leger *et al.* 2009), the black ellipse represents the recently revised range for $R_p = (1.58 \pm 0.10)$ $R_\oplus$ and $M_p = (5.2 \pm 0.8)$ $M_\oplus$ (Bruntt *et al.* 2010).

the Rydberg-Vinet EoS (Vinet *et al.* 2010) are often criticized because of inconsistencies in respect to the physics of the thermodynamic limit (e.g., Stacey & Davis 2004). Hence, we improve previous models (e.g., Valencia *et al.* 2007) by implementing a generalized Rydberg EoS (Stacey 2005), facilitating extrapolation to exceptionally high pressures (Wagner *et al.* 2011).

We adopt a mixing length formulation (Sasaki & Nakazawa 1986) to self-consistently calculate the thermal state of planetary mantles. This is an improvement to our previously proposed model, in which we have used this approach only for the upper mantle to simulate a lithosphere within the uppermost part (Wagner *et al.* 2010). The basic idea behind the mixing length concept is that internally generated heat is primarily transferred by vertical motion of fluid parcels which will entirely loose their individuality after migrating across size-dependent characteristic length scales. Yamagishi & Yanagisawa (2001) demonstrated the feasibility of that concept for modeling of heat transfer within terrestrial planet interiors. The viscosity determining the temperature profile is modeled in the framework of a temperature- and pressure-dependent Arrhenius viscosity law, considering diffusion and dislocation creep as the most important creep mechanisms in silicate aggregates (Ranalli 2001). The pressure-induced reduction of activation volume with depth is approximated as a vacancy in the material (O'Connell 1977). Solidification temperatures for the silicate-dominated mantle are obtained from ab initio molecular dynamics simulations (Belonoshko *et al.* 2005). To construct melting relations under core conditions for pure iron and iron-rich alloys containing impurities, we use a parameterization based on the theory of tricritical phenomena (Aitta 2010) up to 800 GPa and extrapolate according to the well-known Lindemann law (Ross 1969).

3. Results and Discussion

First, we present modeling results as obtained by using CoRoT-7b as case study and calculating self-consistently the thermal structure within planetary mantles and cores. Figure 2 (left-hand side) shows the radial temperature profile of two plausible CoRoT-7b models of 5 $M_\oplus$ that differ in the iron core mass fraction assumed, representing Earth-like (i.e., 32.6 wt.-%) and iron-depleted (i.e., 15.0 wt.-%) bulk compositions. From a fixed surface value of 1270 K (Leger *et al.* 2009), temperature increases rapidly across a thin thermal lithosphere, where heat is transferred by conduction and convection is absent. A mostly adiabatic temperature rise is observed across the convecting mantle

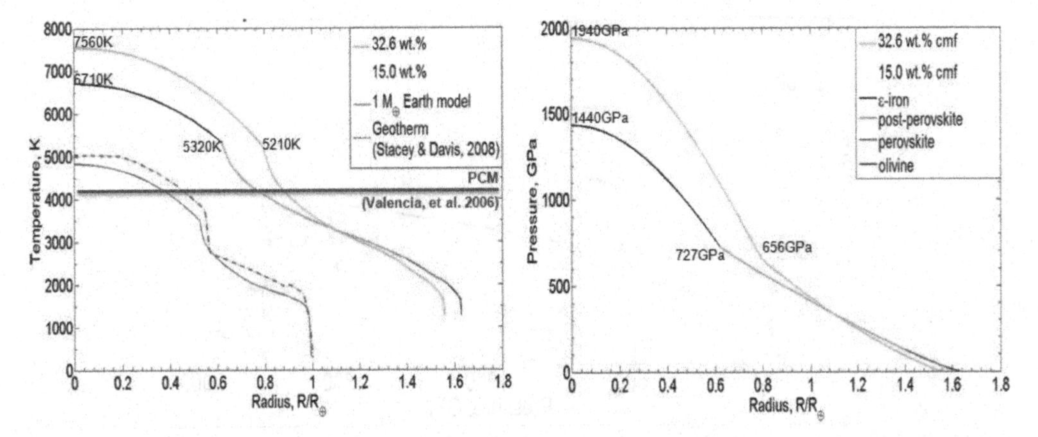

Figure 2. Interior structure of CoRoT-7b corresponding to different iron core mass fractions: (Left) Radial distribution of temperature. An Earth-sized model and a reference geotherm (Stacey & Davis 2008) are shown for comparison. (Right) Radial distribution of hydrostatic pressure.

and underlying core, resulting for the Earth-like and iron-depleted case, respectively, in temperatures of 5210 K and 5320 K at the core-mantle boundary (CMB) and up to 7560 K and 6710 K in the planet's center. Compared to parameterized convection models (e.g., Valencia *et al.* 2006), we find generally higher CMB temperatures. This is caused by a steeper temperature gradient prevalent in the deep interior that is mainly attributed to the pressure-induced increase of viscosity at elevated lower mantle pressures and temperatures. To illustrate environmental conditions in the deep interiors of massive exoplanets, the corresponding pressure profiles for the two CoRoT-7b cases are shown in Fig. 2 right-sided. Whereas CMB pressures of CoRoT-7b are twice as large as the central pressure of the Earth, the central pressure is as much as five times the pressure at the center of the Earth, depending on the planet's iron core mass fraction.

Next, we investigate conditions for the possible existence of liquid cores, which are prerequisite for magnetic field generation. In Figure 3, pressure-temperature (P-T) profiles for two Earth-like CoRoT-7b models are shown together with the corresponding silicate and iron melting relations. It is seen that iron melting temperatures are generally lower than solidus temperatures of $MgSiO_3$ (post-)perovskite under CMB conditions. The temperature profile shown in black corresponds to the Earth-like model case illustrated in Fig. 2 and yields an activation volume of 1.1 cm^3 mol^{-1} at the CMB. Lower mantle and core materials are expected to be solid because of extremely high compression at relatively low temperature. A temperature increase of about 3000 K at the CMB would be required to facilitate melting of the outer core. The temperature profile shown in gray corresponds to the latter case, for which a larger activation volume of 1.8 cm^3 mol^{-1} at the CMB would be required. Hence, the lower mantle becomes stiffer with increasing activation volume and, as a consequence, local temperature will rise until intersection with the iron melting curve. Also notable in that respect is that the melting intervall in the presence of impurities (e.g., siderophile elements like sulfur) would vanish at extremely high pressures (blue curves), according to Aitta (2010)). Due to large pressures within the deep interior of a 5 $M_\oplus$ exoplanet such as CoRoT-7b, the addition of impurities will not modify much the P-T relation compared to that of a pure iron core. This suggests that impurities should only have a minor effect on core melting within massive terrestrial exoplanets (Wagner *et al.* 2011).

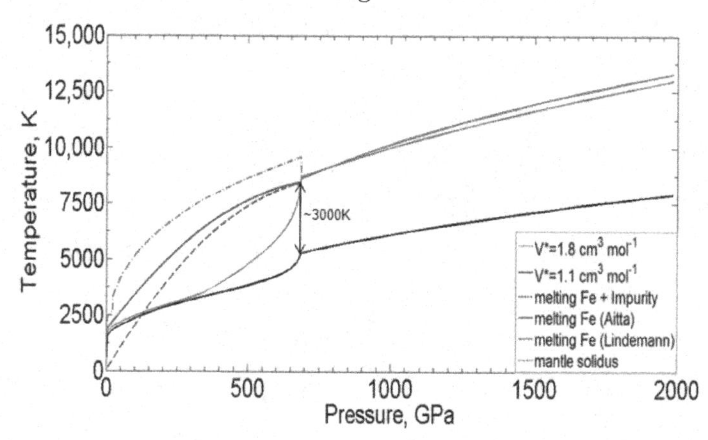

Figure 3. Influence of activation volume V^* and core impurities (e.g., sulfur) on the physical state of matter within a 5 $M_\oplus$ exoplanet.

4. Conclusions

It is concluded that the physical state of the deep interior of massive terrestrial exoplanets is strongly dependent on mantle rheology. Impurities have only a minor effect on core melting, which can be explained conceptually by the large compression of matter under core conditions. Due to the large effect of pressure on melting, a pure iron core is expected to be solid and, therefore, a self-generated magnetic field should be absent on massive terrestrial exoplanets like CoRoT-7b. Nevertheless, liquid cores cannot completely be ruled out, but in that case substantially larger activation volumes (> 1.8 cm^3 mol^{-1}) compared to values predicted by the vacancy approach would be required to initiate core melting. Furthermore, a pressure-induced sluggish convection is prevalent in the lowermost mantle and should influence mantle convection pattern as well as the thermal evolution of massive exoplanets.

Acknowledgement

This research is supported by the Helmholtz Alliance "Planetary Evolution and Life".

References

Aitta, A. 2010, *Phys. Earth Planet. Inter.*, 181, 132
Belonoshko, A. B., Skorodumova, N. V., Rosengren, A., *et al.* 2005, *Phys. Rev. Lett.*, 94, 195701
Birch, F. 1947, *Phys. Rev.*, 71, 809
Bruntt, H., Deleuil, M., Fridlund, M., *et al.* 2010, *A&A*, 519, A51
Hatzes, A., Dvorak, R., Wuchterl, G., *et al.* 2010, *A&A*, 520, A93
Leger, A., Rouan, D., Schneider, J., *et al.* 2009, *A&A*, 506, 287
O'Connell, R. J. 1977, *Tectonophysics*, 38, 119
Pont, F., Aigrain, S., & Zucker, S. 2011, *MNRAS*, 411, 1953
Queloz, D., Bouchy, F., Moutou, C., *et al.* 2009, *A&A*, 506, 303
Ranalli, G. 2001, *J. Geodyn.*, 32, 425
Ross, M. 1969, *Phys. Rev.*, 184, 233
Sasaki, S. & Nakazawa, K. 1986, *JGR*, 91, 9231
Stacey, F. D. & Davis, P. M. 2004, *Phys. Earth Planet. Inter.*, 142, 137
Stacey, F. D. 2005, *Rep. Prog. Phys.*, 68, 341
Stacey, F. D. & Davis, P. M. 2008, *Physics of the Earth*, (Cambridge University Press)

Valencia, D., O'Connell, R. J., & Sasselov, D. D. 2006, *Icarus*, 181, 545
Valencia, D., Sasselov, D. D., & O'Connell, R. J. 2007, *ApJ*, 656, 545
Vinet, P., Rose, J. H., Ferrante, J. *et al.* 1989, *J. Phys.: Condens. Matt.*, 1, 1941
Wagner, F. W., Sohl, F., Rauer, H., *et al.* 2010, *Highlight of Astronomy*, 15, 708
Wagner, F. W., Sohl, F., Hussmann, H., *et al.* 2011, in preparation
Yamagishi, Y. & Yanagisawa, T. 2001, *Frontier Research on Earth Evolution*, 1, 41

The Astrophysics of Planetary Systems: Formation, Structure, and Dynamical Evolution
Proceedings IAU Symposium No. 276, 2010
A. Sozzetti, M. G. Lattanzi & A. P. Boss, eds.

doi:10.1017/S1743921311020187

Exoplanet atmospheres: A theoretical outlook

Sara Seager[1]

[1]Department of Earth, Atmospheric, and Planetary Sciences, Department of Physics, Massachusetts Institute of Technology, Cambridge, Massachusetts, 02139
email, seager@mit.edu

Abstract. With over two dozen exoplanet atmospheres observed today, the field of exoplanet atmospheres is solidly established. The highlights of exoplanet atmosphere studies include: detection of molecular spectral features; constraints on atmospheric vertical temperature structure; detection of day-night temperature gradients; and a new numerical approach to atmosphere temperature and abundance retrieval. As hot Jupiter observations and interpretation are maturing, the next frontier is super Earth atmospheres. Theoretical models of super Earth atmospheres are moving forward with observational hopes pinned on the *James Webb Space Telescope*, scheduled for launch in 2014. Further in the future lies direct imaging attempts to answer the enigmatic and ancient question, "Are we alone?" via atmospheric biosignatures.

Keywords. astrobiology, planetary systems, techniques: photometric

1. Introduction

At the dawn of the first discovery of exoplanets orbiting sun-like stars in the mid-1990s, few believed that observations of exoplanet atmospheres would ever be possible. After the 2002 Hubble Space Telescope detection of a transiting exoplanet atmosphere, many skeptics discounted it as a one-object, one-method success. By 2010, the field was firmly established, with over two dozen exoplanet atmospheres observed today. Hot Jupiters. the type of exoplanet most amenable to study are observed by the dozens. Highlights include, detection of molecular spectral features; observation of day-night temperature gradients; and constraints on vertical atmospheric structure. Atmospheres of giant planets far from their host stars are also being studied with direct imaging. The ultimate exoplanet goal is to answer the enigmatic and ancient question, "Are we alone?" via detection of atmospheric biosignatures. The two paths forward are the near-term focus on transiting super Earths orbiting in the habitable zone of M-dwarfs, and ultimately the space-based direct imaging of true Earth analogs.

2. Past: A Brief History of Exoplanet Atmospheres

The dawn of the discovery of exoplanets orbiting sun-like stars took place in the mid 1990s, with the birth of radial velocity detections. Because of selection effects, many of the exoplanets found in the first few years of discovery orbited extremely close to their host star. Called hot Jupiters, these planets orbit many times closer to their star than Mercury does to our sun. With semi-major axes 0.05 AU, the hot Jupiters are heated externally by their stars to temperatures of 1000-2000 K, or even higher. From the start the high temperature and close stellar proximity of hot Jupiters were recognized as favorable for atmospheric detection (Seager & Sasselov 1998). Even as the number of exoplanet

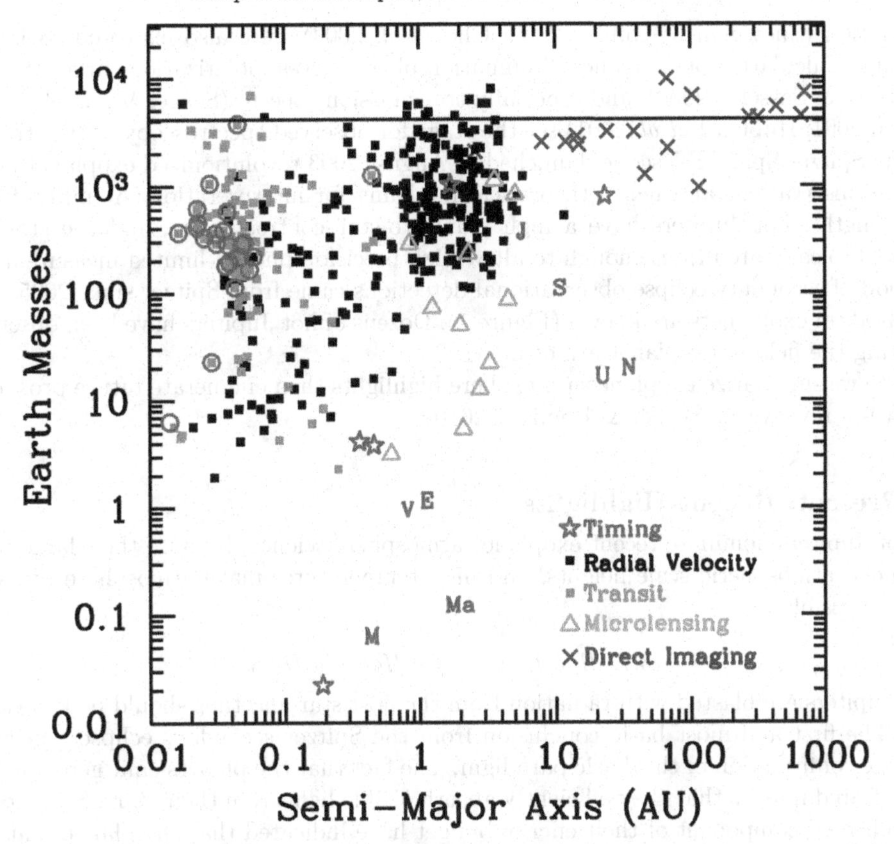

Figure 1. Known planets as of March 2010. Red letters indicate solar system planets. The red circles represent planets with published atmosphere detections. The solid line is the conventional upper mass limit for the definition of a planet. Data taken from http://exoplanet.eu/

detections grew in the late 1990s (just under 30 by the end of the 20th century†), few thought that exoplanet atmospheres could be observed at any time in the foreseeable future.

By the time about seven hot Jupiters were known, the community expected one to transit. With a probability to transit of $R_\star/a$, where $R_\star$ is the stellar radius and a is the semi-major axis, each hot Jupiter has about a 10% chance to transit. Seager & Sasselov (2000) presented transit transmission spectra as a way to detect the atmospheres of hot Jupiters, by way of atomic and molecular transmission spectral features, with a focus on sodium. HD 209458b was found to show transits at the end of 1999 (Charbonneau *et al.* 2000; Henry *et al.* 2000), and the first detection of an exoplanet atmosphere, via atomic sodium, with the Hubble Space Telescope soon followed (Charbonneau *et al.* 2002).

The theory of exoplanet atmospheres was also developing in the 1990s and early 2000s. At that time, theory was leading observation, and observers consulted the model predictions to help define the most promising detection techniques. Most theory papers focused on irradiated hot Jupiters, emphasizing spectral features and 1D temperature/pressure profiles resulting from the intense heating by the host star (Seager & Sasselov 1998; Marley *et al.* 1999; Sudarsky, Burrows, & Pinto 2000; Barman, Hauschildt & Allard 2001. Cloud modeling (Ackerman & Marley 2001; Cooper *et al.* 2003) and atmospheric

† http://exoplanet.eu/catalog.php

circulation (Showman & Guillot 2002; Cho *et al.* 2003) were also expected to be important. Calculation of exoplanet illumination phase curves, polarization curves (Seager, Whitney, & Sasselov 2000), and especially transmission spectra (Seager & Sasselov 2000; Brown 2001; Hubbard *et al.* 2001) set the stage for observed spectroscopy during transit.

The Spitzer Space Telescope, launched in August 2003 revolutionized exoplanet atmosphere observations and hence theoretical modeling for interpretation. At mid-infrared wavelengths, hot Jupiters have a high planet-to-star contrast ratio, and the star and planet typically are bright enough to allow high precision photon-limited measurements. A flood of secondary eclipse observational detections came from Spitzer since 2005. Now hundreds of exoplanets are known (Figure 1). Dozens of hot Jupiters have been observed, creating the field of exoplanet atmospheres.

Here we summarize exoplanet atmosphere highlights then enumerate future prospects. For a full review, see Seager & Deming (2010).

3. Present: Recent Highlights

Hot Jupiters dominate recent exoplanet atmosphere science, because their large radii, extended atmospheric scale heights, and high temperetures make atmosphere measurements possible.

3.1. *Hot Jupiters are Hot and Dark*

Hot Jupiters are blasted with radiation from the host star and thus should be kinetically hot. The first and most basic conclusion from the Spitzer secondary eclipse detections was the confirmation of this basic paradigm. The fact that the planets emit generously in the infrared implies that they efficiently absorb visible light from their stars. Searches for the reflected component of their energy budget have indicated that the planets must be very dark in visible light, with geometric albedos less than about 0.2 (Rowe *et al.* 2008), and likely much lower. Models show that purely gaseous atmospheres lacking reflective clouds will be very dark (Marley *et al.* 1999; Seager, Whitney, & Sasselov 2000).

3.2. *Identification of Atoms and Molecules*

A major achievement for exoplanet atmospheres is the identification of atoms and molecules. In hot Jupiter atmospheres, the atoms and molecules identified are atomic sodium (Na) (e.g., Charbonneau *et al.* 2002), water vapor (H_2O) (e.g., Swain *et al.* 2008; Figure 2), methane (CH_4) (Swain *et al.* 2008), carbon monoxide (CO), and carbon dioxide (CO_2) (e.g., Swain *et al.* 2009a,b; Madhusudhan and Seager 2009). (For a critical discussion of Swain *et al.* 2008, see Gibson *et al.* 2011 and Swain *et al.* this volume.) In addition to molecules, the presence of atmospheric haze has been inferred in HD 189733 via transmission spectra with HST/STIS. While the particle composition has not been identified, the Rayleigh-scattering behavior of the data indicates small particle sizes (Pont *et al.* 2008). A thorough temperature and abundance retrieval method enables statistical constraints on molecular mixing ratios and other atmospheric properties (see Fig. 3 and Madhusudhan & Seager, 2009).

3.3. *Day-Night Temperature Gradients*

Hot Jupiters are theorized to have their rotation synchronized with their orbital motion by tidal forces, a process that should conclude within millions of years (e.g., Guillot *et al.* 1996). Under this tidal-locking condition the planet will have a permanent day side and a permanent night side. Spitzer thermal infrared observations of HD 189733b shows that the planet has only a moderate temperature variation from the day to night side. The

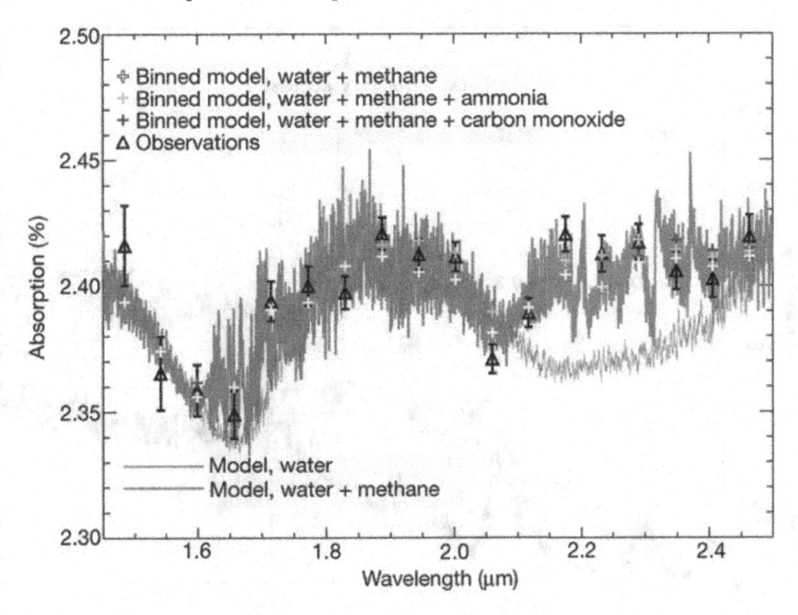

Figure 2. Transmission spectrum of the transiting planet HD 189733. Hubble Space Telescope observations shown by the black triangles. Two different models highlight the presence of methane in the planetary atmosphere. From Swain *et al.* (2008).

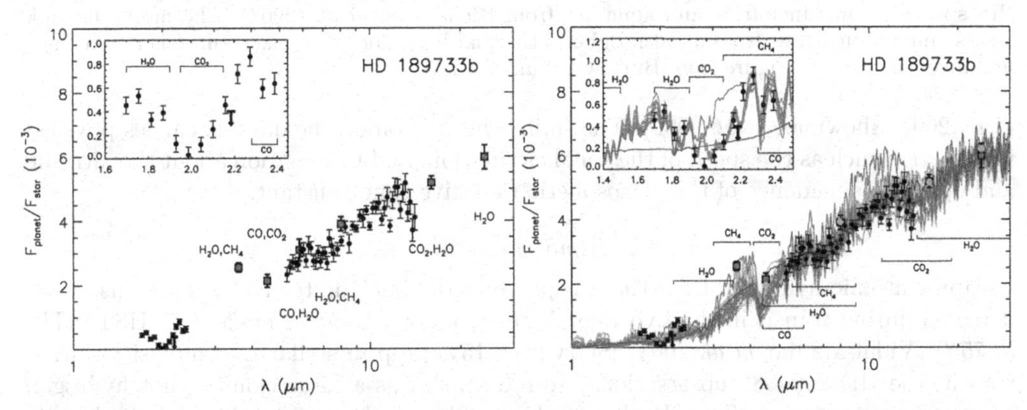

Figure 3. Thermal emission data composite for HD 189733 in secondary eclipse. Data from HST/NICMOS (inset, Swain *et al.* 2009), Spitzer/IRAC (four shortest wavelength red points; Charbonneau *et al.* 2008), Spitzer/IRS-PU (Deming *et al.* 2006), Spitzer/MIPS (Charbonneau *et al.* 2008), Spitzer/IRS (black points from 5 - 13 μm; Grillmair *et al.* 2008). Models shown in the right panel (from Madhusudhan and Seager 2009) illustrate that the best fits to the Spitzer/IRS ((red curve shows fits within the 1.4σ errors, on average; orange 1.7σ, green 2σ, and blue is one best fit model within 1.4σ) and Spitzer photometry (brown curve within 1σ) do not fit the NICMOS data (inset grey curves within 1.4σ) possibly implying variability in the planet atmosphere from data taken at different epochs. For abundance constraints from the different models, see Madhusudan and Seager (2009).

planet shows an 8 μm brightness temperature variation of over 200 K from a minimum brightness temperature of 973±33K to a maximum brightness temperature of 1212±11K (Knutson *et al.* 2007), and a thermal brightness change at 24 μm consistent with the 8 μm data within the errors (Knutson *et al.* 2009). Model interpretation indicates that strong winds have advected the hottest region to the east of the sub-stellar point (Knutson

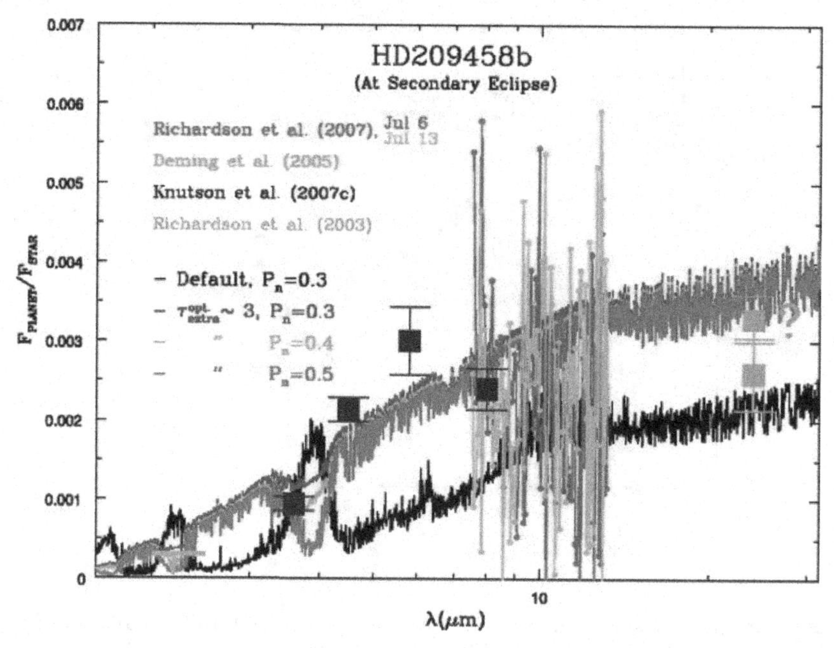

Figure 4. Evidence for an atmospheric thermal inversion for HD 209458b. Spitzer data points from secondary eclipse measurements are shown with brown (IRAC; Knutson *et al.* 2007) and green (Deming *et al.* 2005 and private comm.; the two points are data taken at different times). IRS spectra shown in purple and aqua are from Richardson *et al.* (2007). The model in pink shows emission features from an atmospheric thermal inversion. The black curve is a non-thermal-inversion model. Figure from Burrows *et al.* (2007).

et al. 2007; Showman *et al.* 2009). The shifted hot region on the dayside carries physical information such as the speed of the zonal circulation, and information about the altitude and opacity-dependence of the atmospheric radiative time constant.

3.4. *Atmospheric Escape*

Escaping atomic hydrogen from the exosphere of the hot Jupiter HD 209458b has been detected during transit in the Lyα line. A positive detection was made with HST/STIS (3.75σ) (Vidal-Madjar *et al.* 2003). Showing a 15% drop in stellar Lyα intensity during transit, the HD 209458b observations are interpreted as a large cloud of hot hydrogen surrounding the planet. The cloud extends up to four planetary radii, and the kinetic temperature is as high as tens of thousands of K. Models agree that the implied exospheric heating is likely due to absorption of UV stellar flux, but Jeans escape is not sufficient to account for the hydrogen cloud. The specific origin of the escaping atoms is model-dependent. Escape mechanisms include radiation pressure, charge exchange, and solar wind interaction (see, e.g., Lammer *et al.* 2009 and references therein).

3.5. *Vertical Thermal Inversions*

Evidence for vertical atmospheric thermal inversions in hot Jupiters comes from emission features in place of (or together with) absorption features in the thermal infrared spectrum (for a basic explanation see Seager 2010). Because broad-band photometry does not delineate the structure of molecular spectral bands, the inference of a thermal inversion must rely on models. Spitzer data show that the upper atmospheres of several planets have thermal inversions, if water vapor is present and if abundances are close to solar (e.g., Burrows *et al.* 2008; see Figure 4).

The hot Jupiter temperature inversions are likely created by absorption of stellar irradiance in a high-altitude absorbing layer. Madhusudhan & Seager (2010), however, have found that for many cases existing observations (Spitzer broad-band photometry) are not enough to make robust claims on the presence of thermal inversions.

4. Future Prospects

4.1. *Super Earth Atmospheres*

In exoplanet research the frontier is always the most exciting. In exoplanet atmospheres, the frontier is the field of super Earths. Super Earths are unofficially defined as planets with masses between 1 and 10 Earth masses. The term super Earths should define planets that are rocky in Nature, rather than planets with icy interiors or significant gas envelopes. Because there may be a continuous and overlapping mass range between them, super Earths and "exo-Neptunes" are often discussed together.

The major challenge to studying super Earth atmospheres, is the anticipated wide diversity. This is different from Jupiter and the other solar system gas giants, which have "primitive" atmospheres. That is, Jupiter has retained the gases it formed with, and these gases approximately represent the composition of the sun. The super Earth atmospheres, in contrast, could have a wide range of possibilities for the atmospheric mass and composition. Attempts to evaluate these possibilities used calculations of atmospheres that formed by outgassing during planetary accretion, considering bulk compositions drawn from differentiated and/or primitive solar system meteoritic compositions (Elkins-Tanton & Seager 2008; Schaefer & Fegley 2010). Instead of narrowing down possibilities, this work emphasized the large range of possible atmospheric mass and composition of outgassed super Earths even before consideration of atmospheric escape.

Researchers take different paths for modeling super Earth atmospheres, focusing on different regions of parameter space. One approach is to consider atmospheres similar to Earth, Venus or Mars (or their atmospheres in earlier epochs). Considering the amount of greenhouse gases including CO_2, Selsis *et al.* (2007) and von Bloh *et al.* (2007) both found that Gl 581d is more likely to be habitable (that is with surface temperatures consistent with liquid water) than Gl 581c. Other investigators consider atmospheres that radically depart from the terrestrial planets in our solar system. Water planets, akin to scaled up versions of Jupiter's icy moons, could have up to 50 percent water by mass, with concomitant massive steam atmospheres (Kuchner 2003; Leger *et al.* 2004; Rogers & Seager 2010). In a different approach, Miller-Ricci, Seager, & Sasselov (2009) considered GJ 581c and three possibilities relating to atmospheric hydrogen content. A suggestion of terrestrial planets with sulfur cycles dominating over carbon cycles is described in Kaltenegger & Sasselov (2010). Others have attempted to quantify the atmospheric escape of Earths and super Earths, with little success due to the unknown initial mass and star's activity history (e.g., Lammer *et al.* 2007).

We anticipate the discovery of a handful of rare but highly valuable transiting super Earths in the habitable zones of the brightest low-mass stars. With such targets, we will observe the transiting super Earth atmospheres in the same way we are currently observing transiting hot Jupiters orbiting sun-like stars. NASA's JWST scheduled for launch in 2014, will be capable of observing the absorption signatures of major molecules like water and carbon dioxide. Such observations will require monitoring of multiple transits, often amounting to ~100 hours of JWST observation (Deming *et al.* 2009).

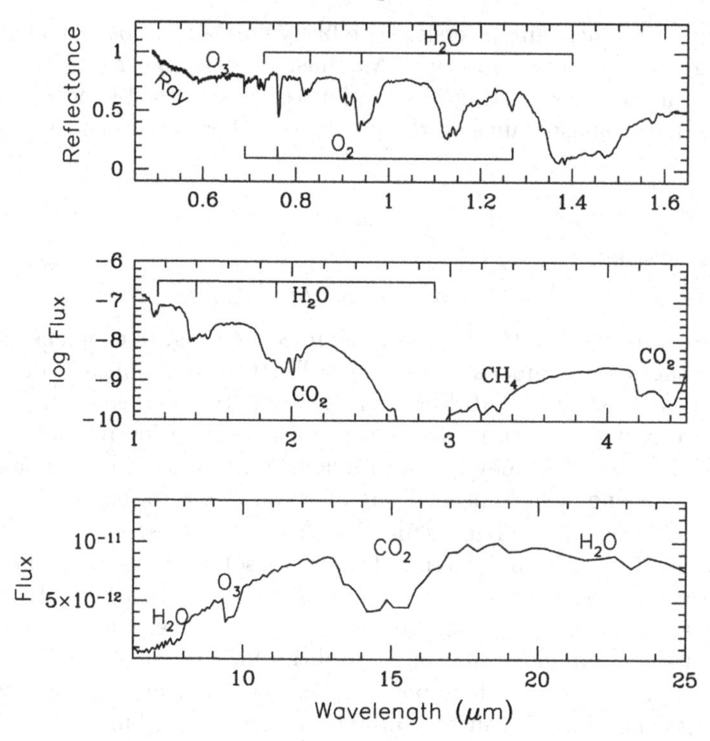

Figure 5. Earth as an exoplanet: Earth's hemispherically averaged spectrum. Top: Earth's visible wavelength spectrum from Earthshine measurements plotted as normalized reflectance (Turnbull *et al.* 2006). Middle: Near-infrared spectrum from NASAs EPOXI mission with flux in units of W m^2 ?m^1 (Robinson *et al.*, 2010). Bottom: Earth's mid-infrared spectrum as observed by Mars Global Surveyor enroute to Mars with flux in units of W m^2 Hz^1 (Pearl and Christensen 1997). Major molecular absorption features are noted including Rayleigh Scattering. Only strongly absorbing, globally mixed molecules are detectable.

4.2. *Earth Analog Atmospheres and Biosignature Gases*

Without question the holy grail of exoplanet research is the discovery of a true Earth analog, an Earth-size, Earth-mass planet in an Earth-like orbit about a sun-like star. We emphasize that discovery of Earth-size or Earth-mass planets-even those in their star's habitable zone-is not the same as identifying a habitable planet. Venus and Earth are both about the same size and mass-and would appear the same to an astrometry, radial-velocity, or transit observation. Yet Venus is completely hostile to life due to the strong greenhouse effect and resulting high surface temperatures (over 700 K), while Earth has the right surface temperature for liquid water oceans and is teeming with life. This is why, in the search for habitable planets, a direct-imaging space-based telescope capable of blocking out the starlight is inevitable.

The main motivation for finding Earth analogs is to search their atmospheric spectra for biosignature gases. An atmospheric biosignature gas is one produced by life. The canonical concept for the search for atmospheric biosignatures is to find an atmosphere severely out of thermochemical redox equilibrium (Lederberg 1965; Lovelock 1965). Indeed Earth's atmosphere has oxygen (a highly oxidized species) and methane (a very reduced species) several orders of magnitude out of thermochemical redox equilibrium.

In practice it could be difficult to detect molecular features from different redox states. The Earth as an exoplanet, for example (Figure 5), has a relatively prominent oxygen

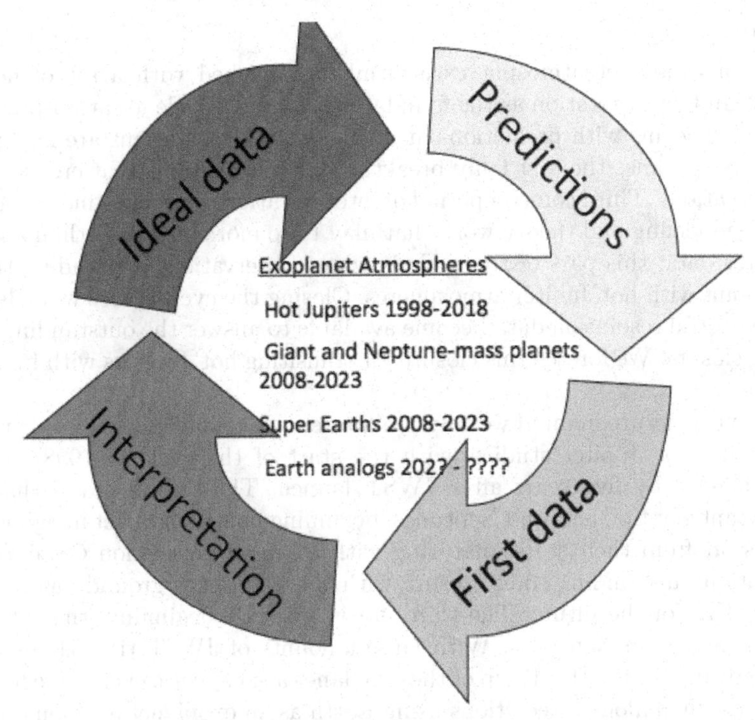

Figure 6. Cycles for exoplanet atmospheres.

absorption feature at 0.76 μm, whereas methane at present-day levels of 1.6 ppm has only extremely weak spectral features. The more realistic atmospheric biosignature gas is a single gas completely out of chemical equilibrium. Earth's example again is oxygen or ozone, about ten orders of magnitude higher than expected from equilibrium chemistry and with no known abiotic production at such high levels. The challenge with a single biosignature outside of the context of redox chemistry becomes one of false positives. To avoid false positives we must look at the whole atmospheric context.

Most biosignatures work to date has focused on mild extensions of exoplanet biosignatures as on Earth (O_2, O_3, N_2O) or early Earth (possibly CH_4) biosignatures. Research forays into biosignature gases that are negligible on Earth but may play a more dominant role on other planets has started. Pilcher (2003) suggested that organosulfur compounds, particularly methanethiol (CH_3SH, the sulfur analog of methanol) could be produced in high enough abundance by bacteria, possibly creating a biosignature on other planets. Pilcher (2003) emphasized a potential ambiguity in interpreting the 9.6 μm O_3 spectral feature since a CH_3SH feature overlaps with it. Segura *et al.* (2005) showed that the Earth-like biosignature gases CH_4, N_2O, and even CH_3Cl have higher concentrations and therefore stronger spectral features on planets orbiting M stars compared to Earth. The reduced UV radiation on quiet M stars enables longer biosignature gas lifetimes and therefore higher concentrations to accumulate. Seager, Schrenk, & Bains (2010) have reviewed Earth-based metabolism to summarize the range of gases and solids produced by life on Earth. A fruitful new area of research will be on which molecules are potential biosignatures and which can be identified as such on super Earth planets different from Earth.

5. Outlook

The field of exoplanet atmospheres is firmly established with a set of hot Jupiter observations and interpretation as the foundation. We see a cycle for atmospheric studies (Figure 6) that begins with predictions at a time when observations are leading theory. Next in the cycle comes the first truly breakthrough observations that enable a flurry of further observations. Third comes a period of interpretation or perhaps more aptly termed retrodiction, modeling and theory work that may raise more questions than answers and beg for better data; this part of the cycle is where observations are leading theory. We are at this point with hot Jupiter atmospheres. Closing the cycle comes when higher S/N and higher spectral resolution data become available to answer the outstanding questions and provide closure. We foresee this closure for transiting hot Jupiters with future JWST observations.

With the cycle picture in mind we envision four eras for exoplanet atmospheric studies. The first is the hot Jupiter studies, with the start of the cycle in 1998 and with at least some closure by five years after JWST launch. The second era is that of more orbitally-distant giant planets and Neptunes, beginning now with direct imaging of young, hot Jupiters far from their stars, maturing with the next generation Gemini and VLT instrumentation, and finding some closure with the very large ground-based telescopes TMT/GMT/ELT of the future. The third era is also just beginning, that of transiting super Earths and mini Neptunes. With large amounts of JWST time the era of super Earths will advance to the third step of the exoplanet atmospheric cycle. The fourth era is that of true Earth analogs. Predictions using Earth as an exoplanet and some extensions are underway, but the first observations will have to await a specialized space telescope that can block out the orders of magnitude brighter starlight.

Thousands of years from now, people will look back and see as one of the most significant, positive accomplishments of our early twenty first century society the first discoveries of exoplanets, and, the human foray into finding and characterizing habitable planets. Exoplanet atmospheres plays a critical component as the hosts of signs of life on other worlds.

References

Ackerman, A. S. & Marley, M. S. 2001, *ApJ*, 556, 872
Barman, T. S., Hauschildt, P. H., & Allard, F. 2001, *ApJ*, 556, 885
Brown, T. M. 2001, *ApJ*, 553, 1006
Burrows, A., Budaj, J., Hubeny, I. 2008, *ApJ*, 678, 1436
Charbonneau, D., Brown, T. M., Latham, D. W., *et al.* 2000, *ApJ*, 529, L45
Charbonneau, D., Brown, T. M., Noyes, R. W., *et al.* 2002, *ApJ*, 568, 377
Charbonneau, D., Knutson, H. A., Barman, T., *et al.* 2008, *ApJ*, 686, 1341
Cho, J. Y.-K., Menou, K., Hansen, B. M. S., *et al.* 2003, *ApJ*, 587, 117
Cooper, C. S., Sudarsky, D., Milsom, J. A., *et al.* 2003, *ApJ*, 586, 1320
Deming, D., Seager, S., Richardson, L. J., *et al.* 2005, *Nature*, 434, 740
Deming, D., Harrington, J., Seager, S., *et al.* 2006, *ApJ*, 644, 560
Deming, D., Seager, S., Winn, J., *et al.* 2009, *PASP*, 121, 952
Elkins-Tanton, L. & Seager, S. 2008, *ApJ*, 685, 1237
Gibson, N. P., Pont, F., & Agrain, S. 2011, *MNRAS*, 411, 2199
Grillmair, C. J., Burrows, A., Charbonneau, D., *et al.* 2008, *Nature*, 456, 767
Guillot, T., Burrows, A., Hubbard, W. B., *et al.* 1996, *ApJ*, 459, L35
Henry, G. W., Marcy, G. W., Butler, R. P., *et al.* 2000, *ApJ*, 529, L41
Hubbard, W. B., Fortney, J. J., Lunine, J. I., *et al.* 2001, *ApJ*, 560, 413
Kaltenegger, L. & Sasselov, D. 2010, *ApJ*, 708, 1162

Knutson, H. A. Charbonneau, D., Noyes R. W., *et al.* 2007, *ApJ*, 655, 564
Knutson, H. A., Charbonneau, D., Cowan, N. B., *et al.* 2009, *ApJ*, 690, 822
Kuchner, M. 2003, *ApJ*, 596, 105
Lammer, H., Lichtenegger, H. I. M., Kulikov, Y. N., *et al.* 2007, *Astrobiology*, 7, 185
Lammer, H., Odert, P., Leitzinger, M., *et al.* 2009, *A&A*, 506, 399
Lederberg, J. 1965, *Nature*, 207, 9
Léger, A., Selsis, F., Sotin, C., *et al.* 2004, *Icarus*, 169, 499
Lovelock, J. E. 1965, *Nature*, 207, 568
Madhusudhan, N. & Seager, S. 2009, *ApJ*, 707, 24
Madhusudhan, N. & Seager, S. 2010, *ApJ*, 725, 261
Marley, M. S., Gelino, C., Stephens, D., *et al.* 1999, *ApJ*, 513, 879
Miller-Ricci, E., Seager, S., & Sasselov, D. 2008, *ApJ*, 690, 1056
Pearl, J. C. & Christensen, P. R. 1997, *JGR* 102, 10875
Pilcher, C. B. 2003, *Astrobiology*, 3, 471
Pont, F., Knutson, H., Gilliland, R. L., *et al.* 2008, *MNRAS*, 385, 109
Richardson, L. J., Deming, D., Horning, K., *et al.* 2007, *Nature*, 445, 892
Robinson, T. D., Meadows, V., & Crisp, D. 2010, *BAAS* 42, 1061
Rogers, L. & Seager, S. 2010, *ApJ*, 716, 1208
Rowe, J. F., Matthews, J. M., Seager, S., *et al.* 2008, *ApJ*, 689, 1345
Schaefer, L., Fegley, B. 2010, *Icarus*, 208, 438
Seager, S. & Sasselov, D. 1998, *ApJ*, 502, L157
Seager, S. & Sasselov, D. 2000, *ApJ*, 537, 916
Seager, S., Whitney, B. A., & Sasselov, D. D. 2000, *ApJ*, 540, 504
Seager, S. & Deming, D. 2010, *ARA&A*, 43, 631
Seager, S. 2010, *Exoplanet Atmospheres, Physical Processes* (Princeton University Press, Princeton)
Seager, S., Schrenk, M., Bains, W., *et al.* 2011, submitted to *Astrobiology*
Segura, A., Kasting, J. F. Meadows, V., *et al.* 2005, *Astrobiology*, 5, 706
Selsis, F., Kasting, J. F., Levrard, B., *et al.* 2007, *A&A*, 476, 1373
Showman, A. P., Fortney, J. J., Lian, Y., *et al.* 2009, *ApJ*, 685, 1324
Showman, A. P. & Guillot, T. 2002, *A&A*, 385, 166
Sudarsky, D., Burrows, A., & Pinto P. 2000, *ApJ*, 538, 885
Swain, M. R., Vasisht, G., & Tinetti, G. 2008, *Nature*, 452, 329
Swain, M. R., Vasisht, G., Tinetti, G., *et al.* 2009a, *ApJ*, 690, L114
Swain, M. R., Tinetti, G., Vasisht, G., *et al.* 2009b, *ApJ*, 704, 1616
Swain, M. R., Deroo, P., & Vasisht, G. 2011, this volume
Turnbull, M. C., Traub, W. A., Jucks, K. W., *et al.* 2006, *ApJ*, 644, 551
Vidal-Madjar, A., Lecavelier des Etangs, A., Desert, J.-M., *et al.* 2003, *Nature*, 422, 143
von Bloh, W., Bounama, C., Cuntz, M., *et al.* 2007, *A&A*, 476, 1365

The Astrophysics of Planetary Systems: Formation, Structure, and Dynamical Evolution
Proceedings IAU Symposium No. 276, 2010
A. Sozzetti, M. G. Lattanzi, & A. P. Boss, eds.

doi:10.1017/S1743921311020199

Exoplanet atmospheres at high spectral resolution: A CRIRES survey of hot-Jupiters

Ignas Snellen[1], Remco de Kok[2], Ernst de Mooij[1], Matteo Brogi[1], Bas Nefs[1] and Simon Albrecht[3]

[1]Leiden Observatory, Leiden University, Postbus 9513, 2300 RA, Leiden, the Netherlands
email: snellen@strw.leidenuniv.nl

[2]SRON, Sorbonnelaan 2, 3584 CA Utrecht, The Netherlands

[3]Department of Physics, and Kavli Institute for Astrophysics and Space Research, Massachusetts Institute of Technology, Cambridge, Massachusetts 02139, USA

Abstract. Recently, we presented the detection of carbon monoxide in the transmission spectrum of extrasolar planet HD209458b, using CRIRES, the Cryogenic high-resolution Infrared Echelle Spectrograph at ESO's Very Large Telescope (VLT). The high spectral resolution observations (R=100,000) provide a wealth of information on the planet's orbit, mass, composition, and even on its atmospheric dynamics. The new observational strategy and data analysis techniques open up a whole world of opportunities. We therefore started an ESO large program using CRIRES to explore these, targeting both transiting and non-transiting planets in carbon monoxide, water vapour, and methane. Observations of the latter molecule will also serve as a test-bed for METIS, the proposed mid-infrared imager and spectrograph for the European Extremely Large Telescope.

Keywords. planetary systems, techniques: spectroscopic

1. Introduction

A planet which crosses the disk of its host star allows its atmospheric properties to be studied in three ways: 1) by transmission spectroscopy, when during a transit starlight filters through a planet's atmosphere leaving a spectral imprint of its atmospheric constituents; 2) by eclipse photometry, when the planet moves behind the star the planet light is temporary blocked, and the contribution from the planet can be determined; and 3) by orbital phase variations, since thoughout the orbit we see varying contributions form the (warm) dayside and (cooler) nightside of the planet, which is seen as minute changes in brightness of the system as function of orbital phase.

Until recently, all the fascinating discoveries using these techniques came from the Hubble and Spitzer space observatories. Transmission spectroscopy and secondary eclipse measurements have yielded detections of absorption signatures from hydrogen, oxygen and carbon atoms in the ultraviolet (Vidal-Madjar *et al.* 2003; 2004), sodium in the optical (Charbonneau *et al.* 2002), and of broadband molecular signatures in the near- and mid-infrared from water, methane, carbon-monoxide and carbon-dioxide (e.g. Tinetti *et al.* 2007; Swain *et al.* 2008; Grillmair *et al.* 2008; Swain *et al.* 2009a,b; Beaulieu *et al.* 2010). The strengths of molecular bands at different wavelengths constrain atmospheric temperature profiles, indicating that some of the hottest planets contain a thermal inversion layer (e.g. Burrows *et al.* 2007). In addition, orbital phase variations have been detected in the mid-infrared showing the day-to-nightside temperature distributions (Knutson *et al.* 2007), work that has been recently extended to the optical by CoRoT and Kepler (Snellen *et al.* 2009; Burucki *et al.* 2009).

For long, ground-based observations did not play any significant role, and it was thought that the disturbing influence of the Earth atmosphere, as emphasized by many unsuccessful attempts, would prevent contributions from ground-based instrumentation to this exciting field of research. However, over recent years, novel observation and data-analysis techniques have been in development, utilizing the many superior qualities of ground- based instrumentation (such as collective power and spectral resolution) over the space telescopes. This has resulted in a string of ground-based detection of transmission features (e.g. Redfield *et al.* 2008; Snellen *et al.* 2008; Sing *et al.* 2011), and optical and near-infrared eclipse measurements (e.g. Sing & Lopez-Morales 2009; de Mooij & Snellen 2009; Croll *et al.* 2010).

2. Molecular absorption in a planet's atmosphere at high spectral resolution

Broad-band space-based observations contain always a level of ambiguity, because molecular signatures overlap, and because high-temperature line lists for some molecular species (e.g. methane) are known to be incomplete. These issues can be circumvented by observing at high spectral resolution and thereby unambiguously identifying the individual molecular absorption lines. We therefore observed one transit of HD209458b with the Cryogenic Infrared Echelle Spectrograph (CRIRES) on the VLT, targeting the rotation-vibration band of carbon monoxide between 2.29 and 2.34 μm at a spectal resolution of R=100,000. By combining the signal of $\sim$50 CO absorption lines in the observed wavelength regime we detected a clear signal from the planet's atmosphere (Snellen *et al.* (2010) – S10). The change in the radial component of the planet's orbital motion is clearly seen, varying between $\pm$15 km/sec over the 3-hr transit, leading for the first time to a determination of the circular velocity of the planet. Combined with the radial velocity variations of the host-star, it results in an assumption-free determination of the mass of the planet and host-star using only Newton's law of gravity (just as for double-line eclipsing binaries).

The detected CO signal seems offset by $\sim$2 km/sec (2σ) with respect to the systemic velocity of the host star. It suggests the presence of a strong wind flowing from the irradiated dayside to the non-irradiated nightside of the planet within the 0.01-0.1 mbar atmospheric pressure range probed by these observations. Such winds may be driven by the large incident heat flux from the star on the dayside. Indeed three-dimensional circulation models indicate that at low pressure (<10 mbar) air should flow from the sub-stellar point towards the antistellar point both along the equator and the poles (Showman *et al.* 2008). The strength of the carbon monoxide signal suggests a CO mixing ratio of 1-3$\times10^{-3}$ in this planet's upper atmosphere. Assuming it is representative for the planet as a whole, and that CO dominates over CH_4, it implies that the C/H ratio in HD209458b is a factor 2-6 higher than that of the parent star, and that HD209458b is substantially enriched in heavy elements, to the level of Jupiter and Saturn in our own solar system.

3. The CRIRES Survey of Hot-Jupiter atmospheres

The results on carbon monoxide open up a range of very exciting possibilies for new observations with CRIRES, and we have recently been awarded a significant amount of CRIRES observing time to fully explore these. These future observations can be divided into three parts, a) transmission spectroscopy, b) dayside observations of transiting planets, and c) dayside observations of non-transiting planets.

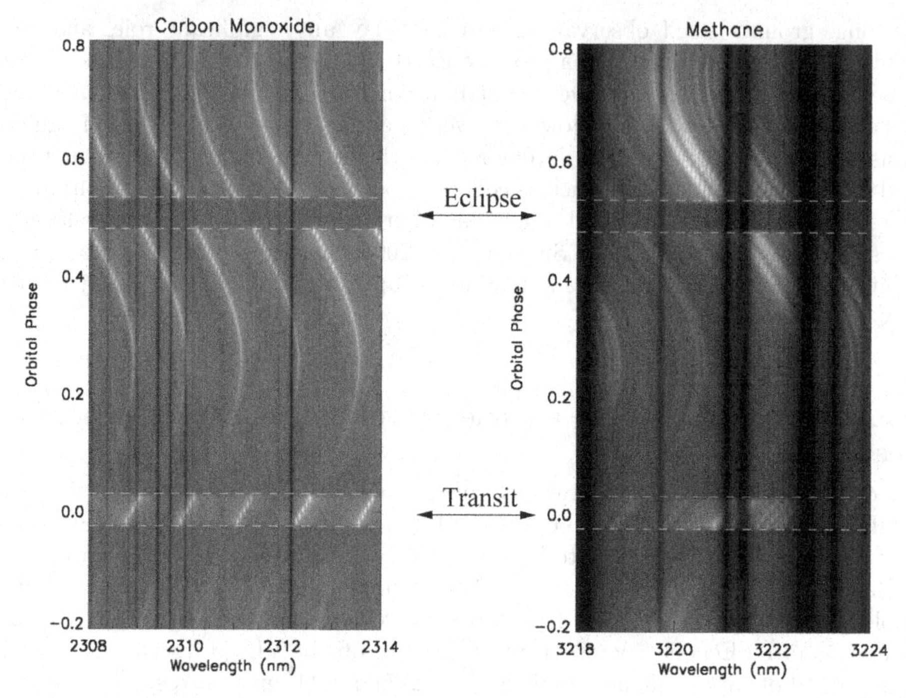

Figure 1. Toy model simulations of ground-based spectra of hot Jupiters as function of planet orbital phase, targeting carbon monoxide (left) and methane (right). The vertical bands are telluric absorption, while the planet signals are shown in emission for clarity, assuming uniform emission from the planet dayside, producing a planet signal equally strong as in transmission. Most of the methane transmission features are blocked by the earth atmosphere, but can be observed around phases ~0.4 and ~0.6.

a) Transmission spectroscopy: A logical next step is to search for a carbon monoxide signal in transmission in the other bright transiting hot Jupiter HD189733b. The CO/CH_4 ratio may be less balanced towards CO, because its effective temperature is lower, and its atmospheric scale-height is also smaller. But these could well be compensated by the factor 2 higher apparent brightness of the system in K-band, and the deeper transit.

b) Dayside spectroscopy of transiting planets: As is apparent from broad-band spectroscopy (e.g. Swain *et al.* 2009), dayside absorption features can be just as strong as during transit. Methane and water are expected to produce strong signals in L-band, but are inaccessible with transmission spectroscopy because the strong telluric absorption blocks out most of any possible signal (see Fig.1). However, when the dayside emission is probed, the orbital motion of the planet is of great help. When observing at a phase of $\theta \sim 0.4$ or $\theta \sim 0.6$, we still observe 90% of the dayside hemisphere, but with the signal blue- or redshifted by 100 km/sec due to the planet's orbital motion - outside the regime where telluric absorption is a problem. This provides some exciting prospects.

Firstly, such observations will constrain the temperature-pressure profiles of the hot Jupiter atmospheres. Since HD209458b is thought to have a thermal inversion layer, its spectral lines may appear in emission, unambigously proving the inversion to be present.

Secondly, the uncertainty in radial velocity of our carbon monoxide detection is about 1 km/sec. This means that we can expect to reach a ~1% uncertainty in the orbital velocity of the planets if we observed them at a red- or blueshift of 100 km/sec, resulting in a mass-determination of the planets and host-stars at a precision of <3%. Not only will these measurements remove any ambiguity in the radii and density of the planets, it will determine these properties of the host stars very accurately, which combined with the effective temperature and metallicity, could result in accurate age determinations for the two systems.

c) Dayside spectroscopy of non-transiting planets: Since for the dayside spectroscopy it is no longer required that the planets transit their host star, these observations can be extended to non-transiting planets. A great benefit is that the brightest hot Jupiter systems visible from Paranal, 51 Peg, tau Boo, and HD179949, are 4–12 times brighter in K-band than HD209458b. A very exciting prospect for these observations is that a measured planet orbital velocity, in combination with the well-known radial velocity variations of its host-star, will give the planet/star mass ratio, and subsequently the orbital inclination and true mass of the planet.

4. Perspectives for the E-ELT

Our planned observation of methane at 3.2 μm will fall within the wavelength range of METIS (Brandl *et al.* 2010), the proposed mid-infrared imager and spectrograph for the European Extremely Large Telescope, and can therefore serve as an interesting test-bed of exoplanet atmosphere observations with the E-ELT. Its baseline design contains a R=100,000 integral field spectrograph ideally suited for this type of work. With its 25x larger collective power, a wealth of new information could be extracted, probing other molecules, structure and variations in absorption features, and/or probing smaller-size planets - possibly down to the super-earth regime.

Acknowledgements

We thank the organisers very much for the extremely stimulating symposium.

References

Beaulieu, J. P., *et al.* 2010, *MNRAS*, 409, 963
Borucki, W. J., *et al.* 2009, *Science*, 325, 709
Brandl, B. R., *et al.* 2010, *Proc. SPIE*, 7735, 77352G1
Burrows, A., Hubeny, I., Budaj, J., Knutson, H. A., & Charbonneau, D. 2007, *ApJL*, 668, L171
Charbonneau, D., Brown, T. M., Noyes, R. W., & Gilliland, R. L. 2002, *ApJ*, 568, 377
Croll, B., Jayawardhana, R., Fortney, J. J., Lafrenière, D., & Albert, L. 2010, *ApJ*, 718, 920
de Mooij, E. J. W. & Snellen, I. A. G. 2009, *A&A*, 493, L35
Grillmair, C. J., *et al.* 2008, *Nature*, 456, 767
Knutson, H. A., *et al.* 2007, *Nature*, 447, 183
Redfield, S., Endl, M., Cochran, W. D., & Koesterke, L. 2008, *ApJL*, 673, L87
Sing, D. K. & López-Morales, M. 2009, *A&A*, 493, L31
Showman, A. P., Cooper, C. S., Fortney, J. J., & Marley, M. S. 2008, *ApJ*, 682, 559
Sing, D. K., *et al.* 2011, *A&A*, 527, A73
Snellen, I. A. G., Albrecht, S., de Mooij, E. J. W., & Le Poole, R. S. 2008, *A&A*, 487, 357
Snellen, I. A. G., de Mooij, E. J. W., & Albrecht, S. 2009, *Nature*, 459, 543
Snellen, I. A. G., de Kok, R. J., de Mooij, E. J. W., & Albrecht, S. 2010, *Nature*, 465, 1049
Swain, M. R., Vasisht, G., & Tinetti, G. 2008, Nature, 452, 329
Swain, M. R., *et al.* 2009, *ApJ*, 704, 1616
Swain, M. R., Vasisht, G., Tinetti, G., Bouwman, J., Chen, P., Yung, Y., Deming, D., & Deroo, P. 2009, *ApJL*, 690, L114
Vidal-Madjar, A., Lecavelier des Etangs, A., Désert, J.-M., Ballester, G. E., Ferlet, R., Hébrard, G., & Mayor, M. 2003, *Nature*, 422, 143
Vidal-Madjar, A., *et al.* 2004, *ApJL*, 604, L69

The Astrophysics of Planetary Systems: Formation, Structure, and Dynamical Evolution
Proceedings IAU Symposium No. 276, 2010
A. Sozzetti, M. G. Lattanzi & A. P. Boss, eds.

doi:10.1017/S1743921311020205

The properties of super-Earth atmospheres

Eliza M. R. Kempton[1]

[1]Department of Astronomy and Astrophysics
University of California, Santa Cruz, CA 95064 USA
email: ekempton@ucolick.org

Abstract. Extrasolar super-Earths likely have a far greater diversity in their atmospheric properties than giant planets. Super-Earths (planets with masses between 1 and 10 $M_\oplus$) lie in an intermediate mass regime between gas/ice giants like Neptune and rocky terrestrial planets like Earth and Venus. While some super-Earths (especially the more massive ones) may retain large amounts of hydrogen either from accretion processes or subsequent surface outgassing, other super-Earths should have atmospheres composed of predominantly heavier molecules, similar to the atmospheres of the rocky planets and moons of our Solar System. Others still may be entirely stripped of their atmospheres and remain as bare rocky cores. Of the two currently known transiting super-Earths one (GJ 1214b) likely falls into the former category with a thick atmosphere, while the other (CoRoT-7b) falls into the latter category with a very thin or nonexistent atmosphere. I review some of the theoretical work on super-Earth atmospheres, and I present methods for determining the bulk composition of a super-Earth atmosphere.

Keywords. planetary systems

1. Introduction

Super-Earths with masses between 1 and 10 $M_\oplus$ represent a fundamentally new class of planets that do not exist in our solar system. These planets lie in the intermediate mass range between the terrestrial planets of our solar system and the gas/ice giant Neptune. A natural question that arises is 'What defines the dividing line between small gas/ice giants ("mini-Neptunes") and scaled-up rock-dominated terrestrial planets?' The division between these two classes of planets may depend on varied parameters such as the history of formation for a particular planet, its mass, and properties of its host star. While terrestrial planets in our solar system have not retained molecular hydrogen in their atmospheres over their lifetimes, super-Earths are predicted to have higher surface gravities, and some of these planets could likely retain massive hydrogen atmospheres.

Recent discoveries of transiting super-Earths give the first constraints on the bulk composition of these planets. However, significant degeneracies exist in the theoretical mass-radius relationship predicted for super-Earths, given that these planets can be composed of a combination of iron, rock, ices, and atmospheric gasses (e.g. Fortney *et al.* 2007). Follow-up observations will therefore be necessary to further constrain the composition of super-Earth planets.

Two transiting super-Earths are currently known. The first, CoRoT-7b (Léger *et al.* 2009) is highly irradiated, and its measured density is consistent with a planet that possesses no significant atmosphere. If the planet does have an atmosphere, it is most likely experiencing atmospheric blow-off and would need to be constantly replenished by surface outgassing (Valencia *et al.* 2010). The second transiting super-Earth, GJ 1214b, (Charbonneau *et al.* 2009) has a low observed density, and fits to the observed mass and radius of this planet imply that it must have a massive atmosphere. Two distinct classes of planets have been proposed that both fit the observed density of GJ 1214b. One is

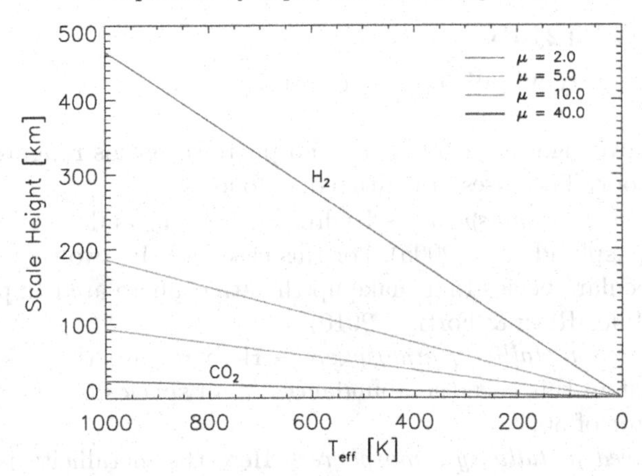

Figure 1. Atmospheric scale height for GJ 1214b as a function of effective temperature for atmospheric compositions ranging from pure hydrogen ($\mu = 2$) to mostly CO_2 ($\mu = 40$). The transmission signature of an exoplanet scales directly proportionally with scale height.

a mostly rock planet with a hydrogen atmosphere that comprises ~1% of the planet's overall mass ("mini-Neptune"). The other possibility is that the planet is composed mostly of water with a small rocky core and a thick water vapor atmosphere ("water-world") (Nettelmann *et al.* 2011; Rogers & Seager 2010).

A key difference between the two classes of planets for GJ 1214b is that they are predicted to posses very different atmospheres that should in turn produce fundamentally different atmospheric signatures. While the mini-Neptune scenario implies an atmosphere composed of mostly hydrogen (and possibly helium) with trace amounts of water, methane, or ammonia, the water-world scenario produces an atmosphere almost entirely composed of water steam. The distinctly different compositions of these two classes of atmospheres will produce spectral features in emission or transmission that are indicative of their overall makeup. Furthermore, the two classes of atmospheres also have atmospheric scale heights that differ from each other by up to an order of magnitude. The equation for scale height is given by:

$$H = \frac{kT}{\mu g} \tag{1.1}$$

where k is Boltzmann's constant, T is the atmospheric temperature, and g is the surface gravity. The mean molecular weight μ is only 2 times that if the hydrogen atom for a hydrogen dominated atmosphere but increases by a factor of 9 if the atmosphere is composed primarily of water vapor (see Fig. 1).

Transmission spectroscopy provides a strong constraint on atmospheric scale height by probing spectral features that arise from absorption of stellar light through the limb of the planet during transit. These spectral features tend to sample layers ranging over several scale heights in the planet's atmosphere, so planets with larger scale heights will produce correspondingly larger spectral features in transmission. For this reason transmission spectroscopy is a promising way to determine the bulk composition of an exoplanet atmosphere and should be useful for breaking the degeneracy between the two classes of planets proposed for GJ 1214b. In what follows we present models of GJ 1214b's atmosphere for a range of possible atmospheres of this planet.

2. Results for GJ 1214b

(See Miller-Ricci & Fortney (2010) for more details.)

For GJ 1214b we investigate six different atmosphere scenarios ranging from hydrogen-rich to hydrogen-poor. The cases are outlined as follows:

(*a*) *Solar composition atmosphere* – Hydrogen, helium, and metals appear in solar abundance ratios (Asplund *et al.* 2005). For this case and the two that follow, the abundances of the molecular species that make up the atmosphere are computed in chemical equilibrium (see Miller-Ricci & Fortney 2010).

(*b*) *30 × Enhanced metallicity atmosphere* – Here we use the base composition of atmosphere (a), but in this case the abundances of all species except for H and He are enhanced by a factor of 30.

(*c*) *50 × Enhanced metallicity atmosphere* – Here the metallicity is enhanced to 50 times relative to solar.

(*d*) *100% water (steam) atmosphere*

(*e*) *50% water, 50% CO_2 atmosphere*

(*f*) *CO_2 atmosphere plus trace gasses* – This model atmosphere is composed of 96.5% CO_2 with other trace gasses appearing in Venusian abundances, notably 3.5% N_2 and 20 ppm H_2O.

For each scenario we produce transmission and secondary eclipse spectra using a 1-D radiative transfer code (see Miller-Ricci *et al.* (2009) and Fortney *et al.* (2008)) to determine the observable signature of GJ 1214b's atmosphere.

Transmission Spectra: The approximate size of spectral features in transmission (relative to the stellar background) is given by

$$\Delta D \approx \frac{20 H R_{pl}}{R_*^2} \tag{2.1}$$

where H is the atmospheric scale height, and we have assumed that spectral features can probe regions of the planet's atmosphere ranging in depth by up to 10 scale heights. For GJ 1214b, if the planet possesses a hydrogen-dominated atmosphere, this corresponds to a value for ΔD of up to 0.3% relative to its M-dwarf host star.

In Fig. 2 we show the results from our spectral modeling of GJ 1214b. Transmission spectra for the three hydrogen-rich atmospheres are dominated by absorption features due to water and methane. These atmospheres have low mean molecular weight and correspondingly large atmospheric scale heights, resulting in transmission features that appear on the level of 0.1-0.3% relative to the host star, which agrees with our back-of-the-envelope calculation above. For the high mean molecular weight atmospheres (cases (d), (e), and (f)) the scale height is an order of magnitude or more smaller, and transmission features are only present at $\sim$ 0.01% relative to the stellar light. For these atmospheres spectral features of water and/or CO_2 are present but would be exceedingly difficult to observe with current instrumentation.

Emission Spectra: Secondary eclipse spectra for GJ 1214b are shown in Fig. 3. We predict secondary eclipse depths of up to 0.3% if redistribution of heat to the night side of the planet is inefficient, and 0.2% if heat circulation is efficient. Unfortunately, shortward of 5 μm at wavelengths that can be probed by the Warm Spitzer mission the secondary eclipse depths are predicted to be quite small – generally less than a few hundred parts per million. This level of precision is unlikely to be attainable with Warm Spitzer and we therefore await mid-infrared instrumentation such as MIRI aboard JWST to measure secondary eclipse spectra for GJ 1214b.

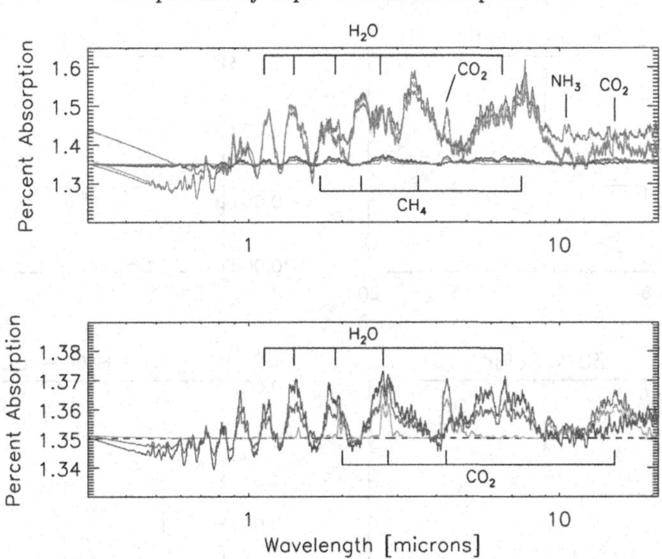

Figure 2. Top: Transmission spectra for atmospheres with differing composition – solar composition (red), 30 × solar metalicity (blue), 50 × solar metalicity (light green), water steam (purple), 50% water - 50% CO_2 (dark green), CO_2 with trace quantities of water (orange). All spectra are for models with efficient day-night circulation and have been normalized to the planet's observed radius of 2.678 $R_\oplus$ in the MEarth bandpass covering 650 - 1050 microns. Bottom: Same as above but zoomed in to show the spectra for the three atmospheres composed of heavier molecules. The dashed black line indicates the radius of a planet with no atmosphere. (From Miller-Ricci & Fortney 2010.)

Longward of 5 mum a whole host of water, methane, CO_2, and ammonia features are present, which will be useful diagnostics of atmospheric chemistry. We note that the spectra for the three hydrogen-dominated atmosphere scenarios ((a), (b), and (c)), all strongly resemble the secondary eclipse spectrum of a "water world" (scenario (d)) despite the very different atmospheric and bulk composition between these two classes of planets. This results from the fact that water is the main source of opacity in each of these atmospheres. For this reason, secondary eclipse spectroscopy may not be useful in breaking the degeneracy between a "water world" and "mini-Neptune" composition of GJ 1214b, and transmission spectroscopy with its strong dependence on scale height is the more decisive method for differentiating between these two classes of models.

Clouds: The presence of clouds can additionally complicate the interpretation of transmission spectra. Clouds impede the transmission of stellar light through a planetary atmosphere below the height where the cloud optical depth equals unity. The overall effect is to flatten out the observed transmission spectrum, since deeper levels in the planet's atmosphere are no longer probed by the incoming stellar light.

We use a toy model to test the effect of clouds at various heights on the transmission spectrum of GJ 1214b. Specifically, we model the cloud deck by cutting off all transmission of stellar light below the height of the cloud deck. We examine clouds ranging in height (in terms of pressure) from 300 mbar to 1 mbar for a solar composition atmosphere (shown in Fig. 4). We note that high clouds in the planet's atmosphere can considerably alter the appearance of the transmission spectrum, and a solar composition atmosphere with very high clouds may be indistinguishable from a high mean molecular weight water (steam) atmosphere over large wavelength ranges.

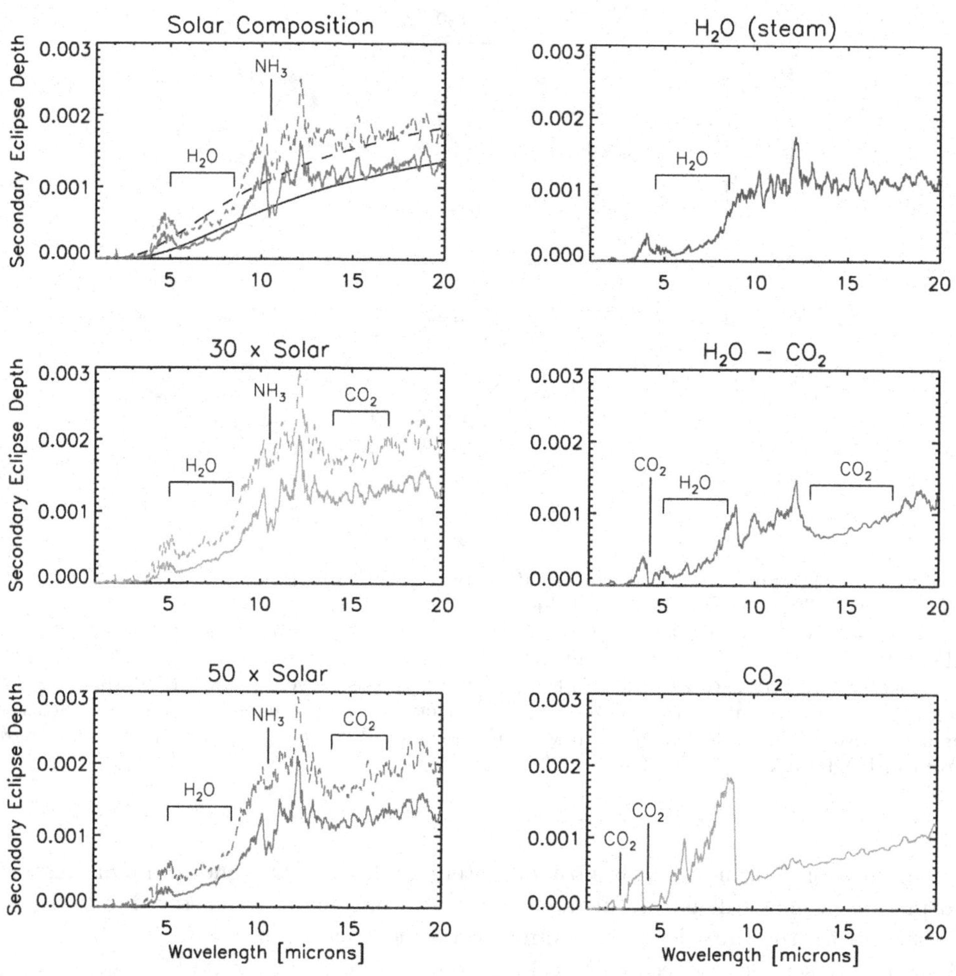

Figure 3. The contrast ratio between the day-side emission from GJ 1214b and the emitted light from its M-dwarf host star, plotted as a function of wavelength for 6 different possible atmospheric compositions. In the top left panel the black lines denote the contrast ratios that would be expected if the planet and star both emitted as blackbodies, with planetary T_{eff} of 555 K (solid) and 660 K (dashed). Dashed lines are spectra for models with inefficient day-night heat redistribution. Solid lines denote models with efficient heat circulation.(From Miller-Ricci & Fortney 2010.)

3. Conclusions

Of the two currently known super-Earths, GJ 1214b is far more likely to have a substantial permanent atmosphere. This planet is also conveniently suited to follow-up observations with current ground-based and spaced-based instrumentation. In transmission, GJ 1214b should vary in its observed transit depth by 0.1-0.3% as a function of wavelength *if the atmosphere is hydrogen-dominated.* If variations in the transit depth are observed to take place on this level, then GJ 1214b's atmosphere is unequivocally composed of predominantly hydrogen. If transit depth variations are not present, then interpretation is more complex, since the planet's atmosphere could be composed of higher mean molecular weight material like water vapor, but could also be hydrogen-rich but with high clouds or haze present.

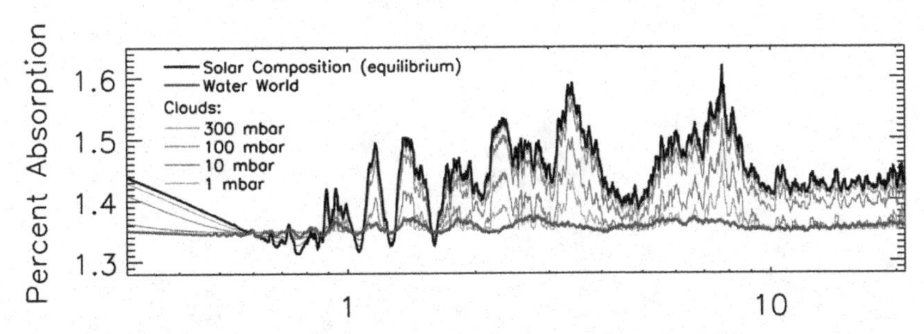

Figure 4. Toy model showing the approximate effect that clouds at differing heights would have on the transmission spectrum of GJ 1214b for solar composition. All transmission of stellar light is cut off at the height of the cloud deck in this simple model. Very high clouds considerably flatten out the transmission spectrum rendering a solar composition atmosphere virtually indistinguishable from a thick steam atmosphere at most wavelengths.

The planet GJ 1214b holds a special place in the family of super-Earths. Because this planet orbits a small M-dwarf and because the planet is low density and most likely possesses an atmosphere with a large scale height, GJ 1214b is uniquely suited to follow-up observations aimed at characterizing its atmosphere. In general, most super-Earths will not be nearly so easily followed up by characterization efforts. Most super-Earths will exhibit transmission and emission signatures at the level of only 1-100 parts-per-million relative to their host stars (see Miller-Ricci *et al.* 2009). This means that follow-up efforts will only be possible with next generation observing facilities like JWST (Deming *et al.* 2009) or any of the ground-based extremely large telescopes. Once these facilities come online, the effort can begin to study the overall diversity of super-Earth atmospheres. Models such as the ones presented here for GJ 1214b are instructive in illustrating the potential observable signatures of these different types of atmospheres.

References

Asplund, M., Grevesse, N., & Sauval, A. J. 2005, in Astronomical Society of the Pacific Conference Series, Vol. 336, Cosmic Abundances as Records of Stellar Evolution and Nucleosynthesis, ed. T. G. Barnes, III & F. N. Bash, 25–+

Charbonneau, D., *et al.* 2009, *Nature*, 462, 891

Deming, D., *et al.* 2009, *PASP*, 121, 952

Fortney, J. J., Lodders, K., Marley, M. S., & Freedman, R. S. 2008, *ApJ*, 678, 1419

Fortney, J. J., Marley, M. S., & Barnes, J. W. 2007, *ApJ*, 659, 1661

Léger, A., *et al.* 2009, *A&A*, 506, 287

Miller-Ricci, E., Seager, S., & Sasselov, D. 2009, *ApJ*, 690, 1056

Miller-Ricci, E. & Fortney, J. J. 2010, *ApJ* (Letters), 716, L74

Nettelmann, N., Fortney, J. J., Kramm, U., & Redmer, R. 2011, *ApJ*, 733, id.2

Rogers, L. A. & Seager, S. 2010, *ApJ*, 716, 1208

Valencia, D., Ikoma, M., Guillot, T., & Nettelmann, N. 2010, *A&A*, 516, A20

PART 3:
INTERACTIONS

The Astrophysics of Planetary Systems: Formation, Structure, and Dynamical Evolution
Proceedings IAU Symposium No. 276, 2010
A. Sozzetti, M. G. Lattanzi & A. P. Boss, eds.

doi:10.1017/S1743921311020217

The diverse origin of exoplanets' eccentricities & inclinations

Eric B. Ford[1]

[1]University of Florida, 211 Bryant Space Science Center, Gainesville, FL 32611-2055, USA
email: eford@astro.ufl.edu

Abstract. Radial velocity surveys have discovered over 400 exoplanets. While measuring eccentricities of low-mass planets remains a challenge, giant exoplanets display a broad range of orbital eccentricities. Recently, spectroscopic measurements during transit have demonstrated that the short-period giant planets ("hot-Jupiters") also display a broad range of orbital inclinations (relative to the rotation axis of the host star). Both properties pose a challenge for simple disk migration models and suggest that late-stage orbital evolution can play an important role in determining the final architecture of planetary systems. One possible formation mechanism for the inclined hot-Jupiters is some form of eccentricity excitation (e.g., planet scattering, secular perturbations due to a distant planet or wide binary) followed tidal circularization. The planet scattering hypothesis also makes predictions for the population of planets at large separations. Recent discoveries of planets on wide orbits via direct imaging and highly anticipated results from upcoming direct imaging campaigns are poised to provide a new type of constraint on planet formation. This proceedings describes recent progress in understanding the formation of giant exoplanets.

Keywords. planets and satellites: formation

1. Introduction

Radial velocity surveys have discovered dozens of systems with multiple giant planets (Wright *et al.* 2011 and references therein). These can be roughly assigned into the following categories: 1) systems in or near a mean motion resonance (MMR), 2) systems with significant secular interaction (but not near a strong MMR), and 3) hierarchical systems where no significant interactions are expected based on the known planets (and assuming inclinations are not extreme). Early results from HARPS (Mayor & Udry 2008) and NASA's *Kepler* mission (Borucki *et al.* 2011) suggest that systems of multiple low-mass (i.e., less than Neptune) planets may differ from those of giant planets. In particular, there appears to be a population of systems with multiple low-mass planets that are closely-spaced, but non-resonant (Lissauer *et al.* 2011a, 201b).

Of course, many (most?) stars with only a single known exoplanet may harbor additional planets that have yet to be detected, perhaps due to their low mass and/or long orbital period (for RV surveys) or their orbital inclination (for transit surveys). Even if future observations exclude additional planets today, there may have once been additional planets that have been ejected, collided with with other planets or been swallowed by the host star. Thus, we interpret the hundreds of systems with a single known planet in the context of the formation and orbital evolution models developed for explaining multiple-planet systems.

2. Eccentricity Distribution

The distribution of orbital eccentricities provides an important constraint for planet formation models (Ford & Rasio 2008). The precision of eccentricity measurements varies widely, depending primarily on the ratio of the Doppler amplitude to the measurement precision and the number of Doppler observations. Precision can also be adversely affected by the presence of multiple planets and/or poor phase coverage, particularly in cases where the orbital period is comparable to or greater than the timespan of precise Doppler observations.

Unfortunately, characterizing the eccentricity distribution of a population of exoplanets is further complicated by measurement biases (Shen & Turner 2008; Zakamska *et al.* 2011). Population analyses suggest that Doppler-detected giant planets can be modeled by two populations: 1) a set of low-eccentricity planets (~20-30%) and a second population with a broad eccentricity distribution (e.g., Rayleigh(0.3), ~70-80%; Wang & Ford 2011).

Complications due to large measurement uncertainties and biases for "small" eccentricities become more significant as one pushes towards Neptune-mass and smaller planets. Transit durations are proportional to $e \sin \omega$ and occultations durations to $e \cos \omega$. When the occultation can be well-measured (e.g., Spitzer, ground, Kepler), the combination provides an accurate eccentricity that is robust to the effects of additional planets (Colón *et al.* 2009). For planets where occultation observations are not practical, it is still possible to characterize the eccentricity distribution of a population of transiting planets, provided accurate stellar properties (Ford *et al.* 2008; Moorhead *et al.* 2011).

3. Eccentricity & Inclination Excitation

Several mechanisms have been proposed to explain the broad range of eccentricities among giant exoplanets. In most cases, these mechanisms have implications for the orbital inclinations. For example, planets formed via disk instability would be expected to have significant primordial eccentricities. The most promising location for disk instabilities for form planets is at large distances from the host star. If the protoplanetary disk is sufficiently warped, then these planets could also have significant inclinations relative to interior planets (and presumably the stellar rotation).

Planet-disk interactions during migration have also been proposed to excite the eccentricities. For isolated planets, this disk perturbations could excite the eccentricity and inclination of the most massive planets (greater than $\sim 10 M_{\rm Jup}$ for solar-mass host). However, most planets are not sufficiently massive. In cases, where there are multiple giant planets, disk migration leading to trapping (or passing through) MMRs can excite eccentricities and inclinations (Lee & Thommes 2009).

For a system of multiple planets, the current eccentricities represent just a snapshot of the range of values visited over a secular timescale (Veras & Ford 2009a; Veras & Ford 2010). Secular perturbations lead to the exchange of angular momentum between bodies in a bound hierarchical system. If all bodies start on circular and coplanar orbits, then there is no angular momentum deficit to be exchanged. Thus, secular evolution of planetary systems can sculpt systems, but does not provide a mechanism for exciting eccentricities in the first place. For sufficiently widely separated binary stars, protoplanetary disks are rarely aligned, suggesting that a substantial fraction of planets around one star in a wide binary could have their eccentricities excited by the "Kozai effect." While this inevitably affects some systems, Monte Carlo simulations show that this mechanism under-predicts the abundance of planets with intermediate eccentricities relative to nearly circular or highly eccentric orbits (Takeda *et al.* 2008). Thus, perturbations from binary stars are not sufficient to explain the whole population of exoplanet eccentricities. For planetary systems with a distant and massive planet, angular momentum conservation implies that a small change in the eccentricity of the outer planet can lead to large eccentricities of interior and less-massive planets (Wu & Lithwick 2011).

Finally, densely packed planetary systems can lead to planet-planet scattering in systems with multiple massive planets. Large eccentricities are generated when one planet scatters a comparable mass planets so that it is effectively removed from the system, either due to being ejected from the system, falling into the host star, or being perturbed by another body (e.g., Rasio & Ford 1996; Weidenschilling & Marzari 1996). N-body simulations show that this mechanism produces a broad distribution of both eccentricities and inclinations (Chatterjee *et al.* 2008; Nagasawa *et al.* 2008; Juric & Tremaine 2008). However, the correlation is statistical and individual system may have a significant eccentricity but low inclinations, or vice versa. Similar interactions in systems with more disparate mass ratios can produce smaller eccentricities. For low-mass planets and small separations, interactions typically result in collisions rather than ejections (Ford & Rasio 2008). In these cases, the final orbits typically have small eccentricities and inclinations, unless at least one planet had already acquired a large eccentricity or pericenter. Of course, the situation is complicated if the planet scattering occurs while there is still significant gas present (Matsumura *et al.* 2010).

4. Implications of Inclined Transiting Planets

The planet scattering model makes several predictions that have been discussed previously (e.g., Ford & Rasio 2008; Juric & Tremaine 2008). Recently, spectroscopic observations during transit (i.e., Rossitter-McLaughlin measurements) provide evidence for a substantial population of hot-Jupiters on highly-inclined, nearly-polar or even retrograde orbits (measured relative to the current stellar spis axis; Fabrycky & Winn 2009; Morton & Johnson 2011; Triaud *et al.* 2010; Winn *et al.* 2010). The highly-inclined population includes some planets that remain highly eccentric (e.g., HD 80606; Winn *et al.* 2009) and some that are nearly circular today. Eccentricity damping is expected to proceed more rapidly than inclination damping (e.g., Hut 1982). Thus, these observations strongly suggest that many hot-Jupiters were formed by eccentricity and inclination excitation followed by tidal damping (Rasio & Ford 1996). The eccentricity and inclination excitation could be due to either planet scattering or secular evolution by a distant massive body (either a star, i.e., "Kozai-effect" or one or more massive planets with significant initial angular momentum deficit; Chatterjee *et al.* this volume, Naoz *et al.* this volume; Wu & Lithwick 2011).

While these mechanisms appear the most natural candidates, researchers are also considering alternative mechanisms, including perturbations due to stellar encounters (e.g., Malmberg *et al.* 2011; Payne *et al.* 2011; Boley *et al.* this volume) or torques on the star (Lai *et al.* 2011). Future observations of Rossiter effect in planets at large separations (e.g., Borucki *et al.* 2011) and the relative inclinations among multiple planet systems (Payne *et al.* 2010; Ragozzine & Holman 2010; Ford *et al.* 2011; Lissauer *et al.* 2011b) can be expected to help determine the significance of each mechanism.

5. Implications of Planet at Wide Separations

The planet scattering model also makes predictions for the frequency of planets in very-long period orbits around young stars and even an abundance of free floating planets

(Scharf & Menou 2009; Veras *et al.* 2009b). As researchers refine these models, they can be tested by upcoming observations from direct imaging campaigns (e.g. Bonavita *et al.*, this volume) and microlensing results (Beaulieu *et al.* this volume).

Acknowledgements

EBF and this review were supported by NASA Origins of Solar Systems grant NNX09AB35G.

References

Beaulieu, J.-P., *et al.* 2011, this volume
Boley, A., *et al.* 2011, this volume
Borucki, W. J., *et al.* 2011, *ApJ*, 736, id.19
Chatterjee, S., Ford, E. B., Matsumura, S., & Rasio, F. A. 2008, *ApJ*, 686, 580
Chatterjee, S., Ford, E. B., & Rasio, F. A. 2011, this volume (*arXiv:1012.0584*)
Colón, K. D. & Ford, E. B. 2009, *ApJ*, 703, 1086
Fabrycky, D. C. & Winn, J. N. 2009, *ApJ*, 696, 1230
Ford, E. B., Quinn, S. N., & Veras, D. 2008, *ApJ*, 678, 1407
Ford, E. B. & Rasio, F. A. 2008, *ApJ*, 686, 621
Ford, E. B., *et al.* 2011, submitted to *ApJ* (*arXiv:1102.0544*)
Hut, P. 1982, *A&A*, 110, 37
Bonavita, M., *et al.* 2011, this volume
Jurić, M. & Tremaine, S. 2008, *ApJ*, 686, 603
Lai, D., Foucart, F., & Lin, D. N. C. 2011, *MNRAS*, 412, 2790
Lee, M. H. & Thommes, E. W. 2009, *ApJ*, 702, 1662
Lissauer, J. J., *et al.* 2011a, *Nature*, 470, 53
Lissauer, J. J., *et al.* 2011b, submitted to *ApJ*, *arXiv:1102.0543*
Malmberg, D., Davies, M. B., & Heggie, D. C. 2011, *MNRAS*, 411, 859
Matsumura, S., Thommes, E. W., Chatterjee, S., & Rasio, F. A. 2010, *ApJ*, 714, 194
Mayor, M. & Udry, S. 2008, Physica Scripta Volume T, 130, 014010
Moorhead, A. V., *et al.* 2011, submitted to *ApJ*, *arXiv:1102.0547*
Morton, T. D. & Johnson, J. A. 2011, *ApJ*, 729, id.138
Nagasawa, M., Ida, S., & Bessho, T. 2008, *ApJ*, 678, 498
Naoz, S., *et al.* 2011, this volume
Payne, M. J., Boley, A. C., & Ford, E. B. 2011, in Detection and Dynamics of Transiting Exoplanets, St. Michel l'Observatoire, France, Edited by F. Bouchy, R. Díaz, & C. Moutou; *EPJ Web of Conferences*, 11, 4005
Payne, M. J., Ford, E. B., & Veras, D. 2010, *ApJL*, 712, L86
Ragozzine, D. & Holman, M. J. 2010, *arXiv:1006.3727*
Rasio, F. A. & Ford, E. B. 1996, Science, 274, 954
Scharf, C. & Menou, K. 2009, *ApJL*, 693, L113
Shen, Y. & Turner, E. L. 2008, *ApJ*, 685, 553
Takeda, G., Kita, R., & Rasio, F. A. 2008, *ApJ*, 683, 1063
Triaud, A. H. M. J., *et al.* 2010, *A&A*, 524, A25
Veras, D. & Ford, E. B. 2009a, *ApJL*, 690, L1
Veras, D., Crepp, J. R., & Ford, E. B. 2009b, *ApJ*, 696, 1600
Veras, D. & Ford, E. B. 2010, *ApJ*, 715, 803
Wang, J. & Ford, E. B. 2011, submitted to *MNRAS*
Weidenschilling, S. J. & Marzari, F. 1996, *Nature*, 384, 619
Winn, J. N., Fabrycky, D., Albrecht, S., & Johnson, J. A. 2010, *ApJL*, 718, L145
Winn, J. N., *et al.* 2009, *ApJ*, 703, 2091
Wright, J. T., *et al.* 2011, *ApJ*, 730, 93
Wu, Y. & Lithwick, Y. 2011, *ApJ*, 735, id.109
Zakamska, N. L., Pan, M., & Ford, E. B. 2011, *MNRAS*, 410, 1895

The Astrophysics of Planetary Systems: Formation, Structure, and Dynamical Evolution
Proceedings IAU Symposium No. 276, 2010
A. Sozzetti, M. G. Lattanzi, & A. P. Boss, eds.

doi:10.1017/S1743921311020229

How planet–planet scattering can create high-inclination as well as long-period orbits

Sourav Chatterjee[1], Eric B. Ford[1] and Frederic A. Rasio[2]
[1]University of Florida, 211 Bryant Space Science Center, Florida, USA
email: s.chatterjee@astro.ufl.edu, eford@astro.ufl.edu
[2]CIERA, Northwestern University, Evanston, IL 60208, USA
email: rasio@northwestern.edu

Abstract. Recent observations have revealed two new classes of planetary orbits. Rossiter-Mclaughlin (RM) measurements have revealed hot Jupiters in high-obliquity orbits. In addition, direct-imaging has discovered giant planets at large (~ 100 AU) separations via direct-imaging technique. Simple-minded disk-migration scenarios are inconsistent with the high-inclination (and even retrograde) orbits as seen in recent RM measurements. Furthermore, forming giant planets at large semi-major axis (a) may be challenging in the core-accretion paradigm. We perform many N-body simulations to explore the two above-mentioned orbital architectures. Planet–planet scattering in a multi-planet system can naturally excite orbital inclinations. Planets can also get scattered to large distances. Large-a planetary orbits created from planet–planet scattering are expected to have high eccentricities (e). Theoretical models predict that the observed long-period planets, such as Fomalhaut-b have moderate $e \approx 0.3$. Interestingly, these are also in systems with disks. We find that if a massive-enough outer disk is present, a scattered planet may be circularized at large a via dynamical friction from the disk and repeated scattering of the disk particles.

Keywords. scattering, methods: n-body simulations, methods: numerical, planetary systems

1. Introduction

The 15 years since the discovery of the first exoplanet around a Solar-like star (Mayor & Queloz 1995) have seen a revolution in our understanding of planet formation and evolution. Observations and theoretical modeling have worked hand-in-hand to discover and explain new architectures of planetary orbits. It is now well known that many exoplanets have large e compared to our Solar system planetary orbits, indicating an active dynamical history (e.g., Chatterjee *et al.* 2008; Jurić & Tremaine 2008; Nagasawa *et al.* 2008).

Recent RM measurements of many transiting planets are putting further constraints on theoretical models of various planet formation and evolution scenarios (e.g., Triaud *et al.* 2010; Winn *et al.* 2010; Morton & Johnson 2011). Indeed, recent measurements find a large population of highly inclined planetary orbits, and even some retrograde orbits (see contributions from Winn *et al.* and Triaud *et al.* in this volume). Disk–planet migration models generally predict alignment between the planetary orbital angular momentum and the stellar spin axis from an aligned protoplanetary disk. Thus high-inclination orbits suggest dynamical evolution to be important in shaping the exoplanet architectures. Alternatively, inclined orbits might also result if the inner portion of the protoplanetary disk itself had been misaligned (Lai *et al.* 2011, see also Lai et al. in this volume).

Recent high-contrast imaging has revealed another class of planets—giant planets at very large a ($\gtrsim 50$ AU; e.g., Kalas *et al.* 2008; Marois *et al.* 2008; Ducourant *et al.* 2008;

Ireland *et al.* 2011). Timescale considerations for the core-accretion model of planet formation indicates that forming these planets in situ may be hard (e.g., Levison & Stewart 2001; Dodson-Robinson *et al.* 2009). Instead, we consider formation at closer orbital separations followed by planet–planet scattering to launch planets in large-a orbits from dynamically active systems (e.g. Chatterjee *et al.* 2008; Jurić & Tremaine 2008; Nagasawa *et al.* 2008). However, these simulations predict that these orbits typically also have high $e \gtrsim 0.6$. Interestingly, some of the observed large-a systems also have disks (e.g. Kalas *et al.* 2008) and dynamical modeling of these disks indicates moderate values of $e \approx 0.3$. We have started to explore the possibility that a planet launched into a large-a and high-e orbit can later be circularized near its apocenter if the planet enters a debris disk during its apocenter excursion.

In Section 2 we summarize our numerical setup and present results for expected inclinations from planet–planet scattering. In Section 3 we discuss how planet–planet scattering followed by circularization due to a residual disk may create moderate-e giant planets at large a. In Section 4 we conclude.

2. Orbital inclinations from planet–planet scattering

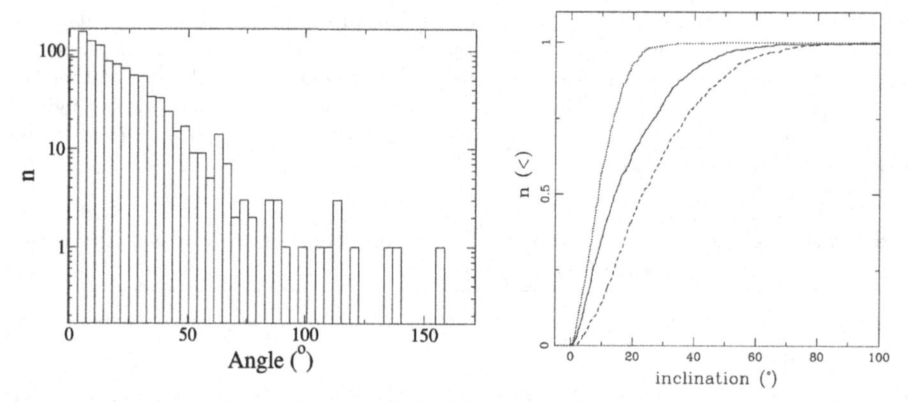

Figure 1. Distribution of orbital inclinations with respect to the initial invariable plane. *Left:* Histogram for the final inner-planet orbit. In our simulations about 2% of final inner planets are in retrograde orbits. *Right:* Cumulative histograms for the final inner (solid), and outer (dotted) planets. The relative inclinations between the inner and outer orbits (dashed) are also shown. About 20% of our simulated systems with two giant planets at the end have relative inclination angles $\geqslant 40°$ and could later go through Kozai-type oscillations.

We have simulated 3 giant planets with masses between 0.4–1.2 Jupitar-mass (M_J) around a Solar mass star. The initial innermost planet is placed at 3 AU, and the other 2 planets are placed with planet–planet separation of 4.4 Hill radii. The initial a's were chosen to avoid Mean-motion resonances. The initial e for the planetary orbits are chosen uniformly between 0–0.1. Initial i is chosen uniformly between $0°$–$10°$ with respect to the initial innermost orbital plane. All phase angles are assigned random values in the full range. Each of the above configurations is integrated using the hybrid integrator of MERCURY6.2 (Chambers 1999) for 10^7 yr. If the energy conservation is poorer than 10^{-3} then the full integration is repeated using the Burlisch-Stoeer (B-S) integrator (see Chatterjee *et al.* 2008 for more details).

Figure 1 shows the distributions of the final inner- and outer-planet orbital inclinations with respect to the initial invariable plane, as well as the relative inclinations between the orbits. We find that planet–planet scattering is very efficient at exciting the inclination of planetary orbits. These high inclinations are not mere reflection of the initial inclinations. There is in fact no correlation between the initial and final inclinations. Starting from only moderate inclinations planet–planet scattering can create large inclinations extending all the way to retrograde orbits (although only about 2% in these simulations). The relative inclinations between the planetary orbits in systems where two giant planets remain bound are also high (median $\approx 25\,^\circ$). Note that the median value is incidentally in good agreement with the recent relative inclination measurement for υ–Andromidae (c & d, $29.9\,^\circ \pm 1\,^\circ$; McArthur *et al.* 2010). In our simulations about 20% systems show relative inclinations angles $\geqslant 40\,^\circ$, making them potentially Kozai active.

3. Long-period planets

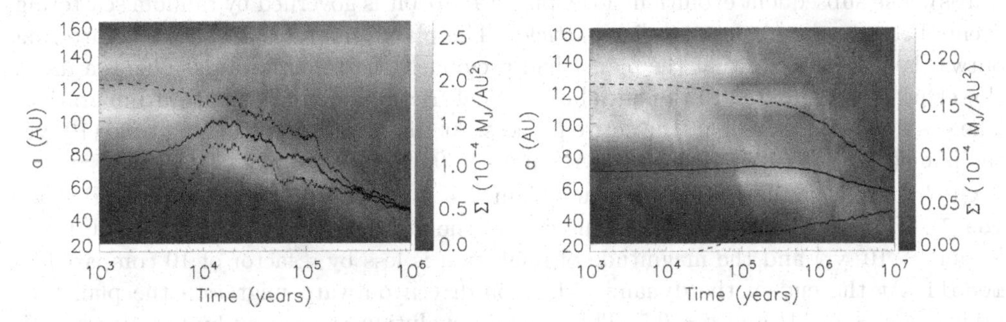

Figure 2. Evolution of the planetary orbit and surface density of the disk. *Left:* Evolution for the higher density disk `model1`. *Right:* Evolution for the lower density disk `model2`. The planets migrate outwards at first due to dynamical friction from the disk. The later inward migration of the planets is due to random outward scattering events of the disk particles by the planet. Planetary e decreases throughout the entire evolution.

Planet–planet scattering naturally creates large-a orbits. In our simulations with 3 giant planets we find outer planets with a up to about 150 AU. Here we explore the following scenario. A massive disk of rocky material (possibly relic from the planet-formation process) remains at large separation from the star. This disk can remain relatively undisturbed for some time while the planets remain much closer to the star. At some point in the evolution, planet–planet scattering launches one of the planets into a large-a, and high-e orbit so that the giant planet enters the disk near apocenter. The Keplerian velocity of the planet near apocenter is less than that of the material in the disk as long as the e of the planet is higher than the typical e of the disk material. Thus, while in the disk the planet experiences a force due to dynamical friction from the disk in the direction of the planet's orbital velocity, increasing the planet's orbital energy and angular momentum. As a result, the planet's pericenter is raised as the orbit is circularized. The amount of migration and the timescale for circularization are directly proportional to the disk density.

Using the B-S integrator in MERCURY6.2 we simulate the evolution of a giant planet orbit initially in a large-a and high-e orbit (possibly created via a planet–planet scattering event). The initial a, and e of the planetary orbit are 70 AU, and 0.7, respectively. Note that, in this case our $t = 0$ is after a planet–planet scattering event in a multi-giant planetary system that has launched a giant planet into this orbit. A disk of material is distributed initially between 90–150 AU with a uniform surface density. The mass in the disk is represented in our simulations by 10^3 equal-mass particles. The disk particles interact with the planet but not with each other. Initial e and inclination of the disk material are chosen uniformly between 0–0.3 and $0\,°$–$4\,°$, respectively. The planetary orbit is assumed to be aligned with the mid-plane of the disk (as a first step). We use two models varying the initial disk surface densities (Σ) keeping everything else fixed. Our `model1` and `model2` have $\Sigma = 10^{-4}$ and $10^{-5}\,M_J\,AU^{-2}$, respectively.

Figure 2 shows the evolution of the planetary orbit as well as the surface density contours of the disk. For `model1` (Figure 2, left), the planet first migrates outwards from 70 AU to $\approx$ 100 AU in $\approx 10^4$ yr. Note that the outward migration happens via circularization of the planetary orbit near the planet's apocenter. At the end of this migration the planet is in an orbit with $a = 100$ AU and $e = 0.14$. The outward migration is halted when the planet's intrusion severely depletes the disk. Until then the planet's evolution is dominated by dynamical friction arising from the disk. During this process the planet scatters a part of the disk material inwards (seen as a strip of over-density extending inwards). The subsequent evolution of the planetary orbit is governed by random scattering events between the planet and disk particles. The planet migrates inwards by scattering outwards disk particles that the planet had previously scattered inwards. This is a much slower process compared to the initial dynamical-friction dominated outward migration. The e is damped throughout the entire process. At the integration stopping time (10^6 yr) the planet is in an orbit with $a = 44$ AU and $e = 0.02$.

Qualitatively similar behavior is noted in the evolution of the planetary orbit for `model2`. However, in this case the timescale of the outward migration is about 10 times longer ($\sim 10^5$ yr) and the magnitude of migration is less by a factor of 10 compared to `model1`. At the end of the dynamical-friction-driven outward migration the planetary orbit has $a = 74$ AU and $e = 0.5$. This stage of evolution is followed by the planet–disk particle scattering stage during which the planet migrates inwards. At the integration stopping time (10^7 yr) for `model2` the planetary orbit has $a = 57$ AU and $e = 0.2$.

In both cases at the end the planet is (or will be) left in a large-a, and modest-e orbit. In addition, the total mass in the disk is reduced dramatically. The intrusion of the giant planet also excites the e and inclination of the disk material. Furthermore, a low density disk may remain outside the planetary orbit which may then, over time, grind to create a dust ring via collisions, similar to the one observed in Fomalhaut-b.

4. Discussion

Using numerical simulations we have studied how planet–planet scattering can naturally create high-inclination orbits. We have studied whether giant planets can be launched into large-a, but modest-e orbits, similar in architecture to Fomalhaut-b, via planet–planet scattering. We find that planet–planet scattering can naturally create many large-a orbits, however, these orbits are also expected to have high e (e.g., Chatterjee *et al.* 2008). Nevertheless, if a massive outer disk exists and the scattered giant planet near its apocenter enters the disk, dynamical friction from the disk can raise the planet's a until the planet reduces the disk density significantly via scattering. Then the planet can migrate inwards via scattering some of the remaining disk particles outwards. During the whole process the e of the planetary orbit reduces. We plan to further study this process in detail exploring different disk masses, densities, as well as varying planet masses. We thank the SOC and LOC for arranging this excellent conference and the opportunity to present these results.

References

Chambers, J. E. 1999, *MNRAS*, 304, 793
Chatterjee, S., Ford, E. B., Matsumura, S., & Rasio, F. A. 2008, *ApJ*, 686, 580
Dodson-Robinson, S. E., Veras, D., Ford, E. B., & Beichman, C. A. 2009, *ApJ*, 707, 79
Ducourant, C., *et al.* 2008, *A&A*, 477, L1
Ireland, M. J., *et al.* 2011, *ApJ*, 726, id.113
Jurić, M. & Tremaine, S. 2008, *ApJ*, 686, 603
Kalas, P., *et al.* 2008, *Science*, 322, 1345
Lai, D., Foucart, F., & Lin, D. N. C. 2011, *MNRAS*, 412, 2790
Levison, H. F. & Stewart, G. R. 2001, *Icarus*, 153, 224
Marois, C., *et al.* 2008, *Science*, 322, 1348
Mayor, M. & Queloz, D. 1995, *Nature*, 378, 355
McArthur, B. E., *et al.* 2010, *ApJ*, 715, 1203
Morton, T. D. & Johnson, J. A. 2011, *ApJ*, 729, id.138
Nagasawa, M., Ida, S., & Bessho, T. 2008, *ApJ*, 678, 498
Triaud, A. H. M. J., *et al.* 2010, *A&A*, 524, A25
Winn, J. N., Fabrycky, D., Albrecht, S., & Johnson, J. A. 2010, *ApJ*, 718, L145

The Astrophysics of Planetary Systems: Formation, Structure, and Dynamical Evolution
Proceedings IAU Symposium No. 276, 2010
A. Sozzetti, M. G. Lattanzi, & A. P. Boss, eds.

doi:10.1017/S1743921311020230

The Rossiter-McLaughlin effect for exoplanets

Joshua N. Winn[1]

[1]Department of Physics, and Kavli Institute for Astrophysics and Space Research, Massachusetts Institute of Technology
email: `jwinn@mit.edu`

Abstract. There are now more than 35 stars with transiting planets for which the stellar obliquity—or more precisely its sky projection—has been measured, via the eponymous effect of Rossiter and McLaughlin. The history of these measurements is intriguing. For 8 years a case was gradually building that the orbits of hot Jupiters are always well-aligned with the rotation of their parent stars. Then in a sudden reversal, many misaligned systems were found, and it now seems that even retrograde systems are not uncommon. I review the measurement technique underlying these discoveries, the patterns that have emerged from the data, and the implications for theories of planet formation and migration.

Keywords. stars: rotation, planetary systems, planetary systems: formation

---

1. Introduction

When was the first announcement of the detection of a planetary transit? About 10 years ago, the transits of HD 209458b were reported by two different groups (Henry *et al.* 2000, Charbonneau *et al.* 2000). However, planetary transits have a much longer history. A transit of Venus was first witnessed in 1639, by Jeremiah Horrocks. And even before that, in 1611, Christoph Scheiner reported the detection of a system of close-in planets transiting the Sun (see, e.g., Casanovas 1997).

Scheiner was wrong, and was eventually convinced he was seeing dark spots in the Sun's atmosphere, rather than transiting planets. He then charted the trajectories of sunspots over the years, and thereby discovered that the Sun's equatorial plane is not perfectly aligned with the plane of Earth's orbit: the solar obliquity is about 7°.

Now that we know of many other stars harboring planets, it would be interesting to know whether this degree of alignment is common, or unusual. Even though planets are thought to have formed on well-aligned circular orbits, there are reasons to expect occasional misalignments. Many exoplanets are observed to have eccentric orbits, and whatever process excited their eccentricities may also have excited their inclinations. Furthermore, for close-in planets, there are various "migration" scenarios for bringing the planets inward from beyond the snow line, which make differing predictions about spin-orbit alignment. Migration through tidal interactions with a protoplanetary disk would damp any initial inclination (see, e.g., Marzari & Nelson 2009). In contrast, planet-planet scattering would amplify any initial inclinations (e.g., Chatterjee *et al.* 2008), and the Kozai effect due a companion star or distant planet can produce drastic misalignments (Fabrycky & Tremaine 2007).

Scheiner's method cannot be used for exoplanets, at least not until our telescopes can resolve the disks of planet-hosting stars. Instead, the stellar obliquity can be assessed by observing a phenomenon known as the Rossiter-McLaughlin (RM) effect. During a transit, part of the rotating stellar surface is hidden, weakening the corresponding velocity components of the stellar absorption lines. When the blueshifted (approaching) half of the star is blocked, the spectrum appears slightly redshifted. Likewise, when the redshifted (receding) half of the star is blocked, the spectrum is blueshifted. The result is an "anomalous Doppler shift" that varies throughout the eclipse (see Figures 1 and 2). By monitoring the stellar line profiles throughout eclipses, we can measure the angle between the sky projections of the orbital and spin axes. This phenomenon was first predicted by Holt (1893) and observed by Rossiter (1924) and McLaughlin (1924).

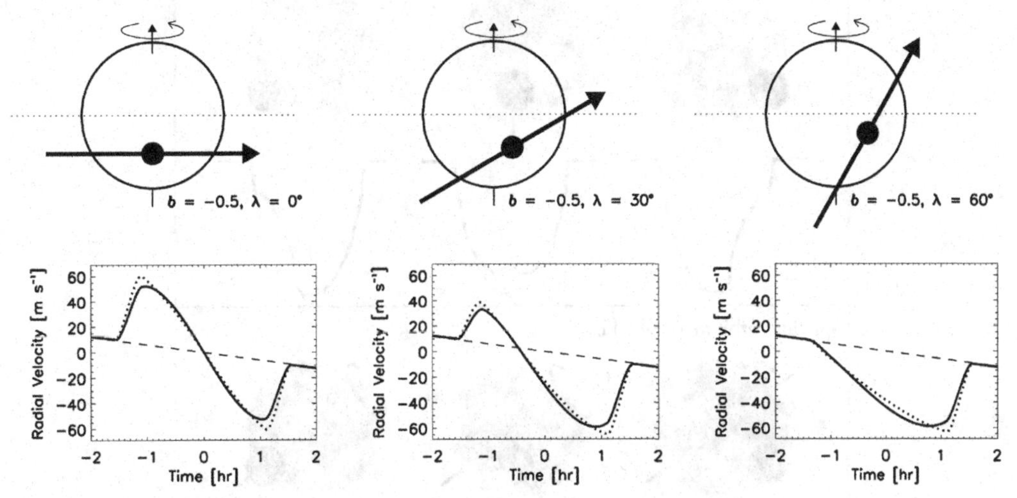

Figure 1. The RM effect as an anomalous Doppler shift. Top: three transit geometries that produce identical light curves, but differ in spin-orbit alignment. Bottom: corresponding radial velocity signals. Good spin-orbit alignment (left) produces a symmetric "redshift-then-blueshift" signal, a 30° tilt (middle) produces an asymmetric signal, and a 60° tilt (right) produces a blueshift throughout the transit. From Gaudi & Winn (2007).

2. Observations of the RM effect

The exoplanetary RM effect was first detected by Queloz *et al.* (2000), who found the orbit of the "hot Jupiter" HD 209458b to be prograde. Further measurements revealed a diversity of orbits, including some that are well-aligned with the equatorial planes of their parent stars, some that are misaligned by more than 30°, and even some that are apparently retrograde. Examples from each of these categories are shown in Figure 3.

A noteworthy case is HD 80606b, with its monstrous orbital eccentricity of 0.93. Wu & Murray (2003) proposed that the orbit had been shaped by the Kozai effect due to a companion, and Fabrycky & Tremaine (2007) realized that this theory predicted a large spin-orbit misalignment. This prediction was eventually confirmed, owing to the amazingly good fortune that the planet's orbit is viewed close enough to edge-on to exhibit transits, and a large community effort to observe the photometric and spectroscopic transits. Hébrard *et al.* (2010) have performed the most definitive analysis.

Another important accomplishment was the measurement by Triaud *et al.* (2010) of the RM effect of six systems, half of which showed evidence for retrograde orbits. This suggested that misaligned orbits may be common, rather than exceptional, and increased the sample size by enough to allow meaningful searches for patterns.

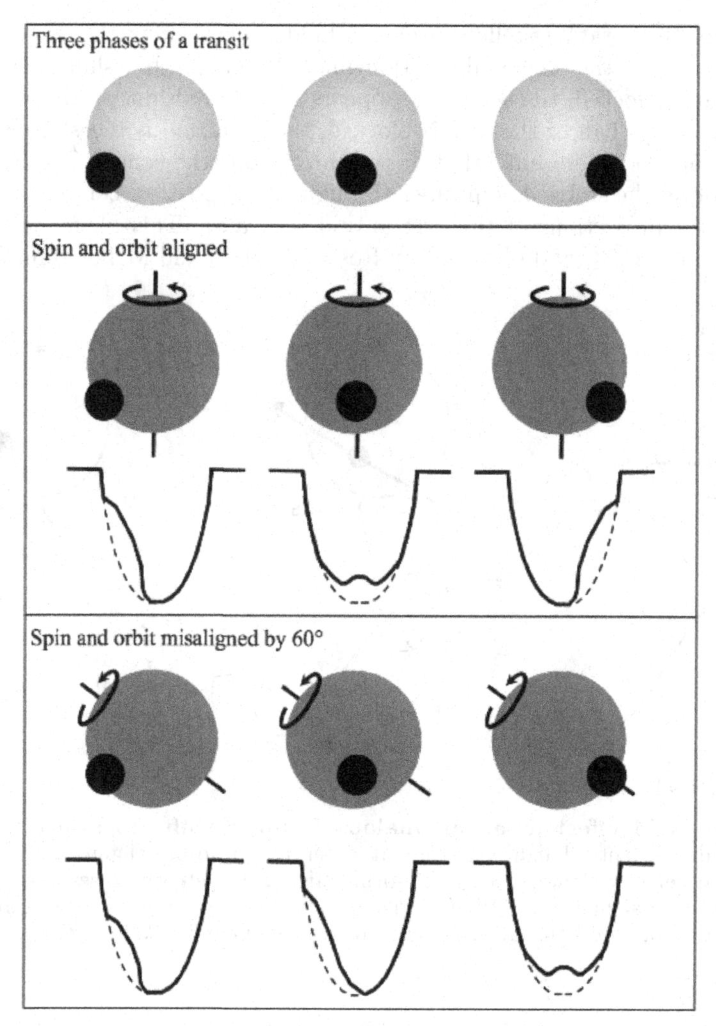

Figure 2. The RM effect as a spectral distortion. Top: three successive phases of a transit. Middle and bottom panels: same, but showing the line-of-sight rotation velocity of the photosphere as a gradient, as well as the distorted spectral line profile. Middle: a low obliquity. The spectral distortion moves from the blue side to the red side. Bottom: a projected obliquity of 60°. In this case the obscuration by the planet weakens the blue wing throughout most of the transit. By observing and modeling the time-varying spectral distortion throughout the transit, one can measure the projected obliquity of the star (see, e.g., Collier Cameron *et al.* 2010).

One possible pattern that has emerged is that stars with high obliquities are preferentially those with the highest effective temperatures or (nearly equivalently) the largest masses. This finding, illustrated in Figure 4, could help to explain the otherwise puzzling history of RM observations: the first 10 published analyses were all consistent with good alignment, while the next 20 showed a much wider range of obliquities. The explanation could be that cooler stars were examined earlier, as they allow for better RV precision and greater ease of detecting planets.

The physical reason for this pattern is not clear. It could be a signal that planet migration is different for low-mass stars than for high-mass stars. Or, perhaps *all* close-in planets begin with a wide range of obliquities, but tidal dissipation causes some of them

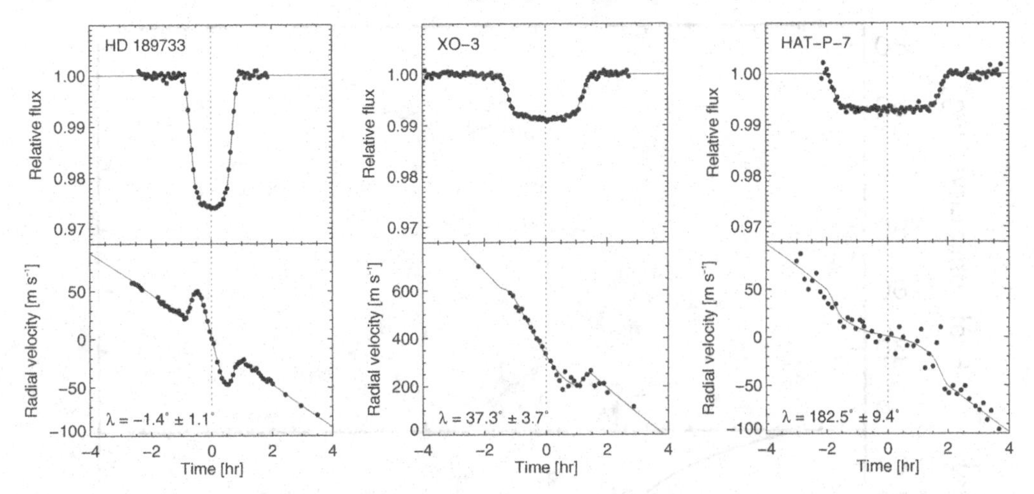

Figure 3. Examples of RM data. The top panels show transit photometry, and the bottom panels show the apparent radial velocity of the star, including both orbital motion and the anomalous Doppler shift. The left panels show a well-aligned system, the middle panels show a misaligned system, and the right panels show a system for which the stellar and orbital "north poles" are nearly antiparallel on the sky. From Winn *et al.* (2006; 2009a,b).

to realign with the orbit, and this occurs preferentially for low-mass stars because of their larger outer convective zones. In support of the tidal hypothesis, the few cool stars with high obliquities are also those with the longest orbital periods, leading to exceptionally weak tidal effects. A serious problem with this hypothesis is that tidal reorientation of the star should be accompanied by orbital decay, leading to engulfment of the planet. To reduce the angular momentum that the orbit must surrender to the star, one could invoke core-envelope decoupling, but it is unclear why the coupling would be so weak, and it is not observed in the Sun. For further discussion, see Winn *et al.* (2010a).

Regardless of the explanation, recent observations have strengthened the evidence for the pattern. The hot stars XO-4 and HAT-P-14 were found to be misaligned (Narita *et al.* 2010, Winn *et al.* 2011), while the cool star HAT-P-4 is well-aligned (Winn *et al.* 2011). Another interesting case is HAT-P-11b, whose orbit is grossly misaligned despite the host star's low effective temperature (Winn *et al.* 2010b, Hirano *et al.* 2011). This might be attributed to the system's unusually weak tidal coupling, due to the planet's relatively low mass and long orbital period. Thus it could be another telling exception to the rule that hot stars have high obliquities: it implicates tidal evolution as the reason for low obliquities among cool stars with more massive planets in tighter orbits (Winn *et al.* 2010a).

To illustrate this point, Figure 5 shows the RM results as a function of a dimensionless parameter that characterizes the expected timescale for tidal dissipation,

$$\tau \equiv \left(\frac{M_\star}{M_p}\right)\left(\frac{a}{R_\star}\right)^6 \frac{(1-e^2)^{9/2}}{1+3e^2+(3/8)e^4}, \tag{2.1}$$

where $M_\star$ and M_p are the stellar and planetary masses, a is the orbital semimajor axis, $R_\star$ is the stellar radius, and e is the orbital eccentricity. Because the specific mechanisms for tidal dissipation are not understood (i.e. the Q values are unknown) we cannot give the absolute timescales, but we may rank the systems according to τ. "Hot" and "cold" stars are plotted separately, since they are hypothesized to have very different Q values.

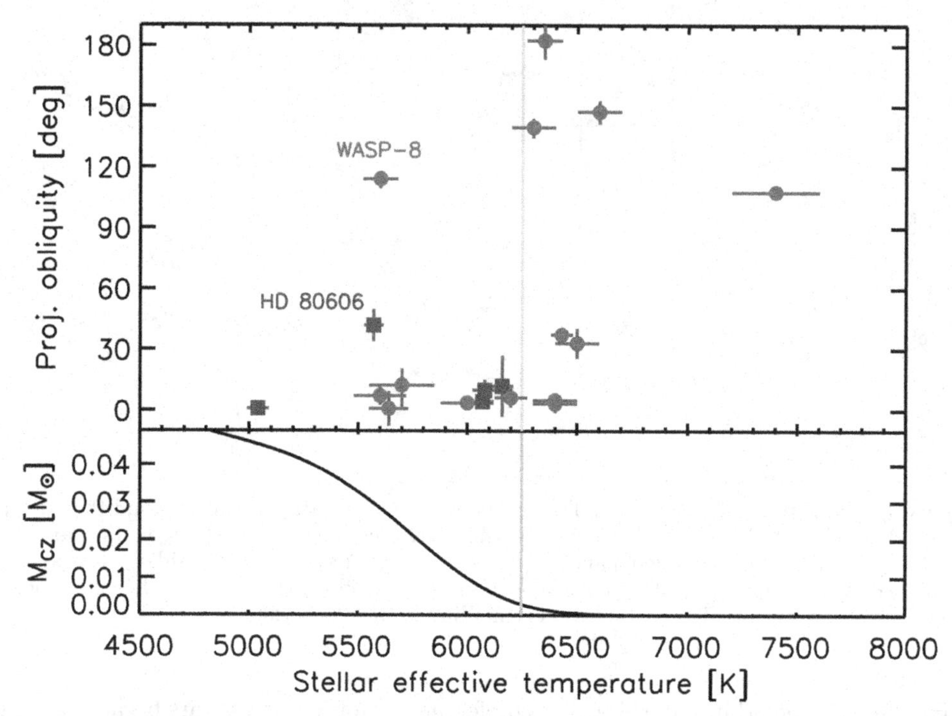

Figure 4. Misaligned systems have hotter stars. Top: the projected obliquity is plotted against the effective temperature of the host star. A transition from mainly aligned to mainly misaligned seems to occur at $T_{\rm eff} \approx 6250$ K. The two strongest exceptions (labeled) are also the systems with the weakest expected tidal interactions. Squares indicate systems discovered by RV surveys, while circles indicate systems found in photometric transit surveys. Bottom: the mass of the convective zone of a main-sequence star as a function of $T_{\rm eff}$, from Pinsonneault *et al.* (2001). It is suggestive that 6250 K is approximately the temperature at which the mass of the convective zone becomes negligible. From Winn *et al.* (2010a).

Figure 5 shows that the hot stars have a wide range of obliquities, and the cold stars generally have low obliquities except for those with the weakest tidal interactions.

In addition, Schlaufman (2010) presented evidence that hot stars have high obliquities, based on a completely different technique. His idea was to compare the $v \sin i$ distribution of stars with transiting planets with those of random stars of the same mass and evolutionary state. To the extent that the transit stars have rotation axies parallel to the orbital axes, they will have systematically larger $v \sin i$ values, because $\sin i = 1$ for the transiting systems while $0 < \sin i < 1$ for random stars. As shown in Figure 6, he found that more massive stars have higher obliquities than lower-mass stars, with the transition at around 1.2 $M_\odot$, consistent with the RM results.

3. Implications for planet migration

What is this collection of measurements telling us? The prevalence of misaligned orbits has been marshalled as evidence against the "standard model" for planet migration, in which disk-planet interactions cause the planet to spiral inward. Instead the results suggest that many close-in giant planets arrived at their current locations through gravitational perturbations from other massive bodies, followed by tidal dissipation (Triaud

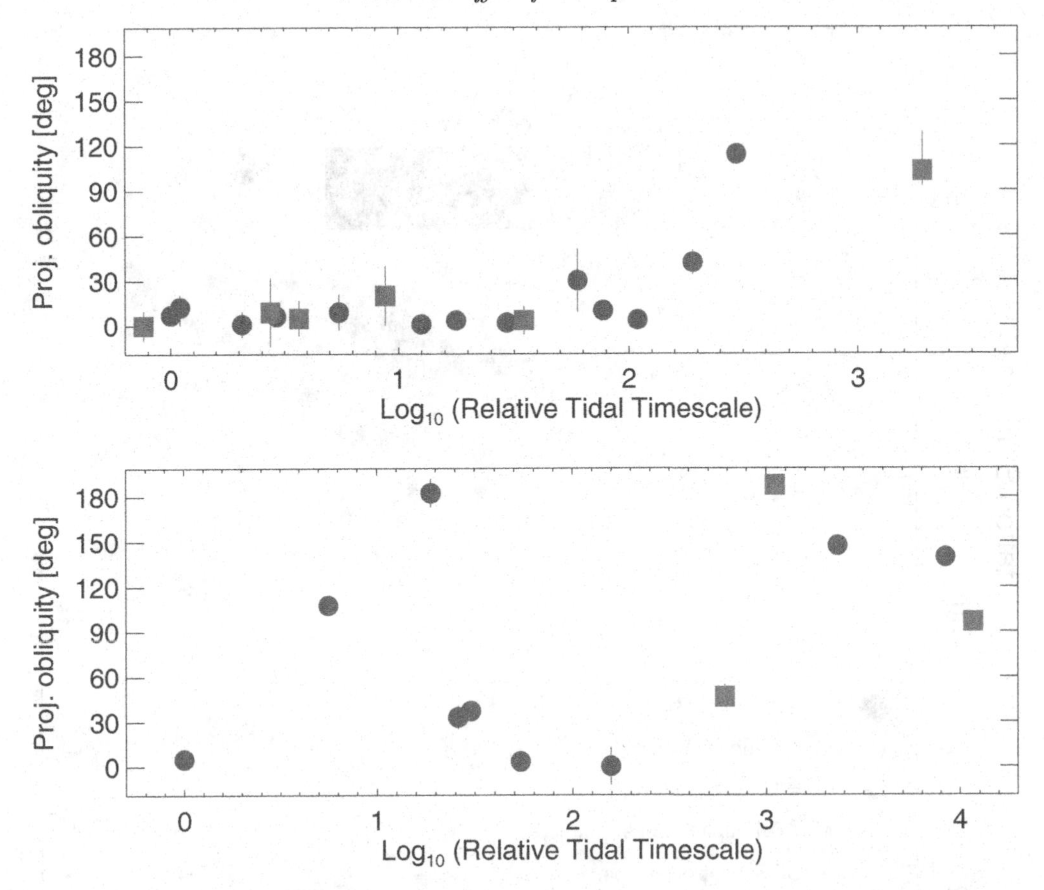

Figure 5. Are low obliquities the result of tidal dissipation? The sky-projected stellar obliquity as a function of the parameter defined in Eqn. (2.1), which is expected to be roughly proportional to the timescale for tidal dissipation. According to the hypothesis of Winn *et al.* (2010b), hot stars have a wide range of obliquities, and cold stars have low obliquities except for those with the longest tidal timescales. This was borne out by 9 observations conducted after the hypothesis was proposed (square symbols).

et al. 2010; Winn *et al.* 2010a; Matsumura, Peale, & Rasio 2010). To appreciate the strengths and weaknesses of this argument, it is useful to frame it as a series of premises:

(*a*) Hot Jupiters formed beyond the snow line and migrated inward.
(*b*) The orbit and spin were initially aligned.
(*c*) The two migration theories are disk-planet interactions and few-body dynamics.
(*d*) Disk-planet interactions damp inclinations.
(*e*) Few-body dynamics excite inclinations.†
(*f*) Nothing else excites inclinations.
(*g*) Many hot Jupiters have high inclinations.

From these premises it follows that the misaligned hot Jupiters migrated via few-body dynamics. Furthermore, it is possible that *all* hot Jupiters migrated this way, if tides are responsible for the low observed inclinations.

Some authors have already challenged these premises. For example, it is possible that protoplanetary disks are frequently misaligned with the rotation of their host stars (Bate,

† "Few-body dynamics" is meant here to include some combination of planet-planet interactions and the Kozai effect of a distant companion.

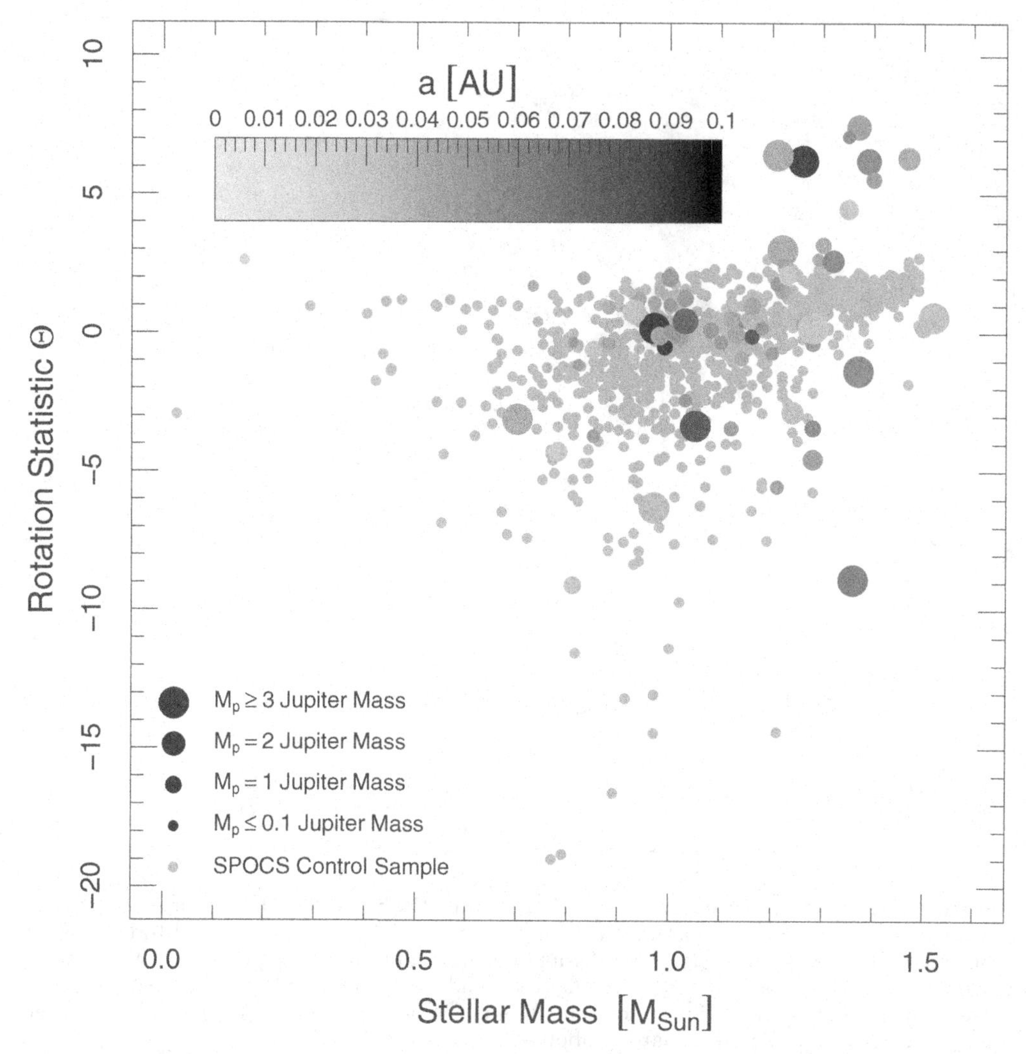

Figure 6. Independent evidence that hot (massive) stars have high obliquities, from Schlaufman (2010). The rotation statistic θ quantifies the degree to which the measured $v \sin i$ is smaller than the expected $v \sin i$ for a star of the given mass and age. Large dots are transit hosts, and small gray dots are random field stars. The transit hosts with high θ, in the upper right corner, represent stars with unusually low $v \sin i$, suggesting $\sin i$ is small and therefore that the stellar rotation is misaligned with the planetary orbit. All the high-θ systems involve massive stars ($> 1.2\ M_{\odot}$).

Lodato, & Pringle 2010; Lai, Foucart, & Lin 2010). Tests of these ideas are possible with RM observations of multiple-planet systems (Fabrycky 2009, Ragozzine & Holman 2010), and of binary stars (Albrecht *et al.* 2009, 2011).

It is important to remember that this argument focuses exclusively on hot Jupiters. Disk migration has become vulnerable as an explanation for those planets, but is still viable (and without a good alternative) for explaining the "medium-period" giant planets, well within the snow line but too distant for tidal effects to be important. The mean-motion resonances that are occasionally observed among such planets are evidence for disk migration. The priority for future observations of the RM effect is to explore a more diverse sample of planets, including rocky planets, long-period planets, and multiple-planet systems.

Acknowledgements

I am grateful to Alessandro Sozzetti and the organizing committees, for the opportunity to participate in this important meeting and visit the fascinating city of Torino. I thank Simon Albrecht for comments on the draft of this contribution, and Kevin Schlaufman for providing Figure 6. Finally, I extend my gratitude to the colleagues with whom I have worked closely on this topic: Yasushi Suto, Ed Turner, Geoff Marcy, Bob Noyes, John Johnson, Norio Narita, Dan Fabrycky, Andrew Howard, Simon Albrecht, Teruyuki Hirano, and Roberto Sanchis Ojeda.

References

Albrecht, S., *et al.* 2009, *Nature*, 461, 373
Albrecht, S., *et al.* 2011, *ApJ*, 726, A68
Bate, M., Lodato, G., & Pringle, J. E. 2010, *MNRAS*, 401, 1505
Casanovas, J. 1997, in: Advances in the Physics of Sunspots, B. Schmieder, J.C. del Toro Iniesta, & M. Vazquez (eds.), *ASP Conf. Ser.*, 118, 3 (San Francisco: ASP)
Charbonneau, D., *et al.* 2000, *ApJ* (Letters), 529, L45
Chatterjee, S., Ford, E., Matsumura, S., & Rasio, F. 2008, *ApJ*, 686, 580
Collier Cameron, A., *et al.* 2010, *MNRAS*, 403, 151
Fabrycky, D. & Tremaine, S. 2007, *ApJ*, 669, 1298
Gaudi, B. & Winn, J. 2007, *ApJ*, 655, 550
Hébrard, G., *et al.* 2010, *A&A*, 516, 95
Henry, G., *et al.* 2000, *ApJ* (Letters), 529, 41
Hirano, T., *et al.* 2011, *PASJ*, 63, 531
Holt, J. R. 1893, *A&A*, 12, 646
Lai, D., Foucart, F., & Lin, D. 2011, *MNRAS*, 412, 2790
Marzari, F. & Nelson, A. 2009, *ApJ*, 705, 1575
Matsumura, S., Peale, S. J., & Rasio, F. 2010, *ApJ*, 725, 1995
McLaughlin, D. 1924, *ApJ*, 60, 22
Narita, N., *et al.* 2010, *PASJ*, 62, L61
Queloz, D., *et al.* 2000, *A&A*, 359, L13
Ragozzine, D. & Holman, M. 2010, *ApJ*, submitted [arxiv:1006.3727]
Rossiter, R. 1924, *ApJ*, 60, 15
Schlaufman, K. 2010, *ApJ*, 719, 602
Triaud, A., *et al.* 2010, *A&A*, 524, 25
Winn, J., *et al.* 2006, *ApJ* (Letters), 653, 69
Winn, J., *et al.* 2009a, *ApJ*, 700, 302
Winn, J., *et al.* 2009b, *ApJ* (Letters), 703, 99
Winn, J., *et al.* 2010a, *ApJ* (Letters), 718, 145
Winn, J., *et al.* 2010b, *ApJ* (Letters), 723, 223
Winn, J., *et al.* 2011, *AJ*, 141, id.63
Wu, Y. & Murray, N. 2003, *ApJ*, 589, 605

The Astrophysics of Planetary Systems: Formation, Structure, and Dynamical Evolution
Proceedings IAU Symposium No. 276, 2010
A. Sozzetti, M.G. Lattanzi & A.P. Boss, eds.

doi:10.1017/S1743921311020242

Tidal evolution of star-planet systems

Rosemary A. Mardling[1]

[1] School of Mathematical Sciences, Monash University, Victoria 3800, Australia
email: rosemary.mardling@monash.edu

Abstract. The equilibrium tide model in the weak friction approximation is used by the binary star and exoplanet communities to study the tidal evolution of short-period systems, however, each uses a slightly different approach which potentially leads to different conclusions about the timescales on which various processes occur. Here we present an overview of these two approaches, and show that for short-period planets the circularization timescales they predict differ by at most a factor of a few. A discussion of the timescales for orbital decay, spin-orbit synchronization and spin-orbit alignment is also presented.

Keywords. methods: analytical, celestial mechanics, planets and satellites: dynamical evolution and stability, planets and satellites: formation, planetary systems

---

1. Two-body tidal evolution

The effects on an orbit due to tidal distortion and rotation can be divided into two categories: those associated with the viscosity of the fluid (and rheology in the case of a rigid component), and those associated with the non-spherical shape of the distorted body. The latter give rise to apsidal advance of the orbit in its plane in the case that the orbit is non-circular, and to precession of the orbital plane and the spin axes of the rotating bodies about the direction of the total angular momentum of the system in the case that the spin axes are not aligned with the orbit normal. They do not affect the semimajor axis or the eccentricity of the orbit, and occur on a timescale which is generally much shorter than those due to tidal dissipation. In contrast, viscous dissipation in the tidally distorted bodies results in changes to all the orbital elements (except the longitude of periastron) as well as the rotation rate, and is the focus of the so-called *equilibrium tide* model.

The equilibrium tide model in the weak friction approximation was devised by Darwin (1879) for the purpose of studying tides in planets and their satellites, and in particular the evolution of the orbital elements due to their resulting non-sphericity and orientation. The shape of the tidal bulge is obtained by assuming hydrostatic equilibrium within the distorted body, that is, the fluid's internal pressure gradients and viscous stresses are exactly balanced by the gravitational force exerted both by the fluid body itself and the companion, implying there is no relative motion of neighbouring fluid elements and hence no viscous heat loss. In fact, as long as the orbit is non-circular (the gravitational field of the companion varies in time), and/or the fluid body rotates at a rate different to the mean motion (fluid elements vary their distance to the companion at a rate equal to the difference between the rotation rate and the mean motion, even if the orbit is circular), and/or the rotation axis points in a different direction to the orbit normal (again the gravitational field varies), the assumption of hydrostatic equilibrium is only approximate. However, one can study the effect of the tidal dissipation on the orbital evolution without studying the fluid motion itself, simply by realizing that dissipation within the fluid will cause its response to be out of phase with the tidal forcing, this producing a misalignment between the tidal bulge and the line of centres and consequently a spin-orbit torque. The misalignment angle ϵ is directly related to the dissipation rate in the fluid via the tidal quality factor or Q-value† such that $2\epsilon = Q^{-1}$, and if one assumes that the height of the tide is unaffected by friction (this is the *weak-friction* approximation), one can calculate the rates of change of the orbital elements by considering the motion of a companion moving in the potential of the obtained figure of equilibrium, the latter slightly rotated with respect to the line of centres. Note that a nice feature of Darwin's model is that in spite of the presence of friction, one can write down a *potential* and thus make use of Lagrange's planetary equations for the rates of change of the elements.

Darwin's original treatment (and most subsequent treatments) involves a Fourier expansion of the tidal potential, and it is the treatment of each individual Fourier component which distinguishes the two equilibrium tide models in common use; the so-called *constant time-lag* model which Darwin himself employed assumes that the lag angle of (and hence the rate of dissipation in) a particular tidal component is proportional to the rate at which it is forced, and that the constant of proportionality (the lag time) is common to all tidal components, while the *constant lag angle* model (Goldreich & Soter 1966) assumes that all tidal components lag behind (or ahead of) the line of centres by the same angle. Darwin's prescription has been improved over the years by various authors, and adopted by the binary star community as appropriate for studying tides in stars. In contrast, the planet community has adopted the prescription of Goldreich & Soter (1966) for solid planets and satellites, one which appeals to the experimental fact that for various Earthly solid materials, the lag angle is independent of the forcing frequency for a wide range of forcing frequencies. Moreover, the binary star community refer to the *apsidal motion constant*, k_A, a quantity reflecting the degree of central concentration of a *fluid* body, while the planet community refer to the *Love number*, k_L, a quantity associated with the degree of rigidity and other properties of a homogeneous *solid* body. In fact, they are both (apart from a factor of two) the ratio of the potential associated with the tidal bulge to the tidal potential produced by the perturber at the surface of the distorted object, and are such that $k_L = 2k_A$.

The choice of prescription for dissipation in a tidally distorted planet - constant lag time or constant lag angle - turns out to vary the tidal circularization timescale by no more than a factor of a few, at least in the simple equilibrium tide model as we show below. Moreover, contrary to the situation to date, it seems more reasonable to use the Darwin prescription in the case of gas giants and Neptune-like ice giants, given that tidal energy is probably mostly stored and dissipated in the dense atmosphere of the latter. For example, it may be that this factor of a few is all that is needed to explain the significantly non-circular orbit of the short-period hot Neptune GJ436b (Butler *et al.* 2004).

Perhaps the most succinct and elegant derivation of the equations governing the variation of the orbital elements due to tidal friction in stars is given by Hut (1981) who follows Darwin in assuming a constant time lag and weak friction, but unlike Darwin and most others, uses energy and angular momentum arguments instead of a Fourier decomposition of the tidal field. He shows that, in this approximation, one is able to write down closed-form expressions for the rates of change of the elements and the rotation frequency.‡ In contrast, the Goldreich & Soter (1966) constant lag angle theory is correct to first order only in the eccentricity, and does not display explicit dependence

† See Goldreich (1963, p260) for a derivation of the relationship between the lag angle of the Q-value, as well as Mardling (2010) for a comparison of the Darwin and Goldreich analyses including a derivation of the Fourier components of the tidal bulge.

‡ Note that Hut's expression for the rate of change of spin obliquity is valid for small values

on the spin frequencies. In particular, the sign of the rate of change of eccentricity is decided by the sign of the dominant contributing tidal component, this being the one with forcing frequency $2\Omega - 3n$,¶ where Ω and n are the spin frequency and the mean motion respectively (Goldreich 1963, p261). As such the relevant forcing frequency for a synchronously rotating body ($\Omega = n$) in a non-circular orbit is the mean motion itself (since $2\Omega - 3n = -n$ in this case).|| This makes sense since a particle in the tidal bulge, although always lying (approximately) at a constant angle away from the line of centres, nonetheless experiences a gravitational field which varies in strength with a frequency equal to the mean motion as its distance to the companion varies over the orbit.

To first order in the eccentricity, the Hut (1981) expression for the rate of change of eccentricity for a synchronously rotating system is given by

$$\frac{1}{e}\frac{de}{dt} = -\frac{21}{4}\frac{k_L}{T}\left(\frac{M_2}{M_1}\right)\left(1+\frac{M_2}{M_1}\right)\left(\frac{R_1}{a}\right)^8, \tag{1.1}$$

while the Goldreich & Soter (1966) expression is†

$$\frac{1}{e}\frac{de}{dt} = -\frac{21}{2}\frac{nk_L}{Q}\left(\frac{M_2}{M_1}\right)\left(\frac{R_1}{a}\right)^5, \tag{1.2}$$

where n is the mean motion, R_1 and M_1 are the radius and mass of the tidally distorted body, M_2 is the mass of the tide-raising body, T is a damping timescale which is related to the time lag τ via $T^{-1} = (GM_1/R_1^3)\tau$, and Q is the tidal quality factor. These two expressions agree to this order in eccentricity if we take $\frac{1}{2}n\tau = Q^{-1}$, consistent with the mean motion being the relevant forcing frequency for a synchronously rotating body.‡

In order to use the constant time lag model to study the tidal evolution of hot Jupiters, while at the same time use information we have about our own Jupiter, we might assume that, instead of the Q-value itself, it is τ which is common to Jupiter and other gas giants of similar mass, although in reality this quantity can be expected to vary from one system to the next according to the planet's density and chemical composition. Since we expect all hot Jupiters to be essentially synchronously rotating,¶ we can use the relation $Q^{-1} \propto n\tau$ to estimate the Q-value of a planet whose orbital period is P via $Q/Q_J = P/P_{Io}$, where Q_J is Jupiter's Q-value and $P_{Io} = 1.77$ d is Io's orbital period. Lainey *et al.* (2009) have estimated a value of $Q/k_L = 9.1 \times 10^4$ for Jupiter by astrometrically measuring Io's acceleration. Thus, for example, a hot Jupiter with the same Love number as Jupiter's but with a period of 4 days will have a Q-value which is 2.3 times that for Jupiter. Note, however, that if such a hot Jupiter is inflated its Love number will be reduced (as is also the case if the planet has a substantial core), and this will further increase the circularization timescale.

of this quantity, while Eggleton, Kiseleva and Hut (1998) derive close-form expressions for the rates of change of all three Euler angles associated with rotation.

¶ In a frame rotating at the same rate as the tidally distorted body; recall that what is relevant is the relative motion of neighbouring fluid parcels.

|| It is interesting to note that the Darwin/Hut expression for the rate of change of the eccentricity is proportional to $11\Omega - 18n$; compare this to $2\Omega - 3n$.

† Note that Goldreich & Soter (1966) employ a modified Q-value involving the Love number and defined by $Q' = \frac{3}{2}Q/k_L$.

‡ Note that in order to study the case of synchronous rotation, Goldreich (1963) assumes a constant time lag for two of the tidal components and Goldreich & Soter (1966) go on to assume a constant lag angle for the remainder, thus employing a curious mixture of both formulations.

¶ In fact, for finite eccentricity they will be rotating slightly sub-synchronously (Hut, 1981, p131).

2. Timescales

The minimum-energy state of a binary system is one in which the orbit is circular, and both bodies spin synchronously with the mean motion and their spin axes are aligned with the orbit normal. The semimajor axis may increase or decrease during this process, depending on the net transfer of angular momentum. For example, while spin synchronization of a planet by a star may have little effect on the orbit because the planet has only a small portion of the angular momentum budget, spin synchronization of a star by a planet tends to shrink the orbit, except in the case where the star spins faster than the mean motion. For mature systems comprised of a hot Jupiter and a solar-type star, this is never the case. In fact such systems tend to be unstable to orbital decay because there is not enough angular momentum in the orbit to spin up the star (Rasio *et al.* 1996), and this is exacerbated if the star is losing angular momentum via a stellar wind. On the other hand, if the star has a convective envelope which is sufficiently decoupled from the core, the synchronization process may only involve this layer and can be achieved before the planet fills its Roche lobe.

For a planet-star system, the timescale for circularization of the orbit, τ_{circ}, is given by contributions from tides raised on the planet by the star and tides raised on the star by the planet, and is such that

$$\tau_{circ}^{-1} = \tau_{ep}^{-1} + \tau_{e*}^{-1}, \tag{2.1}$$

where

$$\tau_{ep} = 0.16 \left(\frac{k_p}{0.3}\right)^{-1} \left(\frac{Q_p}{10^5}\right) \left(\frac{M_p}{M_J}\right) \left(\frac{M_*}{M_\odot}\right)^{-1} \left(\frac{R_p}{R_J}\right)^{-5} \left(\frac{a}{0.04\,\mathrm{AU}}\right)^5 \mathrm{Gyr}, \tag{2.2}$$

is the contribution to orbital circularization from tides raised in the planet and

$$\tau_{e*} = \left(\frac{Q_*}{Q_p}\right) \left(\frac{k_p}{k_*}\right) \left(\frac{M_*}{M_p}\right)^2 \left(\frac{R_p}{R_*}\right)^5 \tau_{ep} \tag{2.3}$$

is the contribution from tides raised in the star. Here the various symbols have their obvious meaning. For a $1M_J$ planet orbiting a $1M_\odot$ star, $\tau_{e*} \simeq 10(Q_*/Q_p)(k_p/k_*)\tau_{ep} \gg \tau_{ep}$ for reasonable values of stellar and planetary Q-values and Love numbers so that, in this case, it is tides raised in the planet which govern the circularization of the orbit.

In the case where no angular momentum is lost from the system, one can write down timescales for spin-orbit synchronization of the star and planet, $\tau_{\Omega*}$ and $\tau_{\Omega p}$, their spin-orbit alignment, τ_{i*} and τ_{ip}, and orbit decay or expansion, τ_a, all in terms of τ_{ep} and τ_{e*} (Hut 1981). These are

$$\tau_{\Omega_*} = \frac{7}{3(\alpha_* - 3)}\tau_{e*}, \qquad \tau_{\Omega_p} = \frac{7}{3(\alpha_p - 3)}\tau_{ep}, \tag{2.4}$$

$$\tau_{i_*} = \frac{7}{(\alpha_* + 1)}\tau_{e*}, \qquad \tau_{i_p} = \frac{7}{(\alpha_p + 1)}\tau_{ep}, \tag{2.5}$$

and

$$\tau_a^{-1} \simeq \tau_{\Omega_*}^{-1}, \tag{2.6}$$

where α_* is the ratio of the orbital angular momentum to the body's spin angular momentum *at equilibrium*, that is, when the star is in the spin-synchronous state, and α_p is the planet analogue.† For a Jupiter-mass planet orbiting at a distance of 0.04 AU from

† The quantity α_* is determined by insisting on conservation of angular momentum and solving for the semimajor axis corresponding to $\Omega_* = n$ and similarly for α_p. Since the timescales

a solar-mass star which has a rotation period of 20 days, $\alpha_p \simeq 27000$ so that $\tau_{\Omega_p} \ll \tau_{circ}$ and $\tau_{i_p} \ll \tau_{circ}$, while there is no stable synchronous state for the star. In fact there is no such state for the star for systems with orbital periods less than 12.7 days for the case where the whole star is spun up. In general, there is no stable state when $\alpha_* < 3$ (Councelman 1973).

In the case that the star's convective envelope is decoupled from its core, α_* should be scaled by $(M_*/M_{env})(R_*/R_{env})^2$, where M_{env} and R_{env} are the mass and depth of the convective zone, in which case there will be stable solutions for systems with smaller orbital periods. It has been suggested by Winn *et al.* (2010) that high stellar obliquities are initially generic to all hot Jupiters, and that the convective envelopes of late-type stars align on timescales shorter than the age of the system while the radiative envelopes of more massive stars are unable to do so.

3. Conclusions

Tidal interactions between exoplanets and their host stars are proving to be play an extremely important role in our knowledge and understanding of the origin, internal structure and dynamical evolution of short-period planets and their companions. While the eccentricities of most short-period exoplanets are more or less consistent with expectations based on our knowledge of the Q-values of the Solar System planets, we still have a long way to go before we fully understand tidal dissipation in the wide variety of circumstances exoplanets present us. In the meantime, the simple equilibrium tide model serves us well in our quest to understand the complex dynamics of single and multiplanet systems.

References

Butler, R. P., Vogt, S. S., Marcy, G. W., *et al.* 2004, *ApJ*, 617, 580
Councelman, C. C. 1973, *ApJ*, 180, 307
Darwin, G. H. 1879, *Phil. Trans. Roy. Astr. Soc.*, 170, 1
Eggleton, P. P., Kiseleva, L. G., & Hut, P. 1998, *ApJ*, 499, 853
Goldreich, P. 1963, *MNRAS*, 126, 257
Goldreich, P. & Soter, S. 1966, *Icarus*, 5, 375
Hut, P. 1981, *A&A*, 99, 126
Lainey, V., Arlot, J. E., Karatekin, Ö., *et al.* 2009, *Nature*, 459, 957
Mardling, R. A. 2010, http://online.kitp.ucsb.edu/online/exoplanets10/mardling/
Rasio, F. A., Tout, C. A., Lubow, S. H., *et al.* 1996, *ApJ*, 470, 1187
Winn, J. N., Fabrycky, D., Albrecht, S., *et al.* 2010, *ApJL*, 718, L145

for spin-orbit synchronization of the star and the planet are so different, one can ignore the planet's angular momentum when calculating α_*, and similarly one can ignore the star's angular momentum when solving for α_p.

The Astrophysics of Planetary Systems: Formation, Structure, and Dynamical Evolution
Proceedings IAU Symposium No. 276, 2010
A. Sozzetti, M. G. Lattanzi, & A. P. Boss, eds.

doi:10.1017/S1743921311020254

Revisiting the eccentricities of hot Jupiters

Nawal Husnoo[1], Frédéric Pont[1], Tsevi Mazeh[2], Daniel Fabrycky[3], Guillaume Hébrard[4,5] and Claire Moutou[6]

[1]School of Physics, University of Exeter, Exeter, EX4 4QL, UK
email: nawal@astro.ex.ac.uk
[2]School of Physics and Astronomy, Tel Aviv University, Tel Aviv 69978, Israel
[3]Harvard-Smithsonian Centre for Astrophysics, Garden Street, Cambridge MA
[4]Institut d'Astrophysique de Paris, UMR7095 CNRS, Université Pierre & Marie Curie, 98bis boulevard Arago, 75014 Paris, France
[5]Observatoire de Haute-Provence, CNRS/OAMP, 04870 Saint-Michel-l'Observatoire, France
[6]Laboratoire d'Astrophysique de Marseille, Université de Provence, CNRS(UMR 6110), 38 rue Frédéric Joliot Curie, 13388 & Marseille cedex 13,France

Abstract. Most short period transiting exoplanets have circular orbits, as expected from an estimation of the circularisation timescale using classical tidal theory. Interestingly, a small number of short period transiting exoplanets seem to have orbits with a small eccentricity. Such systems are valuable as they may indicate that some key physics is missing from formation and evolution models. We have analysed the results of a campaign of radial velocity measurements of known transiting planets with the SOPHIE and HARPS spectrographs using Bayesian methods and obtained new constraints on the orbital elements of 12 known transiting exoplanets. We also reanalysed the radial velocity data for another 42 transiting systems and show that some of the eccentric orbits reported in the Literature are compatible with a circular orbit. As a result, we show that the systems with circular and eccentric orbits are clearly separated on a plot of the planetary mass versus orbital period. We also show that planets following the trend where heavier hot Jupiters have shorter orbital periods (the "mass-period relation" of hot Jupiters), also tend to have circular orbits, with no confirmed exception to this rule so far.

Keywords. planetary systems

1. Introduction

Transiting planets are the only exoplanets for which we can precisely measure the orbital and physical parameters relevant in constraining planet formation and migration models. Today, over one hundred transiting planets are known, most of which orbit their parent stars at a distance smaller than 0.1 AU. These close-in planets are expected to experience strong tidal forces, and their orbits are expected to circularise early in their lifetimes (Lin, Bodenheimer & Richardson 1996).

It is therefore surprising that a significant number of transiting systems with very short periods, such as WASP-12 and WASP-18 (see Figure 1), appear to have small but measurable eccentricities at the few percent level from radial velocity data and transit photometry. For example, WASP-12b was reported by Hebb *et al.* (2009) to have an eccentricity of $e = 0.049 \pm 0.015$, while orbiting at only 3.1 stellar radii. WASP-18 was reported by Hellier *et al.* (2009) to have $e = 0.0092 \pm 0.0028$ and this was refined by Triaud *et al.* (2010) to $e = 0.0088 \pm 0.0012$. This is unexpected from tidal arguments, since the tidal circularisation timescales for many of these systems are considerably shorter than the age of the systems. As a consequence, a lot of theoretical effort has been spent explaining these systems on a case-by-case basis.

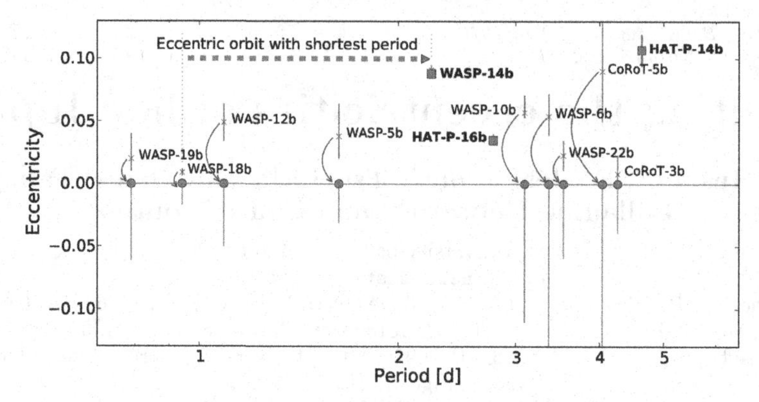

Figure 1. Plot of eccentricity against orbital period. Transiting planets with a reported eccentricity in the Literature ($P < 5$ and $e < 0.12$) are represented by crosses. A square (labelled in **bold**) represents a planet that has a confirmed eccentricity according to our study. Those planets for which we find no evidence for eccentricity are represented by large circles plotted at $e = 0$, with the 95% upper limit on the eccentricity shown on the negative e-axis for clarity.

In addition to the combination of radial velocity measurements and transit photometry, one can constrain one projected component of the eccentricity, $e\cos\omega$, where ω is the argument of periastron, by observing the secondary eclipse as the planet passes behind the parent star. A circular orbit would give a value of $e\cos\omega = 0$, i.e. an eclipse phase of $\phi = 0.5$. López-Morales *et al.* (2010) observed WASP-12 in the z'-band from the ground and obtained a result of $e\cos\omega = 0.0156 \pm 0.0035$, corresponding to an eclipse phase $\phi = 0.510 \pm 0.002$. This is a 4.5-σ departure from a circular orbit, but they later revised their estimate to $e\cos\omega = 0.016^{+0.011}_{-0.009}$, in an update to the paper before publication. Campo *et al.* (2011) observed the secondary eclipse of WASP-12 using *Spitzer* and derived an eclipse phase of $\phi = 0.5012 \pm 0.0006$ (3.6 and 5.8 µm) and 0.5007 ± 0.0007 (4.5 and 8.0 µm). This suggests that one projected component of the eccentricity is consistent with zero. Using new radial velocity data from SOPHIE, we showed that the WASP-12 system probably had a circular orbit (Husnoo *et al.* 2010, *in Press*). We suggested that the original detection could have been caused by instrumental systematics. This can be seen in Figure 2 (left). When a Keplerian orbit is substracted from the radial velocity measurements, the scatter in the residuals is larger than the formal uncertainties in the measurements. When there are few radial velocity measurements, or when most of the measurements are taken on a single night, correlated noise may lead to a spurious detection of eccentricity. In the case of WASP-4 and WASP-5 (Figure 2, right, for WASP-5), the residuals from substracting a Keplerian orbit from our new data showed a clear signal, which is due to stellar activity (an anti-correlation is observed between the line bisector span and the residuals). If unaccounted for, this signal can inflate the eccentricy to a few percent at the few sigma level. In a similar way, Christian *et al.* (2009) suggested that WASP-10 was eccentric at 6% at the 14-σ level, but recently Maciejewski *et al.* (2011) reanalysed this system and concluded that the eccentricity detection was related to stellar activity, and if this is taken into account, the data is consistent with a circular orbit.

2. Analysis

In this study, we take a new look at the orbital eccentricity of 64 known transiting systems. We add new precise radial velocity data using HARPS and SOPHIE for 12

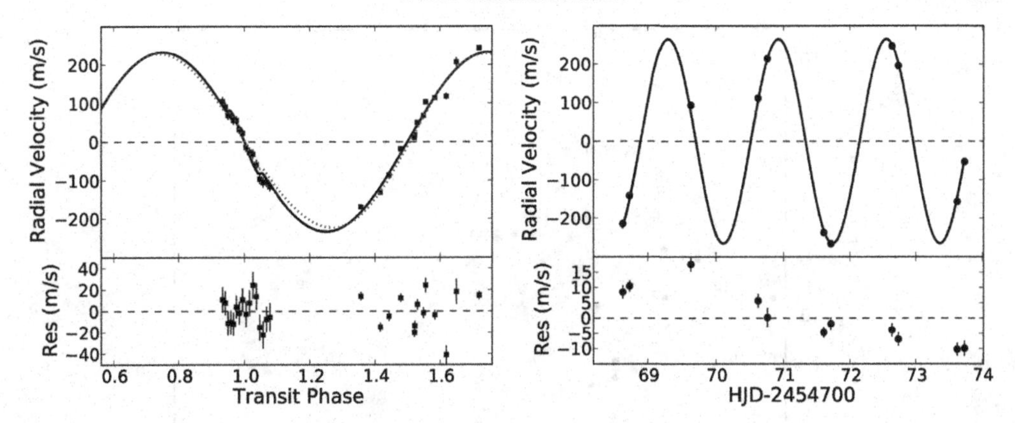

Figure 2. Left: Plot of radial velocity data from SOPHIE against transit phase for WASP-12. An eccentric orbit (dotted line) and a circular orbit (solid line) are overplotted. Note that the scatter of the residuals is larger than the formal uncertainties in the radial velocity measurements. **Right:** Plot of radial velocity against time for WASP-5. Note the signal in the residuals, due to stellar activity, which could mimic the presence of a second companion in this system.

systems and reanalyse the radial velocity data in the literature for another 42 systems. In addition to these 54 systems, we also include the eccentricities of 8 systems from the literature — HD 189733 and HD 209458 which have well measured eccentricities consistent with zero (see for example Agol *et al.* (2010) and Laughlin *et al.* (2005)) and 5 that have precisely measured eccentricities larger than 0.1 (CoRoT-10, HAT-P-15, HD 17156, HD 80606 and XO-3). We also take the eccentricity of the multiplanet system HAT-P-13 from the literature.

For the 54 systems that we analysed, we also collected the relevant orbital constraints from photometry from the original papers. This included the period and mid transit time, but also the secondary eclipse constraint in the form of the eclipse phase ϕ or the projected component of the eccentricity, $e \cos \omega$ where available (see Winn *et al.* 2005, for the relation). For each system in this study (Husnoo *et al.* 2011, *in prep*), we include a realistic treatment of the uncertainties, by penalising data taken on the same night in a fashion similar to Pont, Zucker & Queloz (2006).

Even if all observational uncertainties including correlated noise are taken into account, the value of eccentricity derived from fitting a Keplerian orbit to radial velocity data will always be overestimated. This bias is well known in the literature, and was first pointed out by Lucy & Sweeney (1971) in relation to short period binaries. The reason is that eccentricity is a positive definite value, and for large observational uncertainties and small eccentricities, any noise will inflate its value. This bias has been studied for exoplanets by Shen & Turner (2006) and more recently by Zakamska *et al.* (2010).

3. Results and conclusion

Our results are shown in Figure 1 and Figure 3. In Figure 1, we compare the measured eccentricities of 12 transiting systems from the Literature with our own results. We confirm the eccentricities of 3 of these, and find no evidence for eccentricity in the remaining 9. We therefore displace WASP-18b and WASP-12b as the shortest period objects with a confirmed eccentricity, and replace them by WASP-14b and HAT-P-16b.

Our ensemble results are shown in Figure 3, where we have plotted the planetary mass M_p against the orbital period P. We have deemed systems with precise eccentricities

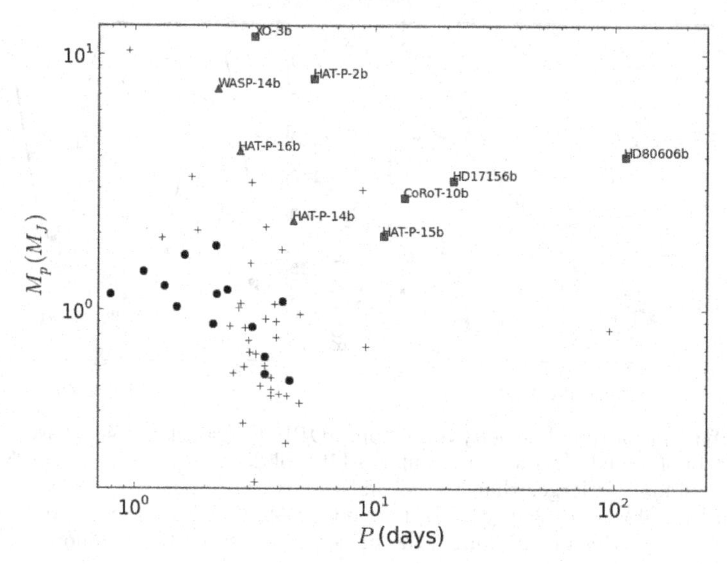

Figure 3. Plot of planetary mass M_p against orbital period P. Squares denote eccentric systems, and circle denote systems with no evidence for eccentricity. The triangles represent the special case of systems with small but precisely measured eccentricies ($e = 3$–10%, at the 10-σ level or more). The crosses denote systems for which the eccentricity is unknown.

$e > 0.1$ as eccentric, and systems with small eccentricities that are not consistent with $e > 0.1$, even at the 3-σ level as compatible circular. An interesting feature in Figure 3 is the mass-period relation. First pointed out by Mazeh *et al.* (2005), this shows that heavier hot Jupiters tend to orbit at shorter periods. Here, we notice that planets following this trend also have circular orbits, with no confirmed eccentric orbit in this region of parameter space. If one considers tides raised on the planet by the star, light planets at short period are expected to circularise quickly, while heavy planets at longer periods are harder to circularise. We thus expect to see planets on circular orbits on the lower left of Figure 3 while we expect to see planets on eccentric orbits on the upper right. This is indeed what we see in Figure 3, and we consider it to be a signature of tidal evolution.

References

Agol, E., Cowan, N. B., Knutson, H. A., *et al.* 2010, *ApJ*, 721, 1861
Campo, C. J., Harrington, J., Hardy, R. A., *et al.* 2011, *ApJ*, 727, id.125
Christian, D. J., Gibson, N. P., Simpson, E. K., *et al.* 2009, *MNRAS*, 392, 1585
Hebb, L., Collier-Cameron, A., Loeillet, B., *et al.* 2009, *ApJ*, 693, 1920
Hebb, L., Collier-Cameron, A., Triaud, A. H. M. J., *et al.* 2010, *ApJ*, 708, 224
Hellier, C., Anderson, D. R., Cameron, A. C., *et al.* 2009, *Nature*, 460, 1098
Husnoo, N., *et al.* 2011a, *MNRAS*, 413, 2500
Husnoo, N., *et al.* 2011b, *in prep*
Laughlin, G., and Marcy, G. W., Vogt, S. S., *et al.* 2005, *ApJL*, 629, L121
Lin, D. N. C., Bodenheimer, P., & Richardson, D. C. 1996, *Nature*, 380, 606
López-Morales, M., Coughlin, J. L., Sing, D., *et al.* 2010, *ApJL*, 716, L36
Lucy, L. B. & Sweeney, M. A. 1971, *AJ*, 76, 544
Maciejewski, G., Dimitrov, D., Neuhaeuser, R., *et al.* 2011, *MNRAS*, 411, 1204
Mazeh, T., Zucker, S., & Pont, F. 2005, *MNRAS*, 356, 955

Nymeyer, S., Harrington, J., Hardy, R. A., *et al.* 2010, *arXiv*:1005.1017
Pont, F., Zucker, S., & Queloz, D. 2006, *MNRAS*, 373, 231
Shen, Y. & Turner, E. L. 2006, *ApJ*, 685, 553
Triaud, A. H. M. J.., Collier-Cameron, A., & Queloz, D. 2010, *A&A*, 524, A25
Winn, J. N., Noyes, R. W., Holman, M. J., *et al.* 2005, *ApJ*, 631, 1215
Zakamska, N. L., Pan, M., & Ford, E. B. 2010, *MNRAS*, 1566

The Astrophysics of Planetary Systems: Formation, Structure, and Dynamical Evolution
Proceedings IAU Symposium No. 276, 2010
A. Sozzetti, M. G. Lattanzi & A. P. Boss, eds.

doi:10.1017/S1743921311020266

Uncertainties in tidal theory: Implications for bloated hot Jupiters

Jérémy Leconte[1], Gilles Chabrier[1] and Isabelle Baraffe[1,2]

[1]École Normale Supérieure de Lyon, 46 allée d'Italie, F-69364 Lyon cedex 07, France; Université Lyon 1, Villeurbanne, F-69622, France; CNRS, UMR 5574, Centre de Recherche Astrophysique de Lyon;
(jeremy.leconte, chabrier, ibaraffe@ens-lyon.fr)

[2]School of Physics, University of Exeter, Stocker Road, Exeter EX4 4PE, UK

Abstract. Thanks to the combination of transit photometry and radial velocity doppler measurements, we are now able to constrain theoretical models of the structure and evolution of objects in the whole mass range between icy giants and stars, including the giant planet/brown dwarf overlapping mass regime (Leconte *et al.* 2009). In the giant planet mass range, the significant fraction of planets showing a larger radius than predicted by the models suggests that a missing physical mechanism which is either injecting energy in the deep convective zone or reducing the net outward thermal flux is taking place in these objects. Several possibilities have been suggested for such a mechanism:

• downward transport of kinetic energy originating from strong winds generated at the planet's surface (Showman & Guillot 2002),
• enhanced opacity sources in hot-Jupiter atmospheres (Burrows *et al.* 2007),
• ohmic dissipation in the ionized atmosphere (Batygin & Stevenson 2010),
• (inefficient) layered or oscillatory convection in the planet's interior (Chabrier & Baraffe 2007),
• Tidal heating due to circularization of the orbit, as originally suggested by Bodenheimer, Lin & Mardling (2001).

Here we first review the differences between current models of tidal evolution and their uncertainties. We then revisit the viability of the tidal heating hypothesis using a tidal model which treats properly the highly eccentric and misaligned orbits commonly encountered in exoplanetary systems. We stress again that the low order expansions in eccentricity often used in constant phase lag tidal models (i.e. constant Q) necessarily yields incorrect results as soon as the (present or initial) eccentricity exceeds ~ 0.2, as can be rigorously demonstrated from Kepler's equations.

Keywords. planets and satellites: general

1. Uncertainties in tidal theory

Tidal heating has been suggested by several authors to explain the anomalously large radius of some giant close-in observed exoplanets (Bodenheimer *et al.* 2001; Jackson, Greenberg & Barnes 2008; Miller, Fortney & Jackson 2009; Ibgui, Spiegel & Burrows 2011). The basic scenario consists in a planet left on a wide, very eccentric orbit by an early event during its formation. The orbit then slowly decays due to tidal dissipation, leading to a circularization on a timescale of a few Gyr's. This *slow* circularization, however, is due to the fact that all these authors use a tidal model based on a *quasi circular approximation*, truncated at 2nd order in eccentricity.

This quasi circular approximation, developed initially to study the tidal evolution of the *solar system* planets (Goldreich & Soter 1966; Ferraz-Mello, Rodríguez & Hussmann

2008), which have negligible eccentricities, is valid only in this very limit, $e \ll 1$. In the context of exoplanetary systems, current high eccentricities are common and initial high eccentricities are very likely, as inferred from non-transiting planets observed by radial velocity. Therefore, this quasi circular approximation is no longer correct, as demonstrated in Eggleton *et al.* (1998), Wisdom (2008) and Leconte *et al.* (2010) and summarized below.

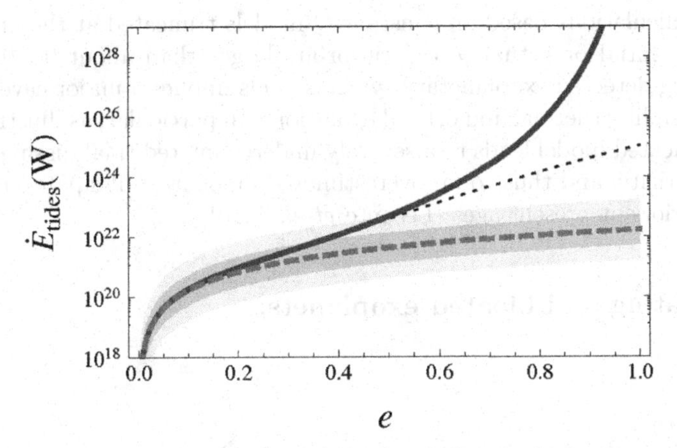

Figure 1. Tidal energy dissipation rate in a pseudo-synchronized planet as a function of the eccentricity calculated with the complete formula (*curve*), with the e^2-truncated formula (*dash*) and to e^{10} (*Dotted*). The ratio of the two curves only depends on the eccentricity and not on the system's parameters. The inner (outer) shaded area shows the uncertainty in the heating when allowing the dissipation parameter to vary within one (two) order of magnitude. The actual values were derived using HD 209 458 b parameters.

Present analytical theories for tidal interaction are all based on the equilibrium tide and weak friction approximation, since no adequate theory for dynamical tides presently exists. These theories differ in two ways

- (i) their parametrization of the dissipative processes. The most common prescriptions are either a constant phase lag (constant-Q) model or a constant viscosity or time lag (constant-Δt) model.
- (ii) their mathematical treatment of the geometry of the orbits: perturbative developments around the coplanar/circular keplerian orbits or closed formulae, valid for any eccentricity.

While these two sources of differences between tidal models are completely different by nature, they are often, erroneously, mixed together. Indeed, only the constant time lag model, because of the linear dependence of the phase lag upon the time lag in this model, allows the calculations to be carried out in terms of closed formulae for *any* eccentricity. High order calculations in eccentricity in the framework of the constant-Q model are very cumbersome (see Ferraz-Mello *et al.* 2008).

As demonstrated by Wisdom (2008) and Leconte *et al.* (2010), even though large uncertainties remain on the quantification of the dissipative processes themselves, the discrepancies arising from the differences in the treatment of the orbital geometry at moderate to high eccentricities ($e \gtrsim 0.2$-0.3) can become dominant *by orders of magnitude*. This is summarized on Fig. 1 which compares the tidal heating given by the constant time lag model of Leconte *et al.* (2010) (solid curve) and by the quasi circular approximation of Peale & Cassen (1978) (dashed curve). In comparison, the inner (outer) shaded area illustrate the impact of the uncertainty in the heating when allowing the tidal dissipation parameter to vary by one (two) order of magnitude. For $e > 0.4$, we see that high order terms in e yield a contribution which is larger than the uncertainty in the quantification of the dissipative processes. Such a behavior at high eccentricity is well understood in the context of celestial mechanics and is due to the slow convergence of elliptical expansion series (Danjon 1980; Cottereau, Aleshkina & Souchay 2010). Using the admitedly large uncertainty in tidal dissipative processes as an argument to use necessarily wrong orbital equations is thus by no means justifiable.

Therefore, calculations based on constant-Q models truncated at the order e^2 cannot be applied to (initial or actual) eccentric orbits larger than about 0.2-0.3, a common situation among detected exoplanetary systems. This implies a major caveat in previous calculations coupling thermal and orbital evolutions. In particular, as illustrated in Fig. 1, using a e^2-truncated model leads to a severely underestimated tidal dissipation timescale at large eccentricity, and thus to an overestimated amount of dissipated tidal energy in exoplanet interiors at present ages (Leconte *et al.* 2010).

2. Tidal heating and bloated exoplanets.

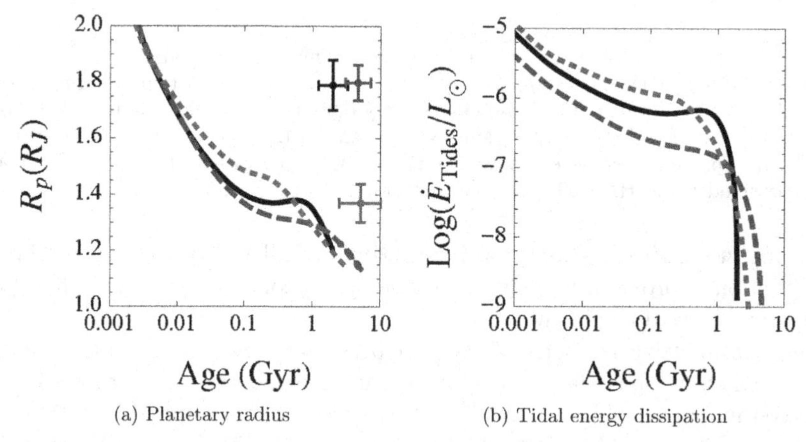

(a) Planetary radius (b) Tidal energy dissipation

Figure 2. Evolutionary tracks for WASP-12 b (solid), TrES-4 b (dashed) and WASP-4 b (dotted) that lead to the best agreement with the observed orbital parameters for these systems. These runs assume $Q'_{0,\mathrm{p}} = 10^6$ and $Q'_{0,\star} = 10^6$. Tidal dissipation is not sufficient to sustain the large radii observed for these planets.

Revisiting the viability of the tidal heating hypothesis to explain the anomalously large Hot Jupiter radii † with the Hut complete tidal model, Leconte *et al.* (2010) showed that, although tidal friction indeed provides a non negligible contribution to the planet heat content and can possibly explain *some* transiting system large radius, the tidal heating hypothesis fails to explain the radii of extremely bloated planets like - among others - HD 209 458 b, TrES-4 b, WASP-4 b or WASP-12 b, as illustrated on Fig 2. These conclusions have been confirmed recently by Hansen (2010). The main reason is the early circularization of the orbit of these systems, as mentioned above, which thus yields insufficient heating at late epochs. Note that these conclusions rely on the assumption of

† Numerical values of the radius anomaly of transiting planets as defined in Leconte *et al.* 2010 can be found at *perso.ens-lyon.fr/jeremy.leconte/JLSite/JLsite/Exoplanets_Simulations.html.*

a genuine two body system. The presence of a third body able to excite eccentricity in a massive giant planet for several gigayears would provide an other explanation. Accurate observations are necessary to support or exclude the existence of such undetected close low-mass or distant massive companions.

Acknowledgements

This work was supported by the Constellation european network MRTN-CT-2006-035890, the french ANR "Magnetic Protostars and Planets" (MAPP) project and the "Programme National de Planétologie" (PNP) of CNRS/INSU. We acknowledge the use of the *www.exoplanet.eu* database. J.L. wishes to thank L. Cottereau for insightful discussions concerning current developments in Celestial Mechanics.

References

Batygin, K. & Stevenson, D. J. 2010, *ApJ*, 714, L238
Bodenheimer, P., Lin, D. N. C., & Mardling, R. A. 2001, *ApJ*, 548, 466
Burrows, A., Hubeny, I., Budaj, J., & Hubbard, W. B. 2007, *ApJ*, 661, 502
Chabrier, G. & Baraffe, I. 2007, *ApJ*, 661, L81
Cottereau, L., Aleshkina, E., & Souchay, J. 2010, *A&A*, 523, A87
Danjon, A. 1980, *Astronomie generale. Astronomie spherique et elements de mecanique celeste*, ed. Danjon, A.
Eggleton, P. P., Kiseleva, L. G., & Hut, P. 1998, *ApJ*, 499, 853
Ferraz-Mello, S., Rodríguez, A., & Hussmann, H. 2008, *Celestial Mechanics and Dynamical Astronomy*, 101, 171
Goldreich, P. & Soter, S. 1966, *Icarus*, 5, 375
Hansen, B. M. S. 2010, *ApJ*, 723, 285
Ibgui, L., Spiegel, D. S., & Burrows, A. 2011, *ApJ*, 727, id.75
Jackson, B., Greenberg, R., & Barnes, R. 2008, *ApJ*, 681, 1631
Leconte, J., Baraffe, I., Chabrier, G., Barman, T., & Levrard, B. 2009, *A&A*, 506, 385
Leconte, J., Chabrier, G., Baraffe, I., & Levrard, B. 2010, *A&A*, 516, A64+
Miller, N., Fortney, J. J., & Jackson, B. 2009, *ApJ*, 702, 1413
Peale, S. J. & Cassen, P. 1978, *Icarus*, 36, 245
Showman, A. P. & Guillot, T. 2002, *A&A*, 385, 166
Wisdom, J. 2008, *Icarus*, 193, 637

The Astrophysics of Planetary Systems: Formation, Structure, and Dynamical Evolution
Proceedings IAU Symposium No. 276, 2010
A. Sozzetti, M. G. Lattanzi, & A. P. Boss, eds.

doi:10.1017/S1743921311020278

Tidal dynamics of transiting exoplanets

Daniel C. Fabrycky[1]

[1]UCO/Lick University of California Santa Cruz, CA 95064
email: daniel.fabrycky@gmail.com

Abstract. Transits give us the mass, radius, and orbital properties of the planet, all of which inform dynamical theories. Two properties of the hot Jupiters suggest they had a dramatic origin via tidal damping from high eccentricity. First, the tidally circularized planets (in the 1-4 day pile-up) lie along a relation or boundary in the mass-period plane. This observation may implicate a tidal damping process regulated by planetary radius inflation and Roche lobe overflow, early in the planets' lives. Second, the host stars of many planets have spins misaligned from the planets' orbits. This observation was not expected a priori from the conventional disk migration theory, and it was a boon for the alternative theories of planet-planet scattering and Kozai cycles, accompanied by tidal friction, which predicted it. Now we are faced with a curious observation that the misalignment angle depends on the stellar temperature. It may mean that the tide raised on the stars realigns them, the final result being the tidal consumption of hot Jupiters.

Keywords. planetary systems: formation, eclipses

---

1. Introduction

Two of the central mysteries of extrasolar gas giants is that they have small periods and large eccentricities, relative to the giants of the solar system (see Figure 1). A host of physical processes may account for these differences, but the most ubiquitous and robust mechanisms seem to be migration in a gas disk to shorten orbital periods (Goldreich & Tremaine 1980; Ward 1997; Lubow & Ida 2010) and chaotic interactions among planets to increase orbital eccentricities (Jurić & Tremaine 2008; Chatterjee *et al.* 2008; Ford & Rasio 2008). At first glance then, it may seem that the orbital evolution of hot Jupiters — a class of planets with typical periods 1-4 days and nearly circular orbits, and usually no known companion planets (Wright *et al.* 2009) — is dominated by migration in the gas disk with not by dynamical interactions. However, orbits with periapses as close to the star as these would tend to circularize via tidal dissipation. Therefore, it is possible that hot Jupiters actually arise from dramatic eccentricity excitation without much migration in the disk, after which the orbits circularize tidally (Rasio & Ford 1996; Wu & Murray 2003; Matsumura *et al.* 2010). These two vastly different evolutionary paths are illustrated on Figure 1. The latter mechanism has a sharp distance dependence — if it happens at all, it must result in a characteristic orbital period, i.e. it could lead to the observed 3-day pile-up (Wu *et al.* 2007), as illustrated by Figure 2.

Transiting planets offer some new possibilities for resolving the origin of hot Jupiters because the orbital and physical properties can be precisely measured. Furthermore, unlike in the case of radial-velocity discoveries which yields $m \sin i$ of the planet, the true mass of transiting planets can be measured. These measurements have revealed an intriguing trend among the hot Jupiters: planets with periods ~ 1 day are more massive than their ~ 3 day brethren. Section 2 is devoted to this statistical result and a potential explanation. Another property of the orbit, its alignment relative to the star's spin, is only measurable via transits, using the Rossiter-McLaughlin effect (Winn, these proceedings). Thus transits offer a new property for discerning planetary migration, and this is discussed in section 3.

2. Circular hot Jupiters follow a mass-period relation

After only a few transiting planets were discovered, Mazeh *et al.* (2005) boldly identified a relationship between the planetary masses and orbital periods: for the hot Jupiters, they were inversely correlated. As more transiting planets were discovered, it came to be characterized more as a lower limit to mass as a function of orbital period, rather than a relation. That is, massive exoplanets were found to transit and were far off the relation. These same planets are typically more eccentric, being less prone to tidal dissipation in the planet, which causes circularization. After separating the planets into those that are robustly eccentric and those that are consistent with circular, Pont *et al.* (2011) (see Husnoo *et al.* these proceedings) found that the circular orbits still follow an anti-correlation in mass versus period.

This observation suggests that being drawn on to the mass-period relation might be considered a fundamental aspect of hot Jupiters, that some property of their origin establishes circular orbits on the mass-period relation. Tidal dissipation in the planet can indeed bring planets to circular orbits, and the period does shrink as the orbit circularizes, but the end point of standard tidal dissipation does not depend on the planet's mass. However, it has been recognized (e.g. Gu *et al.* 2003; Gaudi *et al.* 2005) that hot Jupiters have orbital energies greatly exceeding their binding energy, and that this can lead to planetary inflation, mass loss, and even tidal disruption. In particular, if a planet dissipates from a very high eccentricity orbit (much larger semi-major axis a), then the tidal energy input is:

$$E_{\rm t} = -E_{\rm orb} = GM_{\star}M_p/(2a), \tag{2.1}$$

whereas the binding energy of the planet is:

$$E_{\rm b} = kGM_p^2/R_p, \tag{2.2}$$

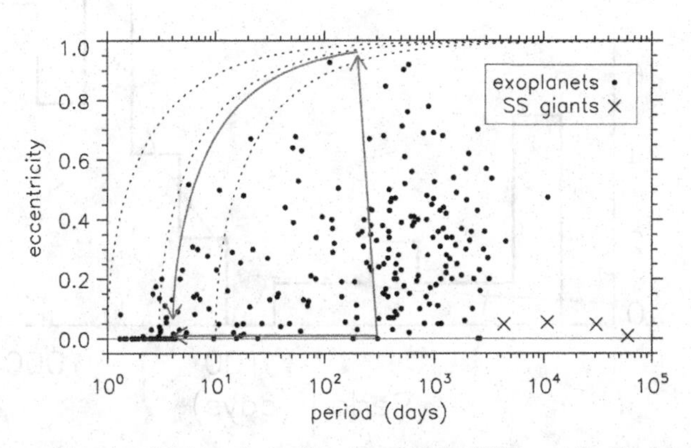

Figure 1. Periods and eccentricities of exoplanets, which differ markedly from the giants of the solar system. Broadly speaking, two distinct pathways for forming hot Jupiters are shown by the arrows. Lines of constant angular momentum (dotted) show the paths of exoplanets that are evolving by tides in the planet alone, with negligible angular momentum transferred to the planet's spin. The large population of tidally circularized hot Jupiters is readily apparent.

where at an age of $10^6 - 10^7$ Myr, $k \approx 1$ and $R_p = 1.3 - 1.6\ R_{\rm Jup}$ (Burrows *et al.* 2000; Marley *et al.* 2007). Thus, overflow is energetically possible ($E_{\rm t} \gtrsim E_{\rm b}$) if

$$M_p \lesssim 19.2 M_{\rm Jup} (M_\star/M_\odot)^{2/3} (P/{\rm day})^{-2/3}. \quad (2.3)$$

Curiously, this mass-period limit lies parallel (although much higher in mass than) the mass-period relation of Mazeh *et al.* (2005).

Energetically, then, all the circularized hot Jupiters might have gone through an epoch of tidal inflation and/or overflow. If the initial perhaps separation is at the tidal radius or below ($\lesssim 2.4(m_\star/m_p)^{1/3} R_p$) then the planet will be ripped apart on one or a few orbital periods (Faber *et al.* 2005). If it starts slightly further out, the radius inflation still meets the Roche surface with unstable results (Guillochon *et al.* 2011).

However, if the initial periapse is distant enough, this process could be rather gentle, tearing off only a very small fraction of the mass per orbital period. The lionshare of the mass will be lost through the (instantaneous) inner Lagrangian point, towards the star. The gas would form an accretion disk around the star, transferring the mass to the star and transporting angular momentum back out, ultimately back to the planet. Therefore, even as the planet loses mass, it will nearly conserve its angular momentum. The result is likely capable of drawing the planet to the mass-period relation. In particular, lower mass gas giants are more susceptible to tidal dissipation and radius inflation, and they will thus lose a larger fraction of their mass, and gain more *specific* angular momentum through mass transfer.

3. Spin-orbit misalignment

Most mechanisms that can set up such extreme eccentricities, such that the hypothesis of the previous section is applicable, are also likely to excite inclinations. Let us first suppose that a planetary system begins aligned with its host star's spin, as both orientations are set by the protoplanetary accretion disk (but see below). Disk migration would presumably maintain this alignment. In contrast, the mechanisms of Kozai cycles

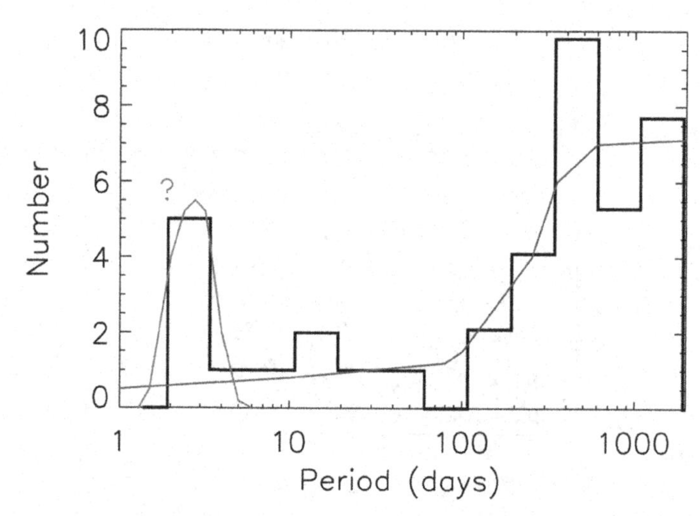

Figure 2. Empirical period distribution of gas-giant ($m \sin i > 0.3\ {\rm M_{Jup}}$) exoplanets (Cumming *et al.* 2008). There is a clear probability enhancement at short orbital periods: the hot Jupiters are piled up at 3 days. Perhaps disk migration creates a smooth component that populates from ~ 1 AU to the stellar surface with declining efficiency, whereas tidal dissipation from high eccentricity creates the pile-up.

due to a binary companion (Fabrycky & Tremaine 2007; Wu *et al.* 2007) and planet-planet scattering (Nagasawa *et al.* 2008) have been shown to yield wide distributions of spin-orbit misalignment. Dramatically, these mechanisms predict that some hot Jupiters should be on retrograde orbits, that is, the stellar obliquity should be > 90 degrees.

Therefore the second probe of migration specific to transit measurements is the angle between planetary orbits and stellar spins, observable during transits through the spectroscopic anomaly known as the Rossiter-McLaughlin effect (Queloz *et al.* 2000; Gaudi & Winn 2007; Fabrycky & Winn 2009, Winn in this volume). Over the past two years, a large statistical sample has been built up by very busy observers (e.g. Triaud *et al.* 2010; Winn *et al.* 2010). Both a large population of aligned systems, as well as large prograde inclinations and retrograde inclinations, have all been observed. In fact, the fraction of highly inclined and retrograde systems are in rough agreement with the fraction predicted by the mechanisms forming hot Jupiters via high eccentricities (Morton & Johnson 2011). However, one difficulty with this hypothesis is that it does not leave enough very nearly aligned systems (Fabrycky & Winn 2009) – this "core" of well-aligned systems is a robust feature, despite the deluge of misaligned systems.

Another curious fact is that the hot stars (defined as $T_{eff} > 6250$ K) are much more often to have misaligned planets (Winn *et al.* 2010). In fact, just these stars are completely consistent with strong scattering prediction (Morton & Johnson 2011). Perhaps *all* hot Jupiters arrive via high eccentricity and inclination, and the cool stars are somehow able to damp those inclinations to quite small values. In particular, tidal dissipation in the cool stars, which have more dynamically significant convective envelopes, seems a likely damping mechanism. Moreover, among the cool stars, there are some misalignments, and these are the systems for which the planets raise the weakest tide on their stars. See (Winn *et al.* 2010) and Winn in this volume.

Despite all these empirical encouragements, the theory of how tides can damp inclinations seems problematic. When one tries to construct a tidal model of tides raised on stars, one immediately finds that the large inertia of stars means that a typical hot Jupiter cannot torque its star into spin-orbit synchrony and survive the process — it will surrender all its orbital angular momentum as it spins up the star (Rasio *et al.* 1996; Levrard *et al.* 2009; Jackson *et al.* 2009; Taylor 2010). In fact, the inclination damping timescale and planetary consumption timescale rely on similar physics, so one expects that moderate-mass exoplanets would only reorient their stars as they plunge in (Barker & Ogilvie 2009).

What is needed is a mechanism that allows the star to reorient without spinning up its bulk, otherwise it would rob the planet of too much angular momentum. Thus Winn *et al.* (2010) proposed the following resolution: the spin rate of the interior radiative zone decouples from the outer convective zone. The outer parts of the star, which are most strongly affected by tides from the planet, are free to respond to it. They are also free to be torqued down by the magnetic wind, assuring that the surface rotation does not violate the observation that hot-Jupiter hosts are not particularly fast surface rotators. This concept of differential rotation within solar-type stars is not new; it is a natural consequence of magnetic wind torque preferentially slowing the outer parts of the star (Pinsonneault *et al.* 1989), and it has received some empirical support through observations of rotation periods in young stellar clusters (Irwin & Bouvier 2009). Several mechanisms act to restore near-solid-body rotation: hydrodynamic instabilities (Goldreich & Schubert 1967), g-modes (Zahn *et al.* 1997), and magnetic torques. One or all of which have apparently coupled the radiative interior and the convective zone of the Sun (Howe 2009). On that coupling timescale, the hot Jupiter would interact with the whole star and tidally spiral in. So if this picture is correct, it implies that most of the hot Jupiters with spin-orbit alignment with cool stars are in a transient state, and they will meet a spectacular end through tidal consumption.

Finally, we must address whether spin-orbit misalignments are *proof* that some hot Jupiters arrive via high eccentricities. On the contrary: after the discovery of several retrograde planets, the assumptions of spin-orbit alignment resulting from disk migration were questioned by several authors. Bate *et al.* (2010) suggested that the host star simply does not need to be aligned with the disk that formed the planets, as its spin orientation represents the whole build-up of mass, accreted from many directions within the birth cloud, whereas the planets are probably built from only the last material that was accreted into a disk. Lai *et al.* (2011) and Foucart & Lai (2011), on the other hand, described a model of protostellar accretion in which the magnetic interaction between the star and the disk can amplify the inclination difference between them. If either of these mechanisms, or additional mechanisms, can explain the correlation between spin-orbit misalignment and stellar temperature, then disk migration may regain its full appeal as the mechanism that produces most hot Jupiters.

4. Conclusions

Transiting planets have offered interesting new windows into the origin of hot Jupiters, the enigmatic set of gas-giant planets parked at few-day periods from their stars. Continued increases of the samples of hot Jupiters, and more investigation into trends and correlations of spin-orbit alignment, are mandatory before these issues will be completely resolved. It is quite likely that hot Jupiters come from some combination of the evolutionary paths shown in Figure 1, and the relative contribution is what needs to be determined. But some universal properties of hot Jupiters, e.g. the mass-period correlation and misalignments preferentially around hot stars, may indicate that future theoretical work may be able to interpret these observational trends within a comprehensive and singular framework.

References

Barker, A. J. & Ogilvie, G. I. 2009, *MNRAS*, 395, 2268
Bate, M. R., Lodato, G., & Pringle, J. E. 2010, *MNRAS*, 401, 1505
Burrows, A., Guillot, T., Hubbard, W. B., Marley, M. S., Saumon, D., Lunine, J. I., & Sudarsky, D. 2000, *ApJL*, 534, L97
Chatterjee, S., Ford, E. B., Matsumura, S., & Rasio, F. A. 2008, *ApJ*, 686, 580
Cumming, A., Butler, R. P., Marcy, G. W., Vogt, S. S., Wright, J. T., & Fischer, D. A. 2008, *PASP*, 120, 531
Faber, J. A., Rasio, F. A., & Willems, B. 2005, *Icarus*, 175, 248
Fabrycky, D. & Tremaine, S. 2007, *ApJ*, 669, 1298
Fabrycky, D. C. & Winn, J. N. 2009, *ApJ*, 696, 1230
Ford, E. B. & Rasio, F. A. 2008, *ApJ*, 686, 621
Foucart, F. & Lai, D. 2011, *MNRAS*, 234
Gaudi, B. S., Seager, S., & Mallen-Ornelas, G. 2005, *ApJ*, 623, 472
Goldreich, P. & Schubert, G. 1967, *ApJ*, 150, 571
Gaudi, B. S. & Winn, J. N. 2007, *ApJ*, 655, 550
Goldreich, P. & Tremaine, S. 1980, *ApJ*, 241, 425
Gu, P.-G., Lin, D. N. C., & Bodenheimer, P. H. 2003, *ApJ*, 588, 509
Guillochon, J., Ramirez-Ruiz, E., & Lin, D. N. C. 2011, *ApJ*, 732, id.74
Howe, R. 2009, *Living Reviews in Solar Physics*, 6, 1
Irwin, J. & Bouvier, J. 2009, in IAU Symposium, Vol. 258, ed. E. E. Mamajek, D. R. Soderblom, & R. F. G. Wyse, 363

Jackson, B., Barnes, R., & Greenberg, R. 2009, *ApJ*, 698, 1357
Jurić, M. & Tremaine, S. 2008, *ApJ*, 686, 603
Lai, D., Foucart, F., & Lin, D. N. C. 2011, *MNRAS*, 231
Lubow, S. H. & Ida, S. 2010, Planet Migration, in EXOPLANETS, pg. 347-371, ed. S. Seager, University of Arizona Press
Levrard, B., Winisdoerffer, C., & Chabrier, G. 2009, *ApJL*, 692, L9
Marley, M. S., Fortney, J. J., Hubickyj, O., Bodenheimer, P., & Lissauer, J. J. 2007, *ApJ*, 655, 541
Matsumura, S., Peale, S. J., & Rasio, F. A. 2010, *ApJ*, 725, 1995
Mazeh, T., Zucker, S., & Pont, F. 2005, *MNRAS*, 356, 955
Morton, T. D. & Johnson, J. A. 2011, *ApJ*, 729, 138
Nagasawa, M., Ida, S., & Bessho, T. 2008, *ApJ*, 678, 498
Pinsonneault, M. H., Kawaler, S. D., Sofia, S., & Demarque, P. 1989, *ApJ*, 338, 424
Pont, F., Husnoo, N., Mazeh, T., & Fabrycky, D. 2011, *MNRAS*, 414, 1278
Queloz, D., Eggenberger, A., Mayor, M., Perrier, C., Beuzit, J. L., Naef, D., Sivan, J. P., & Udry, S. 2000, *A&A*, 359, L13
Rasio, F. A., Tout, C. A., Lubow, S. H., & Livio, M. 1996, *ApJ*, 470, 1187
Rasio, F. A. & Ford, E. B. 1996, *Science*, 274, 954
Taylor, S. F. 2010, preprint *arXiv:1009.4221*
Triaud, A. H. M. J., Collier Cameron, A., Queloz, D., Anderson, D. R., Gillon, M., Hebb, L., Hellier, C., Loeillet, B., Maxted, P. F. L., Mayor, M., Pepe, F., Pollacco, D., Ségransan, D., Smalley, B., Udry, S., West, R. G., & Wheatley, P. J. 2010, *A&A*, 524, A25+
Ward, W. R. 1997, *ApJL*, 482, L211+
Winn, J. N., Fabrycky, D., Albrecht, S., & Johnson, J. A. 2010, *ApJL*, 718, L145
Wright, J. T., Upadhyay, S., Marcy, G. W., Fischer, D. A., Ford, E. B., & Johnson, J. A. 2009, *ApJ*, 693, 1084
Wu, Y. & Murray, N. 2003, *ApJ*, 589, 605
Wu, Y., Murray, N. W., & Ramsahai, J. M. 2007, *ApJ*, 670, 820
Zahn, J.-P., Talon, S., & Matias, J. 1997, *A&A*, 322, 320

The Astrophysics of Planetary Systems: Formation, Structure, and Dynamical Evolution
Proceedings IAU Symposium No. 276, 2010
A. Sozzetti, M. G. Lattanzi & A. P. Boss, eds.

doi:10.1017/S174392131102028X

Spin-orbit angles: A probe to evolution

Amaury H. M. J. Triaud[1], Didier Queloz[1] and Andrew Collier Cameron[2]

[1]Observatoire Astronomique de l'Université de Genève, Chemin des Maillettes 51, CH-1290 Sauverny, Switzerland
email: Amaury.Triaud@unige.ch

[2]SUPA, School of Physics & Astronomy, University of St Andrews, North Haugh, KY16 9SS, St Andrews, Fife, Scotland, UK

Abstract. We will present our campaign to estimate the projected spin-orbit angle for transiting hot Jupiters, obtained via observations of the Rossiter-McLaughlin effect. Combining our results to those of other teams we show what the current distribution in projected spin-orbit angle is, quickly reminding what interpretation we make of it. Finally we will show early results from a campaign that we initiated, surveying the Rossiter-McLaughlin effect on transiting SB1 intended to provide a comparison sample to the transiting planet's results.

Keywords. planetary systems: formation, binaries: eclipsing, techniques: spectroscopic

1. Context

Prior to the discovery of the first extrasolar planet found by Mayor & Queloz (1995), it was usually expected that giant planets would not be found much closer to their parent star than Jupiter is. This discovery called for a migrating mechanism to carry a gas giant from its birth location to the currently observed one. Lin *et al.* (1996) using work from Goldreich & Tremaine (1980) presented the idea that planets ought to migrate via exchange of angular momentum with the protoplanetary disc. An alternative explanation was proposed by Rasio & Ford (1996) whereby in an unstable multiplanetary system, planets would gradually scatter each others, some arriving on close-in orbits.

The discovery of 51 Peg b and the other hot Jupiters altered the generally accepted view of well ordered, hierarchical systems such as our own towards more diverse systems. Indeed, with masses ranging over several orders of magnitude, very different radii, large eccentricities and a curious period distribution, planets have demonstrated an enormous variety in their observed parameters.

More recently a new observable appeared: the projected spin-orbit angle β (also called λ) which is obtained via the Rossiter-McLaughlin effect. First theorised by Holt (1893) as a way to determine stellar rotation, it was observed by Rossiter (1924) and McLaughlin (1924) on eclipsing binaries. It was first observed for a planet by Queloz *et al.* (2000) for HD 209458b and on an increasing number of transiting hot Jupiters since, most systems appearing aligned with their star (see Josh Winn's contribution and citations within).

2. The HARPS Rossiter-McLaughlin survey

The WASP consortium has two instruments located in La Palma, Canary Islands, Spain, and in Sutherland, South Africa. The aim is to discovery transiting planets (see contribution by Andrew Collier Cameron). WASP produces candidates which are then confirmed on the 193cm at OHP, the NOT at La Palma, and by the CORALIE spectrograph on the 1.2m Swiss Euler Telescope at La Silla, Chile. After confirming some of those candidates as being planets we used the HARPS spectrograph, mounted on the ESO 3.6m, also at La Silla, to refine the orbits but also to get high signal to noise and high cadence observations during the transits of the WASP planets.

Our first batch of 8 observations was published in three papers: Gillon *et al.* (2009), Queloz *et al.* (2010) and Triaud *et al.* (2010). On these eight we found that four of our targets were retrograde: WASP-2b, 8b, 15b and 17b (see Fig. 1), an unusually large number compared to other results in the literature. We attempted a statistical analysis of all the Rossiter-McLaughlin effect known then: 26, trying to see which of the published theoretical predictions could be compared to the observations (see Triaud *et al.* (2010)).

On the cumulative distribution of projected spin orbit angles, compared with predictions by Fabrycky & Tremaine (2007) and by Nagasawa *et al.* (2008), the two predictions displaying the largest range of angles, we see that predictions by Nagasawa *et al.* (2008) fit the best (Fig. 2). These authors used the Kozai mechanism caused by an outer binary with tidal friction to produce their results while Nagasawa *et al.* (2008) used planet-planet scattering, initiating Kozai cycles, and tidal friction. Both used dynamical events followed by tidal friction.

We concluded that standard disc migration, not expected to produce large angles, could not alone produce the observations, that another mechanism ought to be present, the Kozai mechanism fitting best the observations.

These results depended on an assumption: that the probability for a planet to have a certain β appeared independent of other parameters for all planets, that we had only one distribution. Winn *et al.* (2010) showed it was not so by plotting the projected spin-orbit angle in function of the effective temperature of the star (see Josh Winn's contribution).

3. New results

Since, we continued observing with HARPS, while other colleagues continued their observations. Combining announcements made during the conference and those that we have in stock we can confirm what Winn *et al.* (2010) had remarked: planets around stars colder than 6250 K tend to be more aligned than planets around hotter stars. In fact the combined observations show a large lack of aligned systems for planets around

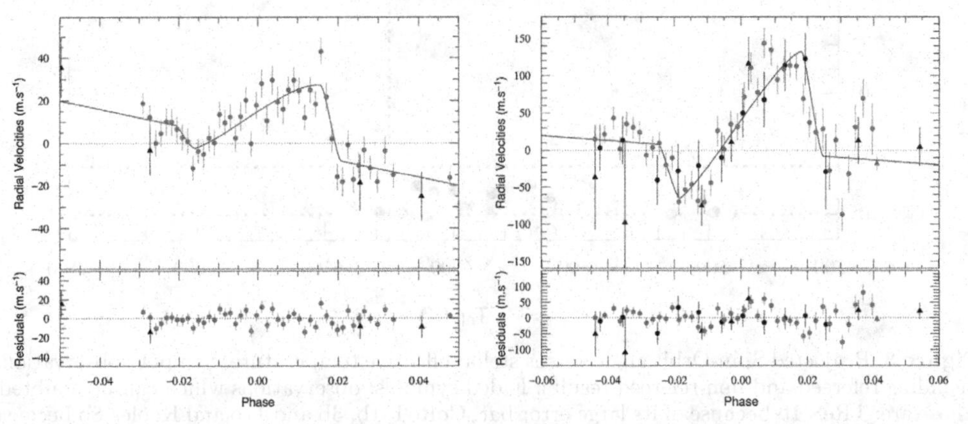

Figure 1. The R-M effect for WASP-15b and WASP-17b and residuals as appearing in Triaud *et al.* (2010). Circles are HARPS observations, triangles are CORALIE observations.

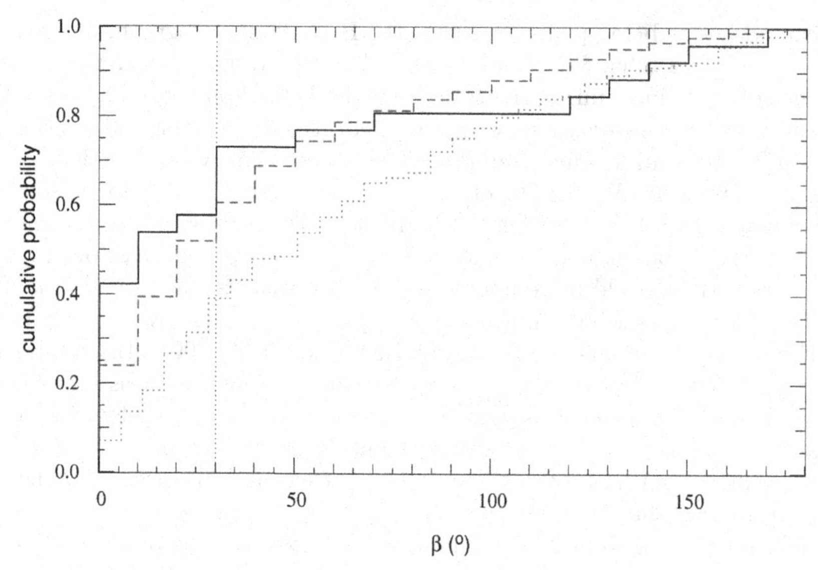

Figure 2. Cumulative distribution of observations (plain). Theoretical predictions by Fabrycky & Tremaine (2007) (dashed). Theoretical predictions by Nagasawa *et al.* (2008) (dotted).

hotter stars compared to those around colder one (Figure 3). Schlaufman (2010) using another method, independently comes to similar conclusions.

From explanations given in Winn *et al.* (2010) 6250 K is the temperature at which the stellar outer convective envelope disappears. Planets around stars colder than this

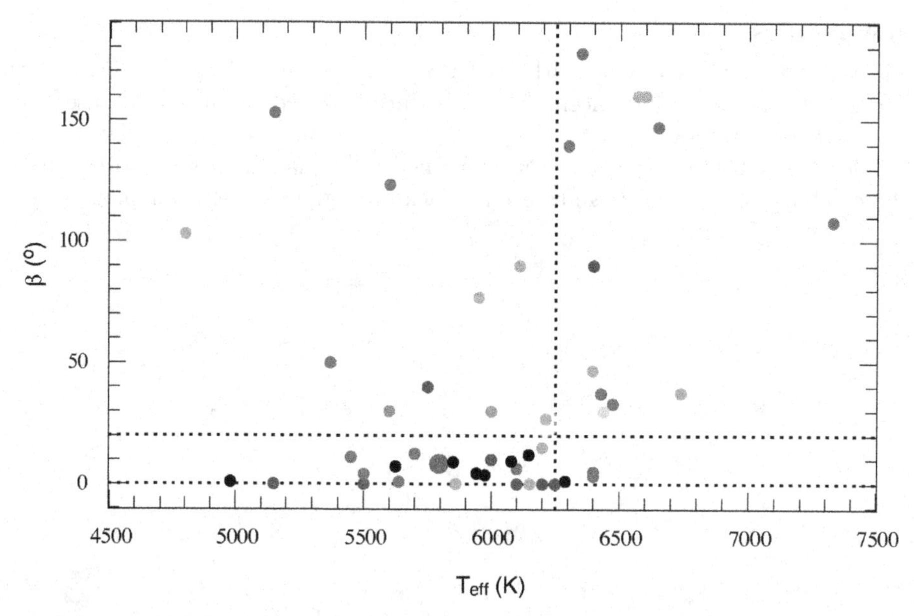

Figure 3. Projected Spin-Orbit angle versus stellar effective temperature. Symbols: observations including refereed and non refereed results. Faded symbols: observations which can be doubted. Those are: TReS-1b because of its large error bar, CoRoT-1b, 3b and 11b and Kepler-8b because notably of the bad sampling due to the faintness of those targets. One of our new observations also suffer from some systematic effect. Numbers likely to change.

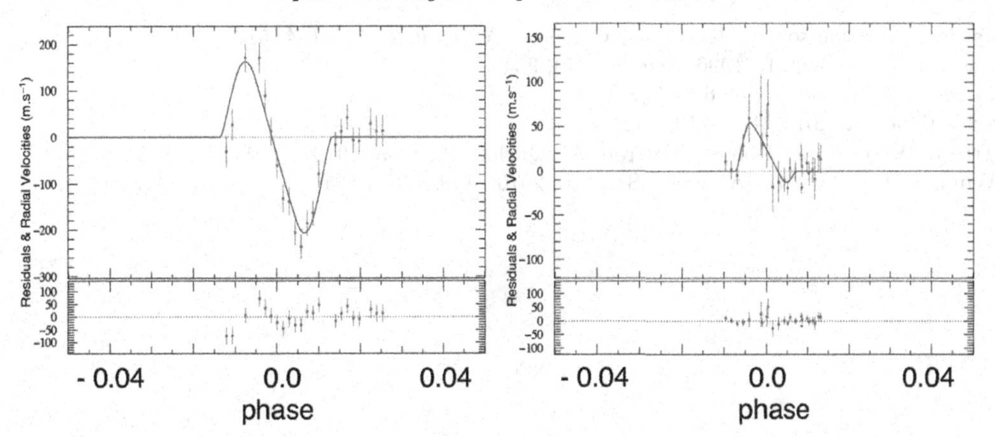

Figure 4. Rossiter-McLaughlin effect for two transiting M dwarfs. On left, the M-dwarf transits a 6600 K star, on a 8.1 day, eccentric period. On right, the M-dwarf is around a 5300 K star, on a nearly circular 6.8 day period. While the first is clearly aligned, the second is 3σ away from $\beta = 0$. Triaud *et al.* in prep.

would realign the convective envelope and therefore would appear aligned; planets around hotter stars retain their original spin-orbit angle. An analysis of the angle of those planets could then be used and compared to theory.

Among our new results we confirm that WASP-7b, observed with the CORALIE spectrograph, is severely misaligned with a quasi polar orbit, but do not confirm the detection of the Rossiter-McLaughlin on WASP-16b (Josh Winn's contribution) where our HARPS results show no deviation from the Keplerian orbit.

4. Comparison sample with Low Mass Binaries

In its search for planets, the WASP consortium found a plethora of transiting M-Dwarfs. These objects, interesting in their own right also provide a good opportunity to have a comparison sample. We now know planets can be aligned and can be misaligned. What about stars? Very few quantitative measurements exist in the literature; only one binary system is known to be misaligned Albrecht *et al.* (2009). We have embarked in a survey to measure the spin-orbit angle for these low mass binary stars, behaving in transit on a first approximation as a planet would: a small dark disc on a bright surface. Being SB1, the radial velocities are extracted exactly like in the planet case. We observe therefore the Rossiter-McLaughlin effect. This is the first survey of its kind. We show here two examples from our observations with CORALIE in figure 4.

References

Albrecht, S., Reffert, S., Snellen, I. A. G., & Winn, J. 2006, *Nature*, 461, 373
Fabrycky, D. & Tremaine, S. 2007, *ApJ*, 669, 1298
Gillon, M., Anderson D. R. & Triaud, A. H. M.. J., *et al.* 2009, *A&A*, 501, 785
Goldreich, P. & Tremaine, S. 1980, *ApJ*, 241, 425
Holt, J. R. 1893, *A&A*, XII, 646
Lin, D. N. C., Bodenheimer, P., & Richardon D. C. 1996, *Nature*, 380, 606
Mayor, M. & Queloz, D. 1995, *Nature*, 378, 355
McLaughlin, D. B. 1924, *ApJ*, 60, 22
Nagasawa, M., Ida, S., & Bessho, T. 2008, *ApJ*, 678, 498
Queloz D., Eggenberger, A., Mayor, M., *et al.* 2000, *A&A*, 359, L13

Queloz, D., Anderson, D. R., Collier Cameron, A., *et al.* 2010, *A&A*, 517, L1
Rasio, F. A. & Ford, E. 1996, *Science*, 274, 954
Rossiter, R. A. 1924, *ApJ*, 60, 15
Schlaufman, K. 2010, *ApJ*, 719, 602
Triaud, A. H. M.. J., Collier Cameron, A., Queloz, D., *et al.* 2010, *A&A*, 524, A25
Winn, J., Fabrycky, D., Albrecht, S., *et al.* 2010, *ApJ*, 178, L145

The Astrophysics of Planetary Systems: Formation, Structure, and Dynamical Evolution
Proceedings IAU Symposium No. 276, 2010
A. Sozzetti, M. G. Lattanzi & A. P. Boss, eds.

doi:10.1017/S1743921311020291

The origin of retrograde hot Jupiters

Smadar Naoz[1], Will M. Farr[1], Yoram Lithwick[1], Frederic A. Rasio[1] and Jean Teyssandier[1]

[1]Center for Interdisciplinary Exploration and Research in Astrophysics (CIERA), Northwestern University, Evanston, IL 60208, USA
email: snaoz@northwestern.edu

Abstract. Many hot Jupiters are observed to be misaligned with respect to the rotation axis of the star (as measured through the Rossiter–McLaughlin effect) and some (about ~ 25%) even appear to be in retrograde orbits. We show that the presence of an additional, moderately inclined and eccentric massive planet in the system can naturally explain close, inclined, eccentric, and even retrograde orbits. We have derived a complete and accurate treatment of the secular dynamics including both the key octupole-order effects and tidal friction. The flow of angular momentum from the inner orbit to the orbit of the perturber can lead to both high eccentricities and inclinations, and even flip the inner orbit. In our treatment the component of the inner orbit's angular momentum perpendicular to the stellar equatorial plane can change sign; a brief excursion to very high eccentricity during the chaotic evolution of the inner orbit can then lead to rapid "tidal capture," forming a retrograde hot Jupiter. Previous treatments of the secular dynamics focusing on stellar-mass perturbers would not allow for such an outcome, since in that limit the component of the inner orbit's angular momentum perpendicular to the stellar equatorial plane is strictly conserved. Thus, the inclination of the planet's orbit could not change from prograde to retrograde.

Keywords. planets and satellites: formation, planetary systems: formation

1. Introduction

The search for extrasolar planets has led to the surprising discovery of many Jupiter-like planets in very close proximity to their host star (e.g., Nelson 2001), the so-called "hot Jupiters" (hereafter HJ). Even more surprisingly, many of these hot Jupiters have orbits that are eccentric or highly inclined with respect to the rotation axis of the star (as measured through the Rossiter–McLaughlin effect (Gaudi & Winn 2007)) and some (8 out of 32) appear to be in *retrograde* orbits (Triaud *et al.* 2010). How they get so close to the star in such orbits remains an open question. Slow migration though a protoplanetary disk (e.g., Lin & Papaloizou 1986 and Masset & Papaloizou 2003) would produce orbits with low eccentricities and inclinations. However, this mechanism cannot explain the recent observations. Some models, such as those of Fabrycky & Tremaine (2007) and Wu *et al.* (2007) invoke a companion star in the system, which perturbs the inner orbit and can produce increases in eccentricity and inclination but not retrograde orbits, i.e., the inclination of the inner orbit with respect to the total angular momentum cannot change sign.

2. Secular interaction between two giant planets

We define the orientation of the inner orbit so that a prograde (retrograde) orbit has $i_1 < 90°$ ($i_1 > 90°$), where i_1 is the inclination of the inner orbit with respect to the *total* angular momentum, assumed parallel to the stellar rotation axis. We note that the directly observed parameter is actually the *projected* angle between the spin axis of the star and the planet's angular momentum. This angle can be above 90° even if i_1 is prograde (i.e., below, but close to 90°; see Fabrycky & Tremaine 2007, and also Fig. 1(c)). However, for simplicity we focus here on the physical angle i_1 (see section 4 for discussion and comparison with observations).

We assume a hierarchical configuration, where the star and Jupiter-like planet form a relative tight binary, and with the outer planet in a much wider orbit than the inner one. In the secular approximation the orbits may change shape and orientation but the semi-major axes (SMA) are strictly conserved in the absence of tidal dissipation (e.g., Ford *et al.* 2000; Eggleton *et al.* 1998). In particular, the Kozai-Lidov mechanism (Kozai 1962; Lidov 1962) produces large-amplitude oscillations of the eccentricity and inclination when the initial relative inclination between the inner and outer orbits is sufficiently large ($40° \lesssim i \lesssim 140°$, for initial eccentricity of the inner orbit close to zero). We have derived the secular evolution equations to octupole order using Hamiltonian perturbation theory (e.g., Kozai 1962; Harrington 1969; Krymolowski & Mazeh 1999; Ford *et al.* 2000). The octupole-level terms can give rise to fluctuations in the eccentricity maxima to arbitrarily high values (see Krymolowski & Mazeh 1999; Ford *et al.* 2000; Blaes *et al.* 2002), in contrast to the regular evolution in the quadrupole potential, where the amplitude of eccentricity oscillations is constant (e.g., Fabrycky & Tremaine 2007; Wu *et al.* 2007; Mazeh & Shaham 1979).

Unlike previous derivations of "Kozai-type" evolution, our treatment allows for changes in the z-components of the orbital angular momenta (i.e., the components along the total angular momentum) $L_{z,1}$ and $L_{z,2}$ (see Naoz *et al.* 2010, 2011 for more details). Many previous studies of secular perturbations in hierarchical triples considered a stellar-mass perturber, for which $L_{z,1}$ is very nearly constant (e.g., Mazeh & Shaham 1979; Fabrycky & Tremaine 2007; Wu *et al.* 2007; Perets & Naoz 2009). Moreover, the assumption that $L_{z,1}$ is constant has been built into previous derivations. However, this assumption is only valid as long as $L_2 \gg L_1$, which is not the case in comparable-mass systems (with two planets). Unfortunately, an immediate consequence of this assumption is that a prograde orbit can never be turned into a retrograde orbit.

3. Tidal friction

We adopt the tidal evolution equations of Eggleton, Kiseleva & Hut (1998), which are based on the equilibrium tide model of Hut (1981). The complete equations can be found in Fabrycky & Tremaine (2007, eqs A1–A5). Following their approach (see their eq. A10) we set the tidal quality factors $Q_{1,2} \propto P_{\rm in}$ [see also Hansen 2010, eq. (11)]. This means that the viscous times of the star and planet remain constant; the representative values we adopt here are 5 yr for the star and 1.5 yr for the planet, which correspond to $Q_\star = 5.5 \times 10^6$ and $Q_J = 5.8 \times 10^6$, respectively, for a 1-day period.

In figure 1 we show the secular dynamical evolution of a two-planet system with tidal dissipation. The inner planet becomes retrograde at 82 Myr, and remains retrograde after circularizing into a hot Jupiter. During each excursion to very high eccentricity for the inner orbit [marked with vertical lines in panels (b) and (d)], tidal dissipation becomes significant. Eventually the inner planet is tidally captured by the star and its orbit becomes decoupled from the outer body. After this point the orbital angular momenta remain nearly constant. The final SMA for the inner planet is at 0.024 AU, typical of a hot Jupiter. We monitor the pericenter distance of the inner planet to ensure that it always remains outside the Roche limit (Matsumura *et al.* 2010).

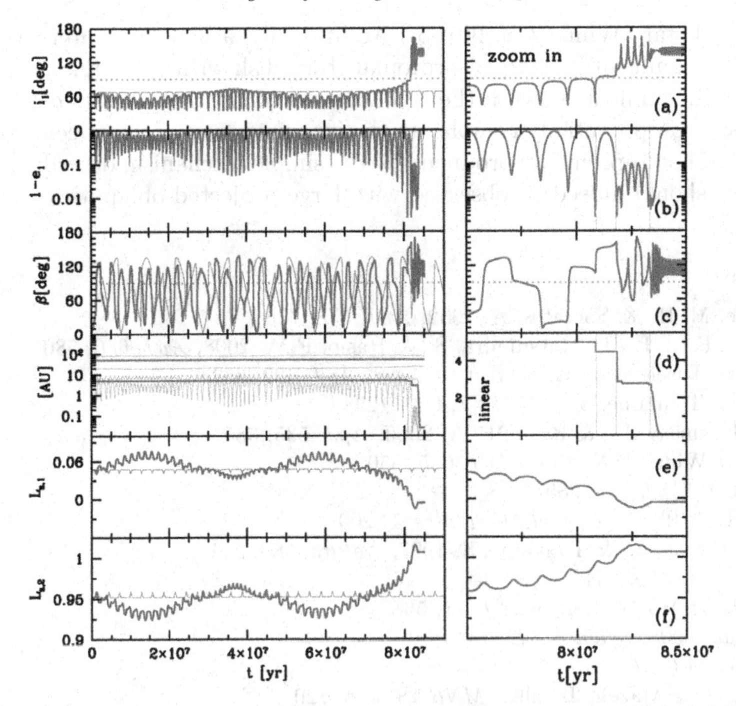

Figure 1. Dynamical evolution of a representative two-planet system with tidal dissipation included. Here the star has mass $1\,M_{\odot}$, the inner planet $1\,M_J$ and the outer planet $3\,M_J$. The inner orbit has SMA $a_1 = 5\,\mathrm{AU}$ and the outer orbit has SMA $a_2 = 51\,\mathrm{AU}$. The initial eccentricities are $e_1 = 0.001$ and $e_2 = 0.6$ and the initial relative inclination $i = 74.5°$. We show: (a) the inner orbit's inclination (i_1); (b) the eccentricity of the inner orbit (as $1 - e_1$); (c) the spin orbit angle β (d) the SMA, peri-, and apo-center distances for the inner orbit and the peri- and apo-center distance for the outer orbit; and, in (e) and (f) the z-components of the inner and outer orbit's angular momentum, normalized to the total angular momentum. The initial mutual inclination of 74.5° corresponds to inner- and outer-orbit inclinations of 67.6° and 6.9°, respectively. The thin curves shows the evolution in the quadrupole approximation (but including tidal friction), demonstrating that the octupole-order effects lead to a qualitatively different behavior.

4. Discussion - Comparison to observations

We showed that retrograde HJ orbits can be the result of a dynamical evolution in the presence of an additional moderately inclined and eccentric planet in the system. As mention above the observable parameter from the Rossiter–McLaughlin effect is the *projected* angle between the star's spin and the orbital angular momentum (the projected obliquity, Gaudi & Winn 2007). In Fig. 1 we show both the inclination i_1 [panel (a)] and also the angle between the star's spin axis and the planet's orbit β [panel (c)]. We note that, in contrast to the quadrupole order result, the spin orbit angle continues to oscillate at the same rate as the inclination, after the HJ is formed. However, the planet still appears to be in retrograde motion (i.e., $\beta > 90°$). Here instead defined retrograde orbits with respect to the total angular momentum, and we focused on the true angle between the orbital angular momentum of the inner planet and the invariable plane.

Projection effects can cause the true inclination and the spin-orbit angle to differ in magnitude, or even sign. Moreover, several mechanisms have been proposed in the literature that could, under certain assumptions, directly affect the spin axis of the star. These mechanisms can re-align the stellar spin axis through tidal interactions with either a slowly spinning star (Matsumura *et al.* 2010) or with the outer convective layer of a sufficiently cold star (Winn *et al.* 2010b). Additionally, a magnetic interaction between the star and a significantly charged protoplanetary disk with negligible accretion could also lead to misalignment between the stellar spin and the disk (Lai *et al.* 2011).

These effects can potentially complicate the interpretation of any specific observation. Nevertheless, if hot Jupiters are produced by the simple mechanism described here, many of their orbits should indeed be observed with large projected obliquities.

References

Blaes, O., Lee, M. H., & Socrates, A. 2002, *ApJ*, 578, 775
Chatterjee, S., Ford, E. B., Matsumura, S., & Rasio, F. A. 2008, *ApJ*, 686, 580
Eggleton, P. P., Kiseleva, L. G., & Hut, P. 1998, *ApJ*, 499, 853
Fabrycky, D. & Tremaine, S. 2007, *ApJ*, 669, 1298
Ford, E. B., Kozinsky, B., & Rasio, F. A. 2000, *ApJ*, 535, 385
Gaudi, B. S. & Winn, J. N. 2007, *ApJ*, 655, 550
Hansen, B. 2010, *ApJ*, 723, 285
Harrington, R. S. 1969, *Celestial Mechanics*, 1, 200
Holman, M., Touma, J., & Tremaine, S. 1997, *Nature*, 386, 254
Hut, P. 1998, *A&A*, 99, 134
Jefferys, W. H. & Moser, J. 1966, *AJ*, 71, 568
Kalas, P., *et al.* 2008, *Science*, 322, 1345
Kozai, Y. 1962, *AJ*, 67, 591
Krymolowski, Y. & Mazeh, T. 1999, *MNRAS*, 304, 720
Lai, D., Foucart, F., & Lin, D. N. C. 2011, *MNRAS*, 412, 2790
Lidov, M. L. 1962, *Planetary and Space Science*, 9, 719
Lin, D. N. C. & Papaloizou, J. 1986, *ApJ*, 309, 846
Marois, C., *et al.* 2008, *Science*, 322, 1348
Masset, F. S. & Papaloizou, J. C. B. 2003, *ApJ*, 588, 494
Matsumura, S., Peale, S. J., & Rasio, F. A. 2010, *ApJ*, 725, 1995
Mazeh, T. & Shaham, J. 1979, *A&A*, 77, 145
Nagasawa, M., Ida, S., & Bessho, T. 2008, *ApJ*, 678, 498
Naoz S., Farr W. M., Lithwick Y., Rasio F. A., & Teyssandier J. 2010, *Nature*, 473, 187
Naoz S., Farr W. M., Lithwick Y., & Rasio, F. A. 2011, in preparation
Nelson, R. P. 2001, *Solar and extra-solar planetary systems*, 577, 35
Perets, H. B. & Naoz, S. 2009, *ApJ*, 699, L17
Pollack, J. B., Hubickyj, O., Bodenheimer, P., Lissauer, J. J., Podolak, M., & Greenzweig, Y. 1996, *Icarus*, 124, 62
Schlaufman, K. C. 2010, *ApJ*, 712 602
Triaud, A. H. M. J., *et al.* 2010, *A&A*, 524, A25
Winn, J. N., Fabrycky, D., Albrecht, S., & Johnson, J. A. 2010, *ApJ*, 718 145
Wu, Y., Murray, N. W., & Ramsahai, J. M. 2007, *ApJ*, 670, 820

The Astrophysics of Planetary Systems: Formation, Structure, and Dynamical Evolution
Proceedings IAU Symposium No. 276, 2010
A. Sozzetti, M. G. Lattanzi, & A. P. Boss, eds.

doi:10.1017/S1743921311020308

Do falling planets cause stellar spin-up?

D. J. A. Brown[1], A. Collier Cameron[1], C. Hall[1] and L. Hebb[1]

[1]SUPA, School of Physics and Astronomy, University of St Andrews, North Haugh, St Andrews, Fife KY16 9SS, UK.
email: djab@st-andrews.ac.uk

Abstract. We investigate the tidal interactions between hot Jupiter extra-solar planets and their host stars in an effort to characterise the effects of such interactions on stellar rotation. We study the WASP-18 and WASP-19 systems, showing that in both cases tidal interactions cause the eventual spiral in of the the planet towards the Roche limit. We find that for both systems this process will cause significant spin up of the host star, independent of the precise value for the tidal quality factors. By fitting tidal evolution models to observed parameters, we are able to determine that WASP-19 b is currently spiralling in, and that it has a very short remaining lifetime ~ 3 Myr.

Keywords. stars: rotation, stars: activity, planetary systems

1. Introduction

As the number of known extra-solar planetary systems rises, focus is increasingly turning to characterisation of the physical, orbital and spectral properties of both the planets and their host stars. The varied interactions between stars and hot Jupiter planets are also coming under scrutiny in an attempt to improve our understanding of the formation and evolution of planetary systems.

Tidal interactions in the context of binary stars have been extensively studied (e.g. Hut 1980), but interest in their effect on planetary systems has intensified only recently. Previous studies have tended to focus on changes in the orbital parameters (e.g. Mardling & Lin 2002; Jackson *et al.* 2008; Barker & Ogilvie 2009; Jackson *et al.* 2009) whilst neglecting the rotation of the two bodies, which were shown to play an important role by Dobbs-Dixon *et al.* (2004). We present work investigating the effects of tidal interactions on the rotation of hot Jupiter host stars, and discuss the applicability of the gyrochronology dating method to hot Jupiter systems.

2. Gyrochronology

It is well documented that stars spin down during their main sequence lifetime through the action of magnetic braking. The standard form of this spin down, $P \propto 1/\sqrt{t}$ (Skumanich 1972), can be derived from $\dot{\Omega} \propto \Omega^3$, the simple stellar wind model of Weber & Davis (1967). This implies that the stellar rotation period can be used as a proxy for stellar age in the absence of any external influence on the stellar rotation.

Gyrochronology, a dating method that takes advantage of this, was developed by Barnes (2007) who showed that there exists a tight, universal relationship between stellar rotation period, age and colour for slowly rotating stars in open clusters. That work has since been used as the basis for studies of a range of stellar clusters (e.g. Collier Cameron *et al.* 2009; Meibom *et al.* 2009), all of which have been shown to follow similar relationships. The existence of such a relation for open clusters is not in doubt, but the applicability of the gyrochronology model to exoplanet host stars has not been examined in great detail.

3. Tidal model

We make use of the tidal equations of Dobbs-Dixon *et al.* (2004), adding the stellar wind model of Weber & Davis (1967) to provide a source magnetic braking. We integrate the equations forward in time from a set of initial conditions, ending the integration when the host star reaches its estimated main sequence lifetime or when the planet reaches the Roche limit. We evaluate the integration through a χ^2 test of the semi-major axis, eccentricity and stellar spin rate, with the addition of a Gaussian prior on the estimated stellar age to give a test statistic C. We take the best fit point in the integration to be the step at which C is a minimum, and take the system age to be the time at that step.

For simplicity we assume that the planet is initially tidally locked, but allow the system to evolve away from this state. We calculate the initial stellar rotation rate by scaling the period-colour relation for the Coma-Berenices cluster (Collier Cameron *et al.* 2009) to the initial time, which is governed by the observed stellar rotation period and estimated age. t_0 must be sufficiently early for all possible evolutionary scenarios to be explorable, but late enough that the demonstrable timescale for convergence of the period-colour relation is accounted for.

Our tidal evolution model is built into both a grid search scheme and a Markov Chain Monte Carlo (MCMC) algorithm, to explore a clearly defined, four-dimensional parameter space in $\log(Q'_s)$, $\log(Q'_p)$, a_0 and e_0. The upper limit on e_0 was defined to avoid the problems with high eccentricity systems discussed by Leconte *et al.* (2010). We assume monotonic shrinking of the planetary orbit with time, and do not account for stellar obliquity, limiting our investigation to systems that are well-aligned. The best solution returned by the grid search is taken to be that which returns the absolute minimum value of the C test statistic, whilst for the MCMC algorithm the best set of initial parameters are taken to be the median values of the respective individual posterior probability distributions.

For further details of our model we refer the reader to Brown *et al.* (2011). Here we present the key results from our investigation into the histories of the WASP-18 (Hellier *et al.* 2009) and WASP-19 (Hebb *et al.* 2010) planetary systems. These systems were selected owing to pre-existing discrepancies between the age estimates obtained from gyrochronology and stellar model fitting.

4. WASP-19

The initial time was set to 0.6 Gyr, approximately the age of the Hyades cluster, to limit computation time and ensure that the period-colour relation of Barnes (2007) had converged. The primary region of interest suggested by the grid search was found to contain unphysical solutions caused by the combination of tidal quality factors in that section of our parameter space. We therefore restricted the $\log(Q'_s)$ parameter space available to the MCMC algorithm to the secondary region of interest found by the grid search.

Integrating the resulting median MCMC solution indicates that WASP-19 A has an age close to 3 Gyr, and that the planet has an extremely short remaining lifetime, of the order of 2 Myr, until it reaches the Roche limit. Figure 4 shows that the planet is in the process of its final spiral in, and that the star has been spun up dramatically in the recent past. We investigated a range of tidal quality factors for the same initial eccentricity and

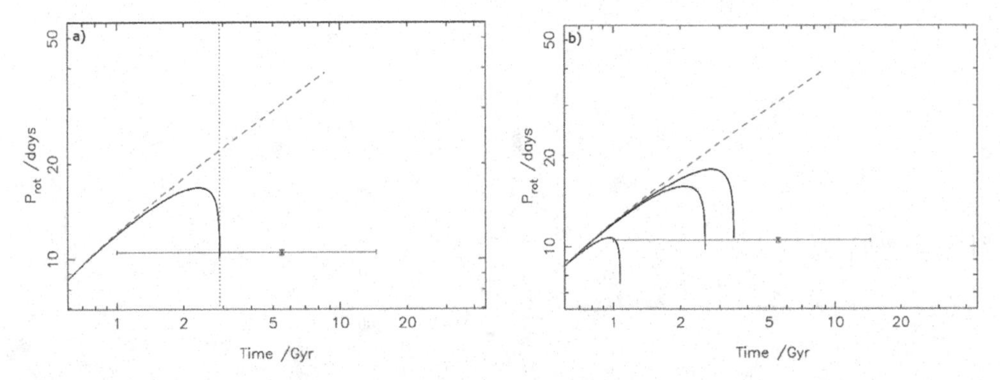

Figure 1. a): Stellar rotation period as a function of time for the median solution obtained by the MCMC algorithm (black, solid line) for the WASP-19 exoplanetary system. Also plotted is the evolution expected from magnetic braking only (red, dashed line). Note the proximity of the observed data (blue datum and error bars) to the end of the evolutionary track, implying a very short remaining lifetime for WASP-19 b. The dotted line denotes the best fit system age derived from the evolutionary track. **b):** Stellar rotation period as a function of time for a range of $\log(Q'_s)$ values consistent with the solution from the MCMC routine. All tracks have the same values of a_0, e_0 and $\log(Q'_p)$. In all cases a short remaining lifetime for the planet is implied, and the system has an age that is inconsistent with the estimate obtained using gyrochronology.

semi-major axis, finding a very short ($< 3\,$Myr) remaining lifetime for WASP-19 b in all cases that were consistent with the observed system parameters. The same evolutionary histories all exhibited a significant, very rapid spin up of WASP-19 A's rotation as the hot Jupiter spiraled in to the Roche limit (figure 4).

The previous lower age limit of 1 Gyr derives from stellar model fitting, and from comparison of the space velocity of WASP-18 A to a simulated population of stars with similar properties (Hebb *et al.* 2010). This robust limit forces all of the evolutionary tracks to fit the observed period at ages that are entirely inconsistent with the gyrochronology estimate, indicating that stellar spin up must have taken place.

5. WASP-18

For the WASP-18 system we set $t_0 = 0.15\,$Gyr, approximately the age of the M35 cluster, to permit the full range of possible solutions to be explored. Integrating the median solution returned by the MCMC algorithm (figure 5) implies that WASP-18 is in a similar situation to WASP-19; a planet on the verge of beginning its final inspiral to the star, a short remaining lifetime for the planet, and a stellar age inconsistent with gyrochronology. This result is somewhat misleading however, as the uncertainty on the age estimate derived from stellar model fitting is such that the star might still be in the magnetic braking only regime of its rotational evolution, and thus gyrochronology might still be applicable.

Figure 5 shows the evolution of the stellar rotation period for a range of $\log(Q'_s)$ values consistent with the MCMC result, and with the same a_0, e_0 and $\log(Q'_p)$. These tracks imply a large range of potential ages for the star, some of which suggest that tidal effects have yet to affect its rotation and that it therefore has a young age consistent with gyrochronology. The evolutionary status of WASP-18 b is therefore considerably less certain than that of WASP-19 b, but it is important to note that all of the tracks show considerable spin-up of the host star towards the end of the planet's lifetime.

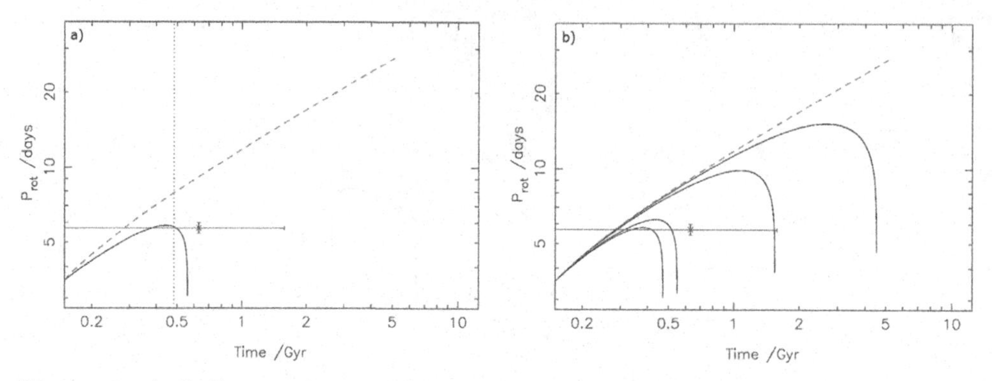

Figure 2. a): Stellar rotation as a function of time for the WASP-18 system. Legend as for figure 1. The track implies that the hot Jupiter is on the verge of spiralling-in, and that the star will be significantly spun up whilst this occurs. **b):** As figure 4. For this system the range of evolutionary histories suggest a large range of possible ages, some of which are consistent with the gyrochronological estimate. In all cases stellar spin up is implied at the end of the planet's lifetime.

6. Conclusions

From our study of the WASP-18 and WASP-19 exoplanetary systems we are able to conclude that tidal interactions between hot Jupiters and their host stars can have a strong effect on rotation of the host star, with the final infall of the planet towards the Roche limit causing notable spin up. This will manifest itself in stellar age estimates, with stars appearing younger as a result of their rotation than implied by their other properties. This signature could potentially be used to detect planet hosting stars.

We also conclude that although gyrochronology is a useful tool, due care should be taken with its application to hot Jupiter systems, as tidal spin up will cause the star to appear much younger than its true age if the technique is applied blindly. This is aptly illustrated by the WASP-19 system, which is certainly older than implied by its rotation. Conversely, the WASP-18 system shows that gyrochronology can give usful indications as to a hot Jupiter host's age, particularly where existing methods appear to give results consistent with the technique.

References

Barker, A. J. & Ogilvie, G. I. 2009, *MNRAS*, 395, 2268
Barnes, S. A. 2007, *ApJ*, 669, 1167
Brown, D. J. A, Collier Cameron, A., Hall, C., Hebb, L., & Smalley, B. 2011, *MNRAS*, 415, 605
Collier Cameron, A., Davidson, V. A., Hebb, L., Skinner, G., *et al.* 2009, *MNRAS*, 400, 451
Dobbs-Dixon I., Lin D. N. C. & Mardling R. A. 2004, *ApJ*, 610, 464
Hebb, L., Collier-Cameron, A., Triaud, A. H. M. J., Lister, T. A., *et al.* 2010, *ApJ*, 708, 224
Hellier, C., Anderson, D. R., Cameron, A. C., Gillon, M., *et al.* 2009, *Nature*, 460, 1098
Hut, P. 1980, *A&A*, 92, 167
Jackson, B., Greenberg, R., & Barnes, R. 2008, *ApJ*, 678, 1396
Jackson, B., Barnes, R., & Greenberg, R. 2009, *ApJ*, 698, 1357
Leconte, J., Chabrier, G., Baraffe, I., & Levrard, B. 2010, *A&A*, 516, A64
Mardling, R. A. & Lin, D. N. C. 2002, *ApJ*, 573, 829
Meibom, S., Mathieu, R. D., & Stassun, K. G. 2009, *ApJ*, 695, 679
Skumanich, A. 1972, *ApJ*, 171, 565
Sozzetti, A., Torres, G., Charbonneau, D., Latham, *et al.* 2007, *ApJ*, 664, 1190
Weber, E. J. & Davis Jr., L. 1967, *ApJ*, 148, 217

The Astrophysics of Planetary Systems: Formation, Structure, and Dynamical Evolution
Proceedings IAU Symposium No. 276, 2010
A. Sozzetti, M. G. Lattanzi & A. P. Boss, eds.

doi:10.1017/S174392131102031X

Orbital migration models under test

Wilhelm Kley[1]

[1]Institut für Astronomie & Astrophysik
Universität Tübingen, Morgenstelle 10, 72076 Tübingen, Germany
email: wilhelm.kley@uni-tuebingen.de

Abstract. Planet-disk interaction predicts a change in the orbital elements of an embedded planet. Through linear and fully hydrodynamical studies it has been found that migration is typically directed inwards. Hence, this migration process gives natural explanation for the presence of the 'hot' planets orbiting close to the parent star, and it plays a mayor role in explaining the formation of resonant planetary systems.

However, standard migration models for locally isothermal disks indicate a too rapid inward migration for small mass planets, and a large number of massive planets are found very far away from the star. Recent studies, including more complete disk physics, have opened up new paths to slow down or even reverse migration. The new findings on migration are discussed and connected to the observational properties of planetary systems.

Keywords. planetary systems: formation, accretion disks, hydrodynamics

---

1. Introduction

The orbital elements of the observed extrasolar planets are distinctly different from the solar system. The major solar system planets have a nearly coplanar configuration and orbits with small eccentricity. In contrast, the exoplanet population displays large eccentricities, and many planets orbit their host star on very tight orbits. Recently, it has been discovered that inclined and retrograde orbits are quite frequent as well, at least for close-in planets. Historically, it was exactly the relatively 'calm' dynamical structure of the solar system that led to the hypothesis that planets form in protoplanetary disks. Within the framework of the sequential accretion scenario planet formation proceeds along a series of substeps, growing from small dust particles all the way to the giant gaseous planets. The discovery of the hot planets, which could not have been formed in-situ due to the hot temperatures and limited mass reservoir, gave rise to the exploration of dynamical processes that are able to change the location of planets in the disk, accompanying the regular formation process.

In the context of moons embedded in the ring system of Saturn it had been noted that disk-satellite interaction can alter the orbital elements of the perturbing moon, in particular its semi-major axis (Goldreich & Tremaine 1980). Later it was recognized that via the very same process, operating between a young embedded planet and the protoplanetary disk, it is possible to bring a planet that has formed at large distances from the star to its close proximity (Ward 1986). Hence, this migration process has provided a natural explanation for the population of hot planets, and their mere existence has been considered as evidence for the migration process. At the same time it was noted that planet-disk interaction may lead to eccentricity as well as inclination damping (Ward 1988; Ward & Hahn 1994).

Another indication for a planetary migration process comes from the high fraction (nearly 20%) of configurations in a low order mean-motion resonance, within the whole sample of multi-planet systems. As the direct formation of such systems seems unlikely, only a dissipative process that changes the energy (semi-major axis) is able to bring planets from their initial non-resonant configuration into resonance. Since resonant capture excites the planetary eccentricities typically to large values in contradiction to the archetypical system GJ 876, it has been inferred that planet-disk interaction should lead to eccentricity damping (Lee & Peale 2002).

However, population synthesis models strongly indicated that the standard migration scenario yields very rapid inward migration rates that appear to be in disagreement with the observations. Additionally, recent observations of close-in planets on eccentric and inclined orbits have questioned the general validity of the migration paradigm to form them. In this review, I will first explain the basic mechanism of migration, present new findings on the migration rate, and then discuss its applicability with respect to the overall planet formation process.

2. Origin of migration

An embedded object disturbs the ambient disk dynamically in two important ways: First it divides the disk into an inner and outer disk separated by a coorbital (horse shoe) region. Secondly, the propagating sound waves that are sheared out by the Keplerian differential rotation generate density waves in the form of spiral arms in the disk. The created structures in the perturbed coorbital region and in the spiral arms back-react on the planet and cause a change in its semi-major axis. Thus, physically speaking, planetary migration is caused by the effect of spiral arms and corotation region. Let us discuss these effects in turn.

Spiral arms: To put it simple, the spiral arms can be considered as density enhancements in the disk that 'pull' gravitationally on the planet. This gives rise to so called Lindblad torques that change the planet's angular momentum. For circular orbits the disk torque exerted on the planet is directly a measure of the speed and direction of migration. The inner spiral forms a leading wave that causes a positive torque, while the outer wave generates a negative contribution. The combined effects of both spirals determine then the sign and magnitude of the total torque. A positive total torque will add angular momentum to the planet and cause outward migration. On the other hand, a negative torque will induce inward migration. It turns out that under typical physical disk conditions the contributions of the inner and outer spiral arm are comparable in magnitude. However, the effect of the outer spiral quite generally wins over the inner one causing the planet to migrate inward.

Corotation region: As viewed in the corotating frame, material within the corotation region performs so called horseshoe orbits. Here, the gas particles upon approaching the planet at the two ends of the horseshoe are periodically shifted from an orbit with a semi-major axis slightly larger than the planetary one to an orbit with slightly smaller value, and vice versa. Hence, at each close approach with the planet there is an exchange of angular momentum between (coorbital) disk material and the planet. The total corotation torque is then obtained by adding the contributions from both ends of the horseshoe. To obtain a net, non-zero torque requires non-vanishing radial gradients of vortensity and entropy across the corotation region (Baruteau & Masset 2008). For an ideal gas without friction or heat diffusion mixing effects within the horseshoe tend to flatten out these gradients yielding a vanishing corotation torque, or so called torque saturation.

2.1. *Type-I migration*

Small mass planets do not alter the global disk structure significantly, in particular they do not open a gap within disk. Hence, the combined effect of Lindblad and corotation torques can be calculated for small planetary masses using a linear analysis. The outcome of such linear, no-gap studies has been termed type-I planet migration. Due to the complexity of considering heat generation and transport in disks these studies have relied nearly exclusively on simplified, locally isothermal disk models. Here, the temperature is assumed to be independent of height and is given by a pre-described function of radius, $T = T(r)$. Typically, it is assumed that the relative scale height H of the disk is a constant, $H/r = const.$, yielding $T \propto r^{-1}$. The total torque Γ_{tot} is given as the sum of Lindblad and corotation torque $\Gamma_{tot} = \Gamma_L + \Gamma_{CR}$. The speed of the induced linear, type-I migration scales inversely with the disk temperature (i.e. disk thickness) as $\propto (H/r)^{-2}$, linear with the planet mass $\propto m_p$, and with the disk mass $\propto m_d$. Linear models have been calculated for flat 2D disks as well as full 3D configurations. The problem of 2D simulations lies in taking into account approximately the neglected vertical stratification of the disk, which is typically done through a smoothing of the gravitational potential near the planet. Additional problems arise when considering radial gradients. Hence, 2D and 3D results may well yield agreeing migration rates at a particular radius but opposite dependence on radial gradients. Additionally, non-linear effects may set in already at a planetary mass of about 10 earth masses. New full 3D, nested grid locally isothermal hydrodynamic simulations of planet-disk interaction give very good agreement with previous 3D linear results (Tanaka & Ward 2004) and yield the following form for the total torque for small mass planets below about 10 M_{Earth} (D'Angelo & Lubow 2010)

$$\Gamma_{tot} = -(1.36 + 0.62\alpha_\Sigma + 0.43\alpha_T) \left(\frac{m_p}{M_*}\right)^2 \left(\frac{H}{r_p}\right)^{-2} \Sigma_p \, r_p^4 \, \Omega_p^2. \tag{2.1}$$

In eq. (2.1) the index p refers to the planet, α_Σ and α_T refer to the radial variation of density and temperature, such that $\Sigma(r) \propto r^{-\alpha_\Sigma}$ and $T(r) \propto r^{-\alpha_T}$.

2.2. *Type-II migration*

For larger planet masses, a gap is opened in the disk because the planet transfers angular momentum to the disk, positive exterior and negative interior to the planet. The depth of the gap that the planet carves out depends for given disk physics (temperature and viscosity) only on the mass of the planet (Fig. 1, left panel). Because the density in the coorbital region is reduced, the corotation torques are strongly affected and are no longer of any importance for larger planet masses. For very large masses even the Lindblad torques are reduced yielding a slowing down of the planet (Fig. 1, right panel). This non-linear regime has been coined the type-II regime of planetary migration; here the drift of the planet is dominated by the disk's viscous evolution. The dip in the migration rate at around $m_p = 10 M_{Earth}$ has been discovered by Masset *et al.* (2006), and can be attributed to a change in the flow structure in the vicinity of the planet due to non-linear effects.

2.3. *Migration in radiative disks*

The migration rates obtained through linear analysis as well as fully non-linear hydrodynamical models have resulted in approximate formulae for the migration speed $\dot{a}$ of a planet (as quoted above for small mass regime, eq. 2.1) that are frequently used in population synthesis models, i.e. growth models of planets that include disk evolution and planet migration in parameterized form as well. These synthesis models can be used to calculate the final location of many planets in the mass/semi-major axis diagram and compare the results statistically with the observed distribution. Through variation of individual parameter of the model, their relative importance can be estimated. The results indicate in particular, that the migration for small mass planets in the type-I regime is

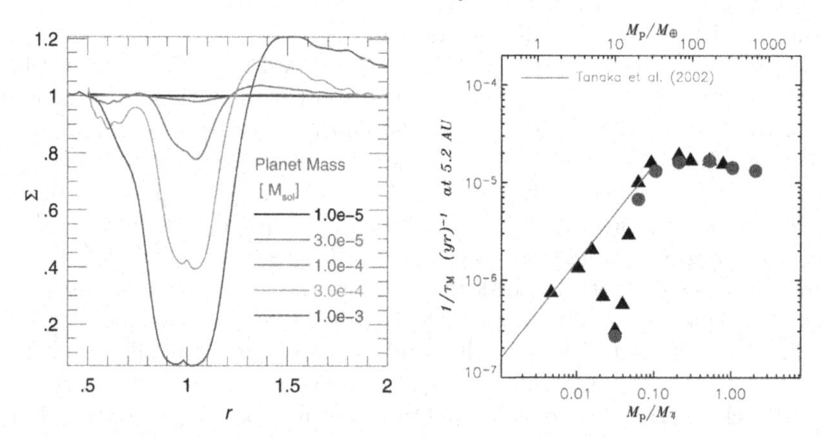

Figure 1. Left: Azimuthally averaged surface density profile of the disk for a variety of planet masses, quoted in solar masses. The density structure is obtained for an isothermal disk with $H/r = 0.05$ using a constant viscosity, equivalent to a $\alpha = 0.004$ at the radius of the planet, as obtained with 2D hydrodynamic simulations. **Right**: The migration rate, quoted in terms of the migration time scale $\tau = a/\dot{a}$ for different planet masses. The results refer to 3D nested grid simulations where the symbols denote different grid layouts. The red solid line denotes the result of Tanaka *et al.* (2002).

by far too fast to account for the observed distribution, the majority of planets would have been lost to the star, as shown for example by Ida & Lin (2008) and Mordasini *et al.* (2009). Only a significant reduction in the type-I migration speed gives satisfactory results. Suggested remedies included: stochastic migration of a planet in a turbulent disk, migration in inviscid self-gravitating disks or nonlinear effects.

Here, we shall concentrate on a very simple and straight forward improvement of the models, that represent a possible solution to the type-I migration problem: the inclusion of more accurate physics. As mentioned above, past modeling relied nearly exclusively on the simplified locally isothermal models, which has the advantage that no energy equation has to be considered. Taking more realistic thermodynamics into account requires the incorporation of a heating and cooling mechanism. The importance of radiative diffusion has first be pointed out by Paardekooper & Mellema (2006), and recent papers quote approximate formulas for viscous and diffusive disks (Masset & Casoli 2010; Paardekooper *et al.* 2011). To demonstrate the effect for two-dimensional flat disks, we show results for a planet-disk simulation where we include viscous heating, local radiative cooling as well diffusive radiative transport in the disk's plane. The energy equation then reads

$$\frac{\partial \Sigma c_v T}{\partial t} + \nabla \cdot (\Sigma c_v T \mathbf{u}) = -p\nabla \cdot \mathbf{u} + D - Q - 2H\nabla \cdot \vec{F} \tag{2.2}$$

where Σ is the surface density, T the midplane temperature, p the pressure, D the viscous dissipation, Q the radiative cooling and $\vec{F}$ the radiative flux in the midplane. Models where the various contributions on the right hand side of eq. (2.2) were selectively switched off and on, have been constructed by Kley & Crida (2008).

The effect of this procedure on the resulting torque is shown in the left panel of Fig.2. The basis for all the models is the same equilibrium disk model constructed using all terms on the rhs. of eq. (2.2) and no planet. Embedding a planet of $20M_{earth}$ yields in the long run positive torques only for the radiative disks, where the maximum effect is given when only viscous heating and local radiative cooling are considered. The inclusion of diffusion in the disk midplane yield a slightly reduced torque. Since the initial state consists of a non-vanishing negative radial entropy gradient, the adiabatic model shows a positive

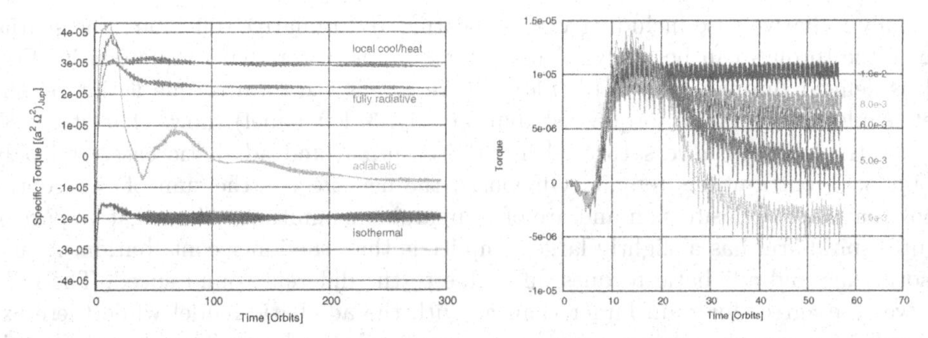

Figure 2. Time evolution of the torque acting on a 20 M_{earth} planet embedded in a disk with 0.01 solar masses with a radial range of 0.4 to 2.5 r_{Jup}. **Left**: Simulations using a constant value of the viscosity. From bottom to top the curves indicate simulations where *i*) no energy equation (isothermal), *ii*) only the first term on the rhs. of eq. (2.2) (adiabatic), *iii*) all terms on the rhs. (fully radiative), and *iv*) all but the last term on the rhs. (heating/cooling), have been used. (after Kley & Crida, 2008) **Right**: The influence of viscosity on the resulting torque for fully radiative simulations using an alpha type viscosity. In the simulations the dissipation has been kept fixed, varying only the value of α in the momentum equation.

torque during the first 20 orbits directly after insertion of the planet. However, in the long run the torque becomes negative as in the isothermal case because the material within the horseshoe region is mixed thoroughly, wiping out the entropy gradient. Hence, the positive corotation torque is obliviated and the negative Lindblad contribution dominates. The adiabatic case does not approach the isothermal result because the corresponding sounds speeds are different. The right hand panel of Fig.2 demonstrates that a non-vanishing value of the viscosity is necessary to maintain torque desaturation. These models have used an α-type viscosity in contrast to the results displayed on the left side. To compare results directly, it has been ensured that the thermal state of the models is identical despite the different value of α used in the angular momentum equation.

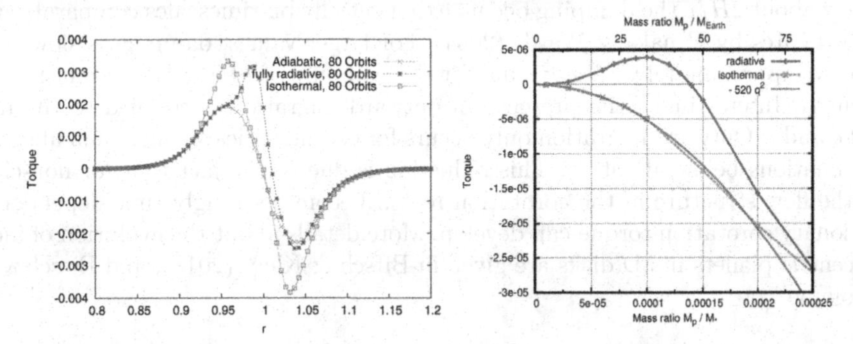

Figure 3. Left: Radial torque density, $\Gamma(r)$, for fully 3D radiation-hydrodynamical simulations with an embedded $20M_{earth}$ planet. The data are shown after equilibrium has been reached. Different assumptions for the thermodynamical state are made. (after Kley *et al.* 2009) **Right**: The total torque acting on the planet for a 2D fully radiative simulation for a variety of planet masses. (after Kley & Crida, 2008)

2.4. *Fully three-dimensional radiative disks*

The previous 2D cases have been repeated using full 3D radiative disk simulations with an identical physical setup. These clearly confirmed the existence of outward migration in the presence of radiative diffusion. To compare isothermal, adiabatic and fully radiative simulations, all start from the same initial state which corresponds to an equilibrium of the fully radiative case including viscous heating and radiative diffusion. The spatial origin of the torques can be analyzed using for example the radial torque density $\Gamma(r)$ which is defined through $\Gamma_{tot} = \int \Gamma(r) dr$, where T_{tot} is the total torque acting on the planet. A plot of the radial torque distribution (Fig. 3, left panel) shows that at $t = 80$ the corotation torques have saturated in the isothermal and adiabatic case, and only the Lindblad contributions remain. Obviously, the net effect is the sum of two contributions that have opposite sign and are of comparable magnitude. The negative part of the outer spiral arm has a slightly larger amplitude than the inner contribution. Again, the isothermal and adiabatic torques differ due to the different sound speed. The fully radiative case agrees for radii larger than a_p with the adiabatic model while there exists a well pronounced torque maximum just inside of the planet. This contribution is responsible for the torque reversal. The right panel of Fig. 3 shows that the strength of this positive corotation effect also scales with the square of the planet mass up to about 20 to 25 M_{earth}. Beyond this mass, gap opening begins and only the Lindblad torques remain, and above 40 M_{earth} planets begin to migrate inwards again. In the full 3D simulations the results are qualitatively the same, the averaging procedure, that is necessary in 2D, leads to some quantitative differences (Kley *et al.* 2009). Interestingly, the full 3D results show even a stronger effect. New population synthesis models based on the modified migration rates indicate better agreement with the observational data set (eg. Mordasini, this volume).

3. Eccentricity and inclination

In addition to a change in semi-major axis, planet-disk interaction will modify the planetary eccentricity (e) and inclination (i) as well. Extending previous isothermal studies by Cresswell *et al.* (2007), fully 3D radiative disk simulations have been performed recently to study the evolution e and i. The results indicate that both are damped for all planetary masses (Bitsch & Kley 2010, and this volume). For small values of e and i that are below about $2H/r$ the damping occurs exponentially on timescales comparable to the linear estimates by Tanaka & Ward (2004). For larger values, damping is slowed down and follows approximately $\dot{e} \propto e^{-2}$, and for the damping of i an identical relation holds, surprisingly. Interestingly, the presence of outward migration is coupled to the magnitude of e and i. Outward migration only occurs for eccentricities smaller than about 0.02, and inclinations below about 4°. This reduction is due to the fact that for non-circular orbits the flow structure in the corotation region becomes strongly time dependent and no stationary corotation torque can develop. More details about the evolution of inclined and eccentric planets in 3D disks are given in Bitsch & Kley (2010), and Bitsch & Kley (this volume).

4. Resonant systems

The mere existence of resonant planetary systems is a strong indication that dissipative mechanisms changing the semi-major axis of planets must have operated, as the likelihood of forming planets in these configurations in situ or later through scattering processes seems to be small. On the other hand, convergent differential migration of a pair of planets will lead under quite general conditions to capture into resonance. The presence of an inner disk in combination with realistic inner boundaries leads to very good agreement with the observations for the best observed system GJ 876, as demonstrated in Crida *et al.* (2008) and shown in Fig. 4. Through transit timing measurements the first double eclipsing system, Kepler 9, has been discovered, and the data clearly indicate a low order

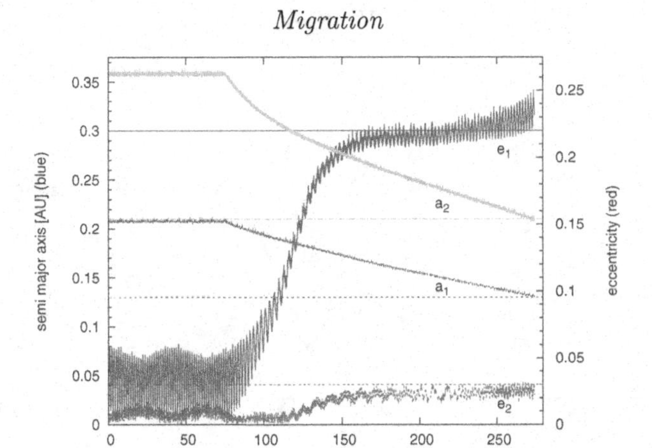

Figure 4. Evolution of the semi-major axis a and eccentricity e of a pair of planets with physical parameter resembling the observed planetary system GJ 876. For the initial 75 years the planets are kept fixed and thereafter they are allowed to migrate freely within the disk. The index 1 refers to the inner and 2 to the outer planet. The horizontal lines refer to the observed values of GJ 876. The evolution has been stopped after the planets have reached the observed distances from the star. (after Crida *et al.* (2008))

mean motion resonance, probably 2:1 (Holman *et al.* 2010). The proximity of the planets to the star and the near coplanarity of the system strongly hints towards a migration scenario for the formation of the system.

The importance of dynamical migration models for systems of planets is indicated by the system HD 45364, where two planets engaged in a 3:2 resonance have been discovered by Correia *et al.* (2009). The inferred orbital parameters for the two planets are semi-major axes of $a_1 = 0.681$AU and $a_2 = 0.897$AU, and eccentricities of $e_1 = 0.168$ and $e_2 = 0.097$, respectively. Fully non-linear hydrodynamical planet-disk models have been constructed for this system by Rein *et al.* (2010). For suitable disk parameter, the planets enter indeed into the 3:2 resonance through a convergent migration process. After the planets have reached their observed semi-major axis, a theoretical RV-curve has been calculated. Surprisingly, even though the hydrodynamically obtained eccentricities ($e_1 = 0.036, e_2 = 0.017$) are quite different, the model fits the observed data point equally well as the published data, see Fig. 5 and Rein *et al.* (2010). This pronounced dynamical difference between the two models that match the observations equally well, can only be resolved with more observational data. Hence, the system HD 45364 is a very good example that it is urgently necessary to have sufficient and better observational data for interacting multi–planet systems to obtain the orbital parameters, that are required to constrain the theoretical models.

5. Summary

We have shown that just by inclusion of more accurate physics, namely radiation transport, it is possible to reduce significantly the otherwise too rapid type I migration. The effect is driven by corotation torques, and the requirement to maintain the horseshoe torques unsaturated is the presence of viscosity and radiative cooling (or diffusion), with timescales of the order of the libration time of the horseshoe material near the separatrix. Including this effect, population synthesis models yield better agreement with the observations. We find that eccentricity and inclination are typically damped by planet-disk interaction for all planet masses.

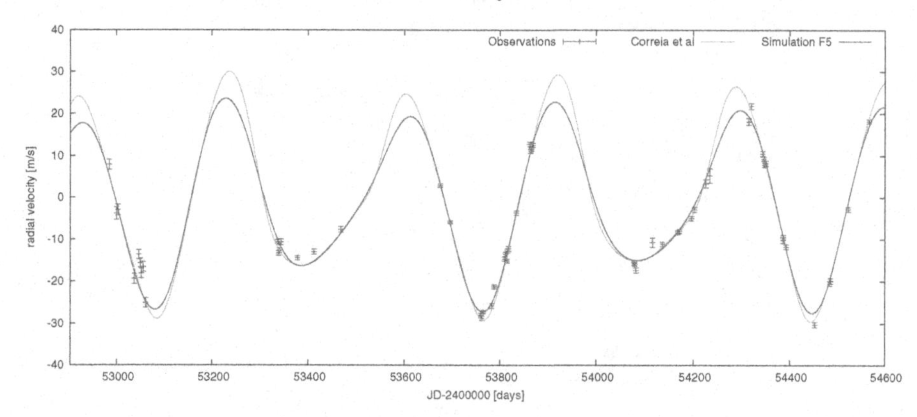

Figure 5. Synthetic radial velocity curves HD 45364 together with observational data points. The light green curve is the fit as given in the discovery paper (Rein *et al.* (2010)). The darker blue curve has been obtained by performing full hydrodynamical simulation of the system. Both curves have comparable χ^2 values.

Differential migration in multi-planet systems frequently result in resonant capture. Upon capture the eccentricity of the planets is strongly increased. For continued migration the systems remain stable only when eccentricity is damped by the disk. We have shown that the formation of the systems GJ 876 and HD 45364 can naturally be explained by planet-disk interaction migration scenarios. For resonant systems where the disk action is reduced in the final stages of the planet formation process the left over configuration may result in an unstable system. The last phase will then be dominated by scattering processes which may pump up the planetary eccentricities and possibly their inclinations to the large values observed.

References

Baruteau, C. & Masset, F. 2008, *ApJ*, 672, 1054
Bitsch, B. & Kley, W. 2010, *A&A*, 523, A30
Correia, A. C. M., *et al.* 2009, *A&A*, 496, 521
Cresswell, P., Dirksen, G., Kley, W., & Nelson, R. P. 2007, *A&A*, 473, 329
Crida, A., Sándor, Z., & Kley, W. 2008, *A&A*, 483, 325
D'Angelo, G. & Lubow, S. H. 2010, *ApJ*, 724, 730
Goldreich, P. & Tremaine, S. 1980, *ApJ*, 241, 425
Holman, M. J., *et al.* 2010, *Science*, 330, 51
Ida, S. & Lin, D. N. C. 2008, *ApJ*, 673, 487
Kley, W., Bitsch, B., & Klahr, H. 2009, *A&A*, 506, 971
Kley, W. & Crida, A. 2008, *A&A*, 487, L9
Lee, M. H. & Peale, S. J. 2002, *ApJ*, 567, 596
Masset, F. S., D'Angelo, G., & Kley, W. 2006, *ApJ*, 652, 730
Masset, F. S. & Casoli, J. 2010, *ApJ*, 723, 1393
Mordasini, C., Alibert, Y., Benz, W., & Naef, D. 2009, *A&A*, 501, 1161
Paardekooper, S.-J. & Mellema, G. 2006, *A&A*, 459, L17
Paardekooper, S.-J., Baruteau, C., & Kley, W. 2011, *MNRAS*, 410, 293
Rein, H., Papaloizou, J. C. B., & Kley, W. 2010, *A&A*, 510, A4
Tanaka, H., Takeuchi, T., & Ward, W. R. 2002, *ApJ*, 565, 1257
Tanaka, H. & Ward, W. R. 2004, *ApJ*, 602, 388
Ward, W. R. 1986, *Icarus*, 67, 164
Ward, W. R. 1988, *Icarus*, 73, 330
Ward, W. R. & Hahn, J. M. 1994, *Icarus*, 110, 95

The Astrophysics of Planetary Systems: Formation, Structure, and Dynamical Evolution
Proceedings IAU Symposium No. 276, 2010
A. Sozzetti, M. G. Lattanzi, & A. P. Boss, eds.

doi:10.1017/S1743921311020321

Direct imaging of massive extrasolar planets

Paul Kalas[1]

[1]Astronomy Department, University of California
Berkeley, CA 94720, USA
email: kalas@berkeley.edul

Abstract. The direct detection of an extrasolar planet can provide accurate measurements of its orbit, mass and composition, greatly improving our understanding of how planets form and evolve. Recent advances in ground-based and space-based imaging techniques have now produced the first direct images of extrasolar planets. Typically these are many-Jupiter-mass planets on wide orbits. Direct imaging therefore probes the outer architecture of planetary systems and it is highly complementary to other techniques sensitive to inner architectures. This brief review summarizes the properties of the currently imaged exoplanets, provides an update on the orbit of Fomalhaut *b*, and highlights the emerging phenomenon of circumplanetary disks.

Keywords. planetary systems, planets and satellites: rings, accretion, accretion disks, interplanetary medium, stars: individual (Fomalhaut)

---

1. Introduction

The direct imaging of extrasolar planets has been one of the late bloomers among the various detection techniques. Direct imaging could of course refer to any method that measures photons that interact with an extrasolar planets. Therefore the thermal infrared light curves that give secondary eclipse light curves, primary eclipse spectroscopy, or some periodic polarization signal might be considered a sort of image for an extrasolar planet (e.g., see Seager & Deming 2010 for a review).

This review covers the scientific findings from recent observations where the planet signal is spatially resolved from the star. Readers interested in understanding the technical problems and the current state-of-the-art may refer to reviews by Duchene (2008) and Oppenheimer & Hinkley (2009).

Table 1 gives a snapshot of the currently observed extrasolar planets arranged in order of increasing heliocentric distance (column 3). I will briefly discuss each of the topics in the table columns, provide an update on Fomalhaut b, and touch on several aspects of circumplanetary material. My approach is to make this information memorable for astronomy students, but with enough detail to make it valuable for more senior scientists.

2. Host and Spectral Type

Probably the most common question asked by scientists listening to talks on directly imaged planets is the following: "Has anyone obtained RV data on these stars?". Columns 1 and 2 in Table 1 provide the answer.

Column 1 is labeled "host" rather than "host star" because several of the currently detected planets appear to be physically associated with either a brown dwarf (2M1207) or a pre-main sequence object (GQ Lup, 1RXJ1609, CT Cha). RV detected planets are for the most part limited to bright, stable, FGKM main-sequence stars, whereas the brown

Table 1. Overview of directly imaged extrasolar planets

Host	SpT	Distance [pc]	Planet	Separation [AU, projected]	Mass [M_J]	Age [Myr]	Reference
Fomalhaut	A4V	7.69	*b*	100	< 3	100-400	Kalas *et al.* (2008)
β Pic	A5V	19.3	*b*	8	7-11	8-20	Lagrange *et al.* (2009)
HR 8799	A5V	39.4±1.0	*b*	68	4-10	30-160	Marois *et al.* (2008)
			c	38	7-13		"
			d	24	7-13		"
			e	14.5	5-13		Marois *et al.* (2010)
AB Pic	K2V	45.5±1.6	*b*	260	11-16	30	Chauvin *et al.* (2005)
2M1207	L2	52.4±1.1	*b*	41	2-10	5-12	Chauvin *et al.* (2004)
GQ Lup	K7	156±50	*b*	100	4-39	< 2	Neuhauser *et al.* (2005)
1RXJ1609	K7	145±20	*b*	330	6-11	4-6	Lafreniere *et al.* (2010)
CT Cha	K7	165±30	*b*	440	11-23	< 4	Schmidt *et al.* (2008)

Notes: For the significance of non-detections, see, e.g., Lafreniere *et al.* 2007; Nielsen & Close 2010; Chauvin *et al.* 2010. Essentially for every host target in Table 1, at least ten more were imaged with a null result.

dwarf and pre-main sequence hosts are both faint and variable. For example, 2M1207b is a common proper motion companion to 2MASSW J1207334-393254 (2M1207A), which is a mid-L dwarf with $M \sim 25\ M_J$. With $m_V = 20.2$ mag (Ducourant *et al.* 2008), 2M1207A is not amenable to RV observations.

The brighter, main-sequence host stars Fomalhaut, Beta Pic and HR 8799 are all A stars ($M \sim 1.7 - 2.9\ M_\odot$). RV techniques are unsuccessful because stellar lines are fewer, shallower and rotationally broadened (100 - 200 km s^{-1}; Galland *et al.* 2005). However, a separate sample of older, "retired" A stars (i.e., subgiants) have slower rotation and more lines, permitting an RV-derived estimate that the exojupiter occurrence rate is relatively high (~26%) for intermediate mass stars (Bowler *et al.* 2010).

In some cases, exoplanets may have been imaged as free floaters in young clusters, and the "host" might be the cluster name. Specifically, the 3 Myr-old σ Orionis cluster at ~400 pc may contain several planet-mass objects (Zapatero Osorio *et al.* 2000), and other young clusters such as IC 348 (Luhman *et al.* 2005) and the Trapezium (Lucas *et al.* 2006) are now targets for finding planet-mass objects via direct imaging and spectroscopy. The free floaters are not represented in Table 1.

These findings of free-floating planets and planets associated with brown dwarfs challenge the notion that planets necessarily form in circumstellar disks in a manner completely different from stars. Moreover, the large separations between the planets and host stars rule out *in situ* formation by core-accretion in a circumstellar disk, with gravitational instability as a more viable mechanism. For a more complete analysis, see recent work by Dodson-Robinson *et al.* (2009), Nero & Bjorkman (2009), & Kratter *et al.* (2010), to name a few. A different class of models assumes a birth site closer to the star with subsequent dynamical transport outward (e.g., Rasio & Ford 1996; Veras *et al.* 2009; Raymond *et al.* 2010).

3. Distance and Separation

A near simultaneous glance at the *Distance* and *Age* columns in Table 1 reveals that direct imaging searches for exoplanets have been successful for the very young ($< 10^6$ yr) star-forming environments at > 140 pc, and for the somewhat older ($10^6 - 10^7$ yr) main sequence stars residing within the local bubble (< 100 pc). The next generation of extreme adaptive optics systems such as the Gemini Planet Imager (GPI; Macintosh *et al.* 2008) sample the host star wavefront at rates exceeding 2 kHz and target stars must therefore have $m_I < 9$ mag (I-band is centered near 0.8 μm). Thus, future ground based imaging experiments will have the greatest impact in discovering new planets orbiting main sequence stars within the local bubble.

The fundamental breakthrough of future imaging experiments is that the inner working angle (IWA) will be reduced (improved) to a few λ / D radius from the star. The IWA specifies the smallest radial distance from a star where a planet-mass object could be detectable. Currently, the β Pic *b* and HR 8799*e* detections represent the smallest achievable IWA's. A near simultaneous glance at the *Distance* and *Separation* columns, and dividing the latter by the former, shows that both β Pic *b* and HR 8799*e* are detected at IWA $\sim 0.4''$ radius. The goal for GPI is IWA $\sim 3.5\ \lambda$ / D. Since the Gemini 8-m telescope has an effective aperture of D=7.77 m, then IWA $= 0.15''$ at 1.6 μm (H-band). Of course the star-to-planet contrast achieved beyond $0.15''$ radius is the other significant metric of planet imaging experiments. The various components of GPI are designed to achieve a goal of $\Delta H > 15$ mag.

Table 1 gives the host-to-exoplanet projected *separations* (ρ), and not the exoplanet deprojected separations (r) in a stellocentric cylindrical or spherical coordinate system, nor the semi-major axes, (a). Moreover, the projected separations are variable due to the exoplanet orbital motion. For β Pic and Fomalhaut, images of light scattering from circumstellar disk grains give line-of-sight *disk* inclinations of $i_d \sim 90°$ (edge-on) and $i_d \sim 24°$, respectively (Smith & Terrile 1984; Kalas *et al.* 2005). Assuming circular ($e = 0.0$) *and* coplanar exoplanet orbits, the host-exoplanet separations correspond to $r \sim 8$ AU for β Pic *b*, and $r \sim 119$ AU for Fomalhaut *b*. However, neither assumption is well-tested for *any* of the exoplanets in Table 1, largely because only a few epochs of astrometry are available. For example, adopting the assumption that Fomalhaut *b* lies in the belt plane, if e=0.12, as deduced from the measured stellocentric belt offset (Kalas *et al.* 2005), then $a \sim$115 AU.

The conversion from apparent separation to semi-major axis is particularly important for multi-planet systems such as HR 8799, where the origin and evolution of the system hinges on dynamical stability analysis (e.g., Fabrycky & Murray-Clay 2010). Not shown in Table 1 are the position angle (PA) of each exoplanet relative to the star, and the PA for the semi-major axis of each debris disk, which are required for the deprojection calculation (here the semi-major axis does not refer to an orbital element; instead it refers to the apparent elliptical morphology of an inclined circle). HR 8799 is surrounded by a debris disk that is resolved at 70 μm with the Spitzer Space Telescope, indicating $i_d < 25°$ (Su *et al.* 2009). If we adopt $i_d = 20°$ and put the major axis along the direction from the star to HR 8799*b*, then for the *b* component we have $\rho = r$, but for HR 8799*c*, which is approximately orthogonal to *b*, the non-face-on inclination translates to $\rho \sim r/cos(20°)$. In other words, r is 6% greater than the projected separation ρ. Reidemeister *et al.* (2009) consider the entire range of position angles, finding that with $i_d = 20°$, position angles of approximately $0° - 90°$ give stable configurations over the age of the system.

From Table 1, β Pic *b* is at the separation most "Jupiter-like" as it resides near the ice-line of the system. Since the inclination and position angle of the disk are well-constrained, it is unlikely that the semi-major axis of β Pic *b* will be shown to be more than 1 AU different from the current estimate of 8 AU. The corresponding $\sim$16 year orbital period means that in their lifetime, the majority of readers will witness a full orbit of β Pic *b*, whereas for Fomalhaut *b* readers need to wait eight centuries.

4. Planet Mass and Age

Stellar age uncertainties may be the most important factor in determining whether or not an object is a planet or a brown dwarf ($M > 13\ M_J$). In the literature one will typically find a paper announcing the discovery of an extrasolar planet through direct imaging, and then at least one follow-up paper discussing the age of the star and the implications for planet masses. For example, Marois *et al.* (2008) estimate the age of HR 8799 as < 160 Myr, which means that all three planets in their manuscript have $M \leqslant 13 M_J$. Moya *et al.* (2010) then describe the age uncertainty in greater detail, and present evidence from astroseismology that supports a likely age near 1 Gyr, implying that the HR 8799 objects have brown dwarf masses. Moro-Martin *et al.* (2010) respond that the system is dynamically stable only for $\leqslant$150 Myr, even if the masses are in the brown dwarf regime, and this timescale is a hard constraint. A new dynamical analysis that adds the fourth planet, HR 8799*e*, suggests that the planet masses must be relatively small, and therefore from the measured luminosities and theoretical cooling curves, the system is younger than 60 Myr (Marois *et al.* 2010).

The analytic estimate for the luminosity evolution of a brown dwarf gives $L \propto M^{2.6}\ t^{-1.3}$ (Burrows & Liebert 1993). Numerical calculations extending down to Jupiter mass planets give $L \propto M^{1.9}\ t^{-1.1}$ (Burrows *et al.* 2003; Fortney *et al.* 2008). A factor of two uncertainty in age corresponds to a 50% uncertainty in mass. Of course there are many more variables that matter, such as the assumptions of chemistry, metallicity, clouds, etc. One significant difference concerns the assumptions of planet formation, as illustrated in Marley *et al.* (2007) and Fortney *et al.* (2008). The scalings above apply to the hot-start planet formation model, but in a cold-start model accreting gas passes through a shock that quickly radiates a large fraction of the energy early in the evolution of the planet. The young exojupiter thereby acquires relatively cool gas, and therefore a low luminosity does not necessarily mean a less massive planet at a given age.

The scientific importance of direct imaging should therefore be obvious. This is the method by which we can empirically anchor theories of planet formation and evolution. Planet masses with direct imaging can be estimated by dynamical modeling, and the mechanism for planet formation and the relevance of various planet atmosphere models are thereby understood.

Several of the exoplanets in Table 1 are highlighted elsewhere in this volume. Below we summarize the tentative new findings concerning Fomalhaut *b*.

5. Fomalhaut *b* Update

At this conference I presented new observations with the Hubble Space Telescope showing the recovery of Fomalhaut *b* in September, 2010. The most significant barrier in producing follow-up observations was that the coronagraphic camera that enabled the Fomalhaut *b* discovery, the Advanced Camera for Surveys High Resolution Channel (ACS/HRC) suffered an electronics failure in January, 2007. The HST servicing mission in May 2009 restored function to the ACS Wide Field Channel (ACS/WFC), but the ACS/HRC was not successfully repaired. Meanwhile, adaptive optics imaging at near-infrared wavelengths with the 10-m Keck telescope, as well as mid-infrared (3.8 μm) imaging with the 8-m Gemini Observatory, lacked sufficient sensitivity to detect Fomalhaut *b* from the ground. We therefore attempted to recover Fomalhaut b using direct imaging with the Wide Field Camera 3 infrared channel (WFC3/IR) using the F110W filter (1.1 μm). The experiment failed due to previously unknown instrumental scattering spread over large azimuth angles and at a radius corresponding to the location of Fomalhaut *b*. We then studied the feasibility of using STIS coronagraphy or direct imaging with the ACS/WFC. Simulations showed that both cameras could be successful, but STIS should achieve a detection more efficiently than ACS/WFC (i.e. with a smaller allocation of telescope orbits). The details of the STIS observing strategy, data reduction and analysis will be presented elsewhere (Kalas *et al.* 2011).

Figure 1 summarizes our current findings. The new STIS observations show a point source near the expected 2010 location of Fomalhaut b with flux 0.4 μJy, in agreement with the 2006 measurement (Kalas *et al.* 2008). The source location matches the expected position in the north-south direction, but lies approximately $0.15''$ west of the expected location for low eccentricity orbits. This discrepancy may indicate that Fomalhaut *b* is on an eccentric orbit ($e \sim 0.67$). However, we also expect uncorrected systematic errors arising from transforming between the ACS and STIS astrometric reference frames. The STIS observations were made in single guide star mode, increasing the uncertainty in the spacecraft roll angle. Also, our geometric distortion solution is derived from

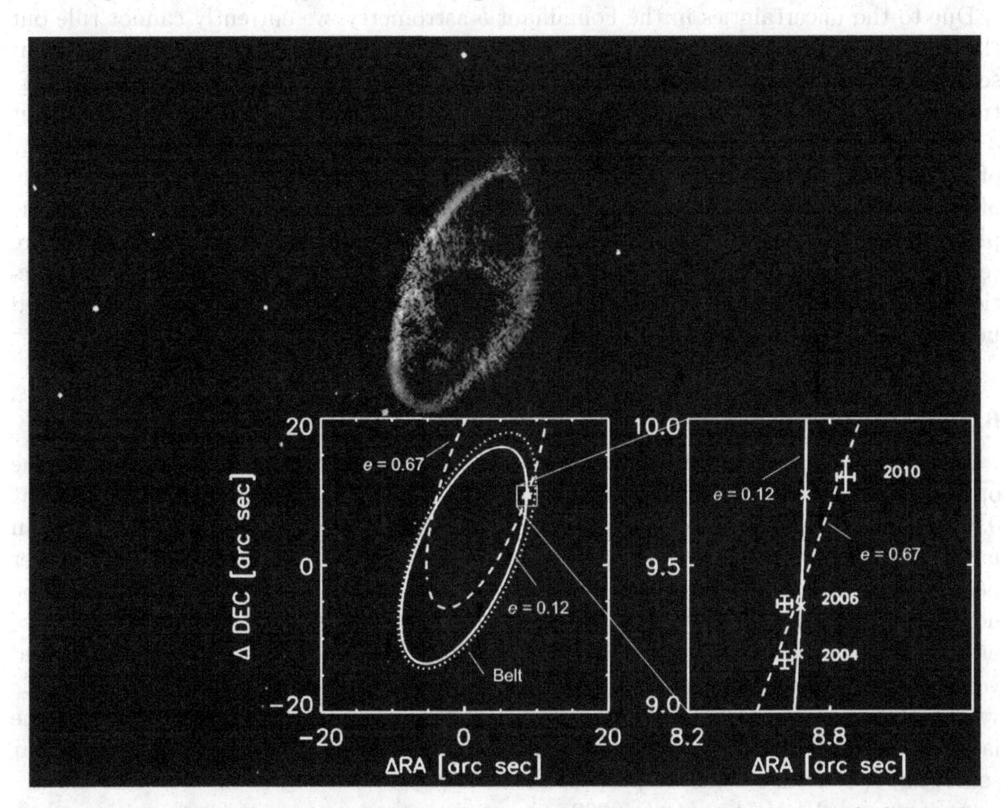

Figure 1. Composite image of the Fomalhaut system constructed from optical observations using three cameras aboard the Hubble Space Telescope (ACS/HRC, WFPC2, and STIS). The central star is occulted by coronagraphic spots. North is up, east is left. The tenuous halo north of the belt is a new feature discovered in recent STIS observations. The geometric offset of the belt $2''$ northward of the star is evident to the eye. The left bottom plot is drawn to the same scale as the image. The dotted line traces the inner edge of the belt, and the solid line traces an orbit nested within the belt with $e = 0.12$. The right bottom plot magnifies our preliminary three-epoch (2004, 2006, & 2010) astrometry and orbital fits. The ACS and STIS astrometric data are the three points with error bars, and the predicted positions are shown with the symbol ($\times$) for the $e = 0.12$ orbit. We are currently analyzing systematic errors due to uncorrected STIS distortion and a roll angle uncertainty due to the single guide star observation in 2010. At face value the observations are consistent with a best-fit bound orbit with $e = 0.67$ that crosses the belt. We cannot rule out an unbound orbit until future observations are conducted. Monte Carlo simulations show that one additional observation in 2012 will exclude unbound orbits.

calibration data obtained *before* the most recent servicing mission. Uncorrected distortions are evident in the calibrated STIS images because diffractions spikes have parabolic shapes. This uncertainty will be resolved when an updated STIS geometric distortion solution is available from new astrometric calibration observations.

Figure 1 also shows the discovery of a faint dust halo that extends northward approximately 4" beyond the previous outer boundary of the belt determined in the ACS/HRC data (Kalas *et al.* 2005). This finding more than doubles the measured extent of Fomalhauts dust belt. Two plausible explanations that are testable with multi-color imaging are: (1) the halo is due to small grains (grain size $a \sim \lambda/2\pi$) driven outward by radiation pressure, in which case the halo should be blue-scattering; or, (2) the eccentric orbit of Fomalhaut *b* dynamically disturbs particles of all sizes, broadening the belt, as shown in dynamical simulations (Chiang *et al.* 2009), in which case the outer halo should share the scattered light color of the main belt.

Due to the uncertainties in the Fomalhaut *b* astrometry, we currently cannot rule out the very intriguing possibility that it has a highly eccentric, belt-crossing orbit. This scenario implies that Fomalhaut *b* may *not* be the object responsible for the secular perturbation that creates the 15 AU stellocentric belt offset (Kalas *et al.* 2005). Fomalhaut *c* therefore remains to be discovered as the perturber. Furthermore, just as the majority of readers will witness β Pic *b* orbit its star over the next 16 years, Fomalhaut *b* may offer the opportunity of witnessing a planet cross into its Kuiper Belt. Depending on the planet mass, belt crossings may need to be relatively fast to preserve the belt structure (e.g. orbital planes inclined relative to each other). Some type of periodic, close interaction with the belt may be consistent with the hypothesis that Fomalhaut *b* is detected because of a large circumplanetary disk that is replenished over the age of the system.

6. Circumplanetary disks

Fomalhaut *b* is unique in Table 1 because it is the only exoplanet detected in the optical. Kalas *et al.* (2008) report detections at 0.6 and 0.8 μm, with non-detections at 0.4, 1.6 and 3.8 μm. The 0.6 μm flux is more than an order of magnitude greater than model atmospheres predict, suggesting non-thermal sources of optical emission. Another possibility is that the optical flux is light reflected from a circumplanetary ring. The non-detection at 0.4 μm does not contradict this scenario due to the low signal-to-noise of observations made in this bandpass. Depending on factors related to the assumed geometry and albedo, the radius of the ring system is $20-40\ R_p$ ($R_p \sim 1.2\ R_J$). This would appear more like a circumplanetary disk than the main rings of Saturn, and the lack of a counterpart in our Solar System made this scenario for Fomalhaut *b* appear somewhat speculative when we proposed it in 2008.

As if on cue, Verbischer *et al.* (2009) reported that a counterpart for the circumplanetary disk hypothesis exists in the Solar System. The surface of Saturn's outer moon Phoebe ($a = 215\ R_p$; $R_p \sim 1.0\ R_J$) is bombarded by micrometeoroids that launch fine dust in orbit around Saturn. The dust spirals inward due to radiation drag, forming a tenuous circumplanetary dust disk that is detected between $128-207\ R_p$. Thus Saturn's largest ring is about five times larger than the ring postulated for Fomalhaut *b*. The estimated optical depth of the Phoebe ring is 10^{-8}, but the inner Solar System at an age of 100 Myr had five to six orders of magnitude more interplanetary debris than the present epoch (Bottke *et al.* 2007 and references therein). Therefore an optically bright Phoebe ring at early times is entirely plausible.

However, Fomalhaut *b* is not the only exoplanet where a puzzling spectrum evokes the existence of a circumplanetary disk. Near-infrared imaging and spectroscopy of 2M1207b give a model-dependent temperature of 1600 K, but the luminosity of the planet is smaller than expected by an order of magnitude (Mohanty *et al.* 2007). An edge-on disk surrounding 2M1207*b* could produce significant gray extinction, or obscures the planet completely such that the flux received at Earth is due to photons scattered from the disk. A resolved example of such an optically thick case is HK Tau B. High angular resolution imaging shows that the source previously thought to be a direct image of a young protostar is actually the scattering surface of a disk, with the protostar completely occulted by the disk midplane (e.g., Koresko 1998). The main consequence of the circumplanetary disk hypothesis for 2M1207*b* is a very large uncertainty in its mass (Table 1).

Naturally the key difference between 2M1207*b* and Fomalhaut *b* is that the latter has a significant source of external illumination from the host star, as in the case of Saturn's rings. For reference, Fomalhaut b is four times farther from its host star as Neptune is from the Sun, but the luminosity of Fomalhaut is a factor of ∼16 greater than the Sun. Therefore, the incident stellar radiation flux on any circumplanetary material surrounding Fomalhaut b is roughly equivalent to the incident flux on Neptune.

One final note is the possibility that circumplanetary dust is produced from a system of irregular satellites, and thus the dust parent bodies reside in a cloud around the planet rather than a flattened disk. Kennedy & Wyatt (2011) model the production of a dust cloud by irregular satellites surrounding Saturn. The resulting cloud has an hourglass morphology with apparent dimensions 2° × 1° (for comparison the Moon's apparent diameter is 0.5°). The maximum radius of the cloud corresponds to approximately 1/2 of a Hill radius [$R_H = a_p(1 - e_p)(M_p/3M_*)^{1/3}$]. If Fomalhaut b has M = 1 M_J, then 1/2 $R_H \sim 3$ AU. However, a dust cloud with diameter 6 AU at the distance to Fomalhaut subtends 0.8″ and would have been resolved with the HST observations. This constraint, as well as other considerations, particularly the collision lifetimes, suggest that Fomalhaut b may have $M < 0.3\ M_J$.

If the circumplanetary disk hypothesis is correct, orbital motion should change the viewing angles such that reflection and extinction vary significantly over time. Unfortunately, unlike Earth and Saturn analogs where the viewing angles change appreciably over months to decades (Arnold & Schneider 2004), the orbital periods of Fomalhaut *b* and 2M1207*b* approach a millennium. Nevertheless, secular variability in the optical to infrared spectra of exoplanets may eventually produce strong evidence for circumplanetary material around these two exoplanets and others.

7. Summary

The big picture is that 1.6 centuries after Neptune was discovered, we are once again finding and studying planets through direct imaging. If one simply tallies the planets in Table 1, we have just entered the domain of having more images of exoplanets than solar system planets. Moreover, two exoplanets may have giant ring systems that give us a notion of what Saturn or Jupiter may have resembled 4.5 billion years ago. Exoplanet imaging thus invokes a very powerful convergence of disciplines. From the study of planet atmospheres, to the theoretical framework of their formation, and proceeding to their dynamical evolution into long-lived systems, observation and theory will ultimately quantify the frequency and architectures of planetary systems.

Acknowledgements

I am grateful to James Graham (Dunlap Institute & UC Berkeley) for the analysis of Fomalhaut b's astrometry and orbit, Michael Fitzgerald (UCLA) for significant work on ground-based observations of Beta Pic and Fomalhaut, as well as Mark Clampin (NASA GSFC) and Matt Mountain (STScI) for enabling the HST follow-up of Fomalhaut b. This work received support from the following: GO-11818 provided by NASA through a grant from STScI under NASA contract NAS5-26555; NSF AST-0909188; and the University of California LFRP-118057.

References

Arnold, L. & Schneider, J. 2004, *A&A*, 420, 1123
Bottke, W. F., Levison, H. F., Nesvorny, D., & Dones, L. 2007, *Icarus*, 190, 203
Bowler, B. P., Johnson, J. A., Marcy, G. W., *et al.* 2010, *ApJ*, 709, 396
Burrows, A. & Liebert, J. 1993, *Rev. Mod. Phys.*, 65, 301
Burrows, A., Sudarsky, D., & Lunine, J. 2003, *ApJ*, 596, 587
Chauvin, G., Lagrange, A.-M., Dumas, C., *et al.* 2004, *A&A*, 438, L25
Chauvin, G., Lagrange, A.-M., Dumas, C., *et al.* 2005, *A&A*, 425, L29
Chauvin, G., Lagrange, A.-M., Bonavita, M., *et al.* 2010, *A&A*, 509, A52
Chiang, E., Kite, E., Kalas, P., Graham, J. R., & Clampin, M. 2009, *ApJ*, 693, 734
Ducourant, C., Teixeira, R., Chauvin, G., *et al.* 2008, *A&A*, 477, L1
Dodson-Robinson, S. E., Veras, D., Ford, E. C., & Beichman, C. A. 2009, *ApJ*, 707, 79
Duchene, G. 2008, *New Astron. Revs*, 52, 117
Fabrycky, D. C. & Murray-Clay, R. A. 2010, *ApJ*, 710, 1408
Fortney, J. J., Marley, M. S., Saumon, D., & Lodders, K. 2008, *ApJ*, 683, 1104
Galland, F., Lagrange, A.-M., Udry, S., *et al.* 2005, *A&A*, 443, 337
Kalas, P., Graham, J. R., & Clampin, M. 2005, *Nature*, 435, 1067
Kalas, P., Graham, J. R., Chiang, E., *et al.* 2008, *Science*, 322, 1345
Kalas, P., Graham, J. R., Fitzgerald, M., & Clampin, M. 2011, *in prep*
Kennedy, G. M. & Wyatt, M. C. 2011, *MNRAS*, 412, 2137
Koresko, C. D. 1998, *ApJ*, 507, L145
Kratter, K. M., Murray-Clay, R. A., & Youdin, A. N. 2010, *ApJ*, 710, 1375
Lafreniere, D., Doyon, R., Marois, C., *et al.* 2007, *ApJ*, 670, 1367
Lafreniere, D., Jayawardhana, R., & van Kerkwijk, M. H. 2010, *ApJ*, 719, 497
Lagrange, A.-M., Gratadour, D., Chauvin, G., *et al.* 2009, *A&A*, 493, L21
Lucas, P. W., Weights, D. J., Roche, P. F., & Riddick, F. C. 2006, *MNRAS*, 373, L60
Luhman, K. L. McLeod, K. K., & Goldenson, N. 2005, *ApJ*, 623, 1141
Macintosh, B. A., Graham, J. R., Palmer, D. W., *et al.* 2008, *SPIE*, 7015, 31
Marley, M. S., Fortney, J. J., Hubickyj, O., Bodenheimer, P., & Lissauer, J. J. 2007, *ApJ*, 655, 541
Marois, C., Macintosh, B., Barman, T., *et al.* 2008, *Science*, 322, 1348
Marois, C., Zuckerman, B., Konopacky, Q. M., *et al.* 2010, *Nature*, 468, 1080
Mohanty, S., Jayawardhana, R., Huelamo, N., & Mamajek, E. 2007, *ApJ*, 657, 1064
Moro-Martin, A., Rieke, G. H., & Su, K. Y. L. 2010, *ApJ*, 721, L199
Moya, A., Amado, P. J., Barrado, D., *et al.* 2010, *MNRAS*, 405, L81
Nero, D. & Bjorkman, J. E. 2009, *ApJ*, 702, L163
Neuhauser, R., Guenther, E. W., Wuchterl, G., *et al.* 2005, *A&A*, 435, L13
Nielsen, E. L. & Close, L. M. 2010, *ApJ*, 717, 878
Oppenheimer, B. R. & Hinkley, S. 2009, *ARAA*, 47, 253
Rasio, A. F. & Ford, E. B. 1996, *Science*, 274, 954
Raymond, S. N., Armitage, P. J., & Gorelick, N. 2010, *ApJ*, 711, 772
Reidemeister, M., Krivov, A. V., Schmidt, T. O. B., *et al.* 2009, *A&A*, 503, 247
Schmidt, T. O. B., Neuhauser, R., Seifahrt, A., *et al.* 2008, *A&A*, 491, 311
Seager, S. & Deming, D. 2010, *ARAA*, 48, 631
Smith, B. A. & Terrile, R. 1984, *Science*, 226, 1421
Su, K. Y. L., Rieke, G. H., Stapelfeldt, K. R., *et al.* 2009, *ApJ*, 705, 314
Veras, D., Crepp, J. R., & Ford, E. B. 2009, *ApJ*, 696, 1600
Verbiscer, A. J., Skrutskie, M. F., & Hamilton, D. P. 2009, *Nature*, 461, 1098
Zapatero Osorio, M. R., Bejar, V. J. S., Martin, E. L., *et al.* 2000, *Science*, 290, 103

The Astrophysics of Planetary Systems: Formation, Structure, and Dynamical Evolution
Proceedings IAU Symposium No. 276, 2010
A. Sozzetti, M. G. Lattanzi & A. P. Boss, eds.

doi:10.1017/S1743921311020333

On the equilibrium rotation of Hot Jupiters in eccentric and excited orbits

Alexandre C. M. Correia[1,2]

[1]Department of Physics, I3N, University of Aveiro,
Campus Universitário de Santiago, 3810-193 Aveiro, Portugal
email: correia@ua.pt

[2]Astronomie et Systèmes Dynamiques, IMCCE-CNRS UMR 8028,
77 Avenue Denfert-Rochereau, 75014 Paris, France

Abstract. Hot-Jupiters are a common sub-class of exoplanets, which are enough close to the star to undergo tidal dissipation. The continuous action of tides modify the rotation of the planets until an equilibrium situation is reached. It is often assumed that synchronous motion is the most probable outcome of tidal evolution, since synchronous rotation is observed for the majority of the satellites in the Solar System. This is true for circular orbits, but when the orbits are eccentric, tidal effects are stronger when the planets are closer to the star, and therefore, the rotation rate tends to equalize the orbital speed rate at the pericenter (which is faster than synchronous rotation). An additional complication arises if the eccentricity is not constant and undergoes periodic perturbations from an external companion. Here we obtain an expression for the equilibrium rotation of Hot-Jupiters undergoing tidal dissipation and planetary perturbations. We show that for these planets, the equilibrium rotation rate is faster than for non-perturbed eccentric orbits.

Keywords. planetary systems, celestial mechanics, methods: analytical

---

1. Introduction

At present, about half of the exoplanets that have been detected are close-in planets with semi-major axis smaller than 0.4 AU (http://exoplanet.eu/). This percentage is biased, since close-in planets are easily detected by radial velocity and transit methods (the two methods with greater success in discovering exoplanets). However, it shows that massive planets are often found within a distance to the star equivalent to the Sun-Mercury distance. As a consequence, these planets undergo significant tidal interactions with the star, resulting that their spins will be slowly modified.

The ultimate stage for tidal evolution corresponds to a low obliquity and synchronous rotation, a configuration where the rotation rate coincides with the orbital mean motion, since the synchronous equilibrium corresponds to the minimum of dissipation of energy. However, when the eccentricity is different from zero some other configurations are possible, such as the 3/2 spin-orbit resonance observed for the planet Mercury (e.g., Correia & Laskar 2004) or the chaotic rotation of Hyperion (Wisdom *et al.* 1984). Indeed, for eccentric orbits the rotation rate of the planet tends to equalize the orbital speed rate at the pericenter, which is faster than synchronous rotation. About 1/3 of the planets with a semi-major axis smaller than 0.4 have an eccentricity above 0.1, so it is important to understand how their equilibrium rotation rates will be modified.

Additional effects may also contribute to the final evolution of the spin, such as atmospheric tides or planetary perturbations. The effect of a dense atmosphere has been studied in Correia *et al.* (2008). Here we will focus on the effect of planetary perturbations on the eccentricity of the close-in planet. A similar problem has been studied for the spin of Mercury (Correia & Laskar 2004), but in that case we also needed to take into account the contribution of the asymmetric equatorial moments of inertia C_{22}, which is responsible for capture in spin-orbit resonances. Here we will look at the behavior of Jupiter and Neptune-like planets for which the second order harmonics of the gravity field are close to zero (e.g. Jacobson 2001), leading to insignificant chances of capture. We thus expect that the final rotation of these planets is controlled uniquely by the equilibrium tide.

2. Equations of motion

We will adopt here a viscous tidal model, with a linear dependence on the tidal frequency, which is suitable to study slow rotations near the equilibrium, and is also valid for any value of the eccentricity (e.g. Mignard 1979). Tidal dissipation and core-mantle friction drive the planet's obliquity (the angle between the equator and the orbital plane) close to zero (e.g. Correia *et al.* 2003), so we will assume zero degree obliquity throughout this study for simplicity. The tidal contribution to the rotation rate, ω, is then given by (e.g. Correia & Laskar 2010):

$$\dot{\omega} = -K \left[\Omega(e)\omega - N(e)n \right], \tag{2.1}$$

with

$$\Omega(e) = \frac{1 + 3e^2 + 3e^4/8}{(1-e^2)^{9/2}}, \tag{2.2}$$

$$N(e) = \frac{1 + 15e^2/2 + 45e^4/8 + 5e^6/16}{(1-e^2)^6}, \tag{2.3}$$

and

$$K = 3n \frac{k_2}{\xi Q} \left(\frac{R}{a} \right)^3 \left(\frac{m_0}{m} \right), \tag{2.4}$$

where ξ is a structure constant, k_2 is the second Love number, Q is the dissipation quality factor, and n, a, R, m, and m_0 are the mean motion, the semi-major axis, the radius of the planet, its mass, and the mass of the star, respectively.

For a non-constant varying eccentricity $e(t)$, the solution of equation (2.1) is given by (Correia & Laskar 2004):

$$\omega(t) = \left(\omega(0) + nK \int_0^t N(e(\tau)) g(\tau) d\tau \right) / g(t), \tag{2.5}$$

with

$$g(t) = \exp(K \int_0^t \Omega(e(\tau))\, d\tau). \tag{2.6}$$

3. Approximations

The eccentricity of a planet that is disturbed by the gravitational perturbations of additional planetary companions can be expressed as a sum of quasi-periodic terms (e.g. Laskar 1988):

$$e(t) = \sum_i e_i \, \mathrm{e}^{\mathrm{i}(\nu_i t + \phi_i)}, \tag{3.1}$$

where e_i is an amplitude, ν_i is the frequency of the perturbation, and ϕ_i is a phase angle. When there is a single companion, its orbit lies in the same plane as the planet (i.e., the system is coplanar), and we take into account only the linear terms of the perturbation, the eccentricity variations can be simply written as:

$$e(t) = e_0 + \Delta e \cos(\nu t + \phi) \ . \tag{3.2}$$

The expressions of $\Omega(e)$ (Eq. 2.2) and $N(e)$ (Eq. 2.3) are functions of e^2. Since $0 \leqslant e \leqslant 1$, we can develop these two functions in power series of e^2 as

$$\Omega(e) = \sum_{j=0}^{+\infty} \Omega_j e^{2j} = 1 + \frac{15}{2} e^2 + \frac{105}{4} e^4 + ..., \tag{3.3}$$

$$N(e) = \sum_{j=0}^{+\infty} N_j e^{2j} = 1 + \frac{27}{2} e^2 + \frac{573}{8} e^4 + ... \tag{3.4}$$

Because $e(t)$ can be expressed as a sum of quasi-periodic terms (Eq. 3.1), so does the functions $\Omega(e)$ and $N(e)$, as:

$$\Omega(e) = \sum_{i,j} \Omega_{ij} e^{\mathrm{i}j(\nu_i t + \phi_i)}, \quad N(e) = \sum_{i,j} N_{ij} e^{\mathrm{i}j(\nu_i t + \phi_i)}. \tag{3.5}$$

Considering the case of a single companion in a coplanar orbit (Eq. 3.2), these expressions can be obtained directly from equations (3.3) and (3.4), with

$$e^{2j} = e_0^{2j} \sum_{k=0}^{2j} \sum_{l=0}^{k} \begin{pmatrix} 2j \\ k \end{pmatrix} \begin{pmatrix} k \\ l \end{pmatrix} \left(\frac{\Delta e}{2e_0} \right)^k \mathrm{e}^{\mathrm{i}(k-2l)(\nu t + \phi)}. \tag{3.6}$$

Henceforward we will restrict our analysis to the case of a single companion in a coplanar orbit, but our conclusions can be easily extended to the general situation given by equations (3.1) and (3.5).

4. Averaging

Periodic varying quantities can be averaged in order to obtain their mean values. This can be particularly interesting when the eccentricity variations occur in a time-scale that is shorter than the time-scale for tidal variations ($\nu \gg K$), which is often the case. Thus,

$$\overline{e(t)} = \frac{2\pi}{\nu} \int_0^{\frac{2\pi}{\nu}} e(t) \, dt = e_0, \tag{4.1}$$

and

$$\begin{aligned} \overline{e^{2j}} &= e_0^{2j} \sum_{k=0}^{2j} \sum_{l=0}^{k} \begin{pmatrix} 2j \\ k \end{pmatrix} \begin{pmatrix} k \\ l \end{pmatrix} \left(\frac{\Delta e}{2e_0} \right)^k \frac{2\pi}{\nu} \int_0^{\frac{2\pi}{\nu}} \mathrm{e}^{\mathrm{i}(k-2l)(\nu t + \phi)} \, dt \\ &= e_0^{2j} \sum_{k=0}^{2j} \sum_{l=0}^{k} \begin{pmatrix} 2j \\ k \end{pmatrix} \begin{pmatrix} k \\ l \end{pmatrix} \left(\frac{\Delta e}{2e_0} \right)^k \delta_{k,2l} \, \mathrm{e}^{\mathrm{i}(k-2l)\phi} \\ &= e_0^{2j} \sum_{l=0}^{j} \begin{pmatrix} 2j \\ 2l \end{pmatrix} \begin{pmatrix} 2l \\ l \end{pmatrix} \left(\frac{\Delta e}{2e_0} \right)^{2l}. \end{aligned} \tag{4.2}$$

Finally, using the above quantity in the expressions of $\Omega(e)$ and $N(e)$ (Eq. 3.3 and 3.4), we get:

$$\bar{\Omega} = \overline{\Omega(e)} = \sum_{j=0}^{+\infty} \sum_{l=0}^{j} \frac{\Omega_j \, e_0^{2j} \, (2j)!}{(2j-2l)! \, (l!)^2} \left(\frac{\Delta e}{2e_0} \right)^{2l}$$
$$= 1 + \frac{15}{2} e_0^2 + \frac{105}{4} e_0^4 + \frac{15}{4} \Delta e^2 + \frac{315}{4} e_0^2 \Delta e^2 + \frac{315}{32} \Delta e^4 + ..., \quad (4.3)$$

$$\bar{N} = \overline{N(e)} = \sum_{j=0}^{+\infty} \sum_{l=0}^{j} \frac{N_j \, e_0^{2j} \, (2j)!}{(2j-2l)! \, (l!)^2} \left(\frac{\Delta e}{2e_0} \right)^{2l}$$
$$= 1 + \frac{27}{2} e_0^2 + \frac{573}{8} e_0^4 + \frac{27}{4} \Delta e^2 + \frac{1719}{8} e_0^2 \Delta e^2 + \frac{1719}{64} \Delta e^4 + ..., \quad (4.4)$$

5. Implications

For a constant eccentricity $e(t) = e_0$, the solution of equation (2.1) becomes

$$\omega(t) = \omega_{e_0} + \omega(0) \mathrm{e}^{-K\Omega(e_0)t}, \quad (5.1)$$

the equilibrium being achieved when $t \gg 1/(K\Omega(e_0))$ for $\omega(t) = \omega_{e_0} = Cte$, where

$$\frac{\omega_{e_0}}{n} = \frac{N(e_0)}{\Omega(e_0)} = 1 + 6e_0^2 + \frac{3}{8} e_0^4 + ... \quad (5.2)$$

This equilibrium is often presented as the equilibrium rotation rate for planets in eccentric orbits, but in fact, this is only true for unperturbed orbits. Since unperturbed eccentricities of close-in planets are quickly damped to zero, contrarily to eccentricities of close-in planets with companions (Mardling 2007), most of the observed close-in planets with some eccentricities probably have companions, and therefore forced eccentricities.

For a non constant eccentricity $e(t)$, the solution of equation (2.1) is given by expression (2.5), which cannot be solved without some approximations. Assuming that $e(t)$ is simply given by expression (3.2), we have for $t \gg 1/(K\bar{\Omega})$ and $\nu/K \gg 1$:

$$g(t) = \exp(K \int_0^t \Omega(e(\tau)) \, d\tau) \approx \exp(K\bar{\Omega}t), \quad (5.3)$$

and

$$\frac{\omega(t)}{n} \approx K \mathrm{e}^{-K\bar{\Omega}t} \int_0^t N(e(\tau)) \mathrm{e}^{K\bar{\Omega}\tau} d\tau. \quad (5.4)$$

Thus, replacing equations (3.4) and (3.6) in the above expression, gives:

$$\frac{\omega(t)}{n} = \sum_{j=0}^{+\infty} N_j e_0^{2j} \sum_{k=0}^{2j} \sum_{l=0}^{k} \binom{2j}{k} \binom{k}{l} \left(\frac{\Delta e}{2e_0} \right)^k \Phi_{kl}(t), \quad (5.5)$$

where

$$\Phi_{kl}(t) = K \mathrm{e}^{-K\bar{\Omega}t} \int_0^t \mathrm{e}^{\mathrm{i}(k-2l)(\nu\tau+\phi)+K\bar{\Omega}\tau} d\tau$$
$$= \frac{K \mathrm{e}^{\mathrm{i}(k-2l)(\nu t+\phi)}}{\mathrm{i}(k-2l)\nu + K\bar{\Omega}} = \frac{\cos[(k-2l)(\nu t+\phi) - \psi_{kl}]}{\sqrt{(k-2l)^2\nu^2/K^2 + \bar{\Omega}^2}}, \quad (5.6)$$

with

$$\tan \psi_{kl} = \frac{(k-2l)\nu}{K\bar{\Omega}}. \quad (5.7)$$

5.1. *average rotation rate*

Both $\omega(t)$ and $\Phi_{kl}(t)$ are periodic functions, with period $2\pi/\nu$. One may then compute their average value,

$$\overline{\Phi_{kl}(t)} = \frac{2\pi}{\nu}\int_0^{\frac{2\pi}{\nu}} \frac{K e^{\mathrm{i}(k-2l)(\nu t+\phi)}}{\mathrm{i}(k-2l)\nu + K\bar{\Omega}}\, dt = \frac{\delta_{k,2l}}{\bar{\Omega}}, \tag{5.8}$$

and

$$\begin{aligned}\frac{\overline{\omega(t)}}{n} &= \sum_{j=0}^{+\infty} N_j e_0^{2j} \sum_{k=0}^{2j}\sum_{l=0}^{k} \binom{2j}{k}\binom{k}{l}\left(\frac{\Delta e}{2e_0}\right)^k \frac{\delta_{k,2l}}{\bar{\Omega}} \\ &= \frac{1}{\bar{\Omega}}\sum_{j=0}^{+\infty} N_j e_0^{2j} \sum_{l=0}^{j} \binom{2j}{2l}\binom{2l}{l}\left(\frac{\Delta e}{2e_0}\right)^{2l} = \frac{\bar{N}}{\bar{\Omega}} \\ &= \frac{N(e_0)}{\Omega(e_0)} + 3\Delta e^2 + \frac{729}{8}e_0^2\Delta e^2 + \frac{369}{64}\Delta e^4 + ...,\end{aligned} \tag{5.9}$$

that is, the average value of the rotation rate $\bar{\omega} = \overline{\omega(t)} \geqslant \omega_{e_0}$ is always faster than the value of the equilibrium rotation rate for the average value of the eccentricity, e_0 (Eq. 5.2). The difference depends on the amplitude of the eccentricity librations Δe. We then conclude that for orbits undergoing eccentricity excitations, the equilibrium rotation rate of the planet is faster than for non-perturbed eccentric orbits.

5.2. *weak tidal dissipation ($\nu \gg K$)*

In general, the time-scale for the eccentricity variations is shorter than the time-scale for the tidal damping ($\nu \gg K$). In this limit situation, we may neglect the contributions from $K\bar{\Omega}$ in expression (5.6), except those from terms with $k = 2l$, which correspond to the non-periodic terms contributing to the average $\bar{\omega}$ (Eq. 5.9). Thus, retaining only the terms of minimal frequency, $k - 2l = 0$ and $k - 2l = 1$, we get

$$\frac{\omega(t)}{n} \approx \frac{\bar{N}}{\bar{\Omega}} + \frac{K}{\nu}\left(\frac{\Delta e}{2e_0}\right) A_1 \cos(\nu t + \phi - \pi/2), \tag{5.10}$$

with

$$A_1 = \sum_{j=0}^{+\infty} N_j e_0^{2j} \sum_{l=0}^{j-1} \binom{2j}{2l+1}\binom{2l+1}{l}\left(\frac{\Delta e}{2e_0}\right)^{2l}. \tag{5.11}$$

We then conclude that the rotation rate of a planet in an eccentric and excited orbit is given by the average value of the rotation rate, $\bar{\omega}$, plus a sinusoidal function similar to the eccentricity (Eq. 3.2), but with smaller amplitude and a phase delay of $\pi/2$. When we include the remaining harmonics the behavior is identical, only the shape of the periodic signal is modified (Eq. 5.6). As the magnitude of tidal effects increase, the amplitude of the periodic variations becomes larger, and the phase delay becomes smaller than $\pi/2$.

Table 1. Initial parameters for the HD 11964 system (Wright *et al.* 2009).

	HD 11964 ($m_0 = 1.1\,M_\odot$)							
parameter (unit)	a (AU)	e	m (M_{Jup})	$P_{\rm rot}$ (day)	R ($\times 10^6$m)	ξ	k_2	Q ($\times 10^3$)
planet (b)	3.16	0.041	0.622	–	–	–	–	–
planet (c)	0.229	0.300	0.0788	2.0	75	0.25	0.50	5

6. Applications

We will now apply the results derived in previous sections to the planet HD 11964 c. The planetary system around HD 11964 is composed of two planets with minimum masses $m_c = 25\,M_\oplus$ (planet c) and $m_b = 0.62\,M_{\mathrm{Jup}}$ (planet b), at $a_c = 0.229$ AU and $a_b = 3.16$ AU, respectively (Wright *et al.* 2009, Table 1). The orbit of the planet c also presents an eccentricity of about 0.3 and is enough close to the star to undergo substantial tidal

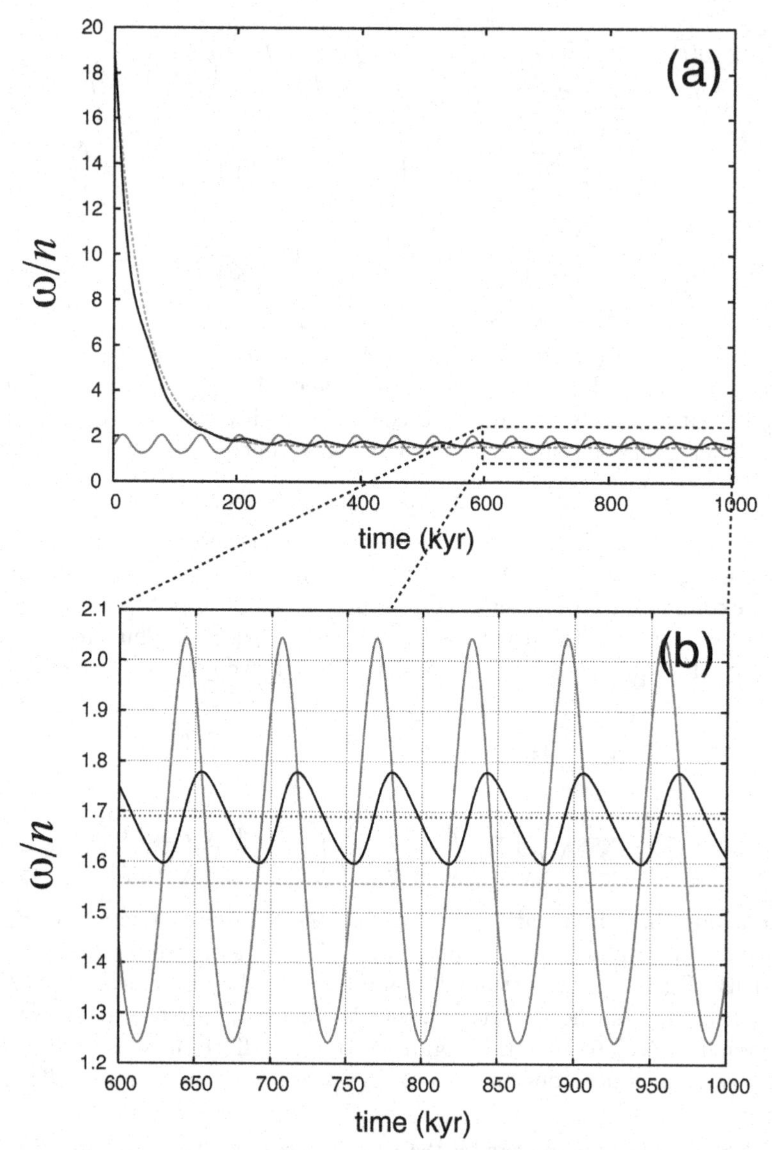

Figure 1. Evolution of the rotation rate with time, starting with the parameters from Table 1, using a model with a varying eccentricity (Eq. 6.1, black solid curve), and a model with constant eccentricity ($e_0 = 0.3$, green dashed curve). We also plot the instantaneous value of the ratio $N(e)/\Omega(e)$ for the varying eccentricity model (red solid curve). We observe that in the case of a constant eccentricity, the rotation rate stabilizes in the equilibrium value $\omega_{e_0}/n = 1.5571$, while in the case of a varying eccentricity, the rotation rate presents a periodic oscillations that follows the equilibrium ratio $N(e)/\Omega(e)$ with a smaller amplitude and a delayed phase angle (Eq.5.10). These oscillations are effectuated around a mean value given by $\bar{\omega}/n = 1.6896$ (blue dotted line).

dissipation. Therefore, this is a good system to illustrate our theoretical results. The average distance between the two planets is relatively high, so if the two orbits are coplanar the eccentricity of the inner body only experiences small oscillations. However, the mutual inclination between the planets is unknown. Veras & Ford (2010) performed extensive n-body simulations for this system and concluded that it is stable for very high mutual inclinations, which can excite significantly the eccentricity of the inner orbit. We will thus assume that the eccentricity of the planet c can be written in the form of equation (3.2) with $\Delta e = 0.1$ and $\nu = 10^{-4}\,\mathrm{yr}^{-1}$:

$$e(t) = 0.3 + 0.1 \sin(10^{-4}\, t_{[\mathrm{yr}]}). \tag{6.1}$$

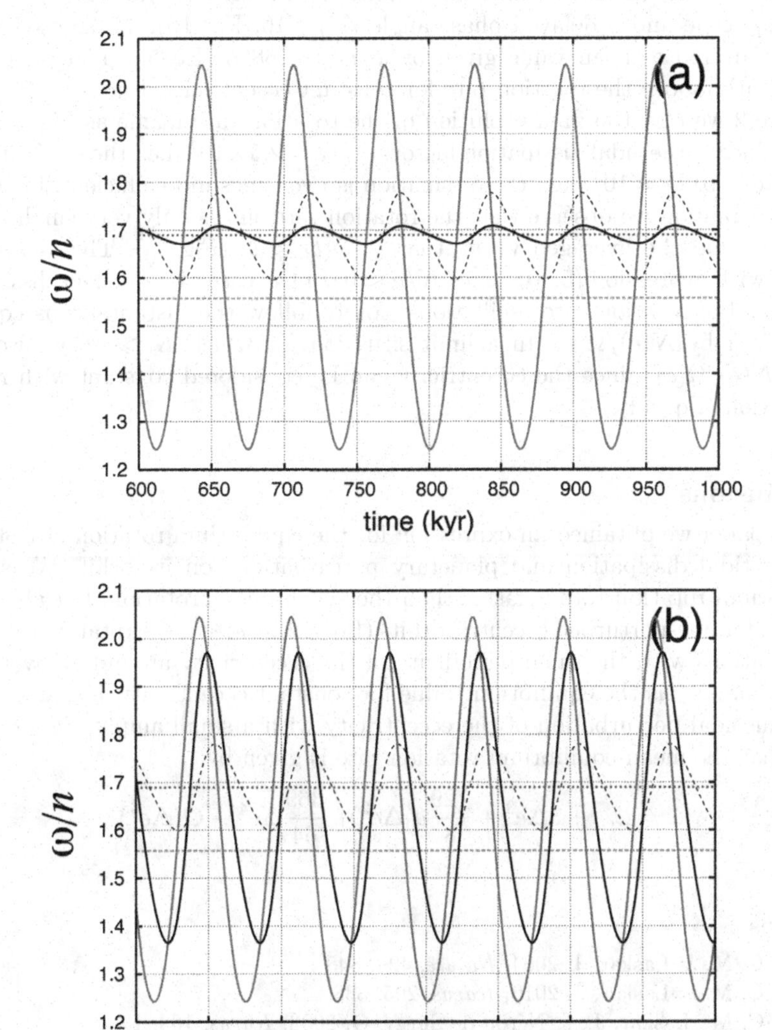

Figure 2. Same as Fig. 1b, but with $Q = 2.5 \times 10^4$ (a) and $Q = 10^3$ (b). In the case of a weaker dissipation (a), we observe that the rotation rate shows only very small amplitude librations around the averaged value $\bar{\omega}/n = 1.6896$ (blue dotted line). In the case of a stronger dissipation (b), we observe that the rotation rate oscillations closely follow the instantaneous equilibrium rotation given by $N(e)/\Omega(e)$ (red solid line).

In Figure 1 we plot the evolution of the rotation rate of the planet, starting with a rotation period of 2 days, that is, $\omega(0)/n \approx 18.9$, and using a dissipation factor identical to Jupiter, $Q = 5 \times 10^3$ (Table 1). We plot the evolution of $\omega(t)/n$ using the varying eccentricity given by the expression (6.1) (Eq. 2.1, black curve), but also using a constant eccentricity $e_0 = 0.3$ (Eq. 5.1, green curve). We observe that the evolution of the rotation rate is identical for two situations during the initial damping (Fig. 1a). We also plot the instantaneous value of the ratio $N(e)/\Omega(e)$ for the varying eccentricity (red curve). After about 0.2 Myr, the rotation rate is fully damped and enter in an equilibrium regime. In the case of a constant eccentricity, the rotation rate stabilizes in the equilibrium value $\omega_{e_0}/n = 1.5571$ (Eq. 5.2), while in the case of a varying eccentricity, the rotation rate presents a periodic oscillations that follows the equilibrium ratio $N(e)/\Omega(e)$ with a smaller amplitude and a delayed phase angle (Eq. 5.10, Fig. 1b). These oscillations are effectuated around a mean value given by $\bar{\omega}/n = 1.6896$ (Eq. 5.9, blue straight line), which is clearly above the rotation rate for constant eccentricity, that is, $\bar{\omega} > \omega_{e_0}$.

In Figure 2 we plot the final evolution of the rotation rate (same as Fig. 1b), but for different values of the tidal dissipation factor: (a) $Q = 2.5 \times 10^4$, i.e., the dissipation is five times weaker; (b) $Q = 10^3$, i.e., the dissipation is five times more efficient. In the case of a weaker dissipation we observe that the rotation rate shows only very small amplitude librations around the averaged value, that is, $\omega(t) \approx \bar{\omega}$ (Fig. 2a). This behavior is in agreement with expression (5.10), since $K/\nu \approx 0$. In the case of a stronger dissipation we observe that the rotation rate oscillations closely follow the instantaneous equilibrium rotation given by $N(e)/\Omega(e)$. In a limit situation for which $K \gg \nu$, we would have $\omega(t)/n \approx N(e)/\Omega(e)$, since the eccentricity can be considered constant with respect to tidal evolution (Eq. 5.2).

7. Conclusions

In these paper we obtained an expression for the equilibrium rotation of Hot-Jupiters undergoing tidal dissipation and planetary perturbations on its orbit. We show that the equilibrium rotation rate is faster than the synchronous rotation, but also than the equilibrium for non-perturbed eccentric orbits (Eq. 5.2). Indeed, the rotation rate presents small oscillations with the same periodicity of the eccentricity around an average value $\bar{\omega} > \omega_{e_0}$, where ω_{e_0} is the equilibrium value for constant eccentricity. In particular, for a regular sinusoidal perturbation of the eccentricity with a small amplitude Δe (Eq. 3.2), we show that the mean equilibrium rotation rate is given by:

$$\bar{\omega} = \omega_{e_0} + 3\Delta e^2 + \frac{729}{8} e_0^2 \Delta e^2 + \frac{369}{64} \Delta e^4 + \mathcal{O}(\Delta e^6) . \quad (7.1)$$

References

Correia, A. C. M. & Laskar, J. 2004, *Nature*, 429, 848
Correia, A. C. M. & Laskar, J. 2010, *Icarus*, 205, 338
Correia, A. C. M., Laskar, J., & Néron de Surgy, O. 2003, *Icarus*, 163, 1
Correia, A. C. M., Levrard, B., & Laskar, J. 2008, *A&A* 488, L63
Jacobson, R. A. 2001, *BAAS*, 33, 1039
Laskar, J. 1988, *A&A*, 198, 341
Mignard, F. 1979, *Moon and Planets*, 20, 301
Veras, D. & Ford, E. B. 2010, *ApJ*, 715, 803
Wisdom, J., Peale, S. J., & Mignard, F. 1984, *Icarus*, 58, 137
Wright, J. T., Upadhyay., S. Marcy, G. W., *et al.* 2009, *ApJ*, 693, 1084

The Astrophysics of Planetary Systems: Formation, Structure, and Dynamical Evolution
Proceedings IAU Symposium No. 276, 2010
A. Sozzetti, M.G. Lattanzi & A.P. Boss, eds.

doi:10.1017/S1743921311020345

Evolution of spin direction of accreting magnetic protostars and spin-orbit misalignment in exoplanetary systems

Dong Lai[1,2], Francois Foucart[1] and Douglas N. C. Lin[3,4]

[1]Center for Space Research, Department of Astronomy, Cornell University, Ithaca, NY 14853
email: dong@astro.cornell.edu

[2]Kavli Institute for Theoretical Physics, University of California, Santa Barbara, CA 93106

[3]Department of Astronomy and Astrophysics, University of California, Santa Cruz, CA 95064

[4]Kavli Institute for Astronomy and Astrophysics, Peking University, Beijing, China

Abstract. Recent observations have shown that in many exoplanetary systems the spin axis of the parent star is misaligned with the planet's orbital axis. These have been used to argue against the scenario that short-period planets migrated to their present-day locations due to tidal interactions with their natal discs. However, this interpretation is based on the assumption that the spins of young stars are parallel to the rotation axes of protostellar discs around them. We show that the interaction between a magnetic star and its circumstellar disc can (although not always) have the effect of pushing the stellar spin axis away from the disc angular momentum axis toward the perpendicular state and even the retrograde state. Planets formed in the disc may therefore have their orbital axes misaligned with the stellar spin axis, even before any additional planet-planet scatterings or Kozai interactions take place. In general, magnetosphere–disc interactions lead to a broad distribution of the spin–orbit angles, with some systems aligned and other systems misaligned.

Keywords. accretion, accretion disks, planetary systems: protoplanetary disks, stars: magnetic fields

1. Introduction

As discussed by others in this conference (J. Winn, A. Triaud), the Rossiter-McLaughlin (RM) effect, an apparent radial velocity anomaly caused by the partial eclipse of a rotating parent star by its transiting planet, can be used to measure the sky-projected stellar obliquity, the angle between the stellar spin axis and the planetary orbital axis. Among the dozens of systems with RM measurements, about 60% have an orbital axis aligned (in sky projection) with the stellar spin, while the other systems show a significant spin-orbit misalignment, including at least 5 with retrograde orbits.

The solar system also provides a clue. Except for Pluto, all planets outside 1 AU lie within 2° of the ecliptic plane, while the Sun's equatorial plane is inclined by 7° with respect to the ecliptic. It is not clear whether the difference between 2° and 7° is significant or needs an explanation.

The process of planetary system formation can be roughly divided into two stages. In the first stage, which lasts a few million years until the dissipation of the gaseous protoplanetary disc, planets are formed and undergo migration due to tidal interaction with the gaseous disc. The second stage, which lasts from when the disc has dissipated to the present, involves dynamical gravitational interactions between multiple planets if they are produced in the first stage in a sufficiently close-packed configuration (e.g., Juric & Tremaine 2007; Chatterjee *et al.* 2008; Nagasawa *et al.* 2008), and/or secular interactions with a distant planet or stellar companion (Wu & Murray 2003; Fabrycky & Tremaine 2007). The eccentricity distribution of exoplanetary systems and the recent observational results on the spin – orbit misalignment indicate that the physical processes in the second stage play an important role in determining the properties of exoplanetary systems. Nevertheless, the importance of the first stage cannot be neglected as it sets the initial condition for the possible evolution in the second stage.

The main message of this talk is that magnetic interaction torque between a protostar and its disk can (but not always) push the stellar spin axis away from the disk axis. This implies that (i) protoplanetary disks do not have to be aligned with the stellar spin, and (ii) before few-body interaction starts (in the second stage), the planet's orbital axis may already be misaligned with the stellar spin. This talk is based on two papers, Lai, Foucart & Lin (2011) and Foucart & Lai (2011), in which more details and references can be found.

2. Physical Origin of the Magnetic Warping Torque

The interaction between a magnetic star and a disk is complex. However, the key physics related to the magnetic torques of interest here can be described robustly in a parametrized manner (see Fig. 1). The stellar magnetic field disrupts the accretion disc at the magnetospheric boundary, where the magnetic and plasma stresses balance:

$$r_{\rm in} = \eta \left[\mu^4/(GM_\star \dot{M}^2)\right]^{1/7}, \tag{2.1}$$

where μ is magnetic moment of the protostar, $\dot{M}$ is the mass accretion rate, and η is a dimensionless constant somewhat less than unity ($\eta \sim 0.5$). Before being disrupted, the disc generally experiences nontrivial magnetic torques from the star. These torques are of two types: (i) A warping torque $\mathbf{N}_w$ which acts in a small interaction region $r_{\rm in} < r < r_{\rm int}$, where some of the stellar field lines are linked to the disc. These field lines are twisted by the differential rotation between the star and the disc, generating a toroidal field $\Delta B_\phi = \mp\zeta B_z^{(s)}$ from the quasi-static vertical field $B_z^{(s)}$ threading the disc, where $\zeta \sim 1$ and the upper/lower sign refers to the value above/below the disc plane. Since the toroidal field from the stellar dipole $B_\phi^{(\mu)}$ is symmetric with respect to the disc plane, the net toroidal field differs above and below the disc plane, giving rise to a vertical force on the disc. While the mean force (averaging over the azimuthal direction) is zero, the uneven distribution of the force induces a net warping torque which tends to push the orientation of the disc angular momentum $\hat{\boldsymbol{l}}$ away from the stellar spin axis $\hat{\boldsymbol{\omega}}_s$. The essential physics of the warping torque can also be understood from the "laboratory" toy model depicted in Fig. 1(b). (ii) A precession torque $\mathbf{N}_p$ which arises from the screening of the azimuthal electric current induced in the highly conducting disc. This results in a difference in the radial component of the net magnetic field above and below the disc plane and therefore in a vertical force on the disc. The resulting precession torque tends to cause $\hat{\boldsymbol{l}}$ to precess around $\hat{\boldsymbol{\omega}}_s$.

The two magnetic torques (per unit area) on the disc can be written as

$$\mathbf{N}_w = -(\Sigma r^2 \Omega)\cos\beta\, \Gamma_w\, \hat{\boldsymbol{l}} \times (\hat{\boldsymbol{\omega}}_s \times \hat{\boldsymbol{l}}), \qquad \mathbf{N}_p = (\Sigma r^2 \Omega)\cos\beta\, \Omega_p\, \hat{\boldsymbol{\omega}}_s \times \hat{\boldsymbol{l}}, \tag{2.2}$$

where $\Sigma(r)$ is the surface density, $\Omega(r)$ the rotation rate of the disc, and $\beta(r)$ is the disc tilt angle (the angle between $\hat{\boldsymbol{l}}(r)$ and the spin axis $\hat{\boldsymbol{\omega}}_s$). The warping rate and precession

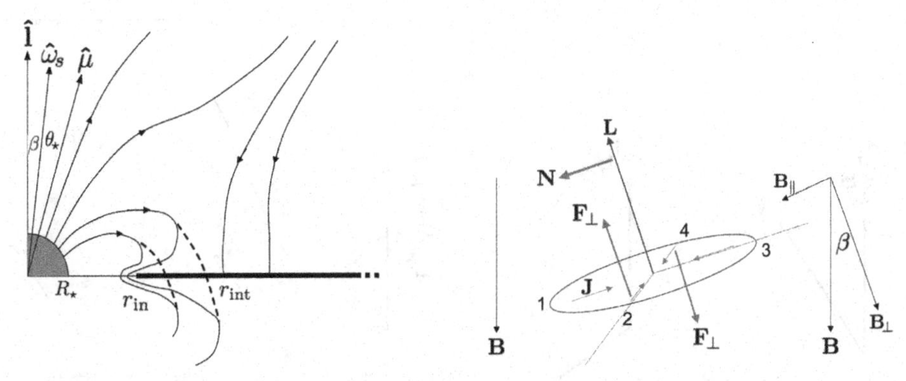

Figure 1. **(a)** A sketch of magnetic field configuration in a star – disc system for nonzero β (the angle between the disc axis and the stellar spin axis) and $\theta_\star$ (the angle between the stellar dipole axis and the spin axis). Part of the stellar magnetic fields (dashed lines) penetrate the disc in the interaction zone between the disc inner radius r_{in} and r_{int} in a cyclic manner, while other field lines are screened out of the disc. The closed field lines are twisted by the differential rotation between the star and the disc, which leads to a magnetic braking torque and a warping torque. The screening current in the disc leads to a precessional torque. **(b)** A toy model for understanding the origin of the warping torque. A tilted rotating metal plate (with angular momentum **L**) in an external magnetic field **B** experiences a vertical magnetic force around region 2 and 4 due to the interaction between the induced current **J** and the external $\mathbf{B}_\parallel$, resulting in a torque **N** which further increases the tilt angle β.

angular frequency at radius r are given by

$$\Gamma_w(r) = \frac{\zeta\mu^2}{4\pi r^7 \Omega(r)\Sigma(r)} \cos^2\theta_\star, \qquad \Omega_p(r) = -\frac{\mu^2}{\pi^2 r^7 \Omega(r)\Sigma(r)D(r)} \sin^2\theta_\star, \tag{2.3}$$

where $\theta_\star$ is the angle between the magnetic dipole axis and the spin axis, and the dimensionless function $D(r)$ is somewhat less than unity.

3. Disk Response to Magnetic Torques and Back-reaction on the Star

In the last section, we showed that the inner region of the disc where magnetic field lines connect star and the disc and where the disc rotates faster than the star ($\Omega > \Omega_s \cos\beta$) experiences a warping torque and a precessional torque. If we imagine dividing the disc into many rings, and if each ring were allowed to behave independent of each other, it would be driven toward a perpendicular state and precess around the spin axis of the central star. Obviously, real protoplanetary discs do not behave as a collection of non-interacting rings: Hydrodynamic stresses (bending waves and viscosity) provide strong couplings between different rings, and the disk will resist the inner disk warping. Detailed calculations by Foucart & Lai (2011) showed that for most reasonable stellar/disc parameters, the steady-state disc warp is rather small because of efficient viscous damping or propagation of bending waves. Thus the inner disk direction is approximate the same as the outer disk direction.

What is happening to the stellar spin? Is there a secular change to the stellar spin direction? Note that there are a hierarchy of timescales in this problem: (1) The dynamical time associated with the spin frequency, disc rotation frequency and the beat frequency $|\omega_s - \Omega|$. This is much shorter than the effects of interest here. (ii) The warping/precession timescale of the inner disc, of order a few months. (iii) The disc warp evolution timescale t_{disc} due to viscosity or bending wave propagation, of order 1000's of years. (iv) The stellar spin evolution timescale. The magnetic misalignment torque on the star is of the

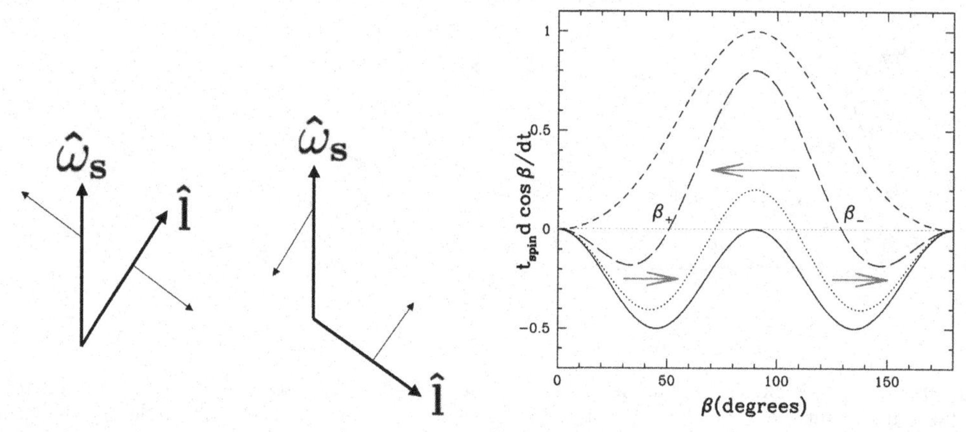

Figure 2. **(a)** A sketch of the effect of the magnetic warping torque. This torque tends to push the disc angular momentum axis $\hat{l}$ toward a state perpendicular to the spin axis When $\hat{l}$ is fixed by the outer disc, the back reaction torque tends to push the spin axis toward being misaligned with $\hat{l}$. **(b)** The rate of change of the stellar inclination angle β for a fixed disc rotation axis. From top to bottom, the four curves correspond to $(\lambda, \tilde{\zeta}) = (1, 0.5)$, $(0.8, 2)$, $(0.2, 2)$ and $(0, 2)$, respectively. The arrows indicate the direction of β evolution.

same order as the fiducial accretion torque $\mathcal{N}_0 = \dot{M}\sqrt{GM_\star r_{\rm in}}$. Assuming the spin angular momentum $J_s = 0.2 M_\star R_\star^2 \omega_s$, we find the spin evolution time

$$t_{\rm spin} = \frac{J_s}{\mathcal{N}_0} = (1.25\,{\rm Myr})\left(\frac{M_\star}{1\,M_\odot}\right)\left(\frac{\dot{M}}{10^{-8}M_\odot {\rm yr}^{-1}}\right)^{-1}\left(\frac{r_{\rm in}}{4R_\star}\right)^{-2}\frac{\omega_s}{\Omega(r_{\rm in})}. \tag{3.1}$$

We are interested in the evolution of the stellar spin direction on timescale of order $t_{\rm spin}$.

We can now consider the back-reaction torque from the disk on the stellar spin. For simplicity, let us assume small warps so that the disk is flat (averaged over the dynamical timescale). If the magnetic torque were the only torque acting on the star, then it is clear (see Fig. 2) that even if initially the stellar spin axis is approximately (but not perfectly) aligned with the disc axis, given enough time, the stellar spin axis will evolve towards the perpendicular state and even the retrograde state.

However, there are other torques acting on the star which counter-act the warping torque. In general, the evolution equation for the spin angular momentum of the star, $J_s\hat{\boldsymbol{\omega}}_s$, can be written in the form

$$\frac{d}{dt}\left(J_s\hat{\boldsymbol{\omega}}_s\right) = \boldsymbol{\mathcal{N}} = \boldsymbol{\mathcal{N}}_l + \boldsymbol{\mathcal{N}}_s + \boldsymbol{\mathcal{N}}_w + \boldsymbol{\mathcal{N}}_p. \tag{3.2}$$

Here $\boldsymbol{\mathcal{N}}_l$ represents the torque component (including the accretion torque) that is aligned with the inner disc axis:

$$\boldsymbol{\mathcal{N}}_l = \lambda \dot{M}(GM_\star r_{\rm in})^{1/2}\,\hat{\boldsymbol{l}}, \tag{3.3}$$

where $\lambda \lesssim 1$ is a dimensionless parameter. The term $\boldsymbol{\mathcal{N}}_s = -|\mathcal{N}_s|\hat{\boldsymbol{\omega}}_s$ represents a spin-down torque carried by a wind/jet from the open field line region of the star. The term $\boldsymbol{\mathcal{N}}_w$ and $\boldsymbol{\mathcal{N}}_p$ represent the back-reactions of the warping and precessional torques, respectively. Note that both $\boldsymbol{\mathcal{N}}_w$ and $\boldsymbol{\mathcal{N}}_p$ are of order $\mu^2/r_{\rm in}^3$, which does not directly depend on $\dot{M}$. But since $r_{\rm in}$ depends on $\dot{M}$, we find that both $\boldsymbol{\mathcal{N}}_w$ and $\boldsymbol{\mathcal{N}}_p$ are of the same order of magnitude as the fiducial accretion torque $\mathcal{N}_0$. The inclination angle of the stellar spin relative to the disc evolves according to the equation

$$\frac{d}{dt}\cos\beta = \frac{\mathcal{N}_0}{J_s}\sin^2\beta\left(\lambda - \tilde{\zeta}\cos^2\beta\right), \qquad \text{with } \tilde{\zeta} = \frac{\zeta'\cos^2\theta_\star}{6\eta^{7/2}}. \tag{3.4}$$

Equation (3.4) is our key result. It reveals the following behavior for the evolution of β (see Fig. 2): (i) For $\lambda = 0$, equation (3.4) describes the effect of the magnetic warping torque acting alone on the star. This torque always pushes the stellar spin toward anti-alignment with $\hat{l}$. (ii) For $\tilde{\zeta}/\lambda < 1$: Regardless of the initial β, the spin always evolves towards alignment. (iii) For $\tilde{\zeta}/\lambda > 1$: There are two possible directions of β evolution, depending on the initial value of β. The condition $d\cos\beta/dt = 0$ yields two "equilibrium" states (β_+ and β_-), given by $\cos\beta_\pm = \pm\sqrt{\lambda/\tilde{\zeta}}$. Of the two equilibria, one is stable (β_+) and the other is unstable. For $\beta(t=0) < \beta_-$, the system will evolve towards a misaligned prograde state β_+; for $\beta(t=0) > \beta_-$, the system will evolve towards the anti-aligned state ($\beta = 180°$).

4. Discussion

Note that there is a 90-degree barrier: Starting from a small angle β, the systen cannot evolve into retrograde if the outer disk orientation is fixed. However, it is possible to produce retrograde systems in two ways: (i) If the outer disk changes direction (e.g., due to external perturbers/flybys); (ii) If the initial condition is retrograde: this may be the case if we consider disk formation in turbulent star-forming clouds, as suggested by Bate *et al.* (2010). But note that even in this scenario, the magnetic misalignment effect discussed above is important – without it, the stellar spin would align with the disk axis on a short timescale. In fact, starting from a random distribution of the initial relative orientation between the spin and disk axes, we can calculate the evolution of the stellar obliquity distribution as a function of time. For "weak" magnetic torques ($\tilde{\zeta}/\lambda < 1$), a distribution peaked at $\beta = 0$ is produced. For "strong" magnetic torques ($\tilde{\zeta}/\lambda > 1$), we produce a bi-modal distribiton: one peak is at β_+, another at $\beta = 180°$ (see Lai *et al.* 2010).

How to test the effects discussed above? One way is to measure the spin-orbit angles for systems with two transiting planets (Kepler-9 is an example). Another is to measure the orientation of the stellar spin and its disk. Recently, Watson *et al.* (2011) carried out such an analysis for several debris disc systems and found no significant difference between $\sin i_\star$ and $\sin i_{\rm disc}$. Note that $i_\star = i_{\rm disc}$ does not necessarily imply alignment between the spin axis and the disc axis. Also, systematic uncertainties in estimating $i_{\rm disc}$ need to be taken into account. For example, for HD 22049 (one of the best cases studied by Watson *et al.*), the disc inclination is consistent with face-on ($i_{\rm disc} \lesssim 25°$; Backman *et al.* 2009). It would be useful to do this for classical T Tauri stars.

References

Backman, D., *et al.* 2009, *ApJ*, 690, 1522
Chatterjee, S., *et al.* 2008, *ApJ*, 686, 580
Fabrycky, C. & Tremaine, S. 2007, *ApJ*, 669,1298
Foucart, F. & Lai, D. 2011, *MNRAS*, 412, 2799
Juric, M. & Tremaine, S. 2008, *ApJ*, 686, 603
Lai, D., Foucart, F., & Lin, D. N. C.. 2011, *MNRAS*, 412, 2790
Nagasawa, M., Ida, S., & Bessho, T. 2008, *ApJ*, 678, 498
Wu, Y. & Murray, N. W. 2003, *ApJ*, 589, 605
Watson, C. A., *et al.* 2011, *MNRAS*, 413, L71

The Astrophysics of Planetary Systems: Formation, Structure, and Dynamical Evolution
Proceedings IAU Symposium No. 276, 2010
A. Sozzetti, M.G. Lattanzi & A.P. Boss, eds.

doi:10.1017/S1743921311020357

Hamiltonian model of capture into mean motion resonance

Alexander J. Mustill[1] and Mark C. Wyatt[1]

[1]Institute of Astronomy, University of Cambridge,
Madingley Road, CB3 0HA, Cambridge, UK
email: ajm233@ast.cam.ac.uk, wyatt@ast.cam.ac.uk

Abstract. Mean motion resonances are a common feature of both our own Solar System and of extrasolar planetary systems. Bodies can be trapped in resonance when their orbital semi-major axes change, for instance when they migrate through a protoplanetary disc. We use a Hamiltonian model to thoroughly investigate the capture behaviour for first and second order resonances. Using this method, all resonances of the same order can be described by one equation, with applications to specific resonances by appropriate scaling. We focus on the limit where one body is a massless test particle and the other a massive planet. We quantify how the the probability of capture into a resonance depends on the relative migration rate of the planet and particle, and the particle's eccentricity. Resonant capture fails for high migration rates, and has decreasing probability for higher eccentricities, although for certain migration rates, capture probability peaks at a finite eccentricity. We also calculate libration amplitudes and the offset of the libration centres for captured particles, and the change in eccentricity if capture does not occur. Libration amplitudes are higher for larger initial eccentricity. The model allows for a complete description of a particle's behaviour as it successively encounters several resonances. The model is applicable to many scenarios, including (i) Planet migration through gas discs trapping other planets or planetesimals in resonances; (ii) Planet migration through a debris disc; (iii) Dust migration through PR drag. The Hamiltonian model will allow quick interpretation of the resonant properties of extrasolar planets and Kuiper Belt Objects, and will allow synthetic images of debris disc structures to be quickly generated, which will be useful for predicting and interpreting disc images made with ALMA, Darwin/TPF or similar missions. Full details can be found in Mustill & Wyatt (2011).

Keywords. celestial mechanics, planets and satellites: general, planetary systems

---

1. Introduction

Mean motion resonances (MMRs) occur when two objects' orbital periods are close to a ratio of two integers, and a particular combination of orbital angles, the resonant argument, is librating. Examples in the Solar System include Neptune and Pluto (3:2 resonance) and the inner Galilean moons of Jupiter (4:2:1 Laplace resonance). There are also now numerous examples of suspected or confirmed MMRs in extrasolar planetary systems (e.g., GJ 876 b and c in a 2:1 resonance, Laughlin & Chambers 2001).

Mean motion resonances also occur between planets and small dust particles, as seen in the Earth's resonant dust ring (Dermott *et al.* 1994). Some extrasolar debris discs, such as Vega, show evidence of non-axisymmetric clumps (Holland *et al.* 1998; Wilner *et al.* 2002), and several authors have attempted to model these as arising from a planet's resonant perturbations (e.g., Kuchner & Holman 2003; Wyatt 2003).

Resonances are common because convergent migration between orbiting bodies can cause them to become captured. There are many mechanisms by which such a semi-major axis change can be driven. Early work looked at the tidal evolution of satellite

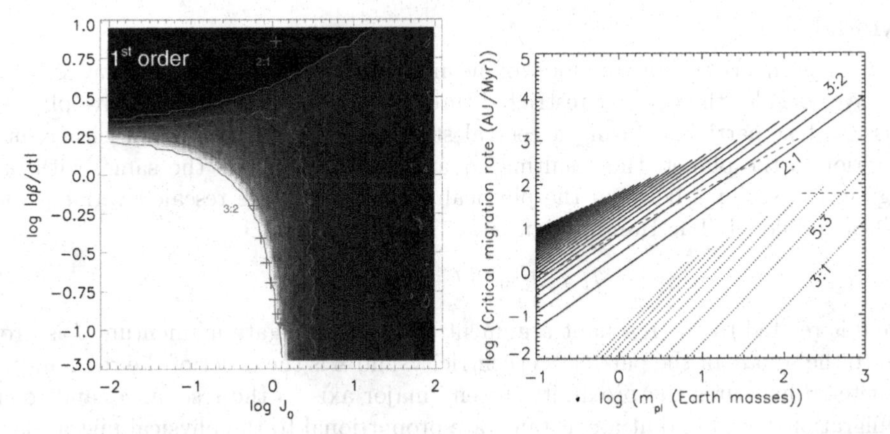

Figure 1. Left: Capture probability for first order resonances, as a function of rescaled eccentricity J_0 and migration rate $\dot{\beta}$. Colour scale and contours show capture probablity, from white (100%) to black (0%). The crosses show the rescaled eccentricity and migration rate corresponding to a particle with Keplerian eccentricity 0.01, migrating at 1 AU Myr^{-1} into exterior resonances with an Earth-mass planet orbiting a Solar-mass star at 1 AU. The 2:1 and 3:2 resonances are labelled. **Right:** Critical migration rate at 1 AU for capture of low-eccentricity particles beign trapped into resonances with a migrating planet. At migration rates faster than this, capture is impossible at low eccentricity. The critical migration rates for first order resonances are shown as solid lines. We show typical migration rates for planets embedded in gas discs at 1 AU (dashed lines). Mass-dependent Type I migration occurs for low mass planets and mass-independent Type II migration for high mass planets.

orbits (Goldreich 1965). In a protoplanetary disc, planets can migrate by tidal interaction with the gas disc (see Chambers 2009, for a recent review), and small planetesimals by aerodynamic drag (Weidenschilling 1977). In a gas-depleted debris disc, planets can migrate by gravitational scattering of planetesimals (Kirsh *et al.* 2009). Interplanetary dust drifts towards the Sun under the influence of Poynting-Robertson (PR) drag (Burns *et al.* 1979).

Resonance capture has been studied by several authors. The regime of adiabatic migration, where the migration timescale is much longer than the resonant argument's libration timescale, has been studied extensively analytically using a Hamiltonian model (e.g., Henrard 1982; Borderies & Goldreich 1984). Rapid migration was studied using full N-body models by Wyatt (2003, henceforth W03) for the case of a planet migrating into a planetesimal disc, and using the Hamiltonian model by Quillen (2006, henceforth Q06) for general migration scenarios. Q06 obtained capture probability as a function of migration rate and eccentricity. We have extended this work, and using the Hamiltonian model we calculate capture probabilities, libration amplitudes and offsets for particles that are captured, and eccentricity jumps for those that pass through the resonance without capture.

The Hamiltonian model we use has several advantages over N-body simulations: (1) it allows some results to be derived analytically; (2) it is faster to integrate numerically than the 3-body problem; (3) all resonances of the same order reduce to a Hamiltonian of the same form, with fewer free parameters than the three-body problem. Once a suite of numerical integrations of the Hamiltonian model is performed, it can be applied to any system, without the need for running a different N-body integration every time the system parameters are changed.

2. Model

We work with a Hamiltonian model of mean motion resonance (e.g., Murray & Dermott 1999). We consider the circular restricted three body problem with a massive planet and a massless test particle orbiting a central star. The test particle orbits either interior or exterior to the planet; the resulting equations of motion are the same, with only a change in the scalings between the physical variables and the rescaled variables of the Hamiltonian model. The Hamiltonian is

$$\mathcal{H} = J^2 + \beta J - J^{k/2} \cos k\theta, \tag{2.1}$$

where θ is related to the resonant argument and the conjugate momentum J is proportional to the square of the particle's eccentricity, and k is the order of the resonance. The parameter β measures the proximity in semi-major axis to the resonance, and to simulate migration β is varied at a constant rate proportional to the physical migration rate. The equations of motion arising from this Hamiltonian were integrated numerically. We varied the initial momentum J_0 and the migration rate $\dot{\beta}$. There are two parameters in the problem: the eccentricity of the particle when it hits the resonance (governed by J_0, the initial value of J), and the particle's or planet's migration rate (governed by $\dot{\beta}$). For each point in $(J_0, \dot{\beta})$ parameter space we integrated 100 trajectories, with the resonant argument chosen from a uniform distribution over $[0, 2\pi)$.

Figure 1 (left) shows capture probabilities for first order resonances as a function of J_0 and $\dot{\beta}$. We see that capture into resonance is guaranteed for small initial eccentricities and migration rates. For low migration rates we are in the well-studied adiabatic regime. For low eccentricities capture is certain since the separatrix forms around the initial trajectory. For high eccentricities the separatrix forms inside the initial orbit and expands to meet it as the migration continues. Capture then is probabilistic, with a probability that decreases as the initial eccentricity increases (Henrard 1982). For low eccentricities, with $J_0 < 1.3$, capture is certain if the migration rate is low and impossible if it is high. In the limit of low eccentricity, the transition occurs at a critical migration rate of $|\dot{\beta}| \approx 2.1$. Certain capture occurs for low migration rates up to $J_0 \approx 1.3$. For higher eccentricities ($J_0 > 1.3$), capture is always probabilistic, with a capture probability that is not strongly dependent of migration rate; however, if migration rate is too high, then capture is still impossible.

We also found the amplitudes and centres of libration of the resonant argument, and the change in eccentricity if a particle is not captured. In the presence of migration, the centre of libration is offset from the centre in the absence of migration by an amount proprtional to the migration rate, almost independent of eccentricity. The amplitude of libration increases slightly with migration rate, and significantly with the particle's initial eccentricity. If a particle is not captured into resonance, its eccentricity is driven down if capture failed due to high eccentricity (in agreement with adiabatic theory; Murray & Dermott 1999), up if capture failed due to fast migration but low eccentricity, and either up or down if both eccentricity and migration rate are high.

In Mustill & Wyatt (2011) we also investigated second-order resonances.

We validated the model against the N-body integrations of W03. The Hamiltonian model generally shows good agreement with the results of N-body integrations.

3. Applications

Using the Hamiltonian model, the result of an encounter with a resonance in many scenarios can be easily determined by rescaling variables and looking up the outcome in the grid of integration results. For example, Figure 1 shows the rescaled eccentricity and migration rate for a particle with Keplerian eccentricity 0.01, migrating at 1 AU Myr^{-1} into exterior resonances with an Earth-mass planet orbiting a Solar-mass star at 1 AU. We see that it is impossible for the particle to be captured into the 2:1 resonance. If migration continues it is almost certain, however, to be captured into the 3:2 resonance, and if it is not it will be captured into the 4:3 resonance. Because determining the fate of a particle is reduced to looking up the outcome in a data table, the fate can be determined much faster than it can by performing N-body integrations. This makes the model very suitable for generating synthetic images of debris discs with many particles (outward migration of the planet and inward migration of the dust can both be handled in the same way), and for determining the outcome of resonant encounters during planet formation in population synthesis models.

3.1. *Debris discs*

In Mustill & Wyatt (2011) we outline how the degree of dynamical excitation of a planetesimal disc can affect the observed structure if a planet migrates into it. More dynamically excited discs will show weaker resonant signatures, since the capture probability is lower and the amplitude of libration higher (see also Reche *et al.* 2008). Future work will explore disc structures in detail using this model.

3.2. *Protoplanetary discs*

In Figure 1 (Right) we show the critical migration rate for capture of particles interior to a migrating planet. Also indicated are typical migration rates. It is straightforward to determine which resonance the planet will capture particles into. For example, a 10 Eart mass planet undergoing Type I migration will capture bodies into the 4:3 resonance, while a 1000 Earth mass planet undergoing Type II migration will capture bodies into the 5:3 second-order resonance. Future work will address the issue of how accurate the Hamiltonian model is at describing the dynaics of two massive planets.

References

Borderies, N. & Goldreich, P. 1984, *Celestial Mechanics*, 32, 127
Burns, J. A., Lamy, P. L., & Soter, S. 1979, *Icarus*, 40, 1
Chambers, J. E. 2009, *AREPS*, 37, 321
Dermott, S. F., Jayaraman, S., Xu, Y. L., Gustafson, B. Å. S., & Liou, J. C. 1994, *Nature*, 369, 719
Goldreich, P. 1965, *MNRAS*, 130, 159
Henrard, J. 1982, *Celestial Mechanics*, 27, 3
Holland, W. S., Greaves, J. S., Zuckerman, B., Webb, R. A., McCarthy, C., Coulson, I. M., Walther, D. M., Dent, W. R. F., Gear, W. K., & Robson, I. 1998, *Nature*, 392, 788
Kirsh, D. R., Duncan, M., Brasser, R., & Levison, H. F. 2009, *Icarus*, 199, 197
Kuchner, M. J. & Holman, M. J. 2003, *ApJ*, 588, 1110
Laughlin, G. & Chambers, J. E. 2001, *ApJL*, 551, L109
Murray, C. D. & Dermott, S. F. 1999, Solar System Dynamics, ed. C. D. Murray & S. F. Dermott
Mustill, A. J. & Wyatt, M. C. 2011, *MNRAS*, 413, 554
Quillen, A. C. 2006, *MNRAS*, 365, 1367
Reche, R., Beust, H., Augereau, J., & Absil, O. 2008, *A&A*, 480, 551
Weidenschilling, S. J. 1977, *MNRAS*, 180, 57
Wilner, D. J., Holman, M. J., Kuchner, M. J., & Ho, P. T. P. 2002, *ApJL*, 569, L115
Wyatt, M. C. 2003, *ApJ*, 598, 1321

The Astrophysics of Planetary Systems: Formation, Structure, and Dynamical Evolution
Proceedings IAU Symposium No. 276, 2010
A. Sozzetti, M.G. Lattanzi & A.P. Boss, eds.

doi:10.1017/S1743921311020369

Turning solar systems into extrasolar planetary systems in stellar clusters

Melvyn B. Davies[1]

[1]Lund Observatory, Box 43, SE-221 00, Lund, Sweden
email: mbd@astro.lu.se

Abstract. Many stars are formed in some form of cluster or association. These environments can have a much higher number density of stars than the field of the galaxy. Such crowded places are hostile environments: a large fraction of initially single stars will undergo close encounters with other stars or exchange into binaries. We describe how such close encounters and exchange encounters will affect the properties of a planetary system around a single star. We define singletons as single stars which have never suffered close encounters with other stars or spent time within a binary system. It may be that planetary systems similar to our own solar system can only survive around singletons. Close encounters or the presence of a stellar companion will perturb the planetary system, leading to strong planet-planet interactions, often leaving planets on tighter and more eccentric orbits. Thus, planetary systems which initially resembled our own solar system may later more closely resemble the observed extrasolar planetary systems.

Keywords. planetary systems, open clusters and associations: general, binaries: general

1. Introduction

As of November, 2010, a little over 500 planets have been discovered orbiting around other stars. A large fraction of these extrasolar planetary systems are different from our own Solar System: planets as massive as Jupiter are found on much tighter, and eccentric orbits compared to the wider, and essentially circular, orbits of our own gas giants. Interactions between Jupiter and Saturn cause their orbits to change slightly, tilting their planes and changing their eccentricities, but the changes are *oscillatory*: their amplitudes do not grow over time or cause the planets to scatter strongly off each other. If left alone, the gas giants of our Solar System will tick for ever. Is this *Solar–System–like* behaviour common, or rare, throughout the entire population of planetary systems? It could be rare, if planetary systems tend to form with more–massive planets arranged more closely together. Planet–planet interactions would then often lead to strong scatterings between planets, ejecting some whilst leaving others on more bound and eccentric orbits. Indeed such a mechanism is found to match the observed distribution of exoplanet orbits (eg Juric & Tremaine 2008). However, there are some certainly some known multiple planet systems beyond our own Solar System where the planets are seen to be on relatively circular orbits (eg see Fig. 11 of Lovis *et al.* 2011). In these systems, migration has almost certainly brought in the massive planets to orbits substantially closer than the case for our own Jupiter. It is possible that a subset of these systems are solar–system–like by my above definition, in that they are stable if left alone.

In this contribution, we consider what would happen to solar–system–like planetary systems within stellar clusters. This is an important question as all stars form in some sort of stellar cluster or association. Indeed there is evidence that the Sun was formed in a cluster containing about 1000 stars (Adams 2010). We consider two processes which may transform these stable planetary systems into unstable ones, where planets can be ejected. Firstly, close, fly-by encounters with other stars will perturb planetary orbits and even, sometimes, directly eject planets during an encounter. Secondly, if a single star encounters a binary system, it may exchange in to the binary replacing one of the original binary components. The stellar companion within a binary may, over time, perturb the orbits of any planets. How often do fly-by encounters and binary companions de-stabilise planetary systems? What fraction of Solar–System–like planetary systems could survive intact?

We introduce here the idea of a *Singleton*, which is defined as

1) a star which has not formed in a binary,

2) a star which has not later spent time within a binary system,

3) a star which has not suffered close encounters with other stars.

Thus, if we consider for now planetary systems formed around only single stars, the question of solar-system survival within a stellar cluster becomes one of singleton fraction. The frequency of encounters will depend in part on how long a particular star spends within a stellar cluster. Some clusters will break up very quickly as gas is driven out by supernovae or winds from massive stars unbinding the system. Other clusters will survive this early phase of mass loss but the stars will escape from the cluster over a few hundred million years or so, driven by the combined effects of two–body scattering and tidal stripping, populating the field of the Galaxy.

Below we first review what can happen to stars, binaries, and planetary systems within stellar clusters. We then consider in more detail the effects of close encounters and stellar companions on planetary systems, concluding with an estimate of the likely fraction of solar systems which will be affected by being formed in a clustered environment.

2. What happens within a cluster

Young stellar clusters contain a few hundred to about one thousand stars within a radius of about 1pc. One can compute the time required for a particular star to undergo an encounter with another star passing within some distance, rmin, which may be approximated by:

$$\tau_{\rm enc} \simeq 3.3 \times 10^7 \, {\rm yr} \left(\frac{100 \ {\rm pc}^{-3}}{n} \right) \left(\frac{v_\infty}{1 \ {\rm km/s}} \right) \left(\frac{10^3 \ {\rm AU}}{r_{\rm min}} \right) \left(\frac{{\rm M}_\odot}{m_{\rm t}} \right). \tag{2.1}$$

Here n is the stellar number density in the cluster, v_∞ is the mean relative speed at infinity of the stars in the cluster, $r_{\rm min}$ is the encounter distance and $m_{\rm t}$ is the total mass of the stars involved in the encounter. The cross–section for an interaction is increased greatly by what is known as gravitational focusing, where stars are deflected towards each other because of their mutual gravitational attraction. This effect is has been included. The terms in the above equation are of order unity for a typical cluster of a few hundred stars. In other words, for the clusters we consider here, a star will have another star pass within 1000 AU every 30 Myr or so. Encounters can therefore play an important role in the evolution of a planetary system formed in stellar clusters.

We have performed N–body simulations of stellar clusters for a range of masses and radii, modelling the evaporation of the clusters in the Galactic tidal field (Malmberg *et al.* 2007). We followed the histories of all stars considering how many of the stars which were initially single had close encounters or exchanged into binaries. Close encounters here are defined as any encounter within 1000 AU. This is also the typical size of the widest binaries that can survive in the stellar clusters we consider. We find that the singleton fraction (ie those stars having no fly-bys and not exchanging into binaries) can be as low as 10–20% for solar–like stars. Most stars will have at least one encounter within 1000 AU with another star, and about 10% will have spent time in binaries, although most of these stars are single again by the end of the simulation as many binaries are broken up during binary-binary encounters.

3. The effects of fly-bys on planetary systems

We consider now the effects of the flybys. We have simulated flyby encounters between intruding single stars of various masses and a solar–mass star with a planetary system (Malmberg, Davies & Heggie 2011). We find that 20–40% of encounters within 100 AU lead to the direct ejection of at least one planet from a system containing the four gas giants of the solar system. In addition, we find a further 20–40% of encounters lead to the ejection of at least one planet within 100 Myr of the encounters as the planetary systems became unstable due to perturbations in the planetary orbits during the flyby.

As planetary systems become unstable, planet–planet scattering leaves some planets on more bound orbits, whilst others are left on wider orbits. These latter planets will be ejected after several scatterings. However, a snap–shot of all planetary systems some time after encounters would reveal a population of planets on very wide, eccentric orbits (100–1000 AU). This is also the case for planetary systems which are unstable from birth (Scharf & Menou 2009; Veras *et al.* 2009). We also find that planets are sometimes grabbed by the intruding star. We find that these planets are typically on orbits of between 10 and 100 AU and a broad range of eccentricities. This planet could destabilise a planetary system around the intruding star.

4. The effects of binary companions on planetary systems

We consider now the situation where a star with a planetary system has had an encounter with a stellar binary and exchanged into the binary. The stellar companion may now perturb the planetary system. If the orbit of the stellar companion is relatively highly inclined with respect to the plane of the planetary system, the Kozai mechanism may operate (Kozai 1962). For a single planet, this will lead to so–called Kozai cycles where the eccentricity of the planet is seen to oscillate. In the case of planetary systems containing multiple planets, the outer planet will undergo Kozai oscillations on the shortest timescale. Providing the orbit of the stellar companion is sufficiently inclined, the orbit of the outer planet may, periodically, cross the orbit of planets which orbit more closely to the host star. This leads to a period of strong scattering between the planets, with some being ejected and others left on more bound, and eccentric, orbits (Malmberg, Davies & Chambers 2007 ; Malmberg & Davies 2009). In some cases, one is left with a single planet orbiting around the star, undergoing Kozai oscillations. In other cases, mutual planet–planet interactions may protect the planetary system from the effects of the Kozai mechanism, providing the timescale for the changes in planetary orbital elements due to planet–planet interactions is shorter than that due to the Kozai mechanism caused by the stellar companion.

In addition, if the orbit of the stellar companion is extremely inclined (ie within a few degrees of 90 degrees) then the Kozai effect will place the planet on extremely eccentric orbits, with maximum eccentricities close to unity. In this case, tidal interactions between the planet and the host star may circularize the orbit, leaving the planet on a much tighter orbit, which may also be rather inclined compared to its initial orbital plane (eg Fabrycky & Tremaine 2007; Wu, Murray & Ramshai 2007). Indeed observations of the so–called Rossiter-McLaughlin Effect in transiting planets suggests that at least some systems are highly inclined (eg Triaud *et al.* 2010).

5. The likely fraction of planetary systems affected

We have performed N–body simulations of stellar clusters to measure the rate of close encounters and exchanges into binary systems. Subsequently, we have performed simulations of close encounters and their subsequent effects on planetary systems to calculate the cross section to significantly alter a planetary system (eg to eject planets). We have also modelled the effects of stellar companions on planetary orbits via the Kozai effect.

We can now combine all of these results to estimate the fraction of planetary systems which are likely to be affected by either flybys or stellar companions within stellar clusters. We consider here only solar–mass stars, having the four gas giants of our own Solar System. Based on the results of our calculations for clusters initially containing 700 stars, we estimate that some 15% of systems will lose at least one planet due to flyby encounters. Stellar companions will account for planetary losses in about 5% of planetary systems. These figures will change with the initial number of stars in the cluster but only relatively slowly (Malmberg, Davies & Heggie 2011). We therefore conclude that stable planetary systems can be made unstable by flybys or stellar companions interestingly often.

We should also recall that here: 1) we have assumed that planetary systems are only formed around single stars, and 2) taken a relatively modest binary fraction within our stellar clusters (one third). Relaxing either of these constraints would increase the number of systems effected within binaries.

References

Adams, F. C. 2010, *ARA&A*, 48, 47
Fabrycky, D. & Tremaine, S. 2007, *ApJ*, 669, 1298
Jurić, M. & Tremaine, S. 2008, *ApJ*, 686, 603
Kozai, Y. 1962, *ApJ*, 67, 591
Lovis, C., *et al.* 2011, *A&A*, 528, A112
Malmberg, D. & Davies, M. B. 2009, *MNRAS*, 394, L26
Malmberg, D., Davies, M. B., & Chambers, J. E. 2007, *MNRAS*, 377, L1
Malmberg, D., de Angeli, F., Davies, M. B., Church, R. P., Mackey, D., & Wilkinson, M. I. 2007, *MNRAS*, 378, 1207
Malmberg, D., Davies, M. B., & Heggie, D. C. 2011, *MNRAS*, 411, 859
Scharf, C. & Menou, K. 2009, *ApJ*, 693, L113
Triaud, A. H. M. J., *et al.* 2010, *A&A*, 524, A25
Veras, D., Crepp, J. R., & Ford, E. B., 2009, *ApJ*, 696, 1600
Wu, Y., Murray, N. W., & Ramshai, J. M., 2007, *ApJ*, 670, 820

PART 4:
THE NEXT DECADE

The Astrophysics of Planetary Systems: Formation, Structure, and Dynamical Evolution
Proceedings IAU Symposium No. 276, 2010
A. Sozzetti, M.G. Lattanzi & A.P. Boss, eds.

doi:10.1017/S1743921311020370

A roadmap towards habitable exoplanets by the Blue Dots Initiative

V. Coudé du Foresto[1], for the Blue Dots participants

[1]Observatoire de Paris – LESIA, Meudon, Paris, France
email: vincent.foresto@obspm.fr

Abstract. This paper is an abridged version of the report produced by the Blue Dots initiative, whose activities include the elaboration of a roadmap towards the spectroscopic characterization of habitable exoplanets. The full version of the Blue Dots report can be downloaded at http://www.blue-dots.net/spip.php?article105. While the roadmap will need to be updated regularly, it is expected that the methodology developed within Blue Dots will provide a durable framework for the elaboration of future revisions.

Keywords. planetary systems

1. Introduction

The Blue Dots initiative was created in 2008 to contribute towards building a community in Europe around the exoplanet theme, and to converge towards a strategy enabling a more coherent approach to Calls for Proposals in ground- and space-based projects. The scope of the initiative is science-oriented and not restricted to a particular detection technique.

The intiative gathers more than 180 scientists, mostly located in Europe, with additional participation from the US, Japan, and India. Participants are organized in different working groups covering the relevant science themes (Targets and their Environments, Formation and Evolution of Planetary Systems, Habitability Criteria, Observation of Planetary Atmospheres) and methods (Single Aperture Imaging, Multiple Aperture Imaging, Microlensing, Modelling Habitable Planets, Radial Velocities, Astrometry, Transits). The complete report and most of the material produced by the initiative can be found on its web site at http://www.blue-dots.net.

2. Elements for a Roadmap

The initial Blue Dots activity consisted of preparing a roadmap towards the detection and characterization of habitable exoplanets, recognizing that this ambitious goal will require several intermediate steps.

To achieve this a framework for discussion had to be created and several underlying principles were applied. First, science questions should drive the roadmap – techniques should be seen as tools to address these questions. Specific missions should be introduced as late as possible in the process if convergence is to be sought. We also felt it was important to clearly identify the points of consensus and the matters of debate. Points of consensus provide an opportunity for the community to send out a common message. Scientific debates are obviously healthy and useful when they can be organized in a way to help clarify the issues.

A roadmap should also find a way to get around the "pathfinder dilemma" which could be expressed this way: it is clear that the goal of spectroscopic characterization of the atmosphere of habitable planets, in search of biomarkers, will ultimately require one or more very ambitious and innovative missions. Those cannot meet the feasibility criteria as currently established by the space agencies, whose current trend is to select missions which are both low-risk and with an immediate science return. And the immediate science return of a more affordable demonstrator (one that would retire the risk on the bigger mission) does not necessarily meet the agency standards.

The frame of discussion required the creation of homogeneous grids that would cover the science that can can/should be achieved, and the methods that can be employed to achieve it. These are presented below.

2.1. *Science Potential Levels*

Following the lines of the step-by-step approach mentioned above, for a given class of objects the **science potential level** (SPL) of a technique is defined by its capacity:

∗ To carry out a statistical study of objects in a given class;
∗∗ To designate targets in the solar neighborhood for spectroscopic follow-up study;
∗∗∗ To carry out a spectroscopic characterization of the object.

2.2. *Target Classes*

These classes are not meant to categorize objects according to their physical nature – rather they group objects of similar detection difficulty. The five classes identified are (by order of increasing detection difficulty):

(*a*) **Hot giant planets**: these planets can be hot either because they are close to their host stars and highly irradiated, or because they are young;

(*b*) **Other giant planets**: these planets include the warm and cold gaseous giants, down to Neptune size;

(*c*) **Hot telluric planets** that might be hot, young or slightly more massive than the Earth (super-Earth), not necessarily located in the habitable zone neither around specific stars;

(*d*) **Telluric planets in the habitable zone of M-type stars**;

(*e*) **Telluric planets in the habitable zone of solar-type stars**.

2.3. *Observing Methods*

The methods employed for detecting and characterizing exoplanets can be classified in six large families, which are represented in the corresponding Blue Dots working groups:

- The **microlensing** method which consists in monitoring the photometry of distant stars in order to detect microlensing events.
- The **transit photometry** method which relies on measuring the relative change of the photometry of the star due to a primary eclipse (the planet transiting in front of the stellar photosphere) or to a secondary eclipse (the planet disappearing behind the star).
- The **radial velocities** method (RV) which relies on measuring the Doppler shift of star spectra with high precision in order to detect the reflex motion due to the presence of one or several planets. This method also includes timing techniques.
- The **astrometry** method whether in narrow-angle or globally consists in measuring the relative position of stars in order to detect the reflex motion due to the presence of orbiting planets.
- The **single aperture imaging** (SAI) technique includes all types of coronagraphic methods, including external occulters as well as imaging techniques using Fresnel lenses, in order to separate the direct light of the planets from the stellar light which is usually hidden.

• The **multiple aperture imaging** (MAI) technique uses interferometric nulling, hypertelescopes, etc. to extract the direct light from the planet and to some extent the stellar light.

It is clearly understood that these methods are in many ways interrelated and complementary, inasmuch as observables obtained by one technique can often be interpreted only with the help of additional information from another technique. The reader is referred to the Blue Dots report for more details on each method, its current state of the art, the prospective performance, and the required R&D efforts needed to achieve them.

2.4. *Project Scales*

Projects related to the different detection methods are listed and discussed in the Blue Dots report. It is recognized that those projects, which can be ground-based instruments or facilities, or spaceborne missions, can be of different scales. These are symbolized by different letter codes:

• **E** code: existing, or already programmed efforts on an existing facility;

• **G** code: projects which are not yet funded but whose effort is equivalent to that of a ground based instrument on an extremely large telescope ($\simeq$30 M euro, 5-10 years);

• **M** code: projects whose effort is comparable to an ESA M-class mission ($\simeq$450 M euro);

• **L** code: flagship projects corresponding to an ESA L-class mission ($\simeq$650 M euro, 10-15 years), or even larger projects (**XL**, $\geqslant$1 G euro) carried out in worldwide collaboration where ESA's participation would be an L-size mission in itself. A typical time frame for such projects is 20+ years.

3. Producing Synthetic Grids

With this approach, it is possible to produce a grid that provides a synthetic view of the potential of each family of detection methods for different classes of exoplanets (Figure 1). Another way to present the information is to produce a timeline grid (Figure 2) which presents the progression in observation capacity for each family of methods, as a function of the increasing size of the projects.

In the context of the goal of Blue Dots (spectroscopic characterization of habitable telluric exoplanets), a roadmap should be represented by the path of least effort from the current state of the art (the E-coded boxes, located mostly in the upper left of the grid of Figure 1) to where lies the goal of Blue Dots (SPL ∗∗∗ in the last two columns, or in the last column if one is only concerned with earth twins around solar twins).

	Planet classes				
Methods	Hot Giant Planets (young or hot)	Other Giant Planets (same as in Solar System)	Hot Terrestrial Planets (hot, young or super-Earth)	Telluric Planet in habitable zone around M-dwarfs	Telluric Planet in habitable zone around solar-type stars
Microlensing	N/A	★ E	★ E	N/A	★ M
Transits	★★★ E	☆☆ G	★★★ M	★★ / ☆☆☆ M	★ E
Radial velocities	★★ E	★★ E	★★ E	★★ G	☆☆ G
Astrometry	★★ L	★★ E	★★ M	N/A	★★ L
V imaging / coronagraphy (SAI)	★★★ E	★★★ G	★★★ L	N/A	★★★ L
IR imaging / nulling (MAI)	★★★ G	N/A	★★★ L	★★★ L	★★★ L

Figure 1. Science potential levels (∗: statistics, ∗∗: identification, ∗∗∗: spectral characterization, N/A when non applicable) of each family of detection methods for different classes of exoplanets.

		Project classes			
Methods	***SPL***	Existing	Ground-based	M-class in space	L-XL class in space
Microlensing	★	Giants	N/A	HZ telluric solar stars	N/A
Transits	★★/☆☆☆	Close giants	Other giants	All other terrestrials	N/A
Radial velocities	★★/☆☆	Giant, close telluric	Habitable telluric	N/A	N/A
Astrometry	★★	Giants	N/A	Young telluric planets	Young + telluric in HZ
V imaging / coronagraphy (SAI)	★★★	Young giants	Far giants	N/A	Telluric planets
IR imaging / nulling (MAI)	★★★	N/A	Hot giants	N/A	All others

Figure 2. Timelines for various families of detection methods.

4. Analysis

One should first note the successes of the discipline and recognize that the E-coded boxes encompass some major achievements: spectroscopic characterization (SPL ***) has already been achieved on one class of exoplanets (hot giants), and a few telluric planets, albeit non habitable, have already been identified (SPL **). If one considers the youth of our field (the first exoplanets around solar type stars were identified just 15 years ago), this is quite remarkable indeed, and if the trend continues it would give all reasons to be optimistic for the future.

One should also note that not every method is relevant for every class of object. This means that a roadmap necessarily has to rely on a portfolio of methods in order achieve the Blue Dots goal. Likewise, the timeline grid (Figure 2) shows that no single method has relevant projects at all scale levels. So, while some methods may be extremely productive now, other techniques need to be developed in order to take the relay when they are needed.

"Least effort" in the roadmap means that priority should be given, at each step towards the Blue Dots goal, to the method which enables to achieve it with the projects of the lowest scale.

Applying this principle leads to the conclusion that :

• SPL * has been undertaken some years ago and results esentially from RV and microlensing surveys. The statistics of giants is quite known within a few AUs (typically < 4) and the frequency of Super Earths is being investigated at short periods. Microlensing yields the statistics in the bulb down to a few Earth masses for separation of a few AU (0.5-5). There is clearly a need to extend this knowledge to longer periods (beyond the snow line) and to lower masses: this can be achieved by pushing radial velocity and direct imaging surveys.

• SPL ** on habitable exoplanets should be carried out preferably from the ground by radial velocity. This is contingent to the acceptation that RV techniques are indeed capable to identify telluric habitable exoplanets. However, an exhaustive approach should involve at some stage an astrometric mission to be more resiliant with respect to the stellar parameters and to identify Earth masses around the nearby stars to be eventually characterized spectroscopically.

• Spectral characterization of telluric planets will certainly require space missions for two main reasons. First, the Earth atmosphere will make difficult (at low resolution) the analysis of other telluric planet atmospheres. Second, performance (stability, contrast) can be met more easily from space. In any case ground based telescopes and in particular ELTs will certainly contribute to this study but at a more modest level. That being said, SPL *** of close-in telluric planets around M stars should be attempted by transit photometry if it is indeed possible. Similarly, direct imaging at short wavelenghts could potentially achieve spectral characterization of telluric planets but for brighter/closer stars and larger separations (around 1-2 AU).

If one of these goals in the last item is achievable, this means that the goal of Blue Dots, at least in a selected sample of targets, can be achieved with a medium term project.

If not, and in any case for the solar type stars and Earth masses, spectroscopic characterization will require a flagship mission which will involve either an infrared interferometer, or an imager in the visible (and obviously preferentially both). It is not possible at this stage to prioritize these two options but we can work on collecting the elements that will help make the decision:

- Pursue the study of the exozodi issue to see how it impacts the detectability of habitable exoplanets in both cases ;
- Comparative system study for the two concepts in order to be able to compare performances/costs with an equivalent maturity level ;
- Pursue the identification of biomarkers and assess their detectability both in the infrared and the visible/near IR range.

Once these three steps are mastered, a flagship mission, possibly with global resources worldwide, in the decade 2020-2030 will follow naturally: the exoplanet community will be sufficiently knowledgeable to devise the best technological approach, to plan a sound observing program and to optimally exploit the data, especially spectroscopy.

5. Conclusion

No roadmap is carved in stone forever, and this is especially true with such a vibrant field as exoplanet science. It is likely to be revised every few years as the technology improves and new (possibly unexpected) results pour in. Blue Dots will continue to provide a forum and a framework to discuss these updates.

Acknowledgements

The material produced by Blue Dots comes from the contribution of all of the Blue Dots participants, and most notably its core team and working group coordinators which include Ignasi Ribas, Hans Zinnecker, Sebastian Wolf, Franck Selsis, Charles Cockell, Lisa Kaltenegger, Giovanna Tinetti, Anthony Boccaletti, Marc Ollivier, Jean-Philippe Beaulieu, Nuno Santos, Damien Segransan, Fabien Malbet, Alessandro Sozzeti, Ewa Szuszkiewicz, Helmut Lammer, Christoph Keller, Gerard van Belle, Szymon Gladysz, Chas Beichman, Hiroshi Shibai, Maxim Khodachenko, Gang Zhao, Abhijit Chakraborty, as well as $\simeq$180 other scientists who participate at some level in the initiative.

The Astrophysics of Planetary Systems: Formation, Structure, and Dynamical Evolution
Proceedings IAU Symposium No. 276, 2010
A. Sozzetti, M.G. Lattanzi & A.P. Boss, eds.

doi:10.1017/S1743921311020382

A European Roadmap for exoplanets

Artie P. Hatzes[1]
and The Exoplanet Roadmap Advisory Team (EPR-AT)

[1]Thüringer Landessternwarte Tautenburg, Sternwarte 5, D-07778, Germany
email: `artie@tls-tautenburg.de`

Abstract. The Exoplanet Roadmap Advisory Team (EPR-AT) was formed by the European Space Agency (ESA) to advise it on the best path for characterizing exoplanets including terrestrial planets. The EPR-AT delivered its report to ESA in August 2010. Here we summarize the findings of this task force.

Keywords. planetary systems

1. Introduction

The study of exoplanets is arguably one of the most exciting and vibrant fields in astronomy. Understanding the process of planet formation and why life developed on our Earth provides answers to fundamental questions important to both scientists and the public.

The European Space Agency (ESA) created an advisory panel, The Exoplanet Roadmap Advisory Team (EPR-AT) to advise it on the best roadmap for the characterization of exoplanets up to and including terrestrial planets. This is no easy task as the field is technologically challenging. The detection and characterization of exoplanets requires a diverse range of astronomical measurements that are pushed to the extreme limits: radial velocity precisions of $0.1 - 1\ \mathrm{m\,s^{-1}}$, astrometric measurements of a few microarcseconds, photometric variations of $\sim 10^{-5}$, and the measurement of contrast ratios of $\sim 10^{-10}$. The needs of the exoplanet community are drivers of new technologies and many of these extreme measurements often require expensive space missions.

The members of the EPR-AT that drafted the roadmap were Artie Hatzes, (Chair, Thüringer Landessternwarte, Germany), Anthony Boccaletti (Observatoire de Meudon, France), Rudolf Dvorak (Institute for Astronomy, University of Vienna, Austria), Giusi Micela (INAF - Osservatorio Astronomico di Palermo, Italy), Alessandro Morbidelli, Observatoire de la Cote d'Azur, France), Andreas Quirrenbach (Landessternwarte, Heidelberg, Germany), Heike Rauer (German Aerospace Center (DLR), Germany), Franck Selsis (Laboratoire d'Astrophysique de Bordeaux (LAB), France), Giovanna Tinetti, University College London, United Kingdom), and Stephane Udry, University of Geneva, Switzerland).

Space only permits us to summarize our report. The entire roadmap can be downloaded at the ESA website at
http://sci.esa.int/science-e/www/object/index.cfm?fobjectid=47855.

2. The Paths to Characterization

The study of exoplanets is rapidly moving from one that was dominated by detections, to one where the characterization of exoplanets are producing the most exciting results. The explosion in transiting planet discoveries have enabled us to measure the mass, radius, mean density, and dominant atmospheric spectral features of exoplanets. Transiting planets have largely driven this shift to characterization. We are at the point where we can start doing comparative exo-planetology: comparing exoplanets not only to each other, but also to the planets of our own solar system. Drawing from solar system planetary studies, exo-planetary science covers three broad themes: 1) Detections: understanding the census and architecture of planetary systems. 2) Characterization of the internal structure: measuring the fundamental exoplanetary parameters of mass and radius and comparing these to structure models. 3) Characterization of the exoplanetary atmospheres: understanding the effective temperature, composition, and presence of possible biosignatures in the atmospheres of exoplanets.

For these reasons an exoplanet roadmap is not a simple path carrying us from point 'A' (the present) to point 'B' (the detection of an exo-earth). Rather it is more like a 3-lane road whose lanes are defined by our themes of detection, the characterization of the internal structure, and the characterization of the atmospheres. Each lane moves at its own pace and with developments, both scientific and technological, providing a burst of speed that is often unpredictable. This 3-lane roadmap is schematized in Figure 1. The "Near-term" is defined roughly as the time frame 2011-2017, "Mid-term" 2015-2022, and "Long-term" beyond 2022.

The "fastlane" of exoplanet studies is clearly detections. In the past 15 years approximately 500 exoplanets have been discovered. The "moderate" lane is now held by characterization studies of the internal structure. This progress has largely been fueled by the explosion in transiting planet discoveries now spearheaded by the CoRoT and Kepler space missions. The mass, radius, and mean density has been characterized for over 100 transiting exoplanets. Initial results could only be obtained on giant gaseous planets, but with CoRoT and Kepler we have now been able to characterize the mass, radius, and density of the first rocky planets. The "slowlane" is defined by atmospheric detections. Chemical species such as CH_4, H_2O, CO and CO_2 have been observed in a handful of exoplanets, all in transiting systems. The progress in the characterizations of exoplanet atmospheres has been hindered by the technological challenges of detecting the tiny ($\approx 10^{-5}$) signal of the exoplanetary atmosphere compared to the stellar signal. But these discoveries hints at the future developments in this area once the characterization of exoplanetary atmospheres have reached full maturity.

3. Key Questions

The goal of this roadmap is to perform comparative planetology of a wide range of planetary systems including planets down to the terrestrial mass regime in the habitable zone of F-M dwarf stars. Although detections are an important step, the community should strive to make those "added value" discoveries, those which will lead to characterization studies.

The key questions we should answer in this journey of comparative exoplanetology are:

(*a*) What is the diversity and architecture of exoplanetary systems as a function of stellar parameters and birth environment?

(*b*) What is the diversity of the internal structure of exoplanets?

(*c*) What is the diversity of exoplanetary atmospheres?

(*d*) What is the origin of the diversity and how do planets form?

(*e*) What are the conditions for planet habitability, how common is exo-life and can we detect the biosignatures.

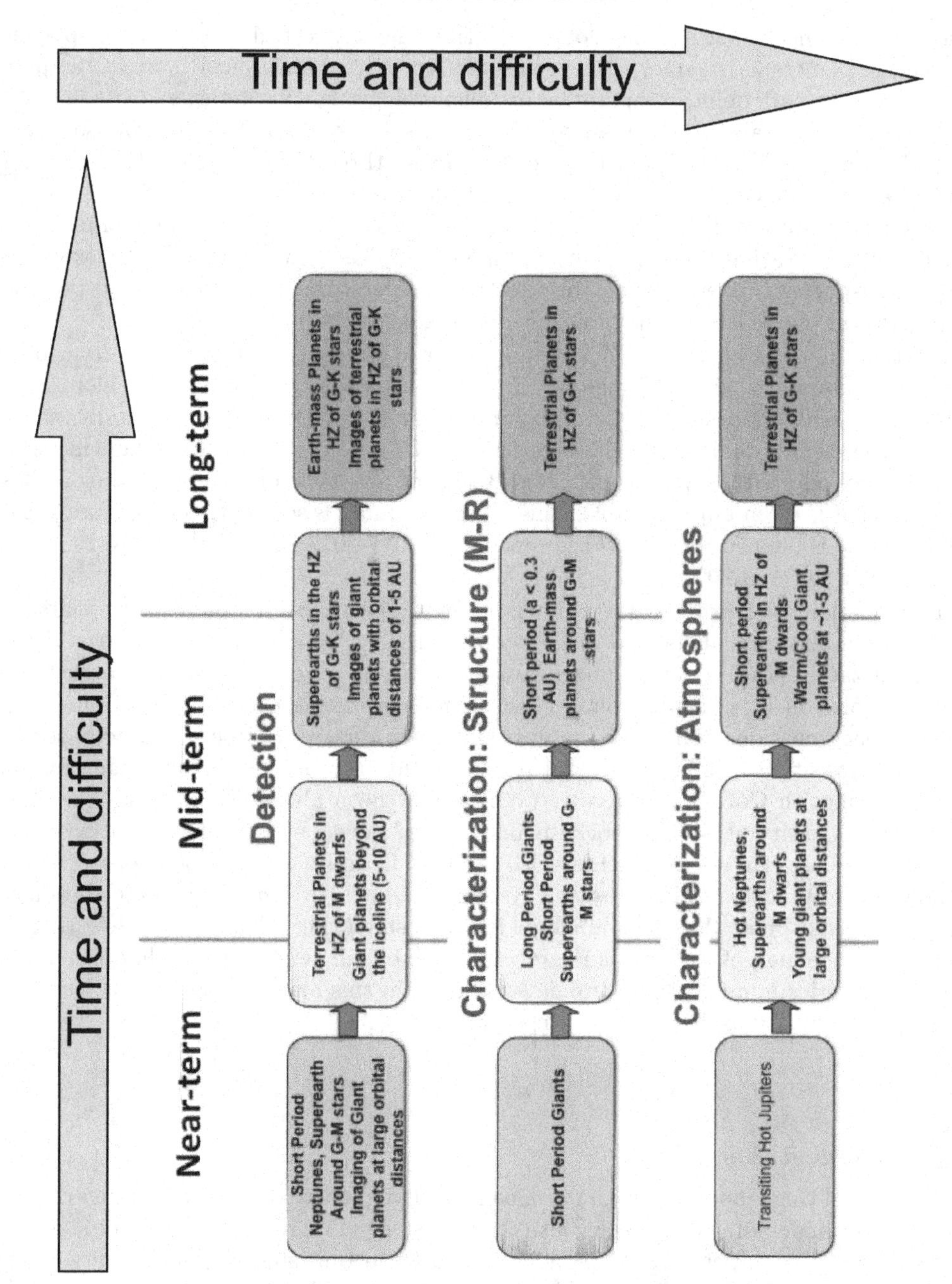

Figure 1. Schematic showing the rough timeline for milestones for our '3-lanes' of detection, characterization of internal structure (mass, radius, mean density) and the characterization of the atmospheres of exoplanets. Near term is approximately 2011-2017, Mid-term 2015-2022, and Long-term beyond 2022. Technological difficulty and time proceeds left to right, and top to bottom.

4. Milestones

As with any roadmap one must have milestones to gauge whether sufficient progress is being made in the field. Some of these milestones include.

Major Goals for Detections:

• *A Complete census of exoplanets.* If we are to understand planet formation we must obtain a representative census of the types of planets that can form around stars. Detections need to push the parameter space to planets of smaller masses and at larger orbital distances, especially around solar-type stars. It also needs to search for planets around host stars with as wide a range in stellar mass as possible and in different evolutionary states. Currently, the census of known extrasolar planets is heavily biased towards solar type stars. To get this complete census one must use a variety of methods: radial velocity measurements, the most successful detection method to date, astrometric measurements which can obtain the true mass of the exoplant, microlensing surveys as these are one of the few ground-based methods that can detect terrestrial planets in the habitable zone of stars, and direct imaging methods which can detect planets at large orbital distances to the parent stars.

• *An Understanding of the Architecture of Planetary Systems.* It is probably a good assumption that all planets are found in planetary systems. However, the multiple planet hosting stars is relatively small - about 10% of the stars known to host at least one exoplanet. The number of known planetary systems must be increased if we are to make progress understanding the architecture of planetary systems.

• *Targets for Characterization* Discovery of extrasolar planets should only be considered as a first step towards the path of characterization. Detection methods should focus on discoveries for which we can characterize the internal structure (mass, radius, and density) and atmospheric composition. Foremost among these discoveries would be terrestrial planets in the habitable zone of stars, in particular stars like our sun.

Major Milestones for Structure Studies

- *The Mass-Radius relationship for Jovian Planets out to 1 AU*
- *The Mass-Radius relationship for Neptunian Planets out to 1 AU*
- *The Mass-Radius relationship for Terrestrial Planets out to 1 AU*

The mass-radius relationship is the fundamental quantity that is needed for comparative planetology of the internal structure of exoplanets. Since this requires a radius determination this relationship can only be derived for transiting planets and proceeds from the the easiest (giant planets) to the most difficult (terrestrial planets). Realistically this relationship can only be derived for planets out to $\approx$ 1 AU due to the low probability for transits.

Major Goals for Atmospheric Studies

- *Composition of Jovian/Neptunian planets out to 0.1 AU*
- *Composition of Jovian/Neptunian planets at greater than 5 AU*
- *Composition Terrestrial Planets out to 0.1 AU*
- *Composition of terrestrial planets out to 1 AU.*
- *The detection of biosignatures*

The most difficult characterization studies is of the stellar atmospheres and is easiest done for transiting giant planets, giant planets at large orbital distances, transiting terrestrial planets, and finally terrestrial planets in the habitable zone of G-K stars.

5. Recommendations

Our general recommendations are divided between "ground-based", those which are best pursued with ground-based facilities, and "space-based", those which can only be accomplished from space.

5.1. *Detections*

Ground-based

(*a*) *Continued Radial Velocity Searches.* The RV has discovered most of the known exoplanets and it will continue to play a dominant role in the forseable future. RV searches will help complete the census of exoplanetary systems out to about 5 AU for G-M stars. It also plays a key role in characterization studies of transiting systems by providing the planet mass. Unfortunately, with the follow up transiting planet candidates, particularly from the CoRoT and Kepler missions, the telescope resources for RV work is overtaxed. The community should ensure that adequate telescope resources are available for carrying out this work, possibly with dedicated 2-4 m telescopes.

(*b*) *Towards a Golden Sample of Stars in the Habitable Zone of G-type stars.* With the demise of SIM-Lite the community must now rely on other methods for providing a sample of stars in the habitable zone of solar-type stars. RV measurements should be able to do this given sufficient telescope resources. RV surveys possibly with dedicated telescopes should focus on a relatively small sample of G-type stars ($\approx$ 50). High cadence observations should enable us to "beat down" the intrinsic stellar noise and to detect the $\approx$ 10 $\mathrm{cm\,s^{-1}}$ amplitude of terrestrial planets. This "Golden Sample" can then serve as the target list for a future flagship mission for characterization studies.

(*c*) *Ground-based microlensing for low mass planets and statistics.* Although microlensing does not provide targets for further characterization studies, it does provide important statistics on the lowest mass planets. These studies can be performed from the ground using a network of modest-sized (1-2 m) telescopes.

(*d*) *RV and transits searches for terrestrial planets in the HZ of M dwarfs targets for characterization studies* M dwarf stars may offer us the first opportunity to study the atmospheres of terrestrial planets in the habitable zone of other stars. However, before we can perform characterization studies, either from space or from the ground, targets should first be found.

(*e*) *Imaging of planets at large orbital radii (Planet Finders)* The next generation of Planet Finders at the Very Large Telescope will enable us to complete the census of planets out to large orbital radii. Because of the large separation between planet and host stars these systems will be the easiest for studying the exoplanetary atmospheres.

Space-based

(*a*) *Use of GAIA to discover a large number of giant exoplanets including determination of mass and orbital inclinations.* GAIA has the potential to discover a large number of giant exoplanets and examine such important phenomena such as the misalignment of planetary systems. This mission should be continued as long as possible to increase time base.

(*b*) *Astrometric detection of terrestrial planets in the HZ of solar type stars.* Astrometric detections are important because they not only may be more effective at discovering terrestrial planets in the habitable zone of G-K stars, but they also provide us with the true mass - *the* fundamental parameter of an exoplanet. Because of this the astrometric mission SIM-Lite played a key role in this roadmap. The U.S. Decadal Review did not support SIM-Lite, therefore there is now a gap in ourroadmap in regards to astrometric measurements. We hope that in the future a “targeted (as opposed to the global) astrometry can achieve the exoplanet goals of SIM-Lite butat significantly reduced cost.

5.2. *Characterization: Structure*

Ground-based

(*a*) *Continue Ground-based transit studies.* Although the space missions CoRoT and Kepler have produced exciting results in the area of transit detections, and hopefully PLATO will do so in the future, the era of ground-based transit searches is not completely over. Ground-based facilities can continue to make contributions by searching for transiting planets in different environments (e.g. stellar clusters) and by searching for photometric transits among the RV-discovered planets, particularly long period planets. Ground-based programs can also do important work by revising the emphemeri of known transiting planets and to search for additional planets via transit timing variations.

(*b*) *Calibration of evolutionary tracks mass and radius of planets detected with imaging at large orbital distances.* Direct imaging detections are important as these probe the planet formation at large orbital distances. This method has the disadvantage in that the mass of the planet is poorly determined and must rely on planet evolutionary tracks. These must be calibrated using planets at large orbital distances for which dynamical masses can be derived.

(*c*) *Ground based support for CoRoT and Kepler and preparation of Ground-based support for GAIA and PLATO.* CoRoT and Kepler have demonstrated that the ground-based follow-up observations in support of transit missions is enormous. PLATO will further tax these limited resources. The community must ensure that the necessary ground-based resources are in place for these important missions.

Space-based

(*a*) *Continuation of CoRoT and Kepler for as long as possible.* The exoplanet community is fortunate that it has two working space missions that are producing breakthrough results. The community should ensure that these continue for as long as possible.

(*b*) *Characterization of exoplanets with PLATO.* An important milestone of this roadmap is the deriving the mass-radius relationship and internal structure for terrestrial planets as these give important clues as to the formation process. Accurate values of the planet radius and mass are needed to distinguish whether the terrestrial exoplanet has a structure similar to the Earth, more like the Moon, or like Mercury. The measurement of the planet radius requires space-based photometry, while the measurement of its mass means that the targets should be relatively bright. *PLATO thus represents a very important, next-step in this roadmap.* PLATO will also play a very important secondary role in providing targets for future studies to detect the atmospheres of these transiting terrestrial planets.

5.3. *Characterization: Atmospheres*

5.3.1. *Ground-based*

(*a*) *Characterization of Giant transiting planets with visible and IR facilities*

(*b*) *Spectroscopy of planets at large orbital radii using very large and extremely large telescopes*

5.3.2. *Space-based*

(*a*) *Effective use of JWST for characterization studies.* The James Webb Space Telescope (JWST) has the capabilities of exoplanet spectral characterization. Although JWST will be a valuable facility for performing characterization studies of exoplanets, it is a general-purpose facility, so it is expected that relatively few targets will be observed. The amount of JWST time available to European scientists will be limited (NIRSPEC is a European contribution and MIRI is a 50-50 US-Europe collaboration). The European community must move fast and organize itself so as to effectively use its share of the time effectively on exoplanet studies. Open Time Key Programs for Exoplanets on JWST, is a concept that worked well with Herschel and should be considered by European scientists in this context.

(*b*) M-class mission in Cosmic Vision 2 to characterize giant planets and if possible superearths. Efforts to characterise the atmospheres of exoplanets should span the entire parameter space from hot, close in planets, to the cooler ones at large distances that are analogous to what is found in our solar system. In the mid-term roadmap preparations for a mission to perform this characterization should be undertaken. We can identify two classes of exoplanet for which spectral characterization can be realistically done in the near term:

1. Combined light mission: This is aimed at the atmospheric characterisation of hot and warm (including the habitable-zone of M-type stars) Giants, Neptunes, and Super-Earths using transit or combined-light spectroscopy. Spitzer, Hubble and large ground-based observatories have already demonstrated the feasibility of such work. While JWST and ground-based facilities will produce similar results for a select sample of targets, the proposed dedicated mission should provide repeated observations of a much larger sample of stars over a broader wavelength coverage.

2. Angular Resolved Detections. These investigations involve the use of high contrast imaging to minimize the light from the host star and to detect directly the light from the exoplanet. In the midterm such spectral characterizations would be for mature giant planets down to maybe super- Earth size at distances > 1 AU from the host star.

(*c*) In the long term (20-25 years) a flagship space mission to characterize terrestrial planets in the HZ of FGK type stars should be flown
Before embarking on such a mission it is essential that the target list be known. This should be a *characterization* mission and not a discovery mission

5.4. *Technology*

There are a number of technology advancements which are also important for this roadmap:

(*a*) *Improved Wavelength Calibration.* This is important for achieving an RV precision of a few $\mathrm{cm\,s^{-1}}$.

(*b*) *Detector Development.* The greatest need is for mid IR detectors in the range 5–20 μm with improved noise level and stability

(*c*) *Improved Deformable Mirrors that can achieve very small wavefront errors.* This is important for angularly resolved detections of exoplanets.

(*d*) *Ongoing developments in cornographic technology*

(*e*) *Free flying and deployable structures.* These are needed if one is to use external occulters to block the light of host stars.

It is hoped that in 20-25 years a flagship space mission is flown to characterize terrestrial planets in the habitable zone of G-K stars. This will require the extraction of the feeble signal of the planet that has an intensity that is 10^{-10} of the star. Before such a mission is flown the community needs to determine which technology: coronography, nulling interferometry, external occulters, etc. has the highest chance for success. This needs to be done in the next decade otherwise the community risks spreading its limited resources over too many technologies.

The Astrophysics of Planetary Systems: Formation, Structure, and Dynamical Evolution
Proceedings IAU Symposium No. 276, 2010
A. Sozzetti, M.G. Lattanzi, & A.P. Boss, eds.

doi:10.1017/S1743921311020394

New Worlds, New Horizons and NASA's approach to the next decade of exoplanet discoveries

Alan P. Boss[1], Douglas M. Hudgins[2] and Wesley A. Traub[3]

[1]Carnegie Institution, 5241 Broad Branch Road NW Washington, DC 20015-1305, USA
email: boss@dtm.ciw.edu
[2]NASA Headquarters, 300 E Street, SW Mail Suite 3W39, Washington, DC 20546, USA
email: Douglas.M.Hudgins@nasa.gov
[3]Jet Propulsion Laboratory, California Institute of Technology, 4800 Oak Grove Drive
M/S 301-355, Pasadena, CA 91109, USA
email: wtraub@jpl.nasa.gov

Abstract. Every ten years the astronomy and astrophysics community in the United States undertakes a survey intended to prioritize plans for major ground- and space-based astronomical facilities for the coming decade. *New Worlds, New Horizons* (NWNH) was released in August 2010 and represents the community's advice to the United States' funding agencies about the top priorities for 2010-2020. Here we focus on the recommendations of NWNH for space-based exoplanet missions to be considered by NASA, and on the plans developed to date for how NASA will respond to the science goals and missions set out for them by NWNH.

Keywords. planetary systems, astrometry, space vehicles, instrumentation: high angular resolution, techniques: interferometric, photometric, radial velocities, spectroscopic

1. Introduction

Beginning in 1964 with the Whitford Report, astronomers and astrophysicists in the United States have prepared detailed reports that prioritize their preferences for major ground- and space-based astronomical facilities for the coming decade. The most recent of these Decadal Survey reports, entitled *New Worlds, New Horizons* (NWNH) was released in August 2010. The chair of this Decadal Survey Committee was Roger Blandford, and as a result NWNH will undoubtedly be known as the Blandford Report in the future (Blandford *et al.* 2010). The other Decadal Surveys issued in the interval between 1964 and 2010 were the Greenstein Report (1972), the Field Report (1982), the Bahcall Report (1991), and the McKee-Taylor Report (2001), all named after their respective chairpersons.

NWNH represents the astronomical community's attempt to speak with a single voice to three of the United States' science funding agencies regarding the top astronomical priorities for 2010-2020: the National Aeronautics and Space Agency (NASA), the National Science Foundation (NSF), and the Department of Energy (DOE). While NASA focuses almost entirely on space-based missions, and NSF on ground-based telescopes, DOE has expressed an interest in both ground- and space-based efforts to understand dark energy. In this brief summary, we focus on the recommendations of NWNH for future space-based exoplanet missions to be considered by NASA, and on the plans developed to date by NASA Headquarters and by NASA's Exoplanet Exploration Program to address these top priority goals.

2. NWNH Decadal Survey

NWNH differed in several important respects compared to the five previous Decadal Surveys. First, determination of the top science goals was performed by a set of Science Frontier Panels covering the major disciplines of astronomy and astrophysics, while a separate set of Program Prioritization Panels considered the wide variety of projects proposed for the next decade and beyond. In the past, both tasks were performed by the same panels. Second, projects that had previously been approved for development by past Decadal Surveys were to be reconsidered anew, unless they were already well underway (e.g., the James Webb Space Telescope – JWST – which was recommended by the McKee-Taylor Report). The Space Interferometry Mission (SIM), intended to detect and determine the masses of Earth-like planets orbiting the closest solar-type stars, first recommended in the 1991 Report, and then recommended again in the 2001 Report, was thus tossed back into the fray in the 2010 Report. Third, all major projects would be subjected to an independent assessment of their cost estimates, in the case of the space-based missions performed by an aerospace corporation with extensive experience in space satellites. Finally, the recommended program had to be consistent with the actual funds expected to become available for new starts in the 2010-2020 decade.

The Blandford Report recommended a number of small, medium, and large projects for deployment both on the ground and in space in 2010-2020. In the context of this particular IAU Symposium, which links exoplanet scientists from around the world, it is also noteworthy that NWNH specifically took note of the increasingly collaborative, international, and interdisciplinary nature of contemporary astronomy, and recommended that the US science funding agencies should consider US participation in international projects, in order to permit access to these new facilities by US scientists, and to maximize the scientific output from the world's major astronomical facilities. Given the fact that the US Decadal Surveys are somewhat out of step with other planning efforts, such as the Cosmic Visions process currently underway in Europe, NWNH recommended that the international astronomical community should meet roughly every five years to share their strategic plans and consider opportunities for collaboration on the largest projects.

In the area of large space-based projects, i.e., those costing over $1B, NWNH ranked the Wide Field InfraRed Survey Telescope (WFIRST) as the top priority, followed by an augmentation to NASA's Explorer Program, the Laser Interferometer Space Antenna (LISA), and the International X-ray Observatory (IXO). WFIRST is intended to be a 1.5-m space infrared telescope with three main science goals, all of equal priority: a search for exoplanets by gravitational microlensing, galactic and extragalactic surveys, and dark energy. The Explorer Program supports the development of relatively low cost (less than $300 million US) space telescopes proposed by US astronomers. LISA would be a joint effort with the European Space Agency (ESA), intended to detect gravitational waves. IXO, also a joint international effort, would be a successor to NASA's current X-ray observatory, Chandra. SIM had been proposed to the 2010 Decadal Survey as a reduced cost mission, dubbed SIM-Lite, yet even with this major concession failed to make the final top priorities of NWNH.

WFIRST was envisioned by NWNH to cost $1.6B, with only a moderate level of risk of meeting its science goals, and to be started in 2013 with the goal of launching in 2020. WFIRST would be coupled with NWNH's top priority for large ground-based projects, the Large Synoptic Survey Telescope (LSST), to be built with the support of NSF and DOE. WFIRST and LSST are interlinked in the area of dark energy studies, where optical observations would be made from the ground by LSST and the corresponding infrared observations would be done by WFIRST.

NWNH noted that ESA is currently considering another space-based dark energy mission, Euclid, which is competing as a part of the current Cosmic Visions process for possible launch in 2017-2018. While Euclid focuses on dark energy, exoplanet surveys by gravitational microlensing are possible as a secondary science objective. NWNH thus stated that a collaborative effort between NASA and ESA might be a possibility, provided that a joint mission addresses all three of NWNH's science goals for WFIRST, makes sense economically, and that the US plays "a leading role in this top-priority mission".

It is interesting to note that NWNH ranked LSST above the Giant Segmented Mirror Telescope (GSMT), a 25- to 30-m ground-based telescope that would be valuable for a variety of exoplanet studies. GSMT had been proposed as the top ground-based major initiative in the 2001 McKee-Taylor Report, but had also been forced to compete again in the 2010 Decadal Survey, where it dropped down a notch in the final prioritized list. LSST moved up several notches in NWNH from its ranking in the 2001 Report. The Terrestrial Planet Finder (TPF), the third-ranked large space mission in the McKee-Taylor Report, disappeared altogether in NWNH, with the possible exception of inclusion in a technology development effort.

NWNH recommended that NASA begin a medium-scale technology development program directed toward the discovery of exoplanets. The New Worlds Technology Development Program is intended to help achieve the objective of discovering and characterizing nearby habitable exoplanets. Before deciding upon a specific concept for a flagship-class mission like TPF, NWNH suggested that the frequency of Earth-like planets be determined first, by a combination of NASA's Kepler Mission (now underway), WFIRST, and ongoing ground-based search programs with a variety of techniques. Further measurements of the level of exozodiacal dust emission and reflection around likely target stars also need to be achieved, as such emission and reflection can stymie direct detection efforts at both optical and infrared wavelengths. A combination of ground-based, sub-orbital, and Explorer-class telescopes was proposed by NWNH to address the exozodiacal light issue. NWNH then recommended that these efforts could lead to an informed choice about the design of a flagship mission between 2015 and 2020. Given that NWNH suggested a 2020 launch date for WFIRST, it is unlikely that WFIRST will inform such a decision, though this role for WFIRST will be obviated by the success of Kepler, whose primary science goal is to determine the frequency of Earth-like planets, which will be accomplished by 2013. NWNH recommended spending between $100 million and $200 million US on this effort over the decade.

While NWNH's top priorities were planned to fit into the expected NASA budget profile for new missions in 2010-2020 that was estimated in the summer of 2009 when the Decadal Survey was well underway, a severe decrease in the total amount of such funds from roughly $4B to $2B in the intervening time period has presented NASA with a major challenge in implementing the NWNH recommendations. NWNH recommended that if NASA's funds were reduced, the top priorities for space should be WFIRST and the Explorer Program augmentation, followed by the New Worlds Technology Development Program, and LISA and IXO Technology Development.

We now turn to a summary of the plans currently envisioned by NASA Headquarters and by the Jet Propulsion Laboratory (JPL), the lead NASA center for the Exoplanet Exploration Program.

3. NASA's Approach to the Next Decade of Exoplanet Discoveries

3.1. *The NWNH Framework for Exoplanet Exploration*

The discussion and recommendations presented in NWNH are arranged according under three broad scientific categories. The first, Cosmic Dawn, is focused on understanding the evolution of the early universe, and the processes that led to the formation of the first stars and galaxies. The second, Physics of the Universe, seeks to exploit the universe as a grand laboratory to deepen our understanding of fundamental physical laws and principles. The third, New Worlds, addresses the quest to search out and characterize extrasolar planets. Not surprisingly, it is the priorities and recommendations that pertain to the New Worlds science theme that will have the most direct impact on NASA's Exoplanet Exploration Program (ExEP) in the coming years.

Within the New Worlds science theme, NWNH articulates three key science objectives for the coming decade. The first is to complete the statistical census of exoplanetary system architectures and develop an unbiased determination of the abundance of rocky, Earth-sized planets on large orbits. Noting that the two techniques responsible for the majority of exoplanet discoveries to date (radial velocity (RV) measurements and transit detections) are strongly biased are toward large planets on small orbits, NWNH advocates an additional, complementary exoplanet detection technique: gravitational microlensing. Gravitational microlensing is a phenomenon wherein the light that we receive from a distant background star is temporarily magnified by the gravitational field of an intervening star as its relative motion carries it in front of the background star. If the intervening star has a planet in orbit around it, the planet can produce detectable substructure in the observed brightness variation. Gravitational microlensing is a very sensitive technique, capable of revealing even sub-Earth-mass exoplanets. More importantly, however, the technique is biased toward the detection of planets on large (1 AU) orbits. Consequently, a microlensing exoplanet census would in some sense mitigate the bias inherent in RV/transit surveys and provide scientists with a more complete picture of exoplanetary system architectures.

The second New Worlds science objective articulated by NWNH is the need to conduct a survey of Earth-mass planets in the habitable zones of stars in the solar neighborhood. Such a survey is desirable because it will provide a target list of promising candidates for a future exoplanet mission capable of direct imaging and spectroscopy of potentially habitable worlds.

Finally, the third New Worlds science objective is the characterization of the exozodiacal dust clouds in exoplanetary systems. Dust in exoplanetary disks will have an important impact on the detectability of planets in those systems. Thus, understanding the amount and distribution of that dust is an important precursor to the design of a future direct detection mission.

In order to achieve the foregoing New Worlds science objectives, and to advance the field of exoplanet exploration in the coming decade, NWNH offers a series of implementation recommendations. These include:

1. Perform a space-based microlensing survey to characterize in detail the statistical properties of habitable terrestrial planets. This is one of the three primary science objectives of the Wide Field Infrared Survey Telescope (WFIRST), the highest priority Large space mission recommended by NWNH.

2. Improve RV measurements on existing ground-based telescopes to locate the prime targets for hosting habitable, terrestrial planets among our closest stellar neighbors and to discover planets as small as 2-3 $M_{\oplus}$ as targets for future spacebased direct detection missions.

3. Use ground-based telescopes or a space-based Explorer mission, to characterize the dust environment around stars like the Sun, so as to gauge the ability of future missions to directly detect Earth-size planets on orbits like that of our own Earth.

4. Develop the technology for a future space mission to study nearby Earth-like planets. The goal of this program, the New Worlds Technology Development Program, is to lay the ground work for a planet-imaging and spectroscopy mission beyond 2020, including precursor science activities. The New Worlds Technology Development Program is the highest priority Medium space project recommended by NWNH.

5. Use JWST to characterize the atmospheric and/or surface composition of planets down to super-Earth masses orbiting the coolest red stars.

6. Carry out a focused program of computation and theory to understand the architectures of planets and disks.

Although the process of developing a program plan to address the recommendations of NWNH is in its very earliest stages, the following sections will summarize some of the plans and activities that are already in place.

3.2. *NWNH Recommendation L-1: The Wide-Field Infrared Survey Telescope*

The highest-priority, Large space mission recommended by NWNH is the Wide-Field Infrared Survey Telescope, or WFIRST. WFIRST is a mission concept for a near-IR space observatory that will conduct wide-field imaging and low-resolution spectroscopy from a vantage point at L2 over a five-year mission lifetime. WFIRST's design will be optimized to support a three-pronged science mission combining Dark Energy science, a microlensing exoplanet census, and other galactic and extragalactic large-area surveys, including a guest observer program. It is worth noting that NWNH emphasized that all three components of the WFIRST science program are of equal importance, and the loss of any one of them would significantly reduce the value of the mission in the eyes of the Decadal Survey Panel.

In the near term, NASA plans to initiate several pre-formulation activities that will begin to lay the foundation for WFIRST. First, NASA has already issued a call for proposals for scientists interested in participating in the WFIRST Science Definition Team (SDT) and plans to announce selections in early 2011. The SDT will include scientists whose specialization spans the complete range of all recommended mission science programs, and will be tasked with articulating the science goals of the mission, and developing an optimized implementation plan. In addition, NASA will also begin technical/engineering support for development of the WFIRST mission concept as resources allow, and will support relevant technology development through its existing programs. Finally, NASA will begin to explore the possibility of interagency and/or international participation in WFIRST, participation which may include representation on the SDT. Beyond these early activities, further development of WFIRST will be influenced by several important budgetary and scientific considerations. On the financial side, NASA's ability to implement WFIRST will be determined almost entirely by the budget profile and schedule for JWST. This is true not only because of the large cost associated with implementing any flagship space mission, but also because NASA's Astrophysics budget is projected to remain essentially flat at ca. $1B through the middle of the coming decade (the current budget horizon). Thus, there will be little funding available to undertake the start of a major new mission until JWST development is complete, and that portion of the budget can be directed to development of the next mission. The science environment in the coming decade will also influence the implementation of WFIRST. This environment will be shaped by such factors as: (a) development of the Large Synoptic Survey Telescope (LSST; NWNH's top-priority, Large ground-based project) and other ground-based facilities by the NSF and the U.S. Department of Energy; (b) ongoing and future investigations by Hubble, Chandra, Spitzer, JWST, etc.; (c) scientific results from future potential Explorer missions; and, (d) missions under development by other nations (e.g., Euclid, PLATO). Consequently, it is likely to be several years before the details of the path that NASA will follow over the coming decade to implement WFIRST will become clear.

3.3. *NWNH Recommendation L-2: Augmentation to NASA's Explorers Program*

The second Large space project recommendation of NWNH is a significant augmentation to the budget for NASA's Explorer Program. The Explorer Program supports competed, PI-led space missions that fall into three categories: (1) Medium-class Explorers (MIDEX), which are cost-capped at ca. $300M; (2) Small Explorers (SMEX), which are cost-capped at ca. $160M; and, (3) Missions of Opportunity (MoO), which are typically $35M. Over the years, Explorer missions have delivered a high level of scientific return on relatively moderate investments, and have provided the capability to respond rapidly to new scientific and technical breakthroughs. Recognizing the proven value of the Explorers Program, NWNH recommended a budget augmentation to the program sufficient to support two Astrophysics MIDEX missions, two Astrophysics SMEX missions, and four Astrophysics MoOs in the coming decade. This represents a doubling of the planned flight rate for the program. As this is a competitive program, there is no guarantee that any exoplanet exploration missions will be selected to fly in the coming decade. However, given the inherently compelling nature of the science, its prominence in NWNH, and the number of highly-competitive, Explorer-class exoplanet mission concepts, prospects for such a mission would appear to be bright.

3.4. *NWNH Recommendation M-1: New Worlds Technology Development*

Few scientific endeavors are as compelling to scientist and non-scientist alike as the search for other habitable, Earth-like worlds. Indeed, it is for just that reason that the New Worlds science theme plays such a prominent role in NWNH. However, simply detecting the presence of habitable, Earth-sized planets around other stars pushes the very limits of current technology; actually isolating the light from such a planet from the overwhelming glare of its parent star so that it can be imaged and its spectrum measured will require significant technological advancements in a number of different areas. Moreover, the optimal architecture by which the necessary degree of starlight-suppression can be achieved is yet unclear. Coronagraphy, external occultation, and interferometry have all been given significant consideration in this regard, and each has its own set of advantages, disadvantages, and challenges. In view of the tremendous challenges yet to be overcome and the tight budgetary constraints it was forced to work within, NWNH concluded that it would not be feasible to execute a direct-detection mission in the coming decade. On the other hand, acknowledging the paradigm-shifting nature of such an endeavor, NWNH recommended a significantly increased investment in exoplanet technology development in the form of its New Worlds Technology Development Program. The goal of that program is to lay the technical and scientific foundations for a space mission capable of imaging and spectroscopy of habitable, rocky planets in the 2020 decade. The importance of this effort within the framework of NWNH is reflected in its placement as the top-priority, Medium class project. The New Worlds Technology Development Program envisioned by NWNH would be implemented through a two-stage process. The first stage calls for stepped-up funding for exoplanet technology development activities spanning all of the candidate starlight suppression techniques (coronagraphy, interferometry, star-shades) through mid-decade. If, at that point, the scientific groundwork and design requirements for a direct-detection mission are sufficiently clear, a technology down-select should be made. Subsequent investments in the latter half of the decade (the second stage of the program) should be increased dramatically and focused on advancing the most promising mission architecture. The goal of the New Worlds Technology Development Program is to develop a mature concept for a flagship mission to conduct imaging and spectroscopy of habitable, terrestrial exoplanets for consideration by the 2020 Decadal Survey. Fortuitously, NASA's existing Astrophysics Strategic Research and Technology (SR&T) portfolio is well-suited to implementing the substance of the New Worlds Technology Development Program. In the current portfolio, exoplanet technology development is funded primarily through two programs:

(a) Astrophysics Research and Analysis (APRA). The APRA program supports (among other things) fundamental research into new exoplanet technologies, i.e. Technology Readiness Levels (TRL) 1-3.

(b) Technology Development for Exoplanet Missions (TDEM). TDEM is a component of NASA's Strategic Astrophysics Technology (SAT) program. It supports the mid-range maturation of exoplanet technologies whose feasibility has already been demonstrated, i.e., TRL 3-6.

It should be noted that mission-enabling, ground-based precursor science activities are included in the scope of TDEM. However, in this context, the term mission-enabling science is carefully defined as science that advances technologies or informs the design of future NASA flight missions. Thus, investigations such as high-precision RV surveys of planetary systems in the solar neighborhood, or characterization of exozodiacal dust disks, would fall within the purview of the program if suitably motivated.

4. Other Exoplanet Exploration Activities in the Coming Decade

4.1. *NWNH Small Initiatives*

Several of the (unprioritized) Small Initiatives recommended by NWNH are also likely to contribute to NASA's Exoplanet Exploration activities in the coming decade. These include:

(a) an augmentation to NASA's Astrophysics Theory Program (ATP) which supports research into the origin and evolution of exoplanetary systems;

(b) an augmentation to NASA's Suborbital Program to increase flight rate, which would create additional opportunities for suborbital exoplanet investigations; and,

(c) definition of a future UV/optical space observatory that could reasonably include exoplanet exploration as a component of its science mission.

4.2. *Ongoing Activities*

Outside of the recommendations of NWNH, NASA is engaged in a number of ongoing exoplanet exploration activities that will likely make important contributions to the field in the coming years. First and foremost is the Kepler mission. At the time of this writing, Kepler has completed about one and a half of its three and a half year prime mission, and the Kepler Science Team is engaged in the prodigious task of analyzing the data and following up on potential exoplanet detections. In addition, the first two quarters of mission data are now available in the public domain and analyses based on those data are eligible for funding under NASA's Astrophysics Data Analysis Program (ADAP). Thus, the number of exoplanets discovered by Kepler will undoubtedly increase dramatically over the next few years, and the mission should yield our first direct measurement of the frequency of Earth-sized planets by the middle of the coming decade. On the ground, NASA continues to support exoplanet exploration investigations at both the W.M. Keck Observatory in Hawaii and the Large Binocular Telescope in Arizona. At Keck, NASA supports high-resolution radial velocity observations with the HIRES instrument for both Kepler follow-up and general exoplanet observations. It should be emphasized that, although a portion of NASA's Keck time is set aside for Kepler follow-up observations by the Kepler Science Team, proposals for Kepler follow-up observations from the wider scientific community in response to NASA's biannual Keck solicitations are welcome and encouraged. NASA support for the Keck interferometer has also been extended into 2012, and options for future support of that facility are currently being discussed with the NSF.

In Arizona, the Large Binocular Telescope is nearing completion, and initial testing of the NASA-funded Large Binocular Telescope Interferometer (LBTI) is already underway. The key science project for the LBTI will be to explore the exozodiacal dust environments of nearby stars at mid-infrared wavelengths, directly addressing one of the New Worlds science objectives articulated in NWNH. In addition to this key science program, once the LBTI is fully operational, NASA plans to make time on this facility available to the scientific community for exoplanet-related and other science investigations on a competitive basis.

Finally, NASA will continue to support fundamental exoplanet research through a suite of competitively selected R&A programs including the previously-mentioned ADAP and ATP programs, as well as other programs such as Origins of Solar Systems and Planetary Atmospheres.

4.3. *Programmatic Impacts of NWNH*

In any prioritization process, some opportunities must be sacrificed so that others can proceed. In this final section, we will examine two prominent exoplanet exploration activities that must be abandoned in deference to the NWNH recommendations.

4.4. *Space Interferometry Mission (SIM-Lite)*

After careful consideration, the Decadal Survey panel made the difficult decision not to include the SIM-Lite mission among its priorities for the coming decade. This decision was driven primarily by two factors. First, the panel concluded that the $1.9B price tag obtained through its own independent cost estimate and the projected 8.5 yr time-to-launch made SIM-Lite uncompetitive in the rapidly changing field of exoplanet science. In addition, the panel found that rapid advances in ground-based observational capabilities had eroded SIM-Lite's importance as a target-finding mission for a future direct-detection mission. Specifically, the panel concluded, the role of target-finding for future direct-detection missions, one not universally accepted as essential, can be done at least partially by pushing ground-based radial-velocity capabilities to a challenging but achievable precision below 10 centimeters per second. As a consequence, NASA's Science Mission Directorate has formally discontinued sponsorship of the SIM-Lite project and directed the project to initiate shut-down activities. Closeout activities will include retention of all SIM hardware by NASA's Exoplanet Exploration Program for potential future use, archiving of all SIM-related technology and design documentation, and reassignment of personnel to new work. Termination activities should be completed by the end of 2010.

4.5. *Participation in ESA's PLATO Mission*

Prior to the release of NWNH, the European Space Agency invited NASA to consider a 20% partnership on two of its Cosmic Visions M-Class mission candidates: the PLAnetary Transits and Occulations of stars (PLATO) mission and the Euclid mission to map the geometry of the dark universe. These opportunities were described in a letter from NASA to the Decadal Survey panel in early April 2010. Subsequent to this invitation, NASA supported the participation of US scientists and engineering teams in ESA's planning and optimization studies for each concept during May/June 2010. However, throughout these activities, NASA emphasized that its participation in either or both of these missions would be contingent upon the recommendations of NWNH.

As discussed above, NWNH emphasized the importance of completing the census of exoplanetary systems with a technique that mitigated the bias inherent in the transit technique, as well as the need to conduct a survey of nearby habitable, Earth-sized planets and determine the exozodiacal dust characteristics of those systems. The PLATO mission does not address either the first or the last of these science objectives. Moreover, the observing strategy currently planned for the PLATO mission (a combination of two long-duration pointings lasting three- and two-years, respectively, followed by a series of shorter pointings of a few months each) is not well suited to the detection of habitable, rocky planets around stars other than late-type dwarfs. Consequently, NASA has concluded that the scientific objectives of the PLATO mission are poorly aligned with the recommendations of NWNH, and has notified ESA that it does not intend to pursue a strategic partnership in the mission, even if it is ultimately selected for flight through the Cosmic Visions process.

5. NASA's Exoplanet Exploration Program

The messages in NWNH from the astronomical community to those of us in its exoplanet subset are quite clear: exoplanet science is an exciting field, exoplanet science deserves a major portion of the WFIRST mission in this decade, and exoplanet science is currently earmarked for a dedicated space mission in the following decade.

The path to a dedicated mission in the 2020s has even been mapped out for us by NWNH: community agreement on a "*mission definition for a space-based planet imaging and spectroscopy mission*" by mid-decade (i.e., 2015), a "*technology down-select*" around that same time, and an augmented "*mission-specific technology program starting mid-decade*" in preparation for "*a mission start in the 2020 decade*". (All quotes in this section are directly from the NWNH report, specifically pages 7-22 and 7-23, italics added.)

NWNH says "*Detecting signatures of biotic activity is within reach in the next 20 years if we lay the foundations this decade for a dedicated space mission in the next.*" The report continues with more specific directions: "*For the direct detection mission itself, candidate starlight suppression techniques (for example, interferometry, coronagraphy, or star shades) should be developed to a level such that mission definition for a space-based planet imaging and spectroscopy mission could start late in the decade in preparation for a mission start early in the 2020 decade.*"

The funding profile for achieving this is hinted at in NWNH, where a low profile for the first half of the decade is suggested, ramping up to a much higher level in the second half. The message here is that we should not expect funding to be much different in the coming 5 years than it has been in the past few years. This is a disappointment to many researchers. However we can also read this as saying that the authors of NWNH feel that we already have in hand a sufficient number of good ideas for imaging and spectroscopy, and that all we have to do is to refine these somewhat, and internally select what we believe is the most promising of the pack.

The NWNH report is not specific on the down-select mechanism, saying only that "*If the scientific groundwork has been laid and the design requirements for an imaging mission have become clear by the second half of this decade, a technology down-select should be made ...*". The report does not say that this should be a formal down-select directed by NASA, although that could be an option.

The exoplanet community might be wise to heed the example of related communities (e.g., far-infrared, x-ray, Mars) by doing the down-select internally, and agreeing to agree on the result, as with the NWNH process itself. By agreeing on a mission, and presenting a united front to the world, we are much more likely to succeed.

To do the opposite, to disagree until the bitter end, and expect a mystical authority figure to make the decision between warring tribes, is to invite disaster.

The obvious and appropriate venue for making a community down-select decision is the Exoplanet Program Analysis Group (ExoPAG). The ExoPAG was created to provide scientific analysis on exoplanet topics to NASA headquarters and to the Exoplanet Exploration Program (ExEP), via the Astrophysics Subcommittee, which is in turn a part of the NASA Advisory Council. Thus the ExoPAG is the single community entity officially designated to provide analysis information on exoplanet matters in the US.

Technically, the ExoPAG membership is the entire exoplanet community. The ExoPAG Executive Council is a 10-member group with a rotating membership, drawn from among volunteer scientists. However since the ExoPAG itself constitutes the entire community, it is expected that there should be wide participation by all interested parties, through the mechanism of ExoPAG meetings (typically along with winter and summer AAS meetings) and through telecons set up for topic-specific discussions.

The ExoPAG chair, James Kasting, has suggested the following tentative schedule for working toward a community down-select by 2015: (a) formulate study groups and assign tasks, by mid-2011; (b) report on exoplanet science goals and instrumentation options for a mission in the 2020s, by mid-2012; (c) report on mission technology, engineering, and verification challenges, mid-2013; (d) present a detailed design reference mission (DRM) for that mission, early 2014. The detailed schedule may change, but the intent will remain, to produce a community down-select by mid-decade.

As we study the candidate missions, what criteria should guide our winnowing process? The President of the AAS, Debra Elmegreen, in the October 2010 AAS Newsletter, lists these attributes of a mission: (a) scientific merit, (b) technical readiness, (c) balance, (d) affordable cost, (e) tolerable risk. We will need all of these attributes to succeed. The question we can ask of each proposed mission concept is, how will it rank in these categories as we enter 2014?

6. Conclusions

While the ambitious space program envisioned by NWNH is unlikely to be realized in its entirety in the coming decade, largely as a result of the severe funding constraints facing NASA's Astrophysics Division, it is clear that NASA intends to follow the guidance provided by NWNH as best it can. Rapid progress in the area of exoplanetary science will continue to be made in 2010-2020 as a result of ongoing ground-based exoplanet discovery and characterization efforts, future space-based observatory-class telescopes such as JWST, and especially by ongoing exoplanet-specific space telescopes, namely Kepler and Europe's CoRoT Mission.

7. Acknowledgments

APB's participation in IAUS276 was supported by grant NNX07AP46G from NASA's Planetary Geology and Geophysics Program. WAT's participation was supported by the NASA's Exoplanet Exploration Program.

References

Blandford, R., *et al.* 2010, *New Worlds, New Horizons in Astronomy and Astrophysics* (Washington, DC: The National Academies Press).

The Astrophysics of Planetary Systems: Formation, Structure, and Dynamical Evolution
Proceedings IAU Symposium No. 276, 2010
A. Sozzetti, M.G. Lattanzi & A.P. Boss, eds.

doi:10.1017/S1743921311020400

The James Webb Space Telescope and its capabilities for exoplanet science

Mark Clampin[1]

[1]NASA goddard Space Flight Center, Greenbelt, MD 20771
email: mark.clampin@nasa.gov

Abstract. The James Webb Space Telescope is a large aperture (6.5 meter), cryogenic space telescope with a suite of near and mid-infrared instruments covering the wavelength range of 0.6 ?m to 28 ?m. JWSTs primary science goal is to detect and characterize the first galaxies. It will also study the assembly of galaxies, star formation, and the formation of evolution of planetary systems. JWSTs instrument complement offers numerous capabilities to study the formation and evolution of exoplanets via direct imaging, high contrast coronagraphic imaging and photometric and spectroscopic observations of transiting exoplanets.

Keywords. planetary systems, instrumentation: high angular resolution, instrumentation: miscellaneous, telescopes, space vehicles

---

1. Introduction

The James Webb Space Telescope (JWST) is a large aperture, space telescope designed to conduct imaging and spectroscopic observing programs over the wavelength range 0.7μm to 29μm (see Figure 1). JWST will be launched into an L2 orbit aboard an Ariane 5 launcher. The Goddard Space Flight Center (GSFC) is the lead for the JWST program, and manages the project for NASA. JWST is an international project partnered with the European Space Agency (ESA), and the Canadian Space Agency (CSA). The prime contractor for JWST is Northrop Grumman Space Technology (NGST).

The JWST observatory is designed to address four major science themes; first light and re-ionization; the assembly of galaxies; the birth of stars and protoplanetary systems; and the formation of planetary systems and origins of life (Gardner 2006). The formation of planetary systems and the origin of life theme seeks to determine the physical and chemical properties of planetary systems including our own, and investigate the potential for life in those systems. In this review we will focus on the capabilities offered by JWST to undertake high contrast direct imaging studies of exoplanets, and characterization studies of exoplanet via observations of transiting systems.

2. JWST's Observatory Design

JWST will be largest telescope built to conduct astrophysical observations in space. An infrared-optimized telescope, JWST is cooled to ~40K to facilitate science observations from 0.7 μm to 29 μm. In order to maximize the scientific lifetime of the observatory, JWST is a passively cooled telescope. Its operational lifetime is thus primarily determined by the propellant required to maintain its orbit around L2 and to conduct momentum unloading of the reaction wheels. The nominal lifetime is 5 years with a 10 year goal. Passive cooling of the telescope is achieved by means of a five layer sunshield, which deploys after launch and keeps the telescope optics shaded from the sun, as shown in

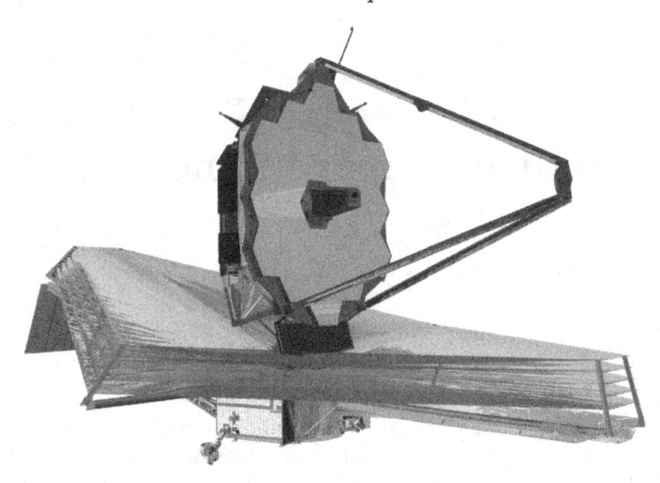

Figure 1. The James Webb Space Telescope

Figure 1. The membranes are made of aluminum coated kapton, and each one is about the size of a tennis court. The membranes interface to the spacecraft bus, which is located on the sun-facing size of the observatory. The spacecraft bus houses electrical systems, attitude control systems, communications systems, propellant tanks, thrusters, and the solar array.

The primary mirror features an 18 segment mirror architecture, with each segment able to make 6 degree of freedom adjustments to optimize the primary mirror figure. The telescope is a three-mirror anstigmat with a collecting area of $\geqslant 25$ m^2. Each primary mirror segment also has a radius of curvature adjustment. JWST will operate at cryogenic temperatures. The telescope optics are made of beryllium since beryllium mirrors are light, mechanically stiff and optically stable over the cryogenic operating temperature range. Excellent progress has been made in the fabrication of JWST's telescope optics. The aft-optics system (AOS), comprising the tertiary and fine steering mirrors is complete. Over half of the flight primary mirror segments have completed polishing and received their gold coating. In Figure 2 we show an image of the first six flight mirrors entering their final acceptance test, which consists of a cryogenic measurement of each mirror's figure.

3. Science Instrument Complement

JWST has a complement of four instruments, the Near Infrared Camera (NIRCam), the Near Infrared Spectrograph (NIRSpec), the Tunable Filter Imager and Fine Guidance Sensor (FGS-TF), the Mid Infrared Instrument (MIRI). The characteristics of the science instruments are summarized schematically in Figure 3 to illustrate their major scientific objectives. The Near Infrared Camera (NIRCam) development team is led by the University of Arizona (Rieke *et al.* 2003). This instrument will be JWSTs primary imager in the wavelength range of 0.6μm to 5μm and is primarily designed for deep, wide field imaging. Required to achieve many of the core science goals, the instrument is particularly well suited to the task of detecting the first luminous sources that formed after the cosmological dark ages. NIRCam is also capable of high-contrast coronagraphic imaging, which will enable observations of debris disks, and searches for giant planets around nearby stars. The NIRCam also fulfills the key role of measuring the wavefront error for the JWST telescope assembly, so that the telescope can be phased. The multi-object Near Infrared Spectrograph (NIRSpec) is provided by the European Space Agency (Jakobsen *et al.* 2009). NIRSPec will serve as the principal multi-object spectrograph in the 0.6μm to 5μm wavelength range. Its ability to obtain simultaneous spectra of more than 100 objects in a 9-square arc-minute field of view at spectral resolutions of $\lambda/\Delta\lambda) = 100$, 1000, and 3000, enables high survey efficiency for a variety of compact sources including primordial galaxies. NIRSPec also offers long-slit spectroscopy and Integral Field Spectroscopy capabilities. The Mid-Infrared Instrument (MIRI), provided by an international collaboration of agencies and the Jet Propulsion Laboratory, will provide broad-band imaging and integral field spectroscopy over the 5μm to 29μm spectrum (Wright *et al.* 2008). This instrument will study the creation of the first heavy elements and will reveal the evolutionary state of high redshift galaxies. It is capable of studying the very early stages of star and planet formation, in regions where all visible light is blocked by dust and most of the emission is radiated at mid-infrared wavelengths. The Tunable Filter Imager (TFI) enables extended objects at high redshift to be imaged in Lyman-α with diffraction limited angular resolution at $\lambda/\Delta\lambda) = 100$ (Doyon *et al.* 2008). This instrument is critical for emission line surveys of primordial galaxies and detailed morphological studies of galaxy nuclei and Galactic nebulae. TFI forms part of the Fine Guidance Sensor (FGS) instrument package, provided by the Canadian Space Agency. The FGS will measure image jitter and provide signals for the Fine Steering Mirror to compensate for that jitter.

JWST was designed to conduct deep, wide-field surveys in the infrared, and so its architecture is not fully optimized for high-contrast imaging. Even so, it offers an impressive suite of capabilities for high contrast imaging in the near and mid-infrared. The 18 segment primary mirror generates additional diffraction structure that must be apodized, compared to that of a monolithic mirror. The design wavefront error of 150 nm at the NIRCam focal plane, is the main driver for the performance of the coronagraphs.

These are both important considerations when making comparisons to fully optimized visible and near-infrared coronagraphs designed for ground-based or space-based applications. However, JWST does offer exceptionally low infrared backgrounds, by virtue of its orbit at L2, and the telescope's cryogenic thermal design. Furthermore, JWST's

Figure 2. Six gold-coated JWST flight mirror segments prior to the start of cryogenic acceptance testing at the X-Ray Calibration Facility, located at the Marshall Space Flight Center.

point spread function will be relatively stable over the 14 day periods between scheduled fine-phasing adjustments. The image stability will be primarily dominated by the effect of thermal drift on mirror alignments. Thus, JWST's wavefront error stability combined with low infrared backgrounds offer unique advantages that make possible a range of observations that will complement other near-term ground-based approaches. JWST's discovery space for high contrast imaging in the thermal infrared is unique because of the low backgrounds, combined with excellent angular resolution offered by its 6.5 meter aperture.

3.1. *JWST's High Contrast Coronagraphs*

The capabilities offered by JWST to provide high contrast imaging of exoplanets, and faint circumstellar structures such as debris disks are summarized in Table 1. NIRCam offers coronagraphic imaging in both its short and long wavelength channels via a set of five occulting masks (Krist *et al.* 2009). This mode is described in detail by Green *et al.* (2005) and Beichman *et al.* (2010). The masks comprise three circular and two wedge shaped masks. The masks are optimized for operating wavelengths of 2.1μm, 3.35μm, 4.3μm and 4.6μ. The predicted contrast at a wavelength of 4.6μm is $6\text{x}10^{-4}$ at an angular radius of 1 asec. Optimum processing using the technique of subtracting a spectrally matched reference star might improve this figure by up to an order of magnitude (Krist 2007).

TFI features a similar coronagraph to NIRCam, however, the observing concept for the instrument is different since TFI is built around a tunable narrowband imager. TFI images in R$\sim$100 increments over most of the 1.6μm - 4.9μm bandpass. TFI coronagraphic observations will employ an analysis technique called Differential Speckle Imaging (Marois *et al.* 2000) to achieve a contrasts up to 10$\times$ greater than NIRCam. TFI also employs a non-redundant mask (NRM) which uses the technique of closure phase imaging to achieve a contrast in the region of 10^{-4}-10^{-5} out to a radius of 0.55 asec (Sivaramakrishnan *et al.* 2009).

MIRI offers four self-contained coronagraphs, three of which are quadrant phase masks and the fourth a traditional Lyot corongraph. Quadrant phase masks are based on the principle of breaking the image plane into four quadrants and applying a π phase-shift to two diagonal quadrants. This produces a deep null in the pupil plane which is then

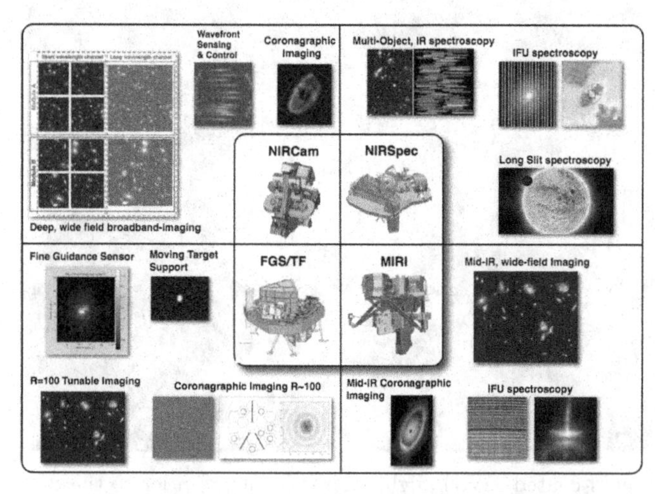

Figure 3. A schematic representation of JWST's science instrument complement and the primary scientific capabilities of each instrument.

passed through a lyot stop and re-imaged. The quadrant phase mask images to $< \lambda/D$ although the contrast is significantly decreased at smaller inner working angles. Each MIRI coronagraph is designed for a specific wavelength, the three quadrant phase masks are assigned to wavelengths of 10.65μm, 11.4μm and 15.5μm, and the Lyot coronagraph to 23μm.

The JWST coronagraph complement provides both near-IR and mid-IR high contrast imaging, making them excellent tools for the study of young, self-luminous planets. Recent modeling by Beichman *et al.* (2010) has shown that JWST should be able to detect 0.2 M_J planets as close as 50 AU. The TFI's non-redundant mask imager and MIRI provide sensitivity to comparable masses in the 10 to 20 AU and $\geqslant$40 AU, regimes respectively. These capabilities will provide the tools to conduct surveys that could constrain planet formation models and planet properties. The low backgrounds offered by JWST also make surveys for planets around M stars especially attractive, as ground-based observations of these relatively faint systems will be limited by the sky background. JWST should be able to detect $\sim$2 M_J planets from at distances of a few AU and beyond (Beichman *et al.* 2010)

Table 1. Overview of JWST's high contrast imaging capabilities. Predicted contrast ratios are presented for modes they they have been modeled.

SI	Mode	λ (μm)	Contrast	Radial Distance
NIRCam	Short λ Lyot Coronagraph	2.0 - 2.3		
NIRCam	Long λ Lyot Coronagraph	2.4 - 5.0	6x10^{-4}	1 asec
TFI	Multi-λ coronagraph	1.6 - 2.5	6x10^{-5}	1 asec
TFI	Multi-λ coronagraph	3.2 - 4.9	6x10^{-5}	1asec
TFI	Non-redundant mask	3.2 - 4.9	10^{-4} - 10^{-5}	$\leqslant$ 0.55 asec
MIRI	Quadrant Phase Coronagraph	10.65		
MIRI	Quadrant Phase Coronagraph	11.4	10^{-4}	1 asec
MIRI	Quadrant Phase Coronagraph	15.5	4x10^{-5}	1 asec
MIRI	Lyot Coronagraph	23		

4. Transiting Exoplanets

The study of transiting exoplanets has provided most of the key data to date on the properties of exoplanets, such as direct estimates of their mass and radius (e.g. Charbonneau 2007), and spectral diagnostics of their atmospheres (e.g. Swain *et al.* 2008). Observations of transiting exoplanets by the Hubble Space Telescope (HST) and Spitzer Space Telescope (SST) have both played lead roles in making demanding, high signal to noise observations of the light curves, and spectra of transiting exoplanets. The launch of JWST will provide new capabilities for the characterization of transiting exoplanets via transit spectroscopy and high precision transit photometry. Spectroscopic characterization of transiting exoplanets demands extremely high signal to noise observations, which JWST will provide by virtue of its large 25 m^2 collecting area and low infrared backgrounds. JWST's orbit around L2 provides it with both excellent sky coverage and long dwell times on targets. JWST offers broad suite of instrumentation which will provide capabilities for both imaging spectroscopic observations of transiting systems. The most important instrument operating modes that address transit photometry and spectroscopy requirements are summarized in Table 2, with descriptions of their application to observations of transiting systems.

Table 2. Overview of JWST's transit science instrument modes

Instrument	λ (μm)	R ($\lambda/\Delta\lambda$)	Comments
NIRCam (Imaging)	0.6 - 2.3 2.4 - 5.0	4, 10 , 100 4, 10 , 100	High precision light curves of primary and secondary eclipses
NIRCam (Defocused Imaging)	0.6 - 2.3 2.4 - 5.0	4, 10 , 100 4, 10 , 100	High precision light curves for bright targets that need to be defocused to avoid rapid saturation of detectors
NIRCam (Spectroscopy)	2.4 - 5.0	1700	Transmission/emission spectroscopy spectroscopy of transiting planets
NIRSpec (Spectroscopy)	1.0 - 5.0	100, 1000 2700	Transmission/emission spectroscopy of transiting planets (1.6 asec ×1.6 asec slit)
TFI (Imaging)	1.6 - 2.6 3.2 - 4.9	100 100	High precision light curves of primary and secondary eclipses
MIRI (Imaging)	5 - 28	100 100	High precision light curves of secondary eclipses
MIRI (Spectroscopy)	5 - 11 5.9 - 7.7 7.4 - 11.8 11.4 - 18.2 17.5 - 28.8	100 3000 3000 3000 3000	Slitless spectroscopy of secondary eclipses Spectroscopy of secondary eclipses: suitable for specific spectral features e.g. CO_2 at 15 μm

4.1. *Imaging*

Both NIRCam, TFI and MIRI offer the opportunity to obtain high precision light curves, a capability that has served as the mainstay of Spitzer transit science. In combination with radial velocity measurements, high precision light curves yield exoplanet mass and radii. High precision light curves are also used to search for unseen companions via transit timing, search for exoplanet moons and rings, and record reflectance and thermal phase variations across the duration of a system's light curve to study atmospheric dynamics. Each of the JWST cameras has the capability to use sub-arrays for detector readout, to increase the dynamic range for brighter targets. In addition to the standard imaging and spectroscopic modes, NIRCam has several instrument modes designed for phasing the primary mirror that also have specific application to the observation of transiting systems. NIRCam's wavefront sensing and control optics includes special lenses that have 4, 8 and 12 waves of defocus to facilitate phase retrieval measurements using instrument optics rather than the secondary mirror. A telescope as large as JWST will be limited to imaging observations of relatively faint stars when measuring photometric transits as the detectors pixels will saturate in even the shortest exposure times. However, the defocus lenses, combined with the use of sub-arrays allows NIRCam to collect to image stars as bright as K$\sim$3 without saturating detector pixels in the minimum exposure time.

NIRCam can employ this capability to obtain high-precision light curves of transiting terrestrial planets e.g. SNR 20-30 for a K=10 star in 6.5 hours (Greene *et al.* 2007).

4.2. *Spectroscopy*

It is in the field of spectroscopic characterization that JWST has the capacity to make major contributions to exoplanet science (Greene *et al.* 2007). Impressive progress has been made with Spitzer (Swain *et al.* 2009)and HST (Charbonneau *et al.* 2002), in obtaining the first spectral diagnostics of exoplanets. Two techniques can be used to probe transiting extrasolar planet atmospheres with JWST. The absorption spectrum of the planet can be measured by detecting the spectral signature imposed on stellar light transmitted through the planet's atmosphere during transit. The emission spectrum of the planet can also be measured during the secondary eclipse. Emission spectra produce potentially larger signals than transmission spectra at infrared wavelengths. However, features in transmission spectra will be present even in the extreme case when the atmospheric temperature profile of the exoplanet is isothermal - which would produce a featureless spectrum in emission. Gas giant planets will present many molecular features (H_2, CO, H_2O, CH_4), strong atomic lines (Na, K), and a spectral shape (due to Rayleigh scattering) that leave distinct imprints on transmission spectra. High quality spectra also probe energy redistribution within the atmospheres. With its large collecting area JWST will be able to conduct detailed comparative studies of gas giant atmospheres and their composition both in transmission and emission, including many of the transiting gas giants discovered by Kepler. JWST will be capable of R=100 to 3000 follow-up spectroscopy of gas giants found by ground and space-based surveys over the 0.7 μm to 10 μm wavelength range. For exoplanets with bright parent stars, it can deliver R 2700 spectra from 1μm - 5 μm wavelength range, and for the first time provide high quality line diagnostic of these exoplanets. In the mid-IR it will be able to deliver R$\sim$100 spectra of gas giants in a single transit.

JWST's large collecting area makes it an obvious choice for characterization studies of intermediate and super earth mass transiting planets. Transiting exoplanets around late-type stars are especially attractive (Charbonneau and Deming 2007) since the relatively small stellar radius yields transit depths that can enable low-resolution spectral characterization of some intermediate and superearth mass exoplanets, in emission and transmission. A recent examples of a candidate system for JWST follow-up is GJ 436b, a hot Neptune, with a mass 0.072 MJ, and a period of 2.6 days, orbiting an M2.5V star. In Figure 4 we show a simulated NIRSpec observation of GJ 436, combining four transits to achieve a R-300 spectrum. The figure demonstrates that JWST will be able to make relatively high precision observations of intermediate mass transiting planets. The simulation includes the effects of detector pixel response functions, the JWST pointing budget and expected detector flat field response.

Recent discoveries of Corot-7b (Leger *et al.* 2009) and GL1214 (Charbonneau *et al.* 2009) have increased interest in the questions of whether JWST will be able to characterize super earths. Deming *et al.* (2009) have addressed this question in detail and find that depending upon the frequency of occurrence and the nature of their atmospheres, JWST can measure the temperature and identify molecular absorptions, such as water and CO_2, of supe rearths orbiting lower main sequence stars. However, significant amounts of observing time will be required for such observations, together with suitable candidates for study. In the coming decade transiting planet surveys focused on late type stars with bright central stars are required to provide targets for observations of super earths with JWST. While JWST will likely not be able to address the question of the

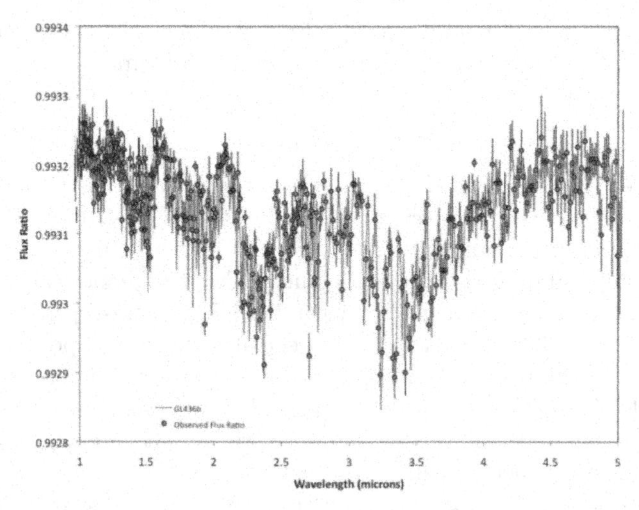

Figure 4. Simulated observation of GJ 436b, a hot neptune, with NIRSpec. The graph shows the simulated observation (filled circles with error bars), plotted over a model spectrum of GL 436B provided Sara Seager. The simulation includes instrumental effects such as the detector pixel response function, and observatory pointing effects based on the JWST pointing error budget

true "earth analog" with a one year period (Traub and Kalteneggar 2009), it will open new discovery space for super earths around late-type stars with short periods.

4.3. *Acknowledgements*

We wish to thank Sara Seager for the GJ436b model spectrum.

References

Beichman, C., *et al.* 2010, *PASP*, 122, 162
Cavarroc, C., *et al.* 2008, *Proc. SPIE*, 7010, 29
Charbonneau, D., *et al.* 2009, *Nature*, 462, 891
Charbonneau, D. & Deming, L. 2007, *arXiv0706.1047C*
Charbonneau, D., Brown, T. M., Noyes, R. W., & Gilliland, R. L. 2002, *ApJ*, 568, 377
Deming, L., *et al.* 2009, *PASP*, 121, 952
Gardner, J. P., *et al.* 2006, *SSRv*, 123, 485
Green, J. J., *et al.* 2005, *Proc. SPIE*, 5905, 185
Greene, T., *et al.* 2007, *Proc. SPIE*, 6693, 15
Wright, G. S., *et al.* 2008, *Proc. SPIE*, 7010, 28
Jakobsen, P., *et al.* 2010, *BAAS*, 215, 396
Krist, J., *et al.* 2009, *Proc. SPIE*, 7440, 31
Krist, J., *et al.* 2007, *Proc. SPIE*, 6693, 12
Leger, A., *et al.* 2009, *A&A*, 506, 287
Makidon, R. B., *et al.* 2008, *Proc. SPIE*, 7010, 22
Marois, C., Doyon, R., Racine, R., & Nadeau, D. 2000, *PASP*, 112, 767
Rieke, M. J., *et al.* 2003, *Proc. SPIE*, 4850, 478
Sivaramakrishnan, A., *et al.* 2009, *Proc. SPIE* 7440, 33
Swain, M. R., *et al.* 2009, *ApJ* 704, 1616
Swain, M. R., *et al.* 2008, *Nature*, 452, 329
Traub, W. A. & Kaltenegger, L. 2009, *ApJ* 98, 519

The Astrophysics of Planetary Systems: Formation, Structure, and Dynamical Evolution
Proceedings IAU Symposium No. 276, 2010
A. Sozzetti, M. G. Lattanzi & A. P. Boss, eds.

doi:10.1017/S1743921311020412

Science with EPICS, the E-ELT planet finder

Raffaele Gratton[1], Markus Kasper[2], Christophe Vérinaud[3], Mariangela Bonavita[1,4] and Hans M. Schmid[5]

[1]INAF-Osservatorio Astronomico di Padova, Italy
email: raffaele.gratton@oapd.inaf.it

[2]European Southern Observatory, Germany

[3]Laboratoire de Astrophysique de Grenoble, France

[4]University of Toronto, Canada

[5]ETH Zurich, Switzerland

Abstract. EPICS is the proposed planet finder for the European Extremely Large Telescope. EPICS is a high contrast imager based on a high performing extreme adaptive optics system, a diffraction suppression module, and two scientific instruments: an Integral Field Spectrograph (IFS) for the near infrared (0.95-1.65 μm), and a differential polarization imager (E-POL). Both these instruments should allow imaging and characterization of planets shining in reflected light, possibly down to Earth-size. A few high interesting science cases are presented.

Keywords. planetary systems, instrumentation: adaptive optics

---

1. EPICS at E-ELT

Direct observations of exosolar planets are among the drivers for the construction of E-ELT, the ESO project for an extremely large telescope (diameter of 42 m). These observations require very high contrasts at very tiny separations from relatively bright sources, and are limited by background and diffraction. From very generic arguments, contrast in this regime is expected to depend on the square of telescope diameter at separations inversely proportional to this parameter. Although actually reaching such performances will be technically very challenging and possibly such a full gain will not be achieved, the gain of E-ELT with respect to existing or designed instrument is expected to be enormous.

EPICS is the instrument proposed for direct imaging and characterization (both in spectroscopy and polarimetry) of exoplanets. To explore its science potential, we notice that GPI and SPHERE are expected to achieve a photon noise limited contrast performance of better than 10^{-7} at 0.5 separation (Marois *et al.* 2008, Beuzit *et al.* 2008) using 8-m class telescopes. Hence a similar instrument at a 42-m telescope should be able to reach a 28 times better contrast at a 5 times smaller angular separation. EPICS should then reach $\sim 10^{-9}$ contrast at 0.1 separation.

Similar to SPHERE, EPICS baseline concept will include an extreme adaptive optics module (XAO), able to provide high Strehl. Due to lack of photons (a XAO system should sample the pupil with a step of ~ 20 cm), the reference source should be bright ($I < 9$). A second module will allow efficient suppression of the central diffraction peak and observations down to an inner working angle (IWA) of a few λ/D (0.03 arcsec, with a goal of 0.02 arcsec). A set of scientific instruments will exploit simultaneous differential imaging techniques. The Integral Field Spectrograph (IFS) will image at Nyquist limit a square field of view (FoV) with a side of 0.8 arcsec, covering the wavelength region 0.95-1.7 μm at low spectral resolution. Additionally, this instrument will provide intermediate

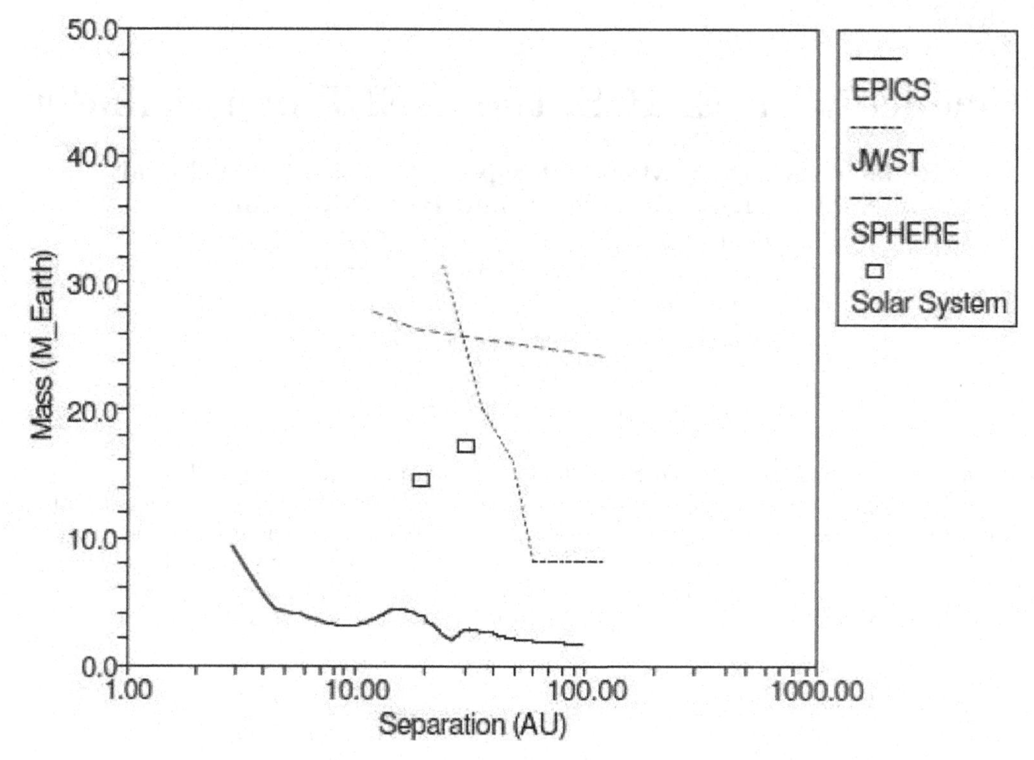

Figure 1. Limiting detectable mass of planets as a function of physical separation around a 10 Myr old G2V star at a distance 120 pc. Observations potentially possible with EPICS and SPHERE are in the J-band; those with JWST are in the L band (data from Green *et al.* 2005). The location of the two outer giant planets of the Solar System (Uranus and Neptune) in this plane is shown for comparison.

($R \sim 4000$) and high ($R \sim 20,000$) spectral resolution for follow up observations of bright planets. A polarization analyser (EPOL) will also be included, based on a concept similar to the ZIMPOL one under development for SPHERE. EPOL will yield high precision (10^{-5}) differential polarimetry over the whole FoV, Nyquist sampled at diffraction limit, over the wavelength range 600-900 nm.

2. Science goals for EPICS

EPICS will be a powerful instrument for detection and characterization of exosolar planets. Its design will be optimized in order to reach very significant goals in a number of areas. In this section we define the main goals of EPICS. This requires a quantitative estimate of the results we might expect from EPICS. At INAF-OAPD we developed a special software tool (MESS: Bonavita *et al.* 2010, in preparation) that allows making such quantitative assessments.

2.1. *Young self-luminous gas planets*

Detection of young self-luminous giant planets allows to determine the initial distribution with mass and separation, to be compared with models of formation and evolution. This is crucial, because these mechanisms are still far from being properly understood. The peak of the distribution of giant planets with separation is expected to lie slightly out of the so-called snow-line, where ices can survive, providing a wealth of material for the formation of large planetary cores: for solar type stars, the snow-line is expected to be at $\sim 3-5$ AU. Exploration of regions even further out (at > 10 AU) provide information on the impact of neighbours on the dynamical evolution of the orbits of already formed planets. These regions are inaccessible with most methods. Direct detection may be very helpful because (i) a single image is enough for describing main characteristics of the whole system; (ii) repeated visits may allow determination or at least constraints on the main orbital parameters; and (iii) possibly coupled with indirect methods, planetary masses can be derived.

For this goal, best data are obtained for star forming regions and young associations, because very young planets are expected to be bright. These regions are typically at distances between 100 and 150 AU, although TW Hya is closer. Note that only a handful of stars in each of these regions are bright enough to be observable with EPICS.

Instruments on 8-10 m class telescope (SPHERE and GPI) should allow detection of a few tens giant planets around a few of these objects (see Figure 1), while JWST can observe farther stars and has a sensitivity to much smaller planet masses but only in the very outer regions of the systems. While this would be by itself of very high interest, the much higher sensitivity of EPICS should allow to observe much fainter planets, that is both less massive and/or older, and moreover to explore with high sensitivity much inner regions, close to the snowline. Figure 1 suggests in fact that Neptune-like planets should be detectable by EPICS down to a few (2-3) AU even if the limiting contrast is only 10^{-6}. EPICS should then allow a complete census of the gaseous planets that form outside of the snowline.

The presence of disks, while providing very important information on their relation with the very young planets, may prevent planet detection. This may occur either because the disk is optically thick or because the disk is so luminous to overcome planet emission. Optically thick disks are observed around very young objects (age ~ 1 Myr or less) in the Orion nebula and elsewhere (see e.g. ODell & Wen 1994). Observation of planets should then be easier around older objects, preferably seen pole-on (it is anyhow dubious that planets can form on such short timescale). Competition with planet emission is very critical for mid-IR instruments on 8-10 m or smaller class telescopes; it is less a problem with EPICS at E-ELT because observations are in the NIR, where disk thermal emission is negligible and the diffraction peak is more concentrated. Still, stellar light scattered by the disk may be a not negligible contribution to the local background, contributing to noise. For instance, in the case of AU Mic (a $\sim$12 Myr old M1V star), the near edge-on disk is ~ 9 mag/arcsec^2 fainter than the star (in the J, H and K band) at 10 AU from the star (Fitzgerald *et al.* 2007). Over the area subtended by a diffraction peak, the disk should then produce a flux equivalent to that of a planet with a contrast of 3×10^{-9}. This is close to the limiting contrast obtained by EPICS.

It should be noticed that young massive planets will be detected at quite high S/N with EPICS. This means that they can be targets of follow-up observations with higher spectral resolution.

2.2. *Mature planets in the solar neighbourhood*

Observation of samples of giant planets in the solar neighbourhood is very important for various reasons: (i) Frequency and mass distribution of giant planets at old ages, once dynamic evolution have cleaned systems from planets in unstable orbits, can be compared with the results obtained in star forming regions and young associations. (ii) These systems may be studied in more detail, even in regions much closer to the central star with respect to the snowline, allowing exploring the habitable zone (HZ) and even inner regions. (iii) These observations are important forerunner for spatial missions for

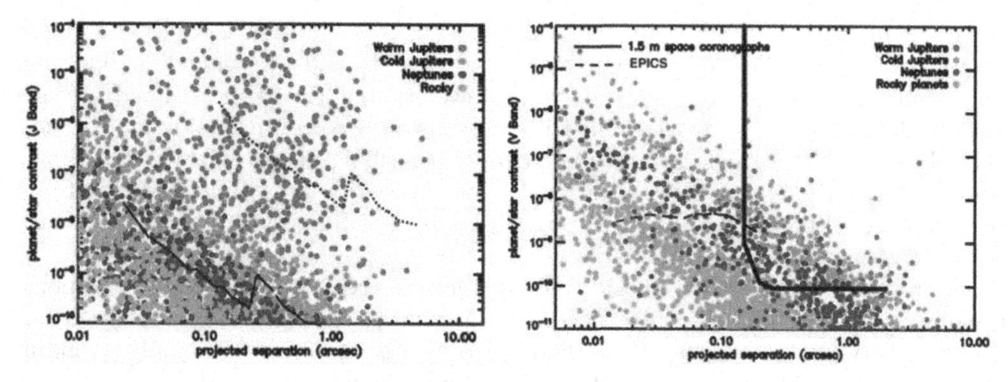

Figure 2. Left panel: Expected simulated planets in the plane contrast versus projected separation, compared with detection limits for EPICS IFS. For comparison, expected detection limits with GPI and SPHERE-ZIMPOL are also plotted (dashed lines). Right panel: the same for EPOL (dashed line). For comparison, expected detection limits with a 1.5 m space coronagraph are also plotted (solid line).

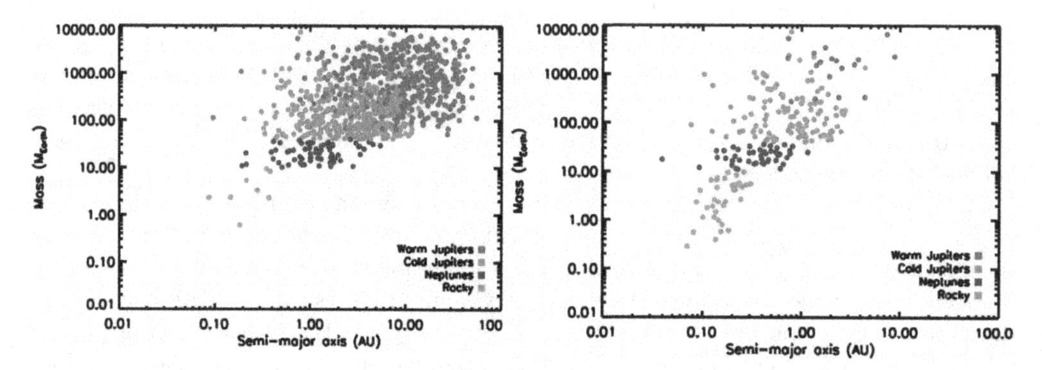

Figure 3. Simulations of planets detectable by the NIR IFS (left panel) and EPOL (right panel) of EPICS, plotted in the semi-major axis versus mass plane.

spectroscopy of Earth-like planets, clarifying which systems are most likely to host rocky planets in the habitable HZ.

Even more interesting would be detection of small mass planets (Neptunes and Super-Earths) that would allow to define the expected frequency of low-mass planets at various separations from the central star. Furthermore, even low resolution and S/N spectra of such objects would allow a first characterization of their atmosphere.

To examine the potentialities of EPICS for these scientific goals, we considered the properties of the planets expected to be detected on both a survey covering a sample of $\sim$600 stars within 20 pc from the Sun and brighter than $I < 9$, and over a sample of 1200 young stars within 100 pc. For each star we randomly choose 5 planets (mass-semi-major axes) from the planetary population predicted by power laws extrapolated from RV surveys (Cumming *et al.* 2008). Luminosity of each planet is estimated taking into account both intrinsic luminosity and reflected light contribution. We then compared the expected contrasts with the curves for limiting detections: a planet is detected if it is above the detection threshold.

We distinguished different planet mass ranges: (i) all the objects with $M_P > 40\ M_{\rm Earth}$ are considered as Jupiter-like planets; (ii) those with $10 < M_P < 40\ M_{\rm Earth}$ are the Neptune-like planets; (iii) and finally the ones with $M_P < 10\ M_{\rm Earth}$ are considered as rocky planets. Figure 2.1 compares the expected distribution of planets in contrast

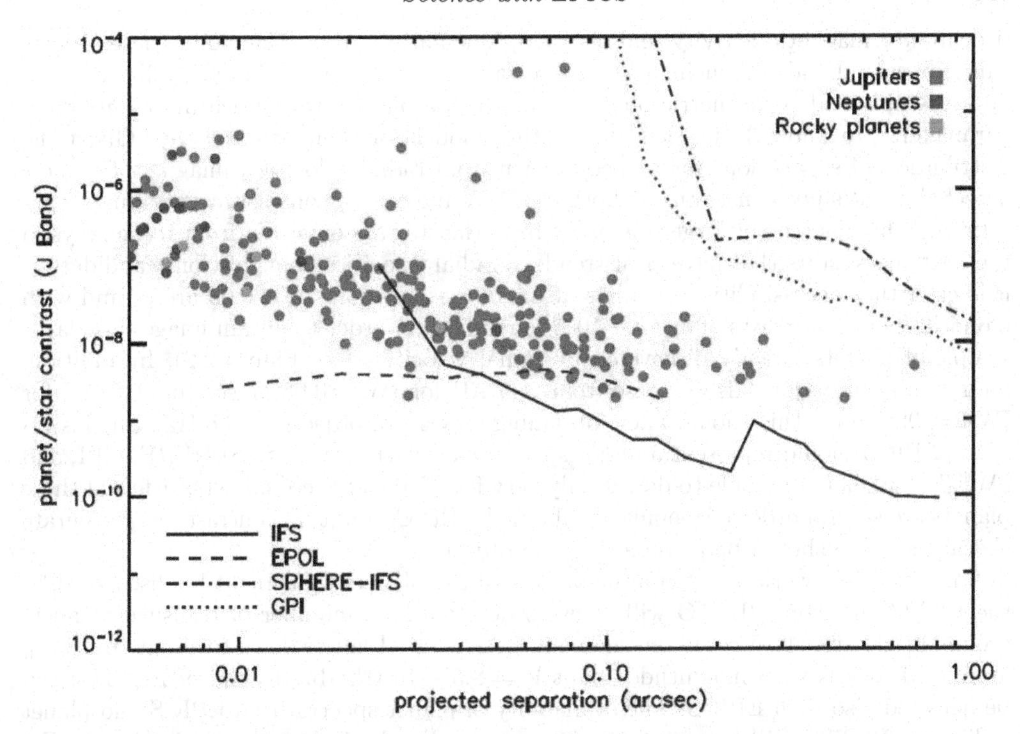

Figure 4. Exosolar planets already discovered with radial velocities in the separation-contrast plane in the J-band. Lines are detection limits with various high contrast instruments.

vs. projected separation plane, against limiting detection curves that might possibly be obtained using EPICS. Figure 2.1 shows the detected planets in the semi-major axis vs. mass plane. Essentially all Jupiter-like planets at projected separation beyond the EPICS IWA will be detected, making up a sample of several hundred objects. A similar survey will also be quite effective in detecting Neptune like planets, with several tens of them detected, in the range of projected separation from the IWA up to about 0.1 arcsec. On the other hand, in our simulations a few rocky planets are detected around very close and bright stars; for the IFS the success rate depends on random fluctuations as well as on the exact real performances of EPICS, while for the polarization analyser the IWA is very critical, and some tens of planets might be detected if this is as small as 0.01 arcsec. At the EPICS IWA (0.02 arcsec at best), most mature Super-Earths have a contrast below 10^{-7}, and are below 10^{-8} at 0.1 arcsec, making them difficult objects at these small separations. We incidentally note that Gl581d is close to the expected detection limit for XAO at E-ELT. While considerable effort can be devoted to the observation of this single object, it is clear that EPICS performances should be put at their limit to allow such an observation.

We finally note that many detections are done even at rather high S/N. This implies that rather accurate spectral information could be gathered from these spectra.

2.3. *Planets discovered by RV, astrometry and transit searches*

In most cases masses for planets detected by EPICS cannot be determined independently of ages. Imaging of planets already detected by RV, transits and/or astrometry would represent a major breakthrough thanks to the availability of dynamical constraints (or even full orbit determination) on the planet masses and on the orbital elements. In most cases, stellar ages are or can be determined rather well exploiting suitable indicators (isochrones, magnetic activity and rotation, kinematics, etc.). Therefore, these objects will represent the ideal benchmarks for the calibration of models for sub-stellar objects. Spectroscopic and polarimetric observation of these planets (for which most important parameters are known) is crucial for testing models of their atmospheres. Given the distribution of separations for planets known from radial velocities, imaging of planets detected by RVs put constrains on both the IWA and on the contrast at very small separations. While planets at larger distances from the star are expected from RV surveys in the next years (stars that have clear trends of radial velocities being obvious candidates), it is clear that accessibility to planets at angular separations of < 0.05 arcsec and with monochromatic contrasts of at least 10^{-8} is required in order to obtain images of a large sample of planets already discovered by radial velocities (see Figure 2.2). In addition, minimum semi-major axis decrease from 1.5 AU for IWA=0.05 arcsec to $\sim$0.8 AU for IWA=0.02 arcsec: this should allow obtaining spectra of objects in the HZ. On this respect, EPICS should represent a major step forward with respect to SPHERE, GPI, and JWST, that should be able to detect only very few if any of these planets. A few of these planets are several order of magnitude above the EPICS limiting contrast: very accurate characterization should be possible for such objects.

An interesting overlap region also exists with the planets that might be discovered by the PLATO mission. PLATO will allow to detect a large number of transiting planets around bright stars that could be studied in close detail. Planets down to about 10 $M_{\rm Earth}$ around M dwarfs with magnitude about $V = 8.5 - 10$ (the bright end of PLATO) can be detected also with EPICS. The availability of planet spectra from EPICS and planet radii from PLATO will be extremely relevant for the physical study of the planets. For G and F stars (and K and M dwarfs as well) planets at separation larger than that accessible to PLATO can be detected, allowing to study the outer planetary system of PLATO targets.

References

Baraffe, I., Chabrier, G., & Barman, T. S. 2003, *A&A*, 402, 701
Beuzit, J.-L., Feldt, M., Doheln, K., *et al.* 2008, *Proc. SPIE*, 7014, 41
Cumming, A., Butler, R. P., Marcy, G. W., *et al.* 2008, *PASP*, 120, 531
Fitzgerald, M. P., Kalas, P. G., Duchêne, G., *et al.* 2007, *ApJ*, 670, 536
Green, J., Beichman, C., Basinger, S., *et al.* 2005, *Proc. SPIE*, 5905, 185
Marois, C., Macintosh, B., Soummer, R., *et al.* 2008, *Proc. SPIE*, 7015, 47
ODell, C. R. & Wen, Z. 1994, *ApJ*, 436, 194

The Astrophysics of Planetary Systems: Formation, Structure, and Dynamical Evolution
Proceedings IAU Symposium No. 276, 2010
A. Sozzetti, M. G. Lattanzi, & A. P. Boss, eds.

doi:10.1017/S1743921311020424

Towards habitable Earths with EUCLID and WFIRST

Jean-Philippe Beaulieu[1,2], David P. Bennett[3], Eamonn Kerins[4] and Matthew Penny[4]

[1]Institut d'Astrophysique de Paris, 98bis Boulevard Arago, 75014 PARIS, France
[2]University College London, Gower street, London WC1E 6BT, UK
email: beaulieu@iap.fr

[3] Department of Physics, University of Notre Dame, Notre Dame, Indiana 46556, USA
email: bennett@nd.edu

[4] Jodrell Bank Centre for Astrophysics, University of Manchester M13 9PL, UK
email: Eamonn.Kerins@manchester.ac.uk

Abstract. The discovery of extrasolar planets is arguably the most exciting development in astrophysics during the past 15 years, rivalled only by the detection of dark energy. Two projects are now at the intersection of the two communities of exoplanet scientists and cosmologists: EUCLID, proposed as an ESA M-class mission; and WFIRST, the top-ranked large space mission for the next decade by the Astro 2010 Decadal Survey report. The missions are to have several important science programs: a dark energy survey using weak lensing, baryon acoustic oscillations, Type Ia supernova, a survey of exoplanetary architectures using microlensing, and different surveys. The WFIRST and EUCLID microlensing planet search programs will provide a statistical census of exoplanets with masses greater than the mass of Mars and orbital separations ranging from 0.5 AU outwards, including free-floating planets. This will include analogs of all Solar System planets except for Mercury, as well as most types of planets predicted by planet formation theories. In combination with Kepler's census of planets in shorter period orbits, EUCLID and WFIRST's planet search programs will provide a complete statistical census of the planets that populate our Galaxy. As of today, EUCLID is proposed to ESA as a M class mission (the result of the selection will be known in october 2011). We are presenting here preliminary results about the expected planet yields. WFIRST has just appointed a Science Definition Team.

Keywords. planets and satellites: formation, planets and satellites: general, planetary systems, planetary systems: formation

1. Microlensing today: Super Earths beyond the snow line

Several different methods have been used to discover exoplanets, including radial velocity, stellar transits, direct imaging, pulsar timing, astrometry, and gravitational microlensing (Mao & Paczynski 1991; Gould & Loeb 1992). Microlensing exploits the light deflection effect according to Einstein's theory of general relativity. So far 12 microlensing exoplanets have been published and whilst this number is relatively modest compared with the radial velocity and transit methods, microlensing probes a part of the parameter space (host separation vs. planet mass) not accessible in the medium term to other methods. The mass distribution of microlensing exoplanets has already revealed that cold super-Earths (at or beyond the snow line and with a mass of around 5 to 15 Earth mass) appear to be common (Beaulieu *et al.* 2006; Gould *et al.* 2006; Sumi *et al.* 2010; Cassan *et al.* 2011). Detections include a scale 1/2 model of our solar system (Gaudi *et al.*

2008; Bennett *et al.* 2010) and several cold Neptunes/Super Earths. It has be shown that the ground-based detection efficiency can, under favourable circumstances, extend down to 1 Earth mass planets (Bennett & Rhie 1996; Batista *et al.* 2009). Microlensing has also provided the first measurement of the frequency of ice and gas giants beyond the snow line. The abundance of such systems is about 7 times higher than closer-in planets probed by the Doppler method. This comparison provides strong evidence that most giant planets do not migrate very far (Gould *et al.* 2010).

Microlensing is currently capable of providing statistics on cool planets of super-Earth mass from the ground. A network of wide-field telescopes, strategically located around the world, could detect planets with mass as low as the Earth. Free-floating planets can also be detected; a significant population of such planets are expected to be ejected during the formation of planetary systems. A wide-field imager network is being implemented now in Chile (OGLE-IV), New Zealand (MOA-2), Tasmania, and the Wise Observatory in Israel. In the near future three additional 1.6m telescopes will be deployed as part of the Korean Microlensing network of Telescopes (PI Han). Microlensing is roughly uniformly sensitive to planets orbiting all types of star, as well as white dwarfs, neutron stars, and black holes. In contrast other detection methods are most sensitive to FGK dwarfs and are only now extending to M dwarfs. Microlensing is therefore an independent and complementary detection method for aiding a comprehensive understanding of the planet formation process. Ground-based microlensing mostly probes exoplanets beyond the snow line, where the favoured core-accretion theory of planet formation predicts a larger number of low-mass exoplanets (Ida and Lin, 2005). The statistics provided by microlensing will enable a critical test of the core accretion model. Exoplanets probed by microlensing are much further away than those probed with other methods and therefore provide an interesting comparison sample with nearby exoplanets. The microlensing road map has been presented in different white papers (ie Gould *et al.* 2007; Beaulieu *et al.* 2008; Gaudi *et al.* 2009; Bennett *et al.* 2010).

2. Space based microlensing and Dark Energy

Ultimately, a comprehensive census of cold planets below Earth masses, including habitable planets, requires a space-based microlensing survey.

Angular resolution is the key to extend sensitivity below a few earth masses
Microlensing relies upon the high density of source and lens stars towards the Galactic bulge to gaurantee the stellar alignments needed to generate microlensing events. But this high star density also means that the bulge main sequence source stars are not generally resolved in ground-based images. This means that the precise photometry needed to detect planets of $\leqslant 1M_{\oplus}$ is not possible from the ground unless the magnification due to the stellar lens is moderately high. This, in turn, implies that ground-based microlensing is only sensitive to terrestrial planets located close to the Einstein ring (at ~2-3 AU). The full sensitivity to terrestrial planets in all orbits from $0.5AU$ to free floating comes only from a space-based survey.

Microlensing from space yields precise star and planet parameters
The high angular resolution and stable point-spread-functions available from space enable a space-based microlensing survey to detect most of the planetary host stars. When combined with the microlensing light curve data, this allows a precise determination of the planet and star properties for most events (Bennett *et al.* 2007a).

The first envisioned project was the Galactic Exoplanet Survey Telescope (GEST) (Bennett & Rhie 2002), which was a wide-field optical telescope that focused on exoplanets, but also had significant weak lensing and supernova programs when proposed to NASA's Midex program. But, the microlensing target fields in the Galactic bulge are most easily observed in the infrared, so the GEST concept was superseded by the wide field infrared Microlensing Planet Finder (MPF) mission concept (Bennett *et al.* 2007b, 2010). In 2008, the Exoplanet Task Force (ExoPTF) released a report (Lunine *et al.* 2008) that evaluated all of the current and proposed methods to find and study exoplanets, and expressed strong support for space-based microlensing. Its finding regarding space-based microlensing states that: *"Space-based microlensing is the optimal approach to providing a true statistical census of planetary systems in the Galaxy, over a range of likely semi-major axes, and can likely be conducted with a Discovery-class mission."* Shortly afterwards the Astro 2010 Decadal Survey ranked as the top priority the WFIRST concept, a wide-field infra-red imager and low resolution spectroscopy on a 1.5m telescope aimed at probing dark energy and exoplanet statistics. It "will open up a new frontier of exoplanet studies by monitoring a large sample of stars in the central bulge of the Milky Way for changes in brightness due to microlensing by intervening solar systems. This census, combined with that made by the Kepler mission, will determine how common Earth-like planets are over a wide range of orbital parameters".

We will briefly describe the status of EUCLID and WFIRST in the next two sections:

2.1. *EUCLID*

The use of cosmic shear to probe dark energy was advanced in Europe in 2006 through the DUNE proposal (Refregier *et al.* 2007), which subsequently developed into EUCLID, based upon a 1.2m Korsch telescope. A 3 month microlensing program is part of the additional science of EUCLID, aiming at low mass telluric planets situated around the snow line (described in the EUCLID yellow book and in Beaulieu *et al.* 2010). The required data for this program is continuous monitoring with a sampling time of around 25 minutes over an area of 1×3 deg^2 centred at the Galactic coordinates $l \simeq 1.1, b \simeq -1.7$ aligned with the Galactic plane so that the long axis of the field is parallel to the Galactic plane. We aim to use both the EUCLID optical and near IR channels, although the efficiency is about twice better on the IR channels.

There is also a proposition to extend the initial 3 month additional science program with a 9 month legacy program which would measure the abundance of cool Earth-mass exoplanets around solar like stars. This program (if accepted) could take place once the dark energy objective will have been reached towards the end of the mission. An attractive possibility would be to have a first survey of 3 months shortly after the launch to guarantee early high-profile results (detection of planets down to the mass of Mars). It is also important to have observations early in the mission life time and after few years to maximize the baseline in order to be able to have better constraints on host masses. With the 9 month legacy program in addition to the 3 months that are currently in the additional science program EUCLID could achieve:

a) a complete census of planets down to Earth mass with separations exceeding 1 AU
b) complementary coverage to Kepler of the planet discovery space
c) sensitivity to planets down to $0.1 M_\oplus$, including all Solar System analogues except for Mercury
d) complete lens solutions for most planet events, allowing direct measurements of the planet and host masses, and distance from the observer.

Currently EUCLID is proposed to ESA as a M-class mission, together with PLATO and SOLAR ORBITER. At most, two out of the three will be selected for launch in 2017-2018. The announcement of the selection process will be made in September 2011.

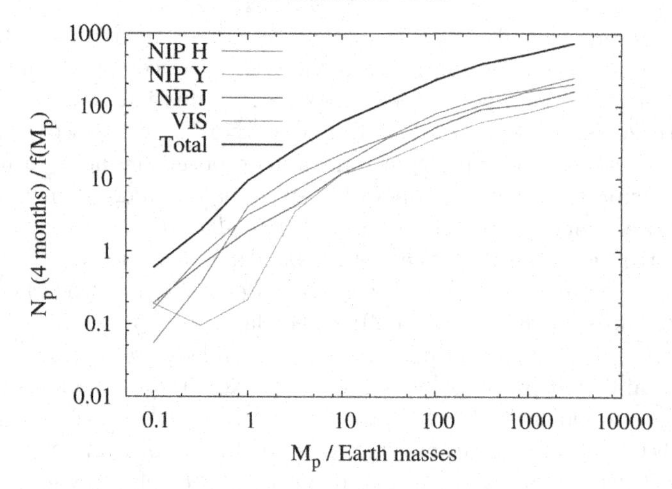

Figure 1. The expected number of planets versus mass for a 4 month program with EUCLID, assuming one planet of mass M_p per host and a logarithmic separation prior out to 30AU (so ignoring free-floating planets). The lines show yields for the H, Y and J EUCLID IR channels, as well as the optical channel (VIS) and the overall yield (black line). The planet yield scales roughly linearly with program duration.

Members of the EUCLID exoplanet Working Group are constructing a detailed simulation of the EUCLID exoplanet catch using the Besancon population synthesis model (Robin *et al.* 2003; Marshall *et al.* 2006). The model includes bulge, disk and spheroid lens and source populations as well as a detailed prescription for the lens and source kinematics. It allows for specific optical and infrared bandpass predictions and also incorporates a calibrated 3D model of the interstellar dust distribution. This model provides a detailed simulation of the spatial and timescale characteristics of the microlensing events (Kerins, Robin & Marshall 2009), allowing the EUCLID field placement to be optimized for maximum exoplanet yield. The model is still under development but initial predictions indicate that even a 3-month EUCLID program should provide useful statistics on the planet abundance down to Earth masses (Figure 1).

2.2. *WFIRST*

A Science Definition Team (SDT) of 19 members has been appointed in December 2010. The WFIRST SDT is charged with developing a mission design, but the "straw-man" design used by the Astro2010 Decadal survey, was the JDEM-Omega design (Gehrels 2010), which was one of three mission concepts that served as the basis for the WFIRST concept. The other two mission concepts were the Microlensing Planet Finder (Bennett *et al.* 2010b) and the Near-Infrared Sky Surveyor (Stern *et al.* 2010). The "straw-man", JDEM-Omega design uses a 1.5m telescope with a wide field-of-view imager that would have three main science programss:
- A microlensing exoplanet survey that would use $\sim$ 500 days of the first five years of the mission with a planet yield equal to half that of MPF.
- A dark energy survey using 2-2.5 years of the first five years of the mission employing baryonic acoustic oscillations, weak lensing and Type Ia supernovae.
- Galactic and and extragalactic infra-red surveys.

One of the specificity of the space-based microlensing surveys and the comparison EUCLID-WFIRST is that the planet yield scales roughly linearly with the focal plane area and the program duration. Therefore, once a design would have been agreed upon for WFIRST it will be straightforward to have the estimates.

References

Batista, V., *et al.* 2009, *A&A*, 508, 467

Beaulieu, J. P., *et al.* 2006, *Nature*, 439, 437

Beaulieu, J. P., *et al.* 2008, White Paper submission to the ESA Exo-Planet Roadmap Advisory Team, *arXiV:0808.0005*

Beaulieu, J. P., *et al.* 2010, *APS Conf. Ser.*, 430, 266

Bennett, D. P. & Rhie, S. 1996, *ApJ*, 472, 660

Bennett, D. P. & Rhie, S. 2002, *ApJ*, 574, 985

Bennett, D. P., Anderson, J., & Gaudi, B. S. 2007a, *ApJ*, 660, 781

Bennett, D. P., *et al.* 2007b, White Paper submitted to the NASA/NSF ExoPlanet Task Force, *arXiV:0704.0454*

Bennett, D. P., *et al.* 2010a, in "RFI Response to The Astronomy and Astrophysics Decadal Survey", *arXiv:1012.4486*

Bennett, D. P., *et al.* 2010b, *ApJ*, 713, 837

Cassan, A., *et al.* 2011, submitted

Gaudi, B. S., *et al.* 2008, *Science*, 319, 927

Gaudi, B. S., *et al.* 2009, in "The Astronomy and Astrophysics Decadal Survey", *arXiv:0903.0880*

Gehrels, N. 2010 in "RFI Response to The Astronomy and Astrophysics Decadal Survey", *arXiv:1008.4936v1*

Gould, A. & Loeb, A. 1992, *ApJ*, 396, 104

Gould, A., *et al.* 2006, *ApJ*, 644, L37

Gould, A., Gaudi, B. S., & Bennett, D. P. 2007, white paper submitted to the NASA/NSF Exoplanet Task Force, *arXiv:0704.0767*

Gould, A., *et al.* 2010, *ApJ*, 720, 1073

Ida, S. & Lin, D. N. C.. 2005, *ApJ*, 626, 1045

Kim, S. L., *et al.* 2010, *Proc. SPIE*, 7733, 77333

Kerins, E., Robin, A. C., & Marshall, D. J. 2009, *MNRAS*, 396, 1202

Lunine, J., *et al.* 2008, in "Exoplanet Task Force Report", *arXiV:0808.2754*

Mao, S. & Paczynski, B. 1991, *ApJ*, 374, L37

Marshall, D. J., Robin, A. C., Reylé, C., Schultheis, M., & Picaud, S. 2006, *A&A*, 453, 635

Refregier, A., *et al.* 2007, "The Dark UNiverse Explorer : proposal to ESA's comic vision", *arXiv:0802.2522v4*

Robin, A. C., Reylé, C., Derrière, S., & Picaud, S. 2003, *A&A*, 409, 523

Stern, D. 2010, in "RFI Response to The Astronomy and Astrophysics Decadal Survey", *arXiv:1008.3563v1*

Sumi, T., *et al.* 2010, *ApJ* 710, 1641

The Astrophysics of Planetary Systems: Formation, Structure, and Dynamical Evolution
Proceedings IAU Symposium No. 276, 2010
A. Sozzetti, M. G. Lattanzi & A. P. Boss, eds.

doi:10.1017/S1743921311020436

The PLATO mission

Heike Rauer[1,2], Claude Catala[3] and the PLATO consortium

[1]Institut für Planetenforschung, DLR,
Rutherfordstr. 2, 12489 Berlin, Germany
email: heike.rauer@dlr.de

[2]Zentrum für Astronomie und Astrophysik, TU Berlin
10623 Berlin, Germany

[3]LESIA, Observatoire de Paris
5 Place Jules Janssen, 92195 Meudon Cedex, France
email: claude.catala@obspm.fr

Abstract. The *PLAnetary Transits and Oscillations of stars* (PLATO) mission is in its definition study phase in the context of ESA's Cosmic Vision 2015-2025 program. PLATO is applying for a launch in 2017/18. Its goal is to detect transiting exoplanets, including terrestrial planets in the habitable zone, and to determine their basic parameters with unprecedented accuracy. In combination with the detailed analysis of the stellar parameters by astroseismology and with ground-based follow-up observations, this will allow characterizing the main properties of exoplanetary systems to a level not achieved before.

Keywords. instrumentation: photometers, techniques: photometric, stars: evolution, stars: fundamental parameters, planetary systems

---

1. Introduction

The *PLAnetary Transits and Oscillations of stars* (PLATO) mission is the next generation space mission to detect transiting extrasolar planets in the solar neighbourhood. PLATO follows the successful CoRoT (CNES) and Kepler (NASA) missions, but aims to detect a sample of planetary systems with terrestrial planets orbiting well-characterized bright stars, including planets in their habitable zone. This new sample of planetary systems will have accurately known basic planet parameters (radius, mass, orbit, age). Planet radii and masses will be derived from highly accurate photometric transit observations with PLATO, combined with a world-wide ground-based follow-up campaign including radial velocity follow-up measurements. Stellar parameters across the HR diagram and for central stars of planetary systems are studied by astroseismology using PLATO's well time-sampled and highly accurate photometric lightcurves. The resulting large sample of such well-known planets orbiting bright stars will provide a breakthrough for numerous further scientific studies.

Examples of the expected scientific impact of PLATO in the field of exoplanets include the study of the planet interior via mass-radius relationships, the orbital evolution of planetary systems, and atmospheric spectral characterization with future spectroscopic telescopes. Furthermore, PLATO has the potential to provide input for new insights into planet evolution. Today, a detailed study of planetary evolution as a fucntion of the age of the system can be made only for the Solar System. We have no information as to whether planets in other planetary systems develop similarly to the well-known planets around the Sun. Therefore, placing our Solar System into a wider context of planetary evolution is currently not feasible with sufficient accuracy because the accuracy of known

Table 1. Number of cool dwarf stars and subgiants in PLATO fields. Cols. 2 and 3: long-term monitoring phase; col. 4: including step-and-stare phase; cols. 5 and 6: comparison with KEPLER; given as a function of noise level (top) and of magnitude (bottom)

	PLATO (4300 deg^2)		20,000 deg^2	KEPLER (100 deg^2)	
noise level (ppm/$sqrt$hr)	nb of stars	m_v	nb of stars	nb of dwarf stars	m_v
27	20,150	9.3 – 10.8	80,400	1,300	11.2
80	292,000	11.6 – 12.9	1,000,000	25,000	13.6
	1,326	8	3,315	30	8
	60,275	11	180,000	1,300	11

ages of extrasolar planets is limited by the only poorly constrained ages of main sequence stars.

PLATO addresses this need for significantly improved stellar parameters, including their ages, by the combination of exoplanet transit detections with astroseismology of their central stars. The accuratly known radii and masses will put meaningful constraints on mass-radius relationships and planet interior. In addition, it has been shown (Kjeldsen *et al.* 2009) that astroseismology with PLATO allows us to derive stellar ages to within 10%. PLATO is therefore expected to provide the basis for a wealth of new discoveries in our understanding of planet and stellar evolution.

PLATO one out of three space missions which are applying for the M2 launch window for medium-sized missions in 2017/18 in the context of ESA's Cosmic Vision program (Stankov *et al.* 2010). At the time of writing, PLATO is in the Definition Phase, in which the spacecraft design will be consolidated and optimized.

2. Mission goals

The prime science goals of PLATO are (Catala *et al.* 2009; Claudi *et al.* 2009): i) The detection and characterization of Earth Analog systems. ii) The search for exoplanets around the brightest stars of solar type at all orbital periods and with all physical sizes. iii) Search for exoplanets around nearby M-type dwarfs with all physical sizes and at all orbital periods, including at orbital distances such that these planets fall within the habitable zones of these very cool stars. iv) Search for and characterization of exoplanets with a wide variety of sizes, masses and orbits around bright stars. v) Full characterization of very bright stars, of all masses and ages, using seismic analysis.

The PLATO mission will detect exoplanets by the transit method and measure the seismic oscillations of the parent stars in order to fully characterize the basic parameters of exoplanetary systems. These goals will be achieved by a long-term, ultra-high precision monitoring in visible photometry of bright dwarfs and subgiants, as well as of a large sample of M dwarfs. The major breakthrough will come from PLATO's strong focus on bright targets, typically with $m_V \leqslant 11$ mag. The PLATO targets will also include a large number of very bright and nearby stars, with $m_V \leqslant 8$ mag, as well as a large sample of cool M dwarfs down to m_V = 15-16 mag.

The PLATO observations will be complemented by a world-wide effort of ground-based follow-up observations, including radial velocity monitoring. In comparison to ongoing space missions for exoplanet detection, spectroscopic follow-up observations to confirm the planet nature of a transit event and to determine the planet mass will be significantly facilitated by the brightness of the PLATO targets.

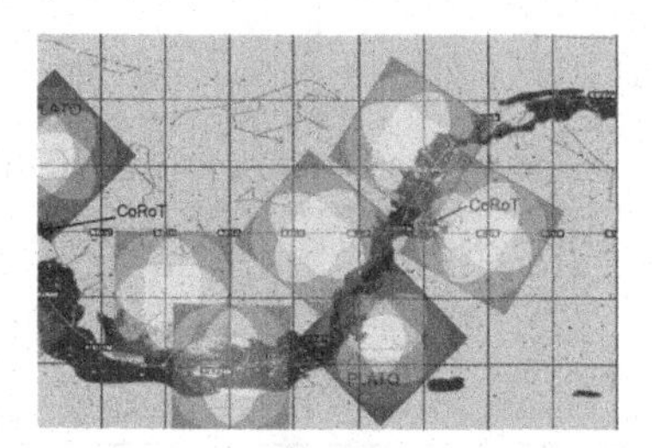

Figure 1. Illustration of sky coverage. Exact field positions will be defined during study phase.

3. The target samples

The prime focus of PLATO is on planets orbiting bright stars. The highest priority target samples, therefore, are about 20,000 dwarfs and sub-giants with mV $\leqslant$ 11 mag. This sample will be monitored with a noise level better than 27 ppm in one hour. This noise level is sufficient to detect Earth-sized exoplanets and to well characterize the central stars by astroseismology. In addition, a sample of at least 1,000 very bright dwarfs/subgiants with mV $\leqslant$ 8 mag will be observed. This sample of very bright stars will provide the prime targets for further spectral characterization of the detected planets, e.g. for atmosphere studies, by spectroscopic telescopes. Finally, more than 245,000 cool dwarfs/subgiants will be observed at a relaxed noise-level (better than 80 ppm in one hr). This sample will still allow detecting transit signals from terrestrial planets around solar-type stars, but radial-velocity confirmation and astroseismology will be difficult due to the faintness of the stars (about 11 mag $<$ V $<$ 13 mag). Nevertheless, this sample will provide a large number of planets covering a wide range of parameter space. The expected approximate number of target stars in the PLATO fields is given in Table 1. To increase the detected number of planets close to the habitable zone of their central star with PLATO, a target sample of more than 10,000 M dwarfs has been added. This target sample will be particularly interesting for further spectroscopic characterizing observations.

It is planned to launch PLATO into an orbit around the Sun-Earth L2 Lagrange point to allow for long-duration observations. Two prime target fields will be monitored for at least 2 - 3 years with high duty cycle, typically above 95%, and high time sampling. In addition, several fields of shorter duration (2-5 months each) will be monitored ("step-and-star" phase) for 2-3 years. In total, about 50 % of the sky will be covered at the end of the mission. The sky coverage is illustrated in Fig. 1. However, the final selection of the exact target field positions will be made later during the PLATO study phases.

4. The instrument concept

The instrument concept of PLATO differs from traditional space observatories (Catala *et al.* 2009). As a result of the required large number of target stars over a wide magnitude range, a multi-telescope approach was chosen (Fig. 2, left). Observations of 32 refractive telescopes are combined to produce highly photometrically accurate lightcurves. Each telescope has a pupil size of about 120 mm and operates in "white" light. They have a read-out cadence of 25 sec, which is, however, binned later to increase the signal-to-noise ratio (SNR) depending on star magnitude (50 sec for the highest priority bright star samples, and better than 600 sec for the fainter star samples). The telescopes are complemented by two so-called "fast" telescopes with higher read-out cadence (2.5 sec) and fixed colour filters. These two telescopes are used to monitor the brightest targets,

Table 2. Instrument characteristics.

pupil diameter per telescope	120 mm
normal telescope field-of-view	$\sim$1100 deg^2 (38.7 diameter)
normal telescope detector	4510^2 pixels, 18 μ m square
fast telescope field-of-view	$\sim$5500 deg^2
fast telescope detector	4510x2255 pixels, 18 μ m square
plate scale	15 arcsec/pixel
total field-of-view (FoV)	overlapping FoV of 2232 deg^2 or 42.4x42.4 deg^2

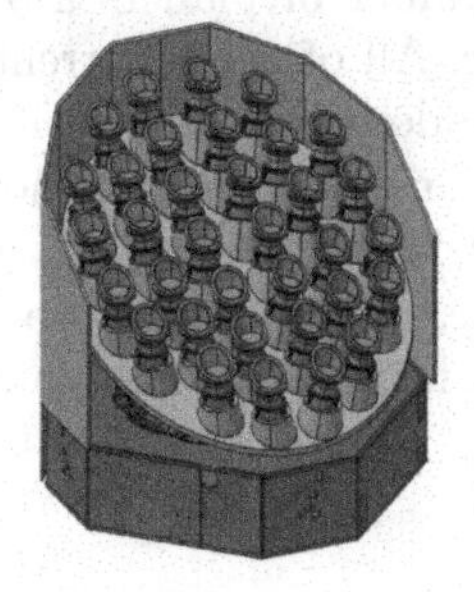

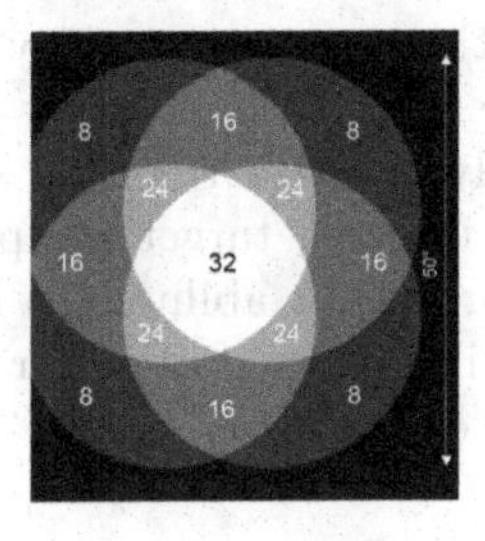

Figure 2. Instrument concept (left) and overlapping viewing directions (right).

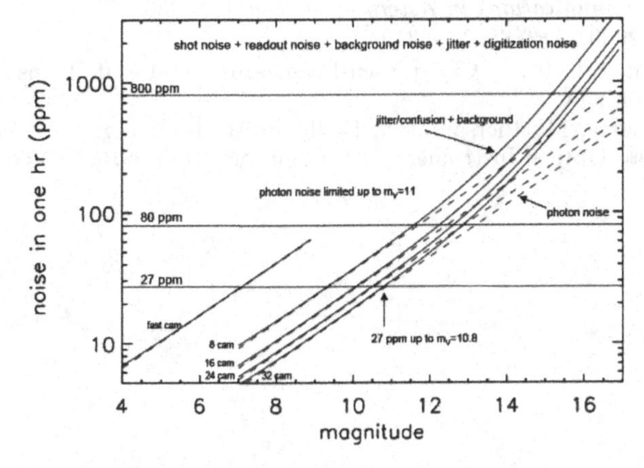

Figure 3. Instrument performance.

but also for satellite fine-pointing. The main foreseen instrument characteristics are given in Table 2.

The 32 "normal" telescopes are combined into 4 groups of 8 telescopes each, pointing into the same viewing direction. The fields of the 4 groups overlap, thereby increasing the area covered in one pointing but keeping a central area where all 32 telescopes are combined (Fig. 2, right). This setting is an optimization of the number of stars at a given noise level and the number of stars at a given magnitude range. Fig. 3 shows the expected noise level for different combinations of telescopes. We note that it is in principle possible to come back to a previous pointing during the "step-and-stare" phase to re-observe interesting targets.

5. Summary

PLATO is the next logical step in the investigation of extrasolar planets. The presently known transiting extrasolar planets have demonstrated the wealth of information that can be gained for these objects, e.g. on their internal structure from their mass and radius, their dynamical history from their orbital parameters (including Rossiter-McLaughlin effect), their atmospheric composition from spectroscopic follow-up. However, the presently known transiting extrasolar planets have also shown where our limitations are to derive conclusive interpretations of what we see, on the observational side as well as on our understanding of planetary systems. Today, the accuracy of mass and radius is limited by our knowledge of the central star. This is in particular true for the most interesting objects: the small, terrestrial planets. Furthermore, up to now a real evolutionary comparison of planets can not be made due to the poorly constraint stellar ages. Since atmospheres can only be detected for a limited number of planets around very bright stars, such investigations are strongly limited so far. All of these current limitations are addressed by the PLATO mission by combining the detection of transiting planets with the detailed analysis of their central stars and stellar evolution in general by astroseismology. In addition, the target sample of PLATO focusses on bright stars which will significantly improve our abilities to investigate these objects in more detail in future. Thus, PLATO will lay the ground for a breakthrough in our understanding of planetary as well as stellar evolution.

References

Catala, C. 2009, *Communications in Asteroseismology*, 158, 330

&Claudi, R., *et al.* 2009, *AP&SS*, 218, 319

Kjeldsen, H., Bedding, T. R., & Christensen-Dalsgaard, J. 2009, in Transiting Planets, *Proc. IAU Symp.*, 253, 309

Stankov, A., Baldesarra, M., Piersanti, O., Fridlund, M., Lindberg, R., & Rando, N. 2010, in Society of Photo-Optical Instrumentation Engineers, *Proc. SPIE*, 7731, 45

The Astrophysics of Planetary Systems: Formation, Structure, and Dynamical Evolution
Proceedings IAU Symposium No. 276, 2010
A. Sozzetti, M. G. Lattanzi & A. P. Boss, eds.

doi:10.1017/S1743921311020448

The science of EChO

Giovanna Tinetti[1] (*UCL*), James Y-K. Cho (*QMUL*), Caitlin A. Griffith (*UoA*), Olivier Grasset (*Un. Nantes*), Lee Grenfell (*DLR*), Tristan Guillot (*Obs. Nice*), Tommi T. Koskinen (*UoA*), Julianne I. Moses (*SSI*), David Pinfield (*UH*), Jonathan Tennyson (*UCL*), Marcell Tessenyi (*UCL*), Robin Wordsworth (*LMD*) **and**
Alan Aylward (*UCL*), Roy van Boekel (*MPIA*), Angioletta Coradini (*INAF/IFSI Roma*), Therese Encrenaz (*LESIA, Obs. Paris*), Ignas Snellen (*Un. Leiden*), Maria R. Zapatero-Osorio (*CAB*) **and**
Jeroen Bouwman (*MPIA*), Vincent Coudé du Foresto (*LESIA, Obs. Paris*), Mercedes Lopez-Morales (*IEEC*), Ingo Mueller-Wodarg (*Imperial College*), Enric Pallé (*IAC*), Franck Selsis (*Un. Bordeaux*), Alessandro Sozzetti (*INAF/OATo*) **and**
Jean-Philippe Beaulieu (*IAP*), Thomas Henning (*MPIA*), Michael Meyer (*ETH*), Giuseppina Micela (*INAF/OAPa*), Ignasi Ribas ((*IEEC*), Daphne Stam (*SRON*), Mark Swain (*JPL*) **and**
Oliver Krause (*MPIA*), Marc Ollivier (*IAS*), Emanuele Pace (*Un. Firenze*), Bruce Swinyard (*UCL*) **and**
Peter A.R. Ade (*Cardiff*), Nick Achilleos (*UCL*), Alberto Adriani (*INAF/IFSI Roma*), Craig B. Agnor (*QMUL*), Cristina Afonso (*MPIA*), Carlos Allende Prieto (*IAC*), Gaspar Bakos (*CfA*), Robert J. Barber (*UCL*), Michael Barlow (*UCL*), Peter Bernath (*Un. York*), Bruno Bézard (*LESIA*), Pascal Bordé (*IAS*), Linda R. Brown (*JPL*), Arnaud Cassan (*IAP*), Céline Cavarroc (*IAS*), Angela Ciaravella *INAF/OAPa*), Charles Cockell *OU*), Athéna Coustenis (*LESIA*), Camilla Danielski (*UCL*), Leen Decin (*IvS*), Remco De Kok (*SRON*), Olivier Demangeon (*IAS*), Pieter Deroo (*JPL*), Peter Doel (*UCL*), Pierre Drossart (*LESIA*), Leigh N. Fletcher (*Oxford*), Matteo Focardi (*Un. Firenze*), Francois Forget (*LMD*), Steve Fossey (*UCL*), Pascal Fouqué (*Obs-MIP*), james Frith (*UH*), Marina Galand (*Imperial College*), Patrick Gaulme (*IAS*), Jonay I. González Hernández (*IAC*), Davide Grassi (*INAF/IFSI Roma*), Matt J. Griffin (*Cardiff*), Ulrich Grözinger (*MPIA*), Manuel Guedel (*Un. Vienna*), Pactrick Guio (*UCL*), Olivier Hainaut (*ESO*), Robert Hargreaves (*Un. York*), Peter H. Hauschildt (*HS*), Kevin Heng (*ETH*), David Heyrovsky (*CU Prague*), Ricardo Hueso (*EHU Bilbao*), Pat Irwin (*Oxford*), Lisa Kaltenegger (*MPIA*), Patrick Kervella (*Paris Obs.*), David Kipping (*UCL*), Geza Kovacs (*Konkoly Obs.*), Antonino La Barbera (*INAF/IASF Palermo*), Helmut Lammer (*Un. Graz*), Emmanuel Lellouch (*LESIA*), Giuseppe Leto (*INAF/OACt*), Mercedes Lopez Morales (*IEEC*), Miguel A. Lopez Valverde (*IAA/CSIC*), Manuel Lopez-Puertas (*IAA-CSIC*), Christophe Lovis (*Obs. Geneve*) Antonio Maggio (*INAF/OAPa*), Jean-Pierre Maillard (*IAP*), Jesus Maldonado Prado (*IEEC*), Jean-Baptiste Marquette (*IAP*), Francisco J. Martin-Torres (*CAB*), Pierre Maxted (*Un. Keele*), Steve Miller (*UCL*), Sergio Molinari (*Un. Firenze*), David Montes (*UCM*), Amaya Moro-Martin (*CAB*), Olivier Mousis (*Obs. Besancon*), napoléon Nguyen Tuong (*LESIA*), Richard Nelson (*QMUL*), Glenn S. Orton (*JPL*), Eric Pantin (*CEA*), Enzo Pascale (*Cardiff*), Stefano Pezzuto (*Un. Firenze*), Ennio Poretti (*INAF/OAMi*), Raman Prinja (*UCL*), Loredana Prisinzano (*INAF/OAPa*), Jean-Michel Réess (*LESIA*), Ansgar Reiners (*IAG*), Benjamin Samuel (*IAS*), Jorge Sanz Forcada (*CAB*), Dimitar Sasselov (*CfA*), Giorgio Savini (*UCL*), Bruno Sicardy (*LESIA*), Alan Smith (*MSSL*), Lars Stixrude (*UCL*), Giovanni Strazzulla (*INAF/OACt*), Gautam Vasisht (*JPL*), Sandrine Vinatier (*LESIA*), Serena Viti (*UCL*), Ingo Waldmann (*UCL*), Glenn J. White (*OU*), Thomas Widemann (*LESIA*), Roger Yelle (*UoA*), Yuk Yung (*Caltech*) **and** Sergey Yurchenko (*UCL*)

[1]University College London, Gower street, London WC1E 6BT, UK
email: `g.tinetti@ucl.ac.uk`

Abstract. The science of extra-solar planets is one of the most rapidly changing areas of astrophysics and since 1995 the number of planets known has increased by almost two orders of magnitude. A combination of ground-based surveys and dedicated space missions has resulted in 560-plus planets being detected, and over 1200 that await confirmation. NASA's Kepler mission has opened up the possibility of discovering Earth-like planets in the habitable zone around some of the 100,000 stars it is surveying during its 3 to 4-year lifetime. The new ESA's Gaia mission is expected to discover thousands of new planets around stars within 200 parsecs of the

Sun. The key challenge now is moving on from discovery, important though that remains, to characterisation: what are these planets actually like, and why are they as they are?

In the past ten years, we have learned how to obtain the first spectra of exoplanets using transit transmission and emission spectroscopy. With the high stability of Spitzer, Hubble, and large ground-based telescopes the spectra of bright close-in massive planets can be obtained and species like water vapour, methane, carbon monoxide and dioxide have been detected. With transit science came the first tangible remote sensing of these planetary bodies and so one can start to extrapolate from what has been learnt from Solar System probes to what one might plan to learn about their faraway siblings. As we learn more about the atmospheres, surfaces and near-surfaces of these remote bodies, we will begin to build up a clearer picture of their construction, history and suitability for life.

The Exoplanet Characterisation Observatory, EChO, will be the first dedicated mission to investigate the physics and chemistry of Exoplanetary Atmospheres. By characterising spectroscopically more bodies in different environments we will take detailed planetology out of the Solar System and into the Galaxy as a whole.

EChO has now been selected by the European Space Agency to be assessed as one of four M3 mission candidates.

Keywords. planets and satellites: formation, planets and satellites: general, planetary systems, planetary systems: formation

---

1. EChO – overview

EChO will provide an unprecedented view of the atmospheres of planets around nearby stars. Those planets will span a range of masses (from gas giants to super-Earths), stellar companions (F, G, K, M) and temperatures (from hot to habitable). EChO will inherit the technology of CoRoT and Kepler to achieve photometric precision at the $10^{-4} - 10^{-5}$ level in the observation of the target star and extend this capability into the mid-infrared.

EChO will observe the atmospheres of planets already discovered by other surveys and facilities. If launched today, EChO would select ~50 targets for atmospheric characterisation out of the 100+ confirmed transiting exoplanets. Most of these targets were discovered by dedicated ground-based transit/radial velocity search programmes (WASP, XO, HAT-P, HARPS, RoPACS etc.). A new generation of transit/radial velocity surveys (NG-WASP, MEarth, APACHE, HARPS-North, ESPRESSO etc.) will provide access to the population of Earth-mass planets orbiting bright late type-stars, e.g. GJ 1214b, 55 Cnc e. In the quest for habitable worlds outside our Solar System, EChO will be able to observe super-Earths in the temperate zone of M dwarfs - not the Earth's and Sun's twins, but rather cousins. Will they present equal opportunities for habitability?

The base-line design for the ESA proposal is a dispersive spectrograph covering continuously the 0.4-16μm spectral range. The spectral resolving power will be adapted to the target brightness, from several tens to several hundreds. The instrument will be mounted behind a 1.2-1.4m class telescope passively cooled. The stability and accuracy of the photometry is critical to the success of EChO and the design of the whole detection chain and satellite will be dedicated to achieving a high degree of photometric stability and repeatability. EChO will be placed in a grand halo orbit around L2. The thermal shield design will be optimised to provide a high degree of visibility of the sky over the year and an ability to repeatedly observe several tens of targets whatever the time of the year (Tinetti *et al.* 2011).

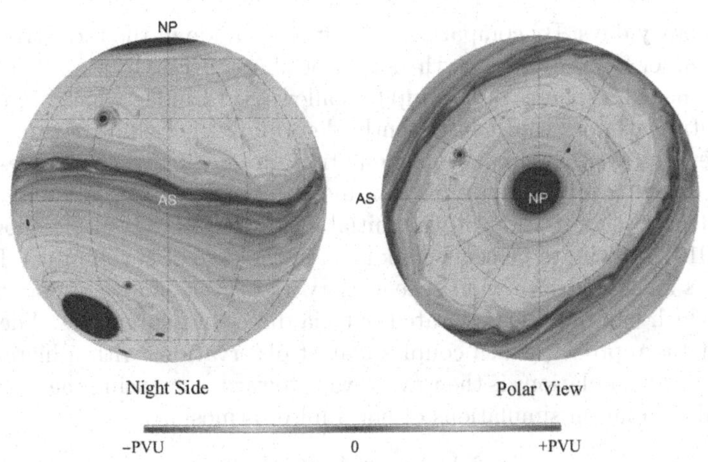

Figure 1. Maps of vorticity in the inertial reference frame and the rotating reference frame from 2-D and 3-D high-resolutions simulations of tidally locked hot Jupiters (Cho *et al.* 2003). We can appreciate the complexity of the flow with structures on an large range of scales.

2. Science return

2.1. *Atmospheric Dynamics of Hot-Jupiters and Hot-Neptunes*

EChO will provide much needed constraints on atmospheric dynamics and circulation models. This is done via careful, repeated observations. The following are the various types of observations that EChO can provide:

- Primary and secondary transits, leading to day and night side information
- Ingress and egress measurements, leading to horizontal/vertical structure information
- Non-transiting planet observations, providing information about the extra-tropics on the planet
- Host stars, providing information about the background and ionisation

Currently, what is lacking is good statistics and times series of observations to assess variability, which is expected to occur on a wide range of scales (Thrastarson & Cho 2010). An iterative approach will be used. First, using plausible vertical temperature profiles from full three-dimensional (3-D) general circulation models, spectra models can give information about the composition and its vertical distribution. The latter will then be inserted as input back into the 3-D models as either initial condition or self-consistently evolved distribution to obtain global temperature and flow distributions. When very high resolution calculations are needed to capture detailed physical or chemical effects, they can be carried out using vertically- or zonally-averaged two-dimensional (2-D) models, as appropriate.

At the cutting edge of the field is whether transient phenomena exist in the light curves and spectra obtained from hot Jupiters, as well as the implications of variability if it exists. Vortices and waves are long-lived, coherent features which should contribute heavily to variability on hot gaseous planets. The variability is expected to be slow and occurs on a large scale, as indicated by three-dimensional simulations. The resulting, computed power spectrum of the temperature field shows that the bulk of the energy is contained in the channel corresponding to a period of about 15 planetary days. The baroclinic instability may also contribute to variability; its basic mechanism is well understood, at least from a terrestrial standpoint. In the case of the hot Jupiter HD 209458b, the gravest (most unstable) mode has a wavenumber of between 2 and 3, while its growth period is about 10 planetary days. By comparison, the gravest mode in the terrestrial atmosphere has a wavenumber of 6 and a growth period of about 2 Earth days. The detection of variability in the atmospheres of hot Jupiters allows us to judge which are the dominant fluid instabilities at work and consequently determine their influence on the observed spectra. In general, general circulation simulations are dealing with a three-dimensional, non-linear problem involving multiple parameters. For example, the outcome of these simulations depend significantly on the initial conditions of the surface flow (Thrastarson & Cho 2010), which are presently unknown in the case of hot Jupiters. Furthermore, the predictions for the surface wind speeds carry an intrinsic range of uncertainty (Heng *et al.* 2011) which can only be calibrated out via direct measurements. The key point is that a pragmatic approach which couples transit observations with a hierarchy of theoretical models and simulations is the way forward towards increasing the predictive power of the general circulation simulations of hot Jupiter atmospheres.

2.2. *Upper Atmosphere*

Within our own solar system, the upper atmospheres of gas giants, both of which have been explored over recent decades both from Earth and from in-situ orbiting satellites, have been found to form regimes of complex interaction between the atmospheric gases, solar radiation, magnetospheres and their plasma population as well as the solar wind. These are regions of particular importance to investigate as they constrain the relative roles of external energy sources, including the magnetosphere/plasma environment, as well as constraining rates of atmospheric gas escape as well as other dynamical processes driven from the deeper atmosphere. In many cases upper atmospheres also feature auroral regions, where energetic particle precipitation deposit energy locally and generate optical emissions which can be observed from Earth, constraining atmospheric gases as well as the magnetic and plasma environments. EChO offers an unprecedented opportunity of expanding this exploration to solar systems outside of our own. We intend with EChO to explore the upper atmospheres of exoplanets, with the aim of addressing the following key science questions:

- What is the thermal structure and energy balance of exoplanet atmospheres? What are the characteristics of stellar forcing? What are the radiative time scales of atmosphere and how important are processes in Local Thermodynamic Equilibrium (LTE) versus those who are in non-LTE.
- What is the composition and vertical distribution of constituents, what chemical processes are active?
- What are the characteristics of the magnetic and plasma environments of exoplanets and how do these interact with the atmospheres?
- What are the rotation rates of exoplanet upper atmospheres?

Over recent years first direct spectroscopic observations have been made of atmospheres of extrasolar planets. Spectra observed during the transit have identified the NaI D lines, the H Ly line and ionised species (CII, SiI) in absorption (Charbonneau *et al.* 2000; Vidal-Madjar *et al.* 2004; Linsky *et al.* 2010). These observations placed first constraints on the structure of extrasolar planet upper-atmosphere. Simulations by Yelle (2004) have shown these observations to be consistent with thermospheric temperatures near 10,000 K, which in turn drive hydrodynamic escape and cooling of the thermosphere by adiabatic expansion. While allowing for detection of unexpected spectral signatures, we intend to specifically investigate amongst other the following lines:

- H_3^+ emission (3.5-4.1 μm). Of particular interest in the study of Gas Giants within our own solar system are emissions of H_3^+ which dominate Gas Giant emissions between 3 and 4 μm. As shown by Miller *et al.* (2006), H_3^+ is a powerful indicator of energy

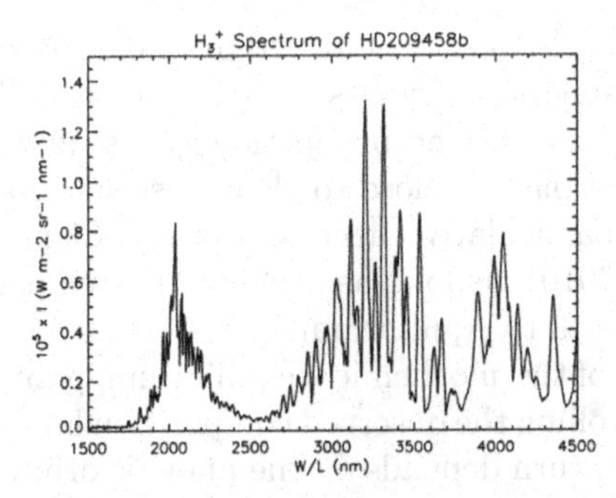

Figure 2. H_3^+ simulated spectrum for hot-Jupiter HD209458b (Koskinen *et al.* 2010). A model of the planet's upper atmosphere (Koskinen & Harris, private communication) was used to calculate the substellar column density of H_3^+ . This model is based on solving the one-dimensional equations of motion for dynamic expansion together with realistic heating rates and photochemistry for an atmosphere composed of hydrogen and helium. The results agree roughly with those of Yelle (2004) and García Muñoz (2007) for the same planet.

inputs into the upper atmosphere of Jupiter, suggesting a possible significance in exoplanet atmospheres as well. Simulations by Yelle (2004) and Koskinen *et al.* (2007) have among other investigated the possible importance of H_3^+ as a constituent and infrared emitter in exoplanet atmospheres. One particular finding of these calculations and those of Yelle (2004) is the fact that close-orbiting extrasolar planets ($R \leqslant 0.2$ AU) may host relatively small abundances only of H_3^+ due to the efficient dissociation of H_2, a parent molecule in the creation path of H_3^+. As a result, the detectability of H_3^+ may depend on the distance of the planet from the star. Fig. 2 shows an example of a simulated emission spectrum of H_3^+ for HD209458b at resolution of R=300, which matches the anticipated EChO resolution in this spectral range.

• CH_4 emission Observations of the auroral regions of Jupiter have given positive detections of CH_4 in emission, which are thought to be generated by energetic particle precipitation which penetrates below the homopause level, reaching stratospheric methane. Therefore, CH_4 can be regarded as a powerful constraint for processes of magnetosphere-atmosphere coupling. Swain *et al.* (2009) identified an unexpected spectral feature near 3.25 μm in the atmosphere of the hot-Jupiter HD 189733 b which was found to be inconsistent with LTE conditions holding at pressures typically sampled by infrared measurements. They proposed this feature to result from non-LTE emissions by CH_4, indicating that non-LTE effects may need to be considered, as is also the case in our solar system for planets Jupiter and Saturn as well as Titan. We intend to specifically address this question with EChO, making use of the improved observing conditions from orbit.

2.3. *The chemistry of Jupiters and Neptunes*

Although it is likely that thermochemical equilibrium prevails in the deeper, hotter regions of the atmospheres of extrasolar giant planets, two main processes can drive the atmosphere out of equilibrium: 1) *transport-induced quenching* and 2) *photochemistry.*

(*a*) In the first process, temperatures in the radiative portion of the exoplanet atmosphere may be cool enough that energy barriers to kinetic reactions are difficult to overcome, so that chemical kinetic time scales can become large. If the vertical transport time scales drop below the chemical kinetic time scales, the mole fractions of some spectroscopically important species may be "quenched" or frozen in at abundances representative of deeper pressure levels (Prinn & Barshay 1977), leading to disequilibrium compositions in the observable regions of the exoplanet atmosphere.

(*b*) In the second process, the energy delivered from the absorption of stellar ultraviolet radiation can excite atmospheric molecules or break chemical bonds, setting off a series of chemical reactions that lead to the production of disequilibrium constituents (Yung & Demore 1999). For giant planets close to their host stars, this disequilibrium photochemical mechanism is a particularly effective process (Liang *et al.* 2003, 2004; Zahnle *et al.* 2009a,b; Line *et al.* 2010), as long as atmospheric temperatures are not so high as to drive the composition back to equilibrium.

The relative importance of thermochemical equilibrium, photochemistry, and transport-induced quenching in controlling the observed composition largely depends on the planet's thermal structure, which in turn depends on the planet's orbital distance and metallicity and the host star's luminosity and stellar type. The host star's chromospheric activity level and the overall UV flux incident on the planet can also affect the photochemistry, but properties like planetary mass or radius play less of a role.

The importance of the thermal structure in controlling chemistry is known. The thermal structures of different Jupiter- or Neptune-mass planets can lie within very different thermochemical equilibrium regimes, affecting not only the equilibrium composition but the effectiveness of disequilibrium processes like photochemistry. A planet like HD 209458b that orbits very close to a bright G0V star is expected to get very hot by planetary standards, which makes it more likely that gas-phase species like TiO, metal sulfides, or Na manage to remain in the gas phase rather than being tied up in condensates (e.g., Hubeny *et al.* 2003; Visscher *et al.* 2006). Silicate cloud formation likely occurs at lower pressures (higher altitudes) on hotter planets, with an increased chance of the stellar radiation interacting with these cloud layers. The thermal profile for HD 209733b lies solidly within the N_2 and CO stability fields (e.g., Lodders & Fegley 2002), making these more photochemically stable molecules the dominant carriers of nitrogen and carbon, thereby reducing the effectiveness of photochemical processes. Moreover, the possible presence of a thermal inversion on the dayside would help drive the chemistry back to equilibrium despite the strong UV flux incident on the planet (Moses *et al.* 2011). Disequilibrium processes on cooler planets like HD 189733b that orbit a fainter K2V star are expected to be more important (Line *et al.* 2010; Moses *et al.* 2011), due to the more sluggish rates of the chemical processes driving the composition back toward equilibrium. Some key molecules like CO, H_2O, and CO_2 may have vertical profiles that remain close to equilibrium predictions on on these cooler "hot Jupiters" like HD 189733b, but transport-induced quenching may allow CH_4 and NH_3 to be much more abundant in the few bar to few mbar region than is expected based on equilibrium, and photochemistry might lead to the production of nitriles like HCN and unsaturated hydrocarbons like C_2H_2 that can affect spectral behavior at visible and infrared wavelengths (Moses *et al.* 2011).

In general, the cooler the exoplanet, the more important that disequilibrium processes are likely to be. This trend is especially true for planets like GJ 436b that orbit close to weaker M stars such that the temperature structure lies within the CH_4 stability field rather than the CO stability field. The carbon-hydrogen bond in CH_4 is much weaker than the carbon-oxygen bond in CO, helping to free up carbon for disequilibrium processes. Complex hydrocarbons and nitriles may be produced on such planets (Zahnle *et al.* 2009b; Moses *et al.* 2011).

2.4. *Super-Earths around M-dwarfs: what should we expect?*

EChO will have the capability to perform transit spectroscopy of Super-Earths near or in the habitable zones of M-dwarf stars. These planets will be of immense scientific interest, as their climates may be comparable to those of the terrestrial planets in our own

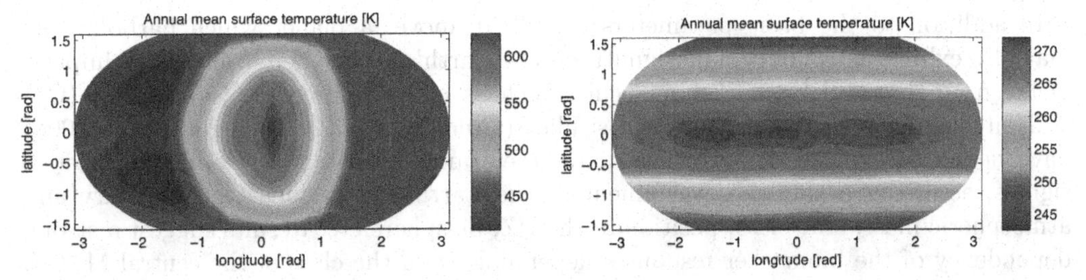

Figure 3. Simulations of the climate of a $R = 1.8R_E$ rocky planet with CO_2-dominated atmosphere around an M-class star of luminosity 0.013 L_s (Wordsworth et al.,). Two cases: hot (orbit 0.05 AU, $T_p \sim 400$ to 650 K) resonance 1:1, cold (orbit 0.22 AU, $T_p \sim 230$ to 280 K) resonance 1:10.

system. In particular, if they are rich in H_2O and have surface temperatures and pressures compatible with liquid water, they may potentially support Earth-like life. In general, the atmospheres of terrestrial exoplanets are expected to depend strongly on details of their formation and subsequent evolution, which means they are more difficult to predict theoretically than gas giants. However, M- dwarfs have some unique features that have already been predicted to make the climates of planets in their habitable zones very different from those in our own Solar System. First, they are relatively faint, so planets must be close in to receive Earth-like amounts of insolation from them. This means that terrestrial exoplanets in M-dwarf habitable zones might be in tidally resonant or locked orbits (see Fig. 3). As in the hot Jupiter case (§6.2), tidal locking can cause super-rotation in the planet's upper atmosphere, with potentially observable consequences. Tidal locking may also have serious consequences for habitability, as volatiles such as H_2O will tend to evaporate on the light side and freeze on the dark side of the planet. In the most extreme cases, the entire atmosphere can even condense out on the dark side. However, modelling has indicated that there are also many scenarios in which locked planets can sustain atmospheres and water cycles. For example, a Super-Earth with a dense atmosphere and a global ocean could efficiently transport heat across its surface and hence maintain a stable climate. One alternative to the scenario of tidal locking is spin-orbit alignment. A planet in a relatively eccentric orbit may escape synchronisation and establish a rotational spin that is some multiple of its orbital period, as happened to the planet Mercury (Correia & Laskar 2004). The climates of terrestrial planets around M-stars will also be altered due to the red-shifted stellar spectra. Red-shifting of the spectrum decreases Rayleigh scattering, so the bond albedos of M-class terrestrial exoplanets should generically be lower than those of planets in the Solar System. This theoretical prediction will be directly testable by EChO through secondary transit measurements in the optical. One side effect of this difference is that greenhouse warming by dense atmospheres becomes more effective than on Earth (Wordsworth *et al.* 2010), which alters the range of orbits for which habitable conditions are possible. Another unusual feature of M-class stars is their increased magnetic activity, which leads to a stronger stellar wind and more stellar flares (Segura *et al.* 2010). Increased stellar wind means increased atmospheric erosion, the consequences of which are still poorly understood for terrestrial exoplanets. The problem of H_2/He escape is a particularly critical one for planets intermediate in mass between the Earth and Neptune, as it ultimately determines the boundary between rocky and ice/gas giants. By studying the atmospheric composition (secondary transit) and probing the scale height through primary transit measurements (the scale height would be noticeably larger for a hydrogen-rich type of atmosphere), EChO will be able to investigate this vital scientific question directly.

In addition to the basic parameters described above, a planet which harbours life may also exhibit astronomical biosignatures The Earth's atmosphere contains an imprint of life from so-called biomarker molecules such as molecular oxygen (O_2), ozone (O_3) and nitrous oxide (N_2O). Theoretical studies (Grenfell *et al.* 2010; Segura *et al.* 2005) have begun to explore the extensive parameter range of potential biomarker spectral signals, assuming a similar development as the Earth and varying e.g. planetary and atmospheric mass, star class, position in the HZ, biosphere etc. Results suggest a strong dependency of the biomarker responses depending upon the class of the central M-star. Care is needed to distinguish true biomarker signals from so-called "false- positives" i.e. cases where planetary atmospheres "mimic" life (Selsis *et al.* 2002) due to inorganic chemical processes producing biomarkers – for example, strong CO_2 photolysis eventually leading to molecular oxygen production. Ozone features a strong infra-red absorption band at 9.6 μ m, easily measurable by EChO, and it may be present in large amounts over a wide range of oxygen concentrations (Segura *et al.* 2003). In this sense, ozone is a good biomarker. However, its photochemistry is complex (WMO, 1998) and is influenced by trace amounts of nitrogen-, chlorine-, and hydrogen-oxides whose abundances are difficult to constrain. Sources of nitrous oxide (N_2O) into Earth's atmosphere (IPCC TAR) are almost exclusively associated with microbial activity. It absorbs mostly in the troposphere with bands at e.g. at 7.8 and 3.9 μm. It is an excellent biomarker from the point of view that inorganic (non-life) production identified so far on the Earth is negligible, implying that false-positives are unlikely. However, its absorption features are weak for typical modern Earth abundances and measurements are extremely challenging. Atmospheres with weak UV-B could favour the build-up of large atmospheric N_2O abundances because its photolytic sink is weak in such cases.

Planets with no atmosphere

We expect that Super-Earths with no or negligible atmosphere would show large variations in intensity as a function of planetary phase. The MIR variability is driven by the difference in day-night surface temperatures. This variability in surface temperature should be relatively high, as a thin atmosphere has a very limited heat capacity to buffer its climate and even out day/night variations.

2.5. *Linking atmospheres and interiors*

The ability of EChO to fully characterise an exoplanetary atmosphere in its composition and thermal structure will provide major improvements for interior models as well. Except for the Earth and the Moon, there is no direct measurements of the deep structure of the planets, as this investigation requires a network of seismometers for terrestrial planets, or techniques similar to the asterosismology for gaseous giants. Nonetheless, the internal structure of planetary bodies in the solar system is, even if not precisely, relatively well understood. Planetary bodies can be split into three main families (Fig. 4) which are: i) the terrestrial planets (or solid planets), ii) the giant planets (or gaseous), and iii) the intermediate planets which are in between the two extreme cases.

The giant planet family

Giant planets are mostly made of hydrogen and helium and are expected to always be in gaseous form (Guillot 2005). Because they play a tremendous role in shaping planetary systems (Tsiganis *et al.* 2005) determining precisely their internal structure and composition is essential to understand how planets form. Contrary to solid planets, they are relatively compressible and the progressive loss of heat acquired during their formation is accompanied with a global contraction. Inferring their internal composition

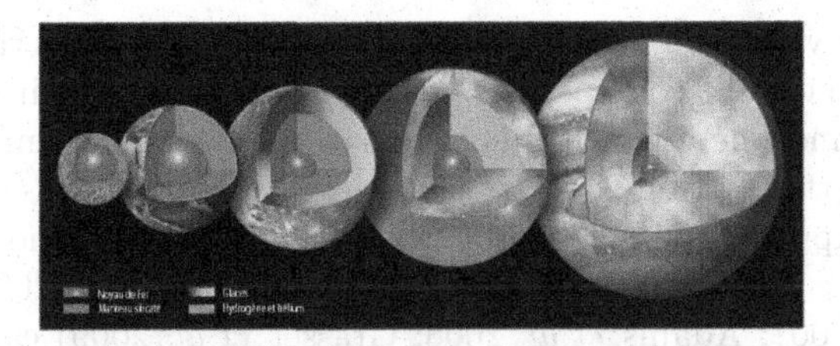

Figure 4. Internal structures of planets (not at scale). The three sub-families on the left are part of the terrestrial family (see text for detail). Giant planets (Jupiter-like) are on the right. Neptune – like planets, are on the fourth position from the left.

thus amounts to understanding how they cool. Fortunately, the dominance of hydrogen and helium implies that the degeneracy in composition (i.e. uncertainty on the mixture of ices/rocks/iron) is much less pronounced than for solid planets, so that the relevant question concerns the amounts and all elements other than hydrogen and helium, i.e. heavy elements, that are present.

The determination of sizes from primary transit measurements and masses from radial measurements have yielded in some cases a constraint on the mass of heavy elements present in the interior that is relatively independent of model hypotheses (Sato *et al.* 2005; Ikoma *et al.* 2006) and otherwise global tendencies showing that this mass is correlated with the metallicity of the parent star (Guillot *et al.* 2006; Burrows *et al.* 2007; Guillot 2008). However, several problems arise. First a large fraction of the known transiting planets are larger than expected, even when considering that they could be coreless hydrogen-helium planets (Bodenheimer *et al.* 2001; Guillot & Showman 2002; Baraffe *et al.* 2003; Guillot *et al.* 2006; Burrows *et al.* 2007; Guillot 2008). There is thus missing physics that is to be identified. Second, we do not know whether these heavy elements are kept inside a central core or distributed inside the planet. This influences how they cool (Guillot 2005; Baraffe *et al.* 2008) and is crucial in the context of formation scenarios (Lissauer & Stevenson 2007). Third, the complex dynamics of the atmosphere of heavily irradiated planets that constitutes the outer boundary condition of evolution models is poorly understood. This has direct consequences for our ability to accurately predict the evolution of these planets (Guillot & Showman 2002; Guillot 2010).

The terrestrial family

Three different sub-families of planets can be considered from left to right in Fig. 4: Mercury-like planets mostly composed of an iron core and a thin layer of silicates, Super-Earth made of an iron core and a thick silicate mantle (such as Venus, Mars and the Earth) and Ocean-planets made of iron, silicates, and water (similar to icy moons of Jupiter and Saturn). Super-Earths are composed of an internal iron-rich core and a thick silicate mantle (lower mantle) covered by a thin layer of low-pressure silicates similar to the upper mantle on Earth, and a very thin liquid layer (like Earth-oceans). Ocean-planets are composed of an iron core, a silicate mantle, and a thick icy layer surrounded by a thin ocean or icy crust at the surface.

For a given mass, one would expect Ocean-like planets have a smaller metallic core and silicate mantles, but also a larger radius than for Earth-like planets because icy materials are lighter than silicates. On the contrary, the radius of a much denser Mercury-like planets is about 80% that of an Earth-like planets (Valencia *et al.* 2007; Grasset *et al.* 2009). Mass - Radius measurements, though, do not give unique solutions. For example, a silicate-rich planet surrounded by a very thick atmosphere could provide the same mass and radius of an ice-rich planet with no atmosphere! (Adams *et al.* 2008). EChO will unravel the ambiguity through primary transit spectroscopic observations in the optical and IR, providing the bulk composition of the atmospheres when they are present. If EChO detects an atmosphere which is not primarily made of helium and hydrogen, thus the planet is most certainly from the terrestrial family, which means that the thickness of the atmosphere is expected to be negligible with respect to the planetary radius. If this is the case, an extensive literature (Léger *et al.* 2004; Valencia *et al.* 2006, 2007; Sotin *et al.* 2007; Seager *et al.* 2007; Adams *et al.* 2008; Grasset *et al.* 2009) can be fully exploited to characterise the inner structure of the new planet.

The intermediate family

Planets in between the gas giants and the small solid terrestrial planets are key to understand the formation of planetary systems. The existence of these intermediate planets close to their star, as found by radial velocity surveys, is already crucial to highlight the shortcomings of theoretical models (Mordasini *et al.* 2009). (i) Standard planet formation scenarios predict that embryos of sufficient mass (typically above 5 Earth masses) should retain some of the primordial hydrogen and helium from the protoplanetary disc. With EChO measurements, we will probe which planets indeed possess a hydrogen helium atmosphere and directly test the conditions of planet formation. (ii) The two only intermediate planets that we can characterise, Uranus and Neptune, are significantly enriched in heavy elements, in the form of methane (Guillot 2005). The reason for this enrichment is unclear: is it due to upward mixing, early or late delivery of planetesimals? EChO will allow these measurements in many planets thereby providing observations that are crucial to constrain these models. (iii) We do not know where to put the limits between solid, liquid and fluid (gaseous) planets. While EChO will not directly measure the phase of a planet as a whole, the determination of its size and of the composition of its atmosphere will be key to determine whether its interior is solid, partially liquid or gaseous.

3. Other science with EChO

While the vast majority of the EChO mission will be dedicated to exoplanet spectroscopy and its design will be fine-tuned for that cause, the ability to do spectroscopy with broad simultaneous wavelength coverage and high sensitivity makes EChO a superb tool to address a host of science cases, in particular:

- Direct spectroscopic characterisation of free-floating (and perhaps in rare cases resolved companion) brown dwarfs and planetary mass objects, with particular focus on constraining surface gravity and composition to compare free-floating planets to models of planets formed through core accretion. In particular, spectroscopic follow-up of L, T, and particularly Y dwarfs from the WISE mission allows confronting models of these very cool objects with observations.
- An important scientific question, is to understand how the elemental abundances of planets follow from the composition and chemistry of the disks in which they formed. The ability to obtain simultaneous visible to mid-IR spectra for variable young stellar objects could make profound contributions to our understanding of how changes in disk accretion and dust attenuation affect disk structure and the evolution of gas and dust composition in planet-forming disks (Ábrahám *et al.* 2009; Banzatti *et al.* 2011).
- Search for extrasolar moons. Exomoons are likely to be rocky bodies and thus offer the same potential of Earths/Super-Earths as possible havens for life. Their discovery would also reap immense new understanding of planet/moon formation. For transiting

planet systems, exomoons can be detected through two principal methods i) transit timing effects ii) exomoon transits (Kipping 2009a,b).

• Rocky transiting planets found with Kepler are unlikely to induce detectable radial velocity signals and thus the only way to confirm their planetary nature is to rule out the probable sources of astrophysical false positives, most pertinently blends (e.g. background eclipsing binaries) that mimic an exoplanet signature in the Kepler bandpass. By measuring the transit depth at multiple wavelengths, such scenarios can be easily excluded.

References

Ábrahám, P., Juhász, A., Dullemond, C. P., Kóspál, Á., van Boekel, R., Bouwman, J., Henning, T., Moór, A., Mosoni, L., Sicilia-Aguilar, A., & Sipos, N. 2009, *Nature*, 459, 224

Adams, E. R., Seager, S., & Elkins-Tanton, L. 2008, *ApJ*, 673, 1160

Banzatti, A., Testi, L., Isella, A., Natta, A., Neri, R., & Wilner, D. J. 2011, *A&A*, 525, A12

Baraffe, I., Chabrier, G., & Barman, T. 2008, *A&A*, 482, 315

Baraffe, I., Chabrier, G., Barman, T. S., Allard, F., & Hauschildt, P. H. 2003, *A&A*, 402, 701

Bodenheimer, P., Lin, D. N. C., & Mardling, R. A. 2001, *ApJ*, 548, 466

Burrows, A., Budaj, J., & Hubeny, I. 2007, *ApJ*, 668, 671

Charbonneau, D., Brown, T., Latham, D., & Mayor, M. 2000, *ApJ*, 529, L45

Cho, J., Menou, K., Hansen, B. M. S., & Seager, S. 2003, *ApJL*, 587, L117

Correia, A. & Laskar, J. 2004, *Nature*, 429, 848

García Muñoz, A. 2007, *Plan. Space Sci.*, 55, 1426

Grasset, O., Schneider, J., & Sotin, C. 2009, *ApJ*, 693, 722

Grenfell, J. L., Rauer, H., Selsis, F., Kaltenegger, L., Beichman, C., Danchi, W., Eiroa, C., Fridlund, M., Henning, T., Herbst, T., Lammer, H., Léger, A., Liseau, R., Lunine, J., Paresce, F., Penny, A., Quirrenbach, A., Röttgering, H., Schneider, J., Stam, D., Tinetti, G., & White, G. J. 2010, *Astrobiology*, 10, 77

Guillot, T. 2005, *Annu. Rev. Earth Plan. Sci.*, 33, 493

—. 2008, *Physica Scripta*, T130, 014023

—. 2010, *A&A*, 520, A27+

Guillot, T., Santos, N. C., Pont, F., Iro, N., Melo, C., & Ribas, I. 2006, *A&A*, 453, L21

Guillot, T. & Showman, A. P. 2002, *A&A*, 385, 156

Heng, K, Menou, K, Phillipps, P. J. 2011, *MNRAS*, 413, 2380

Hubeny, I., Burrows, A., & Sudarsky, D. 2003, *ApJ*, 594, 1011

Ikoma, M., Guillot, T., Genda, H., Tanigawa, T., & Ida, S. 2006, *ApJ*, 650, 1150

Kipping, D. 2009a, *MNRAS*, 392, 181

—. 2009b, *MNRAS*, 396, 1797

Koskinen, T., Aylward, A., & Miller, S. 2007, *Nature*, 450, 845

Koskinen, T., Cho, J.-K., Achilleos, N., & Aylward, A. D. 2010, *ApJ*, 722, 178

Léger, A., Selsis, F., Sotin, C., Guillot, T., Despois, D., Mawet, D., Ollivier, M., Labèque, A., Valette, C., Brachet, F., Chazelas, B., & Lammer, H. 2004, *Icarus*, 169, 499

Liang, M.-C., Parkinson, C. D., Lee, A. Y. T., Yung, Y. L., & Seager, S. 2003, *ApJL*, 596, 247

Liang, M.-C., Seager, S., Parkinson, C. D., Lee, A. Y. T., & Yung, Y. L. 2004, *ApJL*, 605, 61

Line, M. R., Liang, M. C., & Yung, Y. L. 2010, *ApJ*, 717, 496

Linsky, J. L., Yang, H., France, K., Froning, C. S., Green, J. C., Stocke, J. T., & Osterman, S. N. 2010, *ApJ*, 717, 1291

Lissauer, J. J. & Stevenson, D. J. 2007, *Protostars and Planets V*, 591

Lodders, K. & Fegley, B. 2002, *Icarus*, 155, 393

Miller, S., Stallard, T., & Smith, C., *et al.* 2006, Royal Society of London Transactions Series A, 364, 3121

Mordasini, C., Alibert, Y., & Benz, W. 2009, *A&A*, 501, 1139

Moses, J. I., Visscher, C., Fortney, J. J., Lewis, N. K., Showman, A. P., Marley, M. S., Griffith, C. A., & Friedson, A. J. 2011, *ApJ*, 737, id.15

Prinn, R. & Barshay, S. 1977, *Science*, 198, 1031

Sato, B., Fischer, D. A., Henry, G. W., Laughlin, G., Butler, R. P., Marcy, G. W., Vogt, S. S., Bodenheimer, P., Ida, S., Toyota, E., Wolf, A., Valenti, J. A., Boyd, L. J., Johnson, J. A., Wright, J. T., Ammons, M., Robinson, S., Strader, J., McCarthy, C., Tah, K. L., & Minniti, D. 2005, *ApJ*, 633, 465

Seager, S., Kuchner, M., Hier-Majumder, C. A., & Militzer, B. 2007, *ApJ*, 669, 1279

Segura, A., Kasting, J. F., Meadows, V., Cohen, M., Scalo, J., Crisp, D., Butler, R. A. H., & Tinetti, G. 2005, *Astrobiology*, 5, 706

Segura, A., Krelove, K., Kasting, J. F., Sommerlatt, D., Meadows, V., Crisp, D., Cohen, M., & Mlawer, E. 2003, *Astrobiology*, 3, 689

Segura, A., Walkowicz, L. M., Meadows, V., Kasting, J., & Hawley, S. 2010, *Astrobiology*, 10, 751

Selsis, F., Despois, D., & Parisot, J. 2002, *A&A*, 388, 985

Sotin, C., Grasset, O., & Mocquet, A. 2007, *Icarus*, 191, 337

Swain, M. R., Vasisht, G., Tinetti, G., Bouwman, J., Chen, P., Yung, Y., Deming, D., & Deroo, P. 2009, *ApJ*, 690, L114

Thrastarson, H. T. & Cho, J. 2010, *ApJ*, 716, 144

Tinetti, G., *et al.* 2011, *Experimental Astronomy*, submitted

Tsiganis, K., Gomes, R., Morbidelli, A., & Levison, H. F. 2005, *Nature*, 435, 459

Valencia, D., O'Connell, R. J., & Sasselov, D. 2006, *Icarus*, 181, 545

Valencia, D., Sasselov, D. D., & O'Connell, R. J. 2007, *ApJ*, 665, 1413

Vidal-Madjar, A., Désert, J., Lecavelier des Etangs, A., Hébrard, G., Ballester, G. E., Ehrenreich, D., Ferlet, R., McConnell, J. C., Mayor, M., & Parkinson, C. D. 2004, *ApJL*, 604, L69

Visscher, C., Lodders, K., & Fegley, B. 2006, *ApJ*, 648, 1181

Wordsworth, R. D., Forget, F., Selsis, F., Madeleine, J.-B., Millour, E., & Eymet, V. 2010, *A&A*, 522, A22

Yelle, R. V. 2004, *Icarus*, 170, 167

Yung, Y. L. & Demore, W. B., eds. 1999, Photochemistry of planetary atmospheres

Zahnle, K., Marley, M., Freedman, R., Lodders, K., & Fortney, J. 2009a, *ApJ*, 701, L20

Zahnle, K., Marley, M. S., & Fortney, J. J. 2009b, *ApJ* submitted (arXiv:0911.0728)

The Astrophysics of Planetary Systems: Formation, Structure, and Dynamical Evolution
Proceedings IAU Symposium No. 276, 2010
A. Sozzetti, M. G. Lattanzi & A. P. Boss, eds.

doi:10.1017/S174392131102045X

The Gaia astrometric survey for exoplanets in the solar neighborhood

Deborah Busonero[1]
[1]INAF - Osservatorio Astronomico di Torino
Via Osservatorio 20, 10025, Pino Torinese (TO), Italy
email: busonero@oato.inaf.it

Abstract. The Gaia astrometric mission holds the promise for crucial contributions to almost every subject of astrophysics and astronomy, including planetary systems astrophysics. We focus on the potential of the Gaia mission as perfect tool for a complete screening of nearby stars in search for exoplanets. We build our dissertation on the most recent results of the satellites astrometric payload performances and data reduction capabilities. We put the identified capabilities in context by illustrating the outstanding contribution to planetary sciences, in combination with nowadays and next decade exoplanets search programs, as a complement to other indirect and direct methods for the detection and characterization of planetary systems. We conclude by highlighting the crucial improvements in the optimization of the target lists of future dedicated observatory projects.

Keywords. astrometry, space vehicles: instruments, planetary systems

1. Introduction

The Gaia all-sky astrometric survey, due to launch in May 2013, will monitor, during its 5-yr nominal mission lifetime, all point sources (stars, asteroids, quasars, extragalactic supernovae, etc.) in the visual magnitude range 6 - 20 mag, for an amount of about 10^9 objects. The final catalogue is foreseen for 2021.
Using the continuous scanning principle first adopted for Hipparcos, Gaia will determine the five basic astrometric parameters (two positional coordinates α and δ, two proper motion components μ_α and μ_δ, and the parallax ϖ for all objects, with end-of-mission accuracy between 6 μas (at V = 6 mag) and 200 μas (at V = 20 mag) for a G2V star.

Gaia astrometry, complemented by on-board spectrophotometry and (partial) radial velocity information (see de Bruijne *et al.* 2010), will have the precision necessary to quantify the early formation, and subsequent dynamical, chemical and star formation evolution of the Milky Way Galaxy. The broad range of crucial issues in astrophysics that can be addressed by the wealth of the Gaia data is summarized in Perryman (2005).

One of the relevant areas in which the Gaia observations will have great impact is the astrophysics of planetary systems, in particular when seen as a complement to other techniques for planet detection and characterization (e.g., Sozzetti 2009). This paper aims to be the first step towards a clarification about the Gaia astrometric performance and its contribution to the exoplanets research field, taking into account the most recent results on instrumental performances.

2. The Gaia challenge

The Gaia measurement concept requires observations in two lines of sight (LOS), separated by a large angle (named "Basic Angle" (BA) and set to 106.5), continuously scanning the sky along a great circle and completing a rotation in 6 hours, i.e., a scan rate of 60 arcsec/s. The payload is composed of two identical telescopes with primary mirrors of 1.45 x 0.5 m in size and 35 m nominal focal length, feeding a large common focal plane (FP): a CCD mosaic of about one hundred CCDs. The CCD array is divided in three region: the Sky Mapper-Astrometric Field (SM-AF), the Blue Photometer-Red Photometer (BP- RP), the Radial Velocity Spectrometer (RVS) devoted to astrometric, photometric, and spectroscopic measurements, respectively (Fig. 1 left panel). Each CCD operates in Time Delay Integration (TDI) mode matching the projected sidereal velocity (scanning velocity) with the CCD transfer velocity (the clock rate). The measurements repeatedly cover the whole sky, by the composition of rotation, precession, and orbital motion of the satellite. Gaia will collect 80 sky transits for each object on average in 5 years, with a dependency on where the satellite looks, ranging from few tenth to few hundred observations per object along the ecliptic.

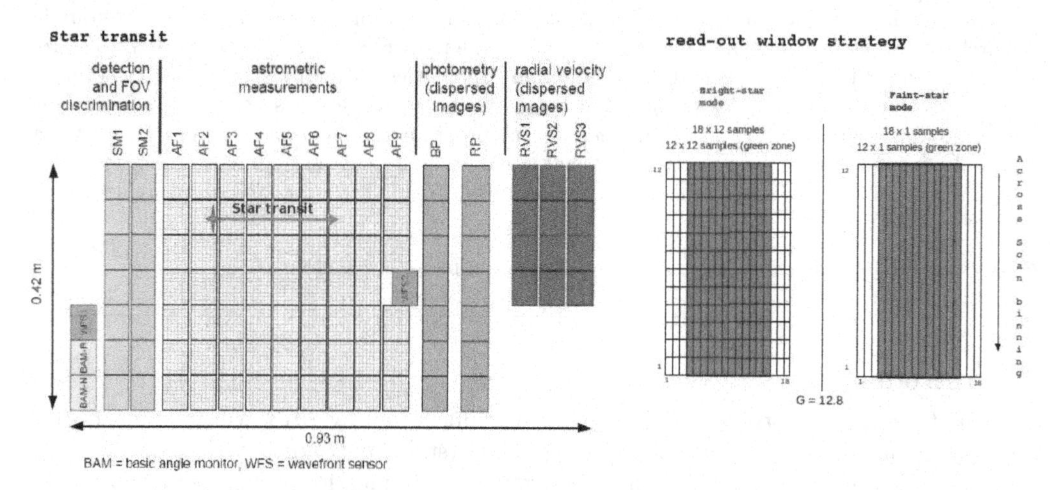

Figure 1. Gaia Focal Plane structure and read-out window strategy.

The instantaneous image is integrated throughout the whole transit over the CCD, as each logical pixel is generated at the leading edge and transferred in step with the corresponding point in object space up to the trailing edge of the device, where the readout process takes place. Readout on the Astrometric Field (AF) is restricted to the regions of interest identified by the SM in a way that for each star a window is selected for read-out according to the adopted windowing strategy (Fig. 1 right panel). The readout mode depends upon the target brightness, in particular using full-bidimensional image readout for stars brighter than G=12.8† and across scan binning (low resolution) to improve upon signal to noise ratio (SNR) and reduce the data volume for stars fainter than G=12.8, so that the output data is a one-dimensional signal. Time of observation and position on the sky are linked by the scan law plus the electro-optical response.

For each transit, a target is observed astrometrically in equal conditions on nine consecutive elementary exposures (the 9 AF CCDs). The independent composition of the elementary exposures provides the transit-level accuracy. In order to meet the end-of-mission accuracy goal, the location error at the level of the elementary exposure has to range from some tenth of as for the brightest stars (G=6-10) to a few mas for G=21.

† Conversion rules from Johnson V to Gaia Astro G magnitude scale depends on star spectral type and for A0V is G=V, B1V is G = V - 0.02, for G2V is G = V 0.19 and for M6V is G = V 2.27

This means that we need to maintain an elementary exposure accuracy on the location ranging from 10^{-4} to 10^{-2} of the along scan pixel size (10 μm).

Gaia, like Hipparcos, is designed to be a self-calibrated instrument (Lindegren 2005), but PSF/LSF calibration, CCD calibration, radiation damage treatment, and transit-level attitude diagnostics indipendent procedures are needed in the data reduction chain (Busonero *et al.* 2010) before starting the astrometric iterative solution (Hobbs 2008).

We don't enter here in the detail of the instrument calibration procedures. We narrow our dissertation down to instrument calibration accuracy and bright objects.

3. Instrumental noise and bright stars

The variation of the instrument response is quite unavoidable so its modeling is crucial to manage the effects that can affect the final micro-arcsecond level astrometric accuracy. Instrumental noise sources are divided in two different classes: random, i.e., photometric errors, and systematic errors. Systematic limitations are imposed by the CCD detectors, e.g., radiation damage (Short *et al.* 2010), gates, deviations from uniformity or from linear pixel response, and by optics imperfections, e.g., optical aberrations and distortions (Busonero *et al.* 2006), pixelization.

A critical item is the treatment of bright objects, which will constitute the bulk of the well-behaved celestial reference points utilized in the core processing and in the sphere recostruction. Operationally the term bright refers to objects of magnitude G<15.8. The search for the best possible centroiding performance is of course critical for such stars. On the other hand, saturation starts at G = 12.8 and will become severe for those objects in the brightest magnitude bin of interest to Gaia (6-10).

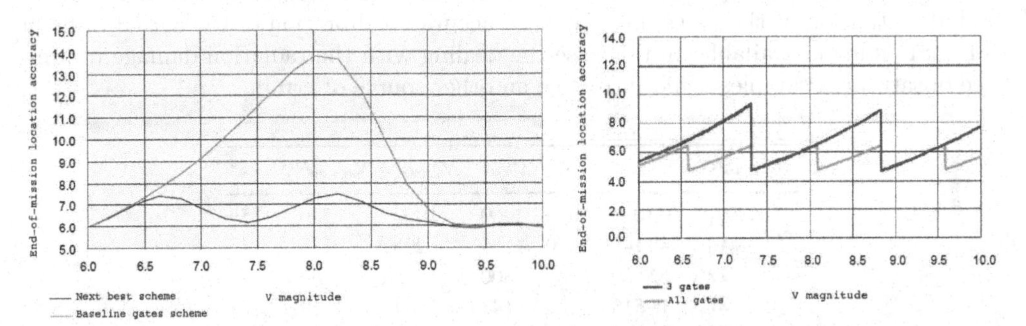

Figure 2. Gaia end-of-mission location accuracy in μas depending on gates activation.(data derived from the "Science performance budget report", GAIA.ASF.RP.SAT.00005)

TDI gate activation is foreseen on board to prevent pixel saturation, i.e., allow the loss of photons to make the brighter stars behave like the fainter ones, thus maintaining similar centroiding performances over the whole bright magnitudes interval. But gates are also sources of instrumental noise since the CCDs have to be calibrated for each activated gate. It means that we need to find a reasonable trade-off among the possibility to avoid the saturation for all the transit objects, the possibility to perform the calibration of the gate catalog within a realistic calibration-block duration (within few weeks), and the limitations imposed by the VPU software which implements for one CCD row one TDI gates table for all CCDs strip and selects the TDI gate activation based on G magnitude, not on AC coordinate. It is a futher possible error source since the saturation level is not uniform over the CCD but has a significant (about 40%) AC variation.

The Industry and the Gaia Scientific Consortium have studied several gates schemes.

We show in Fig. 2 the comparison beetween four different gate schemes and the related performances. From the perspectives of the signal-to-noise of the CCD sample data of bright stars and centroiding precision the optimum solution would clearly be to use all 12 TDI gates available but the calibration needs impose a choice. Required gates calibration time becomes forbidding for a gates activation with more than seven gates as shown in Table 1 .

Mag min	5.70	5.83	6.58	7.33	8.09	8.84	9.59	10.34	11.10	11.47
Mag max	5.83	6.58	7.33	8.09	8.84	9.59	10.34	11.10	11.47	11.95
Gate	3	4	5	6	7	8	9	10	11	12
Calibration duration [weeks]	**136.1**	**16.5**	9.3	5.2	3.0	1.7	0.9	0.7	0.5	0.4

Table 1. Required gates calibration time in weeks for all gates activation.

The actual gates scheme counts six gates (Fig. 2 violet line) and provides an end-of-mission location accuracy for $7 \leqslant V \leqslant 9$ ranging from $7 \mu as$ to $14 \mu as$. A new gates scheme with seven gates is under study (Fig. 2 blue line). It will provide an astrometric accuracy ranging between 6 and 7 μas. In Fig. 3 we show clearly how that degradation of the performances can significantly change the Gaia contribution to the exoplanets survey. We have to keep in mind also that the radiation damage can introduce a similar degradation if not adeguately calibrated.

On the other hand, the saturation of the image core theoretically improves the centroiding performance through improved S/N on the slopes. Significantly better centroid errors can be achieved by dealing directly with saturated images, i.e., without (or very limited) actuation of the gates, but when an accurate calibration of the actual in-flight PSF/LSF is made available. In this case the dealing with the radiation damage in presence of saturated samples could introduce an other source of errors.

$\sigma_{\psi}{}^{a}$ (μas)	$N_{\star}{}^{b}$	$N_{\rm d}{}^{c}$	$N_{\rm m}{}^{d}$	$N_{\rm d,mult}{}^{e}$	$N_{\rm m,mult}{}^{f}$	$N_{\rm copl}{}^{g}$
11	500 000	8000	4000	1000	500	159
16	148 148	2370	1185	296	148	47
22	62 500	1000	500	125	62	19
27	18 519	296	148	37	18	5
60	4000	64	32	8	4	1
100	500	8	4	1	0	0

Figure 3. Gaia worse and better capability for V<13 and d<200*pc* depending on instrument calibration accuracy. (courtesy of Casertano *et al.* (2008))

3.1. *Transit-measurement accuracy*

We pointed out until now how the instrument calibration performances drive the Gaia science case, bringing specific research field like the exoplanets to disappear from the Gaia science case, if the transit-measurement precision degrades significantly,

In Table 2 we list the AF AL scan transit-level location estimation errors in units of μas as function of V and spectra types, taking into account the blue line gates scheme and the possibility of calibrating the image distortion due to the radiation damage so that cleaning the Gaia measurement.

Sp Type	V-G	6.0	7.0	8.0	9.0	10.0	11.0	12.0	13.0	14.0	15.0
B1V	0.03	18,48	29,39	44,40	17,53	14,64	17,33	26,6	31,47	49,87	79,46
A5V	0.02	18,73	29,79	44,93	16,69	14,85	17,51	26,38	31,89	50,54	80,55
F2V	0.08	18,22	28,94	45,02	17,70	14,44	17,46	29,91	29,60	49,14	78,28
G2V	0.16	17,62	28,00	44,68	17,27	13,91	17,41	33,74	28,63	47,53	75,69
K3V	0.33	16,31	25,89	41,27	16,20	17,85	20,73	30,48	31,42	43,94	69,97
M6V	2.18	14,53	16,24	18,17	28,73	45,75	17,84	14,70	21,38	34,2	30,77

Table 2. AF AL scan transit-level location estimation errors in units of μas as function of V and spectral types, for unreddening stars.

4. Contribution to exoplanets discoveries: Conclusion

Gaia will be a great tool for a complete screening of nearby stars in search for exoplanets. Indeed for a position accuracy ranging from 10 to 15 μas it promises the monitoring of hundreds of thousands of FGK stars to $\sim$200 pc, performing a complete census of all stellar types with a dection limit set to $\sim 1 M_J$ planets and $P < 10$ years. About one thousand multiple systems will be measurable, giving relative inclinations and orbits for possibly several hundred systems. The newly discovered single-planets systems will amount to several thousands (Fig. 3).

Such an unbiased, magnitude-limited planet census of hundreds of thousands stars will be essential to study the statistics of planetary systems as a function of stellar properties (the statistics of exoplanets discoveries by astrometry to-date equals zero).

The Gaia exoplanets catalogue strenght lies also in the complementarity with respect to other indirect and direct methods for the detection and characterization of planetary systems, and in the crucial improvements in the optimization of the target lists of future dedicated observatory projects.

Acknowledgement

The Italian participation in the Gaia mission is supported by the Italian Space Agency, under contract ASI I/058/10/0. I thank Alessandro Sozzetti for useful discussions and careful proofreading of this manuscript.

References

Busonero, D., Gai, M., & Lattanzi, M. G. 2010, in: Oschmann, J. M., Jr., Clampin, M. C., MacEwen, H. A. (eds.), *Proc. SPIE*, 7731, 33

Busonero, D., Gai, M., Gardiol, D., Lattanzi, M. G., & Loreggia, D. 2006, *A&A*, 449, 827

Casertano, S., Lattanzi, M. G., Sozzetti, A., Spagna, A., Jancart, S., Morbidelli, R., Pannunzio, R., Pourbaix, D., & Queloz, D. 2008, *A&A*, 482, 699

de Bruijne, J., Kohley, R., & Prusti, T. 2010, in: Oschmann, J. M., Jr., Clampin, M. C., MacEwen, H. A. (eds.), *Proc. SPIE*, 7731, 1

Hobbs, D., Lindegren, L., Holl, B., Lammers, U., & O'Mullane, W. 2008, *Proc. IAU Symp.*, 248, 268

Lindegren, L. & de Bruijne, J. H. J. 2005, P. K. Seidelmann, A. K. B. Monet (eds.), *ASP Conf. Ser.*, 338, 25

Perryman, M. A. C.. 2005, in P. Kenneth Seidelmann and Alice K. B. Monet (eds.), —textitASP Conf. Ser., 338, 3

Short, A., Prod'homme, T., Weiler, M., Brown, S. W., & Brown, A. G. A.. 2010, in: Holland, A. D., Dorn, D, A. (eds.), *Proc. SPIE*, 7742, 12

Sozzetti, A. 2010, in I. F. Corbett, ed., *Highlights of Astronomy*, Proc. XXVIIth IAU General Assembly, 15, 716

The Astrophysics of Planetary Systems: Formation, Structure, and Dynamical Evolution
Proceedings IAU Symposium No. 276, 2010
A. Sozzetti, M. G. Lattanzi & A. P. Boss, eds.

doi:10.1017/S1743921311020461

Super-Earths and life - a fascinating puzzle: Example GJ 581d

Lisa Kaltenegger[1], Antígona Segura[2] and Subhanjoy Mohanty[3]

[1]MPIA, Koenigstuhl 17, 69117 Heidelberg, Germany
also Harvard Smithsonian Center for Astrophysics, 60 Garden st., 02138 MA Cambridge, USA
email: kaltenegger@mpia.de

[2]Instituto de Ciencias Nucleares, Universidad Nacional Autonoma de Mexico, Mexico

[3]Imperial College London, 1010 Blackett Lab., Prince Consort Road, London SW7 2AZ, UK

Abstract. Spurred by the recent large number of radial velocity detections and the discovery of several transiting system and among those two planets, that are consistent with rocky composition, the study of planets orbiting nearby stars has now entered an era of characterizing massive terrestrial planets (aka super-Earths). One prominent question is, if such planets could be habitats. Here we focuss on one particular planet Gl581d. For Earth-like assumptions, we investigate the minimal atmospheric conditions for Gl581d to be potentially habitable at its current position, and if habitability could be remotely detected in its spectra. The model we present here only represents one possible nature an Earth-like composition - of a planet like Gl581d in a wide parameter space. Future observations of atmospheric features of such super-Earths can be used to examine if our concept of habitability and its dependence on the carbonate-silicate cycle is correct, and also assess whether Gl581d is indeed the first detected habitable super-Earth. We will need spectroscopic measurements to probe the atmosphere of such planets to break the degeneracy of mass and radius measurements and characterize a planetary environment.

Keywords. astrobiology, stars: individual (Gl581), planetary systems, Earth, techniques: spectroscopic

---

1. Introduction

The recent large number of radial velocity detections of planets with minimum masses below 15 Earth masses (Mayor *et al.* 2009) and the discovery of the transiting systems CoRoT-7b (Leger *et al.* 2009) and Kepler 10b (Bourucki *et al.* 2011), that have a mean density that is consistent with rocky material, have advanced our study of potential habitats significantly. Here we pick the Gl581 system, that is a particularly striking example among these new discoveries, consisting of an M3V star orbited by a minimum of 4 planets, 3 of which are in the Earth to super-Earth range (Udry *et al.* 2007, Mayor *et al.* 2009). Two super-Earths in this system are located on either edge of the Habitable Zone (HZ). Could these objects be potential habitats? Detailed calculations (Selsis *et al.* 2007, von Bloh *et al.* 2007) show that the inner super-Earth, Gl581c, is too close to its star to maintain water on its surface, if no specific extreme cloud coverage can be invoked. It would thus loose all its water to space, leaving the planet hot and dry. Gl581d, on the other hand, with an updated orbital period of 66.8 days (Mayor *et al.* 2009) was placed within the Habitable Zone of the system with a semi-major axis of 0.22 AU. Its high eccentricity of 0.38 ± 0.09 brings it even closer to its host star and increases the mean flux received by the planet (Williams & Pollard 2002, Spiegel *et al.* 2009, Dressing *et al.* 2010): the mean flux at Gl581d is equivalent to that for a circular orbit around Gl581 with a = 0.20AU. Even so this planet lies within the HZ of its parent star, the specific atmospheric composition needed to maintain liquid water on its surface are not analog to current Earth because of its location on the outer edge of the HZ. The goals of our paper are to explore the atmospheric conditions under which this planet may be habitable (see also Wordsworth *et al.* 2010, von Paris *et al.* 2010, von Bloh *et al.* 2007, Selsis *et al.* 2007) and the remote detectability of such spectral features and biosignatures (see e.g. Kaltenegger *et al.* 2010a, DesMarais *et al.* 2002). For spectroscopic features that could be remotely detected for this planet in transmission and emergent spectra, see also (Kaltenegger *et al.* 2011).

Low mass Main Sequence M dwarfs are the most abundant stars in the galaxy, and representi about 75% of the stellar population. Using this argument, many planets like Gl581d are likely to be found in the near future. Such planets also provid excellent targets for future space missions as shown in this paper, due to their lower contrast ratio to their host star (Scalo *et al.* 2007, Tarter *et al.* 2007, Kaltenegger & Traub 2009, Kaltenegger *et al.* 2010b). This decreased contrast ration should allow us to search for signposts of biological activity in such planetary atmospheres.

2. The star Gl581

Gl581 is an M3 dwarf at a distance of 6.3 pc from the Sun. From the V -band bolometric correction (Delfosse *et al.* 1998), on can infer a bolometric luminosity $L_{Star}= 0.013\ L_{Sun}$ ((Bonfils *et al.* 2005b) and a mass $M_{Star} = 0.31\ M_{Sun}$. We adopt these values unchanged. For this mass, the solar-metallicity evolutionary tracks (Baraffe *et al.* 1998; specifically, the tracks denoted BCAH98 models.) imply a radius $R_{Star} = 0.30 \pm 0.01\ R_{Sun}$ for ages ranging all the way from 150 Myr to 10 Gyr. Similarly, assuming a reasonable age of 5 Gyr suggested by its kinematic properties, intermediate between young disk and old disk, i.e., between 3 and 10 Gyr (Delfosse *et al.* 1998), the radius predicted by these tracks is 0.30 R_{Sun} (consistent with the 0.29 R_{Sun} cited by Bonfils *et al.* 2005b) and recent measurements by van Braun *et al.* (in press). Combined with the empirical luminosity, the latter radius yields an effective temperature $T_{eff} = 3561$K; we thus adopt the rounded-off value $T_{eff} = 3600$K. The magnitude of the stellar chromospheric UV radiation incident on the planet is also very important for calculating the planetary atmospheric properties (Segura *et al.* 2005). With very scant UV observations available for the vast majority of M dwarfs, chromospheric and coronal Hα and X-ray emission have usually been used as proxies for the expected UV emission from these stars. Walkowicz *et al.* (2008) have shown that some M dwarfs with very low Hα and X-ray emission may still have non-negligible near-UV emission, so the use of the former two as a proxy for the latter may not always be accurate. There is very likely no such thing as an M dwarf with no chromosphere, however, with absolutely no indication of activity in Hα or X-rays in Gl581, our conservative approach is to assume a negligible UV as well, instead of speculating. Future direct observations of UV are required to settle the issue.

3. Atmosphere Model Description

In this article we use the term Super-Earths for planets that differ from giant planets (Jupiter- and Neptune-like ones) in that they have a surface: a solid-to-gas or liquid-to-gas phase transition at their upper boundary similar to Earth (see also Valencia *et al.* 2007a, Sasselov et. al. 2008, Sotin *et al.* 2007, Zahnle *et al.* 2007). That surface separates a vast interior reservoir (e.g., Earth's mantle) from an atmosphere with insignificant mass compared to that of the planet. Like on Earth, the atmosphere is fed from the interior reservoir, and its chemical balance is achieved via interactions with the interior (outgassing and burial) and with the parent star (photochemistry and loss to space). These interactions are usually described in terms of geochemical cycles; e.g. on Earth the carbonate-silicate cycle maintains a temperate environment through a feedback cycle of outgassing, rain-out and burial of CO_2.

For Gl581d we adopt a mass and radius of 7 M_{Star} and 1.69 R_{Star} respectively and scale the surface gravity according to the increase in mass and radius (see Valencia *et al.* 2007, Seager *et al.* 2007, Sotin *et al.* 2007). Assuming outgassing and loss rates similar to Earth's, the increase in gravity should proportionally increase the surface pressure. We thus initially set the surface pressure to 2.45 bar in our models, and investigate a range of CO_2 levels (implying e.g. an active carbonate-silicate cycle and resulting CO_2 buildup to stable condition between outgassing and rainout) to explore habitable conditions on the outer edge of the HZ. In our calculations presented in this paper, we concentrate on the case of efficient heat transfer, following recent studies demonstrating that heat transfer should not be limited on a planet with a high density atmosphere even in a synchronous rotation (see Joshi *et al.* 2003; Edson *et al.* 2011. Wordsworth *et al.* 2011, Heng *et al.* 2011). We model atmospheric composition from high oxygen and low CO_2 content to high CO_2 and low oxygen content. These two atmospheric compositions should roughly bracket conditions due to an active carbonate-silicate cycle on a rocky planet in the outer part of the HZ.

To model the atmospheric composition, temperature and spectral features (see Kaltenegger *et al.* 2010 for details on the models), we use Exo-P, a coupled one-dimensional code developed for rocky exoplanets based on the 1D climate (Kasting *et al.* 1984a, 1984b, Haqq-Misra *et al.* 2008), 1D photochemistry (Kasting & Ackerman 1986, Pavlov *et al.* 2000, Kharecha *et al.* 2005, Segura *et al.* 2007) and radiative transfer model based on SAO98 (Traub & Stier 1978, Kaltenegger & Traub 2009) to self consistently calculate the atmosphere and hypothetical spectra of Gl581d, assuming a rocky Earth-analog composition. These codes have been used to calculate HZs around different types of host stars (Kasting *et al.* 1993) as well as model spectra for Earth-analogs around different host stars (Segura *et al.* 2003, 2005, 2007; Kaltenegger & Traub 2009, Kaltenegger & Sasselov 2010) and throughout geological evolution of Earth (Kaltenegger *et al.* 2007).

Atmosphere simulations. In our models we focus on two scenarios for rocky super-Earth atmospheres, $[ModelsA]$ assume a nominal surface pressure of 2.45bar consistent with the increased gravity while $[ModelsB]$ are set to a surface pressure that allows for habitable conditions in our models. $[ModelA1]$ assumes present Earth-like atmospheric composition with 0.21 O_2 and 335 ppmv or 1PAL of CO_2 (PAL = Present Atmospheric Level)).We then increase the amount of CO_2 from 1 PAL $[A1]$ to a mixing ratio of 0.9 $[A2]$ to explore if the surface temperature of the planet would rise above freezing without a surface pressure increase.

Models B explore the minimum amount of CO_2 needed to maintain an average surface temperature above freezing on the planet's surface in an atmosphere with a 0.9 CO_2 mixing ratio with either abiotic $[B1]$ or two different biotic levels of oxygen $[B2andB3]$ (see also Kaltenegger *et al.* 2011).

4. Results

Model Atmospheres. For models A, where the total surface pressure of the planet is scaled to 2.45 bar, we examine two cases within this scenario: the first is an Earth-analog oxygen-rich model $[modelA1]$, which has biotically produced O_2 with a terrestrial mixing ratio of 0.21 and a CO_2 mixing ratio equal to 1 PAL. We assume a CH_4 ougassing of 4.13 x 10^9 molecules s^{-1} cm^{-2}, that scaled to Gl581d surface is 4.93 x 10^9 g yr^{-1}. (Fig. 1)

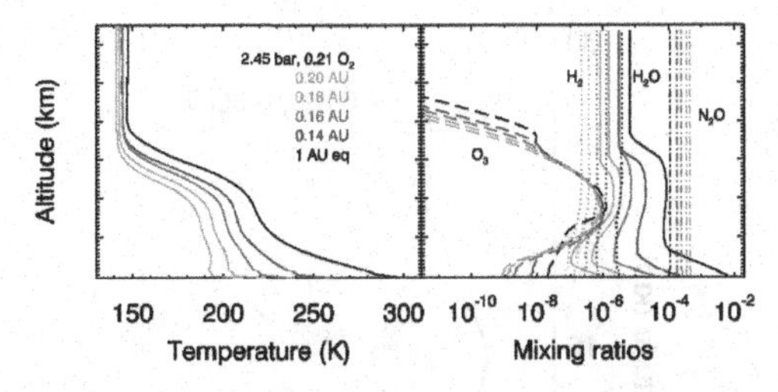

Figure 1. Fig.1: (left) Temperature profile (right) chemical composition for an Earth-like atm. composition for Gl581d-like vs. distance to its host star

shows the atmospheric temperature and mixing ratio profiles as a function of distance from the star for 1 PAL.

Next, we raise the atmospheric CO_2 level in our models to explore what level is actually required in the planet to maintain its mean surface temperature above freezing, and thus provide a habitable environment at its orbit. This in turn sets the surface pressure on the planet. Fig. 3 shows that even a 90% CO_2 (and low, abiotic oxygen) planetary atmosphere yields a temperature of only 237 K well below freezing at Gl581d's 0.2AU nominal orbital radius, if the total surface pressure remains fixed at 2.45 bar (model A2). Increasing the surface pressure to 7.6 bar while maintaining this high level of CO_2 (model B1) increases the surface temperature to an above-freezing value of 275 K (Fig. 2), implying potential habitable conditions. Such a level of atmospheric pressure increase due to CO_2 is reasonable with even a moderate carbonate-silicate cycle or increased outgassing of CO_2.

Note that methane production in these models is kept to abiotic levels, which results in CH_4 concentrations of 1.21 x$10^{-4}$and 1.13x10^{-4} in models $[A2]$ and $[B1]$ respectively. While increasing methane levels can also amplify the greenhouse effect on a planet, we concentrate here on CO_2 levels supplied by the carbonate-silicate cycle to define the outer edge of the HZ, which encapsulates our assumption that it is the carbonate-silicate cycle that regulates habitability.

We use model $[B1]$ as a representative atmosphere for minimal conditions for potential habitability on Gl581d. Having found $[modelB1]$ to yield habitable conditions, we now explore how much biotic oxygen such an atmosphere can support. We examine two such models, both for a 7.6 bar atmosphere: $[modelB2]$, with an O_2 mixing ratio of 10-3 and 90% CO_2; and $[kmodelB3]$, with an O_2 mixing ratio of 10-2 and 90% CO_2. We adopt model $[B2]$ (7.6 bar atmospheric pressure, 90% CO_2 and $10^{-3}O_2$), which provides above-freezing surface conditions for a minimum atmospheric pressure and CO_2 mixing ratio, as our habitable biotic scenario for Gl581d to assess remotely detectable spectral features.

Detectable Spectral Features. In this section we focus on the detectable atmospheric spectral features of a potentially habitable Gl581d. Spectra of terrestrial exoplanets could be obtained in the near future with the same techniques that have successfully provided spectra of Earth (see e.g. Christensen *et al.* 1997, Irioni 2002, Paille *et al.* 2009, Kaltenegger & Traub 2009, Cowan *et al.* 2010) and extrasolar giant planets (EGP) (see e.g. Grillmair *et al.* 2008, Swain *et al.* 2008). The emergent spectra of rocky planets in the HZ are dominated by reflected starlight in the visible to near-IR and thermal emission from the planet in the mid-infrared, while transmission spectra result from starlight that is filtered through the planet's atmosphere. Such spectroscopy provides molecular band strengths of multiple transitions (in absorption or emission) of a few abundant molecules in the planetary atmosphere. We generate synthetic spectra of Gl581d from 0.4 μm to 40μm to explore which indicators of biological activities in the planet's atmosphere may

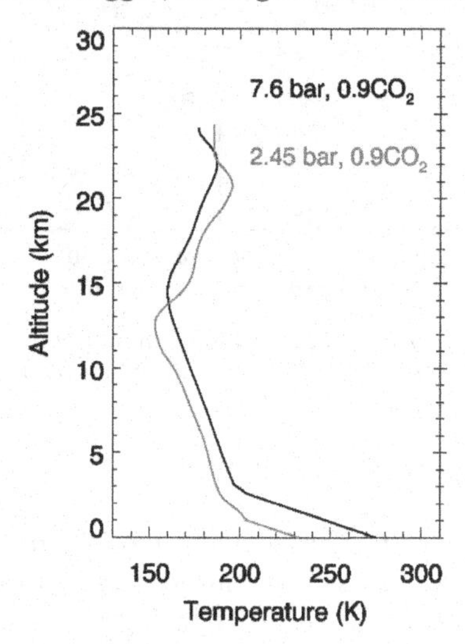

Figure 2. Fig.2: Temperature for a 2.45bar (A2) vs. a 7.6bar(B1) abiotic high CO_2 atmosphere

be observed by future ground- and space-based telescopes such as the Extremely Large Telescope (E-ELT) and the James Webb Space Telescope (JWST).

We use as our template the atmospheric [$modelB2$] (7.6 bar atmosphere, with 90% CO_2 and biotic O_2 with a mixing ratio of 10-3), which is habitable both in the absence of tidal-locking as well as under conditions of relatively efficient heat-transfer in the presence of tidal-locking (see Kaltenegger *et al.* 2011). For simplicity, we restrict ourselves here to spectral features arising from this model in the absence of tidal-locking. The planet is unlikely to be synchronous rotating (see e.g. Leconte *et al.* 2010), or even if it were, the heat transfer should be efficient, given that at least 77% of its surface is likely directly illuminated (see discussion in Selsis *et al.* 2007).

The spectral distribution of M-stars, such as Gl581, generates a different photochemistry on planets orbiting within the HZ of M stars, compared to planets within the HZ of Sun-like stars (Segura *et al.* 2005). In particular, the biogenic gases CH_4, N_2O, and CH_3Cl have substantially longer lifetimes and higher mixing ratios than on Earth, making them potentially observable by space-based telescopes. In addition, the low effective temperatures of M dwarfs yield spectra dominated by molecular absorption bands that redistribute the radiated energy in a distinctly non-black-body fashion. Both effects are crucial to determining the observable spectral features and biosignatures of habitable planets (see e.g. Kalteneger *et al.* 2010a, Kaltenegger *et al.* 2011) around these cool low-mass stars.

The observable quantity to remotely derive what atmospheric features exist in a planet's atmosphere is the contrast ratio versus wavelength, as shown in Fig. 3 for emergent model spectra (for transmission spectra of Gl 581d see Kaltenegger *et al.* 2011). The Sun emits a large fraction of its energy in the visible, a wavelength domain where the atmosphere of a habitable planet is highly reflective, because of the Rayleigh backscattering varying like λ^{-4} and because of the lack of strong H_2O absorption bands. The emission of Gl581, a star with a low effective temperature, peaks in the near-infrared where the contribution of Rayleigh scattering to the albedo becomes negligible and the strong absorption bands of H_2O, CO_2 and CH_4 cause additional absorption of stellar radiation and overall lower the planet's albedo as long as no additional reflective cloud layer forms.

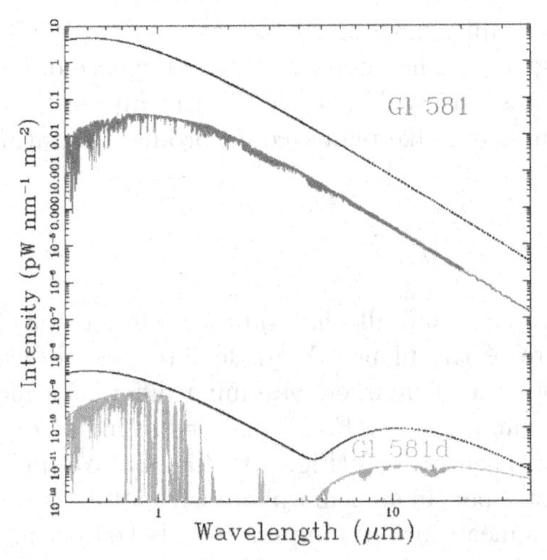

Figure 3. Fig.3: Planet-star contrast ratio for emergent sectra for a clear atmosphere (Model B2). The upper line indicates the levels of the Sun and the Earth for comparison (modified from Kaltenegger *et al.* 2011

For the emergent spectra, the larger surface area of a Super-Earth makes the direct detection and secondary-eclipse detection of its atmospheric features and biosignatures easier than for Earth size planets. Even though Gl581d orbits its host star at a distance less than 1AU, the integrated flux received on top of the planet's atmosphere is half the flux received by Earth. Since Gl581d orbits an M star, it has a lower Bond albedo and thus a smaller part of the stellar light is reflected, making Gl581d appear dimmer in reflected light than an equivalent planet orbiting at 1AU around a sun-like star. Even though the bond albedo decreases from Earth's 0.29 to Gl581d's 0.12, the increased surface area increases the overall reflected flux of the planet by about 15% over Earth's reflected Sunlight (see (Fig. 3). In the MIR the flux increases proportional to the surface area of the planet if the effective temperature is equal. In our model the decrease in effective temperature for Gl581d due to the high CO_2 concentration, reduces its flux in the MIR compared to the Earth significantly.

5. Discussion

One prominent question about planets with masses below 15 Earth masses are, if such planets could be potentially be habitats. Many question regarding this are still open, like if such planets would maintain tectonics in their interior and magnetospheres that can shelter their primary atmosphere. Note that the model we present here only represents one possible nature of a planet like Gl581d in a wide parameter space that includes Mini-Neptunes.

In addition composition of the planet will influence the atmosphere substantially, e.g. the abundance of oxygen found in the abiotic case strongly depends on the oxydation state of the superficial layers and possible oceans of the planet. If e.g. the surface of the planet were highly oxidized, a higher buildup of oxygen would be expected for the same production levels (see e.g. Zahnle *et al.* 2007, Kasting *et al.* 1984b). If e.g. a sulfur cycle were present on a planet, the outer edge of the HZ would shift outwards due to increased greenhouse effect of SO_2 in addition to CO_2 (see Kaltenegger & Sasselov 2009, Domagal-Goldman *et al.* in press,), which would reduce the amount of CO_2 needed to maintain Gl581d's surface temperature above freezing. Note that a planet found in the

HZ is not necessary habitable, since many factors may prevent surface habitability like the lack of water or ingredients necessary for the emergence of life.

This parameter space is very wide and we concentrate on Earth analog models here, to explore the case if an Earth-like planet could produce signatures of habitability that we could remotely detect.

6. Conclusions

We show that Gl581d is potentially habitable, assuming the carbonate-silicate cycle controls the atmosphere of the planet. We calculate the surface temperature and atmosphere including potential biomarkers assuming different atmospheres compositions, high oxygen atmosphere analogous to Earth's as well as high CO_2 atmospheres with and without biotic oxygen concentrations (Fig. 1 to Fig. 2). We find that a minimum CO_2 partial pressure of about 7 bar, in an atmosphere with a total surface pressure of 7.6 bar, are needed to maintain a mean surface temperature above freezing on Gl581d. The model we present here only represents one possible nature - an Earth-like composition - of a planet like Gl581d in a wide parameter space. The surface temperature of a simulated 90% CO_2 and low oxygen planetary atmospheres at 0.2 AU changes from 237 K to 278 K when increasing the surface pressure from 2.45 bar to 7.6 bar surface pressure (see Fig. 2). Such a level of atmospheric pressure increase due to CO_2 is reasonable even with a moderate carbonate-silicate cycle or increased outgassing of CO_2. Additional greenhouse gases like CH_4 and SO_2 as well as clouds assuming a net warming could decrease this amount due to their added warming effect.

Our concept of the habitable zone is based on the carbonate-silicate cycle that should increase the level of CO_2 on the outer part of the HZ. Even so the measurements are hard this concept can be probed by observing detectable atmospheric features by future ground and space- based telescopes like E-ELT and JWST on planets like Gl581d.

Observation of the emergent spectrum could also determine if Gl581d is the first habitable world we have discovered.

Acknowledgements

L.K. acknowledges support from NAI and DFG funding ENP Ka 3142/1-1.

References

Allard, F. *et al.* 2001, *ApJ*, 556, 357
Allen, C. W. 1976, Astrophysical Quantities (London: The Athlone Press)
Baraffe, I., Chabrier, G., Allard, F., & Hauschildt, P. 1998, *A&A*, 337, 403
Barnes, R., Jackson, B., Greenberg, R., & Raymond, S. N. 2009, *ApJ*, 700, L30
Bonfils, X., Forveille, T., Delfosse, X., *et al.* 2005, *A&A*, 443, L15
Chabrier, G., *et al.* 2000, *ApJ*, 542, 464
Christensen, P. R. & Pearl, J. C. 1997, *J. Geophys. Res.*, 102, 10875
Cowan, N. B. *et al.* 2009, *ApJ*, 700, 915
Cox A. N., IV 2000, *Allen's Astrophysical Quantities*, (4th ed; New York: AIP)
Delfosse, X., Forveille, T., Perrier, C., & Mayor, M. 1998, *A&A*, 331, 581
Dole, S. H. 1964, *Habitable Planets for Man*, Blaisdell Publishing, New York.
Dressing, C. D., Spiegel, D. S., Scharf, C. A., Menou, K., & Raymond, S. N. 2010, *ApJ*, 721, 1295
Edson, A., Lee, S., Bannon, P., Kasting, J. F., & Pollard, D., 2011, *Icarus*, 212, 1
Forget, F. & Pierrehumbert, R. T. 1997, *Science*, 278, 1273
Fujii, Y., *et al.* 2011, *ApJ* in press (http://arxiv.org/abs/1102.3625)

Goldblatt, C, Zahnle, K. 2011, *Climate of the Past*, 7, 203
Grillmair, C. 2008, *Nature*, 456, 767
Haberle, R. M., McKay, C., Tyler, D., & Reynolds, R. 1996, in *Circumstellar Habitable Zones*, edited by L.R. Doyle, Travis House, Menlo Park, CA, 29
Haqq-Misra, J. D., Domagal-Goldman, S. D., Kasting, P. J., & Kasting, J. F. 2008, *Astrobiology*, 8, 1127
Hart, M. H. 1978, *Icarus*, 33, 23
Heller, R., *et al.* 2011, *A&A*, 528, A27
Heng, K., Menou, K., & Phillipps, P. J. 2011, *MNRAS*, 413, 2380
Henning, W. G., OConnell, R. J., & Sasselov, D. D. 2009, *ApJ*, 707, 1000
Holland, H. D. 2002, *Geochim. Cosmochim. Acta*, 66, 3811
Houghton, J. T., Meira Filho, L. G., & Callander, B. A., *et al.* 1995, Climate Change 1994: Radiative Forcing of Climate Change and an Evaluation of the IPCC IS92 Emission Scenarios (Cambridge, MA: Cambridge University Press)
Irion, B. *et al.* 2002, *Applied Optics*, 41, 33, 6968
Johnson, J. A. & Apps, K. 2009, *ApJ*, 699, 933
Joshi, M. 2003, *Astrobiology*, 3, 415
Joshi, M. M., Haberle, R. M., & Reynolds, R. T. 1997, *Icarus*, 129, 450
Kaltenegger, L., Traub, W. A., & Jucks, K. W. 2007, *ApJ*, 658, 598
Kaltenegger, L. & Traub, W. 2009, *ApJ*, 698, 519
Kaltenegger L., Sasselov, D. 2009, *ApJ*, 708, 1162
Kaltenegger, L. Selsis, F., *et al.* 2010a, *Astrobiology*, 10, 25
Kaltenegger, L., Eiroa, C., Ribas, I., *et al.* 2010b, *Astrobiology*, 10, 103
Kaltenegger, L., Henning, W. G., & Sasselov, D. D. 2010c, *AJ*, 140, 1370
Kasting, J. F. 1990, *Origins Life Evol. Bios.*, 20, 199
Kasting, J. F., Pollack J. B., Ackerman T. P. 1984a, *Icarus*, 57, 335
Kasting, J. F., Pollack, J. B., & Crisp, D. 1984b, *J. Atmos. Chem*, 1, 403
Kasting, J. F. & Ackerman, T. P. 1986, *Science*, 234, 1383
Kasting, J. F., Whitmire, D. P., & Reynolds, R. T. 1993, *Icarus*, 101, 108
Kasting, J. F. & Catling, D. 2003, *ARA&A*, 41, 429
Kharecha, P. Kasting, J. F., & Siefert, J. L. 2005, *Geobiol.*, 3, 53
Kitzmann D., Patzer, A. B. C., von Paris, P., Godolt, M., Stracke, B., Gebauer, S., Grenfell, J. L., & Rauer, H. 2010, *A&A* 511, A66
Leconte, J., Chabrier, G., Baraffe, I., & Levrard, B. 2010, *A&A*, 516, A64+
Leger, A., *et al.* 2009, *A&A*, 506, 287
Mayor, M., Udry, S., Lovis, C., Pepe, F., Queloz, D., Benz, W., Bertaux, J.-L., Bouchy, F., Mordasini, C., & Segransan, D., 2009, *A&A*, 493, 639
Mischna, M. A., Kasting, J. F., Pavlov, A., & Freedman, R. 2000, *Icarus*, 145, 546
Pavlov, A. A., Kasting, J. F., Brown, L. L., Rages, K. A., & Freedman, R. 2000, *J. Geophys. Res.*, 105, 11981
Pollack, J. B., Kasting, J. F., Richardson, S. M., & Poliakoff, K. 1987, *Icarus*, 71, 203
Rothman, L. S., Jacquemart, D., Barbe, A., *et al.* 2004, *JQSRT*, 96, 139
Rothman, L. S., Gordon, I. E., Barbe, A., *et al.* 2009, *JQSRT*, in press
Sasselov, D. D., Valencia, D., & O'Connell, R. J. 2008, *Phys. Scr.*, T130, 014035
Scalo, J., Kaltenegger, L., Segura, A., *et al.* 2007, *Astrobiology*, 7, 85
Seager, S. *et al.* 2007, *ApJ*, 669, 1279
Segura, A., Krelove, K., Kasting, J. F., Sommerlatt, D., Meadows, V., Crisp, D., Cohen, M., & Mlawer, E. 2003, *Astrobiology*, 3, 689
Segura, A., Kasting, J. F., Meadows, V., Cohen, M., Scalo, J., Crisp, D., Butler, P., & Tinetti, G. 2005, *Astrobiology*, 5, 706
Segura, A., Meadows, V. S., Kasting, J. F., Cohen, M., & Crisp, D. 2007, *A&A*, 472, 665
Selsis, F. 2002, *ESA-SP*, 514, 251
Selsis, F., Kasting, J. F., Levrard, B., Paillet, J., Ribas, I., & Delfosse, X. 2007, *A&A*, 476, 1373
Sotin, C., Grasset, O., & Mocquet, A. 2007, *Icarus*, 191, 337
Spiegel, D. S., Menou, K., & Scharf, C. A. 2009, *ApJ*, 691, 596

Spiegel, D. S., Raymond, S. N., Dressing, C. D., Scharf, C. A., & Mitchell, J. L. 2010, *ApJ*, 721, 1308

Swain, M. R., Vasisht, G., & Tinetti, G., 2008, *Nature*, 452, 329

Tarter, J. C., *et al.* 2007, *Astrobiology*, 7, 66

Traub, W. A. & Stier, M. T. 1978, *ApJ*, 226, 347

Traub, W. A. & Jucks, K. 2002, in Atmospheres in the Solar System: Comparative Aeronomy. Geophysical Monograph 130. Edited by Michael Mendillo, Andrew Nagy, and J.H. Waite. Washington, D.C.: American Geophysical Union, p.369

Udry, S., Bonfils, X., Delfosse, X., *et al.* 2007, *A&A*, 469, L43

Valencia, D., O'Connell, R. J., & Sasselov, D. D. 2006, *Icarus*, 181, 545

Valencia, D., Sasselov, D. D., & OConnell, R. 2007a, *ApJ*, 665, 1413

Valencia, D., O'Connell, R., & Sasselov, D. D. 2007b, *ApJ*, 670, L45

von Bloh, W., Bounama, C., Cuntz, M., & Franck, S. 2007, *A&A*, 476, 1365

von Braun, K., *et al.* 2011, *ApJ*, 729, L26

von Paris, P., Gebauer, S., Godolt, M., *et al.* 2010, *A&A*, 522, A23+

Walker, J. C. G., Hays, P. B., & Kasting, J. F. 1981, *J. Geophys. Res.*, 86, 9776

Walkowicz, L., Johns-Krull, C., & Hawley, S. 2008, *ApJ*, 677, 593

Williams, D. M. & Pollard, D. 2002, *International Journal of Astrobiology*, 1, 61

Wordsworth, R. D., Forget, F., Selsis, F., *et al.* 2010, *A&A*, 522, A22+

Wordsworth, R. D., Forget, F., Selsis, F., *et al.* 2011, *ApJ*, 733, L48

Yung, Y. L. & DeMore, W. B. 1999, *Photochemistry of planetary atmospheres*, Oxford University Press, New York, N.

Zahnle, K., Arndt, N., Cockell, C., *et al.* 2007, *SSRv*, 129, 35

The Astrophysics of Planetary Systems: Formation, Structure, and Dynamical Evolution
Proceedings IAU Symposium No. 276, 2010
A. Sozzetti, M. G. Lattanzi & A. P. Boss, eds.

doi:10.1017/S1743921311020473

Observations and modelling of earth's transmission spectrum through lunar eclipses: A window to transiting exoplanet characterization

E. Pallé[1], Antonio García Muñoz[1], Maria R. Zapatero Osorio[2], Pilar Montañés-Rodríguez[1], Rafael Barrena[1] and Eduardo L. Martín[2]

[1]Instituto de Astrofisica de Canarias, La Laguna, Tenerife, Spain
email: epalle@iac.es

[2]Centro de Astrobiología, CSIC-INTA, Madrid, Spain

Abstract. Recently we were able to retrieve the Earth's transmission spectrum through lunar eclipse observations. This spectrum showed that the depth of most molecular species was stronger than models had anticipated. The presence of other atmospheric signatures, such as atmospheric dimers, were also present in the spectrum. We have been developing a radiative transfer code able to reproduce the Earth's transmission spectra at different depths into the penumbra and umbra, and taking into account transmission, refraction, and multiple scattering. Here we discuss the results to date and the work ahead.

Keywords. astrobiology, radiative transfer, radiation mechanisms: general, planets and satellites: general, Earth

1. Introduction

The observational technique of transmission spectroscopy during extrasolar planet transits allow us to obtain the transmission spectrum of a planet atmosphere. When the transit occurs part of the light that reaches the observer has crossed the planetary atmosphere and contains the signatures of its spectroscopically-active component gaseous species. Based on HST high-precision spectrophotometric observations, Charbonneau *et al.* (2002) detected the absorption from sodium in the atmosphere of HD209458b, while Richardson *et al.* (2007) reported its infrared spectra (7.5-13.2 μm). Using the Spitzer telescope and this same methodology, Tinetti *et al.* (2007) and Swain *et al.* (2008) have recently reported the presence of water and methane, respectively, in the atmosphere of planet HD189733b. Furthermore, the large number of transiting planets expected to be discovered by CoRot and KEPLER and the advent of more sensitive instruments like the JWST, makes this technique extremely promising for the characterization of exoplanetary atmospheres. But what about the detection and characterization of small, rocky planets? To be ready to complete such identifications in the case of terrestrial planets, one requires globally-integrated observations of the Earth and the rest of rocky planets of the solar system. Specially interesting are the observations of the transmitted spectra of their atmospheres. The characterization of the spectral features in our planet's transmission spectra, as if it were observed from a distant star, can be achieved through the earthshine observations during a lunar eclipse, and constitute an important benchmark for future exoplanets studies.

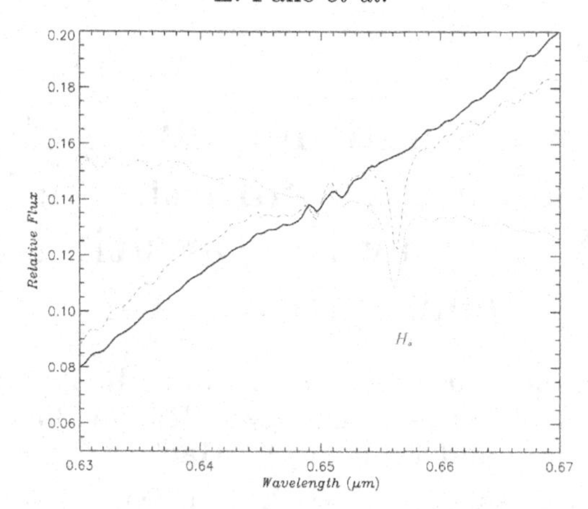

Figure 1. A detail of the umbra spectrum (broke line), the bright Moon spectrum (dotted line) and their ratio spectrum (umbra/bright) around the hydrogen alpha (H_α) line (0.6568 μm). The H_α solar line is a greatw example to illustrate the high S/N of our observations. In this figure, one can see that the H_α is present in the raw spectra of the umbra and the bright Moon, but not in the final transmission spectra.

2. Eclipse observations and modeling

A lunar eclipse is a relatively rare event that occurs when the Moon enters the cone of shadow cast by the Sun-lit Earth. From within the umbra of the eclipse, the planet prevents a direct view of the solar disk. The Moon in umbra, however, is not entirely dark because a fraction of the solar photons that penetrate the terrestrial atmosphere are either refracted or scattered towards the Moon. This fraction, when normalized by the solar irradiance, is a measure of the transmissivity of the Earth's atmosphere, while the spectral features of the Sun, the Moon's albedo and the local telluric atmosphere are canceled (see Figure 1). The fraction is strongly dependent on both the composition of the extended region in the Earth's terminator traversed by the photons and on the photons' wavelength. In the umbra, the pathlength traversed by photons is long, so that weakly absorbing gases yield prominent absorption signatures. In 2009 the first spectrum of a lunar eclipse, with data taken at two ground-based telescopes in La Palma (Spain), was published by Pallé *et al.* (2009). Our 0.36-2.4 μm spectrum of the partial eclipse of 16 August 2008 shows clear evidence of a number of biologically relevant molecules such as O_3, O_2, H_2O, CH_4 and CO_2, and the dimers $O_2 - O_2$ and $O_2 - N_2$ (see Figure 2).

Simulations using the empirical Earth's transmission spectrum, and the stellar spectra for a variety of stellar types, indicate that with the new generation of extremely large telescopes, such as the proposed 42-meter European Extremely Large Telescope(E-ELT), we could be capable of retrieving the transmission spectrum of an Earth-like planet around very cool stars and brown dwarfs ($T_{\rm eff} \lesssim 3100$ K) (Palle *et al.* 2011).

The classical theory of lunar eclipses, nicely summarized by Link (1962), is built on the premise that the sunlight scattered by the gases and particles in the atmosphere contributes negligibly to the brightness of the eclipsed Moon. Our recent modeling effort (García Muñoz and Pallé, 2011) revisits the lunar eclipse theory, extending it to accommodate spectrally-resolved observations and addressing the role of scattered sunlight. Predictions of both direct and diffuse sunlight are produced by integrating the radiative transfer equations over the Earth's disk. We find that omitting scattered sunlight is an acceptable approximation for low and moderate aerosol loadings at visible and longer wavelengths. However, towards the ultraviolet, or at times when the atmosphere contains elevated aerosol amounts, the relative significance of direct and diffuse sunlight may

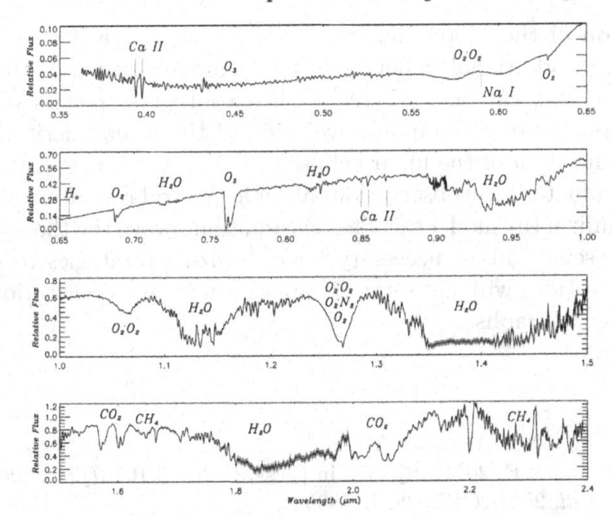

Figure 2. The Earth's transmission spectrum from 0.36 to 2.40 μm. The major atmospheric features of the spectrum are marked. Adapted from Palle *et al.* (2009).

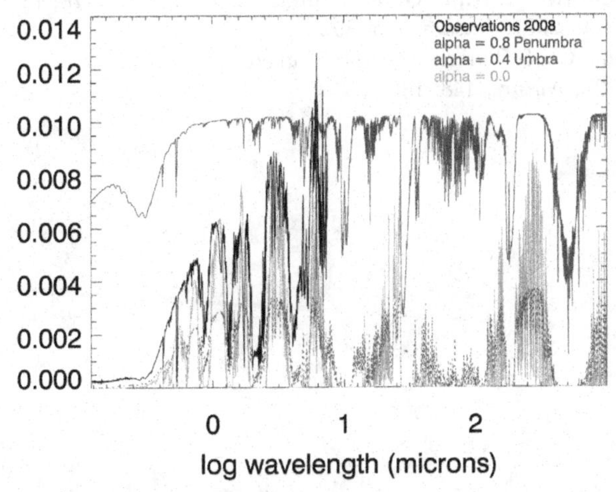

Figure 3. Modeled Moon-reflected radiances for $\Delta\lambda/\lambda$ 1000 at various α angles during the eclipse covering the spectral range from 0.4 to 25 micron. Y scale is in arbitrary units.

reverse (García Muñoz *et al.*, 2011). The conclusion is supported by two distinct features, namely the spectrums tail at short wavelengths and the unequal absorption by the $(O_2)_2$ dimers at two nearby bands. Our findings are consistent with the elevated aerosol loadings reported for several months at high northern latitudes following the 78 August 2008 Kasatochi volcano eruption in Alaska. We also find that lunar eclipse spectra will reveal the presence on enhanced background aerosol levels up to 4 or 5 years after a major volcanic eruption.

3. Future work

As shown in Figure 1, model calculations predict that the structure of the spectrum of Moon-reflected light depends strongly on the phase of the eclipse, with both the total amount of light and its spectral distribution varying (see Figure 3). To quantify this we define a parameter, *alpha*, which is defined as the angle between the center of the umbral disk and the region of the Moon that we are sampling at each time. This dynamical behavior involves an additional difficulty in the characterization of the Moon-reflected light. Such observations can be gathered with 8 to 10 meter class telescopes at very high SNRs, and the complete temporal evolution of the atmospheric signatures can be followed over the duration of the lunar eclipse.

We intend to conduct eclipse observations at high spectral resolution and high temporal cadence to help confirm the model calculations and to improve the theory of lunar eclipses as it stands at present. This is necessary for optimizing strategies to search for Earth analogs. The observations will also serve as input data for new generation planet-hunting high-resolution spectrographs.

References

Charbonneau, D. *et al.* 2002, *ApJ*, 568, 377
García-Muñoz, A. & Pallé, E. 2011, *JQSRT*, in press, DOI: 10.1016/j.jqsrt.2011.03.017
García-Muñoz, A., *et al.* 2011, *GRL*, 38, L14805
Link, F. 1962, *Physics and Astronomy of the Moon*, 560, 161
Pallé, E., *et al.* 2009, *Nature*, 459, 814
Pallé, E., Zapatero-Osorio, M. R., & Garcia-Munoz, A. 2011, *ApJ*, 728, 19
Richardson, L. J., *et al.* 2007, *Nature*, 445, 892
Swain, M. R., Vasisht, G., & Tinetti, G. 2008, *Nature*, 452, 329
Tinetti, G., *et al.* 2007, *Nature*, 448, 169

PART 5:
POSTER PAPERS

Section A:
Planet Formation

The Astrophysics of Planetary Systems: Formation, Structure, and Dynamical Evolution
Proceedings IAU Symposium No. 276, 2010
A. Sozzetti, M. G. Lattanzi & A. P. Boss, eds.

doi:10.1017/S1743921311020485

Characterisation of SPH noise in simulations of protoplanetary discs

Serena E. Arena[1], Jean-François Gonzalez[2] and Elisabeth Crespe[3]

[1]Université de Lyon, Lyon, F-69003, France; Université Lyon 1, Villeurbanne, F-69622, France;
[2]CNRS, UMR 5574, Centre de Recherche Astrophysique de Lyon;
[3]École normale supérieure de Lyon, 46, allée d'Italie, F-69364 Lyon cedex 07, France
email: {Serena.Arena, Jean-Francois.Gonzalez, Elisabeth.Crespe}@ens-lyon.fr

Abstract. The effects of *turbulence* on the dynamics of dust grains in protoplanetary discs is of relevant importance in the study of pre-planetesimal formation. The complex interplay between gas and dust and the modelling of turbulence require *numerical simulations*.

A statistical study of the noise in SPH simulations of gas-only protoplanetary accretion discs is performed in order to determine if it could mimic turbulence and to what extent.

Keywords. planetary systems: protoplanetary disks, turbulence

---

We performed four SPH simulations (see Fig. 1) of the same gas disc changing the two parameters (α, β) in the artificial viscosity prescription of Monaghan & Gingold (1983).

To each realisation of the disc we applied turbulence diagnostics measuring both its magnitude (turbulent viscosity and diffusion) and its structure (power spectrum, structure function and density probability distribution function).

Results: MAGNITUDE of the noise.

Turbulent viscosity and accretion. Turbulent viscosity in accretion discs is often quantified by the parameter $\alpha_{\rm ss}$ (Shakura & Sunyaev 1973) as $\nu_T = \alpha_{\rm ss} c_s H$ (H: scale height of the disc) and it is proportional to the mass accretion rate $\dot{M} \propto \nu_T$. We estimated $\alpha_{\rm ss}$ from: (1) the mass flux: $\alpha_{\rm MF} = -2\sqrt{r}\langle\rho v_r\rangle/(3\langle\rho\rangle)$, (2) the Reynolds stress: $\alpha_{\rm RS} = \langle u_r u_\theta\rangle/\langle P/\rho\rangle$ (u: velocity fluctuations, P: gas pressure), (3) the SPH artificial viscosity: $\alpha_{\rm SPH} = \alpha h/(5H)$, (Meglicki *et al.* 1983, h: SPH smoothing length).

We found that in the simulations the mass accreted onto the star $\dot{M}_{\rm sim}$ increases with both α and β from $7.5 \cdot 10^{-9}$ to $1.3 \cdot 10^{-8}$ $\rm M_\odot yr^{-1}$ remaining consistent with observations ($\dot{M} \approx 10^{-8} \rm M_\odot yr^{-1}$ with $\alpha_{\rm ss} \approx 10^{-2}$, Hartmann *et al.* 1998) for all (α, β) considered.

$\alpha_{\rm SPH}$ increases with α, but it is ten times larger than the expected $\alpha_{\rm ss}$ for high α. In the inner region $\alpha_{\rm MF}$, related to *mean quantities*, also increases with α and in addition with β but with a value closer to the expected $\alpha_{\rm ss}$ for all (α,β). In contrast, $\alpha_{\rm RS}$, related to *fluctuations*, has an opposite behaviour with respect to the other two estimates: we conclude that here fluctuations are not responsible for accretion.

Turbulent diffusion. We found that the diffusive mechanism is correctly represented by large artificial viscosity parameters, for which the turbulent diffusion coefficient $D_T \approx 5 \cdot 10^{-3}$ $c_{\rm S} H$, computed as in Fromang & Papaloizou (2006), is comparable to their value.

Results: STRUCTURE of the noise

Power spectrum. The power spectrum of velocity fluctuations (top right plot in Fig. 1) for scales below the smoothing length h presents a cascade with a slope very close to the Kolmogorov one. For larger scales, $k \in$ (3-16), the cascade is steeper than or close to the Kolmogorov one ($P \approx k^{-5/3}$). In particular, for fluctuations of v_r and v_θ the cascade is: (a) more extended, $k \in (3:10)$, and less steep ($P \approx k^{-2.5}$-$k^{-1.8}$) for large (α,β); (b) less

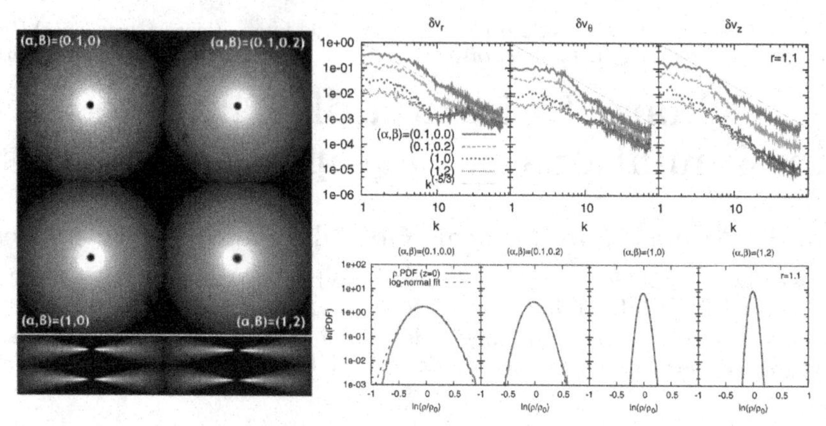

Figure 1. Left: face-on and edge-on view of the four SPH discs around $1M_\odot$ star, sampled by $2 \cdot 10^5$ particles and evolved with the code presented in Barrière-Fouchet *et al.* (2005). Units: 1 $M_\odot$, 100AU, $10^3/2\pi$ yr, in order to have $G = 1$. Initial conditions: $M_{\rm disc} = 0.02 M_{\rm star}$, radial extension from 20AU to 400AU, projected density and sound speed profiles: $\Sigma \propto r^{-3/2}$, $c_s \propto r^{-3/8}$. The disc is let to reach a stationary state and then it is followed for 9 orbits (at 100AU). **Top right**: power spectrum of the three velocity component for the four artificial viscosity combinations. **Bottom right**: density (ρ) probability distribution function (PDF).

extended, $k \in (6:10)$, and steeper ($P \approx k^{-3.5}$-k^{-3}) for small (α,β). For fluctuations of v_z all combinations of (α,β) lead to a similar slope ($P \approx k^{-3.5}$).

Slope of the structure function. The slope ζ_p of the structure function $S(r,s) = \langle |v_i(r) - v_i(r+s)|^p \rangle$ gives information concerning the presence of intermittency. For scales larger than the smoothing length h there is no evidence of intermittency ($\zeta_p \approx 0$). However, for scales close to or smaller than h: low values of (α,β) lead to slopes similar to or slightly larger than the Kolmogorov case ($p/3$); increasing (α, β) significantly decreases ζ_p (similarly to the intermittent case) except for the vertical component of velocity.

Density probability distribution function. It is well described by a log-normal distribution (bottom right plot in Fig. 1), as expected in compressible turbulence. The 3rd and the 4th order moment show deviation from gaussian distribution for small values of (α,β).

Conclusions. Artificial viscosity garantees accretion rates onto the star in the observed range, even for (α,β) as large as $(1,2)$. The properties of the SPH noise depends both qualitatively and quantitatively on (α, β) parameters, in any case no clear sign of intermittency has been observed: (1) Low values lead to a cascade of strong fluctuations, however diffusion is not correctly described. (2) High values lead to a cascade and to a correct description of diffusion, however fluctuations are weak. Since a single (α, β) pair can partially mimic the desired properties of turbulence, additional input/model is necessary. *Acknowledgements.* This research was supported by the Agence Nationale de la Recherche (ANR) of France through contract ANR-07-BLAN-0221.

References

Barrière-Fouchet, L., Gonzalez, J.-F., Murray, J. R., Humble, R. J., & Maddison, S. T. 2005, *A&A*, 443, 185

Fromang, S. & Papaloizou, J. 2006, *A&A*, 452, 751

Hartmann, L., Calvet, N., Gullbring, E., & D'Alessio, P. 1998, *ApJ*, 495, 385

Meglicki, Z., Wickramasinghe, D., & Bicknell, G. V. 1983, *MNRAS*, 264, 691

Monaghan, J. J. & Gingold, R. A. 1983, *Journal of Computational Physics*, 52, 374

Shakura, N. I. & Sunyaev, R. A. 1973, *Proc. IAU Symp.*, 55, 155

The Astrophysics of Planetary Systems: Formation, Structure, and Dynamical Evolution
Proceedings IAU Symposium No. 276, 2010
A. Sozzetti, M. G. Lattanzi & A. P. Boss, eds.

doi:10.1017/S1743921311020497

Porosity models for pre-planetesimals: modified P-α like models and the effect of dissipated energy

Serena E. Arena[1] **and Roland Speith**[2]

[1]Université de Lyon, Lyon, F-69003, France; Université Lyon 1, Villeurbanne, F-69622, France; CNRS, UMR 5574, Centre de Recherche Astrophysique de Lyon;
École normale supérieure de Lyon, 46, allée d'Italie, F-69364 Lyon cedex 07, France
email: serena.arena@ens-lyon.fr

[2]Physikalisches Institut, Universität Tübingen, Auf der Morgenstelle 14, D-72076 Tübingen, Germany
email: speith@pit.physik.uni-tuebingen.de

Abstract. The outcome of collisions between *pre-planetesimals* is important in the theory of planetesimal formation by collisional growth and strongly depends on their *internal structure*. Since pre-planetesimals are *highly porous*, reaching 90% porosity, they could show the so called *anomalous behaviour* (decrease of density during shock compression, e.g. Bolkhovitinov & Khvostov 1978). Due to involved sizes (>dm), laboratory experiments are unfeasible therefore numerical simulations equipped with adequate *porosity models* are necessary.

Here we focus on the P-α model and its variations. We found that they are suitable for applications in the high porosity range only after a modification of the basic equations, that avoids an inconsistency and takes into account the effect of dissipated energy, is performed.

Keywords. equation of state, shock waves, interplanetary medium

---

In the P-α model (Herrmann 1969; Carroll & Holt 1972) the equation of state (EOS) of the porous material is derived from that of its solid matrix (subscript m) by the introduction of the *distension* parameter $\alpha = \rho_m/\rho$, relating the respective density, and the assumption that the specific internal energy of the two components is the same $U = U_m$. The P-α model and its modified ε-α variation have been respectively used by Jutzi *et al.* (2008) and Wünnemann *et al.* (2006) in the related field of impact cratering but only for low and intermediate porosity up to 60%. However at high porosity we found that the first is affected by a singularity and the second cannot model anomalous behaviour of the porous material.

The inconsistent $U = U_m$ assumption.
This assumption implies $dU = dU_m$, but this is valid only in the unrealistic situation where the material maintains a constant porosity (α=const) while it is subject to compression. In fact, applying the First Principle at both the porous material and its matrix we have: $dU = dU_m - \frac{P_m}{\alpha \rho_m} d\alpha$, where the definition of distension α is used and $\delta Q = \delta Q_m$ because the pores in the porous material are assumed to be empty.

The suggested modification.
In order to avoid the singularity, we suggest to derive the internal energy of the *matrix* directly from the First Principle, substituting $U_m = U$ with $dU_m = \frac{P_m}{\rho_m^2} d\rho_m + \delta Q_m$. The term δQ_m represents the heat exchanged by the matrix with its environment and for high porous materials it is mainly due to the *energy dissipated* during plastic deformation.

Application: shock compression.
We use the Hugoniot as energy equation and the simplified Tillotson as matrix EOS.

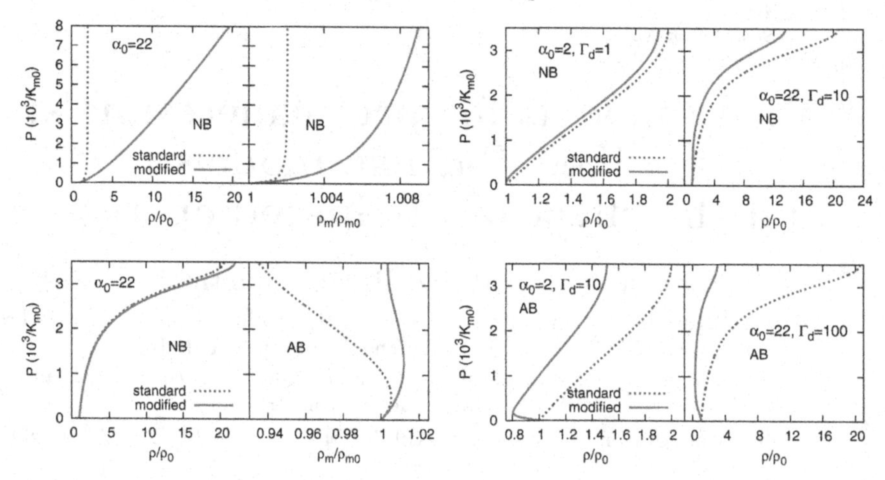

Figure 1. Left: standard (short dashed line) and modified (solid line) ε-α models (top panel) and P-α models (bottom panel) without energy dissipation for high porosity materials. **Right**: standard and modified P-α models with energy dissipation for a low porosity material (left plot in each panel) and high porosity material (right plot in each panel).

The modified P-α model without dissipation. The *singularity*, located at ρ_{sing}/ρ_0=2 in the standard ε-α model (short dashed line in the top left panel in the figure), *disappears* in the modified model (solid line in the same panel). This model predicts *normal behaviour* (NB) for both the porous material and its matrix. The bottom left panel shows that both in the standard and in the modified P-α model the behaviour of the porous material is *normal*, even if the density of the matrix presents a slight anomalous behaviour.
The effects of dissipated energy. The heat exchange is written as: $\Delta Q_m = \Gamma_d \int_{\alpha_0}^{\alpha} M(\alpha')d\alpha'$, where the normalised local dissipation function $M(\alpha)$ is taken to peak at high porosity and the dissipation coefficient Γ_d regulates the magnitude of the dissipation. In the modified version of the P-α model a critical value Γ_{dC} exists for each configuration such that below it the model predicts *normal behaviour* for the porous material independently of the initial porosity (top right panel in the figure) and above it *anomalous behaviour* is predicted for the porous material independently of the initial porosity (bottom right panel in the figure). For modified ε-α models only *normal behaviour* can be reproduced and the dissipated energy has the only quantitative effect of moving the state of full compaction toward higher pressure.

Conclusions. The correction of the inconsistency resolves the singularity problem and the inclusion of dissipated energy allows the description of both normal and anomalous behaviour. Standard P-α like models can be extended to the high porosity range, however the expected features of pre-planetesimals can be better described by the P-α **modified model** because the ε-α model is restricted only to normal behaviour.

Acknowledgements. SEA acknowledges funding by the German Science Foundation DFG under grant SP 646/1-1.

References

Bolkhovitinov, L. G. & Khvostov, Y. B., 1978, *Nature*, 274, 882
Herrmann, W. 1969, *Journal of Applied Physics*, 40, 2490
Carroll, M. & Holt, A. C., 1972, *Journal of Applied Physics*, 43, 759
Jutzi, M., Benz, W., & Michel, P., 2008, *Icarus*, 198, 242
Wünnemann, K., Collins, G. S., & Melosh, H. J., 2006, *Icarus*, 180, 514

The Astrophysics of Planetary Systems: Formation, Structure, and Dynamical Evolution
Proceedings IAU Symposium No. 276, 2010
A. Sozzetti, M. G. Lattanzi & A. P. Boss, eds.

doi:10.1017/S1743921311020503

Stellar companions to exoplanet host stars with Astralux

Carolina Bergfors[1], Wolfgang Brandner[1], Thomas Henning[1] and Sebastian Daemgen[2]

[1] Max-Planck-Institute for Astronomy
Königstuhl 17, 69117 Heidelberg, Germany
email: bergfors@mpia.de

[2] European Southern Observatory
Karl-Schwarzschild Strasse 2, 85748 Garching, Germany

Abstract. A close stellar companion influences the formation of planets in the system. The occurrence of stellar companions and characteristics of the stars and planets in the system provide constraints on the formation processes. We present results from our high-resolution Lucky Imaging survey for binary exoplanet host stars, including the discovery of stellar companion candidates to the transiting planet hosts WASP-12 and HAT-P-8.

Keywords. techniques: high angular resolution, planetary systems, binaries: visual

---

1. Introduction

Today we know of more than 40 planets that belong to a star in a binary or multiple system. While widely separated stellar companions are not expected to affect the formation of planets, the observed binary separation is less than $\sim$ 100 AU for around 20% of the known systems. The presence of a close stellar companion affects planet formation (see, e.g., Nelson 2000; Boss 2006; Kley & Nelson 2008), and the discovery and characterisation of systems like these may provide a way to discriminate between core accretion and disk fragmentation as the dominant formation process.

Transiting exoplanets are in this context especially valuable, since fundamental stellar and planetary properties such as mass, radius and mean density can be derived from the photometric transit in combination with radial velocity measurements.

2. The AstraLux binary exoplanet hosts survey

Resolving close binary systems with separations less than $\sim$ 100 AU in general requires adaptive optics or other high-resolution methods like Lucky Imaging. With the two AstraLux Lucky Imaging instruments at the 2.2 m telescope at Calar Alto (Hormuth *et al.* 2009) and at NTT, La Silla (Hippler *et al.* 2009), we can observe stellar companions to exoplanet host stars at a minimum angular separation of $\sim$ 0.1 arcsec. This corresponds to projected separations of less than 100 AU for $\approx$ 80% of all exoplanets discovered by radial velocity or transit observations today.

So far, $\approx$ 180 exoplanet hosts have been observed in our survey, $\sim$ 30 of which host transiting planets (see Bergfors *et al.*, in prep). Among the observed transiting planet hosts, we find close stellar companion candidates to WASP-12 and HAT-P-8 (Fig. 1). The new observations also include the eastern companion candidate to HAT-P-7 (see Narita *et al.* 2010) and second epoch observations of the companions to WASP-2, TrES-2 and TrES-4 from Daemgen *et al.* (2009).

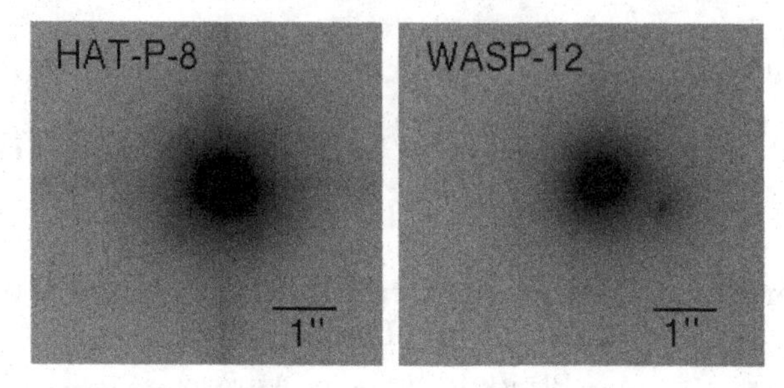

Figure 1. AstraLux Norte z'-band images of the transiting planet hosts HAT-P-8 and WASP-12 and the companion candidates. The images are shown in a logarithmic intensity scale. North is up and east to the left.

Table 1. Properties of planet host star and companion candidate.

Transit host	SpT(A)	SpT(B)	$\Delta z'$	$\Delta i'$	ρ ['']
HAT-P-8	F8 V	M3-M4 V	6.68 ± 0.07	7.34 ± 0.10	1.027 ± 0.011
WASP-12	G0 V	K4-M1 V	3.79 ± 0.10	4.03 ± 0.07	1.047 ± 0.021

3. Binary separation and Safronov number

Hansen & Barman (2007) suggested that hot jupiters could be divided into two classes based on their Safronov number ($\Theta = 0.5 * (v_{esc}/v_{orb})^2$) and equilibrium temperature. Only a small number of transiting planets were known at the time (19), and the significance of the division has been discussed (Fressin, Guillot & Nesta 2009, Southworth 2010). Daemgen *et al.* (2009) found for a somewhat larger sample of 35 transiting exoplanets that the observed gap between groups at $\Theta \sim 0.05$ was still present, and saw a possible correlation between Θ and stellar separation in binary exoplanet hosts where widely separated binaries fell into Class II and closer ones into Class I.

The number of transiting exoplanets is rapidly increasing as a result of new discoveries in large surveys such as e.g. HATNet, SuperWASP, CoRoT and Kepler. There are now more than 100 transiting exoplanets known, and we can investigate the suggested relation for a much larger sample. Not only has the gap between the two classes suggested by Hansen & Barman (2007) become much more narrow, we do no longer see a clear correlation between Θ and binary separation. How stellar companions influence giant planet formation and if it is related to the Safronov number needs to be further investigated.

References

Boss, A. P. 2006, *ApJ*, 641, 1148
Daemgen, S., *et al.* 2009, *A&A*, 498, 567
Fressin, F., Guillot, T., & Nesta, L. 2009, *A&A*, 504, 605
Hansen, B. M. S. & Barman, T. 2007, *ApJ*, 671, 861
Hippler, S., *et al.* 2009, *Msngr*, 137, 14
Hormuth, F., *et al.* 2009, *in AIP Conf. Ser.*, 1094, 935
Kley, W. & Nelson, R. P. 2008, *A&A*, 486, 617
Narita, N., *et al.* 2010, *PASJ*, 62, 779
Nelson, A. F. 2000, *ApJ*, 537, L65
Southworth, J. 2010, *MNRAS*, 408, 1689

The Astrophysics of Planetary Systems: Formation, Structure, and Dynamical Evolution
Proceedings IAU Symposium No. 276, 2010
A. Sozzetti, M. G. Lattanzi & A. P. Boss, eds.

doi:10.1017/S1743921311020515

Disentangling stellar activity and planetary signals

Isabelle Boisse[1,2], François Bouchy[1,3], Guillaume Hébrard[1,3], Xavier Bonfils[4,5], Nuno Santos[2] and Sylvie Vauclair[6]

[1]Institut d'Astrophysique de Paris, Université Pierre et Marie Curie, UMR7095 CNRS, 98bis bd. Arago, 75014 Paris, France, email: iboisse@iap.fr
[2]Centro de Astrofísica, Universidade do Porto, Rua das Estrelas, 4150-762 Porto, Portugal
[3]Observatoire de Haute Provence, CNRS/OAMP, 04870 St Michel l'Observatoire, France
[4]Laboratoire d'Astrophysique de Grenoble, Observatoire de Grenoble, Université Joseph Fourier, CNRS, UMR 5571, 38041, Grenoble Cedex 09, France
[5]Obs. de Genève, Université de Genève, 51 Ch. des Maillettes, 1290 Sauverny, Switzerland
[6]LATT-UMR 5572, CNRS & Université P. Sabatier, 14 Av. E. Belin, 31400 Toulouse, France

Abstract. High-precision radial-velocimetry (RV) is until now the more efficient way to discover planetary systems. Moreover, photometric transit search missions like CoRoT and Kepler, need spectroscopic RV measurements to establish the planetary nature of a transit candidate and to measure the true mass. An active star has on its photosphere dark spots and bright plages rotating with the star. These inhomogeneities of the stellar surface can induce a variation of the measurement of the RV, due to changes in lines shapes and not to a Doppler motion of the star (e.g. Queloz *et al.* 2001; Desort *et al.* 2007; Boisse *et al.* 2009). We study how the Keplerian fit used to search for planets in RV data is confused by spots and we test an approach to subtract RV jitter based on harmonic decomposition of the star rotation. We use simulations of spectroscopic measurements of rotating spotted stars and validate our approach on active stars monitored by high-precision spectrograph HARPS: CoRoT-7 and ι Hor.

Keywords. techniques: radial velocities, planetary systems, stars: activity, stars: individual (CoRoT-7, ι Hor)

---

1. Dark spot simulations

SOAP is a tool that calculates the photometric, RV and line shape modulations induced by one (or more) cool spots on a rotating stellar surface. The RV modulations due to a spot as a function of time for different inclinations i of the star with the line of sight and different spot latitudes lat. These two parameters clearly modify the pattern of the RV modulation. Main peaks are clearly detected in the Lomb-Scargle periodograms at the rotational period of the star P_{rot}, as well as the two-first harmonics $P_{rot}/2$ and $P_{rot}/3$. The periods detected in the periodograms are the same for: 1) stars with different inclinations, 2) spots at different latitudes, 3) spot size and/or temperature varying with time, 4) several spots on the stellar surface.

The purpose is to remove or at least to reduce the stellar activity signals in order to identify a planetary signal hidden in the RV jitter. The Lomb-Scargle periodogram corresponds to sinusoidal decompositions of the data. Three sinusoids with periods fxed at the rotational period P_{rot}, and its two-first harmonics $P_{rot}/2$ and $P_{rot}/3$ reduce the semi- amplitude of the RV jitter by more than 87%.

2. Application to real data

CoRoT-7. Queloz *et al.* (2009) (Q09) reported the intensive campaign carried out with HARPS. RV variations are dominated by the activity of the star. Using our harmonic

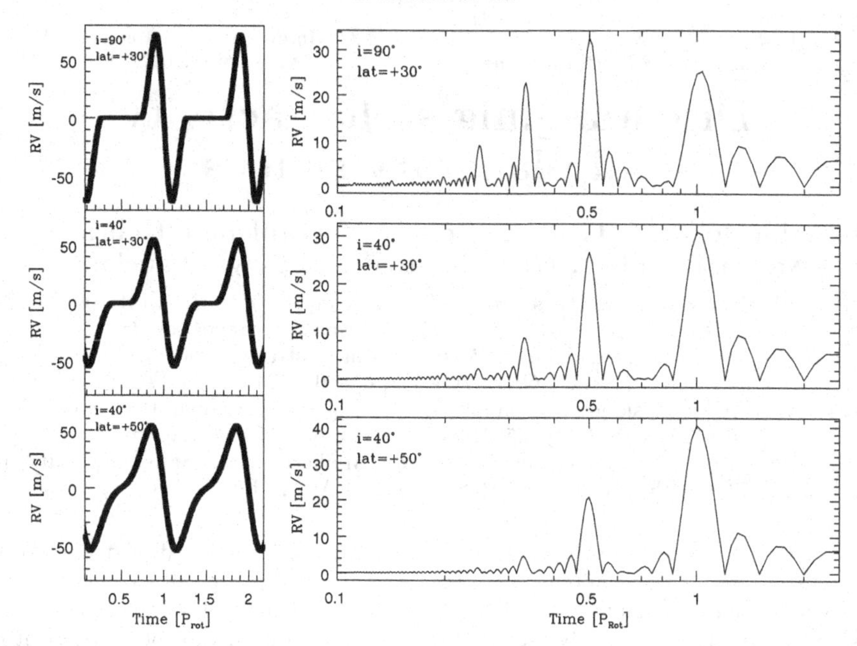

Figure 1. *Left:* RV modulations due to one spot as a function of time (expressed in rotational period unit). At t=0, the dark spot of 1% of the visible stellar surface is in front of the line of sight. The shape of the signal changes with the inclination i of the star and the latitude lat of the spot, labelled in the top left of each panel. *Right:* Lomb-Scargle periodograms of the three RV modulations showed at the left. The fundamental frequency, P_{rot}, and its first harmonics are detected.

filter, we detected in these data the two planets. Then, we fitted simultaneously the stellar activity and planetary system, instead of the method used by Q09 and obtain new values for the masses. We estimate approximately that a systematic noise due to active jitter of 1.5 ms^{-1} must be added quadratically to the error bars. We then find for the masses 5.7±2.5 M_{Earth} for CoRoT-7b agreeing with the value of Q09 and 13.2±4.1 M_{Earth}, slightly higher than the published value, for CoRoT-7c.

ι Hor. Iota Hor is a young G0V star with a long-period exoplanet. We studied 8 nights of HARPS data (Vauclair *et al.* 2008) showing an explicit RV scatter of about 15 m/s probably due to stellar activity. We want to characterize the active jitter and to search for a possible hidden Doppler motion. Based on simulation studies, we fitted 3 sinusoids with period fixed at P_{rot}, $P_{rot}/2$ and $P_{rot}/3$.

We do not detect in the data a short-period companion. Nevertheless, we would like to know if we have subtracted the RV shift due to a companion subtracting the effect of activity. We ran simulations adding RV due to fake planets to the data. We excluded the presence of planet with minimum mass between 6 and 10 M_{Earth} with periods respectively between 0.7 and 2.4 days.

References

Boisse, I., Moutou, C., Vidal-Madjar, A., *et al.* 2009, *A&A*, 495, 959
Desort, M., Lagrange, A.-M., Galland, F., *et al.* 2007, *A&A*, 473, 983
Queloz, D., Henry, G. W., Sivan, J. P., *et al.* 2001, *A&A*, 379, 279
Queloz, D., Bouchy, F., Moutou, C., Hatzes, A., Hébrard, G., *et al.* 2009, *A&A*, 506, 303
Vauclair, S., Laymand, M., Bouchy, F., *et al.* 2008, *A&A*, 428, 5

The Astrophysics of Planetary Systems: Formation, Structure, and Dynamical Evolution
Proceedings IAU Symposium No. 276, 2010
A. Sozzetti, M. G. Lattanzi, & A. P. Boss, eds.

doi:10.1017/S1743921311020527

On the possibility of enrichment and differentiation in gas giants during birth by disk instability

Aaron C. Boley[1] and Richard H. Durisen[2]

[1]Department of Astronomy; University of Florida, 211 Bryant Space Science Center, Gainesville, FL 32611, USA
email:aaron.boley@gmail.com

[2]Department of Astronomy, Indiana University, 727 East 3rd Street, Swain West 319, Bloomington, IN 47405, USA

Abstract. We investigate the coupling between solids and gas during the formation of gas giant planets by disk fragmentation in the outer regions of massive disks. We find that fragments can become differentiated at birth. Even if an entire clump does not survive, differentiation could create solids cores that survive to accrete gaseous envelopes later.

Keywords. planetary systems: protoplanetary disks, planetary systems: formation

---

A gravitationally unstable disk can be very efficient at concentrating solids in spiral arms (Rice *et al.* 2004, 2006). If fragmentation does occur, clumps form from small sections of these arms (Durisen *et al.* 2008; Boley *et al.* 2010), exactly where solids are aerodynamically captured. These circumstances can lead to enrichment and differentiation in fragments at birth. To explore this mechanism, we have run a number of hydrodynamics simulations, exploring different initial conditions, opacities, and particle sizes. In this summary, we highlight the differentiation results from Boley & Durisen (2010), focusing on the simulation called SIM2, which explores 10 cm and 1 km-size particles separately. The disk is the same as SIMD in Boley (2009), which surrounds a 1 $M_\odot$ star and has a disk mass that reaches 0.33 $M_\odot$ through envelope accretion. The re-simulation is begun about 1/2 orbit, measured at the fragmentation radius, before clump formation. Particles are distributed to give a solid-to-gas ratio of approximately 0.01 everywhere.

Rock-size (10 cm) particles show a high degree of concentration in the spiral arms prior to fragmentation (Fig. 1). However, the clump also forms from surrounding solid-depleted gas, diluting the overall enrichment to about 1.4 over the nebular value. As the clump evolves and grows, this ratio is roughly maintained as the clump feeds through solid-enhanced spiral wakes. Although there is only modest enrichment overall, the clump experiences a high degree of differentiation immediately after birth (see Table 1). In contrast, the simulation with only km-size solids shows an overall depletion of heavy elements in the clump because the solids cannot dissipate kinetic energy and contract with the gas. The degree of enrichment and differentiation following fragmentation will depend on the size distribution of solids, which can lead to a variety of outcomes and potential core sizes.

Some clumps may survive to form planets or brown dwarfs directly (Cameron 1978; Boss 1998), while others will be destroyed through, e.g., tidal disruption (Boley *et al.* 2010). In both cases, large cores can form. If the solid core is separated from its nascent clump, it may still undergo a subsequent phase of rapid gas accretion. Because the core is formed rapidly in a clump, we call this scenario *core assist plus gas capture*. For further results and discussions, see Boley & Durisen (2010).

Table 1. Gas and rock mass for the clump that forms in SIM2 with 10 cm and km-size solids. Three different density thresholds are shown, where only mass above the threshold is considered. *Full* represents the entire mass of the clump, here for all gas that has $\rho > 9 \times 10^{-13}$ g cm^{-3} and for $T > 34$ K. *Half* refers to all mass that is above half of the peak density in the clump, $\rho_{\rm max}$, and $\gtrsim 90\%$ is for only mass that is greater than 90% of the peak. This last threshold isolates the core conditions of the clump. The factor $f_{\rm RtoG}$ gives the ratio of rock mass to gas mass, and $f_{\rm enr} \equiv (f_{\rm RtoG} + 0.01)/0.02$ gives the total solids enrichment relative to the average nebula's value. When we write "rocks," we are referring to a mixture of silicates and ices.

SIM2 10cm	Time = 1450 yr	$r \sim 104$ AU	$\rho_{\rm max} = 1.7 \times 10^{-11}$ g cm^{-2}	
Density Threshold	Gas (M_J)	Rocks ($M_\oplus$)	$f_{\rm RtoG}$	$f_{\rm enr}$
Full	8.1	42	0.016	1.3
Half	3.5	38	0.034	2.2
$\gtrsim 90\%$	0.85	32	0.12	6.4
SIM2 km	Time = 1430 yr	$r \sim 106$ AU	$\rho_{\rm max} = 1.3 \times 10^{-11}$ g cm^{-2}	
Full	7.1	7.2	0.0032	0.66
Half	3.7	2.6	0.0022	0.61
$\gtrsim 90\%$	1.3	0.67	0.0016	0.58

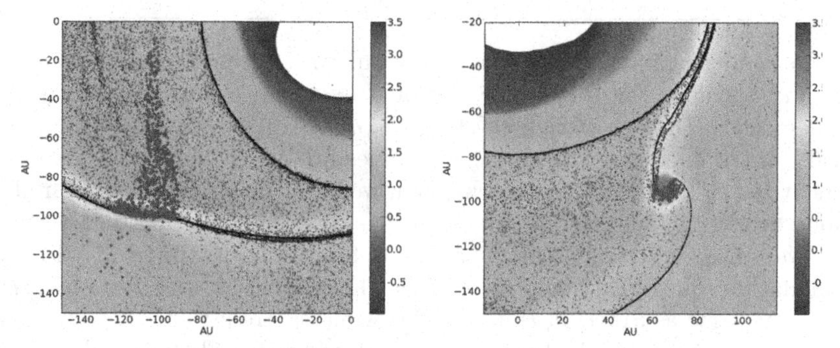

Figure 1. The spiral arm in SIM2 just before fragmentation at 960 yr (left) and after fragmentation at the 1300 yr snapshot (right). Gas density is shown by the colorbar, while solids are given by black dots. The rocks that are considered to be part of the clump at 1300 yr are shown with large, blue crosses. The same particles are shown in the 960 yr snapshot. Most of the mass comes from a 20-30 AU section of the spiral arm, consistent with Durisen *et al.* (2008) and Boley *et al.* (2010). The solids are highly concentrated near the peak gas density, which is slightly offset from the geometric center in projection.

References

Boley, A. C. 2009, *ApJ*, 695, 53L

Boley, A. C., & Durisen, R. H. 2010, *ApJ*, 724, 618

Boley, A. C., Hayfield, T., Mayer, L., & Durisen, R. H. 2010, *Icarus*, 207, 509

Boss, A. P. 1998, *ApJ*, 503, 923

Cameron, A. G. W. 1978, *Moons and the Planets*, 18, 5

Durisen, R. H., Hartquist, T. W., & Pickett, M. K. 2008, *Ap&SS*, 317, 3D

Rice, W. K. M., Lodato, G., Pringle, J. E., Armitage, P. J., & Bonnell, I. A. 2004, *MNRAS*, 355, 543

–. 2006, *MNRAS*, 372, L9

The Astrophysics of Planetary Systems: Formation, Structure, and Dynamical Evolution
Proceedings IAU Symposium No. 276, 2010
A. Sozzetti, M. G. Lattanzi & A. P. Boss, eds.

doi:10.1017/S1743921311020539

Planet candidates from the SARG visual binary survey

Elena Carolo[1], Silvano Desidera[1], Raffaele Gratton[1], Aldo Martinez Fiorenzano[2], Michael Endl[3], Rosario Cosentino[4], Mauro Barbieri[1], Mariangela Bonavita[5], Massimo Cecconi[4], Riccardo Claudi[1], Francesco Marzari[6] and Salvo Scuderi[4]

[1]INAF - Astronomical Observatory of Padova, Italy
email: elena.carolo@unipd.it

[2]INAF - Fundacion Galileo Galilei Santa Cruz de La Palma, Spain

[3]McDonald Observatory, The University of Texas at Austin, Austin, USA

[4]INAF - Astronomical Observatory of Catania, Italy

[5]University of Toronto, Canada

[6]Physics Department, University of Padova, Italy

Abstract. We present preliminary results of the radial velocity survey on going at TNG targeting binaries with similar components.

Keywords. binaries: general, planetary systems

1. Introduction

Determination of the fraction of stars hosting planetary systems of various characteristics is a major tool to test models of planet formation. A large fraction of stars are in binary systems; determination of incidence of planetary system in binary stars is then very interesting.
The SARG radial velocity (RV) survey is devoted to find planets around individual components of wide binary systems with a typical separation of 200 AU and formed by similar main sequence stars (ΔV<1mag). The sample was chosen by Hipparcos catalog (d<100pc). This survey started in September 2000 (Desidera *et al.* 2007) using the Galileo High Resolution Spectrograph (SARG) (Gratton *et al.* 2001) at Telescopio Nazionale Galileo (TNG) and follow up of planets candidates is on-going.

2. Radial velocity analysis

In the last year, some improvements have been implemented in the analysis procedure, e.g. the new version of the AUSTRAL code by Endl (Endl *et al.* 2000). For a better treatment of the shape of the Instrument Profile (IP) the model is set up on a sub-pixel grid. The analytical form of the instrumental profile must be chosen with care. The IP is modeled using a combination of Gaussians. The asymmetries are reproduced using a Gaussian centered at the origin and up to 12 Gaussian satellites placed in the wings. These smaller features are added or subtracted from the main one in order to recreate the asymmetries of the profile. Since the IP varies along the spectrum, this is subdivided into smaller spectral chunks, each of them modeled independently. The modeling process is a multi-parameter χ^2-optimization algorithm.

The modeling of stellar spectra to measure the Doppler shift is similar to the first step, but in this case the stellar spectrum is also present. Even in this case the IP is reconstructed until the model of best fit for each chunk is achieved. The Doppler shift between the iodine and the model provided by the template among the parameters determined for each chunk. The final RV is the weighted mean, and its error is the standard deviation of the mean of the values of each chunk. Tests performed showed that the best number of Gaussians used to fit the IPs is closely related to the S/N of the spectrum beeing analyzed. The final RV precision is about 2-3 m/s for bright stars and 3-10 m/s for the program star (V = 8-10).

3. Abundance analysis and stellar activity

Beside the RV determination, an high precision differential abundance analysis was done on the binary components of the survey, with error of Δ[Fe/H] about 0.02dex (Desidera *et al.* 2004; Desidera *et al.* 2006). For warm stars, with thin convective zone, limits are similar to the quantitiy of meteoritic material accreted by the Sun during its MS lifetime.
Contamination and stellar activity effects have been studied on a sample of the survey through the analysis of line profiles (Martinez Fiorenzano *et al.* 2005).

4. Candidates

The statistical analysis shows us no planets with period shorter then the survey duration and RV semi-amplitude larger than 30 m/s have been found.
We presented a couple of low-amplitude planet candidates: one with probably a Jupiter mass planet in 5yr orbit and a system with two planets with RV amplitude $<$15 m/s.

5. Work in progress

Preliminary statistical analysis shows a trend of lower frequency of planets for tighter binaries and the upper limits simulations indicates for lower frequency of planets in the survey sample then single stars (Bonavita & Desidera 2007).
We are improving a statistical analysis by using bootstrap technique and an orbital parameters grid, taking into account simultaneously the RMS, χ^2 and the Generalised Lomb-Scargle periodogram (Zechmeister & Kürster 2009).

References

Bonavita, M. & Desidera, S. 2007, *A&A*, 468, 721
Desidera, S., Gratton, R. G., Scuderi, S., *et al.* 2004, *A&A*, 420, 683
Desidera, S., Gratton, R. G., Scuderi, S., *et al.* 2006, *A&A*, 454, 581
Desidera, S., Gratton, R. G., Endl, M., *et al.* 2007, *arXiv*, 0705.3141
Endl, M., Kürster, M., & Els, S. 2000, *A&A*, 362, 585
Gratton, R. G., Bonanno, G., Bruno, P., *et al.* 2001, *Experimental Astron.*, 12, 107
Martinez Fiorenzano, A. F., Gratton, R. G., Desidera, S., *et al.* 2005, *A&A*, 442, 775
Zechmeister, M. & Kürster, M. 2009, *A&A*, 496, 577

The Astrophysics of Planetary Systems: Formation, Structure, and Dynamical Evolution
Proceedings IAU Symposium No. 276, 2010
A. Sozzetti, M. G. Lattanzi & A. P. Boss, eds.

doi:10.1017/S1743921311020540

Influence of growth on dust settling and migration in protoplanetary discs

Elisabeth Crespe[1], Jean-Francois Gonzalez[1], Guillaume Laibe[2], Sarah T. Maddison[3] and Laure Fouchet[4]

[1] Université de Lyon, Lyon, F-69003, France; Université Lyon 1, Villeurbanne, F-69622, France; CNRS, UMR 5574, Centre de Recherche Astrophysique de Lyon; École normale supérieure de Lyon, 46, allée d'Italie, F-69364 Lyon cedex 07, France.
email: [elisabeth.crespe;jean-francois.gonzalez]@ens-lyon.fr

[2] Centre for Stellar and Planetary Astrophysics, School of Mathematical Sciences, Monash University, Clayton Vic 3168, Australia.

[3] Centre for Astrophysics and Supercomputing, Swinburne University, PO Box 218, Hawthorn, VIC 3122, Australia.

[4] Physikalisches Institut, Universität Bern, CH-3012 Bern, Switzerland.

Abstract. To form meter-sized pre-planetesimals in protoplanetary discs, dust aggregates have to decouple from the gas at a distance far enough from the central star so they are not accreted. Dust grains are affected by gas drag, which results in a vertical settling towards the mid-plane, followed by radial migration. To have a better understanding of the influence of growth on the dust dynamics, we use a simple grain growth model to determine the dust distribution in observed discs. We implement a constant growth rate into a gas+dust hydrodynamics SPH code and vary the growh rate to study the resulting effect on dust distribution. The growth rate allows us to determine the relative importance between friction and growth.We show that depending on the growth rate, a range of dust distribution can result. For large enough growth rates, grains can decouple from the gas before being accreted onto the central star, thus contributing as planetary building rocks.

Keywords. planetary systems: protoplanetary disks, hydrodynamics, methods: numerical

1. Hydrodynamics SPH Code

We use a 3D, two-fluid (gas+dust) Smoothed Particle Hydrodynamics (SPH) code. The disc has a mass of $0.02M_\odot$ and is made of gas (99%) and dust (1%, ρ_d = 1g.cm^{-3}). It extends from 20AU to 400AU surrounding a $1M_\odot$ star and is vertically isothermal. The initial temperature and density profiles scale as $T \propto r^{-3/4}$ and $\Sigma \propto r^{-3/2}$. The initial state contains 200,000 gas particles in a near-equilibrium disc that relaxes to a stationary state in approximately 8,000 years. Once the disc is relaxed, we add dust particles (size $s_0 = 10\mu$m) superimposed on the gas particles. The disc evolves over 10^5 years.

2. The dust growth model

We implement a constant growth rate model for dust such that $S = S_0 + \gamma\mathcal{T}$, where S represent the grain size, S_0 its initial size, γ is the growth rate and $\mathcal{T}$ represents the time (where capital quantities are dimensionless). The dimensionless growth rate can also be defined as the growth to friction rate:

$$\gamma = \frac{\mathrm{d}S}{\mathrm{d}\mathcal{T}} = \frac{\mathrm{d}(s/s_{\mathrm{opt}})}{\mathrm{d}(t/t_k)} = \frac{\mathrm{d}s/\mathrm{d}t}{s_{\mathrm{opt}}/t_k}, \text{ therefore } \begin{cases} \text{if } \gamma \gg 1, & \text{growth dominates,} \\ \text{if } \gamma \ll 1, & \text{friction dominates,} \end{cases}$$

where s_{opt} is an optimal size for migration and settling processes and t_k is the orbital time.

3. Results

Figure 1 shows the results obtained at the end of the simulation for different values of γ.

• Small values of γ: grains are highly coupled to the turbulent gas and are distributed over the entire disc. The growth is not sufficient to change the regime of evolution. Grains migrate slowly toward the inner disc and can be accreted onto the central star.

• Intermediate values of γ: grains grow slowly as they settle to the mid-plane where they experience a radial drift toward the inner part of the disc. As the growth is moderately efficient, grains can reach larger sizes. Their regime of evolution changes and they can evolve like large grains, decouple from the gas and move on keplerian orbits.

• Large values of γ: growth is very efficient and grains do not have time to feel the gas, decoupling in less than one orbit. They experience little settling and radial drift and are distributed in the whole disc, as they were initially.

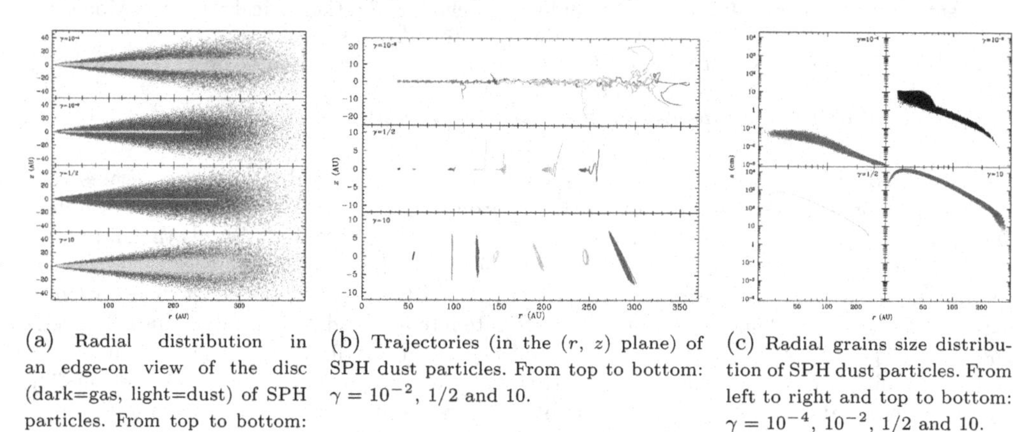

(a) Radial distribution in an edge-on view of the disc (dark=gas, light=dust) of SPH particles. From top to bottom: $\gamma = 10^{-4}$, 10^{-2}, 1/2 and 10.

(b) Trajectories (in the (r, z) plane) of SPH dust particles. From top to bottom: $\gamma = 10^{-2}$, 1/2 and 10.

(c) Radial grains size distribution of SPH dust particles. From left to right and top to bottom: $\gamma = 10^{-4}$, 10^{-2}, 1/2 and 10.

Figure 1. Results of SPH simulations after 10^5 years.

4. Conclusion

This study gives insight on grain size distributions in protoplanetary discs. The results obtained with $\gamma \approx 0.01$ are in good adequation with our study (Laibe *et al.* 2008) of grain growth using the model of Stepinski & Valageas (1997). We conclude that the constant growth rate model gives an approximation of the physical growth process and underlines the fact that we can observe various dust distribution depending on the constant growth rate. SPH simulations validate the analytical results and direct numerical integration performed in Laibe *et al.* (2010a) for non-growing grains and generalized for growing grains in Laibe *et al.* (2010b). *Acknowledgements: This research was partially supported by the Programme National de Physique Stellaire and the Programme National de Planétologie of CNRS/INSU, France, and the Agence Nationale de la Recherche (ANR) of France through contract ANR-07-BLAN-0221. STM acknowledges the support of a Swinburne Special Studies Program.*

References

Barrière-Fouchet, L., Gonzalez, J.-F., Murray, J. R., Humble, R. J., & Maddison, S. T. 2005, *A&A*, 443, 185

Laibe, G., Gonzalez, J.-F., Fouchet, L., & Maddison, S. T. 2008, *A&A*, 487, 265

Laibe, G., Gonzalez, J.-F., & Maddison, S. T. 2010a, *A&A*, under revision after referee's report

Laibe, G., Gonzalez, J.-F., Maddison, S. T., Crespe, E., & Fouchet, L. 2010b, *A&A*, submitted

Stepinski, T. F., & Valageas, P., 1997, *A&A*, 319, 1007

The Astrophysics of Planetary Systems: Formation, Structure, and Dynamical Evolution
Proceedings IAU Symposium No. 276, 2010
A. Sozzetti, M. G. Lattanzi & A. P. Boss, eds.

doi:10.1017/S1743921311020552

3D global simulations of proto-planetary disk with dynamically evolving outer edge of dead zone

Natalia Dzyurkevich[1], Neal J. Turner[2], Willy Kley[3], Hubert Klahr[1] & Thomas Henning[1]

[1]Max-Planck Institute for Astronomy
Königstuhl 17, Heidelberg, D-69117
email: natalia@mpia.de

[2]Jet Propulsion Laboratory, California 91109, USA

[3]University of Tübingen, Auf der Morgenstelle 10, Tübingen, D-72076

Abstract. 3D global MHD simulations of magneto-driven turbulence are performed for the disk of 100 AU with reduced amount of 10μm fluffy dust grains. We use X-ray and cosmic ray ionization, as well as simplified treatment of recombination on dust grains. The ionization of gas and charging of dust grains are dynamically evolving during the simulation, making the zone of high magnetic dissipation ('dead' zone) variable. In our simulations, the jump in MRI-driven turbulent viscosity inside and outside of dead zone is insignificant. We find no hard edge, but rather a smooth transition between active and dead zone. Subsequently, there is no visible pressure bump at outer edge of the dead zone.

Keywords. accretion disks, turbulence, MHD

1. Introduction

Magneto-rotational instability (MRI) is necessary to sustain the observed mass accretion rates in the proto-planetary disks and could govern viscous disk evolution. Due to presence of dust, large part of disk is poorly ionized, leading to a 'dead' zone where MRI turbulence is suppressed. 'Dead' or laminar zone can help making planets in many ways. At the radial transition from active to dead zone we could expect a jump in turbulent stresses, what will lead over time to a density bump and therefore to a trapping of the planetesimals (Johansen *et al.* (2009) and references therein). Usual assumption is that the transition from active to dead zone is one pressure scale height broad. Typically the inverse Elsässer number $\Lambda = 1$ separates dead from active zone, assuming that the fastest growing mode of MRI is dissipated within one orbit. It is known that ideal MRI turbulence is reached only for lower dissipation then $\Lambda > 10$. For dissipations of $0.1 < \Lambda < 10$ the gas is in 'transitional' state, showing still a weak turbulence. If the jump in turbulent viscosity is sharp enough or space separation for $0.1 < \Lambda < 10$ is narrow then trapping of planetesimals is possible. Viscous instability of dead zone may appear when ionization of the gas is treated dynamically. Here, the dead zone may get unstable and broken into rings, each of those could serve as a trap for planets.

2. Model

3D global MHD simulations of magneto-driven turbulence are performed for dynamically ionized disk with reduced amount ($f_{\rm dg} = 0.001$) of fluffy dust grains of 10μm. MHD equations are same as in Dzyurkevich *et al.* (2010). We solve the equations of non-ideal MHD using 3D ZeusMP code. The disk domain is from 5 to 95 AU and includes 8.4 pressure scale heights. Resolution on spherical grid is [256:128:128] for $[r, \Theta, \phi]$. Gas is

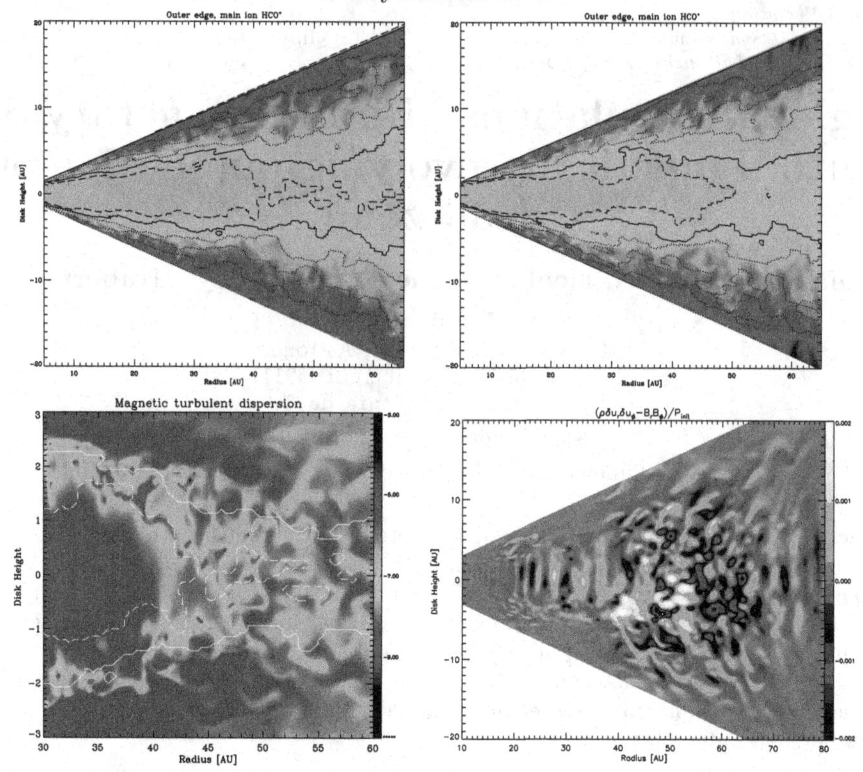

Figure 1. Top: Colors show inverse Elsässer number, decades are notated as solid line for 1, dotted for 10 to 10^3, dashed for 0.1. Location of 'dead' zone edge fluctuates between 40 and 55 AU for time 390 and 450 orbits. Bottom left: Logarithmic turbulent magnetic fields, with white line repeating decades in inverse Elsässer Number. Bottom right: Snap-shop of turbulent stress.

locally isothermal with T(R) constant on cylinders. Gas density is $\rho \propto (r, \Theta)$. Dead zone appears naturally, as we calculate magnetic diffusivity after Okuzumi (2009) and Wardle (2007) for fluffy dust, where $\eta(r, \Theta)$ is space and time dependent and is updated every time-step. Initial magnetic field is a purely azimuthal field, leading to azimuthal MRI.

3. Implications

MRI-free zone is defined within $\Lambda = 0.1$ (not unity) and its edge location can fluctuate over several AU within an orbital time. We find no evidence for viscous instability of dead zone. 'Dead' region with $\Lambda < 0.1$ has significant Reynolds stress, the pillar structures in r, Θ snap-shot of stress are spiral density waves. At midplane, contrast in total stress between 'dead' $\Lambda < 0.1$ and 'MRI-active' $\Lambda > 0.1$ locations is insignificant. - We observe no density bump at $\Lambda = 0.1$ at the length of simulation.

Space separation between dead zone ($\Lambda = 0.1$) and ideal MRI turbulence ($\Lambda = 10$) is stretched over several AU, for given weak magnetic field. Therefore we conclude, that even a longterm formation of a pressure bump would be too shallow to trap dust grains and planetesimals.

References

Dzyurkevich, N., Flock, M., Turner, N. J., Klahr, H., & Henning, Th. 2010, *A&A*, 515, A70
Johansen, A., Youdin, A., & Klahr, H. 2009, *ApJ*, 697, 1269
Okuzumi, S. 2009, *ApJ*, 698, 1122
Wardle, M. 2007, *Ap&SS*, 311, 35

The Astrophysics of Planetary Systems: Formation, Structure, and Dynamical Evolution
Proceedings IAU Symposium No. 276, 2010
A. Sozzetti, M. G. Lattanzi & A. P. Boss, eds.

doi:10.1017/S1743921311020564

Probing the impact of stellar duplicity on the frequency of giant planets: Final results of our VLT/NACO survey

Anne Eggenberger[1], Stéphane Udry[2], Gaël Chauvin[1], Thierry Forveille[1], Jean-Luc Beuzit[1], Anne-Marie Lagrange[1] and Michel Mayor[2]

[1] Université Joseph Fourier – Grenoble 1 / CNRS, Laboratoire d'Astrophysique de Grenoble (UMR 5571), BP 53, 38041 Grenoble Cedex 9, France
email: `Anne.Eggenberger@obs.ujf-grenoble.fr`, `Gael.Chauvin@obs.ujf-grenoble.fr`, `Thierry.Forveille@obs.ujf-grenoble.fr`, `Jean-Luc.Beuzit@obs.ujf-grenoble.fr`, `Anne-Marie.Lagrange@obs.ujf-grenoble.fr`

[2] Observatoire de Genève, Université de Genève, 51 ch. des Maillettes, 1290 Sauverny, Switzerland
email: `Stephane.Udry@unige.ch`, `Michel.Mayor@unige.ch`

Abstract. If it is commonly agreed that the presence of a (moderately) close stellar companion affects the formation and the dynamical evolution of giant planets, the frequency of giant planets residing in binary systems separated by less than 100 AU is unknown. To address this issue, we have conducted with VLT/NACO a systematic adaptive optics search for moderately close stellar companions to 130 nearby solar-type stars. According to the data from Doppler surveys, half of our targets host at least one planetary companion, while the other half show no evidence for short-period giant planets. We present here the final results of our survey, which include a new series of second-epoch measurements to test for common proper motion. The new observations confirm the physical association of two companion candidates and prove the unbound status of many others. These results strengthen our former conclusion that circumstellar giant planets are slightly less frequent in binaries with mean semimajor axes between 35 and 100 AU than in wider systems or around single stars.

Keywords. planetary systems: formation, binaries: visual, techniques: high angular resolution

1. The NACO survey

To probe the impact of stellar duplicity on the frequency of giant planets, we have conducted with VLT/NACO an adaptive optics search for stellar companions to ~60 planet-host stars and to ~70 non-planet-host stars (hereafter control stars). This survey revealed 95 companion candidates near 33 targets (Fig. 1). Using two-epoch astrometry we identified 19 true companions, 2 likely bound objects, and 34 background stars (Eggenberger *et al.* 2007). The companionship of the remaining 40 companion candidates could not be constrained due to the lack of a second-epoch astrometric measurement. Assuming that all but two of these 40 objects were unbound, we showed that giant planets seem slightly less frequent in ~35-100 AU binaries than around single stars (Eggenberger *et al.* 2008).

2. New observations

We recently performed the second-epoch measurements that were missing previously. The new observations show that the companion candidates we detected near the

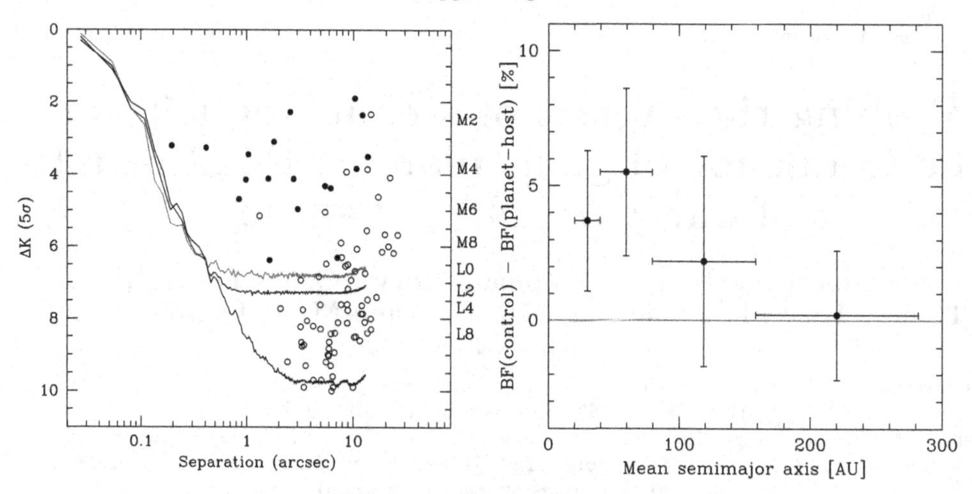

Figure 1. Results of our survey. *Left:* Detections and detection limits. Dots are true companions, circles are unbound objects. The uppermost curve is the detection limit used for the statistical analysis. The two lower curves are median detection limits obtained with the old and with the present detectors of NACO. *Right:* Difference between the binary fraction of control stars and the binary fraction of planet-host stars as a function of binary mean semimajor axis. Vertical error bars are 68% bootstrap confidence intervals. Horizontal error bars represent the bin width.

planet-host stars HD 76700, HD 83443, HD 162020, and HD 330075 are all unrelated background stars. On the other hand, the new data confirm the physical association of the companion candidates to the control stars HD 82241 and HD 134180.

3. Statistical analysis

Figure 1 (right) shows the difference in binary fraction between the control and the planet-host subsamples. According to the updated statistical analysis, the difference in binary fraction is $13.2 \pm 5.1\%$ for semimajor axis below 100 AU, and -1.5 ± 2.9 for semimajor axis between 100 and 200 AU. The positive difference seen for mean semimajor axis $\lesssim$100 AU suggests that giant and intermediate-mass planets are slightly less frequent in moderately close binaries than in wider systems or around single stars. If confirmed with a larger sample, this result would support the idea that the presence of a moderately close stellar companion affects the formation of giant planets, but does not completely stop the process.

Acknowledgements

AE acknowledges support from the Swiss National Science Foundation through a fellowship for advanced researchers.

References

Eggenberger, A., Udry, S., Chauvin, G., Beuzit, J. L., Lagrange, A. M., & Mayor, M. 2008, *ASP-CS*, 398, 179

Eggenberger, A., Udry, S., Chauvin, G., Beuzit, J.-L., Lagrange, A.-M., Ségransan, D., & Mayor, M. 2007, *A&A*, 474, 273

The Astrophysics of Planetary Systems: Formation, Structure, and Dynamical Evolution
Proceedings IAU Symposium No. 276, 2010
A. Sozzetti, M. G. Lattanzi & A. P. Boss., eds.

doi:10.1017/S1743921311020576

Global aspects of the formation of γ Cephei b

Siegfried Eggl[1], Markus Gyergyovits[1] and Elke Pilat-Lohinger[1]
[1]Institute for Astronomy, University of Vienna, 1180 Vienna, Austria
email: siegfried.eggl@univie.ac.at

Abstract. Discoveries of extrasolar planets in tight binaries are of great scientific value since these systems can be used to gain new insights in planetary development processes. Gamma Cephei, one of the most thoroughly investigated double star systems is hosting a Jovian planet at a distance of about 2 AU from its primary, a 1.4 solar-mass K1 III-IV star (Neuhäuser *et al.* 2007; Torres 2007). We comprise aspects of dynamical stability, disc heating processes and different giant planet (GP) formation scenarios in order to gain a better understanding of the open questions that remain in explaining the formation of gamma Cephei b.

Keywords. planetary systems: formation, binaries: general, planetary systems: protoplanetary disks

1. Introduction

A very attractive system from a dynamical point of view, gamma Cep has been a focus of scientific interest ever since the first announcement of the existence of a planet in the system (Hatzes *et al.* 2003). The topics were centered on possible additional planets (Dvorak *et al.* 2003; Haghighipour 2006) as well as different formation scenarios (Thebault *et al.* 2004; Kley & Nelson 2008; Xie & Zhou 2009). Dynamical studies show, that the binaries' highly eccentric orbit ($e \approx 0.4$) and the relatively small separation of the stellar components ($a \approx 20\ AU$) restrict the stable area around the primary gamma Cep A to about 4 AU. When constant gas drag is included in N-Body simulations, Thebault *et al.* (2004) find that a core accretion (CA) model (Pollack *et al.* 1996) is capable of producing a GP of comparable mass in required timescales, but well inside its observed orbit. Besides CA, GP formation through gravitational instability (GI) (Boss 2001) is one of the most widely accepted theories. We are interested whether GI can be considered a viable formation scenario for the GP gamma Cep b.

2. Methods

Apart from surface-density, a protoplanetary disc's temperature is one of the most important parameters in planetary formation. It is vital in all current models, preventing or facilitating GI induced collapses (Kratter *et al.* 2010), as well as granting an effective density boost outside the so called 'Snow Line' (Kennedy & Kenyon 2008). Typically two major heat sources are taken into account, the radiative influence of the star as well as viscous dissipation of gravitational potential energy within the accretion disc itself. In the case of gamma Cephei the cyclic pumping caused by the secondary constitutes another main source of disc heating. In order to gain reliable estimates on the temperature development in the heavily perturbed circumprimary disc, hydrodynamic simulations are essential. We use a modified version of the grid code described in Theis & Orlova (2004) including energy transfer and FARGO transport (Masset 2000). The disc's surface density was chosen, so that the existing GP should be able to form at a distance of 2 AU with respect to the available mass within its feeding zone (Lissauer 1987). This resulted in a density $\rho_0 = 2 \cdot 10^{-9}\ g/cm^3$ equal to the value used in Thebault *et al.* (2004). We also estimated the mass requirements of a CA based planetary core (10 M_{earth}) to form from coagulated dust in the orbital plane which produced a density about three to five times as high featuring a dust to gas ratio of 10^{-2}. The simulation was started with an initial density decay law of $r^{-3/4}$, as well as standard values for viscosity and temperature profiles. In order to separate internal disc heating from stellar heating processes we did not include stellar radiation in our hydrodynamic simulations. Instead we used analytical temperature estimates on the stellar influence for passive flat (Safronov 1972) as well as flared discs (Bell 1999) without photon reprocessing, adapted to the early gamma Cep system. It is important to consider the stars' evolutionary tracks in this respect, putting the primary close to a spectral type of F1.

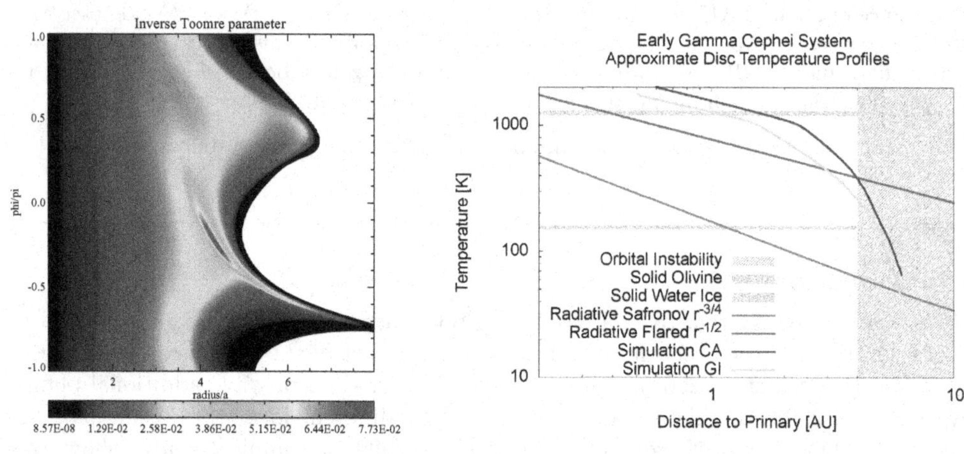

Figure 1. *left:* Snapshot of the inverse Toomre parameter after 100 binary revolutions. Values below 10^{-8} are depicted as white. *right:* Azimuthally averaged temperature profiles of the disc featuring analytical estimates on stellar radiative heating of a passive, flat, optically thick disc, a flared disc, and disc profiles resulting form hydrodynamic simulations without stellar radiative heating for different initial densities. Evaporation temperatures are taken from Pollack *et al.* (1994).

3. Results

We calculated the azimuthally averaged disc temperatures (Figure 1, *right*), the inverse Toomre parameter (Figure 1, *left*) and checked the cooling criterion (Kratter *et al.* 2010) for two non self gravitating disc models around gamma Cep A throughout 100 binary revolutions. After the initial relaxation phase the inverse Toomre parameter did not exceed unity, even-though the cooling criterion was fulfilled in vast regions of the disc, meaning that a GI induced collapse is not possible in this setup. The temperature profiles of the disc are also unfavorable for CA models since the snow line will not be reached inside the dynamically stable region. Our next steps will be inclusion of self-gravity in our calculations, as well as the study of the influence of initial parameters on these results, in order to approach answers to the open questions on the formation of gamma Cep b.

References

Bell, K. R. 1999, *ApJ*, 526, 411
Boss, A. P. 2001, *ApJ*, 563, 367
Dvorak, R., Pilat-Lohinger, E., Funk, B., & Freistetter, F. 2003, *A&AL*, 398, 1

Haghighipour N. 2006, *ApJ*, 644, 543
Hatzes, A. P., Cochran, W. D., Endl, M., McArthur, B., Paulson, D. B., Walker, G. A. H.., Campbell, B., & Yang, S. 2003, *ApJ*, 599, 1383
Kennedy, G. M. & Kenyon, S. J. 2008, *ApJ*,673, 502
Kley, W. & Nelson, R. P. 2008, *A&A*, 486, 617
Kratter, K. M., Murray-Clay, R. A., & Youdin, A. N. 2010, *ApJ*, 710, 1375
Lissauer, J. J. 1987, *Icarus*, 69, 249
Masset, F. 2000, *A&AS*, 141, 165
Neuhäuser, R., Mugrauer, M., Fukagawa, M., Torres, G., & Schmidt, T. 2007, *A&A*, 462, 777
Pollack, J. B., Hollenbach, D., Beckwith, S., Simonelli, D. P., Roush, T., & Fong, W. 1994, *ApJ*, 421, 615
Pollack, J. B., Hubickyj, O., Bodenheimer, P., Lissauer, J. J., Podolak, M., & Greenzweig, Y. 1996, *Icarus*, 124, 62
Safronov, V. S. 1972, *Academy of Sciences USSR, IPST*
Thebault P., Marzari, F., Scholl, H., Turrini, D., & Barbieri, M. 2004, *A&A*, 427, 1097
Theis, Ch. & Orlova, N. 2004, *A&A*, 418, 959
Torres, G. 2007, *ApJ*, 654, 1095
Xie, J. & Zhou, J. 2009, *ApJ*, 698, 2066

The Astrophysics of Planetary Systems: Formation, Structure, and Dynamical Evolution
Proceedings IAU Symposium No. 276, 2010
A. Sozzetti, M. G. Lattanzi & A. P. Boss, eds.

doi:10.1017/S1743921311020588

How common are Earth-Moon planetary systems?

Sebastian Elser[1], Joachim Stadel[1], Ben Moore[1] and Ryuji Morishima[2]

[1]Institute for Theoretical Physics, University of Zurich,
Winterthurerstrasse 190, 8057 Zurich, Switzerland
email: selser@physik.uzh.ch, stadel@physik.uzh.ch, moore@physik.uzh.ch

[2]LASP, University of Colorado,
Colorado 80303-7814, USA
email: ryuji.morishima@lasp.colorado.edu

Abstract. The Earth's comparatively massive moon, formed via a giant impact on the proto-Earth, has played an important role in the development of life on our planet. Here we study how frequently Earth-Moon planetary systems occur. We derive limits on the collision parameters that may guarantee the formation of a circumplanetary disk after a protoplanet collision that could form a satellite. Based on a large set of simulations, we observe potential moon forming impacts and conclude that giant impacts with the required energy and orbital parameters for producing a binary planetary system occur frequently with more than one in ten terrestrial planets hosting a massive moon.

Keywords. planets and satellites: formation

Our main purpose is to explore the giant impact history of terrestrial planets in order to calculate the probability of having a large Moon-like satellite companion. A giant impact between a Mars-size proto-planet and the proto-Earth is the accepted model for the origin of our Moon (e.g. Hartmann & Davis 1975; Cameron & Benz 1991). After its formation, the Moon was much closer and the Earth was rotating more rapidly. The large initial tidal forces created high tidal waves several times per day, possibly promoting the cyclic replication of early biomolecules (Lathe 2004) and profoundly affecting the early evolution of life. Today, the Moon stabilizes the spin axis of our planet (Laskar *et al.* 1993). A stable spin axis and therefore a stable climate on timescales of more than a billion years may be essential to guarantee a suitable environment for, in particular, land-based life.

A large set of N-body simulations, where Earth-like planets in the habitable zone form, provides the background of our study (Morishima *et al.* 2010). This simulations take into account gravitational accretion and hydro-dynamical processes in the planet formation process. The collision parameter space that describes a giant impact is given by the ratio of impactor mass to total mass in the collision, by the impact velocity, by the impact parameter and the total angular momentum. If one does not focus on a strongly constrained system like the Earth-Moon system but just on terrestrial planets of arbitrary mass with satellites that tend to stabilize their spin axis, the parameter space is broad. It becomes difficult to draw strict limits on the parameters because collision simulations for a wider range of impacts were not available for our study. Hence, starting from published Moon-forming SPH simulations by Canup (2008), we derive scaling relations to estimate rough limits on a parameter space that guarantees a circumplanetary debris disk that can form a satellite outside the Roche limit, including obliquity stabilization and tidal evolution.

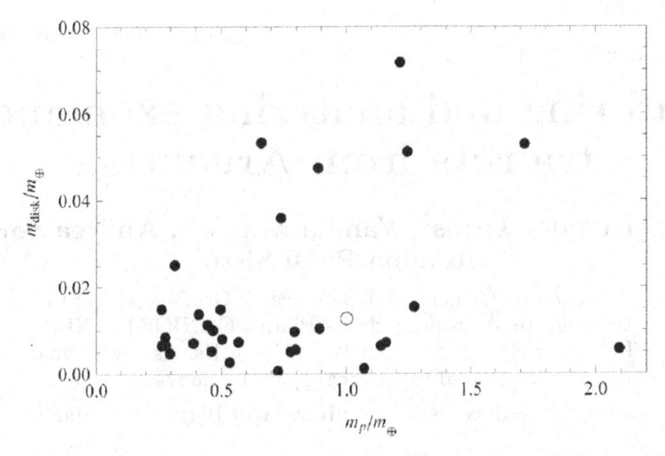

Figure 1. The masses of the final outcomes of the planets for which we identified satellite forming collisions. m_{disk} is the mass of the proto-planetary disk, which is an upper limit on the satellite mass, and m_p is the mass of the planet after the complete core accretion. The circle indicates the position of the Earth-Moon system with the assumption $m_{\text{disk}} = m_{\text{Moon}}$. Since we exclude collisions of the small initial planetesimals in the simulation from consideration for satellite forming events, there are only few small disks.

Under rather restrictive conditions, we identify 31 moon forming events in 64 simulations, the masses of the resulting planet-satellite systems are shown in figure 1. On average, every simulation gives three terrestrial planets with different masses and orbital characteristics. Hence, at least one of six planets has an obliquity stabilizing satellite in its orbit. If we focus on Earth-Moon like systems, where we have a massive planet with a final mass larger than half of an Earth mass, we identify 19 moon forming collisions. Therefore, one in ten planets has a satellite, for which five out of six of those satellites are larger than half a Lunar mass. The mass of a satellite in our study is clearly overestimated since we equate it with the mass of the circumplanetary disk mass, but it shows that there are not only a few small satellites.

Life on planets without a massive stabilizing moon would face sudden and drastic changes in climate, posing a survival challenge that has not existed for life on Earth. Our simulations show that planets with massive moons occur quite frequently.

References

Cameron, A. G. W. & Benz, W. 1991, *Icarus*, 92, 204
Canup, M. R. 2008, *Icarus*, 196, 518
Hartmann, W. K. & Davis, D. R. 1975, *Icarus*, 24, 504
Laskar, J., Joutel, F., & Robutel, P. 1993, *Nature*, 361, 615
Lathe, R. 2004, *Icarus*, 168, 18
Morishima, R., Stadel, J., & Moore, B. 2010, *Icarus*. 207, 517

The Astrophysics of Planetary Systems: Formation, Structure, and Dynamical Evolution
Proceedings IAU Symposium No. 276, 2010
A. Sozzetti, M. G. Lattanzi & A. P. Boss, eds.

doi:10.1017/S174392131102059X

Monitoring and analyzing exoplanetary transits from Argentina

Eduardo Fernández-Lajús[1], Yamila Miguel[1], Andrea Fortier[2] and Romina P. Di Sisto[1]

[1]Facultad de Ciencias Astronómicas y Geofísicas - UNLP,
Instituto de Astrofísica de La Plata - CONICET/UNLP,
Paseo del Bosque S/N, La Plata, Pcia. Bs. As., Agentina
email: eflajus@fcaglp.unlp.edu.ar

[2] Physikalisches Institut, University of Bern, Switzerland

Abstract. Photometric observations of transits can be used to derive physical and orbital parameters of the system, like the planetary and stellar radius, orbital inclination and mean density of the star. Furthermore, monitoring possible periodic variations in transit timing of planets is important, since small changes can be caused by the presence of other planets or moons in the system. On the other hand, long term changes in the transit length can be due to the orbital precession of the planets. For these reasons we started an observational program dedicated to observe transits of known exoplanets with the aim of contributing to a better understanding of these planetary systems. In this work we present our first results obtained using the observational facilities in Argentina including the 2.15 telescope at CASLEO.

Keywords. planetary systems, stars: variables: other, techniques: photometric

1. Introduction

If a planetary System is oriented in space in such a way that the orbital plane is near the observer's visual, the planet will pass periodically in front of the stellar disk producing what is called a transit. Photometric observations of transits can be used to derive physical and orbital parameters of the system, like the planetary and stellar radius, orbital inclination and mean density of the star. Also it is very important to monitoring possible periodic changes in the O-C plots of planets since small changes of the transit moment can be caused by the presence of other planets or moons in the system. On the other hand, long term changes in the transit length can be due to the orbital precesion of the planets. The parameters of a system with transits should be estimated confronting the observations with numerical methods that model the light curves of transits.

Since the discovery of the first exoplanet, known as 51 Peg b by Mayor and Queloz (1995), the number of planets discovered in other stars has grown rapidly. Today we have $\sim$ 500 exoplanets detected of which $\sim$ 100 are transiting planets. However, these numbers change really very quickly and the new observations present constantly new paradigms to be explained by theories of formation of planetary Systems.

In order to contribute to the knowledge and better understanding of the great diversity of these exoplanetary systems, we started an observational program dedicated to observe transits of known exoplanets. We present here our fist results obtained using the observational facilities in Argentina including the 2.15 telescope at CASLEO.

2. Observations

Image acquisition: Digital CCD images were acquired with the 2.15-m "Jorge Sahade" (JS), $f/8.4$ Cassegrain, telescope at Complejo Astronómico El Leoncito, Argentina. We

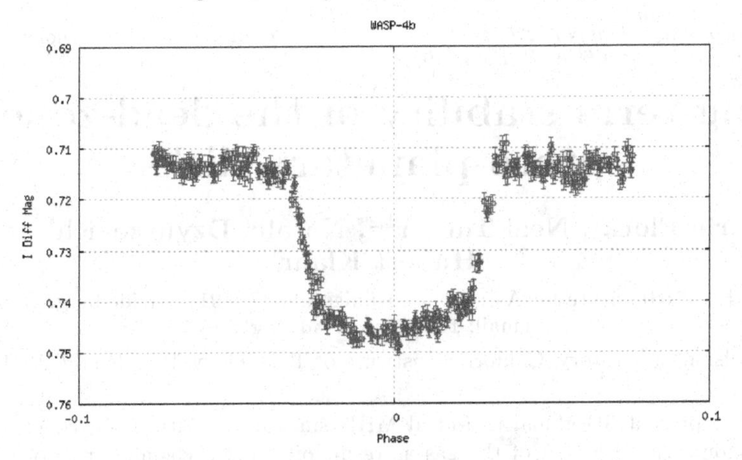

Figure 1. Ligth Curve of WASP-4b.

had two observing runs, with a total of nine usefull nigths during March, 2009 and August, 2010. We used a focal reducer to image a wider field. Image acquisition was performed using a CCD camera attached to the JS telescope, with the broad band filters of the Johnson-Cousins system. The camera is a Versarray 1300B Roper Scientific manufactured by Princeton Instruments. The chip array is 1340 × 1300 pixels (20μm square pixel), the scale being 0.677 $''/px$ (with the focal reducer), with a circular usefull field of $9^{'}$. Bias, dark and flat-field frames are acquired every night to calibrate the science images.

Data reduction: Instrumental magnitudes of each star are determined by means of aperture photometry. An image processing pipeline was written in order to practice real-time aperture photometry to the stars contained in the frames. This pipeline was developed making usage of the IRAF command language and some tasks as those of the DAOPHOT photometry package. The aperture radius was selected for each image series in order to maximize the signal-to-noise ratio. Differential magnitudes of the targets are then obtained using an appropriate comparison star in the field and testing the results with some other check stars. Magnitude errors are calculated considering the poisson noise of the source and background, as well as the detector noise and scintillation noise.

3. Results and conclusions

During both observing runs we could observe a ten transits either partial or complete. In Fig. 1 we show the ligth curve resulting from our I differential photometry of a planetary transit of Wasp-4b as an example of our detections. Some of our results were already published in the exoplanet transit database (http://var2.astro.cz/ETD/). Our first objective was to test the capability of the telescope and site to detect exoplanetary transits, both the limiting magnitude and the transit depth. We observed targets up to $V \sim 14$ mag detecting transit depths up to ~ 0.01 mag. In the coming months and years, we will continue to observe exoplanetary transits in order to obtain parameters of the systems, to detect other perturbing companion planets via the technique of transit timing variations and contribute to the study of exoplanets. We plan to continue using the JS telescope as well as some other facilities in Argentina.

References

Mayor, M. & Queloz, D. 1995, *Nature*, 378, 355

The Astrophysics of Planetary Systems: Formation, Structure, and Dynamical Evolution
Proceedings IAU Symposium No. 276, 2010
A. Sozzetti, M. G. Lattanzi & A. P. Boss, eds.

doi:10.1017/S1743921311020606

Long-term stability of the dead-zone in proto-planetary disks

Mario Flock[1], Neal Turner[1,2], Natalia Dzyurkevich[1] and Hubert Klahr[1]

[1]Max Planck Institute for Astronomy, Königstuhl 17, 69117 Heidelberg, Germany
email: flock@mpia-hd.mpg.de

[2]Jet Propulsion Laboratory, California Institute of Technology, Pasadena, CA 91109, USA

Abstract. We present 3D global non-ideal MHD simulations with a self consistent dynamic evolution of ionization fraction of the gas as result of reduced chemical network. We include X-ray ionization from the star as well as cosmic ray ionization. Based on local gas density and temperature in our chemical network, we determine the magnetic resistivity, which is fed back in MHD simulations. Parameters for dust size and abundance are chosen to have accreting layers and a laminar "dead" mid-plane.

Keywords. accretion disks, MHD, turbulence, planetary systems: formation, instabilities

1. Introduction

Magneto-rotational instability (MRI) is the most prominent process to drive turbulence and angular momentum transport in protoplanetary disks (Balbus and Hawley 1998). To enable MRI to operate in disk, the ionization of gas has to be sufficiently high. Various studies showed that at a certain level of resistivity MRI will be suppressed, mostly due to the small sized dust grains which absorb free electrons (Turner 2010 and references therein). Most studies for MRI in stratified accretion disk rely on local box simulations as well as static ionization profiles of the disk. Until now, no global simulation with a dynamical resistivity profile was calculated. In our work we investigate how far MRI can change the resistivity profile by turbulent motion itself.

2. Model

Numerical setup follows the second order PLUTO configuration in Flock *et al.* (2010) with the HLLD Riemann solver. We use a locally isothermal global disk model with H/R = 0.05; r:3.6 - 6.6 (128 grid cells); θ : ± 5 scale heights (160) and ϕ : 45° (128). The hydrodynamical model follows Dzyurkevich *et al.* (2010). We use a weak vertical net field as seed for MRI. In the dead zone region this field is still visible after 500 local orbits (Fig. 1, right, vectors). For the chemical model (Turner *et al.* 2010), we have chosen dust size and abundance that the dead zone has an extent of ± 2.5 scale heights at the beginning.

3. Result

The upper layers of the disk above 2 scale heights reach after 10 local orbits a quasi steady state. In Fig. 1, right, above 2.5 scale height there are a large scale turbulent structures in all 3 magnetic field components. The dark-violet line shows the Elsässer number (Λ_{el}) of unity around 2.5 scale heights. Radial average of this value stays at 2.5

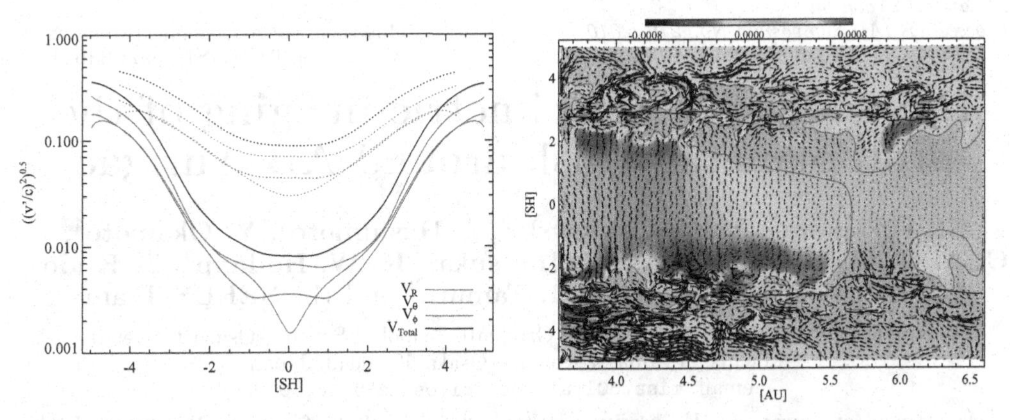

Figure 1. Left: Turbulent RMS velocities over height. Dotted lines present the turbulent RMS velocity for a full ionized and MRI turbulent disk. Solid line present the RMS velocity for our dead-zone run. Right: $r - \theta$ snapshot of the magnetic fields after 500 local orbits. The azimuthal component is plotted as contour color with red lines showing change of sign. Over-plotted are the respective radial and theta component as vector field. The dark-violet line present the Elsässer number of unity.

scale heights, even after 1000 local orbits. In addition we observe a quasi active dead zone region around 2 scale heights but with $\Lambda_{el} < 1$. Here, there is an accumulation of azimuthal magnetic field, antisymmetric for both hemispheres, as well as transport of material outward the dead zone. The midplane of dead zone actually shows a Mach number of 0.01 (Fig. 1, left). In addition we integrate the total mass flux through the vertical boundary at 5 scale height as well as the total radial mass flux at 4 and 6 AU. Even at 5 scale heights the vertical mass transport dominates the radial one by a factor of 3 (integrated until 1000 local orbits).

4. Summary

We do not see separation of the dead zone into rings in our dynamical model (i.e. no viscous instability of dead zone). The dead zone thickness is stable for over thousand of local orbits. In our simulation, evacuation timescale of dead zone is set by the vertical mass outflow. Total integrated mass which escapes the simulation at 5 scale heights is 3 times larger than total radial transport. This value is measured after 1000 orbits of evolution. Turbulent velocities at the midplane in dead zone are around 1% of local sound speed, which is one order of magnitude below the value for ideal MHD case.

References

Balbus, S. A. & Hawley, J. F. 1998, *Reviews of Modern Physics*, 70, 1
Dzyurkevich, N., Flock, M., Turner, N. J., Klahr, H., & Henning, T. 2010, *A&A*, 515, A70
Flock, M., Dzyurkevich, N., Klahr, H., & Mignone, A. 2010, *A&A*, 516, A26
Turner, N. J., Carballido, A., & Sano, T. N. 2010, *ApJ*, 708, 188

The Astrophysics of Planetary Systems: Formation, Structure, and Dynamical Evolution
Proceedings IAU Symposium No. 276, 2010
A. Sozzetti, M. G. Lattanzi & A. P. Boss, eds.

doi:10.1017/S1743921311020618

High-contrast polarimetric imaging of the protoplanetary disk around AB Aurigae

M. Fukagawa[1], J. P. Wisniewski[2], J. Hashimoto[3], Y. Okamoto[4], C. A. Grady[5,6], T. Muto[7], S.-I. Inutsuka[8], K. W. Hodapp[9], T. Kudo[3], M. Momose[4], H. Shibai[1], M. Tamura[3] and the SEEDS Team

[1]Department of Earth and Space Science, Graduate School of Science, Osaka University, 1-1 Machikaneyama, Toyonaka, Osaka 560-0043, Japan
email: misato@iral.ess.sci.osaka-u.ac.jp

[2]Department of Astronomy, University of Washington, Box 351580 Seattle, WA 98195, USA

[3]National Astronomical Observatory of Japan, 2-21-1 Osawa, Mitaka, Tokyo 181-8588, Japan

[4]Institute of Astrophysics and Planetary Sciences, College of Science, Ibaraki University, 2-1-1 Bunkyo, Mito, Ibaraki 310-8512, Japan

[5]Eureka Scientific, 2452 Delmer, Suite 100, Oakland, CA 96002, USA

[6]NASA Goddard Space Flight Center, Code 667, Greenbelt, MD 20771, USA

[7]Department of Earth and Planetary Sciences, Tokyo Institute of Technology, 2-12-1 Oh-okayama, Meguro-ku, Tokyo 152-8551, Japan

[8]Department of Physics, Nagoya University, Furo-cho, Chikusa-ku, Nagoya 464-8602, Japan

[9]Institute for Astronomy, University of Hawaii, 640 N. Aohoku Place, Hilo, HI 96720, USA

Abstract. We present the spatially-resolved polarization measurements for the disk around the Herbig Ae star, AB Aurigae. The images were obtained in J, H, and Ks bands with the coronagraphic camera HiCIAO on the Subaru Telescope. The inner region beyond 30 AU from the star was imaged, which reveals an azimuthal dip, a radial gap at around 80 AU, and complex spiral-like emission in polarized light.

Keywords. planetary systems: protoplanetary disks, planetary systems: formation, techniques: high angular resolution, techniques: polarimetric

1. Protoplanetary Disk around AB Aur

Direct imaging of disks provides valuable information on the distribution of disk material that may be linked to the presence of unseen planets. Imaging has been challenging due to small sizes of disks and high contrast ratios relative to the central stars. However, resolving the structure on ~10 AU or less spatial scales is now possible in near-infrared by employing adaptive optics even from the ground. Moreover, one can effectively overcome the contrast problem by utilizing imaging polarimetry which is a powerful technique to extract scattered light from the disk by suppressing the bright, unpolarized starlight. Polarimetry also allows us to better constrain the disk geometry and to explore the properties of dust such as grain size and composition.

AB Aur is one of the most well-studied young stellar objects, located in the Tau-Aur star-forming region. The star is classified as a Herbig Ae star, and its age is estimated to be ~3 Myr. The previous imaging studies have shown that AB Aur is surrounded by a circumstellar disk with its radius of several hundreds of AU and an outer envelope (Mannings & Sargent 1997; Grady *et al.* 1999; Fukagawa *et al.* 2004). One remarkable feature of the disk is its morphology: the trailing spiral arms have been found in the outer region ($r > 100$ AU). Recently, Oppenheimer *et al.* (2008) and Perrin *et al.* (2009) reported the imaging polarimetry for the inner part ($r > 40$ AU) using the extreme adaptive optics and the *HST*.

2. Imaging Polarimetry with Subaru/HiCIAO

We observed AB Aur with Subaru/HiCIAO in the polarization differential imaging (PDI) mode with adaptive optics (AO188) in October 2009, as part of the ongoing high-contrast imaging survey of exoplanets and disks (SEEDS) (Tamura 2009). The images were obtained in J, H, and Ks bands with the coronagraphic mask of 0.3 arcsec in diameter. The spatial resolution achieved was close to the diffraction limit, 8 AU.

The inner disk ($r \gtrsim 30$ AU) was successfully detected in the polarized light in all the three bands (e.g., Hashimoto *et al. submitted*). The images reveal quite complex nebulosity including an azimuthal depletion at a position angle of ~330° and a radius of ~100 AU. The azimuthal dip was also found in the previous polarimetry (Oppenheimer *et al.* 2008; Perrin *et al.* 2009), and it was suggested that the dip could be attributed to an unseen planet at that location or simply a geometric scattering effect for a smooth inclined disk. Our images are still not conclusive on the existence of planets, but the disk at 100 AU is not uniform at all in azimuthal direction, as an arm-like emission is recognized in the northeast. In addition, the HiCIAO images show a radial gap at ~80 AU clearly seen in the northern direction as well as the inner emission near the mask edge. Note that the location of the outer ring outside of the gap is consistent with the disk wall inferred from the mid-infrared and submillimeter thermal imaging (Honda *et al.* 2010). The observed fine and irregular structure may favor planets in the disk.

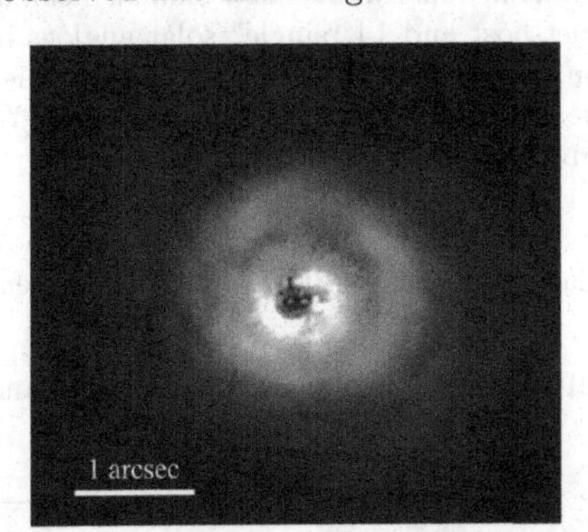

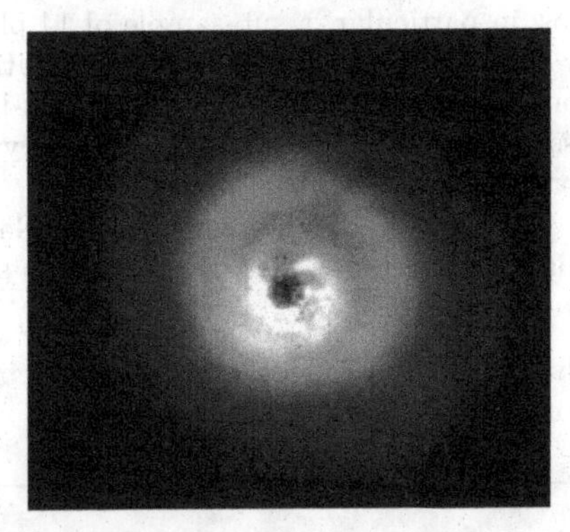

Figure 1. Polarized intensity images for AB Aur in H (*left*) and Ks (*right*) bands. North is up, east is to left.

References

Grady, C. A., Woodgate, B., Bruhweiler, F. C., Boggess, A., Plait, P., Lindler, D. J., Clampin, M., & Kalas, P. 1999, *ApJ*, 523, L151

Fukagawa, M., *et al.* 2004, *ApJ*, 605, L53

Honda, M., *et al.* 2010, *ApJ*, 718, L199

Mannings, V. & Sargent, A. I. 1997, *ApJ*, 490, 792

Oppenheimer, B. R., *et al.* 2008, *ApJ*, 679, 1574

Perrin, M. D., Schneider, G., Duchene, G., Pinte, C., Grady, C. A., Wisniewski, J. P., & Hines, D. C. 2009, *ApJ*, 707, L132

Tamura, M. 2009, *American Institute of Physics Conference Proceedings*, 1158, 11

The Astrophysics of Planetary Systems: Formation, Structure, and Dynamical Evolution
Proceedings IAU Symposium No. 276, 2010
A. Sozzetti, M. G. Lattanzi & A. P. Boss, eds.

doi:10.1017/S174392131102062X

Volatiles and refratories in solar analogs: No terrestial planet connection

Jonay I. González Hernández[1,2], Garik Israelian[1], Nuno C. Santos[3,4], Sergio Sousa[3], Elisa Delgado-Mena[1], Vasco Neves[3] and Stéphane Udry[5]

[1]Instituto de Astrofísica de Canarias, C/ Via Láctea s/n, 38200 La Laguna, Spain
email: jonay@iac.es

[2]Dpto. de Astrofísica y Ciencias de la Atmósfera, Facultad de Ciencias Físicas, Universidad Complutense de Madrid, E-28040 Madrid, Spain

[3]Centro de Astrofísica, Universidade do Porto, Rua das Estrelas, 4150-762 Porto, Portugal

[4]Departamento de Física e Astronomia, Faculdade de Ciências, Universidade do Porto, Portugal

[5]Observatoire Astronomique de l'Université de Genève, 51 Ch. des Maillettes, -Sauverny-Ch1290, Versoix, Switzerland[1]

Abstract. We have analysed very high-quality HARPS and UVES spectra of 95 solar analogs, 24 hosting planets and 71 without detected planets, to search for any possible signature of terrestial planets in the chemical abundances of volatile and refractory elements with respect to the solar abundances.

We demonstrate that stars with and without planets in this sample show similar mean abundance ratios, in particular, a sub-sample of 14 planet-host and 14 "single" solar analogs in the metallicity range $0.14 < [Fe/H] < 0.36$. In addition, two of the planetary systems in this sub-sample, containing each of them a super-Earth-like planet with masses in the range $\sim 7 - 11$ Earth masses, have different volatile-to-refratory abundance ratios to what would be expected from the presence of a terrestial planets.

Finally, we check that after removing the Galactic chemical evolution effects any possible difference in mean abundances, with respect to solar values, of refratory and volatile elements practically dissappears.

Keywords. stars: abundances, stars: fundamental parameters, planetary systems, planetary systems: formation, stars: atmospheres

---

1. Introduction

The discovery of more than 400 exoplanets orbiting solar-type stars by the radial velocity technique have provided a substantial amount of high-quality spectroscopic data (see e.g. Neves *et al.* 2009).

Recently, Meléndez *et al.* (2009) have obtained a clear trend [X/Fe] versus T_C in a sample of 11 solar twins, and claimed (see also Ramírez *et al.* 2009, 2010) that the most likely explanation to this abundance pattern is related to the presence of terrestial planets in the solar planetary system.

Here we summarize the analysis of very high-quality HARPS and UVES spectroscopic data of a sample of 95 solar analogs with and without planets (see González Hernández *et al.* 2010), with a resolving power of $\lambda/\delta\lambda \gtrsim 85{,}000$ and a mean $\langle S/N \rangle \sim 850$.

The stellar parameters and metallicities of the whole sample of stars were computed using the method described in Sousa *et al.* (2008). The chemical abundance derived for each spectral line was computed using the LTE code MOOG (Sneden 1973), and a grid of Kurucz ATLAS9 model atmospheres (Kurucz 1993).

2. Metal-rich solar analogs hosting super-Earth-like planets

We find no substantial differences in the abundance patterns of solar analogs with and without planets. In particular, the slopes of the abundance ratios [X/Fe] versus T_C in two metal-rich stars, HD 1461 and HD 160691, containing each of them one super-Earth-like planet, with 7-11 Earth masses, have the opposite sign to what one would expect if the amount of refractory metals in the atmospheres of planet hosts would depend only on the amount of terrestial planets.

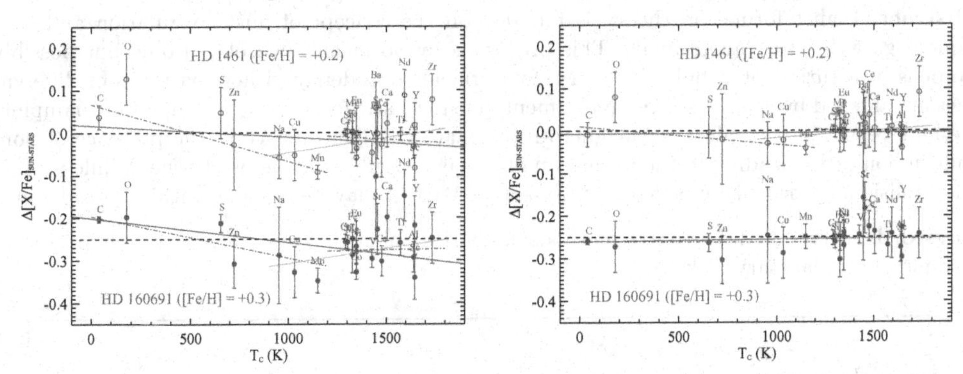

Figure 1. *Left panel*: Abundance differences, $\Delta[X/Fe]_{SUN-STARS}$, between the Sun, and 2 planet hosts with super-Earth-like planets. Linear fits for different T_C ranges to the data points weighted with the error bars are also displayed. We note the different slopes derived when choosing the range $T_C > 1200$ K (dashed-dotted line) as in Meléndez *et al.* (2009) and González Hernández *et al.* (2010), and $T_C > 900$ K (dashed-three-dotted line) as in Ramírez *et al.* (2009, 2010). An arbitrary shift of -0.25 dex has been applied to the abundances of the planet host HD 160691. *Right panel*: Same as left panel of this figure but after correcting each element abundance ratio of each star using a linear fit to the Galactic chemical trend of the corresponding element at the metallicity of each star.

In left panel of Fig. 1 we display the abundances of these two stars and some linear fits for different T_C ranges. The steep positive trend in the linear fit for $T_C > 900$ K is probably affected by chemical evolution effects on Mn, Na and Cu. In right panel of Fig. 1 we have already removed the Galactic chemical evolution effects and both stars do not seem to show any trend. We may conclude that it seems plausible that many of our targets hosts terrestrial planets but this may not affect the volatile-to-refratory abundance ratios in the atmospheres of these stars (see e.g. Udry & Santos 2007).

References

González Hernández, J. I., Israelian, G., Santos, N. C., *et al.* 2010, *ApJ*, 720, 1592
Kurucz, R. L. ATLAS9 Stellar Atmospheres Programs and 2 km s^{-1} Grid, CD-ROM No. 13, Smithsonian Astrophysical Observatory, Cambridge, 1993
Meléndez, J., Asplund, M., Gustafsson, B., & Yong, D. 2009, *ApJ Letters*, 704, L66
Neves, V., Santos, N. C., Sousa, S. G., Correia, A. C. M., & Israelian, G. 2009, *A&A*, 497, 563
Ramírez, I., Meléndez, J., & Asplund, M. 2009, *A&A Letters*, 508, L17
Ramírez, I., Asplund, M., Baumann, P., Meléndez, J., & Bensby, T. 2010, *A&A*, 521, A33
Sousa, S. G., *et al.* 2008, *A&A*, 487, 373
Sneden, C. 1973, *PhD Dissertation*, Univ. of Texas, Austin
Udry, S. & Santos, N. C. 2007, *ARA&A*, 45, 397

The Astrophysics of Planetary Systems: Formation, Structure, and Dynamical Evolution
Proceedings IAU Symposium No. 276, 2010
A. Sozzetti, M. G. Lattanzi & A. P. Boss, eds.

doi:10.1017/S1743921311020631

Planetesimal dynamics in hydromagnetic turbulence

Oliver Gressel[1], Richard P. Nelson[1] & Neal J. Turner[2]

[1]Astronomy Unit, Queen Mary, University of London, Mile End Road, London E1 4NS, UK
[2]Jet Propulsion Laboratory, California Institute of Technology, Pasadena, CA 91109
email: o.gressel@qmul.ac.uk, r.p.nelson@qmul.ac.uk, neal.turner@jpl.nasa.gov

Abstract. Planet formation theory is founded on the concept of dust coagulation and subsequent growth into planetesimals. This process is by no means an isolated one, but possibly happens in a turbulent nebula. It is therefore crucial to understand how particles of different sizes are affected by their gaseous environment via stochastic forcing and aerodynamic damping. We here report on the effects of magneto-rotational (MRI) turbulence in the presence of non-uniform ionisation leading to the formation of a magnetically inactive dead-zone. While we find that collisional growth is impeded by fully-active MRI, it may be possible within a dead-zone.

Keywords. accretion disks, MHD, methods: numerical, planetary systems: formation, planetary systems: protoplanetary disks

---

1. Introduction

Planetesimals are the building-blocks of protoplanets. New approaches to their rapid formation via streaming and gravitational instabilities (e.g. Johansen & Youdin 2007) require a thorough understanding of particle stirring by the gas and the level of sedimentation in stratified disks. When particles grow from metre to kilometre sizes, they aerodynamically decouple from the gas flow. Independent of their mass, they feel a gravitational force from fluctuations in the potential near density waves, and the acquired velocity dispersion governs whether collisional growth eventually becomes disruptive.

2. Results

The global cylindrical disc simulations of Nelson (2005) have demonstrated that density fluctuations from developed MRI turbulence pose a severe limitation to the growth of planetesimals. We have now confirmed this original finding in the framework of local simulations with large enough box sizes (Nelson & Gressel 2010).

2.1. *Stratified discs with dead-zone*

In the following, we give a first report on the extension of this work to stratified discs harbouring a dead-zone due to insufficient ionisation (Gammie 1996). Our stratified dead-zone simulations cover $\pm 5\,{}^{1}/_{2}$ pressure scale heights, and apply a zero-net-flux magnetic configuration with $\beta_{\rm p} = 50$, and an additional weak B_z net field. Since recombination occurs mostly on small grains, we adopt a simplified treatment of the gas-phase reactions. We update the diffusivity, $\eta = \eta(\mathbf{x}, t)$, according to a lookup table derived from the reaction network in model 4 of Ilgner & Nelson (2006), and assuming a dust-to-gas mass ratio of 10^{-3}, and ionisation due to X-rays and cosmic rays (cf. Turner & Drake 2009).

In Figure 1, we show the evolution of the toroidal field $\bar{B}_\phi$ in the presence of a dead-zone. The field exhibits the typical dynamo cycles (cf. references in Gressel 2010) in the MRI active layers. As already seen by Turner & Sano (2008), the toroidal field leaks

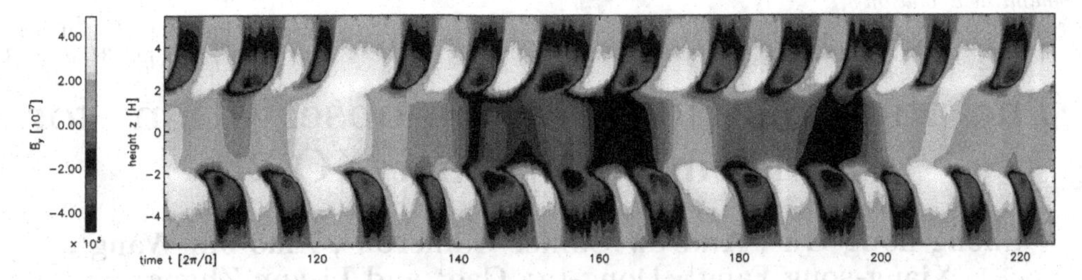

Figure 1. Space-time diagram of the horizontally-averaged toroidal magnetic field $\bar{B}_\phi$.

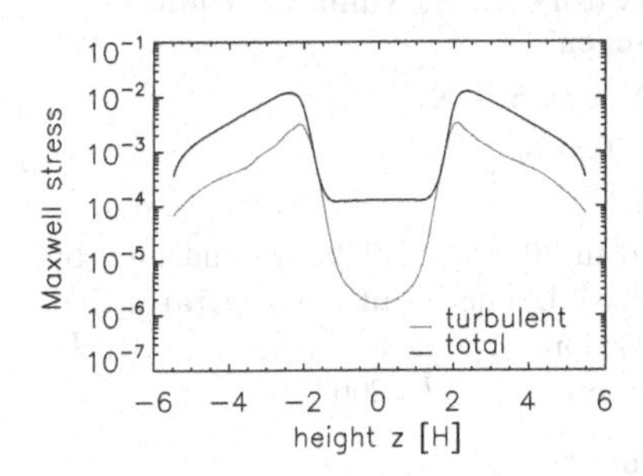

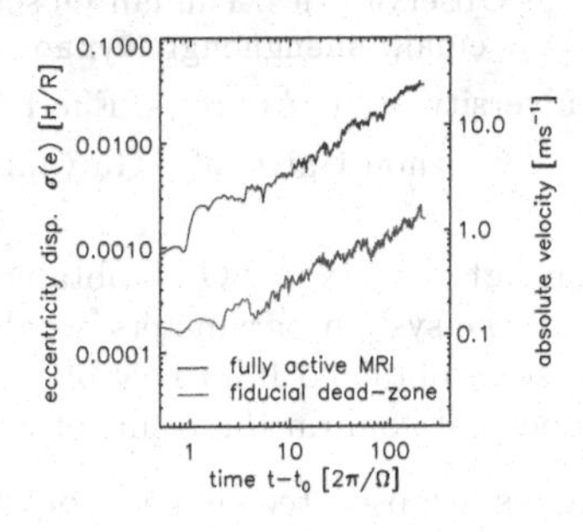

Figure 2. Time-averaged profiles of the total/turbulent Maxwell stress (*left*), and random-walk eccentricity growth (*right*).

into the diffusively dominated midplane region, contributing to the overall stress (see Fig. 2, left panel). Despite the absence of strong turbulence, density fluctuations reach a level of 10-20 percent within the dead-zone. Surprisingly, these fluctuations result in very moderate stochastic torques on the particles. Accordingly, the eccentricity growth is substantially reduced in the case of a dead-zone (Fig. 2, right panel).

2.2. *Conclusions*

Simulations of planetesimals embedded in fully turbulent, non-stratified disc models show that the velocity dispersion of km-sized bodies grows quickly and exceeds the threshold for catastrophic disruption. This raises important questions about the viability of planetesimal accretion in such turbulent discs.

Our simulation of a fully turbulent, vertically stratified disc confirms this basic picture. We find, however, that planetesimals which are embedded in a disc with a dead-zone whose vertical size is approximately two density scale-heights experience a substantially reduced stochastic forcing, being decreased by a factor of 10-20. It thus appears that dead-zones may provide an environment which is conducive to planet formation via planetesimal accretion.

Acknowledgements

This work used the NIRVANA-III code developed by Udo Ziegler at the AIP. All computations were performed on the QMUL HPC facility, purchased under the SRIF initiative.

References

Gammie, C. F. 1996, *ApJ*, 457, 355
Gressel, O. 2010, *MNRAS*, 405, 41
Johansen, A. & Youdin, A. 2007, *ApJ*, 662, 627
Nelson, R. P. 2005, *A&A*, 443, 1067
Nelson, R. P. & Gressel, O. 2010, *MNRAS*, 409, 1392
Ilgner, M. & Nelson, R. P. 2006, *A&A*, 445, 205
Turner, N. J. & Drake, J. F. 2009, *ApJ*, 703, 2152
Turner, N. J. & Sano, T. 2008, *ApJ* (Letters), 679, L131

The Astrophysics of Planetary Systems: Formation, Structure, and Dynamical Evolution
Proceedings IAU Symposium No. 276, 2010
A. Sozzetti, M. G. Lattanzi & A. P. Boss, eds.

doi:10.1017/S1743921311020643

The photometric follow-up observations for transiting exoplanet XO-2b

Sheng-hong Gu[1], Andrew Collier Cameron[2], Xiao-bin Wang[1], Xiang-song Fang[1], Dong-tao Cao[1] and Li-yun Zhang[3]

[1]National Astronomical Observatories/Yunnan Observatory, CAS, Kunming, China
email: shenghonggu@ynao.ac.cn

[2]University of St. Andrews, Fife KY16 9SS, UK

[3]Guizhou University, Guiyang, China

Abstract. Four new transit light curves of XO-2b obtained in 2008 and 2009, are analyzed by using MCMC algorithm, and the system parameters are derived. The result demonstrates that the orbital period of the system obtained from new observations is almost the same as Burke *et al.*'s one (2007), which does not confirm the result of Fernandez *et al.* (2009).

Keywords. planetary systems, eclipses, techniques: photometric

---

1. Introduction

The transiting hot Jupiter XO-2b was discovered by Burke *et al.* (2007), its radius, mass and orbital period are 0.98R_J, 0.57M_J and 2.615857days, respectively, the host star XO-2 has high metallicity and high proper motion. Later, Fernandez *et al.* (2009) observed other six transit events of the system and derived precise radius 0.996R_J and mass 0.565M_J of the exoplanet. They also found that the orbital period of the system changed by 2.5σ. Thus, more observations for its transit events are needed to clarify whether the orbital period of the system is variable.

2. Observations and data reduction

The new observations for transit events of XO-2b were made by using 85cm telescope with 1Kx1K CCD camera (Zhou *et al.* 2009) of Xinglong station, NAOC on Dec.3, 2008 and 1m telescope with 1Kx1K, 2Kx2K CCD cameras of Yunnan Observatory on Jan.19, 2008, Dec.7,12, 2009. In all observations, the R filter was employed. The observed CCD images are reduced by using IRAF package. For the obtained light curves, we remove the systematic errors by using coarse decorrelation method (Collier Cameron *et al.* 2006) and SysRem algorithm (Tamuz *et al.* 2005).

3. Light curve analysis and discussion

In order to get a set of precise system parameters for XO-2, the 4 datasets of transit events are combined in the course of light curve analysis. We model the flux of the transiting system with the parameters $\{T_c, p, \Delta F, t_T, b, M_*\}$ considering the 4-coefficient limb-darkening law of Claret (2000). The basic parameters of the host star are adopted from the recent relative results (Fernandez *et al.* 2009). All observed data points are involved in MCMC (Markov Chain Monte Carlo) analysis to search the optimal parameters $\{T_c, p, \Delta F, t_T, b, M_*\}$ according to the procedure of Collier Cameron *et al.* (2007). The

Table 1. The optimal parameters derived for XO-2 system using the MCMC algorithm.

Transit epoch T_c [HJD]	$2455013.5983^{+0.0003}_{-0.0002}$
Orbital period p	$2.6158553^{+2.2E-06}_{-2.5E-06}$ days
Transit depth ΔF	$0.0136^{+0.0002}_{-0.0002}$ mag.
Transit width t_T	$0.1094^{+0.0005}_{-0.0003}$ days
Impact parameter b	$0.139^{+0.021}_{-0.070}$ R_*
Orbital separation a	$0.0367915^{+1.5E-06}_{-1.1E-06}$ AU
Orbital inclination i	$89.045^{+0.485}_{-0.147}$ degrees
Stellar radius R_*	$0.951^{+0.001}_{-0.003}$ R_{Sun}
Planet radius R_p	$0.945^{+0.006}_{-0.007}$ R_J
Stellar mass M_*	0.971 M_{Sun}
Planet mass M_p	0.565 M_J

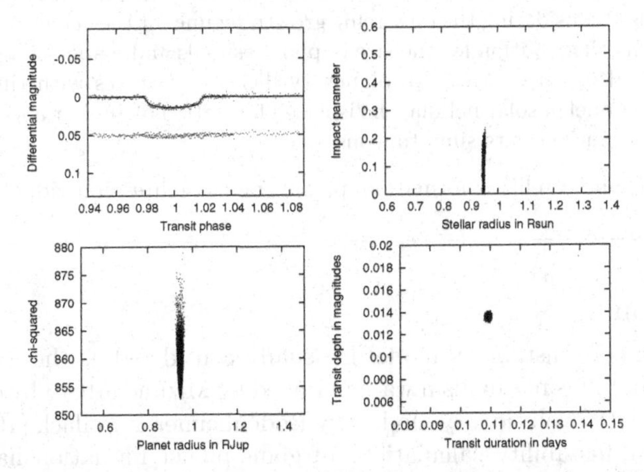

Figure 1. Binned light curve of XO-2b and fitting information.

final result is listed in Table 1, the relative fitting and other information are shown in Fig. 1.

Compared our new result and the previous ones, it can be found that the new orbital period is almost the same as Burke *et al.*'s value (2.615857days). This does not support the suggestion that the orbital period of the system is probably variable, which was given by Fernandez *et al.* (2009). So, it is still necessary to observe more transit events for XO-2b so as to make further investigation on its period behavior.

Acknowledgments

This work is supported by NSFC under grant No.10873031 and Chinese Academy of Sciences under grant KJCX2-YW-T24.

References

Burke, C. J., *et al.* 2007, *ApJ*, 671, 2115
Claret, A. 2000, *A&A*, 363, 1081
Collier Cameron, A., *et al.* 2006, *MNRAS*, 373, 799
Collier Cameron, A., *et al.* 2007, *MNRAS*, 380, 1230
Fernandez, J. M., *et al.* 2009, *AJ*, 137, 4911
Tamuz, O., Mazeh, T.,& Zucker, S. 2005, *MNRAS*, 356, 1466
Zhou, A.-Y., *et al.* 2009, *RAA*, 9, 349

The Astrophysics of Planetary Systems: Formation, Structure, and Dynamical Evolution
Proceedings IAU Symposium No. 276, 2010
A. Sozzetti, M. G. Lattanzi & A. P. Boss, eds.

doi:10.1017/S1743921311020655

Simultaneous formation of Jupiter and Saturn

Octavio M. Guilera [1,2], Adrián Brunini [1,2] and Omar G. Benvenuto[1,2]

[1]Facultad de Cs. Astronómicas y Geofísicas - Universidad Nacional de La Plata, Argentina
[2]Instituto de Astrofísica de La Plata (IALP) - CONICET
email: oguilera@fcaglp.unlp.edu.ar

Abstract. We calculate the simultaneous *in situ* formation of Jupiter and Saturn by the core instability mechanism considering the oligarchic growth regime for the accretion of planetesimals. We consider a density distribution for the size of planetesimals and planetesimals migration. The planets are immersed in a realistic protoplanetary disk that evolves with time. We find that, within the classical model of solar nebula, the isolated formation of Jupiter and Saturn undergoes significant change when it occurs simultaneously.

Keywords. planets and satellites: formation, planets and satellites: individual (Jupiter, Saturn)

1. Introduction

At present, the core instability model is usually considered as the way giant planets formation proceeds. This mechanism was envisaged by Mizuno (1980) by employing static models and later with evolutionary models by Bodenheimer & Pollack (1986) and Pollack *et al.* (1996). Core instability calculations of giant planet formation have been carried out by many groups, e.g., Alibert *et al.* (2005), Hubickyj *et al.* (2005), and Dodson-Robinson *et al.* (2008). Fortier *et al.* (2007, 2009) were the first to consider the oligarchic growth regime for the accretion of planetesimals. However, one usual assumption in detailed simulations of planetary growth is that each planet grows alone in the disk. This would be correct if the population of planetesimals to be accreted by one planet were not appreciably perturbed by the presence of another embryo. At first sight, it may be understood that this is the case if the feeding zone of each planet does not overlap the one corresponding to any other planet. However, this is *not* the case if we include planetesimal migration. This process leads to a net inward motion of planetesimals. A planet will perturb the swarm of planetesimals that may be later accreted by another planet moving along an inner orbit. Moreover, as we show below, even the presence of an inner planet will be able to affect the accretion process of an outer object.

2. Results

In our work Guilera *et al.* (2010), we developed a numerical code to compute the simultaneous formation of giant planets immersed in a protoplanetary disk that evolves with time. We used this code to calculate the *in situ* simultaneous formation of the gaseous giant planets of the solar system. We considered a disk 5 times more massive than the classical "minimum mass solar nebula" of Hayashi (1981). We quantitatively analyzed the effects due to simultaneous formation of Jupiter and Saturn (at its current locations) comparing with the results corresponding to the case of isolated formation. When we refer to isolated formation we mean that we have considered that only one

	Jupiter		Saturn	
	M_c $[M_\oplus]$	t_f $[My]$	M_c $[M_\oplus]$	t_f $[My]$
Isolated formation	28.30	2.20	20.51	6.37
Simultaneous formation	29.68	1.96	3.35(*)	$\gg 15$

Table 1. Comparison between the isolated and simultaneous formation of the solar system gaseous giant planets for a disk 5 times more massive than the Hayashi nebula. Here M_c stands for the final core mass and t_f for the formation time.

planet forms in the disk while it evolves. The results we have obtained are resumed in Table 1.

For the case of the isolated formation, we found that both planets are formed in less than 10 Myr. This is in good agreement with the current observational estimations. We also found that the final core masses were in good agreement with the current theoretical estimations. We remark that we assumed that all the infalling planetesimals reach the core's surface without losing mass on their trajectories throughout the envelope, this meaning that M_c really corresponds to the total heavy element's mass in the interior of the planet (core *plus* solids in the envelope).

Considering the simultaneous formation of both planets we see that Saturn has almost no effect on the formation of Jupiter. However, the opposite is not true: the formation of Jupiter, clearly inhibits the formation of Saturn. The simulation was halted at 15 My. At this time, the embryo of Saturn achieved only a mass of $M_c \sim 3.5$ $M_\oplus$ with a negligible envelope (*).

The inhibition of the formation of Saturn is caused by an eccentricity and inclination excitation of the planetesimals related to Jupiter's perturbations. This excitation causes an increment in the migration velocity of planetesimals at the Saturn's neighborhood when both planets are formed simultaneously. The increment in the migration velocity of planetesimals causes the solid accretion timescale to become longer than planetesimal migration timescales, and the solid accretion rate of Saturn (when it is formed simultaneously with Jupiter) becomes less efficient than for the isolated Saturn formation (see Guilera *et al.* 2010). The most important result is that the rapid formation of Jupiter inhibits -or largely increases- the timescale of Saturn's formation when they grow simultaneously.

References

Alibert, Y., Mordasini, C., Benz, W., & Winisdoerffer, C. 2005, *A&A*, 434, 343
Bodenheimer, P. & Pollack, J. B. 1986, *Icarus*, 67, 391
Dodson-Robinson, S. E., Bodenheimer, P., Laughlin, G., Willacy, K., Turner, N. J., & Beichman, C. A. 2008, *ApJ* (Letters), 688, L99
Fortier, A., Benvenuto, O. G., & Brunini, A. 2007, *A&A*, 473, 311
Fortier, A., Benvenuto, O. G., & Brunini, A. 2009, *A&A*, 500, 1249
Guilera, O. M., Brunini, A., & Benvenuto, O. G. 2010,*A&A*, 521, A50
Hayashi, C. 1981, *Progress of Theoretical Physics Supplement*, 70, 35
Hubickyj, O., Bodenheimer, P., & Lissauer, J. J. 2005, *Icarus*, 179, 415
Mizuno, H. 1980, *Progress of Theoretical Physics*, 64, 544
Pollack, J. B., Hubickyj, O., & Bodenheimer, P. *et al.* 1996, *Icarus*, 124, 62

The Astrophysics of Planetary Systems: Formation, Structure, and Dynamical Evolution
Proceedings IAU Symposium No. 276, 2010
A. Sozzetti, M. G. Lattanzi & A. P. Boss, eds.

doi:10.1017/S1743921311020667

Dead zones and the diversity of exoplanetary systems

Yasuhiro Hasegawa[1] **and Ralph E. Pudritz**[1,2]

[1]Department of Physics and Astronomy McMaster University,
Hamilton ON, L8S 4M1, Canada
email: hasegay@physics.mcmaster.ca

[2]Origins Institute, McMaster University,
Hamilton ON, L8S 4M1, Canada
email: pudritz@physics.mcmaster.ca

Abstract. Planetary migration provides a theoretical basis for the observed diversity of exoplanetary systems. We demonstrate that dust settling - an inescapable feature of disk evolution - gives even more rapid type I migration by up to a factor of about 2 than occurs in disks with fully mixed dust. On the other hand, type II migration becomes slower by a factor of 2 due to dust settling. This even more problematic type I migration can be resolved by the presence of a dead zone; the inner, high density region of a disk which features a low level of turbulence. We show that enhanced dust settling in the dead zone leaves a dusty wall at its outer edge. Back-heating of the dead zone by this wall produces a positive radial gradient for the disk temperature, which acts as a barrier for type I migration.

Keywords. accretion, accretion disks, radiative transfer, turbulence, planetary systems: protoplanetary disks

1. Introduction

Planetary migration is essential in the theory of planet formation in order to understand the observed mass-period relation (Udry & Santos 2007). There are actually two types of migration, distinguished by planetary mass (e.g., Ward 1997). Low-mass planets ($\sim$ several $M_{\oplus}$) undergo the so-called type I migration wherein angular momentum is transferred between planets and the surrounding gaseous disks only at the Lindbald and corotation resonances. In standard disk models, the planets efficiently lose their angular momentum, and hence plunge towards the central star within the disk lifetime ($\sim 1 - 3 \times 10^6$ years). On the other hand, massive planets ($\sim 1 M_J$), known as type II migrators, can open up a gap in the disk due to strong, non-linear resonant torques, and migrate on the viscous evolution timescale of the disk.

The long standing problem of migration is to identify what physical process(es) and/or condition(s) makes type I migration much slower. Considerable effort has been focused on disk properties such as density and temperature gradients to which the tidal torque is sensitive (Hasegawa & Pudritz 2010a).

2. Dust Settling & Rapid Planetary Migration

Dust settling is observationally confirmed in disks around a variety of young stars (e.g., Hasegawa & Pudritz 2010b). Hasegawa & Pudritz (2011) first included the effects of this inescapable aspect of disk evolution on migration, by self-consistently computing the thermal structure of disks. In the computations, the full wavelength dependent, radiative

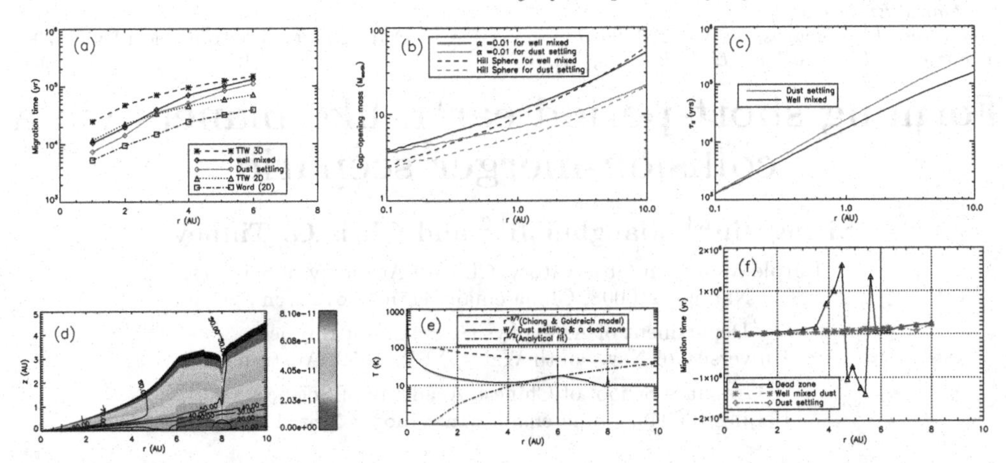

Figure 1. Planetary migration in radiatively heated disks for the well mixed and dust settling cases (in (a)-(c); Adapted from Hasegawa & Pudritz (2011)). In a), the migration time as a function of distance from the central star. In b), the gap-opening mass as a function of distance from the central star. In c), the timescale of type II migration. Planetary migration in radiatively heated disks with dead zones which is 6 AU in size (in (d)-(f); Adapted from Hasegawa & Pudritz 2010a). In d), the dust density distribution with the temperature contours. In e), the temperature structure in the mid-plane region. In f), the migration time as a function of distance from the star.

transfer equation is solved by means of a Monte Carlo method, including dust settling and the gravitational force of a planet.

We show that dust settling results in even more rapid type I migration, by up to a factor of about 2. This arises due to the geometrically flatter shape of the disk which is a consequence of dust settling. On the other hand, dust settling both lowers the gap-opening mass and slows the type II migration rate by about a factor of 2. This can be also understood as a consequence of flatter disk structures. It is obvious that some sort of more robust slowing mechanism is required for even more rapid type I migration.

3. Dead Zones & Outward Migration

Dead Zones change the structure of disks (Hasegawa & Pudritz 2010a). They are the high density, low ionized regions, so that turbulence induced by the magnetorotational instability (MRI) is strongly suppressed (e.g., Gammie 1996). In a dead zone, dust settling is highly enhanced, so that a dusty wall is left at its outer boundary. The wall becomes thermally hot by stellar irradiation, and produces a positive temperature gradient due to its back-heating of the dead zone. We demonstrate that this temperature behavior results in outward migration (Fig. 1). We will address the role of this barrier in the mass-period relation by developing our population synthesis models.

References

Gammie, C. F. 1996, *ApJ*, 457, 355
Hasegawa, Y. & Pudritz, R. E. 2010a, *ApJ*, 710, L167
Hasegawa, Y. & Pudritz, R. E. 2010b, *MNRAS*, 401, 143
Hasegawa, Y. & Pudritz, R. E. 2011, *MNRAS*, 413, 286
Udry, S. & Santos, N. C. 2007, *ARAA*, 45, 397
Ward, W. R. 1997, *Icarus*, 126, 261

The Astrophysics of Planetary Systems: Formation, Structure, and Dynamical Evolution
Proceedings IAU Symposium No. 276, 2010
A. Sozzetti, M. G. Lattanzi & A. P. Boss, eds.

doi:10.1017/S1743921311020679

Forming short-period earth-like planets via a collision-merger scenario

Sheng Jin[1,3], Jianghui Ji[1,2] and Chris G. Tinney[2]

[1]Purple Mountain Observatory, Chinese Academy of Sciences, Nanjing 210008, China email: jijh@pmo.ac.cn

[2]Department of Astrophysics, School of Physics, University of New South Wales, NSW 2052, Australia

[3]Graduate School of Chinese Academy of Sciences, Beijing 100049, China email: qingxiaojin@gmail.com

Abstract. We present a new formation mechanism to produce short-period Earth-like planets in the late stage of planet formation, through a collision-merger scenario. In this scenario, a planetary embryo is directly thrown into a close-in orbit after a collision with another embryo, and then the larger merged body is seized by the central star as a hot Earth-like planet.

Keywords. methods: n-body simulations, celestial mechanics, planetary systems: formation

1. Introduction

It is now widely accepted that short-period planets cannot have formed *in situ*, but rather must have migrated to their current orbits from a formation region much farther from their host star (Lin *et al.* 1996). The formation scenarios for short-period Earth-like planets are also associated with the migration of gas-giant planets (Raymond *et al.* 2006; Terquem & Papaloizou 2007). While in our dynamical simulations for planetesimal evolution in later stage of planet formation, we find a mechanism is revealed by which the collision-merger of planetary embryos can kick terrestrial planets directly into orbits extremely close to their parent stars.

2. Simulation setup and results

Extrasolar planetary systems that harbor pairs of Jupiter-to-Saturn-mass companions are of particular interest to researchers (Gozdziewski 2002; Zhang *et al.* 2010). We have therefore performed 30 simulations to investigate such a system architecture using the MERCURY package (Chambers 1999) for the following two systems:

- **Simulation 1** - two giant planets with initial orbital parameters to emulate the OGLE-06-109L system (Gaudi *et al.* 2008). 500 planetary embryos and planetesimals with total mass $10\,M_\oplus$ were distributed between $0.3\,\mathrm{AU} < a < 5.2\,\mathrm{AU}$ and with $e < 0.02$. Each of the 26 runs evolved over 400 Myr.
- **Simulation 2** - two giant planets with initial orbital parameters to emulate the 47 Uma system (Fischer *et al.* 2002). 648 planetary embryos with total mass of $5.14\,M_\oplus$ were distributed in the region $0.3\,\mathrm{AU} < a < 1.6\,\mathrm{AU}$ with $e < 0.02$. Each of the four runs evolved over 100 Myr.

All simulations exhibit a classical planetary accretion scenario in their late stage formation (Chambers 2001). Figure 1 shows the evolution process of one formed close-in planet in **Simulation 2**. Such a collision-merger mechanism for close-in terrestrial planets happened in $\sim 20\%$ of the total runs. In some cases the short-period planet was

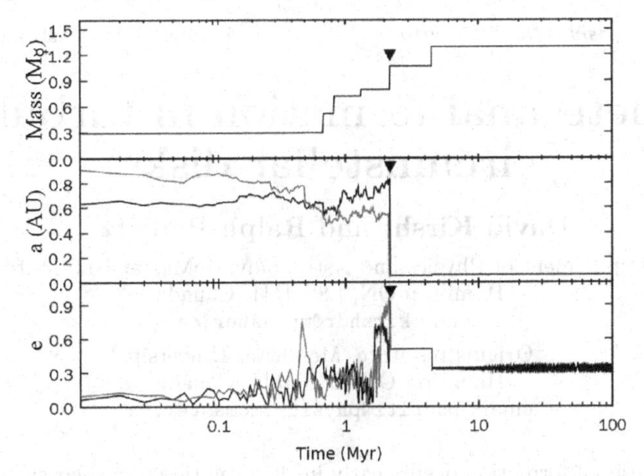

Figure 1. Mass, semi-major axis and eccentricity evolution of the short-period terrestrial planets that emerges from **Simulation 2**

kicked into an inner orbit at a very early stage, and subsequently accreted a majority of the mass available in nearby orbits. This shows that the key role of this mechanism is to throw one body into short-period orbits (Ji *et al.* 2011) at the collision.

3. Discussion

In actual cases collision could have a result that ranges anywhere from merger, like partial fragmentation or complete shattering (Wetherill & Stewart 1993). However, the collision-merger scenario might still be reasonable as it does not require perfect accretion. Rather it depends on the collisions pushing the resultant body inward so that the central star could grasp it. Moreover, the planetesimal disk in the simulations is less massive than that of the Minimum Mass Solar Nebula (Hayashi 1981), which could consist of billions of small bodies, to accordingly increase the probability of this mechanism.

Acknowledgements

This work is financially supported by the National Natural Science Foundation of China (Grants 10973044, 10833001), the Natural Science Foundation of Jiangsu Province, and the Foundation of Minor Planets of Purple Mountain Observatory.

References

Chambers, J. E. 1999, *MNRAS*, 304, 793
Chambers, J. E. 2001, *Icarus*, 152, 205
Fischer, D. A., *et al.* 2002, *ApJ*, 564, 1028
Gaudi, B. S., *et al.* 2008, *Science*, 319, 927
Gozdziewski, K. 2002, *A&A*, 393, 997
Hayashi, C. 1981, *Progress of Theoretical Physics Suppl.*, 70, 35
Ji, J. H., *et al.* 2011, *ApJ*, 727, L5
Lin, D. N. C., Bodenheimer, P., & Richardson, D. C. 1996, *Nature*, 380, 606
Raymond, S. N., *et al.* 2006, *Science*, 313, 1413
Terquem, C. & Papaloizou, J. C. B. 2007, *ApJ*, 654,1110
Wetherill, G. W. & Stewart, G. R. 1993, *Icarus*, 106, 190
Zhang, N., Ji, J. H., & Sun, Z. 2010, *MNRAS*, 405, 2016

The Astrophysics of Planetary Systems: Formation, Structure, and Dynamical Evolution
Proceedings IAU Symposium No. 276, 2010
A. Sozzetti, M. G. Lattanzi & A. P. Boss, eds.

doi:10.1017/S1743921311020680

Planetesimal formation in turbulent circumstellar disks

David Kirsh[1] and Ralph Pudritz[2]

[1]Department of Physics and Astronomy McMaster University, Hamilton ON, L8S 4M1, Canada
email: kirshdr@mcmaster.ca

[2]Origins Institute, McMaster University, Hamilton ON, L8S 4M1, Canada
email: pudritz@physics.mcmaster.ca

Abstract. Planetesimal formation occurs early in the evolution of a solar system, embedded in the circumstellar gas disk, and it is the crucial first step in planet formation. Their growth is difficult beyond boulder size, and likely proceeds via the accumulation of many rocks in turbulence followed by gravitational collapse – a process we are only beginning to understand. We have performed global simulations of the gas disk with embedded particles in the FLASH code. Particles and gas feel drag based on differential velocities and densities. Grains and boulders of various sizes have been investigated, from micron to km, with the goal of understanding where in the disk large planetesimals will tend to form, what sizes will result, and what size ranges of grains will be preferentially incorporated. We have so far simulated particles vertical settling and radial drift under the influence of gas drag, and their accumulations in turbulent clumps.

Keywords. accretion disks, hydrodynamics, turbulence, planetary systems: formation

1. Introduction

The planetesimal is the key to planet formation, and yet their growth is not yet completely understood. During the early phases of core accretion, particles grow by colliding and sticking. However, meter-size particles suffer gas drag and fall into the star faster than they can stick and grow. In fact, collisions at this size range generally lead to fragmentation rather than sticking. These two effects prevent growth beyond this size. Rapid growth to larger sizes is a consequence of accumulation and gravitational collapse within disk gas turbulence. Drag on the gas by the particles helps promote this effect, and gets stronger with accumulation – this runaway effect is termed the streaming instability. Johansen *et al.* 2007 discovered this effect in local periodic shearing box simulations. Here we investigate this process in a global disk model, to determine its efficiency.

2. The Simulations

We utilize the FLASH hydrodynamics 3D grid code, which has adaptive mesh refinement that allows us to apply more resolution in the regions that require it (high density or velocity). Given the low-mass disk setup currently used, the gravity of the system is a background (stellar) potential. The disk is a global model, up to one vertical scale height, with a chosen inner radius cut-off. Accretion through the inner edge sets off chaotic spiral waves. There is a background disk corona in the remaining simulation volume. Both consist of adiabatic gas ($\gamma = 5/3$) in hydrostatic equilibrium with a central gravitational potential. The corona (Ouyed & Pudritz 1997) that surrounds the disk is chosen to be spherically stable. The central potential and the corona are Plummer-smoothed (Dyer

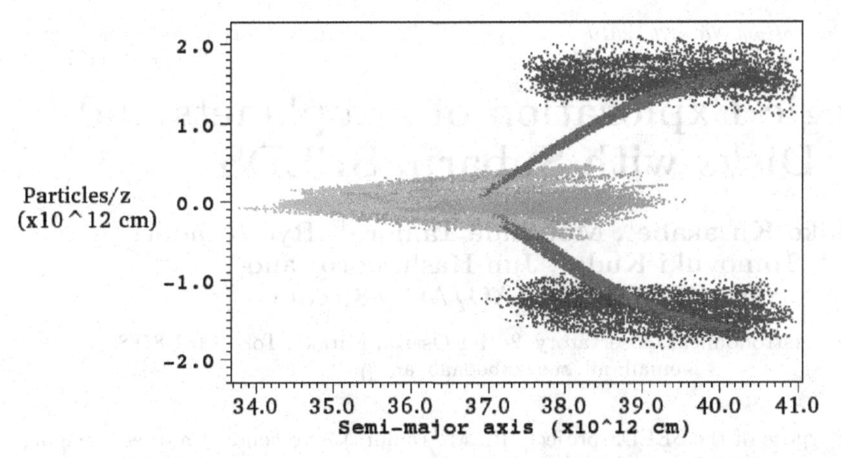

Figure 1. Heights versus semi-major axes for four particle sizes initially placed in two rings of width 0.2AU at 2.65AU (39.6e12cm), at heights of ±0.4AU from the midplane, shown after 5 orbits. The 1mm particles remain suspended as clumps, only falling a fraction of their height. The 1cm particles quickly settle to a thin layer, matching predicted settling rates (Weidenschilling 1977). The 10cm particles fall quickly to the midplane but are still oscillating with a relatively small amplitude. These particles, along with the 1cm size, show the strongest inward drift. The 1m particles oscillate about the midplane, showing only slight reductions in their amplitudes. The curved shape is due to the timescale variation with semi-major axis. Some distortion from the motions of the disk are evident, especially in the smaller sizes.

& Ip 1993) to prevent a too-sharp gradient. The disk density and pressure decrease vertically from the midplane as an exponential, with the scale height given by the ratio of sound to Keplerian speeds. Pressure support leads to slightly sub-Keplerian gas speed throughout the disk, an effect that the particles do not experience – it is this velocity difference that leads to the particles feeling a headwind drag as they orbit through the slower gas.

The Lagrangian particles are added on top of the grid, interacting via gravity and gas drag determined by the density and velocity of the gas in the cells they occupy, as well as their own physical size, density (~ 2 g/cm^3) and velocity. They are superparticles, with a mass that represents swarms of many particles. For models where the solid density is roughly 1/100th the gas density, that mass is divided evenly amongst the particles that are initialized in that cell. They are placed randomly within their cells with Keplerian velocities, and allowed to drift and settle. The drag equations are given by Weidenschilling (1977). Particle drag laws fall into two regimes (Stokes, Epstein) depending on their size compared to the local mean free path of the gas (roughly 10cm for our disk). Figure 1 below shows the details of particle settling and drifting in the gas.

We are continuing to allow the simulations time to run and develop their instabilities. This work will help to show where in the disk planetesimals form, and how efficiently the process occurs.

References

Dyer, C. C. & Ip, P. S. S.. 1993, *ApJ*, 409, 60

Johansen, A., Oishi, J. S., Mac Low, M.-M., Klahr, H., Henning, T., & Youdin, A. 2007a, *Nature*, 448, 1022

Ouyed, R. & Pudritz, R. E. 1997, *ApJ*, 482, 717

Weidenschilling, S. J. 1977, *MNRAS*, 180, 57

The Astrophysics of Planetary Systems: Formation, Structure, and Dynamical Evolution
Proceedings IAU Symposium No. 276, 2010
A. Sozzetti, M. G. Lattanzi & A. P. Boss, eds.

doi:10.1017/S1743921311020692

Strategic Exploration of Exoplanets and Disks with Subaru: SEEDS

Nobuhiko Kusakabe[1], Motohide Tamura[1], Ryo Kandori[1], Tomoyuki Kudo[1], Jun Hashimoto[1] and the SEEDS/HiCIAO/AO188 team

[1]National Astronomical Observatory, 2-21-1 Osawa, Mitaka, Tokyo 181-8588
email: nb.kusakabe@nao.ac.jp

Abstract. The purpose of the SEEDS project (PI: M. Tamura) is to conduct a direct imaging survey, searching for giant planets as well as protoplanetary/debris disks at a few to a few tens of AU regions around 500 nearby solar-type or more massive young stars with the combination of the Subaru 8.2m telescope, the new high-contrast instrument HiCIAO, and the adaptive optics system AO188. After instrument performance verification, the SEEDS survey successfully started in October 2009. We have already detected many companion candidates to be followed-up, and clear and much better detections of disks or details of known disks structures. In this contribution, we will outline our goal, current status, early results, and future instrumentation plans.

Keywords. stars: formation, surveys, circumstellar matter, planetary systems, techniques: high angular resolution

1. Introduction

Since the first detection of exoplanets orbiting normal stars in 1995, many exciting discoveries have been made, but our understanding of planetary systems and their formation is far from complete. As demonstrated with recent successes of direct imaging of planetary-mass objects around Vega-type A stars and a G star, direct imaging approach is indispensable for the detection of such "young" planets, especially planets beyond the snowline (4-40AU), which is complementary to radial velocity or transit searches.

SEEDS is the first Subaru Strategic Programs to conduct a direct imaging survey with the combination of the Subaru 8.2m telescope, the new high-contrast instrument HiCIAO, and the adaptive optics system AO188. The purpose of the SEEDS project is search for giant planets as well as protoplanetary/debris disks at a few to a few tens of AU regions around 500 nearby solar-type or more massive young stars. After instrument performance verification, the SEEDS survey successfully started in October 2009. We have already detected many companion candidates to be followed-up, and clear and much better detections of disks or details of known disks structures. The goals of our survey are to address the following key issues in exoplanet/disk science: (1) the detection and census of exoplanets in the outer circumstellar regions around solar-mass stars and massive stars, (2) the evolution of protoplanetary and debris disks including their morphological diversity, and (3) the link between exoplanets and circumstellar disks. The completeness and uniformity of this systematic survey will provide important statistical, or even useful null, results to be obtained as well as enabling the study of individual objects of particular interest. We will report current status, early results, and future instrumentation plans. A full list of the current SEEDS member (~100 people) can be found on our web site.

2. Early Results

In the first one year, we have already some results and published papers. One of the paper is the direct imaging of a planet candidate around sun like star, GJ 758 (Thalmann *et al.* 2009). GJ 758 is a G9-type star (0.97 M_{sun}), 15.5 pc distant from the sun, and have no radial-velocity planets so far. The newly discovered planet candidate GJ 758 b is estimated to have a mass of 10-20 M_J at a projected distance of 29 AU, resemble to Neptune orbit.

The high contrast imaging of the circumstellar disk of LkCa 15 has revealed the surrounding nebulosity (Thalmann *et al.* 2010). We detect sharp elliptical contours delimiting the nebulosity on the inside as well as the outside, consistent with the shape, size, ellipticity, and orientation of starlight reflected from the far side disk wall, whereas the near side wall is shielded from view by the disk's optically thick bulk. We note that forward scattering of starlight on the disk surface could provide an alternate interpretation of the nebulosity. In either case, this discovery provides confirmation of the disk geometry that has been proposed to explain the SED of such systems, comprising an optically thick disk with an inner truncation radius of ~46 AU enclosing a largely evacuated gap.

On the other hand, polarized intensity image of AB Aur (Hashimoto *et al.*, submitted) is prodicing the sparpest and closest image of the protoplanetary disk, providing the first clear fine structures of the inner disk (<50 AU) regions and providing evidence of an embedded planet.

HAT-P-7 b was reported to have a highly tilted orbit, massive bodies such as giant planets, and a binary star is expected to exist in the the outer region of the system. Our observations have discovered two companion candidates around HAT-P-7 (Narita *et al.* 2009). This paper modeled and constrained the Kozai migration scenario for HAT-P7 b under the existence of a binary star, and found that the Kozai migration scenario was realizable only in a very limited condition, and was not favored if the additional body HAT-P-7 c exist. It conclude that planet-planet scattering is particularly plausible for the migration mechanism of HAT-P-7 b.

3. Future Instruments plan: SCExAO

The SCExAO Project is a project for an upgrade of HiCIAO, that will be installed between Subaru's 188-actuator AO system and HiCIAO, led by Olivier Guyon and Frantz Martinache. The system essentially consists of a 1020-actuator MEMS Deformable Mirror to improve the AO correction, a high-performance PIAA (Phase Induced Amplitude Apodization) Coronagraph as well as aperture masking interferometry capability.

HiCIAO with SCExAO observation will start 2011.

References

Narita, N., *et al.* 2010, *PASJ*, 62, L123
Tamura, M. 2009, *AIPC*, 1158, 11T
Thalmann, C., *et al.* 2010, *ApJ*, 718, L87
Thalmann, C., *et al.* 2009, *ApJ*, 707, L123

The Astrophysics of Planetary Systems: Formation, Structure, and Dynamical Evolution
Proceedings IAU Symposium No. 276, 2010
A. Sozzetti, M. G. Lattanzi & A. P. Boss, eds.

doi:10.1017/S1743921311020709

Non-convergence of the critical cooling timescale for fragmentation of self-gravitating discs

Farzana Meru[1,2] and Matthew R. Bate[1]

[1]School of Physics, University of Exeter, Stocker Road, Exeter, EX4 4QL, UK
email: farzana@astro.ex.ac.uk

[2]Institut für Astronomie und Astrophysik, Universität Tübingen, Auf der Morgenstelle 10, 72076 Tübingen, Germany

Abstract. We carry out a resolution study on the fragmentation boundary of self-gravitating discs. We perform three-dimensional Smoothed Particle Hydrodynamics (SPH) simulations of discs to determine whether the critical value of the cooling timescale in units of the orbital timescale, $\beta_{\rm crit}$, converges with increasing resolution. Using particle numbers ranging from 31,250 to 16 million (the highest resolution simulations to date) we do not find convergence. Instead, fragmentation occurs for longer cooling timescales as the resolution is increased. These results certainly suggest that $\beta_{\rm crit}$ is larger than previously thought. However, the absence of convergence also questions whether or not a critical value exists. In light of these results, we caution against using cooling timescale or gravitational stress arguments to deduce whether gravitational instability may or may not have been the formation mechanism for observed planetary systems.

Keywords. accretion, accretion disks, gravitation, instabilities, hydrodynamics, methods: numerical, planetary systems: formation, planetary systems: protoplanetary disks

1. Introduction

There are two quantities that have historically been used to determine whether a self-gravitating disc is likely to fragment. The first requires the stability parameter, $Q \lesssim 1$ (Toomre 1964), where $Q = \frac{c_{\rm s}\kappa_{\rm ep}}{\pi \Sigma G}$, $c_{\rm s}$ is the sound speed in the disc, $\kappa_{\rm ep}$ is the epicyclic frequency, which for Keplerian discs is $\approx \Omega$, the angular frequency, Σ is the surface mass density and G is the gravitational constant. Gammie (2001) showed the need for fast cooling and suggested that if the cooling timescale can be parameterised as $\beta = t_{\rm cool}\Omega = u\left(\frac{{\rm d}u_{\rm cool}}{{\rm d}t}\right)^{-1}\Omega$, where u is the specific internal energy and ${\rm d}u_{\rm cool}/{\rm d}t$ is the total cooling rate, then fragmentation requires $\beta \lesssim \beta_{\rm crit}$, where the critical cooling timescale, $\beta_{\rm crit} \approx 3$ for a ratio of specific heats, $\gamma = 2$. Rice *et al.* (2005) performed 3D SPH simulations to show the dependence of $\beta_{\rm crit}$ on γ: for discs with $\gamma = 5/3$ and $7/5$, $\beta_{\rm crit} \approx 6 - 7$ and $\approx 12 - 13$, respectively. The above authors show that the cooling condition is equivalent to a maximum gravitational stress that a disc can support without fragmenting. Recently, Meru & Bate (2010) suggested that $\beta_{\rm crit}$ depends on the disc and star conditions.

We carry out a thorough convergence test of the value of $\beta_{\rm crit}$ required for fragmentation using a 3D SPH code. We simulate a $0.1{\rm M}_\odot$ disc surrounding a $1{\rm M}_\odot$ star, spanning a radial range, $0.25 \leqslant R \leqslant 25$ au, using $\gamma = 5/3$. The initial surface mass density and temperature profiles are $\Sigma \propto R^{-1}$ and $T \propto R^{-1/2}$, respectively, and initially, $Q \gtrsim 2$ everywhere. We carry out simulations with this disc setup using 31,250, 250,000, 2 million and 16 million particles. We simulate the discs using various values of β to determine $\beta_{\rm crit}$ at different resolutions.

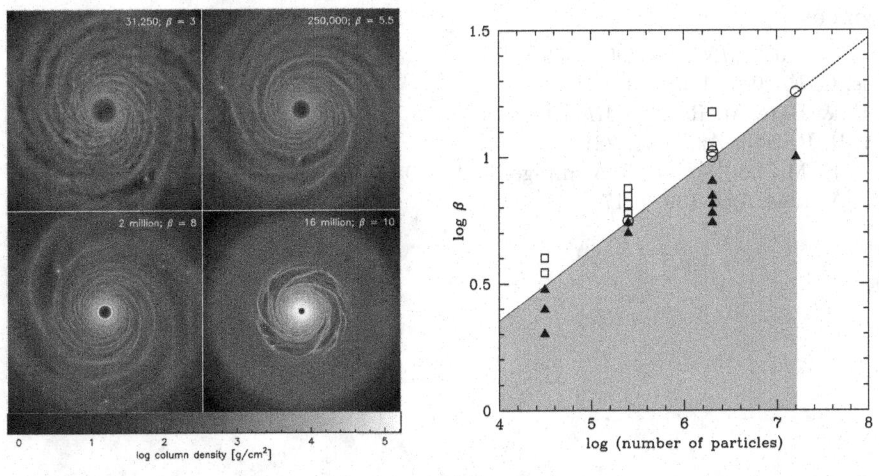

Figure 1. a. Surface mass density rendered images of the fragmenting discs with 31,250, 250,000, 2 million and 16 million particles. At higher resolution, the disc can fragment for larger β. The axes scale from -25 au to 25 au in both directions. **b.** Graph of β against resolution of the non-fragmenting (open squares), fragmenting (solid triangles) and borderline (open circles) simulations (defined as discs that fragment but quickly shear apart with no further fragmentation). The solid black line divides the fragmenting and non-fragmenting cases and the grey region is where fragmentation can occur. The graph shows no evidence of convergence with resolution. The thin dotted line shows how the trend will continue if convergence is not reached with even higher resolution. If convergence can be achieved, the dotted line would follow a flatter profile.

2. Results

Figure 1a shows images of the fragmenting discs at various resolutions. We see that fragmentation occurs for higher values of β as the resolution increases. Figure 1b summarises all the simulations performed. With the data that is available, the dividing line between the fragmenting and non fragmenting cases increases linearly with linear resolution and therefore *convergence has not been reached.*

3. Numerical and observational implications

The lack of convergence certainly shows that the critical value of the cooling timescale is, at the very least, longer than previously thought. However, it also opens up the possibility that there may be no value of β for which such a disc can avoid fragmentation, given sufficient resolution. If this is the case, it suggests that the problem may be ill posed. In other words, it may not be possible for a disc to settle into an equilibrium where there is a balance between heating from gravitational instabilities and a *simple* imposed cooling timescale. This implies that a self-gravitating disc that cools at a rate given by $\frac{\mathrm{d}u}{\mathrm{d}t} = \frac{u\Omega}{\beta}$, and is only heated by internal dissipation due to gravitational instabilities, may not be able to attain a self-regulated state and will always fragment, regardless of the value of β. This re-opens the question of what the criterion for fragmentation of a self-gravitating disc really is and in addition, where in a disc fragmentation can realistically occur. These results cast some serious doubts on previous conclusions concerning fragmentation of self-gravitating discs. In addition, since cooling timescale arguments can be used to determine at what radii in a disc fragmentation can occur (e.g. Rafikov 2009; Clarke 2009), the results presented here need to be considered when making conclusions as to whether observed planetary systems may or may not have formed by gravitational instability.

References

Clarke, C. J. 2009 *MNRAS*, 396, 1066
Gammie, C. F. 2001 *ApJ*, 553, 174
Meru, F., & Bate, M. R. 2010 *MNRAS*, in press
Rafikov, R. R. 2009 *ApJ*, 704, 281
Rice, W. K. M., Lodato, G., & Armitage, P. J. 2005 *MNRAS*, 364, L56
Toomre, A. 1964 *ApJ*, 139, 1217

The Astrophysics of Planetary Systems: Formation, Structure, and Dynamical Evolution
Proceedings IAU Symposium No. 276, 2010
A. Sozzetti, M. G. Lattanzi & A. P. Boss, eds.

doi:10.1017/S1743921311020710

Planetary systems formation and the diversity of extrasolar systems

Yamila Miguel[1,2], Octavio M. Guilera[1,2], and Adrián Brunini[1,2]

[1]Facultad de Cs. Astronómicas y Geofísicas - Universidad Nacional de La Plata, Argentina
[2]Instituto de Astrofísica de La Plata (IALP) - CONICET
email: ymiguel@fcaglp.unlp.edu.ar

Abstract. With the end of answer questions as, how common are planetary systems like our own in the Universe? and What is the diversity of planetary systems that we could find in the universe?, we develop a semi-analytical model for computing planetary systems formation and consider different initial conditions for generating a large sample of planetary systems, which is analysed statistically. We explore the effects in the planetary system architecture of assuming different initial disc profiles and planetary migration rates.

Keywords. planets and satellites: formation, planetary systems: formation

Introduction

The set of planetary systems discovered orbiting around single stars in the solar neighborhood is remarkably diverse and displays a wide range of architectures that reflect the process of planetary formation and are a consecuence of the environment where they were born. Our main objective is to explore the importance of several factors in defining the architecture of a planetary system, explaining the observed diversity of planetary systems. To this end, we explore different gas and solids disc profiles, as well as different planetary migration rates, to find out which factors reproduce the different observed planetary systems and then predict the systems that will be the more common and thus those expected to be find in the Universe.

A Brief Description of the Model

Our model for planetary systems formation is based on our previous works (e.g. Miguel & Brunini 2009; Miguel *et al.* 2010) and its essence is:

Following last protoplanetary discs observations (Andrews *et al.* 2009), we assume that the gas and solid surface density are characterized by a power-law in the inner part of the disc, with an exponent γ which take different values (we assume $\gamma = 0.5$, 1 and 1.5) and an exponential decay in the outer part.

The first initial core is located at the inner edge of the disc, the rest of the cores are separated $10R_H$ each other until the end of the disc.

The cores grow in the oligarchic growth regime (Ida & Makino 1993) and they also grow due to collisions with other embryos. When a core reaches the critical mass, the gas accretion process begins, whose rate was fitted from the results of Fortier *et al.* (2009).

We consider type I and II planetary migration. Type I is very fast and a factor c_{migI} is assumed for delaying it. We assume $c_{migI} = 0.01$, 0.1 and 0, which represent a migration delayed 100, 10 times, and not considered, respectively.

Table 1.

	$\gamma = 0.5$			$\gamma = 1$		
PS Class	$C_{migI} = 0$	$C_{migI} = 0.01$	$C_{migI} = 0.1$	$C_{migI} = 0$	$C_{migI} = 0.01$	$C_{migI} = 0.1$
hot-Jupiter	0.2	1.3	3.6	2.3	10.7	12
solar system	11.3	11.6	6.6	25.1	19.8	12
combined	0	2.2	3.3	0.3	11.8	6.4
rocky	77.2	73.9	75.2	71.4	56.8	67.6
failed	11.3	11	11.3	0.9	0.9	2

	$\gamma = 1.5$		
PS Class	$C_{migI} = 0$	$C_{migI} = 0.01$	$C_{migI} = 0.1$
hot-Jupiter	7.7	13.2	16.9
solar system	22.6	20	9.5
combined	0.8	11.6	2.2
rocky	68.1	54.4	49
failed	0.8	0.7	22.4

Results

We perform 12 simulations, in each one we explore the 3 different gas and solids disc density profiles considered and also the different planetary migration rates. In each simulation 1000 planetary systems are formed, where the initial conditions for our discs are chosen random according to the observations.

Since most of the observed planets are giant planets, we use them for a new planetary systems classification:

hot-Jupiter systems: these planetary systems host planets with masses larger than 15 $M_\oplus$ at a distance less than 1 *au*.

solar systems: these systems harbor giant planets located between 1 and 30 *au*.

combined systems: these systems harbor at least one giant planet within 1 *au* and also in the middle part of the disc.

cold-Jupiter systems: in this case the giant planets are located further from 30 *au*.

rocky systems: these systems have only planets with masses less than 15 $M_\oplus$.

According to this new classification, we analyse statistically the population of planetary systems generated with our model. The table shows the % of different kinds of planetary systems found when assuming different values of γ and different c_{migI}. We exclude the cold-Jupiter systems, since we are unable to form this systems with our model.

As seen rocky systems are the most common in the Universe. We also found that solar systems are not expected to be rare in the Universe, being their formation favoured when $\gamma = 1$ and slow migration rates. Finally, the hot-Jupiter systems are born in discs with $\gamma = 1.5$ and fast migration rates.

References

Andrews, S. M., *et al.* 2009, *ApJ*, 700, 1502
Fortier, A., Benvenuto, O., & Brunini, A. 2009, *A&A*, 500, 1249
Ida, S. & Makino, J. 1993, *Icarus*, 106, 210
Miguel, Y. & Brunini, A. 2009, *MNRAS*, 392, 391
Miguel, Y., Guilera, O. M., & Brunini, A. 2011, *MNRAS*, 412, 2113

The Astrophysics of Planetary Systems: Formation, Structure, and Dynamical Evolution
Proceedings IAU Symposium No. 276, 2010
A. Sozzetti, M. G. Lattanzi & A. P. Boss, eds.

doi:10.1017/S1743921311020722

Metallicity of M dwarfs: The link to exoplanets

V. Neves[1,2], X. Bonfils[2] and N. C. Santos[1,3]

[1]Centro de Astrofísica, Universidade do Porto, Rua das Estrelas, 4150-762 Porto, Portugal
email: vasco.neves@astro.ua.pt

[2]Laboratoire d'Astrophysique, Observatoire de Grenoble, BP 53, F-38041 Grenoble Cédex 9, France

[3]Departamento de Física e Astronomia, Faculdade de Ciências, Universidade do Porto, Portugal

Abstract. The determination of the stellar parameters of M dwarfs is of prime importance in the fields of galactic, stellar and planetary astronomy. M stars are the least studied galactic component regarding their fundamental parameters. Yet, they are the most numerous stars in the galaxy and contribute to most of its total (baryonic) mass. In particular, we are interested in their metallicity in order to study the star-planet connection and to refine the planetary parameters. As a preliminary result we present a test of the metallicity calibrations of Bonfils *et al.* (2005), Johnson & Apps (2009), and Schlaufman & Laughlin (2010) using a new sample of 17 binaries with precise V band photometry.

Keywords. stars: fundamental parameters, planetary systems, stars: late-type, stars: abundances, stars: atmospheres

Preliminary Results

We tested the metallicity calibration of Bonfils *et al.* (2005) (hereafter B05), as well as the calibrations of Johnson & Apps (2009) (hereafter JA09) and Schlaufman & Laughlin (2010) (hereafter SL10), with a sample of 17 M dwarf secondaries with a wide (> 5 arcsec separation) physical FGK companion.

Following B05, three papers with different calibrations were published: JA09, SL10, and Rojas-Ayala *et al.* (2010) (hereafter RA10). Each work claims a calibration with a better precision than the previous ones, and in general, poor V photometry is identified as a serious limitation. In order to address the photometric limitation, only M stars with precise V photometry ($\sigma < 0.04$ mag) were selected. Most stars have V magnitude uncertainties of 0.01 or 0.02 mag. Note that the RA10 calibration was not tested because it requires IR indices that we do not have. This test will be done in the near future.

We found that the metallicity values of our stars (obtained from the FGK primary component) are in reasonable agreement with the [Fe/H] values obtained with all calibrations, as can be seen in Fig. 1. However, our calibrators are found to be more metal poor (on average) than both JA09 and SL10 calibrations.

A better photometry did not improve the dispersion measured around the different calibrations. This means that precision on V photometry may not be the main limitation in the derivation of the [Fe/H] calibration.

Table 1. shows a quantitative comparison between the calibrations. We note that the rms, RMS_P and the R^2_{ap} values were offset-corrected. In general, the calibrations have similar offsets, rms, RMS_P, and correlation coefficients.

Table 1. Comparison of the residuals offset, *rms*, residual mean square (RMS_P), and adjusted square of the multiple correlation coefficient (R^2_{ap}) of the calibrations of Bonfils *et al.* (2005), Johnson & Apps (2009), and Schlaufman & Laughlin (2010) applied to our data. RMS_P and R^2_{ap} definitions were taken from Schlaufman & Laughlin (2010).

Calibration Source + equation	offset (dex)	rms (dex)	RMS_P (dex)	R^2_{ap}
B05 (1) : $[Fe/H] = 0.196 - 1.527M_K + 0.091M^2_K + 1.886(V-K) - 0.142(V-K)^2$	0.05	0.21	0.06	0.18
B05 (2) : $[Fe/H] = -0.149 - 6.508\Delta M, \Delta M = Mass_V - Mass_K$	0.07	0.23	0.06	0.17
JA09 : $[Fe/H] = 0.56\Delta M_K - 0.05, \Delta M_K = MS - M_K$	0.19	0.22	0.05	0.27
SL10 : $[Fe/H] = 0.79\Delta(V-K) - 0.17, \Delta(V-K) = (V-K)_{obs} - (V-K)_{fit}$	0.06	0.22	0.05	0.28

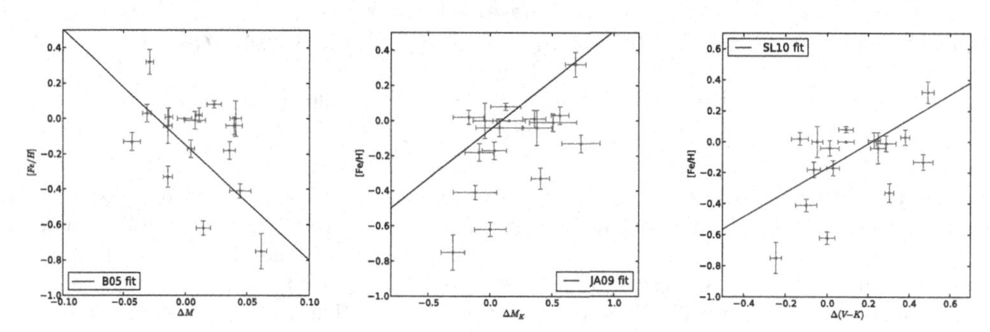

Figure 1. *Left panel:* Plot of metallicity versus the difference between masses calculated from the V- and the K-band Mass-Luminosity equations of Delfosse *et al.* (2000). The black line represents the calibration of B05 (2). *Middle panel:* [Fe/H] versus the difference between the mean value of Mk of M dwarfs and the value of Mk of each star, as defined by Johnson & Apps (2009). The black line represents the calibration of JA09. *Right panel:* [Fe/H] versus the observed difference between V and K magnitudes (V-K) and the fit of the (V-K) corresponding to the horizontal distance (in the V-K, Mk plane) between the mean value of Mk of M dwarfs and the value of Mk of each star, as defined by Schlaufman & Laughlin (2010). The black line represents the calibration of SL10.

Interestingly, the calibration of B05 (1) has the lowest offset and rms. However, the correlation coefficient is a bit lower than the values of JA09 and SL10. The results are inconclusive and require further study.

Acknowledgements: We acknowledge the support by the European Research Council/European Community under the FP7 through Starting Grant agreement number 239953. NCS also acknowledges the support from Fundação para a Ciência e a Tecnologia (FCT) through program Ciência 2007 funded by FCT/MCTES (Portugal) and POPH/FSE (EC), and in the form of grant reference PTDC/CTE-AST/098528/2008. VN would also like to acknowledge the support from FCT in the form of the fellowship SFRH/BD/60688/2009.

References

Delfosse, X., Forveille, T., Ségransan, D., Beuzit, J.-L., Udry, S., Perrier, C., & Mayor, M. 2000 *A&A*, 364, 217

Bonfils, X., Delfosse, X., Udry, S., Santos, N. C., Forveille, T., & Ségransan, D. 2005 *A&A*, 442, 635

Johnson, J. A. & Apps, K. 2009 *ApJ*, 699, 933

Schlaufman, K. C. & Laughlin, G. 2010, *A&A*, 519, A105

Rojas-Ayala, B., Covey, K. R, Muirhead, P. S., & Lloyd, J. P. 2010, *ApJ*, 720, L113

The Astrophysics of Planetary Systems: Formation, Structure, and Dynamical Evolution
Proceedings IAU Symposium No. 276, 2010
A. Sozzetti, M. G. Lattanzi & A. P. Boss, eds.

doi:10.1017/S1743921311020734

The Pennsylvania-Toruń search for planets around evolved stars with HET

Andrzej Niedzielski[1], Alex Wolszczan[2,3], Grzegorz Nowak[1], Paweł Zieliński[1], Monika Adamów[1] and Sara Gettel[2,3]

[1]Toruń Centre for Astronomy, Nicolaus Copernicus University, Gagarina 11, 87-100 Toruń, Poland
email: Andrzej.Niedzielski@astri.uni.torun.pl

[2]Department for Astronomy and Astrophysics, Pennsylvania State University, 525 Davey Laboratory, University Park, PA 16802

[3]Center for Exoplanets and Habitable Worlds, Pennsylvania State University, 525 Davey Laboratory, University Park, PA 16802

Abstract. Searches for planets around giants represent an essential complement to 'traditional' surveys, because they furnish information about properties of planetary systems around stars that are the descendants of the A-F main sequence (MS) stars with masses as high as $\sim 5M_{\odot}$. As the stars evolve off the MS, their effective temperatures and rotation rates decrease to the point that their radial velocity variations can be measured with a few ms^{-1} precision. This offers an excellent opportunity to improve our understanding of the population of planets around stars that are significantly more massive than the Sun, without which it would be difficult to produce abroad, integrated picture of planet formation and evolution. Since 2001, about 30 such objects have been identified, including our five published HET detections (Niedzielski *et al.* 2007; Niedzielski *et al.* 2009a; Niedzielski *et al.* 2009b). Our work has produced the tightest orbit of a planet orbiting a K-giant identified so far (0.6 AU), and the first convincing evidence for a multiplanet system around such as star (Niedzielski *et al.* 2009a). Our most recent discoveries (Niedzielski *et al.* 2009b) have identified new multiplanet systems, including a very intriguing one of two brown dwarf-mass bodies orbiting a $2.8M_{\odot}$, K2 giant. This particular detection challenges the standard interpretation of the so-called brown dwarf desert known to exist in the case of solar-mass stars. Along with discoveries supplied by other groups, our work has substantially added to the emerging evidence that stellar mass positively correlates with masses of substellar companions, all the way from red dwarfs to intermediate-mass stars. We present current status and forthcoming results from the Pennsylvania-Toruń Search for Planets performed with the Hobby-Eberly Telescope (HET) since 2004.

Keywords. planetary systems, stars: low-mass, brown dwarfs

1. Introduction

High precision radial velocities determination for MS stars more massive than $\sim 1.3M_{\odot}$ are not possible due to small number of spectral features present in the spectra of these stars and significant rotational broadening. In search for planetary mass companions to intermediate mass stars (IMS) one may try to use transits or direct imaging. One may also obtain precise RV for intermediate mass stars after they leave the MS and become subgiants or giants cool enough to develop spectra reach in narrow lines. Detailed analysis of the IMS stars with planets shows that relative efficiency of these four approaches differs significantly. In the case of stars more massive that $\sim 1.6M_{\odot}$ RV search for planets around giants is the most efficient way while for stars more massive than $\sim 2.2M_{\odot}$ it is the only way. The Pennsylvania- Torun Search for Plantes (PTPS) is devoted to radial velocity

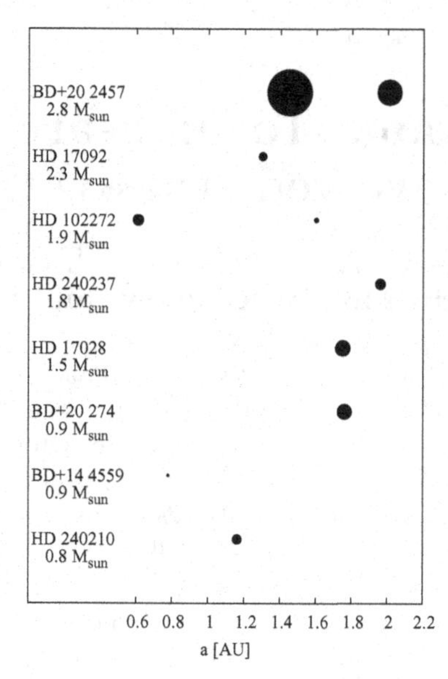

Figure 1. The first 10 substellar-mass companions discovered by PTPS. The symbol size is proportional to mass.

seach for and characterization of planets around stars more massive than the Sun with the Hobby - Eberly Telescope.

2. Observations & the sample

Observations for the Pennsylvania - Torun Planet Search (PTPS) are obtained with the Hobby-Eberly Telescope (HET, Ramsey *et al.* 1998) equipped with the High Resolution Spectrograph (HRS, R=60.000, Tull 1998) in the queue scheduled mode (Shetrone *et al.* 2007). The spectrograph was fed with a 2 arcsec fiber. The spectra consisted of 46 Echelle orders recorded on the 'blue' CCD chip (407-592 nm) and 24 orders on the 'red' one (602-784 nm). Typical signal to noise ratio was 200-250 per resolution element. The basic data reduction was performed in a standard manner using IRAF tasks and scripts. RVs were measured using the standard I_2 cell calibration technique (Butler *et al.* 1996). Details of our survey, the observing procedure, and data analysis have been described in detail elsewhere (Niedzielski *et al.* 2007, Niedzielski & Wolszczan 2008).

The sample of stars observed within PTPS is composed of about 350 giants from the Giant Clump, about 350 giants and subgiants and about 250 evolved MS stars on the upper envelope of the MS (aging dwarfs).

All stars are studied in detail to determine atmospheric parameters (T_{eff}, logg, [Fe/H]) and integrated parameters (mass, luminosity, age). Most of stars included in PTPS have masses ranging between 1 and 3 solar masses.

Results

For about 800 stars conclusive multi epoch observations were gathered. They allow us to subdivide our targets into single stars, stellar binaries and possible substellar-mass companions hosts. The three subsamples do not differ much in frequency of stellar binaries ($\sigma_{RV} > 250ms^{-1}$) which ranges between 21 % for the giants and subgiants sample and

27 % for the aging dwarfs. The frequency of single stars ($\sigma_{RV} < 20ms^{-1}$) varies from 45 % in the clump giant sample to 62% for the giants and subgiants sample. Depending on criteria adopted the substellar-mass companion frequency may reach up to 27 % in the case of the clump giant sample.

In Figure 1 the first 10 substellar-mass companions discovered by PTPS are presented. In addition to published already planets and brown dwarfs we present here also two more planets from Gettel *et al.* (2011) - BD+20 274 b and HD 240237 b and from Nowak *et al.* (2011) - HD 17028 b.

Acknowledgments

We acknowledge the financial support from the Polish Ministry of Science and Higher Education through grants N203 510938 and N203 386237. AW acknowledges support from NASA grant NNX09AB36G. GN is a recipient of a graduate stipend of the Chairman of the Polish Academy of Sciences. We thank the HET resident astronomers and telescope operators for support. The Hobby-Eberly Telescope (HET) is a joint project of the University of Texas at Austin, the Pennsylvania State University, Stanford University, Ludwig-Maximilians-Universität München, and Georg-August-Universität Göttingen. The HET is named in honor of its principal benefactors, William P. Hobby and Robert E. Eberly. IRAF is distributed by the National Optical Astronomy Observatories and operated by the Association of Universities for Research in Astronomy, Inc., under co-operative agreement with the National Science Foundation.

References

Butler, R. P., Marcy, G. W., Williams, E., McCarthy, C., & Dosanjh, P. 1996, *PASP*, 108, 500

Gettel, S. & Wolszczan, A., *et al.* 2011, *in preparation*

Niedzielski, A., Konacki, M., Wolszczan, A., Nowak, G., Maciejewski, G., Gelino, C. R., Shao, M., Shetrone, M., & Ramsey, L. W. 2007, *ApJ*, 669, 1354

Niedzielski, A. & Wolszczan, A. 2008, *Proc. IAU Symp.*, 249, 43

Niedzielski, A., Goździewski, K., Wolszczan, A., Konacki, M., Nowak, G., & Zielinski, P. 2009a, *ApJ*, 693, 276

Niedzielski, A., Nowak, G., Adamów, M., & Wolszczan, A. 2009b, *ApJ*, 707, 768

Nowak, G. & Niedzielski, A. *et al.* 2011, *in preparation*

Ramsey, L. W. & Adams, M. T. *et al.* 1998, *Proc. SPIE*, 3352, 34

Shetrone, M., Cornell, M., Fowler, J., *et al.* 2007, *PASP*, 119, 556

Tull, R. G. 1998, *Proc. SPIE*, 3355, 387

The Astrophysics of Planetary Systems: Formation, Structure, and Dynamical Evolution
Proceedings IAU Symposium No. 276, 2010
A. Sozzetti, M. G. Lattanzi & A. P. Boss , eds.

doi:10.1017/S1743921311020746

A survey of M stars in the field of view of *Kepler* space telescope

Mahmoudreza Oshagh[1], Nader Haghighipour[2] and Nuno C. Santos[1]

[1]Centro de Astrofísica, Faculdade de Ciências, Universidade do Porto, Rua das Estrelas, 4150–762 Porto, Portugal
email: moshagh@astro.up.pt

[2]Institute for Astronomy and NASA Astrobiology Institute, University of Hawaii-Manoa, 2680 Woodlawn Drive, Honolulu, HI 96822,USA

Abstract. M dwarfs constitute more than 70% of the stars in the solar neighborhood. They are cooler and smaller than Sun-like stars and have less-massive disks which suggests that planets around these stars are more likely to be Neptune-size or smaller. The transit depths and transit times of planets around M stars are large and well-matched to the *Kepler* temporal resolution. As a result, M stars have been of particular interest for searching for planets in both radial velocity and transit photometry surveys. We have recently started a project on searching for possible planet-hosting M stars in the publicly available data from *Kepler* space telescope. We have used four criteria, namely, the magnitude, proper motion, H-K_s and J-H colors, and searched for M stars in Q0 and Q1 data sets. We have been able to find 108 M stars among which 54 had not been previously identified among *Kepler*'s targets. We discuss the details of our selection process and present the results.

Keywords. stars: low-mass, planetary systems

1. Introduction

The *Kepler* space telescope is monitoring more than 150,000 stars in a 105 square degree field of view around the constellations Cygnus and Lyra. The data from this telescope provide a great opportunity for identifying potential terrestrial planets around variety of stars. Among these stars, M dwarfs are ideal targets for searching for terrestrial/habitable planets. These stars have the greatest reflex acceleration due to an orbiting planet, their low surface temperatures place their habitable zones at close distances, and their light curves show large decrease when they are transited by a planet. The transit depths and transit times of planets around M stars are also large and well-matched to the *Kepler* temporal resolution. We have surveyed *Kepler*'s publicly available data from quarters Q0 and Q1, and identified more than a hundred M stars. In this paper, we discuss our methodology and present our selection process.

2. Selection Criteria

Traub & Cutri (2008) were the first to make an attempt to identify M stars in *Kepler*'s field of view. They considered stars with K_s smaller than 13.5 mag, and proposed a selection process based on a criterion on H-K_s vs. J-H colors. Since M stars are relatively faint and abundant, Traub & Cutri (2008) suggested that any detected M star is likely to be nearby and therefore likely to have a relatively large proper motion. Using these criteria, these authors estimated that close to 1600 M stars should exist in *Kepler*'s field of view. A recent article by Batalha *et al.* (2010) suggests that the actual number of M stars in *Kepler*'s field of view may be twice as large. Using selection criteria based on

Table 1. Star Counts as a Function of Effective Temperature and Magnitude

Mag	$T_{\rm eff}$ (4500 – 5500)	$T_{\rm eff}$ (3500 – 4500)	Total
$6 < K < 7$	0	0	0
$7 < K < 8$	0	0	0
$8 < K < 9$	0	5	5
$9 < K < 10$	1	3	4
$10 < K < 11$	0	9	9
$11 < K < 12$	0	13	13
$12 < K < 13$	1	13	14
$13 < K < 14$	0	1	1
$14 < K < 15$	0	0	0

a star's surface gravity, effective temperature, and inferred stellar radius, these authors were able to show that approximately 3000 M stars are present in the field of view of *Kepler*.

We used the following criteria to look for M stars in *Kepler*'s publicly available data of quarters Q0 and Q1. These criteria are somewhat similar to those of Traub & Cutri (2008) with the exception that we expanded the requirements on the magnitudes of stars using the ranges suggested by Leggett (1992):

- K magnitude equal to or smaller than 15 mag,
- J-H color in the range of 0.42 to 0.78,
- H-K color in the range of 0.12 to 0.50,
- Proper motion bigger than 0.1.

Our search resulted in identifying 108 M stars. Among these stars, 27 are in both the Q0 and Q1 data sets. When comparing with the sample of M stars presented by Batalha *et al.* (2010), we discovered that there are no data on the effective temperature and surface gravity of 54 of our M stars. As a result, these 54 M stars were not identified by Batalha *et al.* (2010). One example of these M stars is LHS 6343, a member of a M star binary system with a transiting brown dwarf (Johnson *et al.* 2011). Table 1 shows 46 of these 54 M stars that have $\log g > 3.5$. These stars have been binned by magnitude and effective temperature.

This work has been supported by the European Research Council/European Community under the FP7 through a Starting Grant, as well as in the form of grant reference PTDC/CTEAST/098528/2008 funded by Fundação para a Ciência e a Tecnologia (FCT), Portugal. NCS would further like to thank the support from FCT through a Ciência , 2007 contract funded by FCT/MCTES (Portugal) and POPH/FSE (EC). NH acknowledges support from NASA Astrobiology Institute (NAI) under Cooperative Agreement NNA04CC08A at the Institute for Astronomy, University of Hawaii, NAI central, and NASA EXOB grant NNX09AN05G.

References

Batalha, N. M., *et al.* 2010, *ApJ*, 713, L109

Johnson, J. A., *et al.* 2011, *ApJ*, 730, id.79

Leggett, S. K. 1992, *ApJS*, 82, 351

Traub, W. & Cutri, R. 2008, in: Extreme Solar Systems, Eds. D. Fischer, F. A. Rasio, S. E., Thorsett, & A. Wolszczan, *ASP Conf. Ser.*, 398, 475

The Astrophysics of Planetary Systems: Formation, Structure, and Dynamical Evolution
Proceedings IAU Symposium No. 276, 2010
A. Sozzetti, M. G. Lattanzi & A. P. Boss, Eds.

doi:10.1017/S1743921311020758

Observable signatures of dust evolution mechanisms which shape the planet forming regions

Olja Panić[1], Tilman Birnstiel[2], Ruud Visser[3] and Elco van Kampen[1]

[1]European Southern Observatory,
Karl-Schwarzschild Strasse 2, 85748, Garching bei München, Germany
email: opanic@eso.org

[2]Max-Planck-Institut für Astronomie, Königstuhl 17, 69117 Heidelberg, Germany
[3]Leiden Observatory, Leiden University, PO Box 9513, 2300 RA Leiden, The Netherlands

Abstract. Overdensity of dust with respect to the gas in the planet forming regions is a crucial prerequisite to form larger bodies and eventually planets. We use a state-of-the-art code to simulate dust evolution processes in gas-rich circumstellar discs, including the viscous gas evolution. We find significant deviations of the radial distribution of dust from that of the gas as early as 1-2Myr. These deviations are closely related to the efficiency of grain growth. Apparent discrepancies between dust and gas distributions are suggested by the current millimetre interferometer observations, and ALMA will allow us tointerpret any such discrepancies in the context of dust evolution.

Keywords. stars: pre–main-sequence, circumstellar matter, planetary systems: protoplanetary disks, planets and satellites: formation

1. Model description

We use the dust evolution code based on Brauer *et al.* (2008) and Birnstiel *et al.* (2010). The dust growth, fragmentation, radial drift and radial mixing processes are treated in the context of a viscously evolving, irradiated gas disc. We focus on two sets of models, their initial conditions differing only by the assumed limiting relative velocity, v_f, at which the dust grains fragment rather than coagulate upon impact. For low values of v_f, the dust fragments efficiently and remains rather small in size during the evolution of the disc, while for high values of v_f the dust grows to larger sizes. Our two models have $v_f = 1$ m/s (Model A) and 10 m/s (Model B). The main model parameters are the disc radial size set to 250 AU, gas mass of 0.1 $M_\odot$ and dust to gas mass ratio of 1:100. These parameters are fixed at time=0 and vary during the calculation due to aforementioned processes. A more detailed description of the model parameters can be found in Birnstiel *et al.* 2010.

The dust size is an important factor in assessing the effects of radial drift of dust in discs. The choice of fragmentation velocity value, v_f, directly influences the dust size and drift efficiency in our models. Our study examines the consequences of dust evolution, and in particular the radial drift, on the disc structure as a whole, and not only in the planet forming region. The motivation for this is the prospect of directly probing the outer disc regions at high sensitivity with the Atacama Large Millimeter/submillimeter Array in near future.

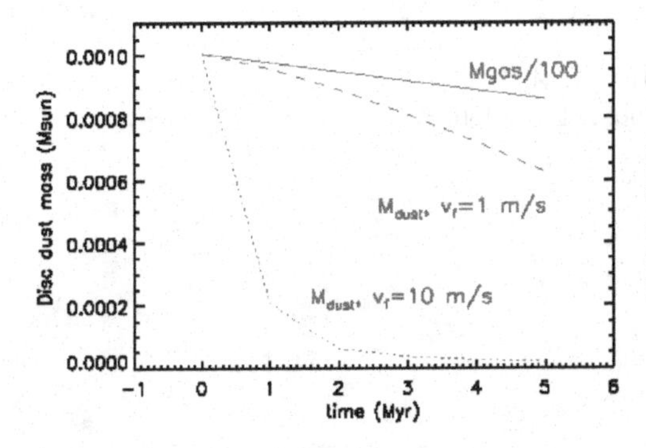

Figure 1. The disc dust mass evolution over time for Model A (dashed line) and Model B (dotted line). The disc gas mass evolution, identical for both Models A and B, is shown scaled by factor 0.01 (solid line) for comparison.

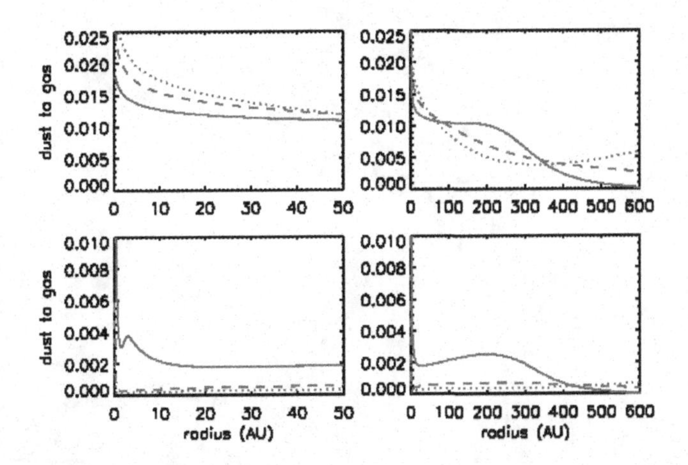

Figure 2. *Upper panels:* The dust to gas mass ratio in Model A at 1 Myr (solid lines), 2 Myr (dashed lines) and 3 Myr (dotted lines). *Lower panels:* Same as in upper panels, for Model B.

2. Results: Dust to gas mass ratio

The influence that dust growth and drift have on disc dust mass is most pronounced in Model B (see Fig. 1). At time of 1 Myr the disc dust mass has decreased by 80% due to efficient radial drift and consequent accretion onto the star. Model A exhibits a dust mass decrease but at a notably slower rate of only 20% over 5 Myr. The gas evolves independently of the dust in our models, including only accretion and viscous evolution. The disc gas mass decreases more gradually than the dust (shown scaled by factor 0.01 in Fig. 1 for comparison).

The overall dust to gas mass ratio of the disc in both our models decreases over time. The most affected are the disc outer regions where most mass resides and where the optimal dust size for drift is small. Figure 2 shows the evolution on the dust to gas mass ratio in Models A (upper panels) and B (lower panels). In Model A the dust is depleted from the outer disc and builds up in the planet forming regions, because the dust remains smaller than the optimal drift size corresponding to the physical properties in these regions. In Model B, the dust grows efficiently and drifts throughout the disc, causing a decrease of the dust to gas mass ratio at all radii.

An increase in the amount of solids in the inner disc is a necessary pre-requisite for planet formation. Our modelling suggests that local processes, such as clumping due to streaming instabilities, are necessary to allow dust to grow to large sizes whilst retaining it in the planet forming regions.

References

Birnstiel, T., Dullemond, C. P., & Brauer, F. 2010, *A&A*, 513, A79
Brauer, F., Dullemond, C. P., & Henning, Th. 2008, *A&A*, 480, 859

The Astrophysics of Planetary Systems: Formation, Structure, and Dynamical Evolution
Proceedings IAU Symposium No. 276, 2010
A. Sozzetti, M. G. Lattanzi & A. P. Boss, eds.

doi:10.1017/S174392131102076X

Wind-shearing in gaseous protoplanetary disks

Hagai B. Perets[1] **and Ruth Murray-Clay**[1]

[1]Harvard-Smithsonian Center for Astrophysics, 60 Garden st. Cambridge MA 02338, US
email: hperets@cfa.harvard.edu

Abstract. One of the first stages of planet formation is the growth of small planetesimals and their accumulation into large planetesimals and planetary embryos. This early stage occurs much before the dispersal of most of the gas from the protoplanetary disk. Due to their different aerodynamic properties, planetesimals of different sizes/shapes experience different drag forces from the gas at these stage. Such differential forces produce a wind-shearing effect between close by, different size planetesimals. For any two planetesimals, a wind-shearing radius can be considered, at which the differential acceleration due to the wind becomes greater than the mutual gravitational pull between the planetesimals. We find that the wind-shearing radius could be much smaller than the *gravitational* shearing radius by the Sun (the Hill radius), i.e. during the gas-phase of the disk wind-shearing could play a more important role than tidal perturbations by the Sun. Here we study the wind-shearing radii for planetesimal pairs of different sizes and compare it with gravitational shearing (drag force vs. gravitational tidal forces). We then discuss the role of wind-shearing for the stability and survival of binary planetesimals, and provide stability criteria for binary planetesimals embedded in a gaseous disk.

Keywords. planets and satellites: dynamical evolution and stability, planets and satellites: formation, planetary systems: protoplanetary disks

---

The interactions between planetesimals play an important role in the evolution of protoplanetary disks and planet formation (Lissauer 1993; Goldreich *et al.* 2004). Most of these interactions and the growth of planetesimals likely occur while the planetesimals are embedded in a gaseous disk. Here we focus on the close interaction between pairs of single planetesimals, i.e. planetesimal-planetesimal-gas interactions.

Planetesimals could vary in size and shape, and therefore have a wide range of aerodynamical properties, which would affect their interaction with the surrounding gas. In particular, planetesimals of different sizes/shapes experience different drag forces from the head wind they encounter in the gaseous disk. The difference between the forces acting on two, different size planetesimals, could effectively change their relative trajectories in respect to their unperturbed motion in the absence of gas (see also Ormel & Klahr 2010). In particular, during an encounter between two different size planetesimals such differential forces result in a wind-shearing (WISH) effect, which could be even stronger than their gravitational interaction. For any two planetesimals, a WISH radius can be considered, at which the acceleration due to aerodynamical wind-shearing becomes greater than the mutual gravitational pull between them. Binary planetesimals can not survive with separation larger than the WISH radius, even if they are gravitationally bound. These issues are discussed in detail in Perets & Murray-Clay (2011).

The differential acceleration between two planetesimal of radii r_b and r_s, due to the wind-shearing effect is given by $\Delta a = |F_D(r_b)/m_b - F_D(r_s)/m_s|$, where $F_D(r)$ is the drag force affecting a planetesimal of radius r.

For small distances, at which the environmental conditions are approximately the same, the differential WISH between any two planetesimals is independent of the

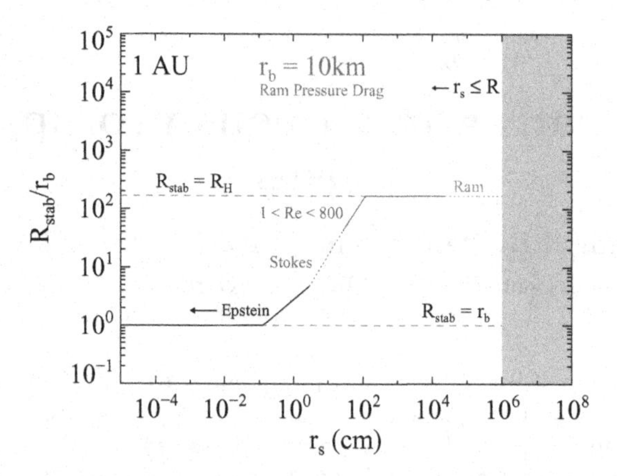

Figure 1. The WISH radius of a 10 km planetesimal and a planetesimal of size r_s, as a function of r_s at 1 AU from the star. The physical size and the Hill radii of the large size planetesimal (lower and upper dashed lines, respectively), are also shown. Also shown are the appropriate drag regimes, which depend on the Reynolds number Re.

distance between them. However, similar to the the Hill radius which arises from the tidal graviataional shearing from the Sun, we can define an important relevant distance scale; the wind-shearing radius. We define this radius as the distance between two planetesimals for which the differential WISH acceleration between them equals their mutual gravitational pull. Beyond this limiting radius even two planetesimals which are gravitationally bound (in the absence of WISH) would be sheared apart by the wind.

We can now derive the WISH radius for any given pair of planetesimals, by equating Δa_{WS} with the gravitational acceleration $a_{grav} = GM/r_{WS}^2$ at the WISH radius (r_{WS}).

The WISH radius can be found for any two planetesimals of arbitrary size, as illustrated in fig. 1. This figure shows the calculated WISH radius as a function of the small planetesimal size. The transition between different drag regimes can be seen in this figure.

Binary planetesimals could be strongly affected by gas drag. In a gas free environment, binary planetesimals are stable as long as their separation is smaller than the Hill radius, whereas wider binaries are destabilized and disrupted by the tidal gravitational shearing from the Sun. However, when gas drag is taken into account, the Hill radius stability limit should be replaced by the WISH radius (as long as $r_{WS} < r_{Hill}$; binaries wider than the Hill radius are always unstable), i.e. we can formulate a stability criteria for binaries embedded in gas: $a_{bin} \leqslant min(r_{Hill}, r_{WS})$. In Fig. 1 the WISH radius represents the limiting separations of binary planetesimals (R_{stab}) or small satellites embedded in a gaseous environment at ahich bound binary planetesimals can exist. The approapariate phase space is delimited by the physical radii, the WISH radii, and the Hill radii.

References

Goldreich, P., Lithwick, Y., & Sari, R. 2004, *ARAA*, 42, 549
Lissauer, J. J. 1993, *ARAA*, 31, 129
Ormel, C. W. & Klahr, H. H. 2010, *A&A*, 520, A43
Perets, H. B. & Murray-Clay, R. A. 2011, *ApJ*, 733, id.56

The Astrophysics of Planetary Systems: Formation, Structure, and Dynamical Evolution
Proceedings IAU Symposium No. 276, 2010
A. Sozzetti, M. G. Lattanzi & A. P. Boss, eds.

doi:10.1017/S1743921311020771

Warm dust around ε Eridani

Martin Reidemeister[1], Alexander V. Krivov[1], Christopher C. Stark[2], Jean-Charles Augereau[3], Torsten Löhne[1] and Sebastian Müller[1]

[1]Astrophysikalisches Institut, Friedrich-Schiller-Universität Jena, Schillergäßchen 2–3, 07745 Jena, Germany
email: martin.reidemeister@astro.uni-jena.de

[2]Department of Physics, University of Maryland, Box 197, 082 Regents Drive, College Park, MD 20742-4111, USA

[3]Laboratoire d'Astrophysique de Grenoble, CNRS UMR 5571, Université Joseph Fourier, Grenoble, France

Abstract. ε Eridani hosts one known inner planet and an outer Kuiper belt analog. Further, Spitzer/IRS measurements indicate that warm dust is present at distances as close as a few AU from the star. Its origin is puzzling, since an "asteroid belt" that could produce this dust would be unstable because of the inner planet. We tested a hypothesis that the observed warm dust is generated by collisions in the outer belt and is transported inward by P-R drag and strong stellar winds. With numerical simulation we investigated how the dust streams from the outer ring into the inner system, and calculated the thermal emission of the dust. We show that the observed warm dust can indeed stem from the outer belt. Our models reproduce the shape and magnitude of the observed SED from mid-IR to sub-mm wavelengths, as well as the Spitzer/MIPS radial brightness profiles.

Keywords. stars: individual (ε Eridani), planetary systems, methods: numerical

1. Introduction

The nearby K2 V star ε Eridani, has a ring of cold dust at ~ 65 AU seen in sub-mm images (Greaves *et al.* 2005), which encompasses an inner disk revealed by Spitzer/MIPS (Backman *et al.* 2009). The star is orbited by an RV planet (Hatzes *et al.* 2000) with $a = 3.4$ AU. Another outer planet may orbit near ~ 40 AU, producing the inner cavity and clumpy structure in the outer ring. The excess emission at $\lambda \gtrsim 15\,\mu$m in a Spitzer/IRS-spectrum indicates that there is warm dust close to the star, at a few AU. Its origin is puzzling, as an inner "asteroid belt" that could produce this dust would be dynamically unstable because of the known inner planet (Brogi *et al.* 2009). Here, we check the possibility that the source of the warm dust is the outer ring, from which dust grains could be transported inward by Poynting-Robertson (P-R) drag and strong stellar winds (30 times the solar wind) (Wood *et al.* 2002).

2. Simulations

With our collisional code *ACE* (*Analysis of Collisional Evolution*) (Krivov *et al.* 2006) we modeled the debris disk beyond 10 AU. This includes the outer ring between 55 and 90 AU and the intermediate region interior to it. We tried different dust compositions, pure astrosilicate (Laor & Draine 1993), as well as mixtures of silicate and water ice (Li & Greenberg 1998). We found that a composition of 30% silicate and 70% ice fits best the far-IR data points. Although the normal optical depth of the disk is rather high ($\tau \sim 10^{-4}$), the disk is transport-dominated, because of the strong stellar wind drag.

To simulate the dust transport further inward from 10 AU through the orbit of the inner planet, we used single-particle numerical integrations. Collisions are not considered in our inner disk model, because they play a minor role in the inner disk. The dust composition was chosen to be pure silicate, because at this distance sublimation of the ice takes effect. We adopted the size distribution from the outer disk simulation and calculated the thermal emission of the dust. The orbital solutions for the inner planet found by Benedict *et al.* (2006) ($e = 0.7$) and Butler *et al.* (2006) ($e = 0.25$) have little influence on the SED. In both cases one can match the Spitzer/IRS spectrum with nearly the same dust mass of $\approx 10^{-7} M_{\oplus}$ for the inner disk.

3. Conclusion

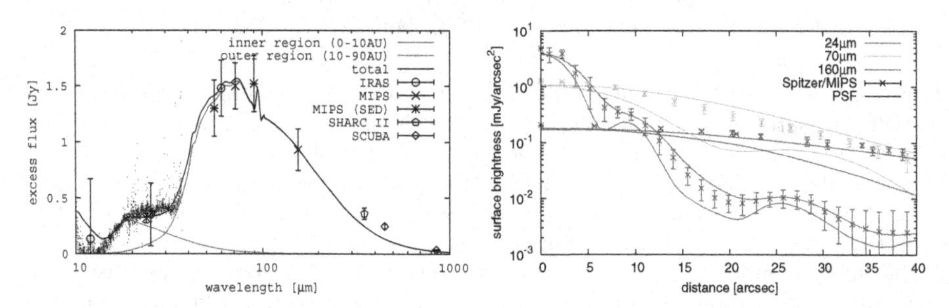

Figure 1. *Left:* SED of the inner and outer dust disk. *Right:* surface brigtness profile compared to Spitzer/MIPS measurements.

Combining the results of the simulations outside and inside 10 AU, we calculated the overall SED and radial brightness profiles at 24, 70 and 160 µm (Fig. 1). The SED is in a reasonable agreement with the available observational data, and it reproduces correctly the shape and the height of the Spitzer/IRS spectrum. Likewise, the surface brightness profiles are consistent with the Spitzer/MIPS data.

The observed warm dust in the ε Eridani system can indeed stem from the outer "Kuiper belt" and be transported inward by P-R and stellar wind drag. The inner planet has little effect on the distribution of dust, so that the planetary orbit could not be constrained. Reasonable agreement between the model of the outer disk and observations can only be achieved by relaxing the assumption of purely silicate dust and assuming a mixture of silicate and water ice.

References

Backman, D., Marengo, M., Stapelfeldt, K., *et al.* 2009, *ApJ*, 690, 1522
Benedict, G. F., McArthur, B. E., Gatewood, G., *et al.* 2006, *AJ*, 132, 2206
Brogi, M., Marzari, F., & Paolicchi, P. 2009, *A&A*, 499, L13
Butler, R. P., Wright, J. T., Marcy, G. W., *et al.* 2006, *ApJ*, 646, 505
Greaves, J. S., Holland, W. S., Wyatt, M. C., *et al.* 2005, *ApJL*, 619, L187
Hatzes, A. P., Cochran, W. D., McArthur, B., *et al.* 2000, *ApJL*, 544, L145
Krivov, A. V., Löhne, T., & Sremčević, M. 2006, *A&A*, 455, 509
Laor, A. & Draine, B. T. 1993, *ApJ*, 402, 441
Li, A. & Greenberg, J. M. 1998, *A&A*, 331, 291
Wood, B. E., Müller, H.-R., Zank, G. P., & Linsky, J. L. 2002, *ApJ*, 574, 412

The Astrophysics of Planetary Systems: Formation, Structure, and Dynamical Evolution
Proceedings IAU Symposium No. 276, 2010
A. Sozzetti, M. G. Lattanzi & A. P. Boss eds.

doi:10.1017/S1743921311020783

A visual guide to planetary microlensing

Leslie A. Rogers[1] and Paul L. Schechter[1]

[1]Department of Physics, Massachusetts Institute of Technology, Cambridge, MA 02139, USA
email: `larogers@mit.edu`, `schech@achernar.mit.edu`

Abstract. The microlensing technique has found 10 exoplanets to date and promises to discover more in the near future. While planetary transit light curves all show a familiar shape, planetary perturbations to microlensing light curves can manifest a wide variety of morphologies. We present a graphical guide that may be useful when understanding microlensing events showing planetary caustic perturbations.

Keywords. gravitational lensing, planetary systems

The microlensing approach to discovering planets relies upon chance alignments of two stars along the line-of-sight to Earth (see, e.g., Bennett (2008) or Gaudi (2010) for a detailed review). During the near-alignments the foreground star acts as a gravitational lens, bending light from the background source star. A point mass foreground lens star produces two images of the source: a positive parity image outside the Einstein radius, and a negative parity image inside the Einstein radius. Together, the two unresolved images result in an overall magnification, A_0, of the source. As the source and lens stars move relative to one another, the overall magnification and image positions vary with time. If the lens star harbors a planet that lies near the path of one of the images, the planet can perturb the magnification, ($A = A_0 + \delta A$), producing an observable signature in the microlensing light curve.

We present a compact yet comprehensive graphic that illustrates the range of possible magnification maps for microlensing events showing planetary caustic perturbations (Figure 1). The diagrams in the middle panels illustrate the lens configurations considered in the surrounding maps. The black circle represents the primary lens star's Einstein ring, while the squares outline the regions spanned by the magnification maps. The center of each magnification map corresponds to a direct-hit, for which the planet falls exactly on one of the unperturbed images of the source lensed by the point-mass primary. Magnification maps on opposite sides of the middle panels correspond to the same direct-hit source position (0.05, 0.45, 0.85 or 1.25 Einstein radii from the lens star); the planet perturbs the positive parity image in the red-outlined maps, and the negative parity image in the blue-outlined maps.

The source-lens-planet configurations at magnification map center are identical between Figures 1 a) and b). In a) the source position varies across the map with the planet fixed, while in b) the planet position varies across the map with the source fixed. Tracing the source trajectory through the source plane maps in a) yields the microlensing light curve. In contrast, the maps in b) reveal the region of the lens plane within which one is sensitive to planets at a given instant in time.

References

Bennett, D. P. 2008, in: J. Mason (ed.), *Exoplanets* (Berlin: Springer), p. 47
Gaudi, B. S. 2010, in: S. Seager (ed.), *Exoplanets* (Tucson, AZ: University of Arizona Press)

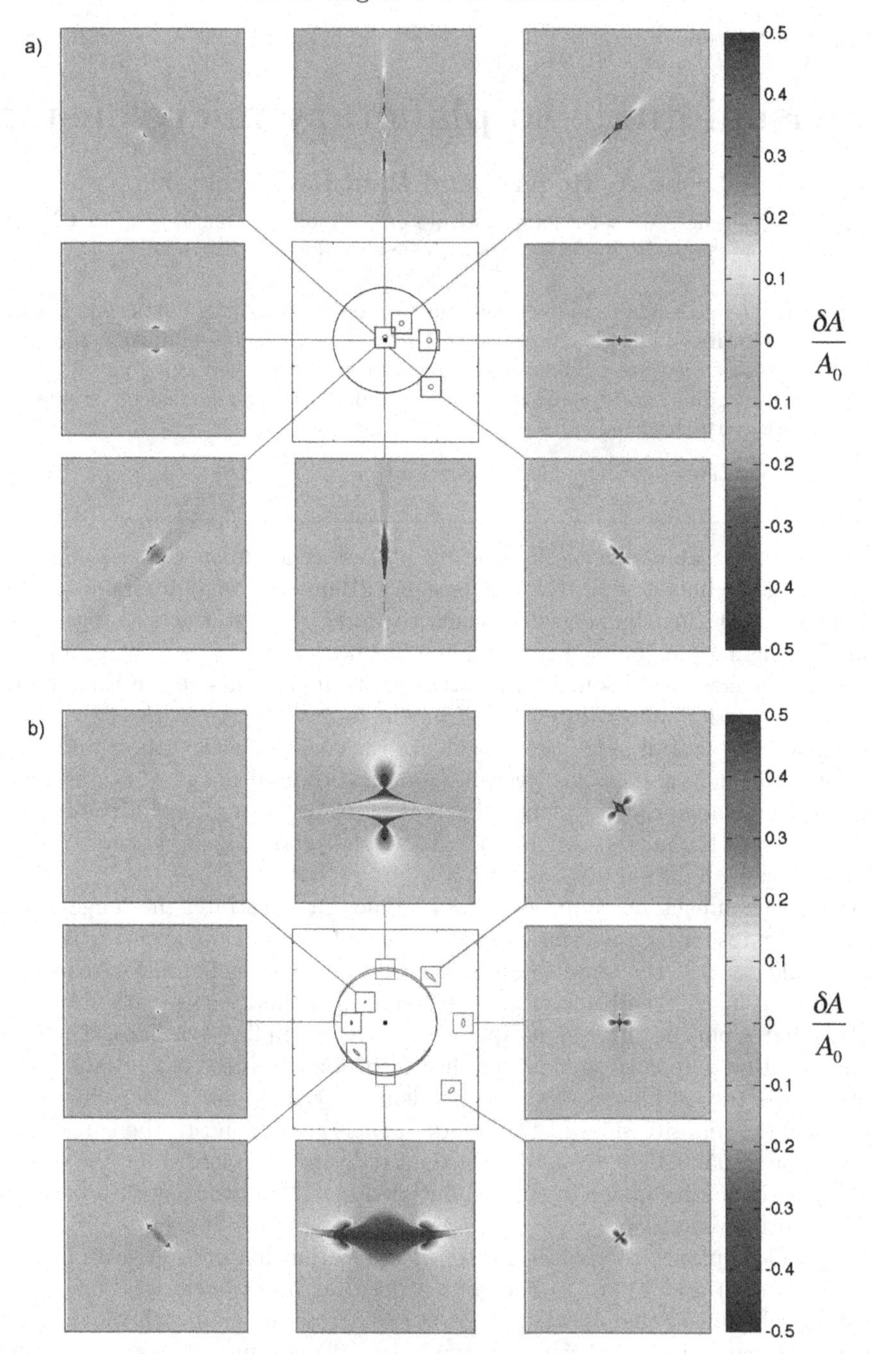

Figure 1. Fractional change in magnification of a point source induced by a planet around the lens star ($M_p = 10^{-4} M_*$) relative to the point mass lens (no planet) case. In the magnification map color scale red and yellow denote magnification increases, shades of blue denote magnification decreases, and green denotes magnifications relatively unchanged by the presence of the planet. In a) we plot how the magnification varies with the position of the source star for a fixed planet position, while in b) we show how the magnification depends on the planet position in the lens plane for fixed source position.

The Astrophysics of Planetary Systems: Formation, Structure and Dynamical Evolution
Proceedings IAU Symposium No. 276, 2010
A. Sozzetti, M. G. Lattanzi & A. P. Boss, eds.

doi:10.1017/S1743921311020795

Exoplanet discovery and characterisation through robotic follow-up of microlensing events: Season 2010 results

Rachel A. Street[1], Yiannis Tsapras[1], Keith Horne[2], Colin Snodgrass[3], Daniel M. Bramich[4], Martin Dominik[2]†, Eric Hawkins[1], Paul Browne[2], Cheongho Han[5], Iain Steele[6], Peter Dodds[2] and Christine Liebig[2]

[1]Las Cumbres Observatory Global Telescope Network, 6740 Cortona Drive, Goleta, CA 93117, U.S.A.,
email: rstreet@lcogt.net
[2]SUPA, St Andrews, School of Physics and Astronomy, North Haugh, St Andrews, Fife, KY16 9SS, U.K.,
[3]Max Planck Institute for Solar System Research, Max-Planck-Str. 2, 37191 Katlenburg-Lindau, Germany,
[4]ESO Headquarters, Karl-Schwarzschild-Str. 2, 85748 Garching bei München, Germany,
[5]Institute for Astrophysics, Department of Physics, Chungbuk National University, Cheongju 361-763, Republic of Korea,
[6]Astrophysics Research Institute, Liverpool John Moores University, Twelve Quays House, Egerton Wharf, Birkenhead, CH41 1LD, U.K.

Abstract. Microlensing searches for planets are sensitive to small, cold exoplanets from 1–6 AU from their host stars and therefore probe an important part of parameter space. Other techniques would require many years of observations, often from space, to detect similar systems. Microlensing events can be characterised from only ground-based observations over a relatively short ($\leqslant$100 d) timescales. LCOGT and SUPA/St Andrews are building a robotic global network of telescopes that will be well suited to follow these events. Here we present preliminary results of the Galactic Bulge observing season 2010 March–October.

Keywords. planetary systems, techniques: photometric

1. Introduction

Our knowledge of cool exoplanets, with orbital separations of 1–few AU, remains scant. Due to the long orbital periods of these planets, many years of radial velocity or photometric (transit) observations are necessary to explore this region of exoplanet parameter space, yet it is of tremendous interest as it includes not only Jovian planets in their formation region beyond the snowline, but also the habitable zone for Sun-like stars. Microlensing offers a way to detect and characterise exoplanets from ~1–6 AU from their host stars from ground-based observations taking only a couple of months.

The survey teams photometrically monitoring the Galactic Bulge, MOA (Hearnshaw *et al.* 2006) and OGLE (Udalski *et al.* 1992), detect >600 microlensing events in progress per year and issue public alerts. Anomalous deviations in a lightcurve from the smooth point-lens-point-source model can indicate that the lensing object may be a binary. Well-sampled lightcurves are necessary to infer the lens and source characteristics

† Royal Society University Research Fellow

(Mao & Paczyński 1991; Gould & Loeb 1992), and to detect planetary anomalies which may have a durations as short as ∼2 hrs.

2. Robotic Telescope Network

As microlensing events do not repeat but span many weeks, a highly responsive, multi-site follow-up network is necessary to monitor them from the ground. Data taken must be reduced as near to real-time as possible to allow early detection of anomalies, and directly influences observation strategy within hours.

We use the 2 m Faulkes-North (Hawai'i) and -South (Australia) and Liverpool (La Palma) robotic telescopes to provide follow-up observations of microlensing events in near real-time response to alerts received.

We have developed a new, adaptive software system called ObsControl, to take advantage of the robotic network's flexible queue-scheduled observing mode. This uses a new command line package, rcstool, to submit requests to the telescope's Java-based observation databases. The system calls on webPLOP (Snodgrass *et al.* 2008, Horne *et al.* 2009) to prioritise the targets and rapidly requests appropriate observations. Data returned are reduced by our fully automated pipeline and the photometry is made publicly available. Target priorities are automatically re-evaluated in the light of new data. We respond to anomaly alerts issued by the ARTEMiS system (Dominik *et al.* 2007), by the follow-up teams and from the surveys. We can request additional observations as necessary, but human authorisation is needed for target-of-opportunity overrides.

3. Season 2010 Results and Future Plans

During our observing season, 2010 March to October, we received 598 event alerts from MOA, of which 197 were selected for observation. In addition, we responded to 30 out of 43 anomaly alerts received. Early modeling work on these anomalies identified 5 candidate lensing planetary systems. MOA-2010-BLG-477, is an example of the sensitivity of this technique to Jovian ($\sim 1.75\,M_J$) planets in cool orbits, where much lower irradiation from the host star means their evolution can be contrasted with that of Hot Jupiters. M-dwarf stars being the most numerous, the lens star is assumed to have a mass of $< M_* > \sim 0.5\,M_\odot$. However, for some events, like MOA-2010-BLG-0073, the lens star mass can be inferred if orbital motion and parallax significantly affect the shape of the lightcurve. Preliminary work suggests a $0.0052 \pm 0.0006\,M_\odot$ ($\sim 5.5\,M_J$) Jovian planet orbiting a $0.0074 \pm 0.008\,M_\odot$ brown dwarf with a projected separation of 0.73 ± 0.07 AU.

In addition to our current facilities, LCOGT and SUPA/St Andrews are building a network of $\geqslant 12 \times 1$ m and $\sim 22 \times 0.4$ m robotic telescopes at $\geqslant 6$ longitudinally-distributed sites in both hemispheres. This will enable us to provide continuous coverage of high priority microlensing events. We anticipate the first of the new telescopes coming online in 2011–2012.

References

Dominik, M., *et al.* 2007, *MNRAS*, 380, 792
Gould, A. & Loeb, A. 1992, *ApJ*, 396, 104
Horne, K., Snodgrass, C., & Tsapras, Y. 2009, *MNRAS*, 396, 2087
Hernshaw, J. B., *et al.* 2006, in: *Proc.9th Asian-Pacific Regional IAU Meeting 2005*, eds. W. Sutantyo and P. Premadi, 272
Mao, S, & Paczyński, B, 1991, *APJ*, 552, 889
Snodgrass, C., *et al.* 2008, in: *Introduction to Microlensing*, Proc. Manchester Microlensing Conference, 56
Udalski, A., *et al.* 1992, *AcA*, 42, 253

The Astrophysics of Planetary Systems: Formation, Structure, and Dynamical Evolution
Proceedings IAU Symposium No. 276, 2010
A. Sozzetti, M. G. Lattanzi, & A. P. Boss, eds.

doi:10.1017/S1743921311020801

The effect of opacity on the evolution of giant planets

Allona Vazan[1], Attay Kovetz[1,2] and Morris Podolak[1]

[1] Department of Geophysics and Planetary Sciences, Sackler Faculty of Exact Sciences
Tel Aviv University, Israel

[2]School of Physics and Astronomy, Sackler Faculty of Exact Sciences
Tel Aviv University, Israel

email: allonava@post.tau.ac.il

Abstract. We use an improved version of the planetary evolution code described in Helled *et al.* (2006) to model the effect of opacity on the evolution of giant planets in the disk instability scenario. We find that changing the opacity law can cause significant changes in the evolutionary path of a protoplanet. Sufficiently high opacities cause oscillatory behavior that delays the final collapse. Peak luminosities just before collapse can exceed $10^{-5}\ L_\odot$.

Keywords. planets and satellites: formation, instabilities

1. Introduction

One of the important parameters in determining the evolutionary path of a protoplanet is the assumed opacity law (e.g. Pollack *et al.* 1994; D'Alessio *et al.* 2001). Microphysical processes can cause the opacity law to differ substantially from the interstellar values. The resulting difference has been shown to have important consequences for the evolutionary track of a protoplanet (Hubickyj *et al.* 2005; Movshovitz *et al.* 2010).

2. The Model

We use a stellar evolution code (Kovetz *et al.* 2009), that was modified to work under planetary conditions (Helled *et al.* 2006). We consider giant planet formation in the disk instability scenario (Boss 2002; Mayer *et al.* 2007; Boley 2009). The initial configuration is a low-density, roughly isentropic, clump of one Jupiter mass, composed of a solar mix of hydrogen and helium. Our standard initial model, kindly provided by P. Bodenheimer (personal communication), has a radius of $101 R_\odot \sim 0.5$ AU. This is roughly equal to the Hill sphere radius.

In order to study the effect of opacity we developed a simple grain opacity code, which assumes a power law size distribution of grains composed of ice and silicates. Although there is no intent to model the details of the chemistry, we do include grain evaporation and the details of Mie scattering for spherical grains. In this way, we can test the importance of such parameters as the grain to gas mass ratio, the ice to silicate mass ratio, and the exponent in the power law size distribution (γ). Gas opacities were taken from tables of opacity kindly supplied by A. Burrows (Sharp & Burrows 2007).

3. Results

For a given grain to gas ratio, the opacity will depend strongly on the ratio of grain size to wavelength, and hence on γ. For low γ most of the mass is in large grains, which have a low surface area to mass ratio. As γ increases the mass is distributed among smaller grains and the opacity increases. For still larger γ the mass is concentrated in the smallest grains, which are inefficient scatterers. Fig. 1 shows the evolutionary track of a clump as a function of γ. For low opacity (low and high γ) the clump cools, becomes dynamically unstable, and collapses to a more compact configuration. For γ in the neighborhood of 4 oscillations are seen which significantly extend the contraction stage, and, in some cases no dynamical collapse is seen. The models that eventually collapse show a markedly

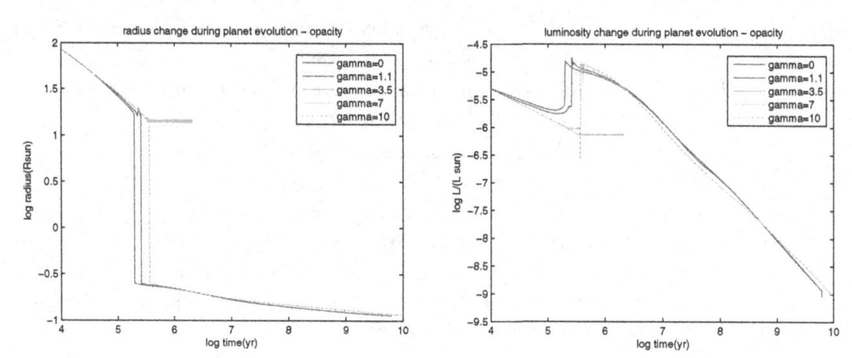

Figure 1. Planetary radius and luminosity as a function of time, for different grain size distributions (γ).

lower opacity at the center. This behavior is reminiscent of the κ-mechanism for stellar pulsations. The actual dynamic collapse, however, appears to be initiated by a radiative core. If the opacity is high enough, there is no such radiative region, and the model appears to settle down, with small oscillations around a quasi-static configuration.

The opacity can be increased, either by raising the mass fraction of grain material, or by concentrating the grain mass at grain sizes where the extinction was most efficient. For sufficiently high values of the opacity, oscillations appear substantially lengthen the contraction time. If the opacity is high enough, the collapse to planetary size could be delayed by millions of years. Delaying the dynamic collapse of an extended protoplanet will allow it to maintain a large cross section for planetesimal capture for a longer time, and this will increase the amount of high-Z material that it can capture and, in turn, increase the grain opacity. A self-consistent calculation of this effect is important, and will be undertaken in future work.

References

Boss, A. P. 2002, *ApJ*, 576, 462

Boley, A. C. 2009, *ApJL*, 695, L53

D'Alessio, P., Calvet, N., & Hartmann, L. 2001, *ApJ*, 553, 321

Helled, R., Podolak, M., & Kovetz, A. 2006, *Icarus*, 185, 64

Hubickyj, O., Bodenheimer, P., & Lissauer, J. J. 2005, *Icarus*, 179, 415

Kovetz, A., Yaron, O., & Prialnik, D. 2009, *MNRAS*, 395, 1857

Mayer, L., Lufkin, G., Quinn, T., & Wadsley, J. 2007, *ApJL*, 661, L77

Movshovitz, N., Bodenheimer, P., Podolak, M., & Lissauer, J. J. 2010, *Icarus*, 209, 616

Pollack, J. B., Hollenbach, D., Beckwith, S., Simonelli, D. P., Roush, T., & Fong, W. 1994, *ApJ*, 421, 615

Sharp, C. M. & Burrows, A. 2007, *ApJS*, 168, 140

The Astrophysics of Planetary Systems: Formation, Structure, and Dynamical Evolution
Proceedings IAU Symposium No. 276, 2010
A. Sozzetti, M. G. Lattanzi & A. P. Boss, eds.

doi:10.1017/S1743921311020813

Formation of massive gas giants on wide orbits

Eduard I. Vorobyov[1,2,3] and Shantanu Basu[3]

[1]Research Institute of Physics, Southern Federal University, Rostov-on-Don, 344090, Russia
[2] The Institute for Computational Astrophysics, Saint Mary's University, Halifax, Canada
email: vorobyov@ap.smu.ca

[3]Dept. of Physics and Astronomy, The University of Western Ontario, London, Canada
email: basu@uwo.ca

Abstract. We present a mechanism for the formation of massive gas giants on wide orbits via disk fragmentation in the embedded phase of star formation. In this phase, protostellar disks undergo radial pulsations which lead to periodic disk compressions and formation of massive fragments on radial distances of the order of 50–300 AU. The fragments that form during the last episode of disk compression near the end of the embedded phase, when torque from spiral arms become weaker, may survive and mature into massive gas giants. This phenomenon can explain the existence of massive exoplanets on wide orbits is such systems as Fomalhaut and HR 8799.

Keywords. instabilities, planetary systems: formation, planetary systems: protoplanetary disks

1. Introduction

The recent discoveries of massive gas giants on wide orbits of the order of 50–300 AU around HR 8799 (Marois *et al.* 2008), Fomalhaut (Kalas *et al.* 2008), and 1RXS J160929.1-210524 (Lafreniére *et al.* 2010) present a new challenge for the currently most favoured core accretion theory for planet formation, whereby giant planet formation is restricted to the inner 5–10 AU. Planet scattering to larger radii is needed to account for the aforementioned systems but this mechanism does not seem to produce stable and low eccentricity orbits (Dodson-Robinson *et al.* 2010).

On the other hand, gravitational fragmentation is expected to occur in the outer disk regions at distances of the order of a few tens or hundreds of AU (e.g., Durisen *et al.* 2007) and thus can naturally explain the in situ formation of wide-orbit extrasolar planets. In this article, we demonstrate that this mechanism is viable and present the results of our numerical hydrodynamics simulations illustrating the formation of a massive gas giant at an orbital distance of 55 AU. We refer the reader to Vorobyov & Basu (2010) for the in-depth discussion of our results.

2. Formation of a massive gas giant

We start our numerical hydrodynamics simulations from a pre-stellar cloud core with mass $M_{\rm core} = 0.9\ M_{\odot}$ and ratio of rotational to gravitational energy $\beta = 1.3 \times 10^{-2}$. The upper rows of images in Figure 1 present the gas surface density (in g cm^{-2}) in the inner 300 AU at different times t after the formation of a protostellar disk. These images demonstrate that fragmentation occurs early in the disk evolution, already at an age of 0.03 Myr. However, at t = 0.14 Myr all fragments are either driven into the star due to exchange of angular momentum with spiral arms or dispersed due to tidal torques.

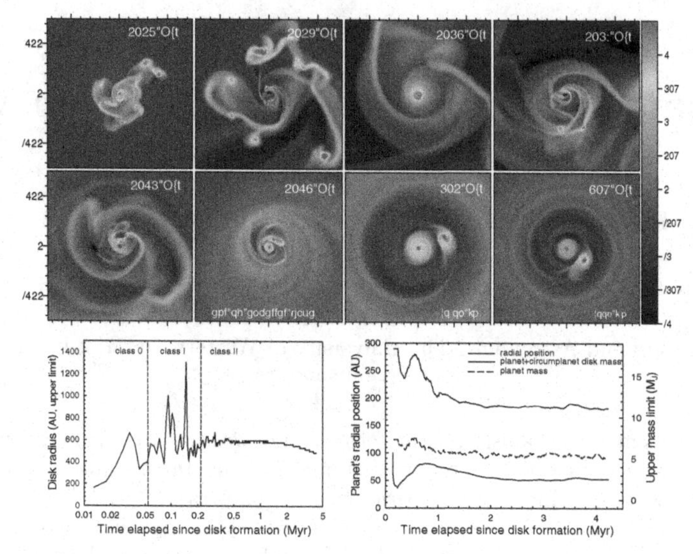

Figure 1. Top two rows. Gas surface density in the inner 300 AU at several distinct times after the disk formation. **Bottom-left.** Disk radius versus time. **Bottom-right.** Planet's radial position (thin solid line), net mass (dashed line), and total (planet+mini-disk) mass (thick solid line) versus time.

The inward radial migration of the fragments deposits angular momentum to the outer parts of the disk, which results in periodic disk expansions as shown by the solid line in the bottom-left panel of the Figure. The last episode of expansion, at $t \approx 0.14$ Myr, is particularly strong and, when followed by disk contraction, it gives birth to a last set of fragments. This happens near the end of the Class I phase (as indicated by the vertical dashed lines) when gravitational and tidal torques from spiral arms are becoming weaker due to diminishing mass infall from the envelope—the major disk destabilizing mechanism. As a result, one of the fragments survives and matures into a massive gas giant at an orbit of ≈ 55 AU (thin solid line in the bottom-right panel). The total (planet plus its mini-disk) mass is about 5 $M_{\rm J}$ (dashed line), while the net planet mass is about 12 $M_{\rm J}$ (thick solid line).

E.I.V. gratefully acknowledges support from an ACEnet Fellowship. Numerical simulations were done on the Atlantic Computational Excellence Network (ACEnet). This project was also supported by RFBR grant 10-02-00278 and by the Ministry of Education grant RNP 2.1.1/1937. S. B. was supported by a Discovery Grant from the Natural Sciences and Engineering Research Council of Canada.

References

Dodson-Robinson, S. E., Veras, D., Ford, E. B., & Beichman, C. A. 2010, *ApJ*, 707, 79

Durisen, R. H., Boss, A. P., Mayer, L., Nelson, A. F., Quinn, T., & Rice, W. K. M. 2007, in *Protostars and Planets V*, eds. B. Reipurth, D. Jewitt, & K. Keil, University of Arizona Press, Tucson, 607

Kalas, P., Graham, J. R., Chiang, E., Fitzgerald, M. P., Clampin, M., Kite, E. S., Stapelfeldt, K., Marois, C., & Krist, J. 2008, *Science*, 322, 1345

Lafreniére, D., Jayawardhana, R., & van Kerkwijk, M. H. 2010, *ApJ*, 719, 497

Marois, C., Macintosh, B., Barman, T., Zuckerman, B., Song, I., Patience, J., Lafreniére, D., & Doyon, R. 2008, *Science*, 322, 1348

Vorobyov, E. I. & Basu, S. 2006, *ApJ*, 650, 956

Vorobyov, E. I. & Basu, S. 2010, *ApJL*, 714, 133

The Astrophysics of Planetary Systems: Formation, Structure, and Dynamical Evolution
Proceedings IAU Symposium No. 276, 2010
A. Sozzetti, M. G. Lattanzi & A. P. Boss, eds.

doi:10.1017/S1743921311020825

The follow-up observations for the transit events of WASP-12b

Xiao-bin Wang[1], Andrew Collier Cameron[2], Sheng-hong Gu[1], Xiang-song Fang[1] and Dong-tao Cao[1]

[1]National Astronomical Observatories/Yunnan Observatory, CAS, Kunming, China
email: wangxb@ynao.ac.cn

[2]University of St. Andrews, Fife KY16 9SS, UK

Abstract. The four new transit light curves of WASP-12b are analyzed by using MCMC simulation so as to derive the system parameters. If the apsidal precession exists in the orbit of WASP-12b, according to the theory of Gimenez & Bastero (1995), the rate of precession is estimated as 0.0076 degrees per cycle in the case of orbit with an inclination of 90°.

Keywords. planetary systems, celestial mechanics, eclipses

1. Introduction

The follow-up observations with high precision for known transiting systems not only can be used to refine planetary parameters, but also to identify the variation of some parameters. Many reasons can result in the variation of parameters. For example, the quadrupole of moment of star, planet and undetectable other planets may produce the precession of line of nodes and apsidal line (Miralda-Escudé 2002). The precession of line of nodes might induce small changes in the shape of transiting light curve, whereas the apsidal precession changes the shape of radial velocity curve. For a very closed-in hot Jupiter (a<0.025AU), the quadrupole of moment of planetary tidal bulge is a main source of apsidal precession (Ragozzine & Wolf 2009). The rate of apsidal precession can be identified in case of orbit with a non-zero eccentricity.

The hot Jupiter WASP-12b discovered by Hebb *et al.* (2009), is close to its host star (a=0.0229AU). The apparent eccentricity is identified by some authors (Hebb *et al.* 2009; Campo *et al.* 2011; Lopez-Morales *et al.* 2010; Husnoo *et al.* 2011). Thus, WASP-12b is an important candidate to measure the apsidal precession.

2. Observations and data reduction

Four transit events of WASP-12b were observed with 1m telescope (2K*2K CCD and R-filter) in Kunming, China on Nov. 9, Dec. 14, 15, 2009 and 2.4m telescope (1340*1300 CCD and R-filter) in Lijiang, China on Mar. 18, 2010. The magnitude of selected objects in CCD frames is measured with APPHOT task of IRAF. 26 and 24 "reference stars" for 1m and 2.4m observations, respectively, are used to simulate systematic errors in photometric data with the methods of Collier Cameron *et al.* (2006) and Tamuz *et al.* (2005). Then these simulated systematic effects are removed from the WASP-12b's photometric data.

Table 1. Part parameters of WASP-12b.

Sidereal period	$1.091412^{+-8.77E-06}_{-1.54E-05}$ days
e	$0.065^{+-0.021}_{-0.022}$

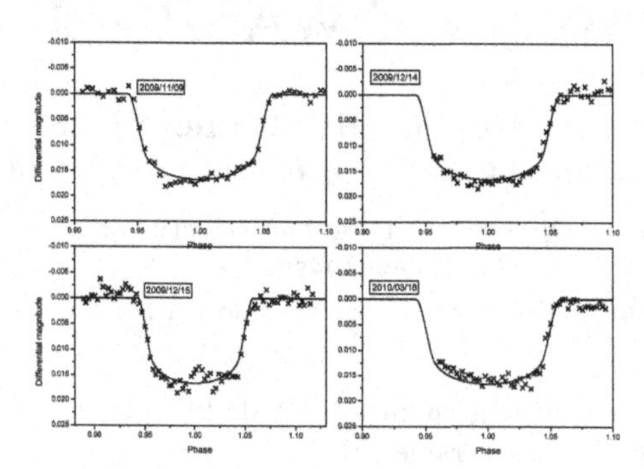

Figure 1. Binned light curves of WASP-12b and relative fitting.

3. Light curve analysis and conclusions

The MCMC (Markov Chain Monte Carlo) routine (Collier Cameron *et al.* 2007) is used to estimate the planetary parameters based on 4 new transit light curves. Firstly, we only analyze 4 new transit light curves of WASP-12b by using the MCMC method. The derived transit period of 1.091437 days is slightly longer than that of Hebb *et al.* (2009). Fig. 1 shows 4 transit light curves and the model data calculated with the best-fitting parameters. Secondly, the observed light curves combined with the radial velocity data of Hebb *et al.* (2009) and Husnoo *et al.* (2011) are analyzed together with MCMC procedure. Table 1 lists the part results. The period of 1.091412 days is different from above one, but is close to the value of Husnoo *et al.* (2011).

We think that two different periods (sidereal period and transit period) may imply the apsidal precession exists in WASP-12b's orbit. Providing an inclination of 90 degrees, we estimate the rate of apsidal precession according to the equation (15) of Gimenez & Bastero (1995) is 0.0076 degrees per sidereal cycle. New accurate photometric and radial velocity observations for WASP-12b are needed for obtaining the more accurate rate of precession.

Acknowledgments

This work is supported by NSFC under grant No.10873031 and Chinese Academy of Sciences under grant KJCX2-YW-T24.

References

Campo, C. J., Harrington, J., Hardy, R. A., *et al.* 2011, *ApJ*, 727, id.125
Collier Cameron, A., *et al.* 2006, *MNRAS*, 373, 799
Collier Cameron, A., *et al.* 2007, *MNRAS*, 380, 1230
Gimenez, A. & Bastero, M. 1995, *ApJ*, 659, 1661
Hebb, L., Collier-Cameron, A., Loeillet, B., *et al.* 2009, *ApJ*, 693, 1920
Husnoo, N., Pont, F., & Hebrard, G., *et al.* 2011, 413, 2500
Lopez-Morales, M., Coughlin, J. L., Sing, D. K., *et al.* 2010, *ApJ*, 716, 36
Miralda-Escude, J. 2002, *ApJ*, 564, 1019
Ragozzine, D. & Wolf, A. S. 2009, *ApJ*, 698, 1778
Tamuz, O., Mazeh, T., & Zucker, S. 2005, *MNRAS*, 356, 1466

The Astrophysics of Planetary Systems: Formation, Structure, and Dynamical Evolution
Proceedings IAU Symposium No. 276, 2010
A. Sozzetti, M. G. Lattanzi & A. P. Boss, eds.

doi:10.1017/S1743921311020837

Modeling of SEDs for substars with disks that have different geometrical and physical parameters

Olga Zakhozhay[1]

[1]Main Astronomical Observatory National Academy of Sciences of Ukraine,
27 Akademika Zabolotnoho St. 03680 Kyiv, Ukraine
email: zkholga@mail.ru

Abstract. The algorithm of spectral energy distribution (SED) calculations for protoplanetary disks and central objects is created. The results of SEDs calculations for substars with protoplanetary disks that have a different ages and inclinational angles are discussed.

Keywords. circumstellar matter, stars: low-mass, brown dwarfs, planetary systems: protoplanetary disks

---

1. Introduction

At the beginning of this century a physical model for substellar (brown dwarf) evolution was created by research group from Astronomical Institution of V.N. Karazin Kharkiv National University (Pisarenko *et al.* 2007). These models describe the evolution of the substars with masses 0.01 – 0.08 solar masses and ages within the range of $1Myr - 10Gyr$.

Based on these investigations the models of disks that would surround such substars were simulated and SEDs were calculated. With this algorithm it is possible to calculate SEDs for substar (or star) with disk that inclined on any angle with taking into account all geometrical particularities that system can have.

The results for the following cases have being received:
- substar masses within the range of $0.01 - 0.08M_{sol}$;
- protoplanetary disks with different inclination angles ($0° - 80°$);
- systems are in the age interval from 1 to $30Myr$;
- substars and protoplanetary disks are radiate like black body based on the Chiang & Goldreich (1997) model;
- distance from Sun to substar equals to $10pc$;
- disk's inner radius equals to central object radius and sublimation radius in the age of $1Myr$.

2. Results

1120 SEDs for systems with all variety of above-mentioned parameters have being received. In this work a case of flat disk without inner hole that surround a substar with mass $0.08M_{sol}$ (as an example) is discussed. On Fig. 1 the evolution of SED's shape and intensity with the age (left panel) and inclination angle (right panel) is shown.

Calculational models were compared with observational data that were received by Scholz *et al.* (2007). On Fig. 2 SED's and observational data for two objects usd15556 and usd155601 are shown. SED's were calculated using parameters that were derived by Scholz *et al.* (2007): for usd155556 - substellar mass $M_{ss} = 0.075M_{sol}$, substellar

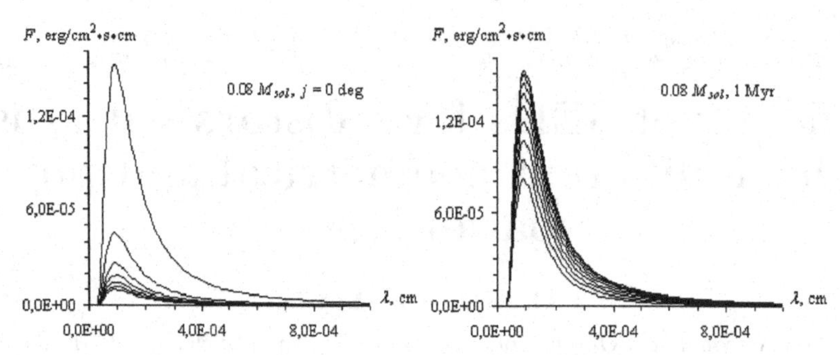

Figure 1. SEDs for systems with substar ($0.08M_{sol}$) and flat disk without inner hole: left panel- different curves show different system ages (from top to bottom): $1Myr, 5Myr, 10Myr, 15Myr, 20Myr, 25Myr$ and $30Myr$; right panel - different curves show different system inclinations (from top to bottom): $0°, 10°, 20°, 30°, 40°, 50°, 60°, 70°$ and $80°$.

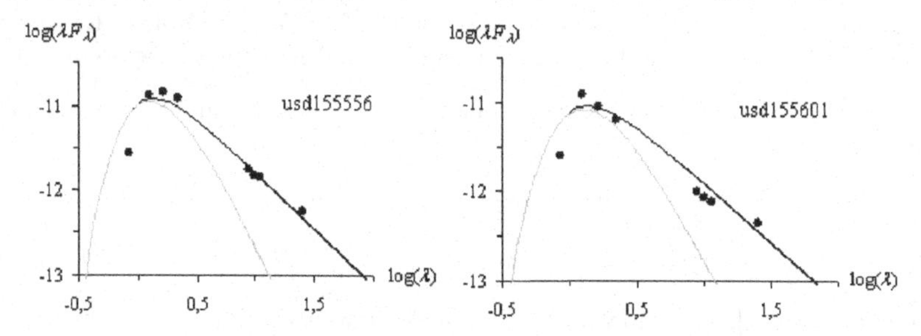

Figure 2. Comparison of observational data (points) with modeled SEDs (solid black line), substellar black body SED is shown with solid grey line

temperature $T_{ss} = 2900K$, substellar radius $R_{ss} = 0.4R_{sol}$; for usd155601 - $M_{ss} = 0.075M_{sol}$, $T_{ss} = 2900K$, $R_{ss} = 0.35R_{sol}$; disks of both objects have inner radius $R_{in} = R_{ss}$, outer radius $R_{out} = 100au$ and inclination angle 20°.

3. Conclusions

- The algorithm of SEDs calculations for substars with disks with and without inner hols, different ages, different geometry and inclination angle has been developed.
- 1120 SEDs for systems with substars within the masses of $(0.01-0.08)M_{sol}$ at the age $1-30Myr$ and flat or flared disks that inclined on different angles have been calculated.
- SEDs that were calculated using our model are in agreement with SEDs that were calculated by Scholz *et al.* (2007).

4. Acknowledgments

Author want to acknowledge International Astronomical Union and Astronomical Institute of Ruhr University (Germany) for the financial support that gives an opportunity to participate in IAU Symposium 276.

References

Pisarenko, A. I., Yatsenko, A. A., & Zakhozhay, V. A. 2007, *Astron. Rep.*, 51, 605
Chiang, E. I. & Goldreich, P. 1997, *ApJ*, 490, 368
Scholz, A., Jayawardhana, R., Wood, K., *et al.* 2007, *ApJ*, 660, 1517

Section B:
Structure & Atmospheres

The Astrophysics of Planetary Systems: Formation, Structure, and Dynamical Evolution
Proceedings IAU Symposium No. 276, 2010
A. Sozzetti, M. G. Lattanzi & A. P. Boss, eds.

doi:10.1017/S1743921311020849

Alien Earth: Glint observations of a remote planet

Richard K. Barry[1] **and L. Drake Deming**[1]

[1]National Aeronautics and Space Administration
Goddard Space Flight Center, Greenbelt, MD
email: Richard.K.Barry@NASA.gov

Abstract. We give a preliminary report on a multi-wavelength study of specular reflections from the oceans and clouds of Earth. We use space-borne observations from a distance sufficient to ensure that light rays reflected from all parts of Earth are closely parallel, as they will be when studying exoplanets. We find that the glint properties of Earth in this far-field vantage point are surprising - in the sense that some of the brightest reflections are not from conventional ocean-glints, but appear to arise from cirrus cloud crystals. The Earth observations discussed here were acquired with the High Resolution Instrument (HRI) - a 0.3 m f/35 telescope on the Deep Impact (DI) spacecraft during the Extrasolar Planet Observation and Characterization (EPOCh) investigation.

Keywords. infrared: planetary systems, astrobiology, astrochemistry, planetary systems, Earth

1. Observations

While EPOCh has as its main scientific goal the observation of transits of exoplanets across the face of their host stars, observations of Earth were also conducted during orbital flyby opportunities. The HRI, used for these observations, has nine filters, seven of which are 100 nm wide, boxcar bandpass filters centered at 350, 450, 550, 650, 750, 850 and 950 nm. Two other filters are uncoated fused silica with no band limiting, transparent longwards of 700 nm. DI also has an NIR spectrometer that covers the wavelength range from 1.05-4.5 microns with a spectral resolution exceeding 200 over the entire pass band [3]. Using the HRI instrument, Earth was observed on five occasions: 2008-Mar-18 18:18 UT, 2008-May-28 20:05 UT, 2008-Jun-4 16:57 UT, 2009-Mar-27 16:19 and 2009-Oct-4 09:37 UT. Each set of observations was conducted over a full 24-hour rotation of Earth and a total of thirteen NIR spectra were taken on two-hour intervals during each observing period. Photometry in the 450, 550, 650 and 850 nm filters was taken every fifteen minutes and every hour for the 350, 750 and 950 nm filters.

2. Data and Discussion

Glints were located to the greatest precision allowed by the data using the images that had been deconvolved with a PSF previously developed using a drizzle process (Barry *et al.* 2010). Once the location of the glint events were noted using the deconvolved images, we constructed spectra of the glints from the un-deconvolved images by centering the 8-pixel patches on the locations found in the deconvolved set. Using the process for extracting glint flux and background flux noted above, the un-deconvolved images were used to report these fluxes for purposes of comparison to the theoretical values. These measured values are given in Table 1.

Table 1. Summary of glint and background blue (450 micron) to IR (850 micron) flux ratios for non-deconvolved Earth observations during the EPOXI science investigation. Individual glints are listed as Glint Event North or South (GEN or GES) together with the event's latitude and longitude. The final column gives the assessment of what is in the 8-pixel photometric aperture for the images deconvolved with the PSF.

Event	Longitude	Latitude	Glint B/IR	Bkg B/IR	Observed
GEN1	-159	40	1.4	1.9	Cloud/Ocean
GEN2	-140	40	1.1	1.8	Ocean/Cloud
GEN3	-121	40	1.2	1.5	Land
GEN4	24	40	1.0	1.3	Aegean/Mediterranean Seas
GEN5	54	40	1.0	1.5	Caspian Sea
GEN6	66	40	1.0	1.3	Land
GEN7	77	40	1.5	1.7	Cloud/Land
GEN8	107	40	1.1	1.4	Land
GEN9	118	40	0.3	1.7	Land/Cloud (Bohia Sea)
GEN10	126	40	1.0	1.7	Seas around Korea
GES1	-170	-45	1.6	1.9	Cloud
GES2	-38	-45	1.1	1.8	Ocean/Cloud
GES3	-31	-45	1.3	1.9	Ocean/Cloud
GES4	11	-45	1.6	1.9	Cloud
GES5	74	-45	2.7	1.9	Cloud
GES6	168	-45	1.0	1.7	Land

Glint Event 1 in the Northern hemisphere (GEN1) together with Glint Events GEN2, GES2, GES3 are unambiguously identified with open ocean and cloud in roughly equal proportion. These are listed in Table 1 together with the comment cloud/ocean or ocean/cloud depending on which hydrological state appeared to dominate the photometric aperture during that observation. GEN3, GEN6, GEN8, and GES6 are all over land with little or no cloud in the photometric aperture. GEN4, GEN5, and GEN10 are all glints from bodies of water that are partially isolated from other, larger bodies of water by land. GEN7 and GEN9 appear to be associated primarily with a combination of land and cloud. Finally, GES1, GES4 and GES5 appear to be glints in photometric apertures dominated by a layer of cloud suggesting a detection of the subsun phenomenon - reflection from flat ice crystals. Referring to Table 1, strong b/I ratio of glint suggests a possible longer passage through atmosphere and consequently more red and IR scattered out by Raleigh scattering. Analysis of these data are ongoing.

References

Barry, R. K., Lindler, D., Deming, L. D., A'Hearn, M. F., Ballard, S., Carcich, B., Charbonneau, D., Christiansen, J., Hewagama, T., McFadden, L., & Wellnitz, D. 2010, *Space Telescopes and Instrumentation 2010*, 7731, 107

The Astrophysics of Planetary Systems: Formation, Structure, and Dynamical Evolution
Proceedings IAU Symposium No. 276, 2010
A. Sozzetti, M. G. Lattanzi & A. P. Boss, eds.

doi:10.1017/S1743921311020850

Modeling giant planets and brown dwarfs

Andreas Becker[1], Nadine Nettelmann[2], Ulrike Kramm[1], Winfried Lorenzen[1], Martin French[1] and Ronald Redmer[1]

[1]Institute of Physics, University of Rostock,
Universitätsplatz 3, 18051, Rostock, Germany
email: andreas.becker@uni-rostock.de

[2]Dept. of Astronomy & Astrophysics, University of California,
CA 95064 Santa Cruz , USA

Abstract. We present new results in modeling the interiors of Giant Planets (GP) and Brown Dwarfs (BD). In general models of the interior rely on equation of state data for planetary materials which have considerable uncertainties in the high-pressure domain. Our calculations are based on *ab initio* equation of state (EOS) data for hydrogen, helium, hydrogen-helium mixtures and water as the representative of all heavier elements or ices using finite-temperature density functional theory molecular dynamics (FT-DFT-MD) simulations. We compare results for the BD Gliese 229B calculated with Saumon-Chabrier-Van Horn EOS (SCVH95) and our EOS data.

Keywords. equation of state, planetary systems, brown dwarfs

1. Introduction

We perform extensive electronic structure calculations for warm dense matter within finite-temperature density functional theory combined with molecular dynamics simulations for the ions (FT-DFT-MD) using the VASP code (Kresse & Furthmüller 1996). We calculate EOS data for hydrogen (Holst *et al.* 2008), helium (Kietzmann *et al.* 2007), hydrogen-helium mixtures (Lorenzen *et al.* 2009) and water (French *et al.* 2009) within a temperature range up to 100,000 K and a pressure range up to 100 Mbar as well as electrical conductivities, pair correlation functions and ion diffusion coefficients. We also determine the high-pressure phase diagrams and the location of metal-nonmetal transitions and of demixing regions.

With this input we model solar and extrasolar Giant Planets (GP) within standard three-layer structure models (Guillot 1999) using a mixture of our EOS, called LM-REOS (Linear Mixing-Rostock Equation Of State). Available observational constraints such as mass M, radius R, rotational period, gravitational moments J_n, the mean helium fraction $\overline{Y}$ which is given by the solar value and the helium fraction at the surface Y_1 are fulfilled by our models. Gravity data like J_2 and J_4 are used to fit metallicities, the size of the core, and the chemical composition within the layers.

Brown Dwarfs are assumed to be fully convective (Stevenson 1991) so that we use an adiabatic one-layer model for the calculations of their interior structure. In our computer code the radius is fitted to the mass by solving the differential equation for hydrostatic equilibrium and for mass conservation in a spherical shell for a given EOS and surface temperature. Typical central conditions for BD's are $p_c \propto 1000$ Mbar, $\rho_c \propto 10^3$ g/cm^3, and $T_c \propto 10^6$ K. We apply our EOS tables down to $\sim$ 100 Mbar and use the Saumon-Chabrier-Van Horn EOS (Saumon *et al.* 1995) for the deep interior. Smooth interpolation is performed between the different EOS tables.

2. Results

We determine the isentrope of Saturn and find that it undergoes H-He demixing for a pressure larger than about 1 Mbar, see Lorenzen *et al.* (2009). This effect has probably a significant impact on its cooling time. We obtain very good agreement for the interior structure of Uranus and Neptune with predictions of dynamo models for their magnetic field made by Stanley & Bloxham (2004) using the phase diagram of water, see Redmer *et al.* 2010. Application of our EOS data to Jupiter restricts the core mass to 1-5 Earth masses (M_E) and the total mass of metals to 30-40 M_E, enriching the inner envelope over 5-9 times the solar value (Nettelmann *et al.* 2008). For the interior of the Neptune-like planet GJ 436b we state that water occurs essentially in the plasma phase or in the superionic phase, but not in an ice phase, see Nettelmann *et al.* (2010).

For the Brown Dwarf Gliese 229B we obtain first models using the observational constraints given in Basri (2000): $R = 1R_{Jup}$, $M = 30M_{Jup}$ and $T_{eff} = 1000K$, see figure 1. We assume that the BD neither has a core nor metals in the envelope. The lines show a difference for the central region in pressure, density and temperature of about 10% depending on the EOS. We obtain a radius 1.5% larger than $1R_{Jup}$ applying LM-REOS and 5% applying SCVH95. Our aim for the future is a full description using *ab initio* EOS data such as LM-REOS.

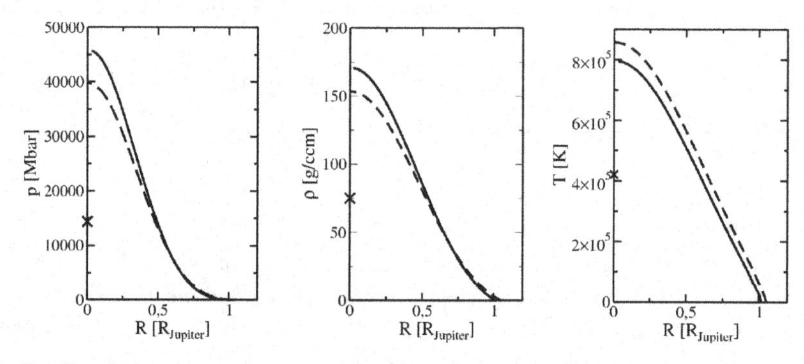

Figure 1. Profiles for pressure, density and temperature through Gliese 229B. Dashed-line profiles are calculated with SCVH95, solid lines with LM-REOS and SCvH95, the crosses represent central values given from Burrows & Liebert (1993).

References

Basri, G. 2000, *Scientific American*, 05/2000, 27

Burrows, A. & Liebert, J. 1993, *Rev. Mod. Phys.*, 65, 301

French, M., Mattson, T. R., Nettelmann, N., & Redmer, R. 2009, *Phys. Rev. B*, 79, 054107

Guillot, T. 1999, *Science*, 286, 27

Holst, B., Redmer, R., & Desjarlais, M. P. 2008, *Phys. Rev. B*, 77, 184201

Kietzmann, A., Redmer, R., Desjarlais, M. P., & Mattson, T. R. 2007, *Phys. Rev. Lett.*, 98, 190602

Kresse, G. & Furthmüller, J. 1996, *Phys. Rev. B*, 54, 11169

Lorenzen, W., Holst, B., & Redmer, R. 2009, *Phys. Rev. Lett.*, 102, 115701

Nettelmann, N., Holst, B., Kietzmann, A., French, M., Blaschke, D., & Redmer, R. 2008, *ApJ*, 683, 1217

Nettelmann, N., Kramm, U., Redmer, R., & Neuhäuser, R. 2010, *A&A*, 523, A26

Redmer, R., Mattson, T. R., Nettelmann, N., & French, M. 2011, *Icarus*, 211, 798

Saumon, D., Chabrier, G., & Van Horn, H. M. 1995, *ApJS*, 99, 713

Stanley, S. & Bloxham, J. 2004, *Nature*, 428, 151

Stevenson, D. J. 1991, *Annu. Rev. Astron. Astrophys.* 29, 163

The Astrophysics of Planetary Systems: Formation, Structure, and Dynamical Evolution
Proceedings IAU Symposium No. 276, 2010
A. Sozzetti, M. G. Lattanzi & A. P. Boss, eds.

doi:10.1017/S1743921311020862

Hot Jupiter secondary eclipses measured by *Kepler*

Brice-Olivier Demory[1] **and Sara Seager**[1]

[1]Department of Earth, Atmospheric and Planetary Sciences, Massachusetts Institute of Technology, Cambridge, USA. email: demory@mit.edu

Abstract. Hot-Jupiters are known to be dark in visible bandpasses, mainly because of the alkali metal absorption lines and TiO and VO molecular absorption bands. The outstanding quality of the *Kepler* mission photometry allows a detection (or non-detection upper limits on) giant planet secondary eclipses at visible wavelengths. We present such measurements on published planets from Kepler Q1 data. We then explore how to disentangle between the planetary thermal emission and the reflected light components that can both contribute to the detected signal in the *Kepler* bandpass. We finally mention how different physical processes can lead to a wide variety of hot-Jupiters albedos.

Keywords. techniques: photometric, eclipses, planetary systems

1. Background and motivation

The reflected light component of the hot-Jupiter population is critical to constrain planetary energy budgets and to explore the upper atmosphere properties. While Jupiter has a geometric albedo of 0.5 in the visible, HD209458b's is surprisingly low : <0.08 (3-σ upper limit, Rowe *et al.* 2008. The planetary to stellar flux ratio of hot Jupiters at visible wavelengths is of the order of 10^{-5}, making it very challenging to measure. After just 1.5 years of operation, *Kepler* has proven to be a facility able to achieve a few parts per million (ppm) photometric precision (e.g. Jenkins *et al.* 2010).

2. Geometric albedo determination

We performed Markov Chain Monte-Carlo analyses on the 6 hot Jupiters that have been observed by *Kepler*, including two previously known exoplanets : TrES-2b and HAT-P-7b. Public data from the first quarter were used to derive the systems parameters.

Results from this preliminary study appear in Table 1 and depict a wide variety of geometric albedos, ranging from 0.06 to 0.35. A single secondary eclipse is also shown on Fig. 1 for HAT-P-7b. Similar analysis has been performed by Kipping & Bakos (2010) and results show good consistency with those presented here.

3. Disentangling thermal emission and reflected light

Hot-Jupiter thermal emission could have a significant contribution to the planetary flux measured in the *Kepler* bandpass. Comparing the range of possible equilibrium temperatures to the brightness temperature corresponding to the secondary eclipse depth allows an estimate of the upper bound for the thermal emission, and thus also an estimate of the reflected light fraction. This simple approach is however challenged by the departure from blackbody radiation of hot Jupiter thermal emission spectra. Moreover, the structure of the atmospheric temperature profile might cause visible band measurements

Table 1. Geometric albedos and equilibrium temperatures (assuming no redistribution) for *Kepler* published giant planets from public data (Q1).

Planet	Geometric Albedo	$T_{eq}[K]$
Kepler 5b	0.21 ± 0.10	1557
Kepler 6b	0.18 ± 0.09	1411
Kepler 7b	0.35 ± 0.11	1370
Kepler 8b	0.21 ± 0.10	1567
TrES-2b	0.06 ± 0.05	1464
HAT-P-7b	0.20 ± 0.03	2085

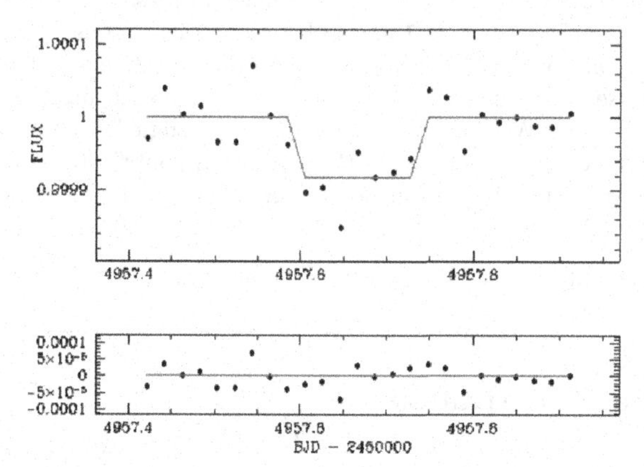

Figure 1. Top : Kepler single long-cadence (30min) secondary eclipse lightcurve of the 2.2-day period hot Jupiter HAT-P-7b with the best fit model superimposed. **Bottom** : residuals of the fit. The occultation depth is 82 ± 12 ppm. Obtained from Kepler Q1 public data.

to probe thermal emission from deep layers of the atmosphere. Brightness temperature estimates at other wavelengths might definitely help in constraining the energy budget and sample the planetary spectral energy distribution.

4. Conclusions

While alkali metal absorption lines and TiO and VO molecular absorption bands are expected to shape the spectrum of hot Jupiters in the *Kepler* bandpass, the planetary thermal emission is an important contributor for the most irradiated hot-Jupiters. Additionally, clouds are expected to form at the intersection of enstatite and iron compound condensation curves with the planetary temperature structure profile (Sudarsky *et al.* 2000). The altitude of iron and enstatite cloud decks in hot Jupiter atmospheres significantly affects the geometric albedo (Seager *et al.* 2000). How representative are giant irradiated planets harboring high altitude reflective clouds and hazes is one of the several points *Kepler* will be able to address, shedding light on the properties of hot Jupiter atmospheres.

References

Jenkins, J. M., Caldwell, D. A., Chandrasekaran, H., & Twicken, J. D., *et al.* 2010, *ApJ*, 713, L120

Kipping, D. M. & Bakos, G. A. 2011, *ApJ*, 730, id.50
Rowe, J. F., Matthews, J. M., Seager, S., Miller-Ricci, E., & Sasselov, D., *et al.* 2008, *ApJ*, 689, 1345
Seager, S., Whitney, B. A., & Sasselov, D. D. 2000, *ApJ*, 540, 504
Sudarsky, D., Burrows, A., & Pinto, P. 2000, *ApJ*, 538, 885

The Astrophysics of Planetary Systems: Formation, Structure, and Dynamical Evolution
Proceedings IAU Symposium No. 276, 2010
A. Sozzetti, M. G. Lattanzi & A. P. Boss, eds.

doi:10.1017/S1743921311020874

A new look at NICMOS transmission spectroscopy: No conclusive evidence for molecular features

Neale P. Gibson[1], Frederic Pont[2] and Suzanne Aigrain[1]

[1]Department of Physics, University of Oxford, email: Neale.Gibson@astro.ox.ac.uk
[2]School of Physics, University of Exeter

Abstract. We present a re-analysis of archival HST/NICMOS transmission spectroscopy of the exoplanet system, HD 189733, from which detections of several molecules have been claimed. As expected, we can replicate the transmission spectrum previously published when we use an identical model for the systematic effects, although the uncertainties are larger as we use a residual permutation algorithm in an effort to account for instrumental systematics. We also find that the transmission spectrum is considerably altered when slightly changing the instrument model, and conclude that the NICMOS transmission spectrum is too dependent on the method used to remove systematics to be considered a robust detection of molecular species, given that there is no physical reason to believe that the baseline flux should be modelled as a linear function of any chosen set of parameters.

Keywords. stars: individual (HD 189733), planetary systems, techniques: spectroscopic

---

Transmission spectroscopy is a powerful technique that can probe the atmospheres of transiting planets for atomic or molecular species (e.g. Seager & Sasselov 2000; Brown 2001), by measuring the wavelength dependence of a planet's observed radius. The size of a planet is determined by the altitude at which the atmosphere becomes opaque to starlight, which may vary due to atomic and molecular absorption.

HST/NICMOS transmission spectroscopy has led to some of the most detailed studies of exoplanet systems to date. However, instrumental systematics are larger than the expected signal due to molecular absorption, and consequently the methods used to remove the instrumental systematics have a considerable effect on the output transmission spectra. Swain, Vasisht, & Tinetti (2008, hereafter SVT08) presented a transmission spectrum of HD 189733, claiming detections of H_2O and CH_4. We present a detailed re-analysis of this and other NICMOS data sets in Gibson, Pont, & Aigrain (2010, hereafter GPA10), and argue that the detection of molecular species is dependent on the choice of instrument model, and therefore cannot be considered as robust. We only briefly summarise our findings here, and refer the reader to GPA10 for further details.

After extracting the raw light curves for each wavelength channel, we first model the systematics using an identical instrument model to that of SVT08. Not surprisingly, we produce a very similar transmission spectrum, although with some disagreement at the blue end and with larger uncertainties. The larger uncertainties are the result of using a residual permutation algorithm, which attempts to take the remaining systematic noise and correlations between the instrument model parameters into account. If we instead evaluate the transit depth using a Levenberg–Marquardt algorithm, we get almost identical uncertainties to those reported in SVT08. However, this method does not properly take all sources of uncertainty into account. We further note that the residual permutation algorithm does not fully account for all sources of uncertainty, in particular those that arise from offsets in flux level between the HST orbits not properly corrected for by the instrument model. Offsets in the in-transit flux level are fitted for by the transit model, and do not appear in the residuals to contribute to the measured uncertainties. Hence the 'true' uncertainties are likely even larger than we reported.

We also produce transmission spectra of HD 189733 using slightly different instrument models to remove the systematics; one using extra quadratic terms in the decorrelation function, another using only two of the three out-of-transit orbits to determine the model, and another excluding the angle parameter from the instrument model (see GPA10 for details). Using different instrument models we produce rather different transmission spectra. Given that there is no physical reason to assume the instrumental systematics should follow a linear function of a chosen set of instrumental parameters, there is no reason to prefer one model over another. We therefore argue that further physical justification for the specific instrument model used in SVT08 (and for other important NICMOS results) is required before these spectra can be considered as robust.

Recently Deroo, Swain, & Vasisht (2010, hereafter DSV10) posted a response to our paper, claiming that our analyses were flawed. They argue that the larger uncertainties we produce in our transmission spectrum are the result of a noisy instrument model, and that the subsequent analyses using varying instrument models are therefore invalid. As stated earlier, the larger uncertainties are due to a different method used to evaluate them. In GPA10, we carefully compared the instrumental parameters used to model the systematics to those provided in SVT10 (supplementary material), and concluded that they show a similar dispersion and amplitude; so why do the plots in DSV10 appear to contradict this? *The answer is that the data marked "Swain et al. (2008)" in DSV10 are not the same data as shown in SVT08* †. The plots in DSV10 are therefore misleading, and their conclusions are based on a misrepresentation of data.

DSV10 also criticises our treatment of the XO-1 data from Tinetti *et al.*(2010), stating that the reason we cannot reproduce the same transmission spectrum is because we omit some parameters from the decorrelation. This is simply re-stating what we already explained in GPA10. The parameters were omitted from the instrument model as these would require extrapolation to the in-transit orbits. Unfortunately, almost no detail is given in Tinetti *et al.*(2010) regarding the data analysis, and we could not compare instrumental parameters and identify the source of the discrepancy.

One of the primary goals of our paper is to encourage open discussion within the exoplanet community about the reliability of methods used to remove systematic noise from this type of dataset, and the robustness of derived results. In our opinion this is a very challenging, unsolved problem. We strenuously contest the claim made by DSV10 that our paper is a confirmation of their results, but will continue to seek a better understanding of the discrepancies between our results and theirs. We are also actively testing new methodologies for the robust characterisation of transmission spectra in the presence of strong systematics, and are in the process of analysing STIS and WFC3 observations of this object, to get a complete transmission spectrum from UV to NIR.

References

Brown, T. M. 2001, *ApJ*, 553, 1006
Deroo, P., Swain, M. R., & Vasisht, G. 2010, *arXiv*:1011.0476v1
Gibson, N. P., Pont, F., & Aigrain, S. 2011, *MNRAS*, 411, 2199
Seager, S. & Sasselov, D. D. 2000, *ApJ*, 537, 916
Swain, M. R., Vasisht, G., & Tinetti, G. 2008, *Nature*, 452, 329
Tinetti, G., Deroo, P., Swain, M. R., *et al.* 2010, *ApJ*, 712, L139

† As an appendix to the astro-ph posting of this article, we provide plots comparing the SVT08, GPA10, and DSV10 datasets.

The Astrophysics of Planetary Systems: Formation, Structure, and Dynamical Evolution
Proceedings IAU Symposium No. 276, 2010
A. Sozzetti, M. G. Lattanzi & A. P. Boss, eds.

doi:10.1017/S1743921311020886

Polarization of the transiting planetary system of the K dwarf HD 189733

Nadia M. Kostogryz[1], **Taras M. Yakobchuk**[1], **Olexandr V. Morozhenko**[1] **and Anatolij P. Vidmachenko**[1]

[1]Main Astronomical Observatory NAS of Ukraine,
27, Zabolotnoho str., Kyiv, 03680, Ukraine
email: kosn@mao.kiev.ua

Abstract. We model the polarization in the system HD 189733 resulting from the planetary transit. This system has a short-period (2.2d) Jupiter-like planet with the radii ratio $R_p/R_* = 0.148$, orbiting at the distance of 0.031 AU around the star.

We calculated the polarization of the system HD189733 to be 0.022% at the limb, which is consistent with the recent observational data. We suggest the shapes of the polarization parameters curves to be used for deriving the planet orbit inclination at the near limb transits as an alternative to standard transit method.

Keywords. polarization, methods: numerical, planetary systems

1. Introduction

During the last 1 - 2 decades a large number of extrasolar planets have been found and substantial effort is devoted to determination of the characteristics of these objects. However, for objects discovered by the most successful radial velocity method, a precise determination of the mass of the companion is not possible without the other techniques being applied (e.g. transits or astrometry). In this sense polarimetry comes as another promising technique (Hough *et al.* 2006; Keller 2006; Schmid *et al.* 2006) for studying the planetary atmospheres by means of scattered host star light, which provides information on their geometry, chemistry, and thermodynamics.

In this study, we model an effect that may be responsible for the polarization increase of the exoplanetary systems by $\sim 2 \times 10^{-4}$ based on Monte Carlo method proposed by Carciofi & Magalhaes (2005).

2. Results and Discussions

HD 189733 is currently the brightest ($m_V = 7.67$ mag) known star to harbor the transiting exoplanet (Bouchy 2005). This fact along with the short planetary period (2.2d) makes it very suitable for polarimetry. Since we are interested in measuring the occultation polarization effect, the ratio of planetary to star radii is also important parameter. The stellar radius of HD 189733 is equal to 0.788 R_* (Baines *et al.* 2009) and the planetary radius is 1.151 R_J (Southworth 2010), which resulting in ratio of 0.148. The panel a) of Figure 1 illustrates the transit configuration of the system. Panel b) shows the time dependence of the stellar flux drop along the planet transiting. The next three panels c), d), e) demonstrate changes of Stokes parameters and polarization degree during occultation. Maximum polarization observed on the stellar limb is $\sim 2 \times 10^{-4}$. This value is very close to that obtained by Berdyugina *et al.* (2008) from direct observations, but authors ascribed this result to the light scattering in the planetary atmosphere. However,

Lucas *et al.* (2009) concluded this effect cannot be explained by reflected light from the planet HD 189733b and explained it by the stellar spots only, given that HD 189733 is an active star. Based on our modelling, we argue that polarization of the system during planet transiting also cannot be rule out. Knowing precisely the moments of ingress and egress, this possibility can be verified by future polarimetric observations.

Until recently the only determination of inclination angle i was based on the shape of the flux drop caused by the planet transit. One of the drawbacks of this method is that at smaller i (i.e. high latitudes transits) flux signatures become weak and hard to detect. In contrast, the linear polarization during the occultation is more sensitive to the inclination angle. In particular, the maximum polarization does not depend on i, being defined by the ratio of planet-to-star radius only. This allows the planet orbit inclination to be determined even for the near limb transits.

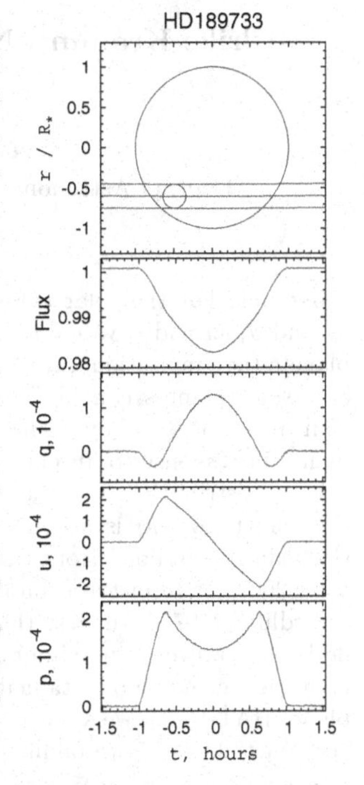

Figure 1. Modeling of the flux and polarization (Stokes $q = Q/I, u = U/I$ and polarization degree) for the planetary transit for HD189733.

3. Conclusions

In this paper we present the results of Monte Carlo simulations of the polarization in the planetary transiting system HD 189733 that ensues from the planet transit over the stellar disk. We derived two polarization maxima at the limb that amount to $\sim$ 0.022%. The maximum polarization for HD189733 is very close to the value obtained by Berdyugina *et al.* (2008). Although they attributed this polarization to scattering of starlight by the planetary atmosphere, this is thought to be most unlikely (Lucas *et al.* 2009) and we suggest the polarization could arise from the planetary transit.

We find that the shapes of the polarization curves at the near limb transits can be used to determine the inclination of the planet orbit, more reliably than from the flux variation observations.

References

Baines, E. K., McAlister, H. A., Brummelaar, Th. A. T, Sturmann, J., Sturmann, L., Turner, N. H., & Ridgway, S. T. 2009, *ApJ*, 701, 154

Berdyugina, S. V., Berdyugin, A. V., Fluri, D. M., & Piirola, V. 2008, *ApJ*, 30, 490

Bouchy, F., Udry, S., Mayor, M., *et al.* 2005, *A&A*, 444, L15

Carciofi, A. C. & Magalhaes, A. M. 2005, *ApJ*, 635, 570

Hough, J. H., Lucas, P. W., Bailey, J. A., Tamura, M., Hirst, E., Harrison, D., & Bartholomew-Biggs, M. 2006, *PASP*, 118, 1302

Keller, C. U. 2006, *Proc. SPIE 6269*,62600T-1

Lucas, P. W., Hough, J. H., Bailey, J. A., Tamura, M., Hirst, E., & Harrison, D. 2009, *MNRAS*, 393, 229

Schmid, H. M., Beuzit, J.-L., Feldt, M., *et al.* 2006, *Proc. IAU Coll.200*, 165

Southworth, J. 2010, *MNRAS*, 408, 1689

The Astrophysics of Planetary Systems: Formation, Structure and Dynamical Evolution
Proceedings IAU Symposium No. 276, 2010
A. Sozzetti, M. G. Lattanzi & A. P. Boss, eds.

doi:10.1017/S1743921311020898

Constraining planetary interiors with the Love number k_2

Ulrike Kramm[1], Nadine Nettelmann[1,2] and Ronald Redmer[1]

[1]Institute of Physics, University of Rostock,
D-18051 Rostock,
email: ulrike.kramm2@uni-rostock.de

[2]Dept. of Astronomy and Astrophysics, University of California Santa Cruz,
CA 95064

Abstract. For the solar sytem giant planets the measurement of the gravitational moments J_2 and J_4 provided valuable information about the interior structure. However, for extrasolar planets the gravitational moments are not accessible. Nevertheless, an additional constraint for extrasolar planets can be obtained from the tidal Love number k_2, which, to first order, is equivalent to J_2. k_2 quantifies the quadrupolic gravity field deformation at the surface of the planet in response to an external perturbing body and depends solely on the planet's internal density distribution. On the other hand, the inverse deduction of the density distribution of the planet from k_2 is non-unique. The Love number k_2 is a potentially observable parameter that can be obtained from tidally induced apsidal precession of close-in planets (Ragozzine & Wolf 2009) or from the orbital parameters of specific two-planet systems in apsidal alignment (Mardling 2007). We find that for a given k_2, a precise value for the core mass cannot be derived. However, a maximum core mass can be inferred which equals the core mass predicted by homogeneous zero metallicity envelope models. Using the example of the extrasolar transiting planet HAT-P-13b we show to what extend planetary models can be constrained by taking into account the tidal Love number k_2.

Keywords. planets and satellites: interiors, planets and satellites: individual (HAT-P-13b), methods: numerical

---

1. Love numbers

Love numbers are planetary parameters that quantify the deformation of the gravity field of a planet in response to an external perturbing body. The perturber of mass M orbits the planet at a distance a and causes a tide raising potential (Zharkov & Trubitsyn 1978) of $W(s) = \sum_{n=2}^{\infty} W_n = (GM/a)\sum_{n=2}^{\infty}(s/a)^n P_n(\cos\theta')$, where s is the radial coordinate of the point under consideration inside the planet, θ' the angle between the planetary mass element at s and the center of mass of M at a, and P_n are Legendre polynomials. The tidally induced mass shift causes a change in the planet's potential, which reads at the planet's surface: $V_n^{\rm ind}(R_{\rm p}) = k_n W_n(R_{\rm p})$, giving a definition of the Love numbers k_n (Love 1911). They solely depend on the radial density distribution of the planet. Of special interest is the Love number k_2: it is a measure for the level of central condensation of an object. A planet of constant density represents maximum homogeneity with $k_2 = 1.5$. In conclusion, k_2 can provide additional constraints for the planet's interior. However, k_2 is not a unique function of the planet's core mass. For three-layer models we find a degeneracy with respect to the density discontinuity in the envelope (Kramm *et al.* 2010).

2. HAT-P-13b

The system HAT-P-13 (Bakos *et al.* 2009) consists of two planets which are assumed to be in apsidal alignment so that the theory described in Mardling (2007) applies and an allowed interval for the Love number k_2 of $0.116 - 0.425$ can be determined, hereafter denoted by Batygin-Bodenheimer-Laughlin (BBL)-Interval (Batygin *et al.* 2009). We constructed models for HAT-P-13b with planet mass $M_{\rm p} = 0.853\,{\rm M_J}$ and planet radius $R_{\rm p} = 1.281\,{\rm R_J}$ within a two-layer model consisting of a rocky core and a H/He/metal-envelope (Saumon *et al.* 1995). We varied the temperatures in the outer atmosphere, which was assumed to be fully adiabatic or isothermal to 1 kbar, and the metallicity in the envelope.

For these models we calculated the core mass and the Love number k_2 (Fig. 1). We conclude that for a given k_2 only a *maximum possible* core mass can be inferred. Assuming the BBL-Interval (Batygin *et al.* 2009) is correct, this maximum possible core mass is $\approx 0.32\,M_{\rm p}$ $(= 87\,{\rm M_\oplus})$. Further constraints could be made if there was more precise knowledge about the envelope's temperature profile and/or metallicity.

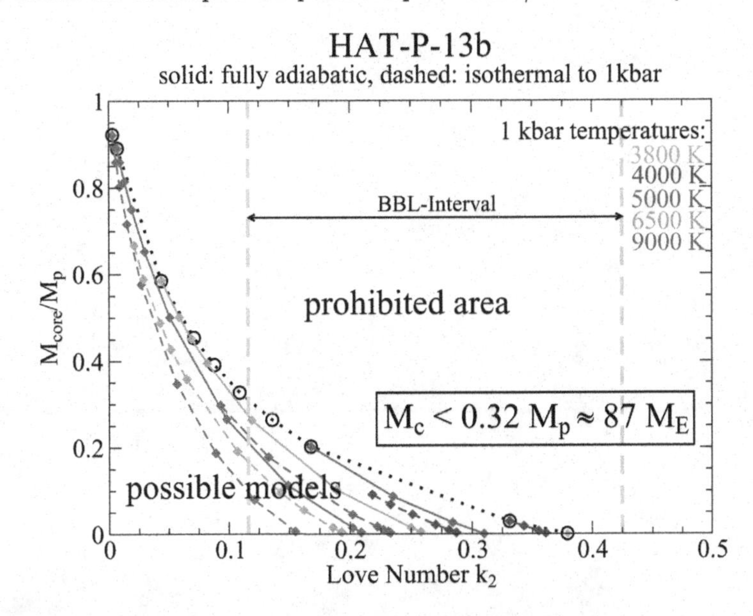

Figure 1. Core masses and Love numbers for several two-layer models of HAT-P-13b. Shown are models with a fully adiabatic (solid) or isothermal to 1 kbar (dashed) envelope for different 1 kbar temperatures (symbol coded, for color version see electronic proceedings). The dotted line consists of models with a fully adiabatic, zero-metallicity envelope. With $k_2 = 0.116 - 0.425$ we deduce a maximum possible core mass of $\approx 87\,{\rm M_\oplus}$.

We also performed calculations for the Love numbers of GJ 436b (Nettelmann *et al.* 2010a, Kramm *et al.* 2010), GJ 1214b (Nettelmann *et al.* 2010b) and Saturn (Kramm *et al.* 2010).

References

Bakos, G. A., Howard, A. W., Noyes, R. W., Hartman, J., *et al.* 2009, *ApJ*, 707,446
Batygin, K., Bodenheimer, P., & Laughlin, G. 2009, *ApJL*, 704, L49
Kramm, U., Nettelmann, N., Redmer, R., & Stevenson, D. J. 2011, *A&A*, 528, A18
Love, A. E. H. 1911, Some problems of geodynamics (Cambridge Univ. Press)
Mardling, R. A. 2007, *MNRAS*, 382, 1768

Nettelmann, N., Kramm, U., Redmer, R., & Neuhäuser, R. 2010, *A&A*, 523, A26
Nettelmann, N., Fortney, J. J., Kramm, U., & Redmer, R. 2011, *ApJ*, 733, id.2
Ragozzine, D. & Wolf, A. S. 2009, *ApJ*, 698, 1778
Saumon, D., Chabrier, G., & van Horn, H. M. 1995, *ApJ*, 99, 713
Zharkov, V. N. & Trubitsyn, V. P. 1978, Physics of planetary interiors (Astronomy and Astrophysics Series, Tucson: Pachart, 1978)

The Astrophysics of Planetary Systems: Formation, Structure, and Dynamical Evolution
Proceedings IAU Symposium No. 276, 2010
A. Sozzetti, M. G. Lattanzi & A. P. Boss, eds.

doi:10.1017/S1743921311020904

Characterization of rocky exoplanets from their infrared phase curve

Anne-Sophie Maurin[1], Franck Selsis[1], Franck Hersant[1] and Marco Delbò[2]

[1] Laboratoire d'Astrophysique de Bordeaux (Université Bordeaux 1), B.P. 89, 33271 Floirac Cedex, France
email: maurin@obs.u-bordeaux1.fr

[2] Laboratoire Cassiopée, Observatoire de la Côte d'Azur, B.P. 4229, 06034 Nice Cedex 4, France

Abstract. During the last few years, observations have yielded an abundant population of short-period planets under 15 Earth masses. Among those, hot terrestrial exoplanets represent a key population to study the survival of dense atmospheres close to their parent star. Thermal emission from exoplanets orbiting low-mass stars will be observable with the next generation of infrared telescopes, in particular the JWST. In order to constrain planetary and atmospheric properties, we have developed models to simulate the variation of the infrared emission along the path of the orbit (IR phase curve) for both airless planets and planets with dense atmospheres. Here, we focus on airless planets and present preliminary results on the influence of orbital elements, planet rotation, surface properties and observation geometry. Then, using simulated noisy phase curves, we test the retrieval of planets' properties and identify the degeneracies.

Keywords. methods: numerical, infrared: planetary systems

1. Infrared phase curve

The light received from a planet varies as the phase seen by the observer changes. At optical wavelengths, variations of the reflected light mainly depend on fraction of the starlit hemisphere in the field of view. At infrared wavelength, however, the phase curve is tightly connected to the temperature distribution on the planet. Infrared phase curves have been observed for short-period giant planets (in transit or not: e.g. Harrington *et al.* 2007; Swain *et al.* 2010), providing constraints on atmospheric circulation. The infrared phase curve of terrestrial planets will be observable provided that the planet is sufficiently hot and the host-star sufficiently cool. Short-period low-mass planets around M star (like for instance Gl581e) satisfy these conditions and are frequent enough to insure the existence of nearby targets.

Phase curves provide a powerful diagnostic for the presence/absence of a dense atmosphere. The temperature distribution, and therefore the thermal emission, of a planet with no atmosphere can be robustly modeled with a simple physical model depending on few parameters: albedo, thermal inertia (Fig. 1a.), rotation, obliquity (Fig. 1b.). In these examples, we assume here a circular orbit and a non synchronous rotation. In the first case, we have a zero-obliquity and we vary the thermal inertia Γ. For high thermal inertias, the heat is redistributed to the night side of the planet, because of the vertical heat diffusion in the subsurface (which reduces the amplitude of the variations) and the hottest region is shifted in longitude relative to the substellar point. A similar effect can be observed for fast rotators planets. The presence of an atmosphere will also make the transport of heat from tropic to pole more efficient. In the Fig 1b., it is the thermal inertia that is fixed, and we vary the obliquity, introducing a seasonal modulation, with a phase difference depending on the observer location relative to the rotation axis.

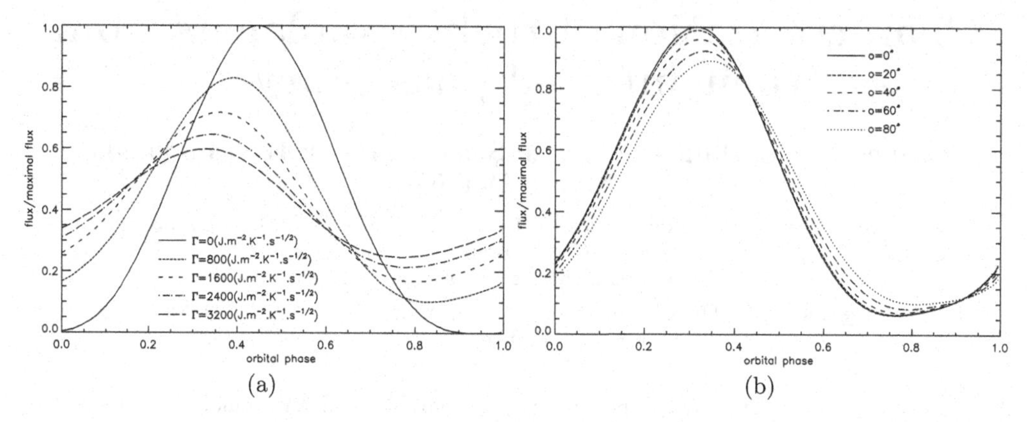

Figure 1. Effect of thermal inertia and obliquity on phase curve. The inclination of the system is 90°(transit configuration). In a. the ratio of orbital by synodic period is 5, in b. it is 2.

2. Retrieval of planet properties

The model we developed computes the variations of surface and subsurface temperature by heat diffusion (Spencer *et al.* 1989), and the resulting infrared phase curve for any set of parameters: orbital elements, planet radius, rotation rate and obliquity, surface properties (thermal inertia, albedo) and observation geometry (inclination of the system). Using this model to produce phase curves, we can retrieve some of the planet properties using a genetic algorithm. When the planet is tidally-locked on a circular orbit, heat diffusion no longer matters and the only parameters to be retrieved are the radius and albedo of the planet and the inclination of the system. For an eccentric orbit, we add the synodic period and the thermal inertia. When the retrieved parameters are unrealistic or vary with the observed wavelengths, this points to the presence of a dense-atmosphere.

If you are interested by a collaboration in using our model, please contact A.S. Maurin.

Acknowledgements

We thank A. De Wit for the using of the genetic algorithm and A. Belu for the helpful discussions we had.

References

Harrington, J., Luszcz, S., Seager, S., *et al.* 2007, *Nature*, 447, 691
Spencer, J. R., Lebofsky, L. A., & Sykes, M. V. 1989, *Icarus*, 78, 337
Swain, M. R., Deroo, P., Griffith, C. A., *et al.* 2010, *Nature*, 463, 637

The Astrophysics of Planetary Systems: Formation, Structure, and Dynamical Evolution
Proceedings IAU Symposium No. 276, 2010
A. Sozzetti, M. G. Lattanzi & A. P. Boss, eds.

doi:10.1017/S1743921311020916

The GROUnd-based Secondary Eclipse project - GROUSE

Ernst de Mooij[1], Remco de Kok[2], Bas Nefs[1], Matteo Brogi[1] and Ignas Snellen[1]

[1]Leiden Observatory, Leiden University, Postbus 9513, 2300 RA, Leiden, The Netherlands;
email: demooij@strw.leidenuniv.nl

[2]SRON Netherlands Institute for Space Research, Sorbonnelaan 2, 3584 CA Utrecht, The Netherlands;

Abstract. Secondary eclipse observations of exoplanets at near-infrared wavelengths are important to constrain the energy budgets of hot-Jupiters, since they probe the radiation from the planet's atmosphere at the peak of the spectral energy distribution. Since this wavelength range is accesible from the ground, we have started the GROUnd-based Secondary Eclipse (GROUSE) project. As part of the GROUSE project, we target a sample of hot-Jupiters at near-infrared and optical wavelengths. Planets include TrES-3b, HAT-P-1, WASP-18b and WASP-33b.

Keywords. techniques: photometric, planetary systems

1. Introduction

Secondary eclipse measurements of transiting extrasolar planets with the *Spitzer* Space Telescope have yielded many detections of thermal exoplanet light (e.g. Charbonneau *et al.* 2005; Deming *et al.* 2005; Knutson *et al.*, 2008). One of the most interesting parts of the planet spectrum (1-3μm) is inaccessible with this satellite. Although the typical planet-to-star contrast ratio in this wavelength range is smaller than at the mid-infrared region probed by *Spitzer*, the near-infrared region is at the peak of the planet's spectral energy distribution and is also a wavelength range where molecular absorption bands can significantly influence the measured spectrum. Since there are several windows in the Earth's atmosphere in the near-infrared, it is possible to use ground-based telescopes to study the properties of the exoplanet atmospheres in this range, which, during the past few years, has led to the detection of the secondary eclipses of several exoplanets (e.g. De Mooij & Snellen 2009; Sing & López-Morales 2009; Croll *et al.* 2010).

2. The GROUSE project

We have started the GROUnd-based Secondary-Eclipse (GROUSE) project to study the optical and near-infrared spectral energy distributions of a sample of hot-Jupiters. For this we use a variety of instruments and telescopes, including LIRIS at the William Herschel Telescope for near-infrared observations, and OSIRIS on the Gran Telescopio Canarias and the WFC on the Isaac Newton Telescope for our optical measurements.

The sample we target can be divided into two parts, the first part contains the hottest and brightest exoplanet systems, such as HAT-P-7b and WASP-18b, for which we should be able to reach a sufficient signal-to-noise ratio to detect their secondary eclipses at both optical and near-infrared wavelengths. The second part of the sample contains planets for which we only expect to be able to measure the the planet's light in K-band, such as HAT-P-1b.

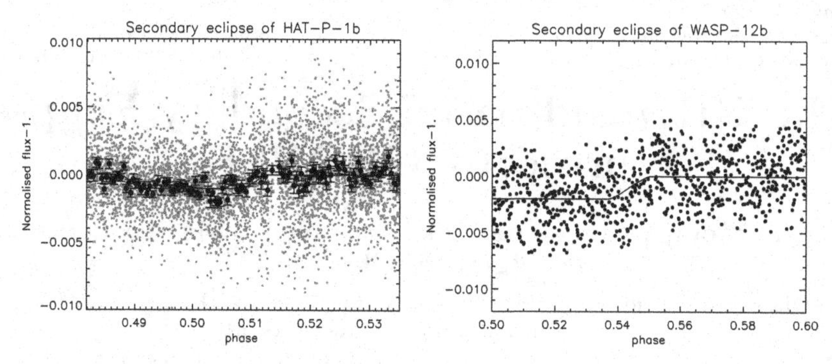

Figure 1. Two examples of secondary eclipse observations from the GROUSE project. Left panel: The Ks-band secondary eclipse of HAT-P-1b (De Mooij *et al.* 2011), the black points show the data binned by 41 frames. Right panel: Partial Ks-band secondary eclipse of WASP-12b observed with the William Herschel Telescope.

As part of this project we have, so far, reported on the detections in Ks-band of the secondary eclipses of TrES-3b (De Mooij & Snellen 2009; Paper I) and HAT-P-1b (De Mooij *et al.* 2011; Paper II). Two examples of secondary eclipse observations are shown in Fig. 1.

Since secondary eclipse observations require very high signal-to-noise ratios, we use a specialised strategy to reduce systematic effects, and to minimize the overheads. Since the best candidate systems are relatively bright (V$\lesssim$12), we defocus the telescope for all our observations. The defocusing has the additional benefit of reducing the sensitivity of the observations to flatfielding errors, but at the cost of an increased noise due to the sky background.

Contrary to typical near-infrared observations, where a dither sequence is used to determine the sky background, we perform our observations in staring mode, which enables us to use guiding without increasing the overheads due to the reacquisition of the guide star. This is very important for the stability of the observations, since guiding enables us to maintain the position of the star within a few pixels for the entire duration of the observation. For the background subtraction, we acquire a set of images on a blank field before and/or after the eclipse.

Since the eclipse depth has to be measured with respect to the out-of-eclipse baseline, we aim to observe the target out-of eclipse for as long as possible. Typically we try to get a baseline that is longer than the eclipse duration, since systematic effects on timescales of the eclipse will then be visible. In this way we manage to obtain eclipse depth uncertainties down to ~0.2 millimag in K-band. This allows us to put tight constraints on the planets' energy budgets, which is especially important now that *Spitzer* has entered its warm mission.

References

Charbonneau, D., *et al.* 2005, *ApJ*, 626, 523

Croll, B., Jayawardhana, R., Fortney, J. J., Lafrenire, D., & Albert, L. 2010, *ApJ*, 718, 920

Deming, D., Seager, S., Richardson, L. J., & Harrington, J. 2005, *Nature*, 434, 740

De Mooij, E. J. W. & Snellen, I. A. G. 2009, *A&A*, 493, L35

De Mooij, E. J. W., De Kok, R. J., Nefs, S. V., & Snellen, I. A. G. 2011, *A&A*, 528, A49

Knutson, H. A., *et al.* 2008, *ApJ*, 673, 526

Sing, D. K. & López-Morales, M. 2009, *A&A*, 493, L31

The Astrophysics of Planetary Systems: Formation, Structure, and Dynamical Evolution
Proceedings IAU Symposium No. 276, 2010
A. Sozzetti, M. G. Lattanzi & A. P. Boss, eds.

doi:10.1017/S1743921311020928

Search and characterization of T-type planetary mass candidates in the σ Orionis cluster

Karla Peña Ramírez[1], Maria R. Zapatero Osorio[2] and Victor J.S. Béjar[1]

[1] Instituto de Astrofísica de Canarias. E-38205 La Laguna, Tenerife. Spain
email: karla@iac.es

[2] Centro de Astrobiología (CSIC-INTA). E-28850 Torrejón de Ardoz, Madrid. Spain

Abstract. We present new photometric and astrometric data available for S Ori 70 and 73, the two T-type planetary-mass member candidates in the σ Orionis cluster ($\sim$3 $\pm$ 2 Myr, d$\sim$ 360 pc). S Ori 70 ($J \sim 19.9$ mag) has a spectral type of T5.5 $\pm$ 1.0 measured from published near-infrared spectra, while no spectroscopic data are available for S Ori 73 ($J \sim 21$ mag). We estimate the spectral type of S Ori 73 by using J, H, and $CH_{4\rm off}$ (λ_c=1.575 μm, $\Delta\lambda$=0.112 μm) photometry and comparing the H-$CH_{4\rm off}$ index of S Ori 73 with the colors of field stars and brown dwarfs of spectral types in the range F to late T. The locations of S Ori 70 and 73 in the J-H vs H-$CH_{4\rm off}$ color-color diagram are consistent with spectral types T8 $\pm$ 1 and T4 $\pm$ 1, respectively. Proper motion measurements of the two sources are larger than the motion of the central σ Ori star, making their cluster membership somehow uncertain.

Keywords. infrared: stars, brown dwarfs, planetary systems, open clusters and associations: individual (σ Orionis)

---

1. Motivation

Knowledge of the low-mass end of the initial mass function (IMF) is crucial to understand the formation mechanisms giving rise to substellar objects. In the σ Orionis cluster ($\sim$3 $\pm$ 2 Myr, d$\sim$ 360 pc), there are only two T-type planetary mass candidates: S Ori 70 ($J \sim 19.9$ mag), which has a spectral type of T5.5 $\pm$ 1.0 measured from near-infrared low-resolution spectra (Zapatero Osorio *et al.* 2002), and S Ori 73 ($J \sim 21$ mag), which has no spectra available so far (Bihain *et al.* 2009). We present new photometric and astrometric data for S Ori 70 and 73 to study the methane nature of the latter and to assess their cluster membership via proper motion analysis.

2. Observational data

Imaging data are summarized in Table 1. HAWK-I and OSIRIS observations were intended to image S Ori 70 and 73, covering areas of $\sim$120 and $\sim$220 arcmin^2, respectively. VISTA data covers the entire region of the σ Orionis cluster. HAWK-I and OSIRIS images were reduced following standard procedures; aperture and point-spread-function instrumental photometry were obtained and calibrated into observed magnitudes using the UKIDSS DR7 database for the near-infrared wavelengths (Lawrence *et al.* 2007) and photometric standard stars from Smith *et al.* (2002) for the optical. The VISTA (science verification) data were reduced by the Cambridge Astronomy Survey Unit (CASU) and only aperture photometry of 2″ in diameter was performed.

Table 1. Log of optical and near infrared observations.

Telescope	Instrument	Filter	Date	Exp.time [s]	Seeing [arcsec]	Completeness [mag]	Limiting [mag]
GTC	OSIRIS	i'	2009 Oct 15	3146	0.80	25.0	26.0
			2010 Jan 11	9360	1.10		
	OSIRIS		2009 Oct 13,14,15	9438	0.70		
			2009 Nov 19	3198	1.10		
VISTA	VIRCAM	Z	2009 Oct 20,21	6084	0.80	22.6	23.2
		Y	2009 Oct 20	1008	0.90	21.0	21.4
		J	2009 Oct 19,20	2112	0.90	21.4	21.8
		H	2009 Oct 20	288	0.90	19.6	20.0
		K_s	2009 Oct 20	288	0.70	18.6	19.1
UT4	HAWK-I	J	2008 Sep 19	160	0.64	22.4	23.4
		H	2008 Dec 8	8410	0.34	22.5	23.4
		$CH_{4\mathrm{off}}$	2009 Feb 24	13500	0.52	22.5	23.1
			2009 Mar 28	13500			
UT4	HAWK-I	J	2008 Oct 27	160	0.75	21.8	22.8
		H	2009 Mar 28	8410	0.52	20.5	21.8
		$CH_{4\mathrm{off}}$	2009 Mar 16	13500	0.57	21.3	22.4

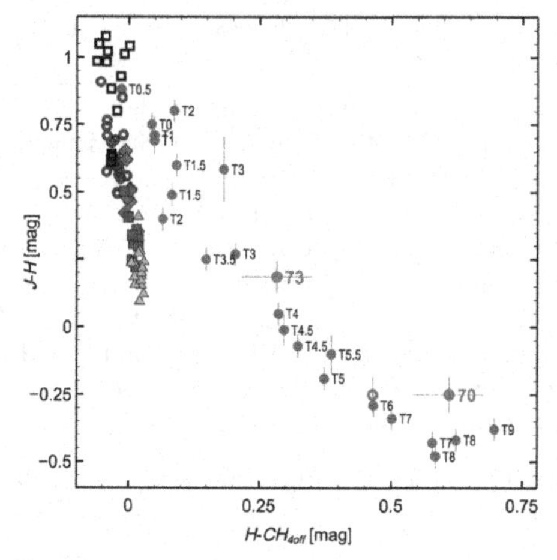

Figure 1. Color-color diagram (including the methane filter) for T (green), L (black), M (magenta), K (olive green), G (blue), and F (yellow) type field sources. The photometry of S Ori 70 and 73 is plotted as red filled circles. The red open circle corresponds to the synthetic methane color of S Ori 70 derived from its spectrum (Zapatero Osorio *et al.* 2002).

3. Results

We estimated the spectral types of S Ori 73 and S Ori 70 by comparing their J-H and H-$CH_{4\mathrm{off}}$ colors with those of field dwarfs of known classification. We derived T4 $\pm$ 1.0 for S Ori 73, and T8 $\pm$ 1.0 for S Ori 70. The "methane" nature of S Ori 73 is thus confirmed. The typing of S Ori 70 is slightly cooler, though still consistent within error bars, than the value previously reported in the literature (T5.5 $\pm$ 1). From our astrometric analysis we derived the following proper motions: $(\mu_\alpha \cos\delta,\ \mu_\delta) = (30.8 \pm 11.0,\ 18.2 \pm 8.0)$ mas yr^{-1} for S Ori 70, and $(\mu_\alpha \cos\delta,\ \mu_\delta) = (43.2 \pm 10.0,\ \text{-}3.7 \pm 7.0)$ mas yr^{-1} for S Ori 73. These values are larger than the Hipparcos proper motion of the cluster central star (σ Ori) by at least 2-σ, making the cluster membership of the two T dwarfs somehow uncertain.

We also carried out a photometric search for additional T-type candidates in the cluster using the HAWK-I (~120 arcmin^2 and completeness magnitude of $J \sim 21.8$ mag), VISTA and OSIRIS data. The selection photometric criteria (i'-$J \geqslant 5$, Z-$J \geqslant 2.5$, J-$H \leqslant 0.5$ and H-$CH_{4\mathrm{off}} \geqslant 0.15$ mag) did not yield any additional candidate with the colors expected for $\geqslant$ T3 dwarfs.

References

Bihain, G., *et al.* 2009, *A&A*, 506, 1169

Zapatero Osorio, M. R., *et al.* 2000, *Science*, 290, 103

Zapatero Osorio, M. R., *et al.* 2002, *ApJ*, 578, 536

Zapatero Osorio, M. R., *et al.* 2008, *A&A*, 477, 895

The Astrophysics of Planetary Systems: Formation, Structure, and Dynamical Evolution
Proceedings IAU Symposium No. 276, 2010
A. Sozzetti, M. G. Lattanzi & A. P. Boss, eds.

doi:10.1017/S174392131102093X

Transmission spectroscopy of the sodium doublet in WASP-17b with the VLT

Patricia L. Wood[1] **and Pierre F. L. Maxted**[1]
[1]Keele University, Staffordshire ST5 5BG, UK
email: plw@astro.keele.ac.uk

Abstract. The detection of sodium absorption during primary transit implies the presence of an atmosphere around an extrasolar planet. WASP-17b (Anderson *et al.* 2010a) is the least dense known planet, with a radius twice that of Jupiter. It orbits an F6-type star, and its low gravity gives its atmosphere a very large scale height. The sodium transit depth is expected to be 4.1 – 5.2 times deeper than for HD 209458b (Seager & Sasselov 2000). We obtained 24 spectra with the GIRAFFE spectrograph on the VLT, 8 during transit. We measured the flux in the sodium doublet at 5889.95 Å and 5895.92 Å using bandpasses 0.75, 1.5, 3.0 and 6.0 Å. We find a transit depth of $0.55 \pm 0.13\%$ at 1.5 Å (4.3σ). WASP-17b therefore has an atmosphere which is depleted in sodium compared to predictions.

Keywords. planetary systems, stars: individual (WASP-17), techniques: spectroscopic

1. Transmission Spectroscopy

When light from a star passes through the atmosphere of an orbiting planet during transit, some wavelengths are absorbed much more strongly than others. The planet will appear larger at these wavelengths, i.e., the transit will be slightly deeper at wavelengths where there is an opacity source in the upper atmosphere. Measurements at wavelengths where the opacity is high, e.g. narrow bandwidths around the NaI lines, probe the upper layers of the atmosphere, whilst wider bandwidths probe the lower atmosphere. Na absorption has so far only been detected in the atmospheres of the two exoplanets – HD 189733b (Redfield *et al.* 2008) and HD 209458b (Snellen *et al.* 2008). As the atmosphere of WASP-17b has a large scale height, the sodium transit depth is predicted to be 4.1 – 5.2 times as large as that for HD 209458b; 0.135% at bandwidth 1.5 Å.

2. Observations

- 24 spectra obtained with VLT GIRAFFE spectrograph
- IFU feed on WASP-17 and comparison star
- 8 spectra in-transit
- 5821 – 6146 Å, 0.48 Å resolution, 0.08 Å per pixel
- Custom-made optimal extraction software
- Telluric lines removed using synthetic telluric absorption spectra

Table 1. Parameters of WASP-17b and HD 209458b for comparison

Planet	Period, d	Mass, M_J	Radius, R_J	T_{eq}, K	H, km	Spec. Type
WASP-17b[1]	3.74	0.49	2.00	1773	2300	F6V
HD 209458b	3.52	0.64	1.35	1300	500	G0V

[1]Parameters taken from Anderson *et al.* 2010b, in prep.

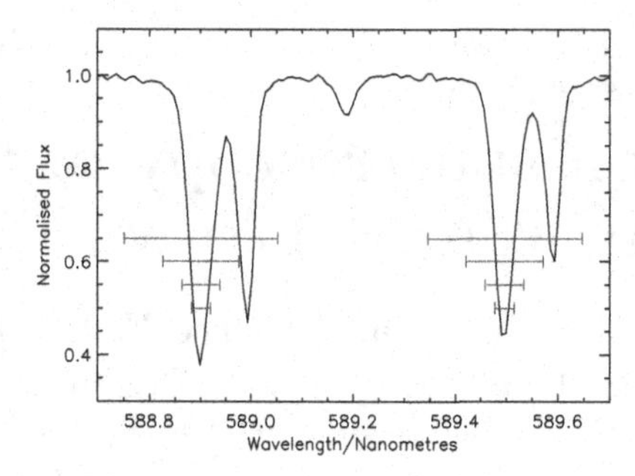

Figure 1. Part of a normalised spectrum of WASP-17 taken with GIRAFFE, showing the Na lines at 5890 Å and 5895 Å. The shallower lines in each pair are interstellar Na absorption features. Measured bandwidths are marked in blue: 0.75 Å, 1.5 Å, 3.0 Å, 6.0 Å.

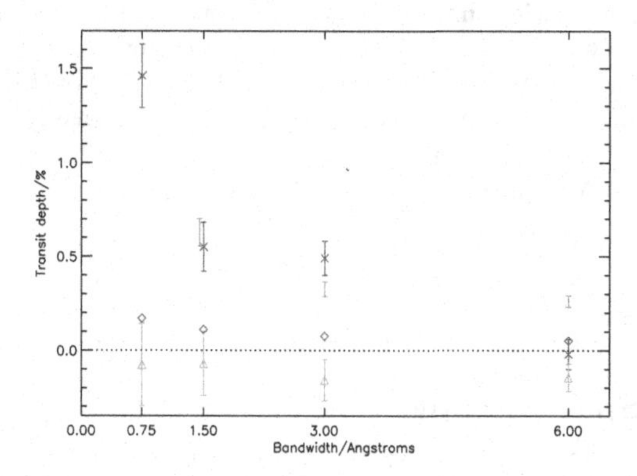

Figure 2. Na transit depths for WASP-17 (blue). In pink are values for HD 2094585 , scaled up by the difference in scale height. Identical measurements on a comparison star observed simultaneously with the same instrument (green triangles) show that the level of systematic errors on these measurements is low. Red diamonds are photon noise.

Table 2. Sodium transit depths for WASP-17b measured at 4 different bandwidths

Bandwidth (Å)	**0.75**	**1.5**	**3.0**	**6.0**
Transit Depth (%)	1.46 ± 0.017	0.55 ± 0.13	0.49 ± 0.09	-0.019 ± 0.076

3. Results and Conclusions

We find that, like HD 209458b, WASP-17b has an atmosphere depleted in sodium compared to predictions, particularly in the 6 Å bandpass that probes the lower depths of the atmosphere. This may be due to photo-ionization, condensation, or the presence of high altitude clouds or haze in the atmosphere. We will now be investigating which is the more likely explanation.

References

Anderson, D. R., *et al.* 2010a, *ApJ*, 709, 159

Anderson, D. R., *et al.* 2010b, *in prep.*

Seager, S. & Sasselov, D. D. 2000, *ApJ*, 540, 504

Redfield, S., Endl, M., Cochran, W. D., & Koesterke, L. 2008, *AJ*, 673, L87

Snellen, I. A. G., Albrecht, S., de Mooji, E. J. W., & Le Poole, R. S. 2008, *A&A*, 487, 357

Section C:
Interactions

The Astrophysics of Planetary Systems: Formation, Structure, and Dynamical Evolution
Proceedings IAU Symposium No. 276, 2010
A. Sozzetti, M. G. Lattanzi & A. P. Boss, eds.

doi:10.1017/S1743921311020941

Two bodies with high eccentricity around the cataclysmic variable QS Vir

Leonardo A. Almeida[1] and Francisco Jablonski[1]

[1]Instituto Nacional de Pesquisas Espaciais/MCT
Avenida dos Astronautas 1758, São José dos Campos, SP, 12227-010, Brazil
email: `leonardo@das.inpe.br`

Abstract. QS Vir is an eclipsing cataclysmic variable with 3.618 hrs orbital period. This system has the interesting characteristics that it does not show mass transfer between the components through the L1 Lagrangian point and shows a complex orbital period variation history. Qian *et al.* (2010) associated the orbital period variations to the presence of a giant planet in the system plus angular momentum loss via magnetic braking. Parsons *et al.* (2010) obtained new eclipse timings and observed that the orbital period variations associated to a hypothetical giant planet disagree with their measurements and concluded that the decrease in orbital period is part of a cyclic variation with period $\sim$ 16 yrs. In this work, we present 28 new eclipse timings of QS Vir and suggest that the orbital period variations can be explained by a model with two circumbinary bodies. The best fitting gives the lower limit to the masses $M_1 \sin(i) \sim 0.0086\ M_\odot$ and $M_2 \sin(i) \sim 0.054\ M_\odot$; orbital periods $P_1 \sim 14.4$ yrs and $P_2 \sim 16.99$ yrs, and eccentricities $e_1 \sim 0.62$ and $e_2 \sim 0.92$ for the two external bodies. Under the assumption of coplanarity among the two external bodies and the inner binary, we obtain a giant planet with $\sim 0.009\ M_\odot$ and a brown dwarf with $\sim 0.056\ M_\odot$ around the eclipsing binary QS Vir.

Keywords. planetary systems, binaries: eclipsing, stars: individual (QS Vir)

1. Introduction

QS Vir is an eclipsing binary consisting of a white dwarf plus a red dwarf that has spectral type M3.5-M4 (O'Donoghue *et al.* 2003). O'Donoghue *et al.* (2003) using the information about the white dwarf spin suggested that QS Vir is a hibernating cataclysmic variable. With orbital period close to the period-gap of the catalysmic variables (CVs), 3.618 hrs, and a secondary close to the transition between stars with a radiative core and completely convective stars, this CV is an interesting target for more detailed studies.

Here, we present 28 new eclipse timings of QS Vir from May to August, 2010. We gathered these to all measurements in the literature and re-analysed the orbital period variation of this system. We suggest that a plausible explanation for the orbital period variations is the presence of two bodies with high-eccentricity around the binary.

2. Analysis of the orbital period variation and discussion

We fit the eclipse timings with the following equation,

$$T_{min} = T_0 + E \times P + \tau_3 + \tau_4, \qquad (2.1)$$

where T_0, E and P are the epoch, the cycle count and the period of the binary, respectively, and τ_3 and τ_4 are the light-time travel effects (LTTEs) (Irwin 1952). Each LTTE includes five parameters: semi-major axis, a, inclination, i, argument of periastron, ω, Keplerian mean motion, n, and epoch of periastron passage, T. We exclude from

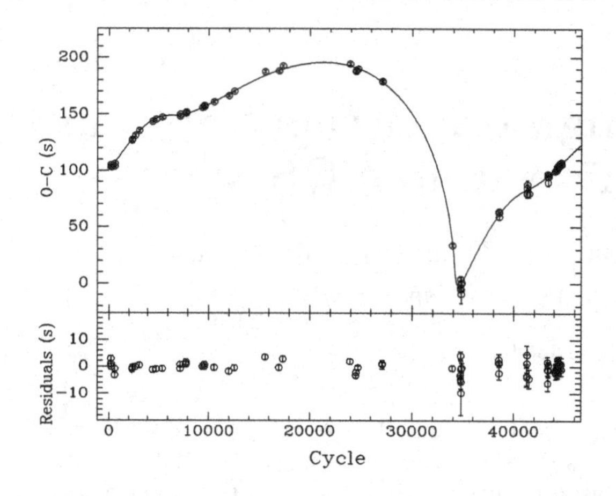

Figure 1: Upper panel: (O−C) diagram of the eclipse timings in QS Vir built with respect to the linear part of the ephemeris in Equation 1. The observed data are presented with open circles and the solid line represents the best fitting including the two LTTEs. Lower panel: The residuals around the fit.

Table 1: Parameters for the linear plus two-LTTEs ephemeris of QS Vir.

Parameter	Value	Unit
Linear ephemeris		
P	$0.150757481 \pm 1 \times 10^{-9}$	days
T_0	$2448689.13995 \pm 2 \times 10^{-5}$	MJD(BTDB)
τ_3 term		
P	14.40 ± 0.08	years
T	2454880 ± 20	MJD(BTDB)
$a \sin i$	0.0446 ± 0.001	AU
e	0.62 ± 0.02	
ω	180.0 ± 2.6	degrees
$f(M)$	$(4.3 \pm 0.3) \times 10^{-7}$	$M_\odot$
τ_4 term		
P	16.99 ± 0.07	years
T	2448689 ± 30	MJD(BTDB)
$a \sin i$	0.320 ± 0.01	AU
e	0.92 ± 0.02	
ω	219 ± 3	degrees
$f(M)$	$(1.1 \pm 0.1) \times 10^{-4}$	$M_\odot$

this analysis the mutual interaction between the external bodies. For the fitting we use the PIKAIA algorithm (Charbonneau 1995) to look for a global solution, followed by a Markov chain Monte Carlo (MCMC) procedure to sample the parameters of Equation 1 around this solution. Figure 1 shows the result of this procedure and Table 1 shows the numerical values with the associated ±68% uncertainties.

Cyclic variations of the orbital period of compact binary systems in time-scales from years to decades can be explained by either the LTTE or the Applegate mechanism. The LTTE is a periodic variation and occurs because the distance from a binary to the observer varies due to gravitational interaction among the inner binary and the external body. The Applegate mechanism was proposed by Applegate (1992) and consists of the coupling between the binary period and changes in the shape of the secondary generated by the quadrupole momentum variation and consequently causing cyclic changes in the binary orbital period. Following the same method used by Brinkworth *et al.* (2006), we obtained that the required energy for the Applegate mechanism is larger than the total radiant energy of the secondary in 1 yr, considering the variation with semi-amplitude $\sim$ 20 s. Thus, both τ_3 and τ_4 terms obtained by us could not be explained by the Applegate mechanism. Therefore, the only explanation for the observed periodic variations of the orbital period in QS Vir is the light-travel time effect by two outer bodies.

References

Applegate, J. H. 1992, *ApJ*, 385, 621
Brinkworth, C. S., *et al.* 2006, *MNRAS*, 365, 287B
Charbonneau, P. 1995, *ApJS*, 101, 309C
Irwin, J. B. 1952, *ApJ*, 116, 211
O'Donoghue, S. G., *et al.* 2003, *MNRAS*, 345, 506
Parsons, S. G., *et al.* 2010, *MNRAS*, 407, 2362P
Qian, S.-B., *et al.* 2010, *MNRAS*, 401L, 34Q

The Astrophysics of Planetary Systems: Formation, Structure, and Dynamical Evolution
Proceedings IAU Symposium No. 276, 2010
A. Sozzetti, M. G. Lattanzi & A. P. Boss, eds.

doi:10.1017/S1743921311020953

Conditions for outward migration

Bertram Bitsch[1] **and Willy Kley**[1]

[1]Computational Physics, University of Tübingen,
72076 Tübingen, Germany
email: bertram.bitsch@uni-tuebingen.de, wilhelm.kley@uni-tuebingen.de

Abstract. The migration of protoplanets in discs not only depends on the thermodynamics of the ambient disc but also on the orbital parameters of the embedded planet. In the fully radiative regime, planets can migrate outward if their orbital parameters fit in a very small range. We simulate planets in fully radiative discs (and isothermal discs - for reference) and determine the influence of the orbital parameters (a, e and i) and the planetary mass on the migration time scale and direction.

Keywords. accretion disks, planetary systems: formation, radiative tranfer, hydrodynamics

1. Introduction

The migration of growing protoplanets depends on the thermodynamics of the ambient disc. Standard modelling, using locally isothermal discs, indicates an inward (type-I) migration in the low planet mass regime. Taking into account non-isothermal effects, recent studies have shown that the direction of the type-I migration can change from inwards to outwards. This change in the direction of migration crucially depends on the properties of the planet's orbit. Eccentricity and inclination influence the migration and can, under certain conditions, restrain the outward migration in non-isothermal discs. In the following we show results of our 3D Simulations of low mass planets, embedded in protoplanetary discs, on circular, eccentric and inclined orbits. All information here is presented in much more detail in the referenced papers.

2. Change of the planetary mass and of the orbital parameters

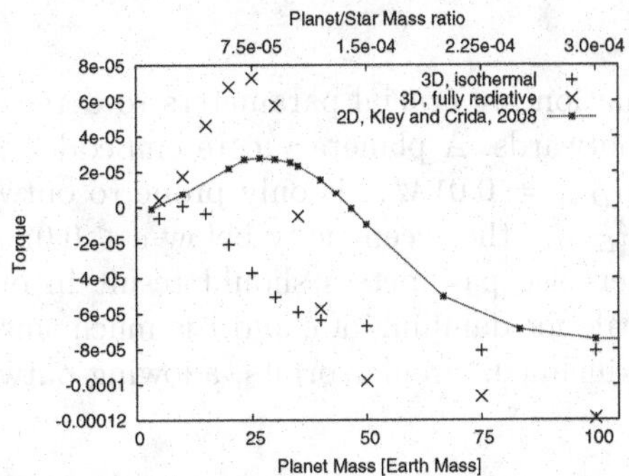

Fig. 1 Torque acting on planets with different masses for 2D and 3D isothermal and fully radiative simulations.

For circular orbits the torque is a direct measurement for the migration, a positive torque indicates outward migration, while a negative torque on the planet results in inward migration. Planets in the fully radiative scheme are prone to outward migration, but only if the planetary core does not exceed a certain threshold for it's mass. In our 3D simulations planetary cores up to $\approx 33 M_{Earth}$ can migrate outward. For 2D simulations the planet's mass can be a little higher (see Fig. 1). Above this limit gap opening diminishes the effect. For more details on this topic, please see Kley *et al.* 2009.

In Fig. 2 we present the theoretical change of the orbital elements of $20M_{Earth}$ planets on fixed orbits. For both thermodynamic cases, initial eccentricity e and initial inclination i will be damped in time, until the planet reaches a circular orbit in the midplane. Only for small eccentricities and inclinations outward migration is possible in the radiative case. The evolutions of a, e and i in time follows the theoretical predicted changes. As soon as the eccentricity is damped below $e \approx 0.02$ the planet can migrate outward.

The threshold for outward migration for inclined planets seems to be $i \approx 4.5°$; lower inclined planets migrate outwards. In the end, eccentricity and inclination have to be damped in order to support outward migration of low mass planets. For more informations on planets on eccentric and inclined orbits, please see Bitsch & Kley 2010 and Bitsch & Kley 2011, respectively.

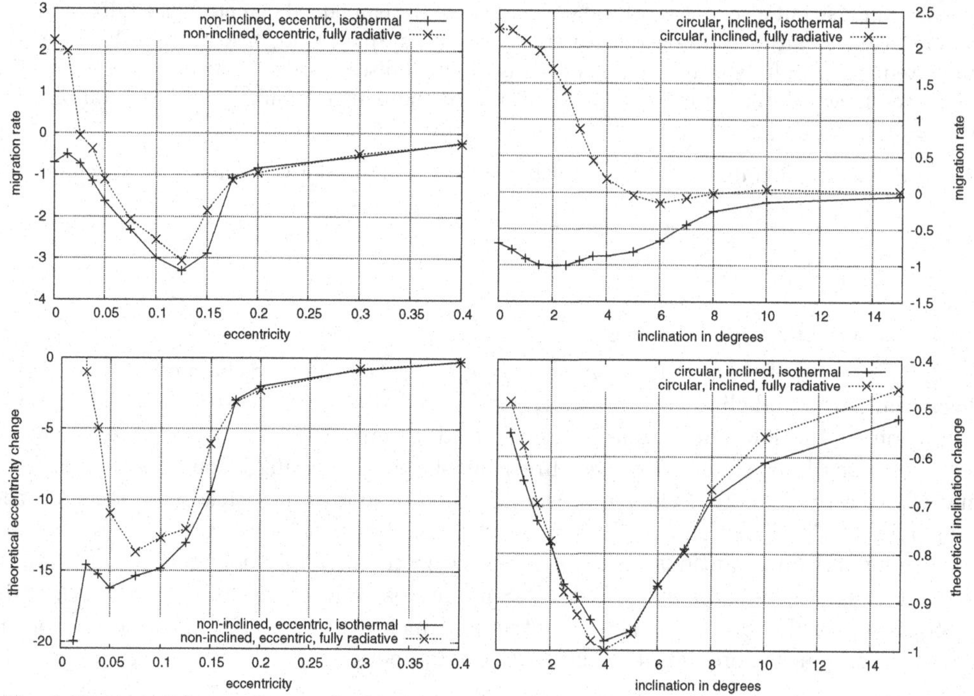

Fig. 2 Theoretical change of a, e and i for $20M_{Earth}$ planets on fixed orbits in isothermal (black '+') and fully radiative (grey-dotted 'x') discs.

3. Conclusions

The planetary mass, eccentricity and inclination are crucial parameters to determine whether the planet will migrate inwards or outwards. A planetary core embedded in a fully radiative disc, with $H/r = 0.037$ and $M_{Disc} = 0.01M_{\odot}$, is only prone to outward migration if the planetary mass is below $33M_{Earth}$, the eccentricity below $e \leqslant 0.02$ and the inclination is lower than $i \approx 4.5°$. Other disc parameters should result in other conditions for outward migration. The timescale for damping of e and i is much smaller than the migration timescale, leading to non-inclined, circular orbits, allowing outward migrating planets.

References

Kley, W., Bitsch, B., & Klahr, H. 2009, *A&A*, 506, 971
Bitsch, B. & Kley, W. 2010, *A&A*, 523, A30
Bitsch, B. & Kley, W. 2011, *A&A*, 530, A41

The Astrophysics of Planetary Systems: Formation, Structure, and Dynamical Evolution
Proceedings IAU Symposium No. 276, 2010
A. Sozzetti, M. G. Lattanzi & A. P. Boss, eds.

doi:10.1017/S1743921311020965

Hot Jupiters and the evolution of stellar angular momentum

Cilia Damiani[1] **and Antonino F. Lanza**[1]

[1]INAF- Osservatorio Astrofisico di Catania,
Via Santa Sofia, 78, 95123, Catania, Italy
email: damiani@oact.inaf.it

Abstract. A close-in massive planet affects the angular momentum of its host star through tidal and magnetic interactions. The transiting planets allow us to study the distribution of the spin and orbital angular momenta in star-planet systems. Considering a sample of about 70 systems, we find that stars having an effective temperature between 6000 and 6700 K and a rotation period shorter than 10 days show a rotation synchronized with the orbit of their hot Jupiters or have a rotation period twice the orbital period of their planets. Such rotational behaviours cannot be explained on the basis of tidal interactions alone. Besides, the gyrochronology relationship for those systems holds if an angular momentum loss rate smaller by about 30 percent than in stars without hot Jupiters is assumed.

Keywords. planetary systems, planet-star interactions, stars: rotation

1. Introduction

Remarkable interactions between a close-in massive planet and its host stars are expected, both as a consequence of tides and reconnection between planetary and stellar magnetic fields. In the present study, updating the work of Lanza (2010), we give observational evidence of such interactions and show that they affect the evolution of stellar angular momentum. We limit ourselves to the sample of transiting systems so that the inclination of the rotation axis can be estimated, assuming that the stellar spin and orbital angular momentum are aligned. As a consequence, we exclude from the study systems for which a notable eccentricity or obliquity have been measured. As of September 2010, this yields 73 systems out of the 97 confirmed transiting planetary systems†.

2. Properties of transiting planetary systems

As a measure of the synchronization of a planet-harbouring star, we adopt the ratio $n/\Omega = P_{\rm rot}/P_{\rm orb}$ between the rotational period of the star and the orbital period of the planet. Here we focus on the relation between n/Ω and the effective temperature of the star, plotted in Fig.1.a. Although there is a large scatter for $T_{\rm eff} < 6000$ K, there is a general trend towards synchronization with increasing effective temperature. For $T_{\rm eff} \geqslant 6000$ K, two subgroups of systems can be identified, those that are close to $n/\Omega = 1$ or to $n/\Omega = 2$ and those showing n/Ω remarkably greater than 2. These two groups appear to be differentiated by their rotation period, with all systems with $P_{\rm rot} \leqslant 10$ days having $n/\Omega < 2$, with the exception of WASP-18 and WASP-24. A Kolmogorov-Smirnov test confirms the statistical difference of the two populations, and their non-uniform distribution in n/Ω, thus confirming the result of Lanza (2010).

† See e.g. http://www.inscience.ch/transits/

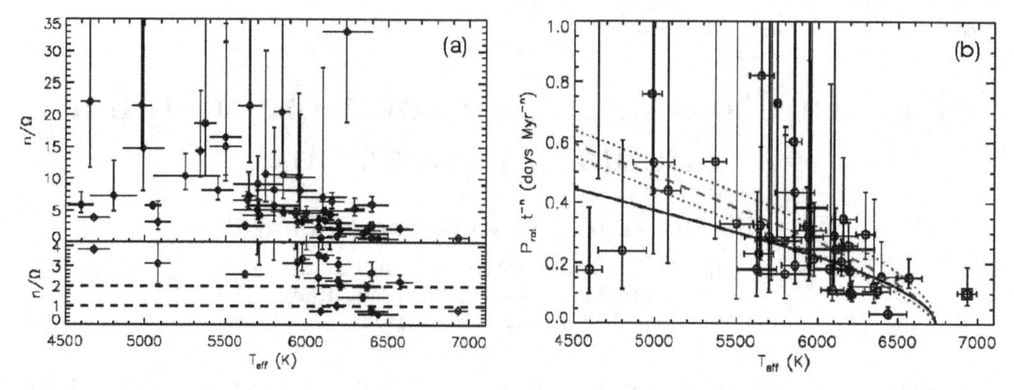

Figure 1. (a) Upper panel: the synchronization parameter n/Ω vs. the effective temperature of the star $T_{\rm eff}$ in transiting planetary systems. Lower panel: an enlargement of the lower portion of the upper panel, to better show the correlation close to $n/\Omega = 1$ and $n/\Omega = 2$. (b) Age-normalized rotation period $P_{\rm rot}t^{-n}$, with n = 0.5189, vs $T_{\rm eff}$, for the systems with a known age estimate of the host star. The relationship found by Barnes (2007) for stars without hot Jupiters is plotted as a dashed blue line, together with its 3σ errors as dotted lines. The solid thick line plots the best one-parameter fit. OGLE-TR-L9 (square symbol) was not included in the study.

According to Barnes (2007), for stars without hot Jupiters, the rotation period $P_{\rm rot}$ in days and the age t in Myr are related to the $B-V$ colour index by the formula $P_{\rm rot}t^{-n} = a\left[(B-V)-0.4\right]^{b}$ with prescribed values for $n, a,$ and b. We plot in Fig.1.b $P_{\rm rot}t^{-n}$ vs $T_{\rm eff}$, converting the $B-V$ index into effective temperature with the calibration by Bessel (1979). The relation by Barnes (2007) is plotted as a dashed blue line, together with its 3σ error range as dotted lines. The goodness of fit is almost null for stars with hot Jupiters. On the other hand, the best fit for a (linked to the angular momentum loss rate) is smaller by a factor 0.7 for our sample of transiting planetary systems (solid line).

This leads to the conclusion that on average, the stars with transiting close-in planets are faster rotators than stars without hot Jupiters, for a given age and temperature. Thus, the angular momentum loss rate of the former is lower than that of the latter (cf. Cohen *et al.* 2010).

3. Conclusions

We have analysed the rotation of stars harbouring transiting hot Jupiters and found a general trend towards synchronization with increasing effective temperature. We have shown that the rotation periods of the planet-hosting stars are, on the average, a factor of 0.7 shorter than those of the stars without planets of the same age. It is thus likely that both tides and magnetic effects are simultaneously at work to shape the distribution of angular momentum and its evolution in stars harbouring hot Jupiters.

Finally, we note that gyrochronology may not be suitable for estimating the age of late-type stars with close-in giant planets, especially if they have $T_{\rm eff} \geqslant 6000$ K and/or are rotating with a period shorter than ≈ 10 days, because their rotational evolution can be remarkably different from that of stars without hot Jupiters.

References

Barnes, S. A. 2007, *ApJ*, 669, 1167

Bessell, M. S. 1979, *PASP*, 91, 589

Cohen, O., *et al.* 2010, *ApJ*, 723, L64

Lanza, A. F. 2010, *A&A*, 512, A77

The Astrophysics of Planetary Systems: Formation, Structure, and Dynamical Evolution
Proceedings IAU Symposium No. 276, 2010
A. Sozzetti, M. G. Lattanzi & A. P. Boss, eds.

doi:10.1017/S1743921311020977

The 2:1 librating-circulating planetary configuration produced by a hybrid scenario

Jianghui Ji[1], Zhao Sun[1,2], Sheng Jin[1,2] and Niu Zhang[1,2]

[1]Purple Mountain Observatory, Chinese Academy of Sciences, Nanjing, 210008, China;
email: jijh@pmo.ac.cn.

[2] Graduate School of Chinese Academy of Sciences, Beijing 100049, China

Abstract. Different migration scenario of two giant planets may play a major role in forming the diverse resonant planetary configurations. The studies on the HD 128311 and HD 73526 systems show that two gas giants are captured in a 2:1 resonance but not in apsidal corotation, because one of the resonant argument circulates over the dynamical evolution. Herein we explore potential scenarios to produce the 2:1 librating-circulating resonance configuration. In the simulations, we find that both colliding or scattering events at early stage of dynamical evolution can induce the configurations trapped into resonance. In this sense, the librating-circulating resonance configuration is more likely to form by a hybrid mechanism of scattering and collision.

Keywords. celestial mechanics, planetary systems: formation

1. Introduction

At present, four resonant pairs of planets (GJ 876, HD 82943, HD 128311 and HD 73526) are believed to be trapped in 2 : 1 mean motion resonance (Marcy *et al.* 2001; Mayor *et al.* 2004; Vogt *et al.* 2005; Tinney *et al.* 2006). Numerical explorations, as well as theoretical analysis, show that the 2 : 1 resonance planetary configuration could be quite diverse as a result of migration in a slightly eccentric disc (Lee 2004; Kley *et al.* 2005). For example, the HD 128311 and HD 73526 systems are stabilized by the 2:1 resonance, but with a librating-circulating resonance configuration, which may be formed either through fast migration or migration with initial eccentricities or via a dynamical scattering event (Tinney *et al.* 2006).

2. Model and Numerical Setup

In the model, we consider the HD 128311 system as the fundamental framework. Here, we change the eccentricities of two giants as our initial eccentricities in order to emulate a slightly earlier state of this system (Zhang *et al.* 2010). The other parameters for the giant planets are the same as in (Sándor & Kley 2006). The modified orbital parameters are dynamically stable at least over 10^6 yr under mutual gravitational interaction. We assume $2-4$ terrestrial planets distributed in the range $0.3-0.75$ AU to explore this evolution. The eccentricities of the terrestrial planets are randomly adopted between 0.1 and 0.3, and the terrestrial bodies have a total mass of $10-15M_{\oplus}$. In each simulation, the terrestrial planets are in coplanar orbits reference to the giant planets. The other angles of the mean anomaly M and the longitudes of periapse ϖ are randomly distributed between 0° and 360° for each orbit.

We set up 120 simulations for two groups and use the hybrid symplectic integrator (Chambers 1999) in MERCURY package to integrate all orbits. In these runs, we consider that the collision and merging occurs when the minimum distance between any of

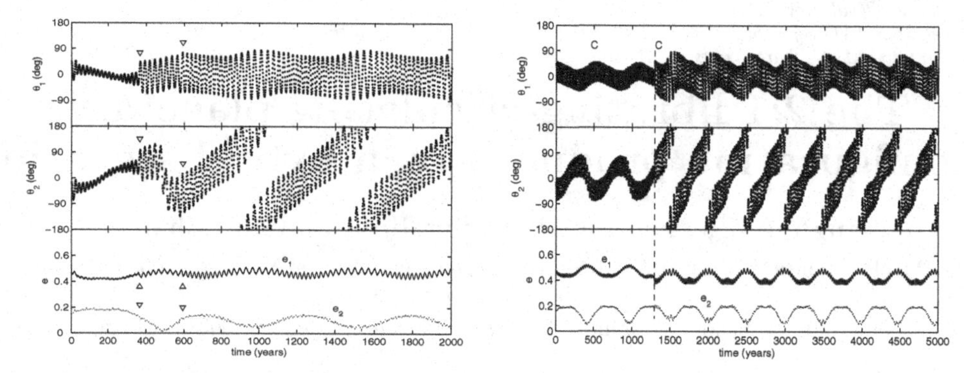

Figure 1. Time evolution for two simulations. *Left panel*: (1a) Evolution of the 2:1 mean motion resonance variables θ_1, θ_2, and the eccentricities of the two giants e_1, e_2, for Run 1. The sign triangles label the time of two terrestrial objects that scattering happens. *Right panel*: (1b) The first upper letter '*C*' denotes a merging event between T1 and T2. And the second collision occurs between T12 and the inner giant.

two objects is equal to or less than the summation of their physical radii. Most of the simulations are run for $10^3 - 10^4$ yr, but a few are extended to a longer integral time.

3. Results

Previous works on the HD 128311 and HD 73526 systems reported that the librating-circulating resonance configuration are produced by scattering of an additional small planet through orbital migration of a sudden stop (Sándor & Kley 2006; Sándor *et al.* 2007). We further show that collision and merger can play a major role in the evolution and both mechanisms can modify the resonance angles of the giants. The configuration engaged in the resonance is altered or destroyed. In the results, there are 16 simulations involved in producing librating-circulating resonance configuration, among which about 56% are shaped by mixed events of planet-planet scattering and merging (or collision). Details of the initial conditions for terrestrial planets may refer to Zhang *et al.* (2010).

We find that if the masses of the scattered terrestrial bodies are low, the changes of the resonance configurations may reinstate after some while. Generally, small planet scattering can increase the amplitude of the resonance angles. It is very likely that continuous scattering events of low-mass planets can alter the resonant configuration. Figure 1(a) shows the result of Run 1. Herein two terrestrial planets with equal masses of $m_{T1} = m_{T2} = 5M_\oplus$ first move about the giants. The 2:1 resonance of two giants is initially in symmetric configuration where $(\theta_1, \theta_2) \approx (0°, 0°)$. At the time the first terrestrial planet T1 is scattered at 366 yr, the amplitudes of the resonance angles become enlarged, but the apsidal corotation is maintained. Until the second terrestrial planet is scattered at 593 yr, soon after the first scattering event, the apsidal corotation is broken up. The eccentricities of two giants e_1 and e_2 fluctuate with large amplitudes, and our outcomes are well consistent with those of (Sándor & Kley 2006; Sándor *et al.* 2007).

Figure 1(b) shows the evolution of Run 2. The first '*C*' denotes a merging event between T1 and T2 at about 500 yr. The second collision occurs between T12 (as a merged larger body) and the inner giant at about 1300 yr. In this case, two resonant arguments θ_1 and θ_2 vary from a librating-librating phase into the librating-circulating state, and both of the eccentricities for two giants are slightly modulated. This shows that a terrestrial planet's colliding with a giant may finally shape librating-circulating configuration of mean motion resonances during the dynamical evolution.

4. Conclusions and Discussions

In the late stage of planetary formation, scattering or colliding among planetesimals and embryos does frequently occur. If two giant planets are initially trapped in a 2:1 symmetric resonance and their eccentricities oscillate with large amplitudes, the collisions arising from the giants and other small bodies may change librating amplitudes of resonant angles over the evolution. If the apsidal corotation is destoyed, the configuration then turns into a librating-circulating state. Obviously, the greater mass of a perturbing terrestrial body may have more significant influence on the commensurable giant planets. In most runs, colliding and scattering events are found to increase or decrease the fluctuation in the amplitude of the resonant angles, even dramatically break up the whole system. In a word, the librating-circulating configuration of mean motion resonance is likely to generate by a hybid mechanism of colliding and scattering.

Acknowledgements

This work is financially supported by the National Natural Science Foundation of China (Grants 10973044, 10833001), the Natural Science Foundation of Jiangsu Province, and the Foundation of Minor Planets of Purple Mountain Observatory.

References

Chambers, J. E. 1999, *MNRAS*, 304, 793
Kley, W., Lee, M. H., Murray, N., & Peale, S. J. 2005, *A&A*, 437, 727
Lee, M. H. 2004, *ApJ*, 611, 517
Marcy, G. W., Butler, R. P., Fischer, D. A., *et al.* 2001, *ApJ*, 556, 296
Mayor, M., Udry, S., Naef, D., *et al.* 2004, *A&A*, 415, 391
Sándor, Z. & Kley, W. 2006, *A&A*, 451, L31
Sándor, Z., Kley, W., & Klagyivik, P. 2007, *A&A*, 472, 981, 28
Tinney, C. G., Butler, R. P., Marcy, G. W., *et al.* 2006, *ApJ*, 647, 594
Vogt, S. S., Butler, R. P., Marcy, G. W., *et al.* 2005, *ApJ*, 632, 638
Zhang, N., Ji, J. & Sun, Z. 2010, *MNRAS*, 405, 2016

The Astrophysics of Planetary Systems: Formation, Structure, and Dynamical Evolution
Proceedings IAU Symposium No. 276, 2010
A. Sozzetti, M. G. Lattanzi & A. P. Boss, eds.

doi:10.1017/S1743921311020989

Evolution of planetary systems in dissipating gas disks

Soko Matsumura[1], Edward W. Thommes[2], Sourav Chatterjee[3,4] and Frederic A. Rasio[3,5]

[1]University of Maryland, MD, USA, email: soko@astro.umd.edu, [2]University of Guelph, Ontario, Canada, [3]Northwestern University, IL, USA, [4]Currently at the University of Florida, FL, USA, [5]Also at Center for Interdisciplinary Exploration and Research in Astrophysics (CIERA), Northwestern University, IL, USA

Abstract. In a recently published paper Matsumura *et al.* (2010) (hereafter M10), we have studied the evolution of three-planet systems in dissipating gas disks by using a hybrid N-body and one-dimensional gas disk evolution code. In this article, we highlight some results which are only briefly mentioned in M10.

Keywords. methods: n-body simulations, planetary systems: protoplanetary disks, planetary systems: formation

---

For the details of initial conditions and the code we use, please refer to M10.

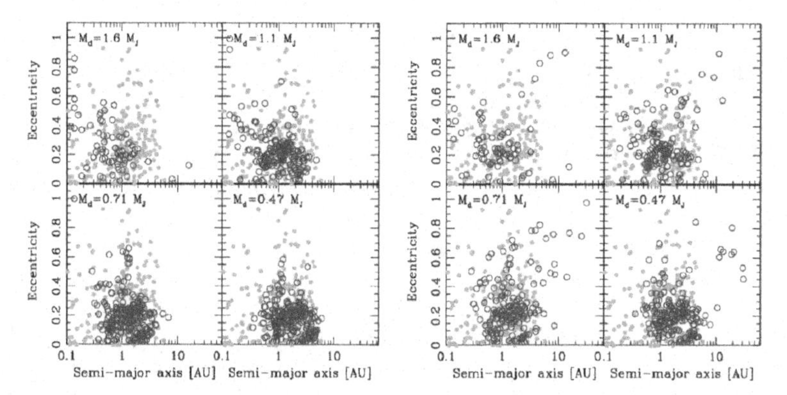

Figure 1. Distribution of observed planets (orange circles) compared with that of our simulations (blue circles) at τ_{GD} (left) and at 100 Myr (right). Evolutions of 100 three-planet systems in various initial disk masses are shown.

In Fig. 1, we compare the distribution of observed planets with that of our simulations at the disk's dissipation time τ_{GD} and at the end of the simulations 100 Myr. We find that the semimajor axis (a) distributions at τ_{GD} and 100 Myr are similar to each other. This supports the expectation that the a distribution of exoplanets is largely determined by planet–disk interactions. Additionally, more massive disks lead to more efficient migration. In contrast to the a distributions, we find that the eccentricity (e) distributions are very different at τ_{GD} and 100 Myr — most planets have $e < 0.2$ at τ_{GD} while the eccentricity distribution is more diverse at 100 Myr, similar to the observed planets. Thus, eccentricity distribution appears to be largely determined by planet–planet interactions after the disk's dissipation. From these results, we can verify the initial conditions used by previous N-body studies without gas disks. Chatterjee *et al.* (2008) and Ford & Rasio (2008) assumed that planets are initially beyond the ice line, on nearly circular orbits.

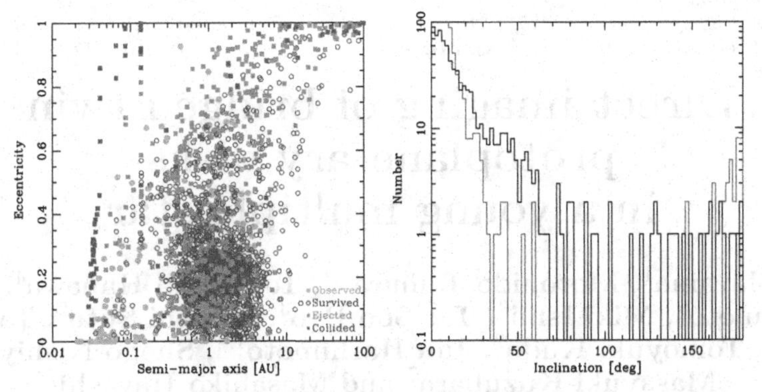

Figure 2. Left: $a - e$ scatter plot of 2000 simulations. Right: corresponding inclination distribution of survived planets (blue) and removed planets (black).

Our results indicate that their initial eccentricity assumption is reasonable, but the semi-major axis distribution may be too conservative. Jurić & Tremaine (2008) studied various initial conditions and found that initially dynamically active systems with a wide range of eccentricity can successfully reproduce the observed eccentricity distribution for $e > 0.2$. We cannot directly compare our results with those of Jurić & Tremaine (2008), because we focus on three-planet systems while they considered up to fifty-planet systems. Having this difference in mind, our results indicate that it is difficult to have dynamically active systems with a wide range of eccentricities at the end of disk dissipation.

In the left panel of Fig. 2, we show the $a - e$ scatter plot for all the simulations at 100 Myr. Also plotted are the final-recorded orbits for planets which are either ejected from the systems, or came too close to the central star (within 0.02 AU, marked with "collided" in the plot). Our simulated results agree well with the observed properties. In the right panel of Fig. 2, we plot the corresponding inclination distribution for survived (blue) and removed (black) planets. Note that most of our planets are on prograde orbits. In fact, only about $\sim 2\%$ of planets are on retrograde orbits. N-body simulations without gas disks done by C08 lead to a similar fraction of retrograde planets (contribution of Chatterjee *et al.* in this volume). These values are much smaller than what is expected from the observations of close-in planets (e.g., Triaud *et al.* 2010). Nagasawa *et al.* (2008) studied evolution of three Jupiter-mass planets with tidal dissipation and without a gas disk, and showed that planet–planet scatterings could also initiate Kozai migration (e.g., Fabrycky & Tremaine 2007; Naoz *et al.* 2010). Their results show a much flatter distribution of orbital inclinations, with more retrograde planets. If planet–star tidal interactions had been included in our simulations, at least some of our "ejected" or "collided" planets could have survived, creating more retrograde planets.

References

Chatterjee, S., Ford, E. B., Matsumura, S., & Rasio, F. A. 2008, *ApJ*, 686, 580
Fabrycky, D. & Tremaine, S. 2007, *ApJ*, 669, 1298
Ford, E. B. & Rasio, F. A. 2008, *ApJ*, 686, 621
Jurić, M. & Tremaine, S. 2008, *ApJ*, 686, 603
Matsumura, S., Thommes, E. W., Chatterjee, S., & Rasio, F. A. 2010, *ApJ*, 714, 194
Nagasawa, M., Ida, S., & Bessho, T. 2008, *ApJ*, 678, 498
Naoz, S., Farr, W. M., Lithwick, Y., Rasio, F. A., & Teyssandier, J. 2011, *Nature*, 473, 187
Triaud, A. H. M. J., Cameron, A. C., Queloz, D., Anderson, D. R., *et al.* 2010, *A&A*, 524, A25

The Astrophysics of Planetary Systems: Formation, Structure, and Dynamical Evolution
Proceedings IAU Symposium No. 276, 2010
A. Sozzetti, M. G. Lattanzi & A. P. Boss, eds.

doi:10.1017/S1743921311020990

Direct imaging of bridged twin protoplanetary disks in a young multiple star

Satoshi Mayama[1], Motohide Tamura[1,2], Tomoyuki Hanawa[4], Tomoaki Matsumoto[5], Miki Ishii[3], Tae-Soo Pyo[3], Hiroshi Suto[2], Takahiro Naoi[2], Tomoyuki Kudo[2], Jun Hashimoto[1,2], Shogo Nishiyama[6], Masayuki Kuzuhara[7] and Masahiko Hayashi[7]

[1]The Graduate University for Advanced Studies (SOKENDAI), Shonan International Village, Hayama-cho, Miura-gun, Kanagawa, 240-0193, Japan
email: mayama_satoshi@soken.ac.jp
[2]National Astronomical Observatory of Japan, 2-21-1, Osawa, Mitaka, Tokyo 181-8588 Japan
[3]Subaru Telescope, National Astronomical Observatory of Japan, 650 North A'ohoku Place, Hilo, Hawaii, 96720, USA
[4]Center for Frontier Science, Chiba University, Inage-ku, Chiba, 263-8522, Japan
[5]Faculty of Humanity and Environment, Hosei University, Fujimi, Chiyoda-ku, Tokyo 102-8160, Japan
[6]Department of Astronomy, Kyoto University, Kitashirakawa-Oiwake-cho, Sakyo-ku, Kyoto, 606-8502, Japan
[7]University of Tokyo, Hongo, Tokyo 113-0033, Japan

Abstract. Studies of the structure and evolution of protoplanetary disks are important for understanding star and planet formation. Here, we present the direct image of an interacting binary protoplanetary system. Both circumprimary and circumsecondary disks are resolved in the near-infrared. There is a bridge of infrared emission connecting the two disks and a long spiral arm extending from the circumprimary disk. Numerical simulations show that the bridge corresponds to gas flow and a shock wave caused by the collision of gas rotating around the primary and secondary stars. Fresh material streams along the spiral arm, consistent with the theoretical scenarios where gas is replenished from a circummultiple reservoir.

Keywords. binaries: general, stars: formation, planetary systems: protoplanetary disks

---

Studies of protoplanetary disks in multiple systems are essential for describing the general processes of star and planet formation because most stars form as multiples (Ghez *et al.* 1993). However, such circummultiple disks and spiral arms in multiple systems have rarely been directly resolved to date. Here, we present the direct image of an interacting binary protoplanetary system (Mayama *et al.* 2010). We investigate the geometry of a young (a stellar age of 4 million years) multiple circumstellar disk system, SR24, to understand its nature based on observations and numerical simulations. SR24 is a hierarchical multiple, located 160 pc away in the Ophiuchus star-forming region.

We obtained an infrared image of SR24 with CIAO (Tamura *et al.* 2000) mounted on the Subaru 8.2-m Telescope. (Fig. 1, left). The emission arises from dust particles mixed with gas in the circumstellar structures scattering the stellar light. Both circumprimary and circumsecondary disks are clearly resolved. Both disks overflow the inner Roche lobes (dotted contours in Fig. 1), suggesting that the material outside the lobes can fall into either of the inner lobes. There is a bridge of infrared emission connecting the two disks and a long spiral arm extending from the circumprimary disk. A spiral arm would suggest that the SR24 system rotates counter clockwise. The orbital period of the binary is

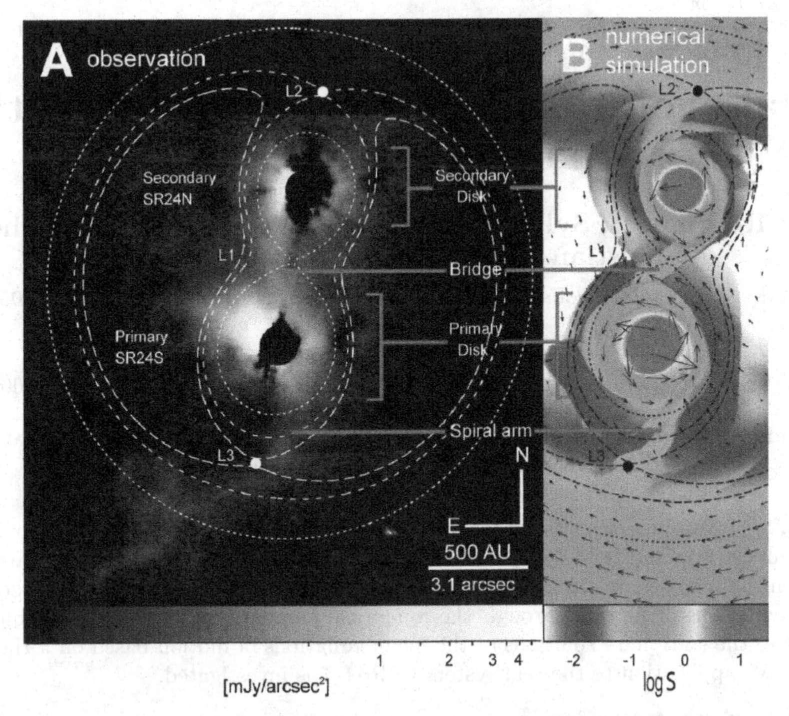

Figure 1. Observed and simulated images of the young multiple star, SR24. (A) H-band (1.6 μm) coronagraphic image of SR24. The total integration time was 1008 s. The PSFs of the final images have sizes of 0.1 arcsecond (FWHM) for the H-band. The inner and outer Roche Lobes are overlaid on the Subaru image as dotted and dashed lines, respectively. L1, L2, and L3 represent the inner Lagrangian point, outer Lagrangian point on the secondary side, and outer Lagrangian point on the primary side, respectively. (B) Snapshot of accretion onto the binary system SR24 based on 2D numerical simulations. The color and arrows denote the surface density distribution and velocity distribution, respectively.

15,000 yr. The arm would also imply replenishment of the twin disk gas from the circumbinary disk.

We performed 2D numerical simulations of accretion from a circumbinary disk to identify the features seen in the coronagraphic image(Fig. 1, right). Although the gas flow was not stationary, especially inside the Roche lobes, the stage of the 2D simulations shown in Fig. 1 shared common features with the observed image. These agreements between observation and simulation suggest that the bridge corresponds to gas flow and a shock wave caused by the collision of gas rotating around the primary and secondary stars. The arm corresponds to a spiral wave excited in the circumbinary disk.

References

Ghez, A. M., Neugebauer, G., & Matthews, K. 1993, *AJ*, 106, 2005

Mayama, S., Tamura, M., Hanawa, T., Matsumoto, T., Ishii, M., Pyo, T. S., Suto, H., Naoi, T., Kudo, T., Hashimoto, J., Nishiyama, S., Kuzuhara, M., & Hayashi, M. 2010, *Science*, 327, 306

Tamura, M., *et al.* 2000, *Proc. SPIE*, 4008, 1153

The Astrophysics of Planetary Systems: Formation, Structure, and Dynamical Evolution
Proceedings IAU Symposium No. 276, 2010
A. Sozzetti, M. G. Lattanzi & A. P. Boss, eds.

doi:10.1017/S1743921311021004

Tidal evolution of a close-in planet with a more massive outer companion

Adrian Rodríguez[1], Sylvio Ferraz-Mello[1], Tatiana A. Michtchenko[1], Cristian Beaugé[2] and Octavio Miloni[3]

[1]Insituto de Astronomia, Geofísica e Ciências Atmosféricas, Rua do Matão 1226, 05508-900 São Paulo, Brazil
email: adrian@astro.iag.usp.br

[2]Observatório Astronómico, Universidad Nacional de Córdoba, Laprida 854, (X5000BGR) Córdoba, Argentina

[3] Facultad de Ciencias Astronómicas y Geofísicas, Universidad Nacional de La Plata, Paseo del Bosque S/N B1900 FWA, La Plata, Argentina

Abstract. We investigate the motion of a two-planet coplanar system under the combined effects of mutual interaction and tidal dissipation. The secular behavior of the system is analyzed using two different approaches, restricting to the case of a more massive outer planet. First, we solve the exact equations of motion through the numerical simulation of the system evolution. We also compute the stationary solutions of the mean equations of motion based on a Hamiltonian formalism. An application to the real system CoRoT-7 is investigated.

Keywords. Celestial mechanics

1. Introduction

The tidal effect produces orbital decay, circularization and spin-orbit synchronization of the orbit of a close-in planet orbiting a slow-rotating star (Goldreich & Soter 1966; Ferraz-Mello *et al.* 2008). When the tidally affected planet has an eccentric companion the inner planet eccentricity is excited, resulting in a rapid migration rate toward the star due to tidal dissipation (Mardling & Lin 2004). Here, we study the coupled tidal-secular evolution of a system with a close-in planet and its outer more massive companion.

2. The model

We work with systems in which the inner planet is a super-Earth with a more massive outer companion. We assume that only the inner planet is deformed under the tides raised by the central star. The reference frame chosen is centered in the star and the motion of the planets occurs in the reference plane (i.e. coplanar motion).

The inner planet evolves under the combined effects of the interaction with the outer planet and tides raised by the star. The tidal force $\mathbf{f}$ acting on the inner mass m_1 is given by Mignard (1979):

$$\mathbf{f} = -3k_1\Delta t_1 \frac{Gm_0^2 R_1^5}{r_1^{10}}[2\mathbf{r}_1(\mathbf{r}_1 \cdot \mathbf{v}_1) + r_1^2(\mathbf{r}_1 \times \mathbf{\Omega}_1 + \mathbf{v}_1)], \tag{2.1}$$

where $\mathbf{v}_1 = \dot{\mathbf{r}}_1$ and Ω_1 is the rotation angular velocity of the inner planet. Δt_1 is the *time lag* and it can be interpreted as a delay in the deformation of the tidally affected body due to its internal viscosity.

3. Secular dynamics of two-planets system

The secular behavior of the system can be also explored through the investigation of the mean dynamics (see Michtchenko & Ferraz-Mello 2001). The *mean equations* of motion of the one degree of freedom system are given by

$$\dot{I}_1 = -\frac{\partial H_{Sec}}{\partial \Delta\varpi}, \qquad \Delta\dot{\varpi} = \frac{\partial H_{Sec}}{\partial I_1}, \tag{3.1}$$

where $\Delta\varpi$ and $I_1 = m_i' \sqrt{\mu_i\, a_i}\,(1 - \sqrt{1-e_1^2})$ are the action-angle variables of the problem, while $H_{Sec} = -(Gm_1m_2/a_2) \times R_{Sec}(L_i, K_i, \Delta\varpi)$ is the secular part of the Hamiltonian and R_{Sec} is the corresponding disturbing function. *Stationary solutions* are given by the roots of Eqns. (3.1), where the second one provides $\Delta\varpi = (0, \pi)$ and are known as Mode I and Mode II, respectively. Periodic solutions of the secular averaged problem are motions around Mode I or Mode II (see Fig. 1). The calculation of stationary solutions can be extended to the case in which dissipation is included. Indeed, tidal dissipation produces orbital decay and thus energy loss due to internal friction. Stationary solutions of Eqns. (3.1) can be found for a range of a_1 values, resulting in a curve in the space (e_1, e_2). This curve is composed by the collection of all equilibrium points for the whole range of a_1, and we refer it as LSE curve (locus of stationary solutions in the space of eccentricities).

4. Application to CoRoT - 7 system

CoRoT-7 planetary system is composed by two short-period super-Earth-like planets orbiting a central Sun-like star of $m_0 = 0.93m_\odot$. The inner and outer planets, CoRoT-7b and CoRoT-7c, are assumed to have masses of $m_1 = 4.8m_\oplus$ and $m_2 = 8.4m_\oplus$, respectively. Their orbital periods are 0.854 and 3.698 days and both orbits are circular (Queloz *et al.* 2009).

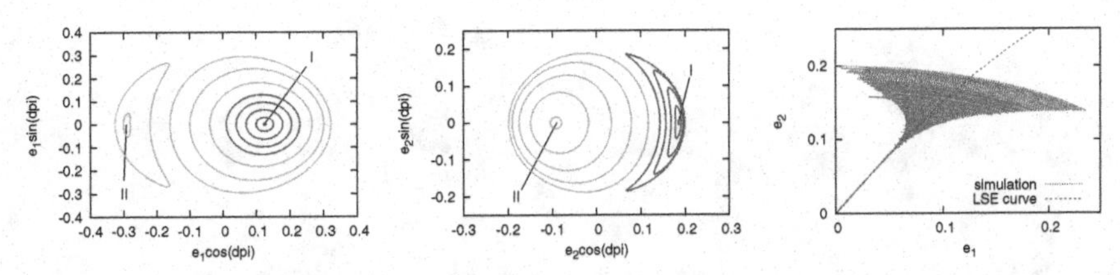

Figure 1. *Left panels*: Secular variation of eccentricities and $\Delta\varpi$ in the CoRoT-7 system for several initial conditions. The locations of equilibrium points are marked by I and II on each plane. *Right panel*:Comparison between LSE curve and the result of numerical simulation for CoRoT-7 system (mode I).

The result of the numerical simulation of the system including tidal dissipation (exact equations) is shown in Fig. 1, and compared with the LSE curve. Initial conditions were $a_1 = 0.0190$ AU, $e_1 = 0$, $e_2 = 0.2$ and $\Delta\varpi = 0$, while $\Delta t = 3$ min, $R_1 = 1.68R_\oplus$. The eccentricities oscillate rapidly with damped amplitudes around the curve of stationary solutions. Note that the LSE curve can be considered as the "mean secular path" of the system, providing thus information about the past and future dynamical behavior.

References

Ferraz-Mello S., Rodríguez A., & Hussmann, H. 2008, *CeMDA*, 101, 171
Goldreich, P. & Soter, S. 1966, *Icarus*, 5, 375
Mardling, R. & Lin, D. N. C. 2004, *ApJ*, 614, 955
Michtchenko, T. A. & Ferraz-Mello, S. 2001, *Icarus*, 149, 357
Mignard, F. 1979, *The Moon and the Planets*, 20, 301
Queloz, D., *et al.* 2009, *A&A*, 506, 303

The Astrophysics of Planetary Systems: Formation, Structure, and Dynamical Evolution
Proceedings IAU Symposium No. 276, 2010
A. Sozzetti, M. G. Lattanzi & A. P. Boss., eds.

doi:10.1017/S1743921311021016

Starspots and spin-orbit alignment in the WASP-4 exoplanetary system

Roberto Sanchis-Ojeda[1], Joshua N. Winn[1], Matthew J. Holman[2], Joshua A. Carter[1,2], David J. Osip[3] and Cesar I. Fuentes[2,4]

[1]Department of Physics, and Kavli Institute for Astrophysics and Space Research, Massachusetts Institute of Technology, Cambridge, MA 02139, USA
e-mail: rsanchis@MIT.EDU

[2]Harvard-Smithsonian Center for Astrophysics, 60 Garden Street, Cambridge, MA 02138, USA

[3]Las Campanas Observatory, Carnegie Observatories, Casilla 601, La Serena, Chile

[4]Department of Physics and Astronomy, Northern Arizona University, PO Box 6010, Flagstaff, AZ 86011

Abstract. We present the photometric analysis of 4 transits of the exoplanet WASP-4b, obtained with the Baade 6.5m telescope, one of the two Magellan telescopes at Las Campanas. The light curves have a photometric precision of 0.5 mmag and a time sampling of 30s. This high precision has allowed us to detect several "spot anomalies": temporary brightenings due to the occultation of a starspot on the transit chord. By analyzing these anomalies we find the sky-projected stellar obliquity to be $\lambda = 1^\circ\,{}^{+12^\circ}_{-14^\circ}$. The small value suggests that the planet migration mechanism preserved the initially low obliquity, or that tidal evolution has realigned the system.

Keywords. stars: spots, planetary systems, techniques: photometric

---

Until now, in the study of transiting planets, starspots have been a nuisance, a factor that limits the precision of our measurements. Most recently, different groups have been able to infer the rotational period of the host star due to the appearance of spots (Dittmann *et al.* 2009; Silva-Valio 2008; Silva-Valio *et al.* 2010). In our project (Sanchis-Ojeda *et al.* 2010, in preparation) we report the observation of four different transits of WASP-4b (Wilson *et al.* 2008), in August and September of 2009 (see Fig. 1). In the figure we have identified two anomalies in the last two transits, which we interpret as the planet crossing in front of the same spot during two different transits. The same sort of spot anomalies can be observed in light curves by Southworth *et al.* (2009). The recurrence of the anomaly at a later phase of the transit favors the configuration where the orbital angular momentum and the axis of rotation of the star are aligned, as in this case, the trajectories of the spot on the surface and the planet would be almost parallel. In this project we quantify this statement, obtaining a value of the sky projected spin-orbit angle of the system with a simple geometric model.

The model only uses the times when the spot events occurred. The parameters are the periods of rotation of the two spots (ours and Southworth's) and the two angles that describe the orientation of the rotation of the star in the sky. In this model, we assume that the spots rotate at a fixed latitude on the surface of the star, with initial conditions fixed by the position of the center of the planet's shadow when the first observation of the spot ocurred. The period of rotation of that particular spot, and the angles λ and i_s, determine completely the position of the spot at any other time. We define a χ^2 function that penalizes models that fail to produce the second anomaly at the appropriate time, or that do produce anomalies during transits when none were observed. We also use the

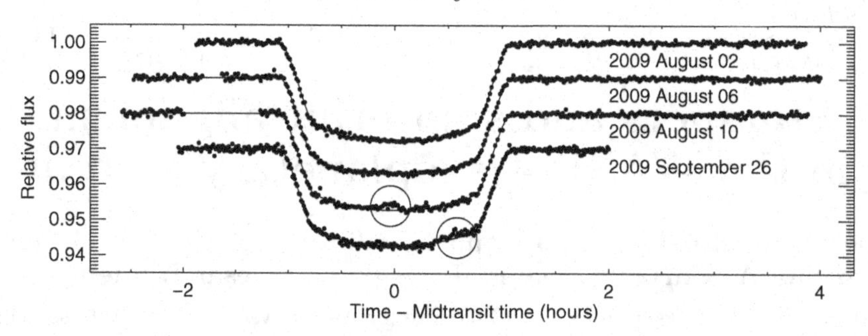

Figure 1. Four transits of WASP-4b, with two probable spot-crossing events highlighted with black circles.

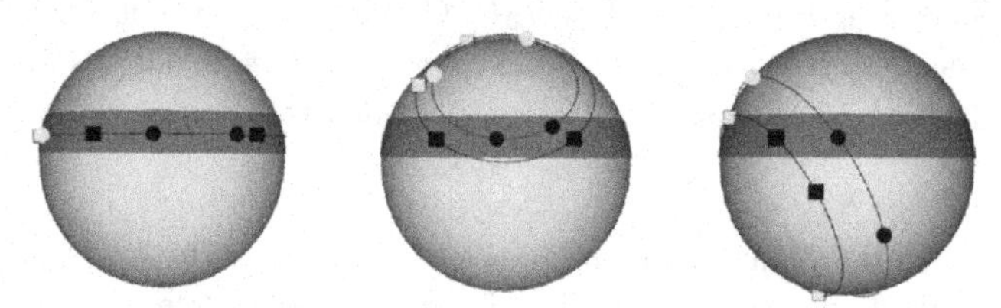

Figure 2. Circles (squares) represent the position of the spot during our (Southworth's) observed transits. Dark marks represent detections whereas light marks represent non-detections. The red dark band represents the transit chord. Left: $\lambda = 0$, $i_s = 90°$, aligned case, allowed. Center: $\lambda = 0$, $i_s = 45°$ case, still allowed. Right: $\lambda = 45°$, $i_s = 90°$ case, not allowed.

constraints $v \sin i_S = 2.14 \pm 0.37$ km/s and $|\lambda| < 90°$ from the study of the Rossiter-McLaughlin effect by Triaud *et al.* (2010).

Four different local minima of χ^2 were found, in which the aligned case ($\lambda = 0$, $i_s = 90°$) is found to be the global minimum. The different allowed cases have very different periods for the spots, and also very different values for i_s, but they all give small values for λ. Fig. 2 shows geometrically why changing i_s is not forbidden, whereas large values of λ are ruled out. Running an MCMC algorithm to explore all possible minima, we are able to obtain a posteriori probability distribution for λ, obtaining a result of $\lambda = 1° {}^{+12°}_{-14°}$.

This method is a new powerful tool to measure the spin-orbit alignment of exoplanetary systems. It works best where the Rossiter-McLaughlin effect does not work, that is, for slowly rotating stars. The precision of the method can be affected by the unknown evolution of the spot, but the results would be highly improved if we could have more consecutive transits, making it possible to constrain also i_s. This is exactly what one will be able to get with the Kepler satellite, which is observing 170,000 stars continuously for 3.5 years. Future research will move in this direction, trying to apply this method to other exoplanetary systems in the Kepler field of view and improving the method at the same time.

References

Dittmann, J. A., Close, L. M., Green, E. M., & Fenwick, M. 2009, *ApJ*, 701, 756
Silva-Valio, A. 2008, *ApJ*, 683, L179
Silva-Valio, A., Lanza, A. F., Alonso, R., & Barge, P. 2010, *A&A*, 510, A25
Southworth, J., *et al.* 2009, *MNRAS*, 399, 287
Triaud, A. H. M. J., *et al.* 2010, *A&A*, 524, A25
Wilson, D. M., *et al.* 2008, *ApJ*, 675, L113

The Astrophysics of Planetary Systems: Formation, Structure, and Dynamical Evolution
Proceedings IAU Symposium No. 276, 2010
A. Sozzetti, M. G. Lattanzi & A. P. Boss, eds.

doi:10.1017/S1743921311021028

Observational signs of planet infall and Roche lobe overflow outward migration

Stuart F. Taylor[1,2,3] and Ing-Guey Jiang[1]

[1]Institute of Astronomy and Department of Physics, National Tsing Hua University, 101 Section 2 Kuang Fu Road, Hsinchu, Taiwan 30013
email: astrostuart@gmail.com, jiang@phys.nthu.edu.tw

[2]Global Telescope Science Group, Los Angeles, California, USA

[3]Eureka Scientific, Inc., Oakland, California, USA

Abstract. Outward migration of planets due to Roche lobe overflow may play an important role in producing the presently observed distribution of planet parameters. We suggest that many of the currently known short period planets may have already migrated into the Roche distance from the star, and then deposited at their current semi-major axes by being migrated outwards due to angular momentum transfer from an episode of Roche lobe overflow (RLO). This RLO outward migration (RLOOM) could be sustained in the region where planetary radius increases with decreasing mass. We are modeling how RLOOM may leave what kind of planet parameter statistics. This modeling seeks to predict what observable signs of RLOOM there may be. Overflow of planetary mass may leave behind characteristic hot dust and gas as well as produce luminous signatures.

Keywords. planetary systems, planetary systems: formation, stars: rotation, stars: statistics, stars: evolution, stars: variables: other

1. Introduction

The pile-up of planet orbits near 3 days (Fig. 1) has been a puzzle from the time there have been enough planet statistics to see patterns in orbital statistics (Santos & Mayor 2003). Also, the radii of many but not all of these planets has been larger than expected (Gu 2010). Finally, some of these planets are reported have nonzero eccentricities despite strong eccentricity damping, though better measurements are needed. It is accepted that these close in planets must have formed further out and migrated inwards to their current position (Pont 2009). Efforts to explain these three mysteries have been based on trying to explain how these patterns could occur as these close in planets migrated inwards. Taylor (2010) and Fabrycky (2010)) suggest that the pile-up may be composed by planets that had migrated all the way to Roche lobe overflow (RLO), from where they had sufficient RLO to migrate *outward* (RLOOM), and that the end of RLOOM may have left planets in the pile-up. Though the role of outward migration following Roche lobe overflow (RLOOM) during the phase when the disk is still present has been studied (Chang & Gu 2010; Gu, Bodenheimer, & Lin 2004; Gu, Lin, & Bodenheimer 2003; Trilling *et al.* 1998), there has been less study of RLOOM during the main sequence phase.

2. Radius-mass relation maintaining outward migration

The mass-radius relation for giant planets turns around due to electron degeneracy such that above a certain temperature-dependent mass, the planet radius begins to decrease with increasing mass (Zapolsky & Salpeter 1969). This creates the potential instability

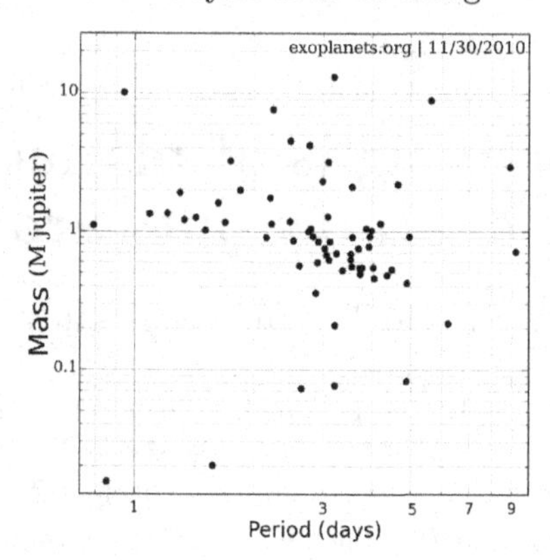

Figure 1. Mass versus period of the transiting planets in the region of the pile-up.

that when RLO removes mass the planet will expand its radius. RLO does not lead to RLOOM for "normal" planets that get smaller upon mass removal, because outward migration quenches the RLO as a planet's Roche radius increases with distance. Population modeling is in progress to see whether this can account for the observed pile-up (Taylor *et al.* in preparation). It must be determined whether the increase in radius due to mass loss is enough to fill the increasing size of the Roche lobe as the planet migrates outward.

If a tidal heating mechanism is necessary for RLOOM to operate sufficiently well to create the pile-up below one Jupiter mass, this may also explain how such planets might be left with a range of inflated radii. Because at the distance of RLO eccentricity is quickly damped out, it has been suggested that perhaps the process of RLOOM allows a means of input of eccentricity (Taylor 2010). Though an increase of RLOOM at periastron would translate into less rather than more eccentricity, work is exploring the range of possible patterns of delayed radius enlargement or other asymetric angular momentum transfer.

The most identifiable signal from past RLO may be the presence of dust and gas. How much depends on how much material is driven away from the star. The potential for a luminous signature from RLO or planet-star collisions has been emphasized (Taylor 2010), but it is also important to model the quantity of infalling dust and gas that may be driven outward to determine whether planetary material may leave an identifiable signal.

We acknowledge the support of National Tsing Hua University and the National Science Council of Taiwan. We made extensive use of the Exoplanet Data Explorer.

References

Chang, S.-H., Gu, P.-G., & Bodenheimer, P. H. 2010, *ApJ*, 708, 1702

Fabrycky, D. 2010, This volume

Pont, F. 2010, *MNRAS*, 396, 1789

Gu, P.-G. 2010, *Nature*, 465, 300

Gu, P.-G., Lin, D. N. C., & Bodenheimer, P. H. 2003, *ApJ*, 588, 509

Gu, P-.G., Bodenheimer, P. H., & Lin, D. N. C. 2004, *ApJ*, 608, 1076

Santos, N. C. & Mayor, M. 2003, in: M. Monteiro (ed.), *The Unsolved Universe: Challenges for the Future, JENAM 2002* (Dordrecht: Kluwer Academic Publishers), p. 15

Taylor, S. F. 2010, submitted to Icarus, *arXiv*:1009.4221

Trilling, D. E., Benz, W., Guillot, T., Lunine, J. I., Hubbard, W. B., & Burrows, A. 1998, *ApJ*, 500, 428

Zapolsky, H. S. & Salpeter, E. E. 1969, *ApJ*, 158, 809

The Astrophysics of Planetary Systems: Formation, Structure, and Dynamical Evolution
Proceedings IAU Symposium No. 276, 2010
A. Sozzetti, M. G. Lattanzi & A. P. Boss, eds.

doi:10.1017/S174392131102103X

3D MHD simulations of planet migration in turbulent stratified disks

Ana Uribe[1], Hubert Klahr[1], Mario Flock[1] and Thomas Henning[1]

[1]Max-Planck-Institut für Astronomie, Königstuhl 17,
D-69117 Heidelberg, Germany
email: uribe@mpia.de

Abstract. We performed 3D MHD numerical simulations of planet migration in stratified disks using the Godunov code PLUTO (Mignone *et al.* 2007). The disk is invaded by turbulence generated by the magnetorotational instability (MRI). We study the migration for planets with different mass to primary mass ratio. The migration of the low-mass planet ($q = M_p/M_s = 10^{-5}$) is dominated by random fluctuations in the torque and there is no defined direction of migration on timescales of 100 orbits. The intermediate-mass planet ($q = M_p/M_s = 10^{-4}$) can experience systematic outwards migration that was sustained for the times we were able to simulate.

Keywords. accretion disks, MHD, turbulence, planetary systems: formation

1. Introduction

The migration of planets in turbulent disks has been studied using MHD simulations in the cylindrical disk approximation by Nelson & Papaloizou (2003) in the first of a series of papers and by Nelson (2005) for low mass protoplanets. Baruteau & Lin (2010) and Laughlin *et al.* (2004) performed HD simulations using a stochastic forcing to model the turbulent fluctuations in the density. Contrary to laminar disk migration, low mass planets migrate stochastically and depending on the amplitude of the artificial forcing, inwards migration can be slowed down for more massive planets. We study the effect of turbulence on migration using global simulations of stratified magnetized disks.

We simulate a global stratified disk in spherical coordinates (r, θ, ϕ), where the computational domain is given by $r \in [1, 10]$, $\theta \in [\pi/2 - 0.3, \pi/2 + 0.3]$ and $\phi \in [0, 2\pi]$, the grid resolution is $(N_r, N_\theta, N_\phi) = (256, 128, 256)$ and the grid is centered in the center of mass. The ratio of the pressure scale height H to the radial coordinate of the disk is taken to be a constant such that $h = H/(r \sin\theta) = 0.07$. Before introducing the planet in the simulations, a weak toroidal magnetic field is imposed on the disk with constant $\beta = 25$. The planet is introduced after the saturation of the MRI. The equations of motion are integrated with a leap frog integrator. The numerical setup follows the setup presented in Flock *et al.* (2010).

2. Results

Figure 1 shows the cumulative torque for simulations with a planet where $q = 10^{-4}$, for different locations in the disk. We see that in both simulations there is an initial stage where the torque is negative, followed by a reversal of the direction of the migration where the torque becomes positive and takes a defined value for the rest of the simulated time. The Hill radius of the planet and the horseshoe region are resolved (by approximately 4 and 7 grid cells per half width for $r_p = 3.3$ and $r_p = 5.0$ respectively). In this case, the

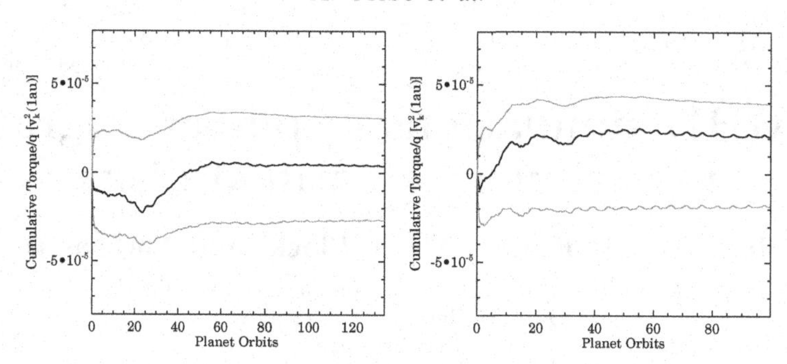

Figure 1. Cumulative torque vs planet orbits for a planet with mass ratio $q = M_p/M_{star} = 10^{-4}$ located at $r_p = 3.3$ (left) and at $r_p = 5.0$ (right). A negative(positive) torque indicates inwards(outwards) migration. The gray lines shows the torque exerted by the disk inside (positive) and outside (negative) of the planet's orbit.

component of the torque originating from the horseshoe region can dominate if there is a mechanism for keeping the corotation torque unsaturated and the local density profile differs from the global profile, possibly increasing outwards, such that the corotation torque can be larger than the Lindblad torque, making the total torque positive. In our simulations the turbulence is responsible for unsaturating the corotation torque and for both simulations we see that the torque reversal ocurrs when the local surface density profile is less steep than the global $\Sigma(r) \sim r^{-0.5}$, being close to flat or increasing outwards. This is consistent with the reversal of migration that has been observed in a certain range of the paramenter space, in disks where the viscosity is modelled as an alpha viscosity (Masset *et al.* 2006a).

For simulations where $q = 10^{-5}$, during the simulated time, migration was dominated by random fluctuations in the torque that can be orders of magnitude larger that what is expected for the value of the Lindblad or corotation torques for this planet mass. This is in agreement with simulations by Nelson (2005) of migration of low mass protoplanets in cylindrical disk models, where stratification is neglected.

3. Discussion and Conclusions

We demonstrate that it is possible for planets of around 30 Earth masses to undergo systematic outwards migration that is sustained up to 100 planet orbits. The question remains about the long term behavior of the torque, and whether this is only a transient behavior lasting for the first few hundred orbits (assuming the same local surface density profile), afterwards saturating and returning to laminar Type I values or fluctuating on longer timescales.

References

Baruteau, C. and Lin, D. N. C. 2010, *ApJ*, 709, 759

Flock, M., Dzyurkevich, N., & Klahr, H. and Mignone, A. 2010, *A&A*, 516, 26

Laughlin, G., Steinacker, A., & Adams, F. C. 2004, *ApJ*, 608, 489

Masset, F. S., D'Angelo, G., & Kley, W. 2006a, *ApJ*, 652, 730

Mignone, A., Bodo, G., Massaglia, S., Matsakos, T., Tesileanu, O., Zanni, C., & Ferrari, A. 2007, *ApJ*, 170, 228

Nelson, R. P. 2005, *A&A*, 443, 1067

Nelson, R. P. & Papaloizou, J. C. B. 2003, *MNRAS*, 339, 993

The Astrophysics of Planetary Systems: Formation, Structure, and Dynamical Evolution
Proceedings IAU Symposium No. 276, 2010
A. Sozzetti, M. G. Lattanzi & A. P. Boss, eds.

doi:10.1017/S1743921311021041

Planetesimal and protoplanet dynamics in a turbulent protoplanetary disk

Chao-Chin Yang[1,2]†, Mordecai-Mark Mac Low[2] and Kristen Menou[3]

[1]Department of Astronomy, University of Illinois,
1002 West Green Street, Urbana, IL 61801, U.S.A

[2]Department of Astrophysics, American Museum of Natural History,
Central Park West at 79th Street, New York, NY 10024, U.S.A

[3]Department of Astronomy, Columbia University,
550 West 120th Street, New York, NY 10027, U.S.A

Abstract. Due to the gravitational influence of density fluctuations driven by magneto-rotational instability in the gas disk, planetesimals and protoplanets undergo diffusive radial migration as well as changes of other orbital properties. The magnitude of the effect on particle orbits has important consequences for planet formation scenarios. We use the local-shearing-box approximation to simulate an ideal, isothermal, magnetized gas disk with vertical density stratification and simultaneously evolve numerous massless particles moving under the gravity of the gas and the host star. Although the results converge with resolution for fixed box dimensions, we find there exists no convergence of the response of the particles to the gravity of the gas against the horizontal box size, up to 16 disk scale heights. This lack of convergence indicate that caution should be exercised when interpreting local-shearing-box models involving gravitational physics of magneto-rotational turbulence.

Keywords. accretion, accretion disks, instabilities, methods: numerical, MHD, planetary systems: formation, planetary systems: protoplanetary disks, turbulence

1. Introduction

It is generally believed that turbulent accretion must operate in protoplanetary gas disks. One of the most promising mechanisms to drive the turbulence is the magneto-rotational instability (MRI; see, e.g., Balbus & Hawley 1998). The resulting density fluctuations are significant enough that the gravitational influence of the turbulent gas makes the embedded planetesimals or protoplanets undergo diffusive radial migration as well as changes of other orbital properties. The magnitude of this effect on particle orbits has important consequences for planet formation scenarios.

To self-consistently measure the influence of the turbulent density fluctuations on particle orbital dynamics, numerical simulations capturing both large-scale and small-scale coherent structures as well as statistical techniques to quantify stochastic particle orbital evolution are required. We adopt the local-shearing-box approximation, which has high-resolving power, can be integrated for long physical timescale, and can simultaneously involve a significant number of particles.

We presented in Yang *et al.* (2009) our simulations of massless particles moving in a local, unstratified, isothermal, ideal, magnetized disk. We found that the response of the particles to the gravity of turbulent gas subject to the MRI is systematically lower than what was previously reported in global disk models. We also found in the same study

† Present address: Department of Astronomy and Astrophysics, University of California, Santa Cruz, 1156 High Street, Santa Cruz, CA 95064, U.S.A.; email: ccyang@ucolick.org

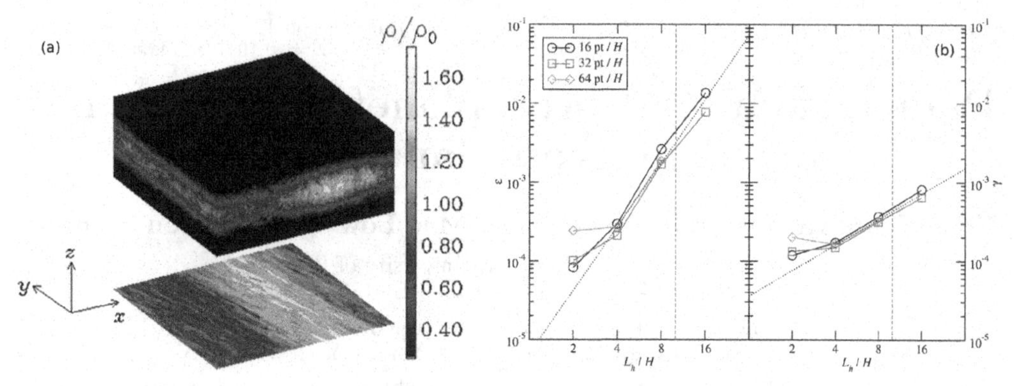

Figure 1. (a) Snapshot of the gas density of a local region in a turbulent protoplanetary disk driven by the magneto-rotational instability. The x, y, and z axes indicate radial, azimuthal, and vertical directions, respectively. The bottom plane shows the slice through the disk mid-plane. The dominance of large-scale structures can be seen; this type of structures are typical in our simulations and have long correlation time compared with orbital period. (b) Dimensionless constants ϵ and γ as a function of horizontal box size L_h. The constant ϵ represents the strength of radial diffusive migration driven by turbulence and was defined by Johnson *et al.* (2006), while γ is related with the strength of eccentricity excitation due to turbulence and was defined by Ida *et al.* (2008). The dotted lines are power-law fits to data points with $L_h \geqslant 4H$. The vertical dashed lines indicate the box size of choice for the case of a disk aspect ratio of $H/R = 0.1$.

that the effect significantly depends on the horizontal box size. In this work, we further consider disks with vertical density stratification by including linearized vertical gravity from the central star.

2. Results and Discussions

We find that no matter what the horizontal box size is, the largest-scale structure in either radial or azimuthal direction dominates the density fluctuations in the gas, up to the largest box we have investigated (Fig. 1a; also see Johansen *et al.* 2009). These large-scale density structures significantly affect the stochastic evolution of particle orbital radius and eccentricity (Fig. 1b). Both the magnitude and the correlation time of the resulting stochastic torques increase with horizontal box size L_h. Consequently, the larger the shearing box, the stronger the response of the particles to the gravity of the turbulent gas. We see no convergence with L_h up to $16H$, where H is the vertical disk scale height.

This lack of convergence in particle dynamics in shearing boxes poses major difficulty in quantifying the influence of the magneto-rotational turbulence on particle orbits. Using heuristic arguments, we suggest that $L_h \sim R$, where R is the distance of the box center to the host star, might be a natural scale of choice for a local model to approach reality. Comparison to recent global disk models conducted by Nelson & Gressel (2010) suggests this criterion offers consistency between local and global disk models.

References

Balbus, S. A. & Hawley, J. F. 1998, *Reviews of Modern Physics*, 70, 1
Ida, S., Guillot, T., & Morbidelli, A. 2008, *ApJ*, 686, 1292
Johansen, A., Youdin, A., & Klahr, H. 2009, *ApJ*, 697, 1269
Johnson, E. T., Goodman, J., & Menou, K. 2006, *ApJ*, 647, 1413
Nelson, R. P. & Gressel, O. 2010, *MNRAS*, 409, 639
Yang, C.-C., Mac Low, M.-M., & Menou, K. 2009, *ApJ*, 707, 1233

Section D:
The Next Decade

The Astrophysics of Planetary Systems: Formation, Structure, and Dynamical Evolution
Proceedings IAU Symposium No. 276, 2010
A. Sozzetti , M. G. Lattanzi & A. P. Boss, eds.

doi:10.1017/S1743921311021053

Exoplanet transit spectro-photometry with SOFIA

Daniel Angerhausen[1], Alfred Krabbe[1] and Hans Zinnecker[1]

[1]Deutsches SOFIA Institut, Universität Stuttgart, Pfaffenwaldring 31, 70569 Stuttgart, Germany
email: angerhausen@irs.uni-stuttgart.de

Abstract. We present the prospects of observing extrasolar planets with the Stratospheric Observatory for Infrared Astronomy (SOFIA). Our analysis shows that optical and near-infrared photometric and spectrophotometric follow-up observations during planetary transits and eclipses will be feasible with SOFIA's instrumentation, especially with the HIPO-FLITECAM optical/NIR instruments. SOFIA has unique advantages in comparison to ground- and space-based observatories in this field of research which will be outlined.

Keywords. planetary systems, telescopes

1. Introduction

The Stratospheric Observatory for Infrared Astronomy (SOFIA) is a NASA 80:20 partnership with the German Space Agency (DLR) to develop a Boeing 747SP airliner fitted with a 2.7-meter reflecting telescope, making it the largest airborne observatory in the world (Gehrz *et al.* 2009). SOFIA has successfully had its first light observation in May 2010. Its first generation of science instruments consists of 7 imagers and spectrographs spanning the entire wavelength range from the optical to the far infrared ($0.3 - 200\mu m$). As warm Spitzer winds down and the future role of HST's infrared capabilities is uncertain, ground-based and SOFIA observations will play a dominant role for exoplanet atmosphere observations in the next decade, before JWST starts operating (see fig. 1).

2. Advantages of SOFIA for exoplanet observations

Even the very close-in extrasolar planets, with distances to their host star of only a few stellar radii, are not much hotter than $T \simeq 2000K$. Therefore the equivalent blackbody emission always peaks in the infrared. Furthermore many important atmospherical properties such as the chemical constituents or the T-P-profiles of planets can be analyzed with IR spectra, observed during transits or eclipses of the planet (e.g. Swain *et al.* 2009). From this perspective, SOFIA operates in the optimal wavelength regime for exoplanet observations.

For short-period close-in transiting planets with transits and eclipses occurring every 2-5 days, the optimal observing schedules for ground-based transit observations are reduced to only few nights per year for a given observing site as the event is best observed close to target culmination and local midnight. The HST, on the other hand, is able to observe transits at many more opportunities but is limited to series of 96 minute on/off-target batches due to its near-earth orbit. This presents a substantial hurdle - in particular for transiting planets with very long orbits (such as HD80606b) and therefore transit durations of more than 6 hours. The analysis of potential flight schedules shows that the mobile platform SOFIA will be able to take off close to the optimal geographic location

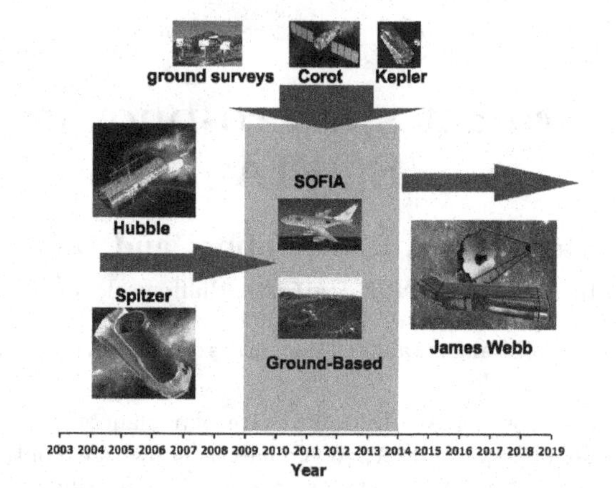

Figure 1. Diagram to explain the perspectives of transit spectro-photometry in the next decade. In the past few years Hubble and Spitzer (left) have been successfully used for spectroscopic characterization of extrasolar planets. As warm Spitzer winds down and the future role of HST's infrared capabilities is uncertain, space-based opportunities are limited until JWST starts operating. For the next years ground-based surveys, CoRoT and Kepler (top) are going to deliver a large number of new targets with interesting opportunities for characterization.The airborne observatory SOFIA will help to follow-up these planets and as testbed for future JWST observations.

for each of those events. The airborne observatory will be able to observe the complete event continuously and with a very stable setup (telescope elevation, airmass etc.) during observing flights of up to 10 hours duration.

The variability of earth's atmospheric transmission as well as the temporal variability of its constituents is the most crucial challenge in ground-based transit observations, in particular when it comes to the spectroscopic analysis of molecular features in the exoplanet's atmosphere that are also present as telluric trace gases. Again, SOFIA will also deal very favorably with these effects since it will be able to fly high enough (12-14 km) to be independent of near surface processes affecting in particular the water and methane lines. The SOFIA telescope is operating at much lower temperatures (240 K) than ground-based telescopes. Therefore thermal background contributions, that are the dominant noise source for transit observations at wavelengths longer than 3 micron, will be significantly reduced (Angerhausen *et al.* 2010).

3. Summary

We demonstrate that SOFIA has a specific and unique phase space for exoplanet research in the next decade. SOFIA operates in the right wavelength regime, above most of the perturbing variation of atmospheric trace gases and can observe rare transient events under optimized conditions. SOFIA will instantaneously be a competitive observatory in the field of state of the art exoplanet astronomy and some science topics are even exclusively observable with SOFIA.

References

Angerhausen, D., Krabbe, A., & Iserlohe, C. 2010, *PASP*, 122, 1020

Gehrz, R. D., Becklin, E. E., de Pater, I., Lester, D. F., Roellig, T. L., & Woodward, C. E. 2009, *AdSR*, 44, 413

Swain, M. R., *et al.* 2009, *ApJ*, 704, 1616

The Astrophysics of Planetary Systems: Formation, Structure, and Dynamical Evolution
Proceedings IAU Symposium No. 276, 2010
A. Sozzetti, M. G. Lattanzi & A. P. Boss, eds.

doi:10.1017/S1743921311021065

Detection of small-size planetary candidates with CoRoT data

Aldo S. Bonomo[1], Pierre-Yves Chabaud[1], Magali Deleuil[1], Claire Moutou[1] and Pascal Bordé[2]

[1]Laboratoire d'Astrophysique de Marseille, 38 rue Frédéric Joliot-Curie, 13388 Marseille Cedex 13, France
email: aldo.bonomo@oamp.fr

[2]Institut d'Astrophysique Spatiale, centre universitaire d'Orsay Bât 120-121, 91405 ORSAY CEDEX

Abstract. With the discovery of CoRoT-7b, the first transiting super-Earth, the CoRoT space mission has shown the capability to detect short-period rocky planets around solar-like stars. By performing a blind test with real CoRoT light curves, we want to establish the detection threshold of small-size planets in CoRoT data. We investigate the main obstacles to the detection of transiting super-Earths in CoRoT data, notably the presence of short-time scale variability and hot pixels.

Keywords. methods: data analysis, techniques: photometric, stars: late-type

1. Introduction

The CoRoT space mission searches for planetary transits by monitoring the optical flux of thousands of stars in several fields of view. It has recently led to the discovery of CoRoT-7b, the first transiting super-Earth (Léger *et al.* 2009). By simulating transits of super-Earths and Neptunes in real CoRoT light curves and searching for them blindly, we want to investigate the capability of CoRoT to detect small-size planets, super-Earths and Neptunes, in short-period orbits ($P < 10$ days).

2. Simulations and data analysis

We chose 500 real light curves of the CoRoT long run LRa01 which lasted 131.5 days. Central transits by super-Earths with orbital period P between 1 and 15 days were simulated in 100 of the aforementioned light curves with visual magnitude $11 < V < 14.5$. In as many light curves with $12 < V < 16$, central transits by Neptune-like planets were inserted. For simplicity, the parent star was assumed in all cases to be a Sun-like star ($R_* = R_\odot$ and $M_* = M_\odot$).

The simulated transits by super-Earths and Neptunes were searched for blindly, i.e. without knowing in advance in which light curves they had been inserted, by means of the LAM transit detection pipeline (Bonomo *et al.* , in preparation). It foresees the following steps: a) a 5-sigma clipping to filter out outliers due to proton impacts during the passage of the satellite at the South Atlantic Anomaly; b) a high-pass sliding median filter to remove stellar variability; c) a Savitzky-Golay low-pass filter to remove high-frequency variations of instrumental origin; d) an automatic detection and correction of the most evident hot pixels; e) the search for transits by means of the BLS algorithm (Kovács *et al.* 2002) with the directional correction.

3. Results and discussion

The percentage of detected super-Earths with $P < 10$ days around stars with $11 < V < 13.5$ is 35 %. 78% of Neptunes with $P < 10$ days and $V < 15$ were correctly found. Figure 1 shows the depth of the simulated transits as a function of their Signal-to-Noise Ratio (SNR). Most of the discovered transits have $SNR \geqslant 11$. This shows that CoRoT is able to detect super-Earths with a SNR much lower than CoRoT-7b (Fig. 2). We had no false positives.

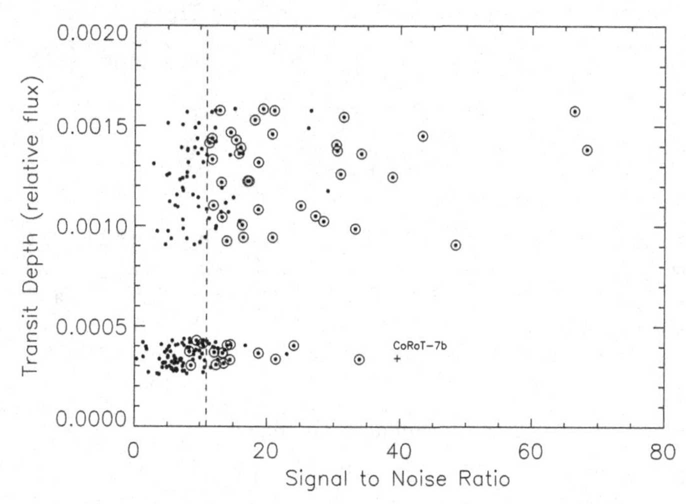

Figure 1. Depth of the simulated transits vs their Signal-to-Noise Ratio. Vertical dashed line: empirical detection limit at $SNR \sim 11$. The cross indicates the position of CoRoT-7b.

All the undiscovered transits with $SNR > 15$ were not detected because of short time-scale (< 1.5 days) variability affecting the light curves in which they had been simulated. The undiscovered transits with $11 < SNR < 15$ went undetected because of short time-scale variability or, more often, the presence of hot pixels. Indeed, the latter hinder the search for shallow planetary transits since, after the filtering of stellar variability, they give rise to artificial dips in the residuals that can be erroneously identified as transits by the detection algorithms.

Starting from the aforementioned results, work is in progress to infer some statistics on the presence of low-mass planets in the CoRoT fields of view. Indeed, it is an intriguing issue to understand if the detection of only one low-mass planet by the CoRoT team (e.g., CoRoT-7b) is compatible with the occurrence rate given by radial velocity surveys (Howard *et al.* 2010; Mayor *et al.* 2009), after taking both the transit probability and the detection rate derived from our blind test into account. Both radial velocity survey (Howard *et al.* 2010; Mayor *et al.* 2009) and theoretical core accretion models (e.g., Mordasini *et al.* 2009) indicate a a pile up of planets in the Neptunian and, even more, super-Earth mass domain.

References

Howard, A. W., Marcy, G. W., Johnson, J. A., *et al.* 2010, *Science*, 330, 653
Kovács, G., Zucker, S., & Mazeh, T. 2002, *A&A*, 391, 369
Léger, A., Rouan, D., Schneider, J., *et al.* 2009, *A&A*, 506, 287
Mayor, M., Udry, S., Lovis, C., *et al.* 2009, *A&A*, 493, 639
Mordasini, C., Alibert, Y., Benz, W., & Naef, D. 2009, *A&A*, 501, 1161

The Astrophysics of Planetary Systems: Formation, Structure, and Dynamical Evolution
Proceedings IAU Symposium No. 276, 2010
A. Sozzetti, M. G. Lattanzi & A. P. Boss, eds.

doi:10.1017/S1743921311021077

A microvariability study of nearby M dwarfs from the Western Italian Alps: Status update

Mario Damasso[1,2], Andrea Bernagozzi[1], Enzo Bertolini[1], Paolo Calcidese[1], Paolo Giacobbe[3], Mario G. Lattanzi[4], Matteo Perdoncin[5], Alessandro Sozzetti[4], Richard Smart[4] and Giorgio Toso[1]

[1]Astronomical Observatory of the Autonomous Region of the Aosta Valley, Loc. Lignan 39, 11020 Nus (Aosta), Italy
[2]Dept. of Astronomy, University of Padova, Vicolo dell'Osservatorio 5, I-35122 Padova, Italy
email: m.damasso@gmail.com
[3]Dept. of Physics, University of Trieste, Via Tiepolo 11, I-34143 Trieste, Italy
[4]INAF - Astronomical Observatory of Torino, Via Osservatorio 20, I-10025 Pino Torinese, Italy
[5]Dept. of Physics, University of Torino, Via Giuria 1, I-10125 Torino, Italy

Abstract. Small ground-based telescopes can effectively be used to look for transiting rocky planets around nearby low-mass M stars, as recently demonstrated for example by the MEarth project. Since December 2009 at the Astronomical Observatory of the Autonomous Region of Aosta Valley (OAVdA) we are monitoring photometrically a sample of red dwarfs with accurate parallax measurements. The primary goal of this 'pilot study' is the characterization of the photometric microvariability of each target over a typical period of approximately 2 months. This is the preparatory step to long-term survey with an array of identical small telescopes, with kick-off in early 2011. Here we discuss the present status of the study, describing the stellar sample, and presenting the most interesting results obtained so far, including the aggressive data analysis devoted to the characterization of the variability properties of the sample and the search for transit-like signals.

Keywords. techniques: photometric, planetary systems

---

1. Introduction

Since December 2009 a photometric survey of a small sample of nearby M dwarfs has been carrying out at the Astronomical Observatory of the Autonomous Region of the Aosta Valley (OAVdA), using less than 1-meter class telescopes. The aim of this 1-year long observing campaign is the monitoring of the microvariability of the stars over a period of some weeks to assess the implications for the detection of small-size transiting planets. The study represents a preparatory step toward a long-term survey of thousands of red dwarfs aimed at the detection of rocky planets. The OAVdA has been chosen as the hosting institution for the campaign on the base of the results of a site characterization study (Damasso *et al.* 2010).

2. Instrumentation and methodology

The instrumental set up used in this study is detailed in Damasso *et al.* (2010) and is composed of a 810, 400 and 250 mm telescope. We selected the targets from a list of nearby M dwarfs with precise parallaxes observed by Smart *et al.* (2010). On average each target has been monitored for 3 hr/night for a maximum period of 2 months. In parallel to the observations the data reduction and analysis pipeline TEEPEE (Transiting

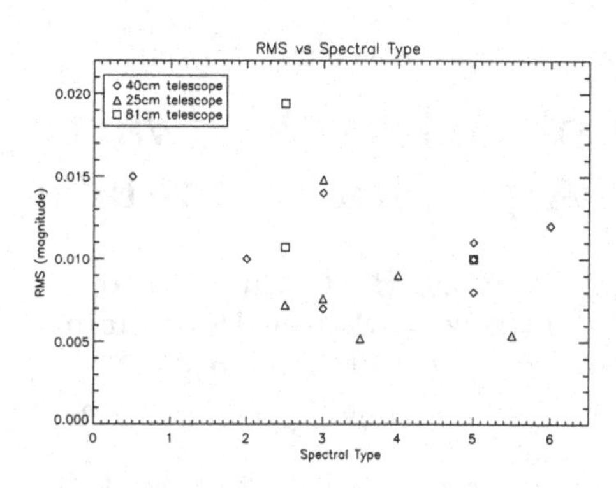

Figure 1. RMS of the target light curves as a function of target spectral subtype (M0-M6).

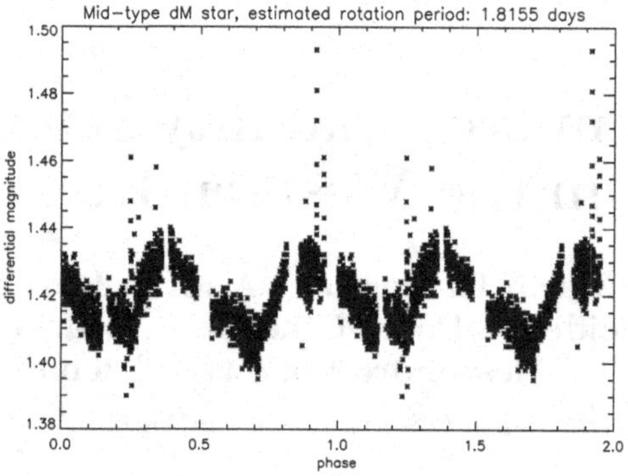

Figure 2. Differential light curve of a M dwarf folded according to the tentative rotation period of 1.8155 days.

ExoplanEts PipElinE; Damasso *et al.* 2010) has been upgraded, e.g. now including the SysRem algorithm to correct systematic effects in the light curves (Tamuz *et al.* 2005).

3. Some preliminary results

Fig. 1 shows the RMS of the light curves of some targets of our sample organized according to their spectral subtype. Each RMS is calculated for the whole period of observation. No correlation is evident from the data. Fig. 2 shows the differential light curve of a mid-type M dwarf which is made up of 19 observing sessions and is folded according the tentative rotation period of 1.8155 days, under the hypothesis that the observed modulations depend on the visibility conditions of active regions or star spots on the stellar disk. Three stellar flares are visible in the data, which were observed in two almost consecutive nights.

4. Toward a long-term automated survey

With kick off in Spring 2011, a long-term automated survey of thousands of nearby M dwarfs will start at the OAVdA, using an array of dedicated 400 mm telescopes. The main aim of the campaign will be the detection of small-size transiting extrasolar planets, designing an Europe-based observational program similar to the US MEarth project which has already discovered the super-Earth GJ 1214b (Charbonneau *et al.* 2009).

Acknowledgements

MD, PC, AB, and GT are supported by grants of the European Union, the Autonomous Region of the Aosta Valley and the Italian Department for Work, Health and Pensions. The OAVdA is supported by the Regional Government of Valle d'Aosta, the Town Municipality of Nus and the Monte Emilius Community.

References

Charbonneau, D., Berta, Z. K., Irwin, J., *et al.* 2009, *Nature*, 462, 891
Damasso, M., Giacobbe, P., Calcidese, P., *et al.* 2010, *PASP*, 122, 895, 1077
Smart, R. L., Ioannidis, G., Jones, H. R. A., *et al.* 2010, *A&A*, 514, A84
Tamuz, O., Mazeh, T., & Zucker, S. 2005, *MNRAS*, 356, 4, 1466

The Astrophysics of Planetary Systems: Formation, Structure, and Dynamical Evolution
Proceedings IAU Symposium No. 276, 2010
A. Sozzetti, M. G. Lattanzi & A. P. Boss, eds.

doi:10.1017/S1743921311021089

Stellar noise and planet detection.
I. Oscillations, granulation and sun-like spots

Xavier Dumusque[1,2], Nuno C. Santos[2], Stéphane Udry[1], Cristophe Lovis[1] and Xavier Bonfils[3]

[1]Observatoire de Genève, 51 ch. des Maillettes, 1290 Sauverny, Switzerland
email: xavier.dumusque@unige.ch

[2]Centro de Astrofísica, Universidade do Porto, Rua das Estrelas, 4150-762 Porto, Portugal

[3]Université J. Fourier (Grenoble 1)/CNRS, Laboratoire d'Astrophysique de Grenoble, France

Abstract. Spectrographs like HARPS can now reach a sub-ms^{-1} precision in radial-velocity (RV) (Pepe & Lovis 2008). At this level of accuracy, we start to be confronted with stellar noise produced by 3 different physical phenomena: oscillations, granulation phenomena (granulation, meso- and super-granulation) and activity. On solar type stars, these 3 types of perturbation can induce ms^{-1} RV variation, but on different time scales: 3 to 15 minutes for oscillations, 15 minutes to 1.5 days for granulation phenomena and 10 to 50 days for activity. The high precision observational strategy used on HARPS, 1 measure per night of 15 minutes, on 10 consecutive days each month, is optimized, due to a long exposure time, to average out the noise coming from oscillations (Dumusque *et al.* 2011a) but not to reduce the noise coming from granulation and activity (Dumusque *et al.* 2011a and Dumusque *et al.* 2011b). The smallest planets found with this strategy (Mayor *et al.* 2009) seems to be at the limit of the actual observational strategy and not at the limit of the instrumental precision. To be able to find Earth mass planets in the habitable zone of solar-type stars (200 days for a K0 dwarf), new observational strategies, averaging out simultaneously all type of stellar noise, are required.

Keywords. planetary systems, stars: activity, stars: oscillations, stars: spots, techniques: radial velocities

---

Generating synthetic radial-velocity measurements with stellar noise

To simulate the effect of new observational strategies, we first have to generate synthetic RV measurements that contain all the considered type of noises: oscillation, granulation phenomena and short-term activity.

For the two first perturbations, which have a typical time scale less than a few days, we use asteroseismology measurements to derive the corresponding noise levels. The technique used consists in calculating the velocity power spectrum density (VPSD) and in fitting it with a function that depend on the different type of noises considered Once the fit is done, we can generate synthetic RVs containing all the considered noises by calculating the inverse fourrier transform of the fitted function (see Dumusque *et al.* 2011a for more details)

Short-term activity noise induced by sun-like spots have a timescale of 10 to 50 days. Asteroseismology measurements, spanning a maximum of 10 days, are too short to address this type of noise whereas longer observations are not precise enough. We therefore need to simulate the RV effect of spots using Sun observations. Starting with simple sunspot properties, such as their number, their position in latitude and longitude, their lifetime, we generate a realistic model of sun-activity using a Poisson process for spot appearance. The RV effect induced by these spots is then calculated using the program

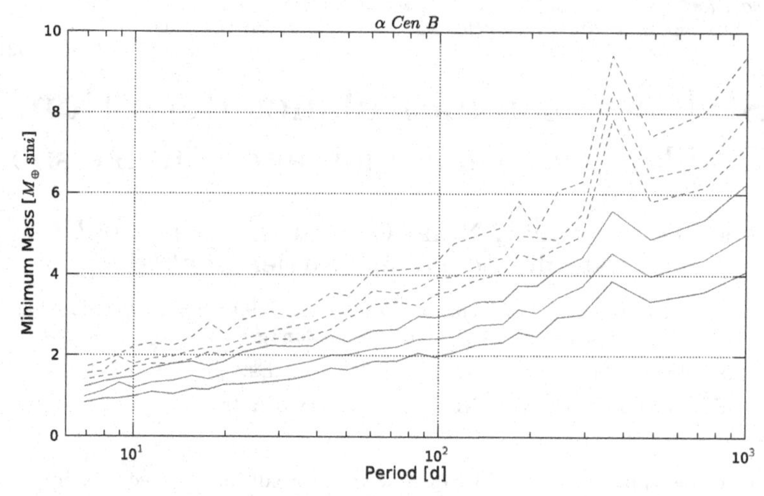

Figure 1. Mass-period detection limits for different strategies (continuous lines: 3N strategy and dashed lines: 1N strategy) and for different activity levels ($\log(R'_{HK}) = -5$ (blue), -4.9 (green) and -4.8 (red) from bottom to top for each strategy).

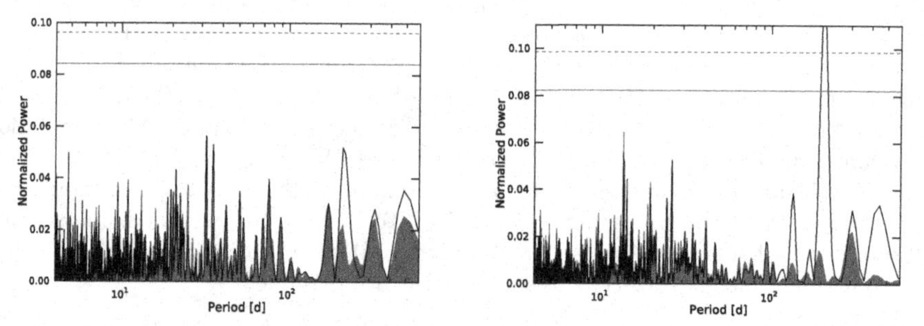

Figure 2. Detection simulation of a $2.5\,M_\oplus$ habitable planet orbiting an early K dwarf (200 days of period), with a star activity level fixed at $\log(R'_{HK}) = -4.9$. *Left :* Periodogram for the 1N strategy with the 1 % FAP (continuous line) and the 0.1 % FAP (dashed line). The red (gray) periodogram corresponds to the noise, and the black, to noise + planet. *Right :* Same but for the 3N strategy.

SOAP (X. Bonfils & N.C. Santos (2011), in prep.), which calculate the RV effect of one spot given its size and position (see Dumusque *et al.* 2011b for more details).

An efficient and affordable observational strategy

Once synthetic RV measurements containing oscillation, granulation phenomena and activity noises are generated, we choose an observational calendar (10 nights per month, 8 months a year on 4 years) and calculate for different observational strategies the planetary mass detection limits using 1 % False Alarm Probability (FAP) in periodograms (see Fig. 1). Depending on the activity level considered, the best tested strategy, 3 measurements per night of 10 minutes each, each 3 nights (3N strategy), could detect planet of 2.5 to $3.5\,M_\oplus$ in the habitable region of early K dwarfs (200 days of period). This strategy gives detection limit 40 % better than the present strategy used on HARPS (1N strategy). In Fig. 2, we can see the detection simulation of a 2.5 $M_\oplus$ habitable planet orbiting an early K dwarf. In this case, the best tested strategy is able to resolve the planet, whereas the present HARPS strategy is not (see Dumusque *et al.* 2011b for more details).

References

Dumusque, X., *et al.* 2011a, *A&A*, 525, A140
Dumusque, X., *et al.* 2011b, *A&A*, 527, A82
Lefebvre, S., *et al.* 1984, *A&A*, 490, 1143
Mayor, M. *et al.* 2009, *A&A*, 507, 487
Pepe, F. & Lovis, C. 2008, *Physica Scripta*, 130

The Astrophysics of Planetary Systems: Formation, Structure, and Dynamical Evolution
Proceedings IAU Symposium No. 276, 2010
A. Sozzetti, M. G. Lattanzi & A. P. Boss, eds.

doi:10.1017/S1743921311021090

Stellar noise and planet detection. II. Radial-velocity noise induced by magnetic cycles

Xavier Dumusque[1,2], Cristophe Lovis[1], Stephane Udry[1] and Nuno C. Santos[2]

[1]Observatoire de Genève, 51 ch. des Maillettes, 1290 Sauverny, Switzerland
email: xavier.dumusque@unige.ch

[2]Centro de Astrofísica, Universidade do Porto, Rua das Estrelas, 4150-762 Porto, Portugal

Abstract. For the 451 stars of the HARPS high precision program, we study correlations between the radial-velocity (RV) variation and other parameters of the Cross Correlated Function (CCF). After a careful target selection, we found a very good correlation between the slope of the RV-activity index (log(R'$_{HK}$)) correlation and the $T_{\rm eff}$ for dwarf stars. This correlation allow us to correct RV from magnetic cycles given the activity index and the $T_{\rm eff}$.

Keywords. planetary systems, stars: activity, techniques: radial velocities

Introduction

The Sun has a 11-years magnetic cycle, during which the activity level (Noyes *et al.* 1984) varies between log(R'$_{HK}$)=-5 at minimum and log(R'$_{HK}$)=-4.75 at maximum. This cycle can also be seen looking at the total number of sunspots or the Sun luminosity variation. In a recent paper, Meunier *et al.* (2010) show a correlation between the Sun RV variation and its magnetic cycle. Amplitudes of tenths of meter per second could be induced by such cycles, hiding signals of long period small-mass planets.

Solar type stars share the property of having an external convective enveloppe. Since upward flows of convection (granules) have a total surface more important than downward flows (intergranules), the stellar spectrum will be blushifted (Dravins 1982; Gray 2009). When the activity level increases, the number of magnetic features (spots and plages) raise up, and since these regions are known to inhibit the convection due to strong magnetic fields, the stellar spectrum will be shifted to the red. Since the activity level is correlated to the number of magnetic features, a long-term variation of the activity level, as it is the case in magnetic cycles, will induce a long-term RV variation.

Target selection

To study only long-term activity noise, we select stars that are measured more than 3 years, with a minimum of 5 bins of 3 months (each bin must contains at least 3 measurements). In addition, only stars with log(R'$_{HK}$) < -4.75 was selected to address solar-like activity. After this selection, we are left with 91 stars out of 451.

To study potential correlations between the activity index and RV variations (see Dumusque *et al.* (2010), in prep.), we first remove all the planets present in the sample. Then, we select stars in the following way (always for the 3 months binned data):

- Correlation between the CCF Full Width at Half Maximum (FWHM) and log(R'$_{HK}$), $R_{FWHM} > 0.5$ or between the CCF BISsector span (BIS) and log(R'$_{HK}$), $R_{BIS} > 0.5$,
- Max(log(R'$_{HK}$)) > -5,
- Peak to peak variation in log(R'$_{HK}$) > 0.05.

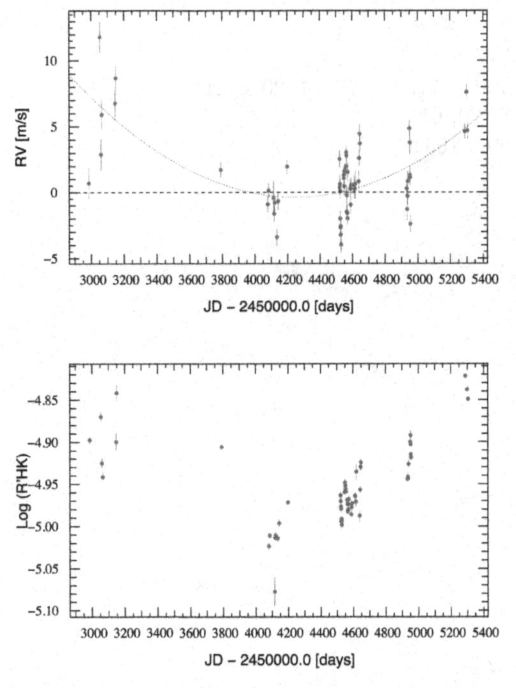

Figure 1. Comparison of the long-term RV and activity index (Log(R'HK)) variation. The Pearson correlation between the 2 parameters is very good, for 3 month bins, R = 0.85.

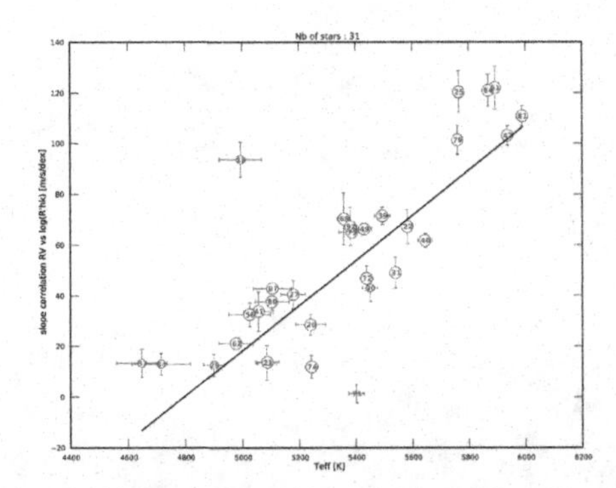

Figure 2. Slope of the correlation RV-log(R'_{HK}) as a function of the $T_{\rm eff}$. The size of the circle surrounding each star scales with R_{RV}, the correlation coefficent between RV and log(R'_{HK}). Small size if $R_{RV} > 0.5$ and large size if $R_{RV} > 0.75$. Green numbers correspond to stars which have $R_{FWHM} > 0.5$ or $R_{BIS} > 0.5$ and red numbers (71 and 23), to stars with no well defined correlation.

After this second selection, only 31 stars are left. However, a nice correlation appears when we plot these stars in a graph representing the slope of the correlation RV-log(R'_{HK}) as a function of the $T_{\rm eff}$ (see Fig 2). The higher the slope, the more the RV will be affected by magnetic activity. Therefore, RVs of early-Kdwarfs are less affected by magnetic cycles than RVs of early-G dwarfs, making early-K dwarfs optimal targets for very small mass planets surveys. In addition, this relation allow us to correct RVs from the magnetic cycle given the activity index and the $T_{\rm eff}$ of the star.

References

Noyes, R., *et al.* 1984, *ApJ*, 279, 763
Meunier, N., Desort, M., & Lagrange, A.-M. 2010, *A&A*, 512, A39
Dravins, D. 1982, *ARAA*, 20, 61
Gray, D. F. 2009, *ApJ*, 697, 1032

The Astrophysics of Planetary Systems: Formation, Structure, and Dynamical Evolution
Proceedings IAU Symposium No. 276, 2010
A. Sozzetti, M. G. Lattanzi & A. P. Boss, eds.

doi:10.1017/S1743921311021107

Space interferometry beyond exoplanetology: Can interdisciplinary collaboration contribute to the future of this technique?

Pavel Gabor[1]

[1]Vatican Observatory, Vatican City
email: p.gabor@jesuit.cz

Abstract. Although a formation-flying space interferometer designed for exoplanet spectroscopy is feasible in principle, the novelty and cost of such an instrument is likely to remain daunting unless the scientific benefits of this technology are demonstrated by intermediary, precursor missions. Such instruments would represent intermediary steps in the real-life testing of the technology, and therefore, by the very reason of being intermediary, they may not have the resolving or collecting power needed for the study of the objects where biomarkers could be hoped to be detected, i.e., exo-Earths in the habitable zone of their stars. This paper examines the potential applications of such intermediary instruments. The direct line of thought focuses on exoplanetology (gas giants, protoplanetary discs, Neptunes, super-Earths, etc.); what we would like to stimulate is an exercise in lateral thinking, looking at what might an intermediary interferometric mission contribute to other fields of astrophysical research (galaxies, supernova precursors, planetary nebulae, molecular clouds, etc.). The paper raises the question of collaboration with astrophysicists studying areas other than exoplanets and its potential gains for the future of space interferometry.

Keywords. instrumentation: interferometers, techniques: high angular resolution, space vehicles: instruments

---

1. Nulling Interferometry in Space

Since its inception in 1993, the proposed *Darwin* nulling interferometer (Léger *et al.* 1996) together with its US counterpart TPFI, has been an important factor in a number of instrumental studies. The science case, as currently presented, is this: The study of extrasolar planets, for the moment, relies on indirect methods. At some point, however, the field will only be able to advance if efficient direct imaging (including spectroscopy) becomes available. The feasibility of the concept has been demonstrated quite satisfactorily by Peters *et al.* (2010) and Martin & Booth (2010). In the eyes of many it was the clear choice for a flagship space mission (cf. Liseau 2010). The funding agencies, however, have not selected this mission for implementation so far. Consequently, if realised, the launch of *Darwin*/TPFI is highly unlikely before 2025.

Apart from a nulling interferometer, other approaches have been under development (the most advanced ones being an occulter and an internal coronograph). Since 2008, roadmapping has been under way both in Europe (ExoPlanetary Roadmap Advisory Team; Blue Dots Initiative; Coudé Du Foresto *et al.* 2010), and in the US (Exoplanet Analysis Committee; Exoplanet Task Force; Traub *et al.* 2010), seeking consensus as to the best strategy in future instrumentation, with special attention to a flagship space mission. The outcome of the process remains unclear while a sense of urgency is growing.

A major issue appears to be a certain reluctance towards space interferometry which is present to varying degree not only among decision makers but also in the scientific community at large, among astrophysicists in general as well as among exoplanetologists and astrobiologists. This attitude can be easily understood, considering that

- even ground-based visible/IR interferometry is very difficult to implement,
- few ground-breaking science results have been obtained using this technique,
- nulling is considerably more demanding than constructive interferometry, and
- implementing this method in space adds another degree of difficulty to the enterprise.

2. A Scientifically Valuable Precursor

In order to overcome this ambivalence, interferometry needs to demonstrate its potential. Schneider (2009) called for a "scientifically valuable precursor". Such an accomplishment could indeed bring about a change of mentality. How to make such a precursor valuable not only technologically but also scientifically?

Historically, the space interferometer community focuses on circumstellar environements . This may be creating an unnecessary dilemma for the precursor. On the one hand, a precursor, by definition, does not achieve the performance of the flagship mission, and therefore – in exoplanetology – it is unlikely to produce major discoveries, nor even can it do cutting-edge science, e.g., as a pathfinder. On the other hand, since the flagship mission's application is exoplanetology, other fields are passed over even when planning a precursor. But beyond exoplanetology it *could* prove itself "scientifically valuable" .

A wider scope of science targets may benefit the cause of a precursor immensely.

(*a*) First, by simple opening more possibilities of achieving inspiring results, and,

(*b*) second, by involving a larger portion of the astrophysical community as a whole.

This applies to space precursors, but also to ground-based (and airborne) instruments.

3. Broadening Horizons

Is there anything a space interferometer could contribute to the field of star formation? In addition to protostellar disks, envelopes, jets, and debris disks, could it discover any new features (let us recall that polar jets were an unexpected phenomenon)? What about molecular clouds or planetary nebulae, and other stellar remnants? Is there anything it could contribute to the study of active galactic nuclei? Or perhaps of quasars? (3C 273 with its jet could be an interesting targets because a detailed map may allow to infer the physical nature of the quasar; Uchiyama *et al.* 2006.) And what about microquasars, black-hole accretion disks, supernova precursors, and other interacting binary systems?

References

Coudé Du Foresto, V. & Blue Dots 2010, in V. Coudé Du Foresto, D. M. Gelino, & I. Ribas (eds.), *ASP Conf. Ser.*, 430, 15

Léger, A., Mariotti, J.-M., Mennesson, B., Ollivier, M., Puget, J.-L., Rouan, D., & Schneider, J. 1996, *Icarus*, 123, 249

Liseau, R. 2010 in V. Coudé Du Foresto, D. M. Gelino, & I. Ribas (eds.), *ASP Conf. Ser.*, 430, 219

Martin, S. R. & Booth, A. J. 2010, *A&A*, 520, A96

Peters, R. D., Lay, O. P., & Lawson, P. R. 2010, *PASP*, 122, 85

Schneider, J. 2009, *ArXiv*, 0906.0068

Traub, W. A., Lawson, P. R., Unwin, S. C., Muterspaugh, M. W., Soummer, R., Danchi, W. C., Hinz, P., Gaudi, B. S., Torres, G., Deming, D., Lazio, J., & Dressler, A. 2010, in V. Coudé Du Foresto, D. M. Gelino, & I. Ribas (eds.), *ASP Conf. Ser.*, 430, 21

Uchiyama, Y., Urry, C. M., Cheung, C. C., Jester, S., Van Duyne, J., Coppi, P., Sambruna, R. M., Takahashi, T., Tavecchio, F., & Maraschi, L. 2006, *ApJ*, 648, 910

The Astrophysics of Planetary Systems: Formation, Structure, and Dynamical Evolution
Proceedings IAU Symposium No. 276, 2010
A. Sozzetti, M. G. Lattanzi & A. P. Boss, eds.

doi:10.1017/S1743921311021119

Gravitation Astrometric Measurement Experiment (GAME)

Mario Gai[1], Alberto Vecchiato[1], Alessandro Sozzetti[1], Sebastiano Ligori[1] and Mario G. Lattanzi[1]

[1] Istituto Nazionale di Astrofisica – Osservatorio Astronomico di Torino, V. Osservatorio 20, I-10025 Pino T.se (TO), Italy
email: gai@oato.inaf.it

Abstract. GAME (Gravitation Astrometric Measurement Experiment) is a mission concept based on astronomical techniques for high precision measurements of interest to Fundamental Physics and cosmology, in particular the γ and β parameters of the Parameterized Post-Newtonian formulation of gravitation theories extending the General Relativity.

High precision astrometry also provides the light deflection induced by the quadrupole moment of Jupiter and Saturn, and, by high precision determination of the orbits of Mercury and high elongation asteroids, the PPN parameter β.

The astrometric and photometric capabilities of GAME may also provide crucial complementary information on a selected set of known exo-planets.

Keywords. gravitation, astrometry, techniques: miscellaneous, space vehicles: instruments

---

1. Introduction: the main science goal of GAME

The main goal of GAME (Gravitation Astrometric Measurement Experiment) is to estimate the γ and β parameters of the Parameterized Post-Newtonian (PPN) formalism, by repeated measurement from space of the star light deflection close to the solar limb. The mission is conceived as a novel rendition, from space and using modern technology, of the experiment by Dyson, Eddington and collaborators conducted during the solar eclipse of 1919, when the gravitational bending of light was measured for the first time. This makes GAME a decisive experiment for the understanding of gravity physics, cosmology and the Universe evolution at a fundamental level.

General Relativity can act as a cosmological attractor for scalar-tensor theories, with expected deviations in the $10^{-5} - 10^{-8}$ range for γ, as from Damour & Nordtvedt (1993). Also, the experimental evidence of an accelerated expansion of the Universe is interpreted as a long range perturbation of the gravity field of the visible matter, generated by Dark Energy, in addition to the measurements, at different scale length, explained with non-barionic Dark Matter (e.g. galaxy rotation curves). An alternative explanation might be provided by a modified version of General Relativity, e.g. in which the curvature invariant R is no longer constant as in Einsteins equations, i.e. the $f(R)$ gravity theories. A $< 10^{-7}$ level determination of γ will provide stringent constraints on acceptable theories, as derived e.g. in Capozziello & Troisi (2005).

The original experiment of Dyson, Eddington and collaborators confirmed the General theory of Relativity at 10^{-1}-level. The same technique was used for several decades after 1919, but its accuracy was not significantly improved, because of fundamental experimental limitations: the short time available for the observations (limited by the eclipse duration, which constrained also the number and brightness of the observed stellar sample), the atmospheric disturbances, and the background from the solar corona. The best

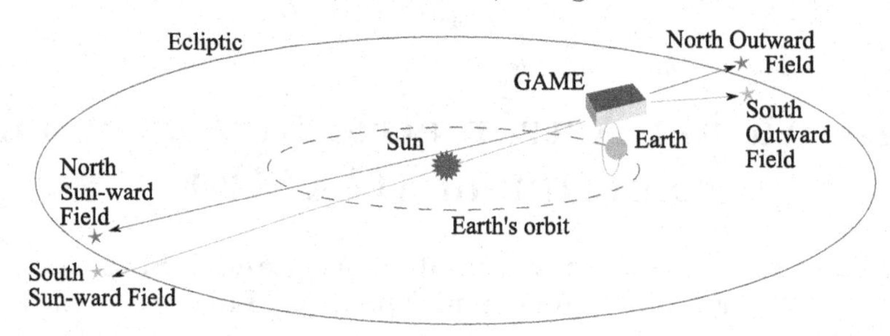

Figure 1. The GAME satellite, observing field pairs either close to, or far away from, the Sun.

current estimate of γ was achieved from the Cassini data, reaching the 10^{-5} level of accuracy, by Bertotti *et al.* (2003), exploiting the derivative of the Shapiro effect.

In the near future, and focusing on astrometric measurements, the most promising effort presently under development is the Gaia mission, which, building on the same principles of Hipparcos, might achieve a level of accuracy of $10^{-6} - 10^{-7}$ during the second half of the next decade, as shown in Vecchiato *et al.* (2003).

GAME, described in a small mission version in Gai *et al.* (2009), merges astrometric and coronagraphic techniques to perform micro-arcsec differential determination of the deflection angle on stellar field pairs, measured in two epochs: close to the Sun and in opposition, after six months (Fig. 1). GAME requires a limited time to fulfil its main science goal, since the γ determination is most conveniently performed on the high stellar density regions of intersection between the Ecliptic plane and the Milky Way disc.

2. Exo-planetary system applications of GAME

GAME will carry out a robust, adaptive additional science program using the available fraction of observing time. Its high-precision, fully differential astrometry, and its high-accuracy, high-cadence photometry, make it an ideal instrument for a program targeted to well-selected, known extrasolar planetary systems, critically deepening our understanding of key issues in exo-planet astrophysics. The goal is four-fold: 1) improving the orbital phase coverage against bad and/or insufficient sampling; 2) looking for planetary companions at all orbital periods; 3) looking for low- as well as high-mass components (Super Earths, Neptunes, and Jupiters); 4) refining the characterization of multiple planet systems (including accurate coplanarity measurements).

The three-fold exo-planet science program is focused on: a) GAME astrometry of nearby, bright stars known to host one or more exoplanets through long-term high-precision RV monitoring; b) GAME follow-up studies of Gaia-detected exoplanets around stars of varied masses, metallicities, and ages (with and without ground-based RV support); c) GAME astro-photometric investigations of planetary systems in which one or more component is known to transit.

References

Damour, T. & Nordtvedt, K. 1993, *Phys. Rev. Lett.* 70, 2217

Capozziello, S. & Troisi, A. 2005, *Phys. Rev. D* 72, 044022

Bertotti, B., Iess, L., & Tortora, P. 2003, *Nature* 425, 374

Vecchiato, A., Lattanzi, M. G., Bucciarelli, B., Crosta, M., de Felice, F., & Gai, M. 2003, *A&A* 399, 337

Gai, M., Vecchiato, A., Ligori, S., Fineschi, S., & Lattanzi, M. G. 2009, *Proc. SPIE* 7438, 74380T

The Astrophysics of Planetary Systems: Formation, Structure, and Dynamical Evolution
Proceedings IAU Symposium No. 276, 2010
A. Sozzetti, M. G. Lattanzi & A. P. Boss, eds.

doi:10.1017/S1743921311021120

Detectability of Earth mass planets with RV techniques around Sun-like stars. The example of the Sun

Anne-Marie Lagrange[1], Nadege Meunier[1] and Morgan Desort[1]

LAOG, Université J Fourier, F-38041 Grenoble Cedex, France
email: lagrange@obs.ujf-grenoble.fr, meunier@obs.ujf-grenoble.fr, desort@obs.ujf-grenoble.fr

Abstract. We present the results of detailed simulations of the RV and astrometric signals expected from the Sun, when taking into account its activity (spots, plages, convection). To do so, we considered all structures (2,000,000) identified on the Sun surface over a full cycle. We show that the Sun activity would prevent the detection of the Earth in the Habitable Zone with RV technics with today or future forthcoming instruments, mainly because of inhomogeneous convection. We also show that the activity-induced signal would be comparatively easier for the astrometric detection of the Earth of similar planets.

Keywords. planetary systems, Sun: activity, stars: activity, techniques: radial velocities, astrometry

---

1. Introduction

It has been known that stellar spots could impact on the detectability of extrasolar planets with radial velocities (RV) technics. The most detailed simulations have shown (Desort *et al.* 2007) that depending on their characteristics (size, location, temperature), and depending on the star properties (inclination), spots can produce RV variations quite similar to those induced by planets. Such spots represent a potential problem and are of growing importance as lower and lower mass planets are being (and will be) searched at both short and long periods with next generation RV instruments. In addition, other contrasted structures (plages, networks) or convection will impact the RV signal, but prior to the present study, no detailed estimation was available. To further investigate, in a realistic case, the impact of activity on the detection of Earth mass planets in the Habitable Zone (HZ), we have performed a complete simulation of the RV variations expected from the Sun, if observed like a star, during one entire, 11 yr solar cycle. The results are presented below. We finally also address the same planet detectability, under similar conditions, with astrometric technics.

2. Impact of activity on planet detection with RV technics

<u>*Spots only (Lagrange et al. 2010a*</u> We considered all (160000) spots with sizes $\geqslant$ 1ppm identified each day on the Sun surface during one cycle (Jan 93-Dec 03) from Debrecen Observatory. Using this map, and assuming that the spots are 550K cooler than the Sun surface, we computed the emerging spectrum and photometric signal that one would measure on the spectrum of the Sun if seen like a star with Teff=5800K and vsini=1.9 km/s. We then analysed the spectra as done in RV searches to extract the resulting RVs, Bissector Velocity Span temporal variations. These variations are then those expected from a solar type star with the same level of activity, seen edge-on. We then added

1ME planet in the HZ as well as instrumental noise (1-5-10 cm/s) corresponding to that expected from the forthcoming RV instruments (Espresso, Codex).

It appears that during the high-activity period, the planet would not or hardly be detected from the corresponding periodograms of the RV variations. During low activity periods, it could be detected with the next generations of RV instruments provided very intensive and long surveys (star observed twice a week over several years).

Impact of solar spots and plages on planet detection (Meunier et al. 2010) We considered then both (20800) spots groups larger than 10 ppm (NOAA data) and bright structures (plages and magnetic network; 1.803.344 structures from SOHO MDI magnetograms) observed between May 96 to Oct 07. The spots and bright structures temperatures were deduced from the comparison between our simulated photometric data and the observed TSI data. The conclusions are similar to the spots-only case.

Impact of solar spots, plages and convection on planet detection (Meunier et al. 2010) We finally considered in addition the impact of inhomogeneous convection. To do so, we assumed that the quiet Sun (ie without spots/bright structures) has a convective blueshift of 200 m/s, and in a first step, that this convective blueshift is canceled within the spots, plages and network. It appears that convection induced RV signal dominates that of the spots and plages over the whole cycle (Table 1). With the knowledge and tools available today, an Earth mass planet orbiting the Sun in the HZ would not be detected with the next generation instruments, even with a daily temporal sampling. A slightly different (e.g. 2/3) attenuation of the convective blueshift instead of a total one would not change the present results.

Table 1. Measured RMS for the shifts along and perpendicular to the equatorial plane (in μas) due to the spots and bright structures. The entire cycle as well as low and high activity periods are considered. Associated rms of the RV (m/s), taking or not convection into account, as well as photometry (relative variations) are also given.

	rms(deltax)	rms(deltay)	rms(RV without conv.)	rms(RV with conv.)	rms(TSI)
all	0.07	0.05	0.33	2.4	3.6e-4
high	0.09	0.06	0.42	1.42	4.5e-4
low	0.02	0.01	0.08	0.44	1.2e-4

3. Astrometric detection

Observables which are not sensitive to convection (photometry, astrometry) represent an attractive solution to search for Earth mass planets in the HZ. Using the same spots+bright structures simulations as described above, we have simulated the resulting astrometric variations. It appears that the Earth astrometric signal (amplitude 0.3 μas) dominates the activity one (rms $\leqslant$ 0.1 μas; Table 1), and that a monthly monitoring of the astrometric position over about 4 years would allow the detection of an Earth in the HZ, provided instrumental precisions of the order of 1. μas per measurement (Lagrange *et al.* 2010b).

References

Desort, M., Lagrange, A.-M., Galland, F, *et al.* 2007, *A&A*, 473, 893
Lagrange, A.-M., Desort, M., & Meunier, N. 2010, *A&A*, 512, A38
Lagrange, A.-M., Meunier, N., Desort, M., & Malbet, F. 2011, *A&A*, 528, L9
Meunier, N., Desort, M., & Lagrange, A.-M. 2010, *A&A*, 512, A39

The Astrophysics of Planetary Systems: Formation, Structure, and Dynamical Evolution
Proceedings IAU Symposium No. 276, 2010
A. Sozzetti, M. G. Lattanzi & A. P. Boss, eds.

doi:10.1017/S1743921311021132

The Italian contribution to the Gaia data processing and archiving

Michele Martino[1], Armando Ciampolini[1], Rosario Messineo[1], Angelo Mulone[1], Enrico Pigozzi[1], Vilma Icardi[1], Filomena Solitro[1], Marco M. Castronuovo[2], Mario G. Lattanzi[3], Roberto Morbidelli[3], Ronald Drimmel[3], Maria Sarasso[3], Deborah Busonero[3], Mariateresa Crosta[3], Daniele Gardiol[3], Alberto Vecchiato[3] and the Italian Gaia Team[3,4,5,6,7,8]

[1] ALTEC S.p.A
email: michele.martino@altecspace.it

[2] ASI Agenzia Spaziale Italiana

[3] INAF - Osservatorio Astronomico di Torino

[4] INAF - Osservatorio Astronomico di Padova

[5] INAF - Osservatorio Astronomico di Bologna

[6] INAF - Osservatorio Astronomico di Roma-Teramo

[7] INAF - Osservatorio Astronomico di Napoli

[8] INAF - Osservatorio Astrofisico di Catania

Abstract. Gaia is an ESA Cornerstone mission, scheduled to be launched in spring 2013, dedicated to precisely measure the positions and motions of over a billion stars in our galaxy: the Milky Way. Gaia Data Processing Center Turin (DPCT), the Italian DPC, is hosted and operated at ALTEC in Turin. The primary objective of DPCT is to provide the infrastructure and operations support to the Astrometric Verification Unit (AVU) activities for CU3 and the Italian participation to the Gaia data processing tasks. DPCT will archive all of the data , produced for and delivered to DPAC as part of the Italian contribution to the activities of CU4, CU5, CU7, and CU8.

Keywords. astrometry, surveys, catalogs

1. Gaia Mission

The Gaia mission aims to obtain astrometric data with a precision two hundred times greater than that of Hipparcos and astrophysical information on brightness in different spectral bands that will allow studying in detail the formation, dynamics, chemistry and evolution of our galaxy. It will also be possible to detect extrasolar planets and observe asteroids, galaxies and quasars. Gaia satellite is completely developed by ESA, including also the scientific instruments. The contribution of member states is devoted to the scientific program. Italy is the third contributor country.

2. Gaia DPCT

Gaia DPCT (see Fig. 1 for one of its constituting elements) is hosted and operated at ALTEC centre in Turin. The main objective of DPCT is to provide the infrastructure (in terms of HW, SW and communications) and operations support to the Astrometric Verification Unit (AVU) activities for CU3 and the Italian participation to the Gaia data processing tasks. DPCT will archive all of the data (photometric calibration data, libraries of spectra, etc) produced for and delivered to DPAC as part of the Italian contribution to the activities of CU4, CU5, CU7, and CU8. AVU is the unit responsible for the development and maintenance of the following CU3 software products: AIM - Astrometric Instrument Model, in charge of processing the Astro data telemetry in order to monitor and analyse the Astro instrument response over the mission lifetime. BAM/AVU - Basic Angle Monitoring software system, in charge of processing the BAM device telemetry in order to monitor and analyse the BA behaviour over time. GSR - Global Sphere Reconstruction, the mathematical and numerical framework that shall be used to verify the global astrometric results produced by AGIS. In addition the DPCT will provide hosting and operations of the GAia Relativistic Experiment on Quadrupole experiment (GAREQ) and of Initial Gaia Source List (IGSL) database. The IGSL DB will be maintained and upgraded throughout the Gaia operational lifetime and will be used to produce upgraded version of the IGSL catalogue to be delivered periodically to ESAC and ingested in the Main DB.

3. Ground Segment

The Gaia mission requires, as a critical element, the construction of a Ground Segment (GS) that shall be operative for all the duration of the mission and, subsequently, up to the achievement of the expected quality before the result publication and distribution to the international scientific community. The GS is composed of six different Data Processing Centres (DPCs) and its implementation and management is the primary task of the European consortium DPAC, responsible for ESA of the reduction of data of the mission.

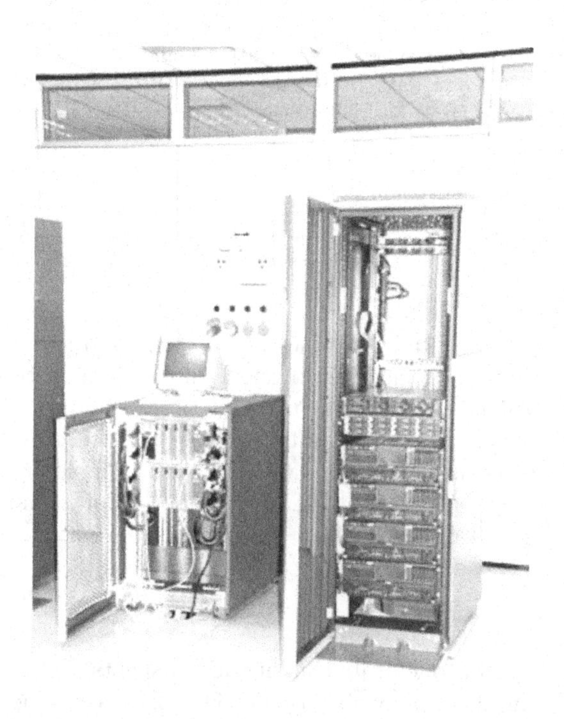

Figure 1. DPCT Development & Test Platform

The six centers are: ESAC/DPCE, CNES/DPCC, Barcelona/DPCB, Cambridge/ DPCI, Geneva/DPCG and Turin/DPCT. The responsibility for reducing the enormous amount of data that will be produced by the mission is entrusted to the DPAC (Data Processing and Analysis Consortium), a consortium of European research institutions created in response to an ESA Announcement of Opportunity. The DPAC is structured into 8 Coordination Units (CU), each responsible for a particular aspect of data processing. INAF-OATO is responsible for the definition, design, development, test and validation of the software for all AVU subsystems, GAREQ and the IGSL that shall be delivered to DPCT. It is also responsible for all of the coordination activities necessary for the development and operations of the DPCT. ALTEC is responsible for the definition, design, implementation, test, validation, and operations of the infrastructure of the DPCT that will be hosted on their premises and of the integration, testing and operations (science operations excluded) of the AVU, GAREQ and IGSL delivered software.

The Astrophysics of Planetary Systems: Formation, Structure, and Dynamical Evolution
Proceedings IAU Symposium No. 276, 2010
A. Sozzetti, M. G. Lattanzi & A. P. Boss, eds.

doi:10.1017/S1743921311021144

The variation of biomarkers in the spectrum of Earthshine

Marie-Eve Naud[1], R. Lamontagne[1] and F. Wesemael[1]

[1]Département de physique, Université de Montréal,
C.P. 6128, Succ. Centre-Ville, Montréal, Qué. H3C 3J7, Canada
email: naud, lamont, wesemael@astro.umontreal.ca

Abstract. We present preliminary results from a study of the variations of the optical spectrum of the Earthshine over a period of a year. Our goal is to follow, on several timescales, the spectral changes of the signature of various biomarkers that are potential indicators of habitability and/or biological activity, and of the Vegetation Red Edge, a feature around 700 nm due to photosynthetic organisms.

Keywords. Earth, planetary systems

---

1. Introduction

The idea of studying the Earth, the only inhabited planet known, as a proxy for habitable exoplanets emerged from the *Galileo* mission in the early 1990s (Sagan *et al.* 1993). Since then, numerous signs of habitability and life were identified in spectroscopic data gathered by spacecrafts. While useful to study the Earth from afar, spacecrafts are so costly that they are generally unsuitable to follow the variations of the Earth's spectrum over a long period. An alternative way to do so is to observe the Earthshine (ES), the weak glow produced by sunlight reflected from the surface of our planet on the otherwise dark side of the lunar disk. This method, already proposed a hundred years ago as a tool to study the Earth, was revived in the last 10 years as exoplanet science bloomed. The ES spectrum, composed of the spectra of the Sun, the Earth and the Moon, can be divided by the Moonshine (MS) spectrum, which only contains contributions from the Moon and the Sun, to isolate the Earth's spectrum. The portion of the Earth reflecting light towards the Moon depends on the lunar phase and on the longitude of the observer. In this contribution, we present the first results of a study of the variations of the ES spectrum over timescales ranging from a fraction of an hour to months.

2. Observations and Data Reduction

The 1.6 m telescope at the Observatoire du Mont-Mégantic (45°27′N, 71°09′W) was used with an optical spectrograph equipped with a 300 l mm^{-1} grating that allowed coverage of the $\sim$ 350-950 nm region. The longitude of the site is such that a few days before new moon, the portion of the Earth contributing to the ES is Africa, Europe and the Atlantic ocean, while a few days after it, it is the Pacific ocean and the Americas. Between 2008 July 28 and 2009 November 23, in the limited periods when the Moon displayed the right illumination, we secured data on 10 different nights, 4 before new moon (phase angle: $-105°$ to $-125°$) and 6 after (phase angle: 90° to 140°). The quality of the data varies appreciably, as it depends on the cloud cover on site and on the illumination of the Moon (10-40 %). On each of these 10 dates, typically 5-10 spectra with

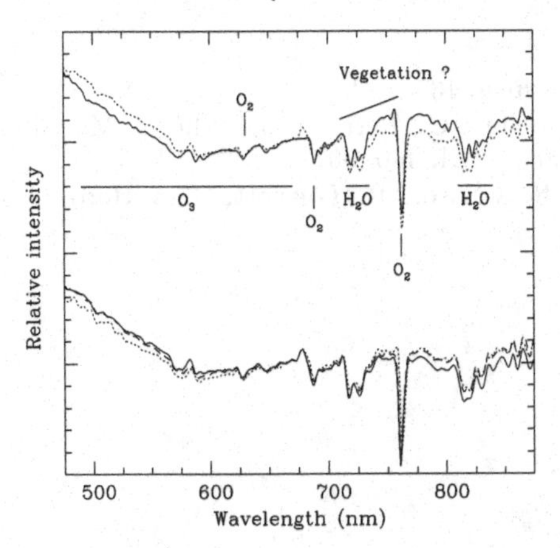

Figure 1. Top panel: The Earth's spectrum on 2008 July 28 (07:50 UT; solid line) and 2009 March 1 (00:00 UT; dotted line); Bottom panel : Changes observed in the Earth's spectrum during the night of 2009 February 28-March 1. The spectra secured at 23:40 (solid line), 00:00 (dotted line) and 00:25 (dashed line) UT are displayed.

the spectrograph slit positioned on the ES and its adjacent sky were taken. MS spectra were taken before and after each ES. Our data reduction procedure closely followed that of Hamdani *et al.* (2006).

3. Preliminary Results

Many atmospheric biomarkers that suggest the presence of life can be seen in our low resolution ($R \sim 1000$) optical spectra, examples of which are shown in Fig. 1: water (720, 820 nm), oxygen (630, 690, and 760 nm) and ozone (around 600 nm) are readily seen. Another feature of interest is the Vegetation Red Edge (VRE), an abrupt rise in the spectrum around 700 nm associated with photosynthetic organisms. This feature is difficult to detect in the disk-averaged spectrum of the Earth because of the restricted area typically covered by vegetation and of the frequent presence of cloud cover. Previous efforts (Arnold 2008) suggest that the VRE should be observed as a few percent increase in reflectivity around 700 nm. Our preliminary analysis suggests that *i)* the VRE is detectable in our data and *ii)* on some nights, we observe a variation of the VRE as vegetation-covered landmasses (esp. Amazonia) appear into view. Further analysis of the VRE and its variation is in progress.

Our data also display time variations of the ES spectrum on different scales. The top panel of Fig. 1 displays sample results secured on nights roughly seven months apart. In this specific case, the Earth phase differed drastically. The bottom panel, in contrast, displays spectra secured within a fraction of an hour of each other. In this case, the Earth phase did not change substantially, but the cloud pattern had time to evolve. Both types of variations could be observed in the spectra of terrestrial exoplanets.

Acknowledgements

M.-E.N. acknowledges the continuing support of the NSERC Canada.

References

Arnold, L. 2008, *Space Sci. Revs*, 135, 323

Hamdani, S., Arnold, L., Foellmi, C., Berthier, J., Billeres, M., Briot, D., François, P., Riaud, P., & Schneider, J. 2006, *A&A*, 460, 607

Sagan, C., Thompson, W. R., Carlson, R., Gurnett, D., & Hord, C. 1993, *Nature*, 365, 715

The Astrophysics of Planetary Systems: Formation, Structure, and Dynamical Evolution
Proceedings IAU Symposium No. 276, 2010
A. Sozzetti, M. G. Lattanzi & A. P. Boss, eds.

doi:10.1017/S1743921311021156

CARMENES: Calar Alto high-Resolution search for M dwarfs with Exo-earths with Near-infrared and optical Echelle Spectrographs

Andreas Quirrenbach[1], Pedro J. Amado[2], José A. Caballero[3], Holger Mandel[1], Reinhard Mundt[4], Ansgar Reiners[5], Ignasi Ribas[6], Miguel A. Sánchez Carrasco[2], Walter Seifert[1] and the CARMENES Consortium[1,2,3,4,5,6,7,8,9,10,11]

[1]Landessternwarte Königstuhl, Heidelberg, Germany
email: A.Quirrenbach@lsw.uni-heidelberg.de
[2]Instituto de Astrofísica de Andalucía, Granada, Spain
[3]Centro de Astrobiología, Madrid, Spain
[4]Max-Planck-Institut für Astronomie, Heidelberg, Germany
[5]Insitut für Astrophysik, Göttingen, Germany
[6]Institut de Ciències de l'Espai, Barcelona, Spain
[7]Instituto de Astrofísica de Canarias, Tenerife, Spain
[8]Thüringer Landessternwarte, Tautenburg, Germany
[9]Hamburger Sternwarte, Hamburg, GermanY
[10]Universidad Complutense de Madrid, Madrid, Spain
[11]Centro Astronómico Hispano-Alemán, Almería, Spain

Abstract. CARMENES (**C**alar **A**lto high-**R**esolution search for **M** dwarfs with **E**xo-earths with **N**ear-infrared and optical **E**chelle **S**pectrographs) is a next-generation instrument for the 3.5 m telescope at the Calar Alto Observatory. CARMENES will conduct a five-year exoplanet survey targeting $\sim$ 300 M stars. The CARMENES instrument consists of two separate fiber-fed spectrographs covering the wavelength range from 0.52 to 1.7 μm at a spectral resolution of $R = 85,000$. The spectrographs are housed in a temperature-stabilized environment in vacuum tanks, to enable a 1 m/s radial velocity precision employing a simultaneous emission-line calibration.

Keywords. instrumentation: spectrographs, techniques: radial velocities, infrared: planetary systems

1. CARMENES Science

he aim of CARMENES is to perform high-precision measurements of stellar radial velocities with long-term stability. The fundamental science objective is to carry out a survey of late-type main sequence stars (with special focus on moderately active stars of spectral type M4V and later) with the goal of detecting low-mass planets in their habitable zones. For stars later than M4–M5 ($M < 0.20\,\mathrm{M}_\odot$), a radial velocity precision of $1\,\mathrm{m\,s}^{-1}$ (per measurement; σ_i) will permit the detection of super-Earths of $5\,\mathrm{M}_\oplus$ and smaller inside the entire width of the habitable zone with $2\sigma_i$ radial-velocity amplitudes (i.e., $K_p = 2\,\mathrm{m\,s}^{-1}$). For a star near the hydrogen-burning limit and a precision of $1\,\mathrm{m\,s}^{-1}$, a planet as small as our own Earth in the habitable zone could be detected. In addition, the habitable zones of all M-type dwarfs can be probed for super-Earths.

The CARMENES survey will be carried out with the 3.5 m telescope on Calar Alto, using at least 600 clear nights in the 2014-2018 time frame. We plan to survey a sample of 300 M-type stars for low-mass planet companions. This will provide sufficient statistics to assess the overall distribution of planets around M dwarfs: frequency, masses, and orbital parameters. The seemingly low occurrence of Jovian planets should be confirmed, and the frequency of ice giants and terrestrial planets should be established along with their typical separations, eccentricities, multiplicities, and dynamics.

2. The CARMENES Spectrographs

In order to identify the wavelength range that is most suitable for the search for radial velocity variations in low-mass stars, we have carried out detailed simulations of the achievable precision of such a measurement (see also Reiners *et al.* 2010). To optimize both the radial-velocity precision and the ability to discriminate between (wavelength-dependent) intrinsic radial-velocity jitter and (wavelength-independent) Keplerian signals, we decided to build two separate spectrographs, which will cover the wavelength ranges from 0.52 to 1.05 μm and 1.0 to 1.7 μm, respectively.

To take full advantage of the information content in a stellar spectrum for radial velocity measurements, the spectrograph resolution has to be matched to the stellar line width. Since many of the potential targets for CARMENES have rotational velocities of 3 km s^{-1} or less, a resolution of up to $\sim$ 100,000 could be usefully exploited. As a compromise between cost, size, and complexity on one hand, and performance on the other hand, the CARMENES spectrographs will have $R \sim 85{,}000$, realized through cross-dispersed echelle formats with white-pupil designs (for details see Quirrenbach *et al.* 2010). The visible-light spectrograph will use a 2048×4096 pixel CCD; the near infrared spectrograph will employ a mosaic of two 2048×2048 pixel detectors. To ensure sufficient stability, both spectrographs will be housed in temperature-stabilized vacuum tanks.

The spectrographs will be coupled to the 3.5 m telescope with optical fibers. Because there is currently no known gas cell that gives a sufficiently dense grid of absorption lines over the whole bands used by CARMENES, the simultaneous emission line method, using ThAr or alternatively UNe lamps, has been adopted as a baseline for the precise wavelength calibration. A stabilized etalon is currently under investigation as a potential alternative.

References

Quirrenbach, A., *et al.* 2010, in: Ground-based and airborne instrumentation for astronomy III, Eds. McLean, I.S., Ramsay, S.K., & Takami, H., *Proc. SPIE*, 7735, 1

Reiners, A., *et al.* 2010,*ApJ*, 710, 432

The Astrophysics of Planetary Systems: Formation, Structure, and Dynamical Evolution
Proceedings IAU Symposium No. 276, 2010
A. Sozzetti, M. G. Lattanzi & A. P. Boss, eds.

doi:10.1017/S1743921311021168

Earth like planets albedo variations versus continental landmass distribution

Esther Sanromá[1] and Enric Pallé[1]

[1]Instituto de Astrofísica de Canarias (IAC),
Vía Láctea s/n 38200, La Laguna, Spain
email: mesr@iac.es

Abstract. By making use of real information about the continental and oceanic surface distribution of the Earth, and cloudiness data from the International Satellite Cloud Climatology Project (ISCCP), we have studied the large-scale cloudiness behavior according to latitude and surface types (ice, water, vegetation and desert). These empirical relationships are used here to reconstruct the possible cloud distribution of historical epochs of the Earth history such as Late Cretaceous (90 My ago) and Late Triasic (230 My ago) when the landmass distribution was different. This information can be used to simulate the photometric variability of these planets according to their different geographical distribution.

Keywords. astrobiology, atmospheric effects, Earth, planets and satellites: general

---

1. Introduction

Over the past few years, advances in astronomy have enabled us to discover more and more planets orbiting around stars other than the Sun, making this planet search one of the most active and exciting fields in astrophysics. It is only a matter of time before we could detect Earth-like planets.

The exploration of our own planet will provide a useful tool to interpret observations and characterize such terrestrial worlds. Because of that, in this poster we present a study about clouds and their effects on the Earth's reflectance. As it is reasonable to expect that the future observed population of planets in the galaxy will exhibit a wide range of planet types and evolutionary stages, we have studied this cloud-reflectance relation not only for the Earth at present day, but also for past epochs in the history of our planet such as the Earth 90 and 230 My ago.

2. Data analysis

To carry out our study, we have made use of real satellite data of the Earth's total cloud amount from the ISCCP (Rossow *et al.* 1996), with the aim of finding a way to modelate the distribution of clouds over the Earth's surface. By making use of real information about the surface properties of the Earth, i.e., the location of deserts, forests and oceans, we have classified the averaged cloud conditions according to different surface types, obtaining empirical relationships between the amount of clouds, surface type and latitude. We have found that these derived cloudiness functions have a very particular shape which is different for each surface type. Therefore, global cloudiness distribution could be empirically traced depending on the latitude and surface type.

These empirical relations have been used to reconstruct the possible cloud distribution of hypothetical planets, with very idealized surface properties (Sanromá *et al.* in preparation), and also the possible cloud distribution of past epochs of the Earth history such

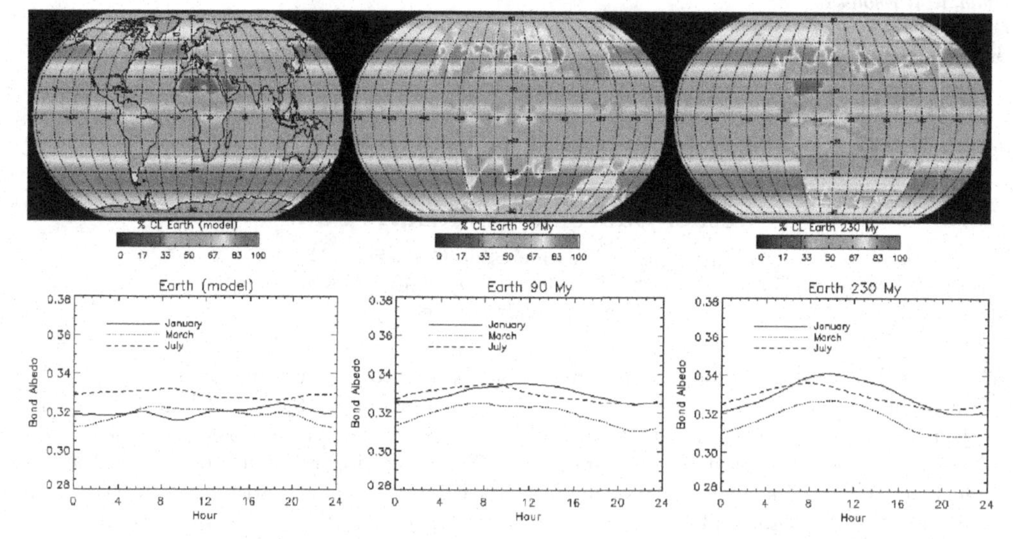

Figure 1. Top figures illustrate the reconstructed cloudiness of the Earth at present, 230 and 90 My ago. Bottom figures show the corresponding Bond Albedo 24 hour variability for three different months.

as Late Cretaceous and Late Triasic. To check our model, we have also reconstructed the cloudiness of the Earth at present obtaining that it reproduces the general features of the Earth's cloudiness distribution.

With these reconstructed cloud maps and using a simple Earth reflectance model (Pallé *et al.* 2003; Pallé *et al.* 2008) we have computed the first-order globally-integrated photometric variability of each epoch. Here we only show the results obtained for the Earth 230, 90 My ago, and at present (Fig. 1). Our results show that the less fractioned the planet's continental surface is, the larger the amplitude of variations is. By comparing the results obtained for the Earth, one can see that historical epochs of the Earth history present larger amplitude of daily variations in reflected light and larger mean albedo than at present.

References

Pallé, E., Ford, E. B., Seager, S., Montañés-Rodríguez, P., & Vazquez, M. 2008 *ApJ*, Vol. 676, 1319

Pallé, E., Goode, P. R., Yurchyshyn, V., Qiu, J., Hickey, J., Montañés Rodriguez, P., Chu, M. C., Kolbe, E., Brown, C. T., & Koonin, S. E. 2003, *J. of Geophys. Res.*, Vol. 108, No. D22, 4710

Rossow, W. B., Walker, A. W., Beuschel, D. E., & Roiter, M. D. 1996, *World Climate Research Programme Rep.* (WMO/TD 737)

The Astrophysics of Planetary Systems: Formation, Structure, and Dynamical Evolution
Proceedings IAU Symposium No. 276, 2010
A. Sozzetti, M. G. Lattanzi & A. P. Boss, eds.

doi:10.1017/S174392131102117X

Resolving blended radial velocities

Alexandre Santerne[1,2], Claire Moutou[1], François Bouchy[2,3] and the CoRoT Exoplanet Science Team

[1]Laboratoire d'Astrophysique de Marseille, Université d'Aix-Marseille & CNRS, 38 rue Frédéric Joliot-Curie, 13388 Marseille cedex 13, France
email: alexandre.santerne@oamp.fr

[2]Observatoire de Haute Provence, Université d'Aix-Marseille & CNRS, 04670 Saint Michel l'Observatoire, France

[3] Institut d'Astrophysique de Paris, UMR7095 CNRS, Université Pierre & Marie Curie, 98bis boulevard Arago, 75014 Paris, France

Abstract. In space, photometric surveys are very efficient to detect small transiting planets or stars which are contaminated by blended eclipsing binaries. We present some simulations compared to radial velocity (RV) observations obtained with the SOPHIE spectrograph (OHP, France) in order to determine the true nature of a brown dwarf candidate revealed by CoRoT: a background eclipsing binary diluted by a foreground star.

Keywords. techniques: radial velocities, techniques: photometric, planetary systems

1. Introduction

CoRoT and Kepler space missions are detecting by high accuracy photometry a large number of transiting planetary candidates, with lots of neptune-like planets or Super-Earth candidates (e.g. CoRoT-7 (Léger *et al.* 2009) or Kepler-9d (Torres *et al.* 2011)). Radial velocity (RV) follow-up is required to discard eclipsing binaries from true planets and characterize the mass of the planet. In the regime of shallow transits, is expected a high frequency of diluted eclipsing binaries (Brown 2003) that could mimic both a transit and also the radial velocity signal of a planet. We perform simulations of precise radial velocity diagnostics to help identifying spectroscopic blends.

2. A transiting brown dwarf candidate

A $\sim 0.3\%$–depth transiting candidate with an orbital period of 2.25 days was detected in the CoRoT light curve. As shown in Fig. 1 (left pannel), the radial velocity follow-up made with the SOPHIE spectrograph (Bouchy *et al.* 2009) mounted on 1.93-m telescope in Observatoire de Haute Provence (France) indicate a clear RV signature compatible with a brown dwarf in phase with CoRoT ephemeris. But, bisector span also showed variations in anti-correlation with velocities (Fig. 1 middle pannel). FWHM variations are seen at both, the orbital period and half the orbital period (Fig. 1 right pannel).

3. Blend simulations

We performed blend simulations of this brown dwarf candidate in order to explain the observed RV diagnostics (RV, bisector and FWHM). We simulated two blended CCF (Cross Correlation Function): the main one without any RV variation, the second one with variations in phase with the CoRoT ephemeris. The width and contrast of the primary CCF is fixed to the observed profile. We find, in this preliminary work,

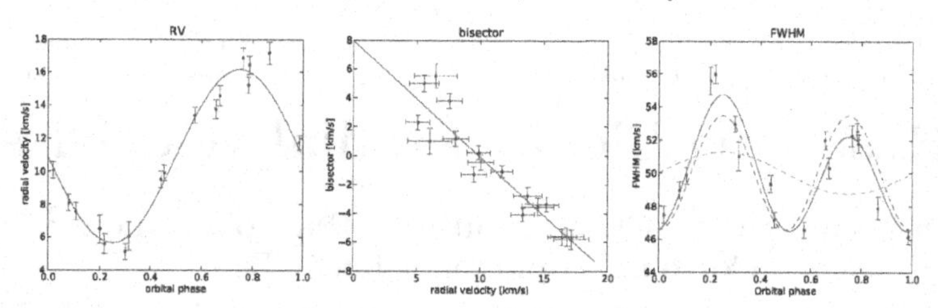

Figure 1. From left to right: phase-folded RV curve of this candidate with the best circular fit, bisector span against RV that show a clear anticorrelation and phase-folded FWHM curve that shows two variations, one at half the orbital period (dashed red line) and one at the orbital period (dashed-light blue line).

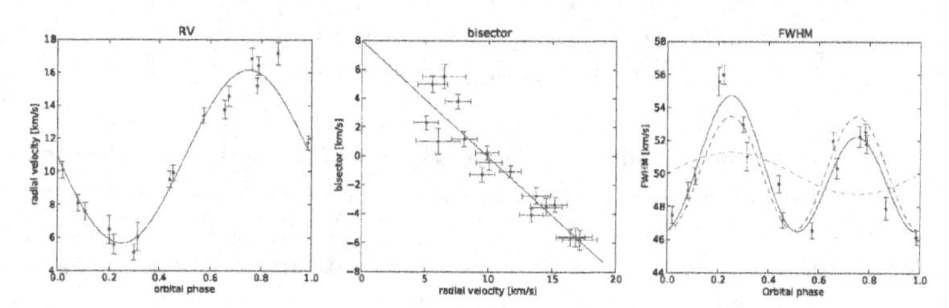

Figure 2. From left to right: simulated CCF of the main and secondary star, phase-folded simulated RV curve, simulated bisector against simulated RV, simulated FWHM as a function of the orbital phase.

that we could qualitatively reproduce RV data (see Fig.2) with a background spectroscopic and eclipsing binary which have the following characteristics: K_{blend} ~25km.s^{-1}, Δ_{flux} ~14%, FWHM$_{blend}$ ~25km.s^{-1}, FWHM$_{target}$ ~50km.s^{-1}, Δ_{v_0} ~100m.s^{-1}.

The negative slope of the bisector seems to be caused by the fact that the secondary star is a lower-rotating star than the primary. FWHM variations at half the orbital period are caused by the secondary star which increases the measured FWHM of the primary at extremal phases, when variations at the orbital period are caused by the slight RV offset between the main star and the binary.

4. Conclusion

Futur space-based surveys, like PLATO or TESS, will find amount of small-transit candidates where part of them will be blended by background stars. This kind of simulations, based on RV CCF diagnostics will be very useful to characterize the observed systems or to discard background blends (Santerne *et al.* in prep). The behaviour of the bisector span with respect to the RV appears as a very sensitive diagnostics of distant spectroscopic binaries. This tools of RV simulations will be completed soon by transit simulations as in Torres *et al.* (2011) and by a robust statistical analysis of best solutions.

References

Bouchy, F., Hébrard, G., Udry, S., *et al.* 2009, *A&A*, 505, 853
Brown, T. M. 2003, *ApJ* (Letters), 593, L125
Léger, A., Rouan, D., Schneider, J., *et al.* 2009, *A&A*, 506, 287
Torres, G., Fressin, F., Batalha, N. M., *et al.* 2011, *ApJ*, 727, id.24

The Astrophysics of Planetary Systems: Formation, Structure, and Dynamical Evolution
Proceedings IAU Symposium No. 276, 2010
A. Sozzetti, M. G. Lattanzi & A. P. Boss, eds.

doi:10.1017/S1743921311021181

Imaging faint companions very close to stars

Eugene Serabyn[1], Dimitri Mawet[1] and Rick Burruss[1]

[1]Jet Propulsion Laboratory, California Institute of Technology, Pasadena, CA 91109, USA
email: gene.serabyn@jpl.nasa.gov

Abstract. A vortex coronagraph on our extreme adaptive optics "well-corrected subaperture" on the Hale telescope has recently allowed the imaging of the triple-planet HR8799 system with a 1.5 m subaperture. Moreover, a faint, low-mass companion to a second star was imaged only one diffraction beam width away from the primary. These results illustrate the potential of the vortex coronagraph, which can enable exoplanet imaging and characterization with smaller telescopes than previously thought.

Keywords. instrumentation: adaptive optics, planetary systems, binaries (including multiple): close

1. Vortex Coronagraphy

The vortex coronagraph provides close to ideal small-angle performance (Guyon *et al.* 2006), allowing observations very close to stars. This small-angle capability allows a significant reduction in the size of potential space telescopes aimed at detecting and characterizing exoplanets. Over the past several years, we have thus implemented a high-contrast vector vortex coronagraph on the 1.5 m well-corrected subaperture of the Palomar Hale telescope, in order to begin to demonstrate its capabilities (Mawet *et al.* 2010). Indeed, even with our rather small subaperture, the observational results to date have been competitive with results obtained at much larger telescopes.

Our well-corrected subaperture had earlier already enabled observations of brown dwarfs and debris disks, using a four-quadrant phase mask coronagraph. With the upgrade to vector vortex masks, and the improvement of our contrast performance to roughly 10^{-5} within $\sim \lambda/D$ of the star (Burruss *et al.* 2010), we have been able to image the triple planet HR8799 system with our small aperture coronagraph (Figure 1; Serabyn *et al.* 2010). Recently, we have also imaged a binary star's faint secondary companion only $\sim 1\lambda/D$ from the primary (Figure 2; Mawet *et al.* in prep.). Thus, both the vortex's small-angle performance, and the utility of small telescopes for exoplanet imaging have begun to be validated. Indeed, a well-corrected subaperture employing a coronagraph capable of reaching small angles is a very promising method for quickly initiating exoplanet observations with a given telescope, especially important as we head toward the era of 30-40 m telescopes, where a well-corrected subaperture exceeding 10 m in diameter can quickly provide state-of-the-art high-contrast observations.

References

Burruss, R. S., Serabyn, E., Mawet, D. P., Roberts, J. E., Hickey, J. P., Rykoski, K., Bikkannavar, S., & Crepp, J. R. 2003, *Proc. SPIE*, 7736, 77365X
Guyon, O., Pluzhnik, E. A., Kuchner, M. J., Collins, B., & Ridgway S. T. 2006, *ApJS*, 167, 81
Mawet, D., Serabyn, E., Liewer, K., Burruss, R., Hickey, J., & Shemo, D. 2010, *ApJ* 709, 53
Serabyn, E., Mawet, D., & Burruss, R. 2010, *Nature*, 464, 1018

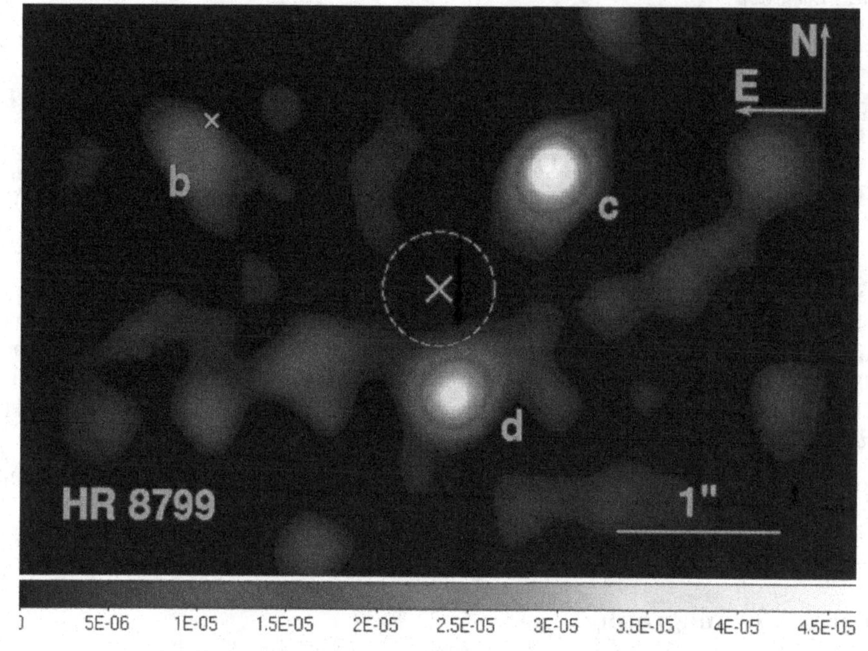

Figure 1. The HR8799 exoplanets imaged with our 1.5 m well-corrected subaperture.

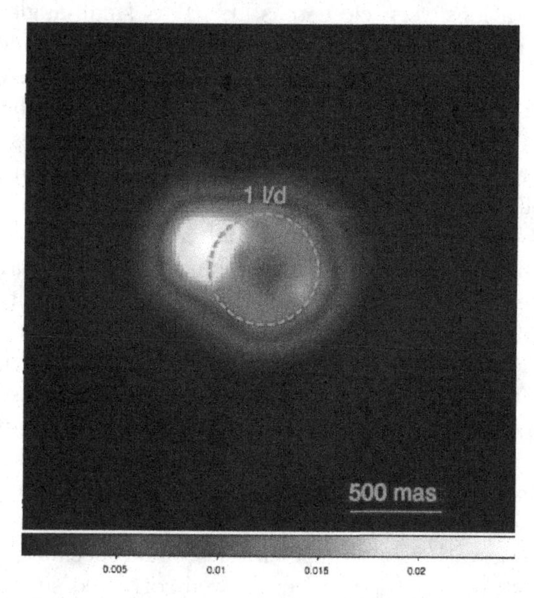

Figure 2. A low-mass binary companion seen at $\sim \lambda/D$ with our vortex coronagraph. The residuals of the primary star appear as a faint ring

The Astrophysics of Planetary Systems: Formation, Structure, and Dynamical Evolution
Proceedings IAU Symposium No. 276, 2010
A. Sozzetti, M. G. Lattanzi & A. P. Boss, eds.

doi:10.1017/S1743921311021193

The LCOGT Network

Avi Shporer[1], Tim Brown[1], Tim Lister[1], Rachel Street[1], Yiannis Tsapras[1], Federica Bianco[1], Benjamin Fulton[1] and Andy Howell[1]

[1]Las Cumbres Observatory Global Telescope Network, 6740 Cortona Drive, Suite 102, Santa Barbara, CA 93117, USA; http://lcogt.net
email: ashporer@lcogt.net

Abstract. Motivated by the increasing need for observational resources for the study of time varying astronomy, the Las Cumbres Observatory Global Telescope (LCOGT) is a private foundation, whose goal is to build a global network of robotic telescopes for scientific research and education. Once completed, the network will become a unique tool, capable of continuous monitoring from both the Northern and Southern Hemispheres. The network currently includes 2×2.0 m telescopes, already making an impact in the field of exoplanet research. In the next few years they will be joined by at least 12×1.0 m and 20×0.4 m telescopes. The increasing amount of LCOGT observational resources in the coming years will be of great service to the astronomical community in general, and the exoplanet community in particular.

Keywords. telescopes, techniques: photometric, planetary systems

---

The LCOGT network currently consists of two 2.0 m telescopes: Faulkes Telescope South (FTS), located at Siding Spring Observatory, Australia, and Faulkes Telescope North (FTN), located on Mt. Haleakala on the Hawaiian island of Maui. Two 0.4 m telescopes are also located within the FTN clamshell dome, and they are currently being commissioned.

In the next 1–2 years LCOGT will deploy telescopes at several sites, as shown in Figure 1. The deployment of the network in the Southern hemisphere will commence first. Construction has already begun at CTIO (Chile) and SAAO (South Africa). In each site we plan to put 3×1.0 m and 4–6×0.4 m telescopes, starting 2011. The third node of our 0.4 m and 1.0 m network in the South will be in Australia, possibly at Siding Spring next to FTS, but we are considering other sites as well. In the North, we are now negotiating a site agreement with the IAC (Canary Islands, Spain) and McDonald Observatory (TX, USA). Our intention is to have an additional Northern node of the network in Asia, where a few possibilities are being investigated.

Another site now being commissioned is the Byrne Observatory at UC's Sedgwick Reserve (BOS) in the Santa Ynez valley, approximately 30 miles from LCOGT's base in Goleta (CA, USA). This site currently has a 0.8 m telescope and will be used for testing new instruments and for education.

The LCOGT science team includes two UCSB faculty members and close to 10 postdocs and project scientists. Within the domain of time-variable astronomy LCOGT focuses primarily on two observational fields: Exoplanets and Supernovae. LCOGT is taking part in most of the important Supernovae surveys, including the Supernova Legacy Survey, Pan-STARRS, the Palomar Transient Factory, and the La Silla/QUEST Supernova search. The study of exoplanets is carried out through observations of microlensing and transiting planets. Microlensing events are being monitored by the LCOGT-based network for the detection of microlensing planets, RoboNet (Tsapras *et al.* 2009).

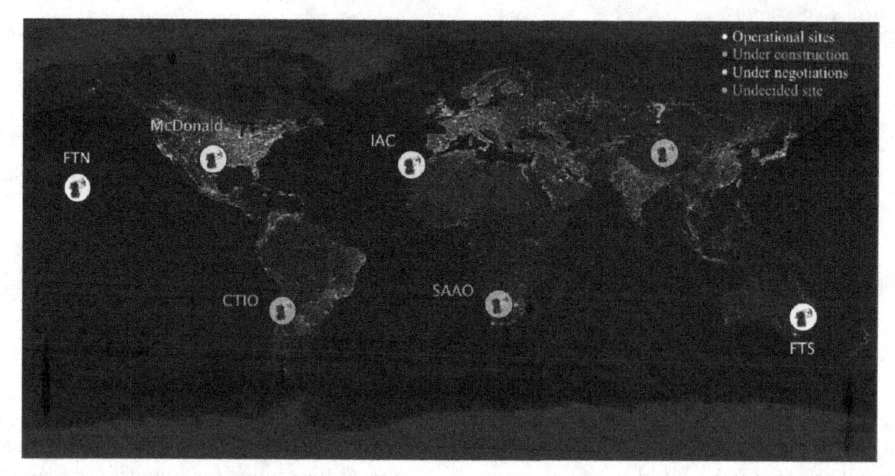

Figure 1. The planned LCOGT network, showing the location and status of the various sites.

LCOGT scientists are collaborating with most of the transiting planet surveys, including WASP, HATNet, TrES, MEarth, CoRoT and Kepler. A large fraction of FTN and FTS telescope time is devoted for observations of planetary transit candidates, in order to resolve candidates blended with nearby stars in survey images and obtain high precision photometry. This is usually done before or contemporaneously with the gathering of high resolution spectroscopic observations at other telescopes. Light curves obtained with LCOGT telescopes were part of the discovery of many of the currently known transiting planets, e.g., WASP-4b (Wilson *et al.* 2008), WASP-24b (Street *et al.* 2010), TrES-3 (O'Donovan *et al.* 2007), HAT-P-24b (Kipping *et al.* 2010) and CoRoT-9b (Deeg *et al.* 2010). LCOGT also participates in follow-up studies of known transiting exoplanets. For example, Hidas *et al.* (2010) and Shporer *et al.* (2010a) describe two follow-up campaigns led by LCOGT researchers where a complete coverage of HD 80606b 12 hour transits was obtained.

In addition, members of the LCOGT science team are involved in the study of other variable stars (e.g., Lister *et al.* 2009, Shporer *et al.* 2010b) and open clusters (e.g., Cieza & Baliber 2007; Fulton & Baliber 2011, in preparation).

A few other projects are currently on-going at LCOGT. Among them is the commissioning of a high-speed camera mounted on FTN, to be used for lucky imaging and observations of Kuiper Belt Object occultations and other short time scale phenomena. We also note here that LCOGT is developing a medium-resolution (R=25,000) fiber-fed spectrograph, for stellar spectroscopy. It will reach a radial velocity accuracy on the order of 100 m/s, making it capable of identifying planetary transit false positives. The spectrograph is expected to be mounted on the Byrne Observatory telescope for on-sky testing during 2011, and will eventually be mounted on the 1.0 m telescopes.

References

Cieza, L. & Baliber, N. 2007, *ApJ*, 671, 605

Deeg, H. J., *et al.* 2010, *Nature*, 464, 384

Tsapras, Y., *et al.* 2009, Astronomische Nachrichten, 330, 4

Hidas, M. G., *et al.* 2010, *MNRAS*, 406, 1146

Kipping, D. M., *et al.* 2010, *ApJ*, 725, 2017

Lister, T., Metcalfe, T., Brown, T., & Street, R. 2009, astro2010 : The Astronomy and Astrophysics Decadal Survey, 2010, 184 (arXiv:0902.2966)

O'Donovan, F. T., *et al.* 2007, *ApJL*, 663, L37
Shporer, A., *et al.* 2010a, *ApJ*, 722, 880
Shporer, A., *et al.* 2010b, *ApJL*, 725, L200
Street, R. A., *et al.* 2010, *ApJ*, 720, 337
Wilson, D. M., *et al.* 2008, *ApJL*, 675, L113

The Astrophysics of Planetary Systems: Formation, Structure, and Dynamical Evolution
Proceedings IAU Symposium No. 276, 2010
A. Sozzetti, M. G. Lattanzi & A. P. Boss, eds.

doi:10.1017/S174392131102120X

Practical suggestions on detecting exomoons in exoplanet transit light curves

Gyula M. Szabó[1,2], A. E. Simon[1], Laszlo L. Kiss[1,3] and Zsolt Regály[1]

[1]Konkoly Observatory of the HAS, Budapest, Hungary
email: szgy@konkoly.hu
[2]Dept. of Experimental Physics, University of Szeged, Hungary
[3]School of Physics, University of Sydney, Australia

Abstract. The number of known transiting exoplanets is rapidly increasing, which has recently inspired significant interest as to whether they can host a detectable moon. Although there has been no such example where the presence of a satellite was proven, several methods have already been investigated for such a detection in the future. All these methods utilize post-processing of the measured light curves, and the presence of the moon is decided by the distribution of a timing parameter. Here we propose a method for the detection of the moon *directly in the raw transit light curves.* When the moon is in transit, it puts its own fingerprint on the intensity variation. In realistic cases, this distortion is too little to be detected in the individual light curves, and must be amplified. Averaging the folded light curve of several transits helps decrease the scatter, but it is not the best approach because it also reduces the signal. The relative position of the moon varies from transit to transit, the moon's wing will appear in different positions on different sides of the planet's transit. Here we show that a careful analysis of the scatter curve of the folded light curves enhances the chance of detecting the exomoons directly.

Keywords. techniques: photometric, planetary systems, planets and satellites: general, eclipses

1. Introduction

We simulated transit light curves of different qualities for exoplanets with moons. Space measurements covering 3 years were represented by Kepler space telescope short cadence and long cadence samplings and the bootstrap noise of non-variable stars. Model moons had 0.7, 0.8, 0.9, 1.0 Earth-radius size, orbited on circular orbit with $P_{moon} = 4.3$ days, while the planet orbited its circular orbit with $P_{planet} = 10$ days period. The planet was a hot Jupiter with 0.7 M_J, 1.0 R_J mass and radius, while the mass of the moon was neglected. The central star was a solar analog, the radius of orbits were $a_{planet} = 0.09$ AU, $a_{moon} = 1.3 \times 10^6$ km, putting the moon to the border of the Hill-sphere.

We concluded that the most efficient tool of detecting the signal of the moon is observing the residual scatter in the folded light curves, after subtracting a boxcar median of all light curves. The scatter peak reflects the tiny light variations that occur because of the changing position of the moon in individual transits. We have deduced from simulations that for a clear detection (false alarm rate $< 1\%$), cut levels in the 4.2–4.4σ range must be chosen. Selecting 4.4σ treshold, 90% of the Earth-sized moons can be discovered in long cadence data, and practically all moons of this size will be discovered in short cadence (detection rate is 99%.) There is still some chance for the discovery of large exomoons with Earth-based quality observations via the scatter peak, in this case the discovery rate is 30% for Earth-sized moons.

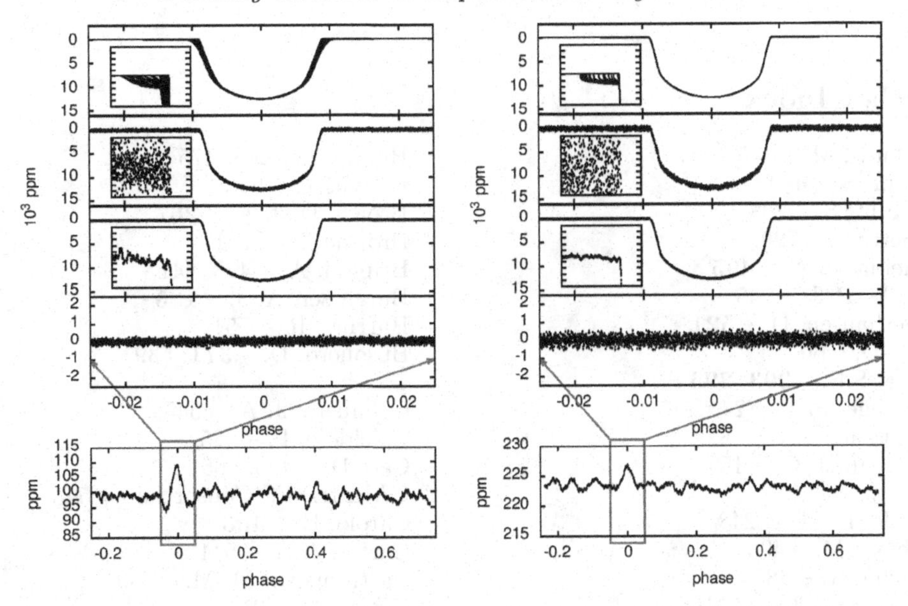

Figure 1. Simulations of 110 transits with exomoon distortions, sampled as Kepler long cadence (left) and short cadence (right) data. Each column shows the input and the noisified light curves, the median filtered data, the residuals to the median and the rms scatter of the residuals. The inserts zoom to the exomoon signal (50 ppm per tick).

2. Practical suggestions for observations

Testing the Scatter Peak from a sequence of light curves is a promising tool for detecting moons directly in the light curves. The successful detection relies on three important conditions:

- All light curves must be stacked in such way that the transit time of the planet exactly coincide in every light curves involved to the analysis.
- Transit observations must include the out-of transit phases immediately before and after the transit of the planet, where the scatter due to the moon is the highest. The wings must be at least as long as the transit duration.
- Trend filtering the light curves must be done such that the tiny brightness variations due to the exomoons shall remain unaffected.

Acknowledgements

This work is supported by the Hungarian OTKA Grants K76816 and MB08C 81013, the "Lendület" Young Researchers' Program of the Hungarian Academy of Sciences and the Hungarian State "Eötvos" Fellowship.

Author Index

Subject Index